Paolo Segneri

The manna of the soul : meditations for each day of the year

Paolo Segneri

The manna of the soul : meditations for each day of the year

ISBN/EAN: 9783742875174

Manufactured in Europe, USA, Canada, Australia, Japa

Cover: Foto ©berggeist007 / pixelio.de

Manufactured and distributed by brebook publishing software
(www.brebook.com)

Paolo Segneri

The manna of the soul : meditations for each day of the year

THE

MANNA OF THE SOUL.

Meditations for every Day of the Year.

BY

FATHER PAUL SEGNERI.

Second Edition.

IN TWO VOLUMES

NEW YORK, CINCINNATI, CHICAGO:

BENZIGER BROTHERS,

Printers to the Holy Apostolic See.

———

1892.

CONTENTS.

JULY.   PAGE

1. WAKEFULNESS IN THE SERVICE OF GOD. —"Love not sleep,
   lest poverty oppress thee; open thy eyes, and be filled with
   bread" (Prov. xx. 13).   1

2. KIND INTEREST IN THE WELFARE OF OTHERS. —"Recover thy
   neighbour according to thy power, and take heed to thyself
   that thou fall not" (Ecclus. xxix. 26).   4

3. THE PROPER OBJECT OF PRAYER. —"Every best gift and every
   perfect gift is from above, coming down from the Father of
   lights, with Whom there is no change or shadow of alteration"
   (St. James i. 17).   8

4. TRUE HONOUR. —"My son, keep thy soul in meekness, and
   give it honour according to its desert" (Ecclus. x. 31).   .   .   12

5. SALUTARY FEAR. —"Pierce Thou my flesh with Thy fear, for
   I am afraid of Thy judgments" (Psalm cxviii. 120).   .   .   15

6. THE PATIENCE WITH WHICH GOD WAITS FOR OUR INVITA-
   TION.—"Behold, I stand at the gate and knock. If any man
   shall hear My voice, and open to Me the door, I will come
   in to him, and will sup with him, and he with Me" (Apoc.
   iii. 20).   19

7. THE READINESS WITH WHICH GOD ACCEPTS OUR INVITA-
   TION.—"If any man shall hear My voice, and open to Me the
   door, I will come in to him, and sup with him, and he with
   Me" (Apoc. iii. 20).   23

8. DISOBEDIENCE A GREAT CRIME.—"It is like the sin of witchcraft to rebel, and like the crime of idolatry to refuse to obey"
   (1 Kings xv. 23).   28

9. MERCY GIVING PLACE TO JUSTICE.—"The jealousy and rage of
   the husband will not spare in the day of revenge, nor will he
   yield to any man's prayers, nor will he accept for satisfaction
   ever so many gifts" (Prov. vi. 34, 35).   32

10. LIFE TOO SHORT FOR THE PURSUIT OF TRIFLES.—"What
    needeth a man to seek those things that are above him, whereas
    he knoweth not what is profitable for him in his life, in all the
    days of his pilgrimage, and the time that passeth like a
    shadow?" (Eccles. vii. 1).   36

11. THE DARK NIGHT OF ETERNITY.—"These are they to whom
    the storm of darkness is reserved for ever" (St. Jude i. 13).   .   40

Contents.

PAGE

12. JEALOUS CARE OF PURITY.—"I made a covenant with my eyes, that I would not so much as think upon a virgin; for what part should God from above have in me, and what inheritance the Almighty from on high?" (Job xxxi. 1). . . . 44

13. CONFORMITY WITH CHRIST CRUCIFIED A MARK OF GOD'S FAVOUR.—"Whom He foreknew He also predestinated to be made conformable to the image of His Son, that He might be the First-born among many brethren" (Romans viii. 29). . 48

14. HAPPINESS IN DEATH.—"Blessed are the dead who die in the Lord. From henceforth now, saith the Spirit, that they may rest from their labours: for their works follow them" (Apoc. xiv. 13). 53

15. REMEMBRANCE OF CHRIST'S PASSION.—"Christ, therefore, having suffered in the flesh, be you also armed with the same thought" (1 St. Peter iv. 1). 59

16. EARNESTNESS IN SERVING GOD.—"Exercise thyself unto godliness; for bodily exercise is profitable to little, but godliness is profitable to all things, having promise of the life that now is, and of that which is to come" (1 Timothy iv. 7, 8). . . 61

17. THE MISERABLE STATE CAUSED BY LONG RESISTANCE TO GRACE.—"A hard heart shall fare ill at the last" (Ecclus. iii. 27). 65

18. THE REWARD OF A COMPASSIONATE HEART.—"Be ye merciful, as your Father also is merciful" (St. Luke vi. 36). . . 69

19. PRAYER OF PROPITIATION.—"Enter not into judgment with Thy servant, O Lord, for in Thy sight no man living shall be justified" (Psalm cxlii. 2). 74

20. THE CROWN OF PERSEVERANCE.—"Be thou faithful unto death, and I will give thee the crown of life" (Apoc. ii. 10) . 79

21. CARE TO GUARD OUR HEARTS.—"Give not place to the devil" (Ephes. iv. 27). 83

22. PERFECT REPENTANCE.—"Thou wert naked, and full of confusion, and I passed by thee, and saw thee; and behold thy time was the time of lovers; and I spread My garment over thee, and covered thy ignominy; and I swore to thee, and I entered into a covenant with thee, said the Lord God, and thou becamest Mine" (Ezech. xvi. 8). 85

23. THE THOUGHT OF ETERNITY.—"I thought upon the days of old, and I had in my mind the eternal years" (Psalm lxxvi. 6). 90

24. THE DREADFUL DOOM OF APOSTASY.—"If we sin wilfully after having received the knowledge of the truth, there is now left no sacrifice for sins: but a certain dreadful expectation of Judgment, and the rage of a fire which shall consume the adversaries" (Hebrews x. 26). 93

25. DILIGENT SELF-RESTRAINT.—"I beseech you, as strangers and pilgrims, to refrain yourselves from carnal desires which war against the soul, having your conversation good among the Gentiles" (1 St. Peter ii. 11). 99

26. VALUE OF THE INTERIOR LIFE.—"The Lord . . . will be silent in His love, He will be joyful over thee in praise" (Sophonias iii. 17). 103

27. PEACE WITH OURSELVES.—"Be at agreement with thy adversary betimes, whilst thou art in the way with him, lest perhaps

the adversary deliver thee to the judge, and the judge deliver thee to the officer, and thou be cast into prison. Amen, I say to thee, thou shalt not go out from thence till thou repay the last farthing " (St. Matt. v. 25). 105

28. THE TWO PRECEPTS OF CHARITY.—"Thou shalt love the Lord thy God with thy whole heart, and with thy whole soul, and with thy whole mind, and with thy whole strength. This is the first commandment. And the second is like to it: Thou shalt love thy neighbour as thyself" (St. Mark xii. 30, 31). . 110

29. THE FIRST PRECEPT OF CHARITY.—"Thou shalt love the Lord thy God with thy whole heart, and with thy whole soul, and with thy whole mind, and with thy whole strength. This is the first commandment" (St. Mark xii. 30). 114

30. THE SECOND PRECEPT OF CHARITY.—"And the second is like to it: Thou shalt love thy neighbour as thyself" (St. Mark xii. 31). 117

31. SEEKING THE GLORY OF GOD.—"How can you believe, who receive glory one from another, and the glory which is from God alone you do not seek " (St. John v. 44). 122

AUGUST.

1. THE LIBERTY OF THE SONS OF GOD.—"A vain man is lifted up into pride, and thinketh himself born free like a wild ass's colt " (Job xi. 12). 128

2. THE WORDS OF ETERNAL LIFE.—"Amen, amen, I say to you: if any man keep My word, he shall not see death for ever " (St. John viii. 51). 133

3. THE RESURRECTION OF THE BODY.—"The hour cometh wherein all that are in the graves shall hear the voice of the Son of God, and they that have done good things shall come forth unto the resurrection of life, but they that have done evil unto the resurrection of judgment" (St. John v. 28, 29). 136

4. LIVING FOR CHRIST.—"The charity of Christ presseth us . . . that they also who live, may not now live to themselves, but unto Him Who died for them" (2 Cor. v. 14, 15). . . . 141

5. DEVOTION TO OUR BLESSED LADY A TREASURE OF LIFE.— "Blessed is the man that heareth me, and that watcheth daily at my gates, and waiteth at the posts of my doors. He that shall find me shall find life, and shall have salvation from the Lord " (Prov. viii. 34, 35). 142

6. THE TRANSFIGURATION.—"This is my Beloved Son, in Whom I am well pleased: hear ye Him " (St. Matt. xvii. 5). . . 146

7. HUMILITY AND CONFIDENCE.—"Be you humbled, therefore, under the mighty hand of God, that He may exalt you in the time of visitation; casting all your care upon Him, for He hath care of you " (1 St. Peter v. 6, 7). 151

8. DANGER OF CARELESSNESS.—"He that contemneth small things shall fall by little and little " (Ecclus. xix. 1). 155

9. LEAVING THE CREATOR FOR HIS CREATURES.—"Be astonished, O ye heavens, at this, and ye gates thereof be very desolate, saith the Lord: for My people have done two evils.

They have forsaken Me, the fountain of living water, and have digged to themselves cisterns, broken cisterns, that can hold no water " (Jerem. ii. 12, 13). 159

10. PERFECT RELIANCE UPON THE FIDELITY OF GOD.—"I suffer, but I am not ashamed : for I know Whom I have believed, and I am certain that He is able to keep that which I have committed to Him against that day " (2 Timothy i. 12).. . . 162

11. KNOWLEDGE OF OUR OWN NOTHINGNESS.—"If any man think himself to be something, whereas he is nothing, he deceiveth himself " (Galat. vi. 3). 168

12. DISENGAGEMENT FROM THE WORLD.—"Be ye like the dove that maketh her nest in the mouth of the hole in the highest place " (Jerem. xlviii. 28). 173

13. CHRIST OUR PATTERN OF CHARITY.—"This is My commandment, that you love one another, as I have loved you" (St. John xv. 12). 175

14. THE DEATH OF THE JUST.—"The souls of the just are in the hand of God, and the torment of death shall not touch them. In the sight of the unwise they seemed to die, and their departure was taken for misery, and their going away from us for utter destruction, but they are in peace " (Wisdom iii. 1—3). . . 180

15. OUR BLESSED LADY HUMBLING HERSELF AND EXALTED.—"Humility goeth before glory " (Prov. xv. 33). . . . 185

16. REPOSE FOR THE WEARY.—"Come to Me, all you that labour and are burdened, and I will refresh you " (St. Matt. xi. 28). . 190

17. BEARING THE YOKE OF CHRIST.—"Take up My yoke upon you, and learn of Me, because I am meek and humble of heart, and you shall find rest to your souls" (St. Matt. xi. 29). . 193

18. PEACE PROMISED TO HUMILITY.—"Learn of Me, because I am meek and humble of heart, and you shall find rest to your souls " (St. Matt. xi. 29). 196

19. A SERVICE LIGHT AND EASY TO THOSE WHO LOVE.—"For My yoke is sweet, and My burden light " (St. Matt. xi. 30). . 200

20. THE WRATH WHICH ENDURETH FOR EVER.—"Thy arrows pass : the voice of Thy thunder in a wheel " (Psalm lxxvi. 19). 204

21. THE FRUIT OF GOOD AND EVIL LIFE.—"What things a man shall sow, those also shall he reap. For he that soweth in his flesh, of the flesh also shall reap corruption. But he that soweth in the spirit, of the spirit shall reap life everlasting " (Galat. vi. 8). 207

22. AN UNFAILING RECOMPENSE.—"And in doing good let us not fail. For in due time we shall reap, not failing " (Galat. vi. 9). 211

23. THE CHEATING PROMISES OF THE WORLD.—"Dreams have deceived many, and they have failed that put their trust in them." (Ecclus. xxxiv. 7). 214

24. FIGHTING IN THE GOOD CAUSE.—"Strive for justice for thy soul, and even unto death fight for justice, and God will overthrow thy enemies for thee " (Ecclus. iv. 33). . . . 218

25. CONTEMPT FOR THE FOOLISH JUDGMENTS OF MEN.—"Fear ye not the reproach of men, and be not afraid of their blasphemies : for the worm shall eat them up as a garment, and the moth shall consume them as wool, but My salvation shall be for ever " (Isaias li. 7, 8). 222

Contents.

PAGE

26. MAKING TRUE PROGRESS.—"Doing the truth in charity, that we may in all things grow up in Him Who is the Head, even Christ" (Ephes. iv. 15). 226

27. THE PORTION OF THE REPROBATE.—They shall go into the lower parts of the earth: they shall be delivered into the hands of the sword, they shall be the portions of foxes" (Psalm lxii. 11). 229

28. THE PROPERTIES OF CELESTIAL FIRE.—"The fire of the Lord is in Sion, and His furnace in Jerusalem" (Isaias xxxi. 9). . 235

29. THE GRACE OF DISCERNMENT.—"If thou wilt separate the precious from the vile, thou shalt be as My mouth" (Jerem. xv. 13). 240

30. A PRIVILEGE TO SHARE IN THE DISGRACE OF THE CROSS.— "Jesus, that He might sanctify the people by His own Blood, suffered without the gate. Let us go forth therefore to Him without the camp, bearing His reproach" (Hebrews xiii. 12). 243

31. THE TERRIBLE PUNISHMENT OF TEPIDITY.—"I know thy works, that thou art neither cold nor hot. I would that thou wert cold or hot. But because thou art lukewarm, and neither cold nor hot, I will begin to vomit thee out of My mouth" (Apoc. iii. 15). 248

SEPTEMBER.

1. THE SCHOOL OF CHRIST.—"Blessed is the man whom Thou shalt instruct, O Lord, and shalt teach him out of Thy law" (Psalm xciii. 12). 254

2. THIS LIFE A TIME OF PROBATION.—"The life of man upon earth is a warfare" (Job vii. 1). 257

3. GOD IN THE SOUL.—"There hath stood one in the midst of you Whom you knew not" (St. John i. 26). 262

4. THE TEMPLES IN WHICH GOD LOVES TO DWELL.—"You are the temple of the living God, as God saith: I will dwell in them, and walk among them, and I will be their God, and they shall be My people" (2 Cor. vi. 16). . . . 265

5. THE SLEEPLESS ENEMY OF MEN.—"Be sober and watch: because your adversary the devil, as a roaring lion, goeth about seeking whom he may devour. Whom resist ye, strong in faith" (1 St. Peter v. 8). 268

6. SPIRITUAL THINGS INVISIBLE TO CARNAL-MINDED MEN.— "The sensual man perceiveth not these things that are of the Spirit of God: for it is foolishness to him, and he cannot understand" (1 Cor. ii. 14). 273

7. CHRIST SUFFERING FOR OUR SAKE.—"Christ suffered for us, leaving you an example, that you should follow His steps" (1 St. Peter ii. 21). 277

8. MARY THE OBJECT OF GOD'S PREDILECTION.—"The Lord possessed me in the beginning of His ways, before He made anything from the beginning" (Prov. viii. 22). . . 280

9. THE EVIL EFFECTS OF ENVY.—"Where envying and contention is, there is inconstancy and every evil work" (St. James iii. 16). 284

10. STRIVING TO MAKE OUR ELECTION SURE.—"I therefore so run, not as at an uncertainty: I so fight, not as one beating the air : but I chastise my body, and bring it into subjection, lest perhaps, when I have preached to others, I myself should become a castaway " (1 Cor. ix. 26).   288

11. THE CHILDREN OF SATAN.—"You are of your father, the devil; and the desires of your father you will do " (St. John viii. 44).   291

12. REPAIRING WASTED YEARS.—"See, therefore, brethren, how you walk circumspectly, not as unwise, but wise : redeeming the time, because the days are evil " (Ephes. v. 15, 16).   . 294

13. THE DISCOMFITURE OF THE PRINCE OF THIS WORLD.— " Now is the judgment of the world : now shall the prince of this world be cast out : and I, if I be lifted up from the earth, will draw all things to Myself" (St. John xii. 31, 32).   . 298

14. THE POWER OF THE CROSS.—"And I, if I be lifted up from the earth, will draw all things to Myself " (St. John xii. 32).   . 301

15. PROCURING THAT GOD MAY BE PRAISED.—"So let your light shine before men, that they may see your good works, and glorify your Father Who is in Heaven " (St. Matt. v. 16).   . 306

16. VAINGLORY TO BE SHUNNED WITH CARE.—"Take heed that you do not your justice before men, to be seen by them, otherwise you shall not have a reward of your Father Who is in Heaven " (St. Matt. vi. 1).   310

17. OUR BLESSED LADY'S PLENITUDE OF GRACE.—"Hail, Mary, full of grace " (St. Luke i. 28).   315

18. CHRIST IN HIS POOR.—"Blessed is he that understandeth concerning the needy and the poor ; the Lord will deliver him in the evil day " (Psalm xl. 1).   319

19. GAINING STRENGTH FROM OPPOSITION.—"Be not overcome by evil ; but overcome evil by good " (Romans xii. 21).   . 324

20. LIFE AND DEATH.—"If you live according to the flesh, you shall die ; but if by the spirit you mortify the deeds of the flesh, you shall live " (Romans viii. 13).   .   .   . 328

21. VOCATION A FREE GIFT.—"The Spirit breatheth where He will, and thou hearest His voice, but thou knowest not whence He cometh and whither He goeth : so is every one that is born of the Spirit " (St. John iii. 8).   331

22. GOD BEARING PATIENTLY WITH THE INSOLENCE OF SINNERS.— "Thou hast made Me to serve with thy sins, thou hast wearied Me with thy iniquities " (Isaias xliii. 24).   .   .   . 335

23. VICES OF THE TONGUE.—"If any man think himself to be religious, not bridling his tongue, but deceiving his own heart, this man's religion is vain " (St. James i. 26).   .   . 339

24. THE REPENTANT SINNER RECEIVED WITH A GREAT WELCOME. —"I say to you, that there shall be joy before the angels of God upon one sinner doing penance more than upon ninety-nine just who need not penance " (St. Luke xv. 7, 10).   . 342

25. THE STRENGTH WHICH IS DERIVED FROM OBEDIENCE.—"An obedient man shall speak of victory" (Prov. xxi. 28).   . 345

26. ACCEPTING CORRECTION BEFORE IT IS TOO LATE.—"Be instructed, O Jerusalem, lest My soul depart from thee " (Jerem. vi. 8).   349

27. TRUE CONVERSION OF HEART.—"Be not conformed to this world, but be reformed in the newness of your mind, that you may prove what is the good, and the acceptable, and the perfect will of God" (Romans xii. 2). 352

28. PERSEVERANCE IN GOOD PURPOSES.—"No man putting his hand to the plough, and looking back, is fit for the Kingdom of God" (St. Luke ix. 62). 356

29. HUMILITY TRIUMPHING OVER PRIDE.—"He hath showed might in His arm, He hath scattered the proud in the conceit of their heart. He hath put down the mighty from their seat, and hath exalted the humble" (St. Luke i. 51, 52) . . . 360

30. WHOLESOME DISTRUST OF OURSELVES.—"I sat alone, because Thou hast filled me with threats" (Jerem. xv. 17). . . . 364

OCTOBER.

1. SCRIPTURE.—"We have the more firm prophetical word, whereunto you do well to attend, as to a light that shineth in a dark place, until the day dawn, and the day-star arise in your hearts" (2 St. Peter i. 19). 367

2. THE HOLY ANGEL GUARDIANS.—"For He hath given His angels charge over thee, to keep thee in all thy ways; in their hands they shall bear thee up, lest thou dash thy foot against a stone" (Psalm xc. 11). 371

3. VIOLENCE AND THE KINGDOM.—"The Kingdom of Heaven suffereth violence, and the violent bear it away (St. Matt. xi. 12). 376

4. LOSING ALL FOR CHRIST.—"The things that were gain to me, the same I have counted loss for Christ. Furthermore, I count all things to be but loss, for the excellent knowledge of Jesus Christ my Lord, for Whom I have suffered the loss of all things, and count them but as dung, that I may gain Christ" (Philipp. iii. 7). 381

5. THE MOTE AND THE BEAM.—"Why seest thou the mote in thy brother's eye, but the beam that is in thine own eye thou considerest not" (St. Luke vi. 41). 385

6. WATCHFULNESS.—"I will stand upon my watch, and fix my foot upon the tower, and I will watch to see what will be said to me, and what I may answer to him that reproveth me" (Habacuc ii. 1). 389

7. CHRIST THE VINE.—"I am the Vine; you are the branches: he that abideth in Me, and I in him, the same beareth much fruit; for without Me you can do nothing" (St. John xv. 5). . 393

8. THE BRANCH CUT OFF.—"If any one abide not in Me he shall be cast forth as a branch, and shall wither, and they shall gather him up, and cast him into the fire, and he burneth" (St. John xv. 6). 398

9. TEMPTATIONS.—They have said to thy soul, Bow down, that we may go over: and thou hast laid thy body as the ground, and as a way to them that went over" (Isaias li. 23). . 403

10. UNCERTAINTY OF OUR STATE.—"Man knoweth not whether he be worthy of love or hatred: but all things are kept uncertain for the time to come" (Eccles. ix. 1, 2). 405

PAGE

11. PREPARATION FOR TEMPTATION.—"Son, when thou comest to the service of God, stand in justice and in fear, and prepare thy soul for temptation" (Ecclus. ii. 1). 408

12. DEPTH OF HEART.—"Man shall come to a deep heart, and God shall be exalted" (Psalm lxiii. 7, 8). 414

13. CHRISTIAN SUFFERING.—"Let none of you suffer as a murderer, or a thief, or a railer, or a coveter of other men's things. But if, as a Christian, let him not be ashamed, but let him glorify God in this name" (1 St. Peter iv. 15, 16). 418

14. FEAR OF THE SAINTS.—"For I have always feared God as waves swelling over me, and His weight I was not able to bear" (Job xxxi. 23). 422

15. MEDITATION.—"I will cry like a young swallow, I will meditate like a dove" (Isaias xxxviii. 14). 426

16. CONDITIONS OF PRAYER.—"Thus, therefore, shall you pray: Our Father Who art in Heaven," &c. (St. Matt. vi. 9). . . 429

17. OUR FATHER.—"Father." 434

18. OUR FATHER.—"Our Father." 439

19. OUR FATHER IN HEAVEN.—"Who art in Heaven." . . 443

20. HALLOWING GOD'S NAME.—"Hallowed be Thy Name." . 448

21. GOD'S KINGDOM.—"Thy Kingdom come." 453

22. THE WILL OF GOD.—"Thy will be done on earth as it is in Heaven." 459

23. DAILY BREAD.—"Give us this day our daily bread." . . . 464

24. FORGIVENESS OF DEBTS.—"And forgive us our debts." . 470

25. FORGIVENESS OF OTHERS.—"As we also forgive our debtors." . 474

26. LEADING INTO TEMPTATION.—"And lead us not into temptation." 479

27. DELIVERANCE FROM EVIL.—"But deliver us from evil." . 484

28. THE "PATER NOSTER."—"Thus, therefore, shall you pray: Our Father," &c. 489

29. EATING WITH OUR LORD.—"You are they who have continued with Me in My temptations, and I dispose to you as My Father hath disposed to Me, a Kingdom, that you may eat and drink at My table in My Kingdom" (St. Luke xxii. 28, 29). . . 494

30. THE ANGER OF MAN.—"Let every man be slow to anger. For the anger of man worketh not the justice of God" (St. James i. 19, 20). 499

31. THE ANGER OF MAN.—"The anger of man worketh not the justice of God" (St. James i. 20). 503

NOVEMBER.

1. THE HOUSE OF GOD.—"Blessed are they that dwell in Thy house, O Lord; they shall praise Thee for ever and ever" (Psalm lxxxiii. 5). 507

2. PRAYER FOR THE DEAD.—"It is a holy and wholesome thought to pray for the dead, that they may be loosed from sins" (2 Mach. xii. 46). 510

3. THE BEGINNING OF WISDOM.—"The fear of the Lord is the beginning of wisdom" (Psalm xc. 10). 515

4. APOSTOLICAL CONFIDENCE.—"I can do all things in Him Who strengtheneth me" (Philipp. iv. 13). 519

PAGE

5. HYPOCRISY.—" Dissemblers and crafty men provoke the wrath of God, neither shall they cry when they are bound" (Job xxxvi. 13). 523
6. THE LAW OF LIBERTY.—"He that hath looked into the perfect law of liberty and hath continued therein, not becoming a forgetful hearer, but a doer of the work, this man shall be blessed in his deed" (St. James i. 25). 527
7. THE FIRST BEATITUDE.—" Blessed are the poor in spirit, for theirs is the Kingdom of Heaven" (St. Matt. v. 3). 531
8. THE SECOND BEATITUDE. —" Blessed are the meek, for they shall possess the land" (St. Matt. v. 4). 535
9. THE THIRD BEATITUDE.—"Blessed are they that mourn, for they shall be comforted" (St. Matt. v. 5). 539
10. THE FOURTH BEATITUDE.—" Blessed are they that hunger and thirst after justice, for they shall have their fill" (St. Matt. v. 6). 544
11. THE FIFTH BEATITUDE.—" Blessed are the merciful, for they shall obtain mercy" (St. Matt. v. 7). 548
12. THE SIXTH BEATITUDE.—"Blessed are the clean of heart, for they shall see God" (St. Matt. v. 8). 552
13. THE SEVENTH BEATITUDE.—"Blessed are the peaceful (*pacifici*), for they shall be called the children of God" (St. Matt. v. 9). 557
14. THE EIGHTH BEATITUDE.—" Blessed are they that suffer persecution for justice' sake: for theirs is the Kingdom of Heaven" (St. Matt. v. 10). 561
15. PROGRESS BY GOD'S AID.—"Blessed is the man whose help is from Thee : in his heart he hath disposed to ascend by steps, in the vale of tears, in the place which he hath set" (Psalm lxxxiii. 6, 7). 564
16. THE SEARCHING OF JERUSALEM.—"And it shall come to pass at that time that I will search Jerusalem with lamps, and will visit upon the men that are settled on their lees, that say in their hearts, The Lord will not do good, nor will He do evil" (Sophonias i. 12). 569
17. GLORYING IN INFIRMITY.—"Gladly will I glory in my infirmities, that the power of Christ may dwell in me" (2 Cor. xii. 9). 573
18. THE WINNING OF PEACE. — "Turn away from evil, and do good, seek after peace and pursue it" (Psalm xxxiii. 15). 576
19. THE CROSS FOR ALL.—"And He said to all : If any man will come after Me, let him deny himself, and take up his cross daily, and follow Me" (St. Luke ix. 23). 579
20. FAITHFULNESS IN LITTLE THINGS.—"He that is faithful in that which is least, is faithful also in that which is greater; and he that is unjust in that which is little, is unjust also in that which is greater" (St. Luke xvi. 10). 583
21. OUR LADY AS THE DAWN.—"Who is she that cometh forth as the morning rising, fair as the moon, bright as the sun, terrible as an army set in array?" (Cant. vi. 9). 586
22. WASHING THE HEART.—"Wash thy heart from wickedness, O Jerusalem, that thou mayest be saved : how long shall hurtful thoughts abide in thee?" (Jerem. iv. 14). 590
23. INVITATION TO BLISS.—"Then shall the King say to them that shall be on His right hand: Come, ye blessed of My Father, possess you the Kingdom prepared for you from the foundation of the world (St. Matt. xxv. 34). 592

PAGE

24. Dismissal to Hell.—"Then He shall say to them also that shall be on His left hand : Depart from Me, you cursed, into everlasting fire " (St. Matt. xxv. 41). 597
25. Divine Wisdom.—"Wisdom will not enter into a malicious soul, nor dwell in a body subject to sins " (Wisdom i. 4). . . 600
26. Grounds of Encouragement.—"The Lord is my helper : I will not fear what man can do unto me " (Psalm cxvii. 6). . 604
27. Judgment of God.—"When I shall take a time, I shall judge justices " (Psalm lxxiv. 2). 608
28. The Evil of Hell.—"I will heap evils upon them, and will spend my arrows among them " (Deut. xxxii. 23). . . 611
29. Diligence of Holy Fear.—"He that feareth God neglecteth nothing" (Eccles. vii. 19). 614
30. The Tree of Life.—"She is a tree of life to them that lay hold on her ; and he that shall retain her is blessed" (Prov.iii.18). 618

December.

1. Preparation for Prayer.—"Before prayer prepare thy soul, and be not as a man that tempteth God " (Ecclus. xviii. 23). . 622
2. Labour for Salvation.—"Brethren, labour the more, that by good works you may make sure your calling and election. For, doing these things, you shall not sin at any time " (2 St. Peter i. 10). 625
3. The Successors of the Apostles.—"As arrows in the hand of the mighty, so the children of them that have been shaken " (Psalm cxxvi. 4). 629
4. Pains of Hell proportionate to Sins.—"As much as she hath glorified herself, and lived in delicacies, so much torment and sorrow give ye to her " (Apoc. xviii. 7). 634
5. How to gain Wisdom.—"If any of you want wisdom, let him ask of God, Who giveth to all men abundantly, and upbraideth not, and it shall be given him. But let him ask in faith, nothing wavering " (St. James i. 5, 6). 637
6. The Searchings of the Spirit.—"The Spirit searcheth all things, yea, the deep things of God " (1 Cor. ii. 10). . . 640
7. Acceptance of Persons.—"I will not accept the person of man, and I will not level God with man : for I know not how long I shall continue, and whether, after a while, my Maker may take me away " (Job xxxii. 21, 22). 643
8. Our Lady the House of God.—"Wisdom hath built herself a house, she hath hewn her out seven pillars " (Prov. ix. 1). 646
9. Trusting in Man.—"Cursed be the man that trusteth in man, and maketh flesh his arm, and whose heart departeth from the Lord " (Jerem. xvii. 5). 649
10. The Testimonies of God.—"I have been delighted in the way of Thy testimonies as in all riches " (Psalm cxviii. 14). . 653
11. Attachment to Ease.—"Moab hath been fruitful from his youth, and hath rested upon his lees, and hath not been poured out from vessel to vessel, nor hath gone into captivity ; therefore his taste hath remained in him, and his scent is not changed " (Jerem. xlviii. 11). 657

12. WASHING FROM INIQUITY. — "Wash me yet more from my iniquity, and cleanse me from my sin : for I know my iniquity, and my sin is always before me" (Psalm l. 4). 659

13. MEDITATING TRUTH. — "My mouth shall meditate truth, and my lips shall hate wickedness" (Prov. viii. 7). 664

14. SACRIFICING ALL FOR GOD.—"None of us liveth to himself, and no man dieth to himself : for whether we live, we live unto the Lord, or whether we die, we die unto the Lord ; therefore, whether we live, or whether we die, we are the Lord's" (Romans xiv. 7, 8). 668

15. SPIRITUAL RICHES.—"The riches of salvation, wisdom, and knowledge ; the fear of the Lord is his treasure" (Isaias xxxiii. 6). 671

16. THE THIEF AND THE WATCHERS.—"If thou shalt not watch, I will come to thee as a thief, and thou shalt not know at what hour I will come to thee" (Apoc. iii. 3). 674

17. SINNERS IN HELL.—"They are laid in Hell like sheep : death shall feed upon them" (Psalm xlviii. 15). 676

18. MARKS OF GOD'S CHILDREN. — "Whosoever are led by the Spirit of God, they are the sons of God" (Romans viii. 14). . 678

19. RESCUE FROM THE EVIL WAY.—"He who causeth a sinner to be converted from the error of his way, shall save his soul from death, and shall cover a multitude of sins" (St. James v. 20). . 682

20. THE FRUIT OF THE VIRGIN EARTH.—"Drop down dew, ye heavens, from above, and let the clouds rain the Just ; let the earth be opened, and bud forth a Saviour" (Isaias xlv. 8). . 686

21. BLESSINGS OF FAITH AND OF SIGHT.—"Blessed are they that have not seen, and have believed" (St. John xx. 29). . . 689

22. THE EMBRACE OF PRAYER.—"Who shall give Thee to me for my Brother, sucking the breasts of my Mother, that I may find Thee without, and kiss Thee, and now no man may despise me?" (Cant. viii. 1). 693

23. GOING TO THE THRONE OF GRACE.—"Let us go with confidence to the throne of grace, that we may obtain mercy, and find grace in seasonable aid" (Hebrews iv. 16). 697

24. FAITHFULNESS IN FRIENDSHIP.—"Keep fidelity with a friend in his poverty, that in his prosperity also thou mayest rejoice" (Ecclus. xxii. 28). 700

25. THE APPEARING OF THE GRACE OF GOD.—"For the grace of God our Saviour hath appeared to all men, instructing us that, denying ungodliness and worldly desires, we should live soberly, justly, and godly in this world, looking for the blessed hope and coming of the glory of the great God and our Saviour Jesus Christ" (Titus ii. 11—13). 703

26. THE BOWELS OF MERCY.—"Put ye on, therefore, as the elect of God, holy and beloved, the bowels of mercy, benignity, humility, modesty, patience ; bearing with one another, and forgiving one another, if any have a complaint against another : even as the Lord hath forgiven you, so you also" (Coloss. iii. 12, 13). 707

27. THE EAGLE OF CONTEMPLATION.—"Will the eagle mount up at thy command, and make her nest in high places? She abideth among the rocks, and dwelleth among cragged flints

PAGE

and stony hills where there is no access. From whence she looketh for the prey, and her eyes behold afar off. Her young ones shall suck up blood, and wheresoever the carcass shall be, she is immediately there " (Job xxxix. 27—30). . . . 710

28. THE CUP TO BE DRUNK.—" Behold, they whose judgment was not to drink of the cup shall certainly drink ; and shalt thou come off as innocent ? Thou shalt not come off as innocent, but drinking thou shalt drink " (Jerem. xlix. 12). 716

29. THE OLD LAW AND THE NEW.—" God, Who at sundry times and in divers manners spoke in times past to the fathers by the Prophets, last of all in these days hath spoken to us by His Son, Whom He hath appointed Heir of all things, by Whom also He made the world " (Hebrews i. 1). 718

30. SACRED SCRIPTURE.—" Thy testimonies are wonderful ; therefore my soul hath sought them " (Psalm cxviii. 129). . . 722

31. THE MOST BLESSED TRINITY.—" For of Him, and by Him, and in Him are all things ; to Him be glory for ever. Amen " (Romans xi. 36). 726

JULY.

FIRST DAY.

Love not sleep, lest poverty oppress thee ; open thy eyes, and be filled with bread (Prov. xx. 13).

I. Consider first, God's jealous care lest His people should love sleep in the wilderness. Therefore He chose to supply them with manna, not only day by day, but at so early an hour that it melted in the first rays of the sun ; so that whoever was not careful to go out to gather it in the morning dawn, was obliged to go fasting that day. And why was this? That we may understand how in this earthly pilgrimage of ours, we must not allow ourselves to be overpowered by sleep, but rather to shake it off early, so as to provide ourselves with the heavenly refreshment that we need in so toilsome a journey. This refreshment is that which we receive in prayer, a thing which is always pleasing to God, but especially when addressed to Him very early in the morning. And lest any one should think that this interpretation is more pious than sound, listen to what was specially appointed by God concerning the manna which He gave to His people : "That which could not be destroyed by fire, being warmed with a little sunbeam, presently melted away ; that it might be known to all that we ought to prevent the sun to bless Thee, and adore Thee at the dawning of the light."[1] And it would seem that the Wise Man meant to allude to this good pleasure of God in this passage also, where he said : "Love not sleep, lest poverty oppress thee ; open thy eyes, and be filled with bread." For, doubtless, he well knew what we see happen every day, namely, that he who does not rise early to pray, either does not pray at all, or else very carelessly. What care dost thou take in this matter? Whenever sleep tempts thee to stay in bed too long,

[1] Wisdom xvi. 27, 28.

B

repeat to thyself these words of the Wise Man, and thou wilt
find them urge thee to spring at once from that couch where
heaviness, not necessity, keeps thee lying : "The words of the
wise are as goads, and as nails deeply fastened in ;" they are
"goads" to spur thee on to good, and "nails" to withhold
thee from evil.

II. Consider secondly, that when the Wise Man says,
"Love not sleep," it is understood that he does not forbid
a proper, but an excessive amount of sleep. And this he has
every reason to forbid ; for sleep brings idleness, idleness
sloth, sloth negligence, and negligence poverty : and this is
a chain of evils which are so closely and inseparably linked
together that the Wise Man, for the sake of greater brevity,
passes from the first to the last, and says at once : "Love not
sleep, lest poverty oppress thee." And what is the poverty
which it brings in the case we are considering ? It is a
miserable poverty of soul—the worst of all. For if thou
sufferest thyself to be overcome by sleep in the morning, either
thou neglectest thy usual prayer altogether, or at any rate it
is so careless and so scanty that thou gainest thence no strength
for doing well, in which, after all, true riches consist. Observe,
therefore, that it is not said "lest hunger," but "lest poverty
oppress thee," because the man who is not nourished with
bodily food feels oppressed with hunger, but not so he who is
not nourished with spiritual food, which is that which is
especially spoken of here. Such a man rather loses the sense
of hunger, but finds himself oppressed with poverty : for when
he desires to do any good thing, he has no fund to draw
upon ; he yields to every suggestion of the devil, however
slight ; he cannot endure a small injury, he cannot bear a
small misfortune, he cannot withstand any of the trials which
are of daily occurrence : "I am smitten as grass, and my
heart is withered ; because I forgot to eat my bread."[1] Who
is there that forgets to take bodily food? The less any one
takes, the more he remembers that he has to take it, because
he is hungry in proportion. It is spiritual food which men
forget to take, because if they go long without doing so, they
lose the sense of hunger. And so, at last, they are without
vigour, like dry grass : this, indeed, is the poverty which
oppresses, which robs thee of all strength.

III. Consider thirdly, how suitable are the words which
the Wise Man adds: "Open thy eyes, and be filled with

[1] Psalm ci. 5.

bread." Thou canst open the eyes both of thy body and of
thy soul: of thy body, by driving sleep from them; of thy
mind, by fixing them on the contemplation of the truths which
thou hast proposed for thy meditation the evening before.
And by so doing, enjoy the bread with which Jesus feeds
souls in the wilderness of this world: "Be filled with bread."
This bread is of two kinds: one, which nourishes the understanding, and the other the will. The first consists in the
lights which a man receives directly from God in prayer, or
which he seeks for himself; the second in the affections he
forms: and who can say which of the two is sweeter? Do
not, therefore, let the word "bread" take away thy relish,
for it is the Heavenly Bread which is spoken of. Dost thou
suppose that it is bread such as ours, without savour or
sweetness? Surely not: on the contrary, it is that of which
the manna was but a figure, and which, therefore, far surpasses
it in the multiplicity of flavours which it contains in itself:
"Thou gavest them bread from Heaven, having in it all that
is delicious."[1] Do not think, then, that when the Wise Man
here bids thee "be filled with bread," he means thee to be
filled with dry bread; for he well knew that in the Psalms
prayer is compared to a banquet: "Let the just feast before
God."[2] He means that thou art to be filled with the delight
which the soul enjoys both in knowing and in loving its God.
These are not, like worldly delights, vain and deceitful, but
substantial, and therefore they are expressed by the name of
bread, rather than of any other kind of food, to denote the
especial strength which they impart to the soul: "Bread
strengthens man's heart."[3] And what food can the world
possibly offer to compare with this bread on which man feeds
when conversing with God? Earthly food gives a superficial
pleasure which only touches the palate; but this food imparts
a deep delight which reaches to the heart: "Thy words were
found, and I did eat them, and Thy word was to me a joy
and gladness of my heart."[4] A "joy," inasmuch as it rejoices
the understanding; and "gladness," inasmuch as it gives
delight to the will, for these are the two faculties which are
comprehended in the one word "heart." And dost thou not
know how poisonous all the dainties of the world are? They
are like unwholesome food, which the more it pleases thee
by the small amount of sweetness it causes so long as it rests

[1] Wisdom xvi. 20.
[2] Psalm lxvii. 4.   [3] Psalm ciii. 15.   [4] Jerem. xv. 16.

on the palate, the more it troubles thee by the great bitterness it afterwards produces in the stomach. Whereas heavenly food is both delicious and wholesome; and, therefore, this is another reason for calling it bread, to show that it is a safe and salutary food, which is suitable even for the sick. And besides, it is known to every one that, in Scriptural language, the word "bread" is not confined to one particular kind of food, as in ours, but that it includes all. And therefore it is here used to signify both the intellectual lights and the affections on which thou feedest in the blessed banquet here spoken of. Leave, therefore, to the world, all its dainties, to be offered in abundance to those who desire them. Do thou choose this bread, which God gives, and be filled with it: never, however, wilt thou be satisfied with it, for thou wilt always desire it more and more.

SECOND DAY.

THE VISITATION OF OUR LADY.

Recover thy neighbour according to thy power, and take heed to thyself that thou fall not (Ecclus. xxix. 26).

I. Consider first, the many obligations which bind thee to the God Who elected thee to glory from all eternity, created, and preserved thee, allowed thee to be born in the bosom of Christendom; Who has waited for thy penitence, extended forgiveness to thee, and gone so far as to die for thee on the bitter Cross. Unless thou hast the heart of a wild beast, thou oughtest by every right to be consumed with the desire of making Him some grateful return. But what canst thou do? He is rich in all things, in need of nothing, great and glorious. In what way is it possible to show thy gratitude? In this way, by doing for Him what the Blessed Virgin did on this day, that is, by winning to Him as many souls as possible. For thou oughtest to think that being so rich in Himself, He has made over to those who are most poor and needy the claim which He had a right to make on thee. In an especial manner has He done this on behalf of those souls who are in danger of being lost for want of some one to help them. If then thou wouldst have Him declare Himself satisfied, do for the servants what it is out of thy power to do for the Master. This is the example which Mary gave thee on this her feast. As soon as she knew herself so highly favoured as to be raised to the dignity of Mother of God, what did she do in return? Did she seclude herself in her chamber to sing canticles of praise? Not at all: she immediately crossed the hills of Juda to co-operate with her Divine Son in saving souls. She went to visit her cousin Elizabeth—not out of ceremony or merely to congratulate her, or from an idle curiosity to see the truth of what the Archangel had told her, but to take the opportunity of giving back to God the little Forerunner of whom He had been robbed by the great thief of Hell. If thou art a true child of Mary, show it by treading in her footsteps. Think, therefore, that on this day she says to thee with her own lips these beautiful words of the Preacher, in which he gives thee not merely the order, but the rule of what thou art bound to do: "Recover thy neighbour according to thy power, and take heed to thyself that thou fall not."

II. Consider secondly, who it is that thou art to "recover." It is "thy neighbour," that is to say, the neighbour whom thou art also bound to love as thyself: "Thou shalt love thy neighbour as thyself." For if there were no other motive to incite thee to succour him, is not this enough? Thou art bound by the law of charity to feel thy neighbour's ills as thine own, to "weep with them that weep."[1] But those which thou shouldst feel most keenly are his spiritual ills, because, on the one hand, they are those which are the greatest that can affect him, and on the other, they are those which trouble him the least, and against which he is least on his guard. He allows himself to be taken straight to Hell, like a miserable slave, without making the slightest resistance: "My young men are gone into captivity;"[2] not "are led," but "are gone." All the more need is there then of some one to hasten to avert their ruin. A sick person helps himself by calling in a physician to cure him, a hungry man helps himself by appealing to some one to give him food, a thirsty man by asking for drink, a man in need of clothing by finding some one to cover him from the cold; whereas a sinner, not only does not help himself by finding some one to ransom him from his wretched captivity, but very often disdains such aid: "But they have thought to cast away my price."[3] If then

[1] Romans xii. 15.   [2] Lament. i. 18.   [3] Psalm lxi. 5.

thou art bound to aid thy neighbour even in those corporal necessities which he endeavours to relieve himself, how much more in the spiritual ones of which he does not feel the importance.

III. Consider thirdly, that if thou art bound to "recover" thy neighbour, then art thou bound to recover him out of the hands of one who has seized upon him. And who is it that has done this? It is the devil. He it is who has dared to make him a slave. Is it right then to allow such a robber to keep with impunity what he has wickedly got into his possession? Justice and charity alike forbid it: justice, because it is not right that this insolent robber should thus daily defy the God Who cast him down to Hell, as though his power to empty Heaven were greater than that of Christ Himself to fill it; and charity, because he is not a common robber, who enslaves souls from avarice, like the corsairs of Barbary. He does so out of the animosity, fury, and undying hatred which he bears towards them, so that thou must believe that if he makes slaves of them, it is in order to keep them in everlasting torments: "His heart shall be set to destroy."[1] If therefore thou art moved to compassion on seeing thy neighbour carried away on board ship as a slave to Algiers, where, after all, it is possible to effect his ransom by a sum of money, how is it that thou art not moved to compassion when thou seest him going as a slave into Hell, from whence there is no possibility of ransoming him? "There is no one to redeem."[2]

IV. Consider fourthly, that if thou art to recover thy neighbour, thou art to do so for One Who has lost him. Who is this? It is Jesus. Oh, how much has He done to win back for Himself the souls which thou seest now in the possession of His enemy! He came down from Heaven to earth, He gave His labour, His sweat, and all His Blood for them, and after all He has to see them from His Cross going away into perdition! It is this which most of all should move thee to go to their aid, the thought that these souls which thou savest, debased and abominable as they are, are the souls for which the Son of God died upon the Cross: "For whom Christ hath died."[3] See then how noble a work thou performest when thou recoverest thy neighbour from the diabolical slavery which holds him captive! Thou co-operatest with Jesus Christ in the redemption of mankind, which is the

[1] Isaias x. 7.　　[2] Psalm vii. 3.　　[3] 1 Cor. viii. 11.

greatest work which has ever been done in the world ; thou
art His companion, His coadjutor. How is it possible then
adequately to express the amount of grace which thou wilt
thus gain ? "We are God's coadjutors."[1]

V. Consider fifthly, that thou mayest be as much deterred
by the thought of thy own weakness from this work of
ransoming thy neighbour from the hands of his enemy, as
thou art incited to undertake it by the motives we have been
dwelling upon, as though the requisite ability were wanting
in thee. But see how, in order to cut off so cold an excuse,
the same voice which bids thee "recover thy neighbour,"
immediately adds, "according to thy power." Thou art not
able to denounce sinners from the pulpit like many zealous
preachers, nor to go into the streets and woods in search of
them : but what of that? Do what thou art able to do
according to thy state of life, thy knowledge, and thy strength.
But what is there that thou wilt not be able to do if only thou
hast true zeal? For zeal is love, and love is very ingenious
in doing good. See how ingenious it was in the Blessed
Virgin who, under the appearance of a common act of courtesy,
so dexterously found a way of delivering a soul from sin.
It is not only by means of the loud thunder of sermons that
souls are saved, they are saved also by a very faint whisper :
"I will whistle for them, and I will gather them together ;"[2]
they are saved by private admonitions, by individual rebukes,
by alms given in time to preserve them from evil, by prayer
and penance and tears, by the offering of daily sacrifices, and,
if nothing else is possible, by good example. All that is
wanting is a real determination to work "according to thy
strength, that is, in a manly way, thinking, meditating, studying the matter, and then, no matter who thou art, thou wilt
be able to do great good to thy neighbour in a short time :
"For the Kingdom of God is not in speech, but in power."[3]

VI. Consider sixthly, that at the same time that thou art
told to labour as much as possible for the salvation of thy
neighbour, there is added this loving admonition, to take care
meanwhile not to lose thyself : "And take heed to thyself
that thou fall not." Who was so safe from all danger of sin
as the Blessed Virgin Mary, who was impeccable? And yet
remark with what care, what speed, what anxiety she travelled
through that hill country of Juda, as though even she feared
the dangers of the way. She "went with haste," although

[1] 1 Cor. iii. 9.   [2] Zach. x. 8.   [3] 1 Cor, iv. 20.

her defence against robbers was the God she bore in her womb. And what should thy care be when thou art so inclined to evil? If thou wouldst really devote thyself to the salvation of thy neighbour, thou wilt often have to drag him out of pits so deep and slippery, that there is great risk merely in approaching them. Listen therefore to these words, "Take heed to thyself that thou fall not," and the word translated "fall" is not *cadas*, which might also refer to the act of one who purposely threw himself in, but *incidas*, which can only refer to one who falls indeed, but against his will. It is not enough to approach the pit with a sincere intention of helping others: it is necessary also to proceed with much care and wise precautions, lest when thou stretchest forth thy hand to pull thy neighbour out, his strength to drag thee in should be the greater: "They shall be turned to thee, and thou shalt not be turned to them."[1]

THIRD DAY.

Every best gift and every perfect gift is from above, coming from the Father of lights, with Whom there is no change nor shadow of alteration (St. James i. 17).

I. Consider first, that what St. James here has especially in view is to incite thee to ask of God everything which thou hast most need of, since it is from Him that it all comes. "Every best gift and every perfect gift is from above." All good that can come to thee from God comes from Him, either as He is the Author of nature or the Author of grace. If thou regardest Him as the Author of nature, it is from Him that "every best gift" (*datum*) comes. If thou regardest Him as the Author of grace, it is from Him that "every perfect gift" (*donum*) comes. The good which is of nature is called *datum*, because, although it also was a gift in principle, it nevertheless has some proportions to the person who receives it. The good which is of grace is called *donum*, because there is no proportion of any kind whatever; it is altogether gratuitous: "Otherwise grace is no more grace."[2] The adjective "best" is here applied to the former (*datum optimum*), because there are three degrees of natural good:

[1] Jerem. xv. 19.   [2] Romans xi. 6.

existence, life, and understanding. Existence is good, but we have this in common with stones; life is better, but we have it in common with animals; understanding is best, and is proper to man. It is this which thou shouldst ask of God, since it is the most excellent gift which He can give thee as the Author of nature. Ask Him to give thee a right under-standing of things: "Give me wisdom," or "understanding" (*intelligentiam*), for good works depend in great measure on good understanding: "He would not understand that he might do well."[1] And so thou seest that it is of this "best" only that mention is here made: "Every best gift." Of those gifts which are good and better nothing is said; because, as to the former, namely, existence, it is not a thing which can be asked; and as to the latter, namely, life, it is not a thing which ought to be asked. Whether thy life is to be longer or shorter should be left entirely to God. To the second kind of gift the adjective "perfect" is applied (*donum perfectum*), because the good which we receive from God as the Author of grace contains in itself four degrees: elevation, redemption, justification, and glorification. You see at once how good it was for us to be raised to a supernatural state. Still better was redemption, for what would it avail us after our fall to be raised to such a state if Christ had not saved us by His own Blood? Justification is best, for what will it avail us to be redeemed by Christ if we are not just? Glorifi-cation is perfect, for of what avail is it to be just if we are not crowned through our perseverance? This it is which thou shouldst especially ask of God without ceasing, final perse-verance, for it is the most excellent gift that He can give thee as the Author of grace. Elevation and redemption cannot be asked, and the Apostle takes for granted that thou already hast the gift of justification, when he urges thee to ask for the last and perfect gift; for how canst thou ask of God that thou mayest persevere to the end in the state of justice unless thou art already in that state? And this shows then the nature of the good which thou art to ask of God; there is the "best gift," which is to understand rightly everything which can help thee to do well; and there is the "perfect gift," which is to persevere in well-doing to the end, by a continual increase of grace.

II. Consider secondly, that "every good gift and every perfect gift" which is possessed by every man on earth in

[1] Psalm xxxv. 4.

very truth comes from God alone: it "is from above." For what canst thou do of thyself? Nothing whatever. What great need, then, thou hast to ask everything of God! Thou art bound to ask the "best gift," that is, understanding, because, although God has given thee the faculty of the intellect, He has reserved to Himself the act of understanding: "The inspiration of the Almighty giveth understanding."[1] Still more art thou bound to ask for the "perfect gift," that is, final perseverance, because, although in giving thee the grace to be just, God has given thee the power to persevere, that is to say, justifying grace, He has not given thee the act, that is to say, perseverance. This, as St. Augustine shows in his treatise, *De bono perseverantiæ*, requires another grace distinct from justifying grace, that, I mean, by which God leads thee gently step by step till death, taking out of thy way all the obstacles which might make thee fall from the exalted state thou art in, and inciting thee to good, encouraging, strengthening, and protecting thee: so that thou mayest see that this is a grace containing in itself many others, one which it is impossible to merit, at least by condignity, like faith; but yet which can be obtained by continual prayer, the object of which is to gain from the mercy of God that which we cannot in any way claim of His justice: "For it is not for our justifications that we present our prayers before Thy face, but for the multitude of Thy tender mercies."[2] Thou seest, then, how truly both "every best" and "every perfect gift" is "from above, coming down from the Father of lights." And so God is here especially called "Father of lights," because it belongs to Him, as the Father of natural lights, to give the "best gift," that of understanding; and it belongs to Him, as the Father of supernatural lights, to give the "perfect gift," that of perseverance, for this perseverance comes from the grace which especially consists in good thoughts. And as the sun not only gives light but heat, and not only heat but strength, so, too, does God (a far better "Father of lights" than the sun) by His most holy grace. Not only does He enlighten thy understanding, but He inflames thy will; not only does He influence thy will, but He gives thee strength, so that thou mayest both know, and choose, and have power easily to do that good which thou art bound to do to the end of thy life; and this is in truth the "perfect gift."

[1] Job xxxii. 8. [2] Daniel ix. 18.

III. Consider thirdly, that having seen the nature of that which thou hast to ask of God, and also the necessity thou art under of asking it, it remains for thee to see how easily thou mayest obtain what thou askest, so as to be the more encouraged to do so. What in the world is easier than to obtain light from the sun, "the Father of lights"? And this, thou hast heard, is the title of God. Indeed, He is a far more excellent Sun than that which thou beholdest with thy bodily eyes. For although that material sun never suffers any change in itself, but is always alike the source of unexhausted light, yet is it subject to vicissitudes in its effects: now it rises, now it sets, at one time it is farther from us, at another nearer to us, sometimes higher, sometimes lower in the heavens; and so thou art not always able to receive from it in an equal measure the light thou desirest. Not so the Divine Sun. Not only does He know no change in Himself, for He is "the self-same,"[1] but neither does He know any alteration. This is the meaning of these words, "With Whom there is no change nor shadow of alteration," that is to say, with Whom not only is there, as in the material sun, no change, but neither is there the shadow of alteration, of which there is a great deal in that sun, which may, therefore, be called "another, yet the same," but not "the self-same." It is true that in the Divine Sun also there is a "shadow," and that, too, very often; but it is not a "shadow of alteration," not a diminution of light, which is occasioned, as in the material sun, by the variations to which it is subject during its course, but one which arises solely from the clouds which pass before it, which are our frequent acts of ingratitude: "Thou hast set a cloud before thee, that our prayer may not pass through."[2] Thou seest, then, that the obstacle which intercepts the light comes, not from the Sun, but from thyself. It is thou who placest this cloud before thyself: "thou hast a cloud before thee," not "before Him," for not only is the shadow not "in Him," it is not even "with Him," it is "with thee." Take away the cloud by ceasing to be ungrateful to God for the benefits which thou art continually receiving from Him, and thou shalt see how easily thou wilt obtain everything.

IV. Consider fourthly, that it is said that "every best gift and every perfect gift is from above, coming down from the Father of lights." Remark the word "coming down," not

---

[1] Hebrews i. 12. [2] Lament. iii. 44.

"falling," but "coming down," for the good which is from Heaven does not fall at random, as foolish men think, but comes down with much deliberation : thus it descends gradually, which is the proper sense of the word. The rain falls from heaven, the light of the sun does not indeed fall in the same way as the rain, still it does fall, at least apparently ; it does not descend, because it does not come step by step, but all at once. Not so the Divine light, which comes down, like the angels who visited the sleeping Jacob in his famous vision, by a ladder ; for according as thou correspondest with the first favour granted thee by God, which is the first step, God will go on to grant thee a second and a third, and so on, by degrees. Thou wouldst, therefore, be greatly mistaken in thinking thou couldst receive it all at once : "Every best gift and every perfect gift is from above," but "coming down" (*descendens*). And from this thou shouldst draw the conclusion that there are two things which help thee to obtain from God the favours which thou askest in prayer : the first is not to be ungrateful for the favours He has granted thee, for this is placing a cloud before the Sun by thy own act ; the second is not only to be ungrateful for these favours, but to be actually grateful by showing positive correspondence with grace, for this is the ladder by which these favours come down, step by step.

FOURTH DAY.

ST. ELIZABETH.

My son, keep thy soul in meekness, and give it honour according to its desert (Ecclus. x. 31).

I. Consider first, what this honour is which is due to the soul. It is to see that it commands, instead of obeying. It is an honour justly due to it, since to this it was born, to command as a queen, not to obey as a slave : "Thy lust shall be under thee, and thou shalt have dominion over it."[1] See, therefore, how greatly thou wrongest it by making it continually subject to the flesh, and to the vilest works of the flesh, to gluttony, dissipation, sloth, and even lust. "My son," do

[1] Genesis iv. 7.

not act thus, "keep thy soul" in its right place, which is that
of a sovereign, and so "give it honour according to its desert."

II. Consider secondly, that the honour here spoken of is
interior honour. There is also that which is exterior, and this
consists in preferring the soul, as it merits, to everything which
is of less value than itself, that is to say, everything perishable,
for what is perishable comes to an end, and the soul is
immortal. Give it, therefore, "honour according to its desert,"
by valuing it more than the vain friendship of men, more than
reputation, more than worldly goods, more even than the life
of thy body which is so dear to thee. This is its "desert."
"All that a man hath he will give for his life."[1] How is it
possible, then, that any one should be so bold as to sell it to
his enemy so lightly? "My son," do not act thus, for thou
wouldst repent thy bargain : "keep thy soul" from him who
would deprive thee of it, as though it were a thing of no
value, and so "give it honour according to its desert."

III. Consider thirdly, that the greatest honour that can be
given to the soul is not to give it the sceptre of a queen, nor
to prefer it to all the transitory goods of the world ; for this is
an honour which every one, no matter how imperfect, is bound
to give it. There is another honour which belongs to the
more perfect to pay, namely, to make it enjoy God even in
this world : "And to Him my soul shall live."[2] This is the
end for which it was created : why, then, wouldst thou, at all
events, make it defer this enjoyment till a future life ? Rather
give it, even in the present life, as much as possible, by occupying thyself with prayer, by thinking of God, speaking of God,
conversing as much as possible with God : "We shall live in
His sight."[3] How high an honour wilt thou give thy soul by
doing this ! Moreover, it is an honour which will make it
easy for thee to keep up the other two : for the man who
converses much with God despises his senses, and so there is
no danger of his making his soul their slave ; and he who
converses much with God despises also whatever is perishable,
and so there is no danger of his ever preferring anything of
the sort to the value of his soul : therefore, "My son, keep thy
soul" by interior recollection, and do not let it roam along
every road like a mean servant, a "wandering daughter,"[4] but
"give it honour according to its desert."

IV. Consider fourthly, that the means which the Wise
Man recommends for easily attaining to all this is to be meek :

[1] Job ii. 4.   [2] Psalm xxi. 31.   [3] Osee vi. 3.   [4] Jerem. xxxi. 22.

" Keep thy soul in meekness." And this is not wonderful, for there is nothing which is a greater obstacle to the reverence due to thy soul than a tendency to anger. And the reason is, that anger disturbs the understanding, and when it is violent, clouds and darkens it : and how is it possible in such a state to value thy soul duly? This esteem of the soul is not acquired by means of what the exterior senses suggest, for these miserable senses will urge thee, on the contrary, to slight it ; it is acquired by following the dictates of the understanding. See, then, how necessary it is always to keep the understanding clear ! And this is done by meekness, which checks in time all possible ebullitions of anger. Hence this disposition, namely meekness, is the most requisite for hearing the truth—" Be meek to hear the word "[1]—because it is that which is the most suitable for knowing it ; so much so, that according to St. Augustine, those persons are meek who "do not contradict the word of truth," and this, because it is the meek who, more than all others, have an unclouded understanding, and so, other things being equal, they know the truth better than others. This, then, will be the effect of meekness : it will keep the understanding clear and calm, so as to judge of things rightly. And if this is so, see how important it is to study to restrain anger : " My son, keep thy soul in meekness." If thou givest way to anger, thy understanding is troubled, for there is nothing which troubles it so much : " My eye is troubled through indignation,"[2] and in this condition, not only wilt thou not give honour to thy soul "according to its desert," but thou wilt rob it of that honour, and expose it to the risk of the utmost disgrace.

V. Consider fifthly, what a wonderful example of everything which the Wise Man here enjoins is given thee by the glorious Saint whom we especially honour on this day. Who can sufficiently express how perfectly she always gave to her soul the honour which was its due? If thou considerest her as a virgin, how did she surpass others in giving it the first honour, making it absolutely mistress of the rebellious body ; as a wife she excelled as to the second honour, preferring it, though her state was that of royalty, surrounded with all that could flatter her, to everything perishable. Then, in the state of widowhood, how did she excel in paying her soul the third honour, making it her study to give it the enjoyment of God in a contemplation which was not only daily but continual.

[1] Ecclus. v. 13.   [2] Psalm vi. 8.

And all this she attained mainly by her wonderful meekness. This is the virtue for which she was most conspicuous, for not only did she possess it herself, but she had the gift of inspiring others with it, so great was the power she had of appeasing angry spirits. Think, therefore, that she is looking down on thee from Heaven, and saying to thee with the lips of a loving mother : " My son, keep thy soul in meekness," as I did, who am now rewarded by so great a glory, "and give it honour according to its desert."

FIFTH DAY.

Pierce Thou my flesh with Thy fear, for I am afraid of Thy judgments (Psalm cxviii. 120).

I. Consider first, what perhaps seems astonishing to thee, namely, that one who is afraid, and who even admits that he has known this fear for a long time, should ask of God the gift of fear : " Pierce Thou my flesh with Thy fear, for I am afraid of Thy judgments." But thou wilt cease to wonder at this if thou observest that not only ought we to ask of God what we do not possess, but also what we do possess, because we are at every moment in great danger of losing it unless He confirms our possession of it. Besides, what is it which David here asks? It is that the fear which he had in his soul might also pass into his body ; and so, not only did he ask for what he had, but for something which he had not. For indeed he had the superior part of himself well under subjection to God, but not so the inferior, or rather the animal part, which very often made fierce war upon him. Therefore he desired that the fear of God which he had in his heart might not stop there, but be communicated by a strong movement of the heart even to the body, so as to make it cold and insensible to those motions of repugnance and rebellion which belong to it. It was this, then, which he meant to ask when he said, " Pierce Thou my flesh with Thy fear "—the subjection of his concupiscence. This is the teaching of St. Augustine. Happy, indeed, wilt thou be if thou attainest so far. In any case, this should be thy desire ; and therefore thou shouldst continually pray to God that He would pierce with this holy fear thy thoughts, thy tongue, thy eyes, thy ears, thy whole self, so that the arrogance of thy senses may no longer trouble thee, at least to any great extent. A piercing which is material passes from the body to the soul; that which is spiritual from the soul to the body; and so in process of time holy men come to have even the flesh crucified: "They have crucified their flesh,"[1] making it to be dead, or at least mortified. But when is it that they attain to this? After they have first crucified their souls by making them obedient to God. Thou complainest that thy flesh becomes more and more masterful. How can it be otherwise? If thou dost not yet fear God, not even in thy soul, if thou art of an easy conscience, bold, arrogant, and careless as to thy true advantage, how canst thou possibly attain to fearing Him in the flesh, that part of man which is the last to be conquered? No one has a right to say to God, "Pierce Thou my flesh with Thy fear," if he is not able to assign truly, as a reason for obtaining that gift, the same which David pleaded: "For I am afraid of Thy judgments." It is not a grace which is bestowed on those who are beginners in the ways of God.

II. Consider secondly, that the meaning of Divine judgments in the Sacred Scriptures is frequently the commandments of God: "If they walk not in My judgments;"[2] "I have not declined from Thy judgments;"[3] "Thou hast despised all them that fall off from Thy judgments."[4] This being so, why was it that David so earnestly begged of God to repress the motions of the rebellious flesh? "Pierce Thou my flesh with Thy fear." Because, otherwise, he feared to transgress: "For I am afraid of Thy judgments," that is, "I am afraid of departing from Thy judgments." And we may believe that he left the word "departing" (*discedere*) to be understood, instead of expressing it, because the fear he felt made him shrink from doing so. And what is to be said of thee, who, on the contrary, art so ready to promise what it is so difficult to persevere in doing? The interior conflict of which he was conscious prevented such a one as David from making sure of obtaining this, and thou art ready to believe that it is within thy grasp. Oh, how greatly oughtest thou to fear a concupiscence which is so easily disordered as thine! So long as it exists thou art in continual danger of being, in the end, conquered by sin. And if that should happen, what would it avail thee to have fought against it with great courage,

----

[1] Galat. v. 24.
[2] Psalm lxxxviii. 31.
[3] Psalm cxviii. 102.
[4] Psalm cxviii. 118.

or even to have vanquished it, up to that moment? Thou wilt win the crown by perseverance alone. David overthrew the giant with stone cast from a sling, which struck him in the forehead; but it was neither the stone nor the sling that he hung up on the walls of the Temple, but the sword which belonged to the giant, because it was with it that he completed his victory over the enemy by cutting off his head.

III. Consider thirdly, that in the Scriptures, by the Divine judgments are also meant those inscrutable counsels of God by which He governs the universe: "Thy judgments are a great deep."[1] Some of these regard His mercy, others His justice. Of the former class are those imperceptible counsels by which God sometimes follows a sinner, whom He sees not only flying from Him, but even ill-using Him: "Saul, Saul, why persecutest thou Me?"[2] and of the latter, those by which He abandons him after the first sin, letting him go on from bad to worse: "How long wilt thou mourn for Saul, whom I have rejected from reigning over Israel?"[3] In the passage which we are meditating, the Psalmist certainly is not speaking of those judgments which belong to the mercy of God, for these are to be admired, not feared. He is speaking of those which belong to His justice, and therefore he declares to God how greatly he fears them: "For I am afraid of Thy judgments." Although, indeed, if we observe closely, David does not say absolutely that he fears the judgments of God, but rather that he fears because of those judgments; for he does not say, *Judicia tua timui*, as it was in some ancient readings, but *A judiciis tuis timui*, as is the right reading. He feared some unexpected attack from his senses which he might not be able to resist manfully, in consequence of which he might be allowed, by the mysterious counsels of God, to be lost. And certainly the danger thou, too, art in of falling into some very great sin does not threaten thee from the judgments of God, for they do not decree evil to any one; the utmost they do is to allow it; the danger is from thyself, who art so prone to wickedness. It is true that, because of those judgments, there is very great reason for thee to fear thyself, especially if thou art under the dominion of thy senses, for they frequently allow even the saints to fall very shamefully in many ways, but especially as to sensuality. Think of the falls of Victorinus the Hermit, Guarino, James, Macarius, and thou wilt be full of terror. All of these, indeed, had the grace to repent in the

---

[1] Psalm xxxv. 7. [2] Acts ix. 4. [3] 1 Kings xvi. 1.

end, but how many are there who have not had that grace! How is it, then, that thou dost not cry to God every day in great fear, "Pierce Thou my flesh with Thy fear, for I am afraid of Thy judgments."

IV. Consider fourthly, that by the Divine judgments in Scripture are meant those accurate judgments which God will form of every one of us on our leaving this life: "The Lord shall be known when He executeth judgments."[1] Who can express the rigour with which God will proceed in this matter, not passing over a single thought, word, or work, not the very least, but examining it to see if it is in accordance with the laws of right? And thinking of this, too, David asks of God to subdue under him once for all the motions of that concupiscence which is so prone to evil: "Pierce my flesh with Thy fear," for knowing the severity with which everything will then be judged, he fears to be led astray into some irregular gratification, which after having been thought little of by him as a fault into which he was surprised, should be judged to be deliberate at God's tribunal: "I am afraid of Thy judgments," —of what? of every thought, word, and work, however small— "I feared all my works, knowing that Thou didst not spare the offender."[2] Now, then, what can be thought of those who have so bold a conscience as to be confident of repelling, and completely repelling, every evil suggestion, even the strongest? For such persons, most of all, it is useful to meditate on the severity of these Divine judgments of which we are speaking, in order that they may no longer have so dangerously easy a conscience. It is quite certain that poison can never be fatal to the body till it touches the heart, neither can an evil suggestion be fatal to the soul so long as the will does not consent to it. But then, just as poison reaches the heart more quickly in those animals which have large than in those which have small veins, so does an evil suggestion more quickly gain consent in those men whose conscience is what is called large. True safety is not in presumption, but in fear. And nothing is more calculated to inspire this in the case we are considering than pondering on those rigorous judgments of God which await us at our death. The man who keeps these always before his eyes will not only very easily succeed in avoiding evil in time of temptation, but he will also do that which is good: "I have kept the ways of the Lord," which is doing good, "and have not done wickedly against my God,"

[1] Psalm ix. 17.   [2] Job ix. 28.

which is avoiding evil, "for" (and this is the reason) "all His judgments are in my sight."[1]

V. Consider fifthly, that although it has been so often said to thee here that thou oughtest most fervently to beg of God this mastery over the flesh which is so necessary—"Pierce Thou my flesh with Thy fear "—yet thou art not to conclude from this that the utmost possible co-operation on thy part is not requisite for obtaining it, as though everything came from God. Everything does indeed come from God, but not exclusively; it must also come from thyself. Dost thou not beg of God every day to give thee thy daily bread? And yet thou, on thy part, dost not leave off sowing, reaping, and using all the means which most conduce to the providing it. So, too, shouldst thou beg of God to crucify thy rebellious flesh—"Pierce Thou my flesh with Thy fear;"—and every day do thy utmost to crucify it thyself. In this King David is thy example, who did not leave to God this holy work of piercing the flesh in such a way as not also to take the hammer in his own hands, and to afflict his body now with watchings which he says he anticipated—"My eyes prevented the watches "[2]—now with hair-shirts, and ashes, and fastings, and other austerities, which he practised so long that they wrought a visible alteration: "My knees are weakened through fasting, and my flesh is changed for oil."[3]

SIXTH DAY.

Behold, I stand at the gate and knock. If any man shall hear My voice, and open to Me the door, I will come in to him, and will sup with him, and he with Me (Apoc. iii. 20).

I. Consider first, who this great Person is Who says: "Behold, I stand." It is the King of Glory. And yet, what is He doing? He is standing at the door of a sinner. I say of a sinner, because if He had come to visit a just man, He would be, not at the door, but in the house. Is not this a very great wonder? When was it ever heard that a prince went uninvited, to the house of so base, so abject, so vile a man as this miserable sinner is in respect of God? And yet it is certain that God is standing there uninvited; for if He

[1] Psalm xvii. 22.   [2] Psalm lxxvi. 5.   [3] Psalm cviii. 24.

had been invited He would, at least, have found the door open, and there would have been no need to say, "Behold, I stand at the gate and knock." And then, when a prince does choose to go to any such house, he first sends messengers, officers, and servants to put everything in proper order for the reception of a person of his dignity; and at last he comes in person. Not so God. He stands knocking all alone: "I stand;" not any others, but "I," and certainly He stands there without having sent any messenger beforehand. Had He done so, it is not possible that He would be obliged to knock without even being sure of gaining admittance. But He says, "I stand and knock," like a man still waiting for an answer. Lastly, even supposing there was a prince willing to come and knock himself, he would not choose to stay so long knocking. If the door were not immediately opened, he would turn and go away in anger. Not so God. He implies that He stood there knocking for a long time: "Behold, I stand at the gate and knock." He would not say "Behold," if He had but just come. And then He adds, "I stand." If He were sitting, or walking, or occupying Himself in any way to beguile the tedious time of waiting, we might better understand it; but he says, "I stand." There He stands on His feet, unmoved, unwearied, though with great inconvenience, and even great disgrace in the eyes of those who see Him waiting on such a threshold. Such is the unspeakable excess of the love of God in order to gain admittance into the heart even of one who is a rebel against Him. Do thou behold this excess of love with a very bewilderment of wonder, and think how base thy conduct has been if thou hast ever made this great God wait in this shameful manner, in order to gain admittance into thy heart by His exciting grace: "Behold, I stand at the gate and knock."

II. Consider secondly, that in order to be admitted into a house that is locked, persons sometimes call without knocking, sometimes knock without calling, and sometimes both call and knock: but they always knock before calling, in order to draw attention to their voice when they call. So also does God. First He knocks: "I stand at the gate and knock;" afterwards He speaks, and therefore He adds: "If any man shall hear My voice." His call is undoubtedly the inspiration of grace; but what is the knocking? It is the remorse which He arouses in the conscience. Thou knowest that the sound of knocking is more unpleasant than that of a voice calling; so too is this remorse: it is literally a knocking at the heart, which forcibly recalls to the sinner the danger of the miserable state he is in, and the object of which is to rouse him to listen to the voice of his Lord which follows it, and which sweetly invites him to open his heart to feel compunction for his sins, to confess them, to declare everything, to resolve firmly to change his way of life. If thou askest why God acts in this manner, when He could, without all this trouble, enter at once and take possession even of a rebellious heart, the only answer to be given is that He does so because He chooses to do so. It is not His pleasure to take forcible possession: "With great favour," or rather "reverence" (*reverentia*), "Thou disposest of us." Thou art master of thy free-will; He leaves it inviolate, that so He may receive from thee a reception which will do Him honour. Moreover, when He comes to visit thee, is it not to bestow on thee a very great benefit? And how can that be done by force? Benefits are not conferred on unwilling persons. The officer of justice who comes to inflict punishment forces his way, breaks open the door like a thunder-bolt if thou dost not open to him; but the benefactor who comes to give thee a rich present would have thee open the door to him gladly by thy own choice, as thou openest to admit the sun. However this may be, God has done enough and far more than enough to gain admittance; if He does not gain it, certainly there is nothing more to be done on His side. Thou canst not complain of Him by saying that He is far from thee: "The Lord is far from the wicked;"[1] if He is far from thee, it is by thy choice; thou hast but to open the door, and He would be very near. Listen to Him telling thee that He is on the very threshold: "I stand at the gate;" not "near," but "at the gate," so close is He to thee. He is, indeed, far from thee as to justifying grace, but He is near thee as to the desire which He has of imparting it to thee; He is near thee by motions, by invitations, by calls, all of which are the grace by which He arouses thee.

III. Consider thirdly, that in order to show the entire freedom which He leaves thee, He says in express terms: "If any man shall hear My voice and open to Me the door, I will come in to him." It is not enough to hear, the door must be opened. But why does He say: "If any man shall

---

[1] Prov. xv. 29.

hear"? The power of hearing is not like that of opening.
The hand is free to open or not to open, at pleasure; but the
ear is not free to hear or not to hear. It is what is called an
involuntary power. Granted: but what does that prove?
Thou knowest the saying: "None so deaf as those who will
not hear." "Who is deaf but he to whom I have sent My
messengers?"[1] And observe that when some one makes a
noise at the door of our house, it is impossible not to hear it
at first. But then we can at our pleasure either pay attention,
so as to hear better who it is, or pay none; either go to the
door or not; either keep quiet or drown the sound we have
heard by making a greater. All this applies to the present
case. When, for example, God calls thee to leave the world,
in which thou art living in a state of almost continual sin, it
is impossible that thou shouldst not hear His voice the first
time; but it rests with thee to pay the more attention in thy
understanding in order to hear better what it is that God
requires of thee: "I will hear what the Lord God will speak
in me."[2] It rests with thee to draw near to the door of thy
heart by the application of thy affections, by retirement,
recollection, detachment, not only exterior but interior, from
those inmates of the house who distract thee. "Approach
thou rather, and hear,"[3] as it was said to Moses. It rests
with thee to keep silence and to quit for a time the other
affairs which keep thee so constantly occupied: "To wait
with silence for the salvation of God."[4] But if thou doest
none of these things, if thou dost not attend, or approach,
but on the contrary purposely makest a noise to avoid hearing;
if, as soon as thou hast heard God's first call, thou hastenest
to the company of thy friends who raise their voices loudly,
jesting, talking and laughing, in order to drive out of thy
head what they call glowing fancies, whose fault is it if thou
hearest no more? "They would not hearken," this is the
first fault; "and they turned away the shoulder to depart,"
this is the second; "and they stopped their ears not to hear,"[5]
this is the third. Observe, therefore, that our Lord does not
say, "If any man shall hear My knock," but "My voice;"
because it is more difficult not to hear a knock than a voice.
It is difficult not to listen to remorse of conscience, although
some persons do at last become deaf even to it. But it is
easier not to hear a voice, which is in its nature a much gentler

---

[1] Isaias xlii. 19.   [2] Psalm lxxxiv. 9.   [3] Deut. v. 27.
[4] Lament. iii. 26.   [5] Zach. vii. 11.

sound. And therefore it is possible that a Divine inspiration may come without observation : "Now, there was a word spoken to me in private, and my ears by stealth as it were received the veins of its whisper."[1] Do thou attend carefully to all that God requires of thee: "Speak, Lord, for Thy servant heareth."[2] Willingness to hear is the preparatory disposition for willingness to open : it is an inchoate consent. Do not let Him be wearied by waiting in vain ; because one who desires to enter a house does not always knock, or always call, but does so at intervals of various length, at one time knocking or calling more, at another less loudly ; there is no rule in the matter. And if our Lord is so good and gracious as never altogether to go away from the door of thy heart, even though He is so unworthily treated, yet He will knock less frequently and call in a less audible voice.

And here we will leave the sinner who remains in his sinfulness, although God by His preventing grace has the will to render him not only penitent, but to lead him on to further proficiency and even to perfection, as will be shown in the explanation of the remaining portion, which could not well be included in this meditation.

SEVENTH DAY.

If any man shall hear My voice, and open to Me the door, I will come in to him and sup with him, and he with Me (Apoc. iii. 20).

I. Consider first, that even when the resolution has been made to open the door to Him Who calls—which is the state in which we left the sinner at the close of the last meditation —there still remain some difficulties to be overcome before the act is accomplished. He has to put himself to inconvenience, to rouse himself, to make haste to shake off that slowness in well-doing which is natural to him. But who would not gladly overcome them for the sake of the great result to be attained by doing so, which is to receive such a Guest as God in the house? "If any man will open to Me the door, I will come in to him." What a wonderful word is that *Mihi!* The greatest difficulty, then, is at the door

[1] Job iv. 12.   [2] 1 Kings iii. 10.

which closes the entrance. What is this door? It is sin. This is the barrier which has so long prevented God from entering the heart, and which is removed by a firm resolution of change of life; when this is made, the door is opened. There are some, indeed, who not only keep this door locked, but guarded by bars, bolts, and chains. Such persons have to go through hard work, I mean such as are entangled in habits of sin, hedged in by evil practices, or laden with heavy restitution in the matter of reputation, property, and the like. St. Augustine at one time desired to open his door; he struggled, toiled, and laboured, without after all succeeding in doing so. He answered God in his heart, that he would open it to Him, but not just yet: *Sed non modo.* O fatal power of habit! And how is it with thee? If thou hast still difficulty in opening, pray to God to help thee to do so; for although it is true that He so respects thy free-will, that He would have thee consent to open the door with thy own hand, yet no sooner dost thou make what effort thou canst on thy side, than He will bring so strong a pressure to bear on the door from without, that thy opening and His entering will be simultaneous: "If any man will open to Me the door, I will come in to him." See how great is thy Lord's desire to enter! He will not allow so much as a second to intervene between thy opening and His entrance.

II. Consider secondly, that He would have every right to wait for thee to go out into the road to meet Him, and conduct Him into thy house, an act of courtesy always shown to guests of distinction. But He does not do so; no sooner is the way open than He enters: "I will come in to him." To open is the sinner's business; to enter is entirely God's act, and so it is wholly performed by Him. He will not linger a moment on the threshold, because He comes, not to ask, as one who is in need of something; He comes to give as a Benefactor, to converse as a Friend, to console as a Comforter, to counsel as a Guide, to heal as a Physician, to instruct as a Teacher, and therefore He enters at once: "If any man shall open to Me the door, I will come in to him." Strangers, even after the door has been opened to them, wait for a word from the person who has opened it; but not so those who are on intimate terms. Thou seest, therefore, from this, that as soon as the sinner is justified by hearty contrition, the moment the obstacle is removed, he is at once the friend of God, even though, up to that time, he may have been one

of His worst enemies. How dear, then, should this holy
contrition be to thee, which so quickly gives thee such a
Friend! All that thou hast to do is to open, and He is thy
own : "If any man shall open to Me the door, I will come in
to him." Not merely "I will come in," but "I will come
in to him;" for He does not come to thee out of a desire to
be in thy house. He comes out of a far tenderer desire. He
comes to thee out of love for thee : He comes to clasp thee
in His arms, to embrace thee, to lavish treasures on thee;
and no sooner does He come than He would have thee make
use of Him without delay : "I will come in to him."

III. Consider thirdly, that there are two main reasons for
this speedy entrance. One regards God, and has been already
mentioned; it is His great desire to abide with men: "My
delights are to be with the children of men."[1] The other
regards man; and it is this : God would not have man stop
short, so to say, on the threshold of his conversion, content
with mere acts of hatred of sin, of abhorrence, detestation,
abomination, and sorrow, although these are very salutary;
but He would have him go on at once to practise works of
piety, progress, and perfection worthy of one who has received
into his house so great a Guest as God. And so, thou seest
that no sooner has He entered than mention is made of
preparing a supper; nay, why do I say of preparing it?
rather of partaking of supper, as though it were all ready:
"I will come in to him, and will sup with him, and he with
Me," to show how ready we ought to be to do good works, if
we really wish to please Him. And, if we consider well, this
supper is an amazing honour; for what is man, that God
should not only visit him, as it was once said, but even sup
in his house? Nay, this is saying too little. I should have
said, sup with him, for God says with His own lips: "I will
sup with him," not *apud illum*, but *cum illo*. What thinkest
thou of this marvellous condescension? There have been
instances of a king while walking or hunting in the woods,
going to rest himself in the cottage of some poor shepherd,
and consenting to accept from him a humble present of
flowers, or strawberries, or chestnuts, which he offered in his
simple hospitality. But when was the king ever known to sit
down with him at his poor table? He would be more likely
to allow the shepherd to sit down at his table. But our Lord
does both the one and the other: "I will sup with him, and

[1] Prov. viii. 31.

he with Me." These words plainly show that two tables are spoken of, one which is spread by God for man, the other by man for God : otherwise, it would have been enough to say either "I will sup with him," or "he shall sup with Me." But by distinguishing between the two, it is certain that there are two banquets given by man to God, and by God to man in turn ; as though these two were persons of equal rank. Does not the mere thought of this plunge thee in a very ecstasy of wonder? "My Beloved to me, and I to Him."[1]

IV. Consider fourthly, what the first supper is, which is prepared by man for God. It consists of the poor food which is his. But what can a man give who has but just passed from the state of sin to that of grace? Nothing surely but "fruit worthy of penance."[2] And this is meat which is most welcome to the Guest he entertains : not meat which supports Him as it does man ; but which is a refreshment, nay, a delight so great to Him, that on earth He even prefers it to that which He offers to man ; and so, as thou seest, He first sits with man at the table spread by man for Him, and afterwards entertains man at His own table : "I will sup with him, and he with Me," not "he shall sup with Me, and I with him." Neither should this surprise thee. The meat which God here receives from man are acts of virtue : the meat which man receives here from God are the spiritual consolations, the sweetness, the delight with which He compensates him for what he suffers. Now, undoubtedly, God takes greater pleasure in the acts of virtue which He receives from man than He takes on earth in the gifts which He bestows on man. Although, indeed, He may also mean thee to understand by giving this preference to the supper prepared by man, that in the same measure with which thou offerest food to God shalt thou be fed by Him. If thou givest Him an abundant banquet, thou too shalt be abundantly feasted by God ; if, on the other hand, it is poor and scanty, thy treatment will be poor and scanty in return. Consider how it was with those saints who laboured greatly for God ; they were not able to contain the raptures with which He filled their souls ; they were compelled to cry out : No more, "it is enough, Lord, it is enough." If, then, thou receivest but slight refreshment from thy Lord, it is that the refreshment thou givest Him is also slight : "The Lord will reward me according to my justice."[3]

[1] Cant. ii. 16. [2] St. Matt. iii. 8. [3] 2 Kings xxii. 21.

V. Consider fifthly, the refreshment which man gives to God is a figure of the state of progress, that which God gives to man is a figure of the state of perfection ; not that in each of these states there is not both the refreshment of man by God with spiritual consolations, and of God by man with acts of virtue, but because in the state of progress it is the part contributed by man which preponderates, and in the state of perfection the part contributed by God. In the former state there is more of labour than of enjoyment, and so man is said to be the giver of the feast ; in the latter state there is more of enjoyment than of labour, and so God is the giver. Now, we all know that the state of progress always comes before that of perfection, and therefore it is that the supper given by man to God precedes, as thou seest, that given by God to man : " I will sup with him, and he with Me," not " he shall sup with Me, and I with him." If, therefore, at any time, thou expectest great consolations from God before thou hast practised great acts of virtue for God, thou art expecting to arrive at perfection before making progress ; and this is a confusion of the right order : first comes "I will sup with him," and afterwards "he with Me."

VI. Consider sixthly, that both these repasts which are here mentioned are taken not by daylight, but by candlelight, by which I mean the light of faith ; they belong, not to day, but to night, and therefore they are called suppers : "I will sup." They are partaken of by means of that perception of faith which is given to man in this low valley of darkness, where, although it is most true that God reveals Himself from time to time with a greater degree of clearness, yet it is always far different from that with which He reveals Himself by the light of glory to the blessed in Heaven. And yet, there too, it is a supper which will be given to man in the light of glory : "Blessed are they that are called to the marriage-supper of the Lamb."[1] But the reason of this being a supper is not that it will be eaten by night, for there in Heaven it will be everlasting day ; but is that on account of which every supper is so called, that is, because it will be the last meal, after which there will never be any other. So delicious will it be that no one will ever wish to change it. It is that supper which will take place at last after all the labours endured in this sorrowful world are for ever at an end ; and therefore also, it is the supper which God promises in return to the just man

[1] Apoc. xix. 9.

when He says : "I will sup with him, and he with Me." He
does not promise him a feast which, although it is eaten by
candlelight, is not the last, because so far from removing
hunger, it increases it ; He promises him one which will be
the last, because it absolutely removes hunger, but it is eaten
in the light of day. Oh, happy wilt thou be, my reader, if it
is thy lot to sit down at that blessed supper ! Remember,
then, that before thy Lord prepares the table for thee, thou,
according to the measure of thy feeble powers, must prepare
it for Him : "I will sup with him, and he with Me ;" otherwise, not only wilt thou not partake of His supper, of either
kind, but no sooner will He have entered thy heart by thy
conversion, than He will depart from it because He does not
find there the refreshment of thy good works which He so
greatly desires to receive from thee. It is for this that He
comes : "I will come in to him and sup with him, and he
with Me." If thou neglectest these good works, to give
thyself to sleep, to amusement, to the idleness habitual to
thee, all thy conversion will be at an end, and at length the
Lord Whom thou hast insulted will depart from thee, just as
a guest would do whom thou hadst received into thy house,
and to whom thou hadst never offered any refreshment.

EIGHTH DAY.

*It is like the sin of witchcraft to rebel, and like the crime of idolatry
to refuse to obey* (1 Kings xv. 23).

I. Consider first, that real obedience does not consist in
merely doing what a Superior orders ; thou must go further
than this, and do it precisely because He does order it. If
thou doest what thou art bidden because it is to thy taste, or
out of the hope of a reward or the fear of punishment, thou
art not yet really obedient, because if these motives were
wanting thou wouldst cease to obey. Thou art really obedient
when thou submittest to thy Superior not only by the material
act, but also with thy will, so as to will what he wills, and
simply for the reason that he wills it. This, therefore, is the
reason why God here expresses the refusal to obey by the
words, *nolle acquiescere.* He does not say "to refuse to
perform" (*nolle exequi*), but to "refuse to consent" (*nolle*

acquiescere), because obedience consists in that consent of the will which has attained perfection when the will of the subject goes so far as to rest in the will of the Superior, as in its centre. But this consent of the will, necessary as it is, is extremely difficult to acquire when the understanding is rebellious. And, therefore, in order to obey well thou oughtest first to endeavour to persuade thyself that thy Superior does well in commanding what he commands. If, on the contrary, thou seekest after reasons for believing that he does ill, thou art guilty of a serious error, because by so doing thou makest thyself indisposed to obey him. And this it is which is here called "to rebel." He does not rebel who, when he has heard the order, humbly represents to his Superior any difficulties which he sees, but he who, after having represented them, persists in maintaining his own opinion, contradicting and disputing, and trying to bend his Superior's judgment to his own. Now, it is in order to make thee understand how great an evil this is, that God here says that "it is like the sin of witchcraft to rebel, and like the crime of idolatry to refuse to obey." "To rebel" belongs to the understanding ; "to refuse to obey" belongs to the will. These words are, if I am not mistaken, the most terrible of all the denunciations hurled in the Sacred Scriptures against the disobedient. Tremble, then, merely at hearing them, and examine thyself seriously in order to see clearly whether, under any circumstances, thou rebellest against thy Superior, and rebellest in such a way as in the end to refuse to obey.

II. Consider secondly, why it is said that to rebel, in other words, to oppose thy own judgment to that of thy Superior, is a vice like that of one who practises magic arts : "It is like the sin of witchcraft to rebel." The reason is that it is beyond a doubt that by following the judgment of thy Superior in everything in which there is not evident sin, it is impossible but that thou shouldst please God ; not so if thou followest thy own judgment, for even as to actions which are praise-worthy in themselves, such as fasting, taking the discipline, hearing Mass, and the like, if they are done at thy own choice, it is very possible that under those circumstances thou pleasest God less by doing them than thou wouldst please Him by doing other different actions, and that at thy death He may say to thee, as He said to the unhappy Jews, "Who required these things at your hands?"[1] But when thou followest the

[1] Isaias i. 12.

judgment of thy Superior the contrary is the case. For the best thing that thou canst ever do, under any circumstances, is to do what thou art bidden. If thou observest well thou wilt see that obedience does a work like grafting: it inserts the will of God into the will of man, and thus enables the latter, which is of itself a wild stock, to produce fruit which it never could have borne in its natural unassisted state. Now, to leave what is certain for what is uncertain is to take one's self to divination, and therefore it is well said in the case we are considering: "It is like the sin of witchcraft to rebel." In following thy own judgment thou mayest possibly be acting rightly, but the contrary is equally possible; whereas in following the judgment of thy Superior thou art always sure of doing so. How thinkest thou, then? Is it a small sin to act like one who practises witchcraft rather than to act like a wise man? This was the conduct of Saul when, after defeating the Amalekites, he thought it much better to spare some fat beasts, "that they might be sacrificed to the Lord," than to kill them all, as Samuel had commanded, on which occasion the latter addressed him in the words thou art meditating: "It is like the sin of witchcraft to rebel." This is a kind of "witchcraft" of which thou, too, art guilty when thou knowest that thy Superior judges a certain place, occupation, office, or way of life to be the best for thee, and yet rebellest in thy judgment, and obstinately persistest in thinking the contrary. "All that resist Him shall be confounded."[1]

III. Consider thirdly, why it is said that to refuse to obey is a wickedness like that of idolatry: "It is like the crime of idolatry to refuse to obey." Thou wilt understand this by carefully considering what the disobedient man, as a disobedient man, makes his object. The sensual man, as a sensual man, makes it his object to give reins to his sensuality; the miser, as a miser, to heap up riches; the ambitious man, to push his fortunes. The disobedient man's object is to have his own way, and what is this but the desire of making a god of his own will? To be the first law of all the operations thou hast to perform is an attribute so exclusively belonging to God that it cannot possibly appertain to any besides, unless God has imparted it to him. He has, indeed, imparted it to thy Superiors as regards thyself, but it is precisely for this reason that He says, "He that heareth you heareth Me."[2] When, therefore, thou wouldst rob any one of them of this

---

[1] Isaias xlv. 25.   [2] St. Luke x. 16.

attribute in order to transfer it to thy free-will, how does this differ from the conduct of idolaters who gave at their choice, now to the beasts of the wood, now to stones or plants, that Name which is, by its nature, incommunicable? "They gave the incommunicable Name to stones and wood."[1] Nay, in a certain sense, thou doest what is even worse, because, whereas idolaters gave only the name of God to stocks and stones, thou actually givest His authority to thy will by making it the law of thy conduct to which thou payest honour. Saul acted as an idolater when, in defiance of Samuel's prohibition, he chose to take his own way, and to spare from the general slaughter of the Amalekites their King Agag : to keep that part of the spoil which he wished to keep, and to burn that which he wished to burn, and therefore Samuel told him that " it is like the crime of idolatry to refuse to obey." And dost not thou, too, act as an idolater, when thou adorest thy own will so as to render to it a Divine worship, which is, in other words, making it thy first law? Such conduct is setting altar against altar, or, rather, it is casting down from the altar thy Superior's will, which thou art bound to reverence on earth precisely as that of God, in order to put thy own will in its place.

IV. Consider fourthly, that if witchcraft is a great sin, idolatry is a far greater, and so, whereas the former is called a sin, "the sin of witchcraft," the latter is a crime, "the crime of idolatry." And there is the same proportion in our subject. Doubtless to rebel against a Superior, to argue, and dispute, and maintain a contrary opinion to his is a sin, and a serious sin, because it is preferring what is doubtful to what is infallible : "It is like the sin of witchcraft to rebel." But "to refuse to obey" passes all bounds, for this is to venture to subject to one's own will that of him who is in the place of God. Is it not a grievous disorder for thy Superior to have to do as thou wishest, rather than that thou shouldst do as he wishes? It is but right for thee to say to him as Saul said when he was struck down on the road to Damascus : "What wilt thou have me to do?"[2] And yet thou wouldst have him, on the contrary, say to thee, as our Lord said to the blind man of Jericho : "What wilt thou that I do to thee?"[3] Take good heed, for this will of thine is but an empty idol, and if thou worshippest it, it is in reality the devil that thou adorest in it, who will surely send thee to perdition. If thou wouldst

---

[1] Wisdom xiv. 21.   [2] Acts ix. 6.   [3] St. Luke xviii. 41.

be saved, take care to detest this abominable idolatry : "turn away from thy own will,"[1] throw down the idol, trample on it, destroy it, do not pay it the slightest respect, and render without reserve to the will of thy Superior the homage which is due from thee, of making it thy first law upon earth.

———

NINTH DAY.

The jealousy and rage of the husband will not spare in the day of revenge, nor will he yield to any man's prayers, nor will he accept for satisfaction ever so many gifts (Prov. vi. 34, 35).

I. Consider first, that no indignation of heart can be imagined equal to that of a noble prince who, returning unexpectedly some night from a distant country, finds his wife faithless to him, and in the company of a strange lover. What would be his anger, wrath, and fury! And how much would this anger be increased if that lover should happen to be the greatest enemy the prince had in the world! And, further still, how would it be increased if the wife had been a maiden, of noble birth indeed, but reduced to extreme indigence, servitude, and slavery, from which she had been redeemed by the prince, and if, moreover, he had rescued her from the hands of the very villain whose mistress she afterwards became, and rescued her for no reason but to raise her from her miserable condition to a royal station! And, lastly, how greatly would it increase the prince's fury to know certainly that she had not been forcibly won, but actually bribed by that wicked lover! Then, indeed, it would reach such a height that it would suffer no delay, deeming it but a poor revenge to strike the ungrateful adulteress with the sword, to smite her, cut her to pieces, slay her, and tear her heart from her breast with his own hand. Of what avail would be entreaties, promises, or tears? It would be no time for heeding them : "The jealousy and rage of the husband will not spare in the day of revenge, nor will he yield to any man's prayers, nor will he accept for satisfaction ever so many gifts." For it is not rage only which inspires revenge in this case, it is, still more, jealousy : "jealousy and rage." The jealousy inflames the rage, and the rage increases the jealousy. And

[1] Ecclus. xviii. 30.

this may perhaps be the reason why the Wise Man did not say, as he might easily have done, "The jealousy and rage of the husband . . . will not yield (*acquiescent*), . . . nor will they accept (*suscipient*)"—thus speaking of the two affections as distinct—but he said, *Non acquiescet—non suscipiet*, and so these verbs, which in the English translation are made to agree with the pronoun "he" (*i.e.*, the husband), may very likely refer to these two affections considered as one ; because, in fact, they are not two, but by their combination there results a mixed affection of equal rage and jealousy, so violent that it can only be felt, not expressed. And now, if Divine things may be, not indeed perfectly expressed, but faintly outlined by human things, do thou imagine thy soul to be represented by the wife here spoken of, Christ by the husband, and the devil by the wicked lover, and make a very close application of the figure by asking thyself, What will be the conduct of this mighty Prince when, coming back from that far country, whither He went "to receive for Himself a kingdom, and to return,"[1] He surprises my soul in the night of utter darkness in the arms of one who is His mortal enemy, a renegade, and a rebel, and that because he falsely promised her those delights which she fancied were not given her by her Spouse. Can I think that I can find any way to appease Him ? It is too late : "The jealousy and rage of the husband will not spare in the day of revenge."

II. Consider secondly, what this day is which is here called "the day of revenge." It is the Day of Judgment, and so it is written in the Septuagint. It is the day both of the Particular and the General Judgment, the first being that of private, the last that of public vengeance : "These are the days of vengeance."[2] And they are both thus styled, because the Spouse is then firmly resolved to take vengeance, that is, to vindicate His injured honour. In men such a resolution is wrong ; and why ? Because it springs from vice, not from virtue—"The anger of man worketh not the justice of God "[3]—since virtue requires them never to cease to pardon the injuries they receive, but to practise constantly towards others those offices of kindness, forgiveness, and charity which they would desire to be constantly done to them : "As you would that men should do to you, do you also to them in like manner."[4] But in God this is most just, because the

---

[1] St. Luke : ix. 12.
[2] St. Luke xxi. 22.   [3] St. James i. 20.   [4] St. Luke vi. 31.

rule does not hold with regard to Him. It is impossible for Him to be in circumstances requiring kindness, forgiveness, and charity, and neither is He therefore bound by the law of doing good in return, to practise these offices to any one. When He does this, it is because it is His pleasure to do it. Hence it is that when He takes vengeance in anger, in other words, when He repairs the injury done to His honour, He does an action which is not only virtuous, but necessary. He "worketh justice," seeing that it is possible for Him to allow injuries to be done to Himself in order to teach us not to be so solicitous concerning exterior honour; but it is not possible for Him to leave them unpunished, because He is the Sovereign Ruler, and, as such, He is bound to chastise not only the wrongs of others but His own. If He does not avenge them now He must do so at another time, and that time will be "the day of revenge." Only think how He now refrains from taking vengeance. Thou hast ample evidence of this in thy own soul, which up to this time has so often been faithless to this perfect Spouse, and yet He seems not to notice it. Nay, more than this, He sends His messengers to her, saying by their mouth: "Thou hast prostituted thyself to many lovers, nevertheless return to Me, saith the Lord, and I will receive thee."[1] And so, if thou considerest closely, the Wise Man does not here say absolutely that "the jealousy and rage of the husband will not spare," but only, "will not spare in the day of revenge." Ah, how ready He is to spare now!

III. Consider thirdly, why it is that this Prince, this Spouse of thy soul, acts so leniently at present. It is because at present He is supposed to be in a distant country: "He went into a far country to receive for Himself a kingdom."[2] And thus, as thou seest, He very often acts as though He were really ignorant of what is being done in the world; He appears not to see or hear, so that His spouses are foolish enough to believe that He is really not at home: "My husband is not at home, he is gone a very long journey."[3] Therefore they are bolder in sinning. But take heed, for at length, however far off He seems to be now, He will show Himself to be present: "He went into a far country to receive for Himself a kingdom, and to return." And how will it be if, as may easily happen, He arrives unexpectedly, and surprises thee in the act of infidelity to Him? Oh, what will be thy

[1] Jerem. iii. 1.　　[2] St. Luke xix. 12.　　[3] Prov. vii. 19.

shame, anguish, and terror. But it will all be in vain: "Behold, I come against thee, saith the Lord, and I will discover thy shame to thy face,"[1] so that the faithless wife will be utterly unable to deny her infamous treachery, though she would fain do so. What wonder, then, if that miserable one is inexorably punished? She was discovered by him in her guilt; there is no escape. And so thou must know that, for the souls of men, that day is called "the day of revenge," on which Christ surprises them in the guilt of infidelity. Now, because He has gone "into a far country," He seems rather to know the injuries done Him, as one knows a thing by hearsay, than to perceive them of Himself, and therefore He does not, as yet, give an irrevocable sentence. But then, if we may so express it, He will see these outrages against His honour as being actually perpetrated before His very eyes, and then, therefore, this fatal "day of revenge will have arrived at last." Hence it is that whenever our Lord spoke of either the Particular or the General Judgement He always made use of this expression of "coming." "Behold I come quickly," and the like. So much so, that in the Gospels the Day of Judgment and the day of His coming are one and the same thing, to teach us that as soon as our Lord has come there is no longer any hope of pardon for one who is found sinning: "The jealousy and rage of the husband will not spare in the day of revenge," or, in other words, "in the day of His coming." If, then, this is so, do not wait till He comes, but bid thy soul at once discard every unlawful love, be filled with compunction, and lead a new life. Otherwise, if surprised in sin, all is over; and remember that it is the habit of the Bridegroom to arrive unexpectedly: "At midnight there was a cry made, Behold the Bridegroom cometh."[2]

IV. Consider fourthly, that if this day is the day of vengeance for so glorious a Spouse, it is impossible but that all the prayers of the faithless soul will then be powerless with Him. Nay, why do I say the prayers of the faithless soul? All the prayers of all the world will be equally vain: "He will not yield to any man's prayers." So that, though all the saints should throw themselves on their knees on that day, to ask pardon for thy soul, they could not obtain it: "I will take vengeance, and no man shall resist Me."[3] There is but one way in which man can resist God—by prayer; when, therefore, God says that no man shall resist Him in that day, it is because no

---

[1] Nahum iii. 5.   [2] St. Matt. xxv. 6.   [3] Isaias xlvii. 3.

man's prayers will be of any avail. Nor is this all; no matter who may plead for that soul, or offer for it abundant alms, fastings, hair-shirts, or disciplines, the Spouse, to Whom such gifts were once so pleasing, will now have none of them: "Nor will He accept for satisfaction ever so many gifts." No, though all the saints were to unite in offering to return to the earth for no other purpose than to make satisfaction for that unhappy soul, He would never accept it to all eternity. And why is this? Thou hast already heard the reason : because it is "the day of revenge." But now, jealous as this Spouse is of His honour, see how little is needed to appease Him. A sigh, a prayer, a single act of contrition ; whereas then, all the united riches of Paradise would not suffice : " Riches shall not profit in the day of revenge."[1] And the only reason that can be given for this is that "the jealousy and rage of the husband will not spare in the day of revenge."

TENTH DAY.

What needeth a man to seek those things that are above him, whereas he knoweth not what is profitable for him in his life, in all the days of his pilgrimage, and the time that passeth like a shadow? (Eccles. vii. 1).

I. Consider first, that in this passage the Wise Man condemns all ambitious and avaricious persons, and all who, after the common custom of the world, seek things superior to the state of life in which God has placed them ; for such people "seek those things which are above" them, or literally "greater than themselves" (*majora se*). And why greater than themselves? Is it because they seek after things superior to their condition? No: for this is not sufficient to make things greater than themselves, seeing that sometimes many of their number deserve a happier and more elevated condition than that which they occupy. They seek after things greater than themselves, because they seek after things superior to their capacity. What do I mean by this? All those persons who aim at raising the condition they are in, because they are not satisfied with it, undoubtedly seek after things future ; and therefore they seek after things superior to their capacity. For how do they know whether this acquisition will be hurtful

[1] Prov. xi. 4.

or profitable to them? This is only known to God, Who has fixed in His mind the order of their predestination. It may be that the office, the dignity, the money, the alliance after which they are straining so eagerly, may bring with it their everlasting ruin; they are foolish, therefore, to seek after it so anxiously, and so the Wise Man here mocks at such persons when he says: "What needeth a man to seek those things that are above him, whereas he knoweth not what is profitable for him in his life, in all the days of his pilgrimage?" The words of the original text are "to seek future things," but the translator of our version wrote "things greater than himself" (*majora se*), because all others are superior to human capacity; God alone knows what that way is by which we are to be saved at last. Therefore, just as a traveller, born an exile in some distant country, does not know by what road to go to his native land, and must let himself be guided by one who does, so, too, must thou let thyself be guided by God, or him who occupies the place of God to thee on earth: thy Bishop in the Church, thy Superior in the monastery, thy spiritual Father in the world. This is the right rule. How great then is thy folly, not only in choosing to be governed by thy own caprice, but in raising thyself, encircling thyself in defiance of God, weaving, as it were, the web of thy life in defiance of God Himself! Be satisfied with the state in which God has placed thee, or, if at any time it is necessary to seek for anything beyond it, do so under the sure guidance of obedience. Then thou wilt be safe: "He that keepeth the commandment shall find no evil."[1] Not only does he find none now, because he acts in a holy manner, but neither "shall he find" any in future, because he acts in a safe manner. Though thou shouldst live a hundred years, there would never be a time when thou wouldst suffer through obedience. Oh, what Divine encouragement is this!

II. Consider secondly, that even though no harm were to come to thee by the acquisition of those posts which thou spendest thy life in striving after, there would still be harm in seeking after them. And the reason is, that this seeking after them turns away thy thoughts from that which is really important, that is, the business of thy eternal salvation; and how many doubts are there about that business! Thou art ignorant what things most tend to ensure it, "what is profitable" for thee; and consequently, thy one only study should be con-

[1] Eccles. viii. 5.

cerning a matter so important. This is the point on which all thy anxiety, all thy thoughts, all thy words, all thy desires ought to be centred, while, on the contrary, by seeking after things in the world which are difficult of attainment, such as lofty position, thou necessarily must neglect what is important for what is unimportant. Is it not better to devote to the good of thy soul the pains and anxiety which thou didst so eagerly devote to things of this world? "What needeth a man to seek those things that are above him, whereas he knoweth not what is profitable for him in his life, in all the days of his pilgrimage?" A traveller does not waste his time on the road in useless employments; he concerns himself only with his object, which is to get safe to his country. This is what he thinks of, speaks of, inquires after; he does not trouble himself at all about other things. On the journey he is contented with an ordinary lodging, ordinary clothing, ordinary food, with attendance even less than ordinary; and why? because there is no time to lose, he is a traveller. And thou, too, art a traveller on earth, thou knowest this, and thy country is Heaven. Spend thy time then in learning what it is important to learn, what is the surest way of reaching it. As to the rest : "What needeth a man to seek those things that are above him, whereas he knoweth not what is profitable for him in his life?" If he did know "what is profitable," that is, what is profitable for the attainment of everlasting salvation, then, indeed, there would be less harm in employing his time in other things; but as he does not know it, he should take pains to learn it. This only is important : "But one thing is necessary."[1] And this is why the Wise Man does not say, "What advantage is it for a man to seek those things that are above him," but "what needeth a man to seek" them? Because he takes for granted that a man has no business on earth but that which is necessary, namely, to work out his everlasting salvation.

III. Consider thirdly, that besides all this, there is the additional consideration that the time is short, for it is not a question of ages, but merely of days, which make up the sum of thy life. How, then, canst thou waste it in seeking after anything but that which ought to be sought for? In a besieged city men buy water, because it is scarce, at a high price : so too is it with corn in a time of famine, hay in a time of drought, iron in time of scarcity; and all goods, in the same

[1] St. Luke x. 42.

way, however common, rise to a high price if they are scarce. How is it, then, that time alone does not rise in value with thee, when it is, in itself, so precious? Listen to what narrow limits human life is confined: "The number of days," for so it stands in the Vulgate, *Numero dierum.* Do not imagine, therefore, that at the last thou canst beg of God to add to this number, for it is so called because it is fixed; and this is why the Wise Man did not here say, as he might have done, *Diebus peregrinationis*—"The days of his pilgrimage," but *Numero dierum*—"The number of the days," in order to teach thee that it is in vain to hope to increase it. All things have their appointed number with God, and so is it with the days of thy life: "O Lord, make me know the number of my days, that I may know what is wanting to me."[1] Do thou, therefore, spend them all in this one thing, in travelling by a safe road to thy native land, both because they are few in number, and because their number is fixed. It is exactly thus that a traveller does who has to reach his country within a short space of time appointed to him, under pain of losing his inheritance. How such a man strives to be on the right side, so as to arrive too soon rather than too late!

IV. Consider fourthly, that not only is this time, as has just been said, short and fixed, but it is also swift as a shadow. Wherefore, after saying, "The number of the days of his pilgrimage," the Wise Man concludes with these words: "and the time that passeth like a shadow." But why does he say that time passes like a shadow, and not rather like a courier along the plain, who never slackens speed, or a ship in the sea, a bird in the air, an arrow shot from the bow? Wouldst thou know why? It is because it passes with the greatest rapidity, and yet thou thinkest that it pauses; and this is peculiar to a shadow. The courier who speeds on his way with tidings of victory passes swiftly indeed, but it is evident to those who see him that he goes quickly; the ship, the bird, the arrow are swift, but thou art aware of it, and so of other things. Not so the shadow on the dial. Watch it as closely as thou pleasest, thou wilt never perceive that it moves, so insensible is its motion. Just so is it with time: it "passeth like a shadow;" like that shadow which marks its passage, and to which its own motion answers. And so, although it passes with wonderful rapidity, it passes in such a manner that thou perceivest that it is past before perceiving that it is passing.

[1] Psalm xxxviii. 5.

Who, therefore, can say how great a danger there is of losing it unless thou takest heed? There is, then, all the more necessity for taking heed? There is nothing that a traveller holds more precious than time, especially if he is in straits. He steals it from sleep, from social intercourse, from sightseeing, from his meals, and that simply because he is a traveller. So, too, art thou; thou art a traveller, and moreover a traveller who art journeying to a goal, which if thou dost not reach within the time appointed by God for thy salvation, thou art lost, there is no more possibility of reaching it to all eternity. Think then whether there is any time to be lost on earth as is done by so many ambitious and avaricious persons, so many who are wholly given up to pushing their way in this miserable world, which they know all the while that they are but travelling through. And if this is so, then thou seest the meaning of the words which thou art meditating: "What needeth a man to seek those things that are above him, whereas he knoweth not what is profitable for him in his life?" a life which is described in these terms, "The number of the days of his pilgrimage, and the time that passeth like a shadow."

ELEVENTH DAY.

These are they to whom the storm of darkness is reserved for ever
(St. Jude i. 13).

I. Consider first, that one of the grievous torments which the damned will suffer in the abyss of Hell will be that of darkness. And not only will this darkness be palpable, as was that of Egypt, but stormy; and so this blessed Apostle says of those unhappy ones: "These are they to whom the storm of darkness is reserved for ever." This darkness will also be of two kinds, exterior and interior, the former belonging to the pain of sense, the latter to the pain of loss. Pray God to give thee light to understand, and so to fear both.

II. Consider secondly, the exterior darkness of which Christ so often made mention in the Gospels: "They shall be cast out into exterior darkness;"[1] "Cast him into the exterior darkness."[2] Not that the exterior is more full of

[1] St. Matt. viii. 12.   [2] St. Matt. xxii. 13.

torment than the interior darkness, but because it is more sensible. This darkness of Hell will proceed from three causes : the place, the position, and the matter. First, it will be caused by the place in which the damned will dwell. For in representing Hell to thy mind, thou must imagine a vast abyss in the deepest centre of the earth, "in the earth" (so that the damned may be at the greatest possible distance from the blessed), in which, as in a close-shut grave, there is no chink through which light can penetrate, because of the weight of earth covering it above, below, and on every side : "And they went down alive into Hell, the ground closing upon them."[1] Secondly, it will be caused by the position in which the damned will be. For in this their grave they will all lie, after the Judgment Day, just as dead bodies lie in theirs when the plague is raging, crowded, piled up, heaped together ; so that, as they will never be able to stir or move, or to open their lips to relieve themselves by uttering an articulate syllable or sound—according to the interpretation of the words : "The wicked shall be silent in darkness "[2]—so neither will they be able to raise their eyelids to try whether they can see anything. So tremendous will be the weight of God's anger, when at the last it shall put under its feet all its enemies together and trample on them : "I have trampled on them in My indignation."[3] In the third place, it will be caused by the thick smoke of brimstone in which the damned will be always enveloped. For it is certain that there is, at the bottom of this infernal pit, a great lake of burning sulphur—"A pool of fire burning with brimstone "[4]—and this fire, being dark and lurid, and not of a kind to give light, will send up volumes of terrible flames mingled with dense smoke, which will keep incessantly rising : "The smoke thereof shall go up for ever."[5] This will indeed be a "storm of darkness," for when that smoke finds no way of issue above, it will descend again, and in its descent come into violent contact with that which is rising up, and so, meeting and mingling together on all sides, it will so darken that vast cavern, that even if every other hindrance to seeing, both of place and position, were removed, still those wretched lost ones would not be able to glance around them without being blinded. Think for an instant, what would be thy lot, if thou, too, wert in this condition ; and tremble at the signs of the

[1] Numbers xvi. 33.   [2] 1 Kings ii. 9.
[3] Isaias lxiii. 3.   [4] Apoc. xix. 20.   [5] Isaias xxxiv. 10.

storm, even while thanking God, that, so far, it has not over-taken thee.

III. Consider thirdly, that the interior darkness is, beyond a doubt, more terrible than the exterior, although less easily apprehended by our minds. This will surround the soul of every one who is lost, just as the other darkness sur-rounds his body. And it will be caused by the absence of all Divine light: "Woe unto us, for the day is declined." For them that sun has ceased to shine which here is so beneficial to all: there there are no more lights, or inspira-tions, or visitations of mercy, only of punishment. Secondly, it will proceed from the extreme severity of the torments, which will prevent those who endure them from ever being able to consider, discern, or think of anything but, as in a stupor of suffering, of the torment which weighs upon them without intermission: "My heart failed, darkness amazed me."[1] Thirdly, it will be caused by the passions which will not only overwhelm their will, but their understanding also. Here, too, will arise a storm, the "storm of darkness." For if a violent passion of anger is of itself sufficient to darken the understanding of even a wise man—"My eye is dim, through indignation"[2]—how will it be with the damned, who will be always consumed with implacable anger and fury against God? The effect of this will be that, although they know that they are rightly punished, they will persist in cursing Him as though He were very unjust; that they will despise His grace and hate His glory; that their pride will not allow them to humble themselves before Him, although they know how deeply they are humbled. Alas for him who is in so fearful a storm! What shouldst thou do to escape it? Thou shouldst be exceedingly grieved that God meets with such ill-treatment in that thick darkness, since even there He ought to be greatly honoured; for it is certain that no less praise is due to Him in Hell on account of the justice He exercises, than is rendered Him in Heaven for the mercy which causes its joy.

IV. Consider fourthly, that here on earth storms are generally brief in proportion to their violence. Not so that storm which the reprobate will be exposed to in Hell. Therefore, lest thou shouldst suppose from these words of the holy Apostle that "a storm of darkness is reserved" to these miserable ones, that this storm, violent as it may be, is

----

[1] Isaias xxi. 4.　　　[2] Job xvii. 7.

of short duration, he adds in plain words that it will not only be a storm, but an everlasting one : "To whom the storm of darkness is reserved for ever." If, then, this storm would be so terrible, even if it were to last but for an hour, what will it be when it will never cease through all eternity? "He shall never see light."[1] If thou passest one sleepless night how weary thou art of the darkness which, after all, the dawn will so soon chase away! What will it be, then, when there is no hope of any dawn, and when the night of suffering is passed, not on a soft bed, but in flames? There indeed it may be said : "We looked for light, and behold darkness."[2] For night will follow night, another, and another, and yet another ; but the last night will never come. So that, even if there were nothing else to deprive the damned of all hope of escape from that deep abyss of Hell, the mere fact of their being always encompassed by such thick darkness would be enough to show that they can never leave it to all eternity.

V. Consider fifthly, that the Apostle says that this great "storm of darkness" is not only prepared, but "reserved" for these miserable beings. Things may be prepared for a person who has no claim of any kind to them, but, strictly speaking, those things only are "reserved" for a person which belong to him : "Save me (*servate*) the boy, Absalom."[3] See, too, how justly this darkness is reserved for sinners, who, although surrounded by so clear and brilliant a light as that of the Gospel, yet purposely shut their eyes that they might not see it, and preferred their own foolish code of honour to the teaching of Jesus Christ Himself : they "loved darkness rather than the light."[4] And what art thou doing? Art thou, too, a lover of darkness? Take good heed ; for, if so, thou art also a lover of thine own perdition. For these two things are so closely connected, that very often damnation is described merely by the word darkness : "Alms will not suffer the soul to go into darkness."[5]

[1] Psalm xlviii. 20.   [2] Isaias lix. 9.
[3] 2 Kings xviii. 5.   [4] St. John iii. 19.   [5] Tobias iv. 11.

TWELFTH DAY.

I made a covenant with my eyes, that I would not so much as think upon a virgin; for what part should God from above have in me, and what inheritance the Almighty from on high? (Job xxxi. 1).

I. Consider first, that in the words here quoted it was holy Job's intention to show the firm resolution he had made to abstain from impure looks, so as not to incur the danger of damnation. But, this being the case, why was he not satisfied with saying that he had so resolved, without making use of the word "covenant"? *Fœdus*, which is the word here translated "covenant," has two meanings : sometimes it means a league, sometimes a truce. The first meaning is clearly out of the question in this passage, for holy Job did not intend to make an agreement with his eyes to see, but not to see. It must, therefore, mean truce, that is, cessation. But why use this expression ? For three reasons : first, to teach thee that the eyes which thou holdest so dear are thy worst enemies. A truce, unlike an alliance, is not made with friends but with foes ; and it is made, moreover, when some great harm is feared from them if their career of triumph is not checked as soon as possible. And, oh ! what great harm is it in the power of thine eyes to cause thee, if they are not restrained in time ! They are able to reduce thy soul to the lowest slavery of which it is capable, that, namely, of lust : "Holofernes was caught by his eyes."[1] The second reason is, to show thee that not only are thine eyes thy worst, but also thy leading enemies. A truce is not made with the soldiers of an army, but with the commanders ; and such are the eyes. It is these which give ingress into thy heart to that body of soldiers who overcome thee, I mean, thy thoughts. Do thou, therefore, make a truce with thy eyes. If thou wouldst make one with thy thoughts and not thine eyes, thou art acting precisely like one who makes a truce with the private soldiers, not with the generals of an army. Thirdly, it is to teach thee never entirely to trust thine eyes. When peace is made with the enemy, every one trusts him as a friend, but not so when it is merely a truce ; in that case the soldiers are kept at their posts, and ammunition in the country, much as when war was going on. Never mayest thou make entire peace with thine eyes till they are closed in death, it can only be a truce ; and therefore thou

[1] Judith x. 17.

Eyes — trust them not — they are traitors

shouldst never trust them entirely, even when they seem to give thee no trouble. They are traitors. They may, indeed, promise thee a lasting peace, but this is not true; they will soon break it; tell them, therefore, that thou wilt never make with them a peace which will oblige thee to lay down thy arms. These are the three reasons which led holy Job, when he wished to declare his resolution to restrain his looks, to prefer this expression to any other: *Pepigi fœdus.*

II. Consider secondly, that it might seem as if, in speaking of a truce which consists in ceasing to look, Job should rather have said, "I made a covenant with my eyes that I would not look," rather than "that I would not think upon." For, true as it is that most frequently it is the eyes which give admittance into the mind to the thoughts which are the body of the army, yet they never do so except by means of looks which are, so to speak, their scouts, their spies, their guides; and so it would seem that, in a covenant of such importance, Job ought to include, not thoughts only, but looks; nay, looks first, then thoughts. Who can doubt it? He included without expressing them, because he judged it superfluous; it was an understood thing. If the whole of the army, which has to cease from all acts of hostility, is included, certainly there are also included the scouts, spies, and guides who always precede it, even though they may not be mentioned in express terms. It follows, therefore, that in saying "that I would not think," he also said "that I would not look." Unless, indeed, we prefer to think that he chose to say "that I would not think," instead of "that I would not look," because he considered that there was no difference between thinking and looking, that they are one and the same thing. And oh! how certain it is that no sooner is the look given than the thought has insinuated itself! They always go together—"If my heart hath followed my eyes"[1]—so that it was indifferent whether he said "that I would not think," or "that I would not look." Nevertheless, Job preferred the former expression in order to show the kind of look of which he was speaking, namely, fixed looks. It is evident that there is no question of an accidental look; for how is it possible for any commander to prevent every one of his soldiers from attempting any act of hostility in time of truce? it is sufficient that he forbids it. But all such looks as are called voluntary are here intended; and because it was of these that he meant to speak, he preferred to

[1] Job xxxi. 7.

say "that I would not think," rather than "that I would
not look." When the mind thinks purposely of anything,
we say that it sees it ; and so, on the other hand, when the
eyes purposely look at anything, we say that they think of it :
"But thou shalt consider with thy eyes."[1]   And these are,
generally speaking, the looks which injure the soul, voluntary
looks.   What art thou doing, then, who, whenever a dangerous
object happens to come in thy way, pausest to contemplate it ?
Rather shouldst thou instantly turn away thine eyes, because,
so long as the look is accidental, it is merely a look ; but as
soon as it is voluntary, it is no longer a look but a thought :
"I made a covenant with my eyes, that I would not so much
as think upon a virgin."

III. Consider thirdly, that this timidity of Job may perhaps
appear over-scrupulous, since he adds, "upon a virgin."   For,
if he desired to keep himself from hostile looks, that is, from
such as might lead to evil, it ought to be sufficient to restrain
the eyes from looking upon a dissolute woman ; but not upon
a young maiden who is full of reserve, modest and innocent,
and whose countenance expresses the utmost purity.   Oh, how
greatly art thou mistaken !   It is possible that to gaze upon
such a virgin may be as dangerous to thee as to gaze upon a
woman of bad life : "Gaze not upon a maiden," says the
Preacher, "lest her beauty be a stumbling-block to thee."[2]
Observe that this pronoun "her" is not *suo*, but *illius*, because
such a virgin, who, as we said, goes on her way modest,
reserved, and innocent, will never be a stumbling-block to thee
in decore suo, as an evil woman would be, but she may be so
in decore illius, that is to say, thou mayest take scandal from
her, although she does not give it to thee. And what difference
does it make to thee whether the scandal is active or passive ?
The latter is enough to be thy ruin.   Indeed, in the case we
are considering, it is never the active, but always the passive
kind which is thy ruin ; not what is given, but what is taken :
and therefore Job said *de virgine*, not merely *virginem*, because
he would not only look upon her, but upon nothing belonging
to her : *Aliquid de virgine*.   Less than the hair of Judith, the
very shoes she wore were enough to captivate Holofernes—
"Her sandals ravished his eyes"[3]—therefore there must be a
total abstinence from all such looks.   Then the truce is secure,
but not otherwise ; it must include all enemies without exception, not merely such as are declared, but such even as are

only suspected. Those looks which seemed to be without arms will, unless thou art on thy guard, soon draw the sword which they now so carefully conceal, and slay thee without mercy.

IV. Consider fourthly, that unless these looks were enough to bring this destruction upon thee, Job would not have said so plainly, " For what part should God from above have in me, and what inheritance the Almighty from on high ? " But since he does say this, be very sure that there is no possible harm which may not befall thee from such looks as these. They are sufficient, of themselves, to prevent God's having any part in thee whatsoever : " For what part should God from above have in me ? " Because God cannot come from above into such a heart, even by means of His inspirations, which find it so easy generally to enter, though the doors should be shut ; so that if God still has some part in that soul, *deorsum*, as He is the Author of nature, yet He has no longer any *desuper*, as He is the Author of grace.

V. Consider fifthly, that even this is not the whole of the evil ; for it would be less terrible for this sin of lust to rob God of the possession of man in life, if it at least left it to Him after death ; but the worst of it is, that it robs Him of it both in life and in death. And so, after saying, " For what part should God from above have in me ? " Job goes on to say, "and what inheritance the Almighty from on high ? " A man is said to share in the goods of another during his life-time, but he can only inherit them after his death. Now, this vice prevents God's possessing thee even by inheritance, for it is exceedingly likely to make thy death as impenitent as thy life ; and the reason is, that it is one which, more than all others, passes from actual into habitual sin, and so brings thee into a condition in which salvation becomes morally impossible. For the order followed by the eyes in bringing about this total overthrow of the soul, is that which has been already explained to thee in the fourth point. First come looks, these bring together with them thoughts, thoughts induce complacency, complacency consent, and after consent come the acts which complete the work that robs God of the living man. The consequence of these acts is habit, which becomes in time necessity, and this necessity is succeeded by despair of escap-ing from such a state, and after despair comes damnation, which robs God finally of the man after death. See then how, as Job says, God has in thee neither " part " nor " inheritance,"

at least the inheritance which He has in thee is not "from on
high." God is everywhere, in Hell as well as in Heaven—" If
I ascend into Heaven, Thou art there ; if I descend into Hell,
Thou art present "[1]—and therefore, in the case we are considering, it is "the Almighty from beneath " Who has an
inheritance in thee by His justice, but not "the Almighty from
on high " Who has an inheritance in thee by His mercy. And
this is a thing of daily occurrence ; so that this vice is the one
which, more than any other, peoples Hell. This being so,
dost thou not think that Job was fully justified in exclaiming,
" I made a covenant with my eyes, that I would not so much
as think upon a virgin "? He saw how important it was not
to allow his eyes any hostile act, for to do so would be
irreparable ruin : " My eye hath wasted my soul."[2] And if a
man of his lofty virtue feared so greatly, what shouldst thou do
who art so disposed to evil?

THIRTEENTH DAY.

*Whom He foreknew He also predestinated to be made conformable to the
image of His Son, that He might be the First-born amongst many brethren*
(Romans viii. 29).

I. Consider first, the anxiety of some persons to have a sign,
and the surest possible sign, of their predestination. There
is no need to seek far for it, it is here given by the Apostle
—the conformity of the copy with the original : "Whom He
foreknew He predestinated" (and after the word *præscivit* we
should add, as most commentators do, *et hos*)—"Whom He
foreknew these also He predestinated, to be made conformable
to the image of His Son, that He might be the First-born
amongst many brethren." Represent, then, to thyself the
general matter of predestination as happening in the following
way. First of all, the Father established the first of His elect,
namely Jesus, His Son by nature, and predestinated Him to
win for Himself, by virtues practised with much suffering, the
glory of Redeemer. "It behoved Christ to suffer, . . . and
so to enter into His glory."[3] Then, speaking in our human
way, He went on to choose the rest of His elect gradually, but
in such a manner that Jesus should be the Pattern to Whom

[1] Psalm cxxxviii. 8.   [2] Lament. iii. 51.   [3] St. Luke xxiv. 26.

all, as adopted children, were to be conformed, so that whoever should not be willing to be conformed to that Pattern must be excluded from glory, and that all who were so willing should be admitted to it according to the greater or less degree of their conformity.   Now, see the final conclusion of the Apostle : "Whom He foreknew, these also He predestinated." That is, those whom He pre-elected, for this expression is peculiar to the Scriptures, for instance, "To the strangers . . . elect, according to the foreknowledge of God the Father."[1] These therefore He also predestinated, but to what?   "To be made conformable to the image of His Son," to be conformed to the image, that is, to the pattern given them in His Blessed Son.   And thou seest what this pattern was.   Read His life, and thou wilt see what an example He chose to leave thee in it of poverty, humility, obedience, purity, modesty, meekness, and patience, not in one kind of suffering only, but in every kind : "Tempted in all things."[2]   Is thy life like this?   If so, happy art thou, because in that case the copy is like the pattern ; if not, fear and tremble because it is different.

II. Consider secondly, the justice of this proceeding of the Eternal Father.   For if the rest of the elect were to be His adopted children, how right it was that they should be like His Son by nature!   It is our adoption which gives us conformity to the image of our glorious Elder Brother in our heavenly country ; it is just, therefore, that in the way to that country it should give us conformity to the image of the same Elder Brother when suffering, that, "as we have borne the image of the earthly," so we may "bear the image of the heavenly."[3] If thou shouldst choose a contrary course, thou wouldst be an unworthy brother. Do not think it a small thing that He, Who was the Only-Begotten according to His Divine Nature, was willing to receive them as a brother by becoming First-born according to His Human Nature. How, then, canst thou desire a better lot than His?   "Reuben, my first-born, . . . excelling in gifts, greater in command."[4]   Since then, even by right of the lesser title of First-Begotten, Christ was to be "greater in command" in Heaven, He might well have chosen also to be "excelling in gifts" on earth by enjoying those supreme advantages of possession and pre-eminence and ease which were His due according to law. Yet He disdained them all merely that He might save thee ;

1   1 St. Peter i. 1. 2.   2 Hebrews iv. 15.   3 1 Cor. xv. 49.
                          4 Genesis xlix. 3.

and is it possible that thou, whom He has saved, shouldst find it hard to be conformed to Him?

III. Consider thirdly, that there is, therefore, not the least exaggeration in the words of the Sacred Scriptures or of the saints, when they expressly declare that suffering is necessary to salvation. It is these only which our Lord has appointed for attaining that end : "Through many tribulations we must enter into the Kingdom of God."[1] Of course He might have appointed a different way, every one knows that ; but this being the way which He has chosen to appoint, there is no help for it, thou must take courage and follow it. For a prince, if he has entered into a brotherhood of arms with thee, is not satisfied if thou offerest him a literary homage, what he desires is military service ; and if, on the other hand, the friendship he has made with thee is a literary friendship, then he is not satisfied with military, but with literary service. Just in the same way, God has founded His friendship with thee on the life of His Blessed Son, and this it is which He requires of thee. All else that thou canst offer Him is like an empty compliment, it will never satisfy Him. Observe, however, that He does not say that we must be made "uniform," but "conformable to the image of His Son." If the former were the case, alas for us ! We may piously believe of the Blessed Virgin, that she, as His Mother, did attain to an exact conformity with the life of her Divine Son ; and so St. Thomas says that she so well represents Him as rather to be His very image than to be made after His image. But this cannot be believed of any one else ; and therefore the Apostle claims nothing for himself, with respect to Christ, but simply to imitate Him : "Be ye followers (imitators) of me, as I also am of Christ."[2] He says, then, "conformable," not "uniform," because conformity admits of degrees, and this is our encouragement. For it is true that the degree of our conformity to Christ in Heaven will be in proportion to the degree of our conformity to Him on earth. And so He will indeed then be the "First-born among many brethren," because just as one brother here is taller, another shorter than the rest, so will it be with the elect. But though this is so, yet they will all love each other as brethren, and therefore every one will rejoice in the superiority of the other as if it were his own. Well will it be for thee if thou art in this blessed number ! But how canst thou hope to be their brother in the inheritance

[1] Acts xiv. 21.   [2] 1 Cor. xi. 1.

if thou refusest to be their brother in the labour? "A brother is proved in distress."[1]

IV. Consider fourthly, how well the Apostle said "to be made conformable," not "to be conformable." And why? Because the latter would not be true. Infants who die immediately after holy Baptism are predestinated; but they are not predestinated to have this conformity to the image of Jesus suffering on earth, although they are predestinated to have the conformity to the image of Jesus glorious in Heaven. But that does not signify; because the obligation is not to have this conformity, but to endeavour to attain it when it can be attained—"to be made," not "to be conformable." See, too, how well he has said "to be made conformable," to teach thee, as one of the elect, that if thou dost not of thyself desire to make thyself conformable to this image, still thou shalt "be made" so. So great is the necessity of suffering laid upon thee by God even when thou wouldst escape it. It is the mark of real predestination. And so, although the Apostle might have said, "He predestinated them to make themselves conformable," yet he did not say so; he said "to be made conformable." And he said it, moreover, without restriction, because there are many who have to concur in producing this conformity. God, by giving thee His holy grace in the troubles which He lets thee suffer; men, by disturbing thee; the devils, by assailing thee; even irrational creatures, by molesting thee; and lastly, thou thyself, by observing how Christ behaved in similar circumstances, and behaving thyself in like manner. This is the right rule: "To be made conformable." That is, not only to make thyself, but also to be made so: "He hath set me as a mark for His arrows."[2]

V. Consider fifthly, the reason which the Apostle assigns for this Divine arrangement; it is, that by this means Christ may gain many brethren, that is, elect: "That He might be the First-born among many brethren." For the greater the number of His brethren, the greater the glory of the First-born: "About Him was the crown of His brethren."[3] At first sight this reason may strike thee as false, and thou mayest think that the inhabitants of Heaven would be more numerous, if it were granted to those who had most enjoyment instead of most suffering. But this is an error. There is no way by which God could have made the attainment of Paradise more common than by making it to be purchased by sufferings, for

[1] Prov. xvii. 17.   [2] Lament. iii. 12.   [3] Ecclus. L. 13.

thorns and briars meet us at every step, thou hast but to put out thy hand to gather them. It is much easier for every one to be poor like Christ than to abound in riches; to humble himself than to be in authority; to obey than to rule; to restrain his appetite than to live in luxury; and so on. Since, then, the Eternal Father has attached the gaining of Paradise to conformity with the life which Christ led, He has attached it to what is in the power of every one; all that is wanted is a firm resolution, and this is a thing which would not be sufficient in the opposite case. Well, therefore, did the Apostle say : "He predestinated to be made conformable to the image of His Son, that He might be the First-born amongst many brethren." For even though, in the opposite case, there should be many to reign with Christ, yet they would not be His brethren, and so He would indeed be the "First-born," but not "amongst many brethren." For what sort of adopted brethren would those be who in no way resembled Him Who is the Son by nature? And how could it be possible for those to be like Him in glory who were not like Him in humiliation?

VI. Consider sixthly, how people do all in their power to avoid suffering, and therefore to avoid being saved : "If you be without chastisement, whereof all are made partakers, then are you bastards, and not sons."[1] And yet it is said that these brethren of Jesus are to be many—"That He may be the First-born amongst many brethren"—and consequently the elect will be numerous. This is certain. "I saw a great multitude which no man could number,"[2] and this shows us the gratitude which we owe to God for having ordained a Purgatory after this life. Were it not so, alas for us! What would become of all those Christians who are so given up to earthly comforts? Who amongst them could be saved? This is why God has, in His infinite mercy, ordained that those of the elect who commit so many imperfections through fear of suffering in this life, shall have their suffering after death. So that Purgatory may be very truly described as a place where those who have not chosen voluntarily to make themselves conformable to the suffering life of the First-born, are made conformable to it. There, by way of mere satisfaction, every one will have to gain that which he would not gain by way of merit. But is it not the height of folly to make up one's mind voluntarily to that abyss of pain, to refuse to merit here, in

---

[1] Hebrews xii. 8.   [2] Apoc. vii. 9.

order to make such hard satisfaction there? How sharp will those sufferings be which bring no merit, but must pay the debt! "Amen, I say to thee, thou shalt not go out from thence till thou repay the last farthing."[1]

VII. Consider seventhly, that thy favourite devotional exercise should be to take in thy hand the crucifix, which is the image that our Redeemer especially chose to leave us of Himself here on earth, and then to consider it in all particulars, and examine how far thy copy is like the original. Alas! how great a difference! Christ stripped of His garments, and thou well clad; Christ in suffering, thou in luxury; Christ forsaken, thou given up to pleasure; Christ in contempt, thou in honour. Is this, thinkest thou, a true copy? Rather, if thou art unable of thyself to choose suffering, shouldst thou pray to God to be pleased to compel thee to suffer. And indeed what but this dost thou pray for when thou prayest Him to bring thee to Paradise? Thou dost but pray, in other words, that He will send thee great suffering; for this is the law: "Whom He foreknew, these He also predestinated to be made conformable to the image of His Son."

FOURTEENTH DAY.

Blessed are the dead who die in the Lord. From henceforth now, saith the Spirit, that they may rest from their labours: for their works follow them (Apoc. xiv. 13).

I. Consider first, who they are who "die in the Lord." They are those who have lived in the Lord; for this is what commonly happens, that every one dies where his place of abode is. It may sometimes accidentally occur that a person dies in a place through which he is passing, but this is but seldom; the rule is that he dies where he has lived. And so the man who lives in sin, dies in sin; and the man who "lives in the Lord," dies "in the Lord." Where art thou living? Consider that thou wilt die where thou art living, if not in the particular sin which thou committest by accident, at least in that which thou committest habitually, the sin of lust or envy, or whatever it may be which may be rightly called thine: "You shall die in your sin."[2]

[1] St. Matt. v. 26.   [2] St. John viii. 21.

II. Consider secondly, what to "die in the Lord" means. It means to die, if not *for* the Lord, like the martyrs, at least *in* the Lord, like His confessors, that is, like those who have served Him faithfully, not merely by living in Him by grace as all the just do, but by living in Him by a special affection of charity. Such persons are rightly said to "die in the Lord," not merely because they die in a state of grace, which is common to all the just who die, but because they die with an entire abandonment of themselves into the bosom of their Lord : they die in His Sacred Side, in His Heart, in the blessed embraces of His arms. How sweet a death to die in God's embrace ! Alas for those who live in the arms of His enemy, as his especial favourites! in his arms they must expect also to die.

III. Consider thirdly, as a proof that this blessed death of which we are speaking does not fall to the lot of all those who die in a state of grace, but only of those whose life has been singularly perfect, God says, "Blessed are the dead who die in the Lord." It seems beyond a doubt that two different deaths are here spoken of, one of which comes after the other : how, otherwise, can the dead be said to die? It is the living who die, not those who are already dead. Yet this is what is said here : "Blessed are the dead who die in the Lord ;" and surely there is a mystery in the words, for if there is not a single superfluous particle in all the Sacred Scriptures, least of all can this be so in the Apocalypse, where it is expressly threatened to strike out of the Book of Life the name not only of the man who shall contradict a single word as incorrect, but of him also who shall cancel a single word as superfluous : "If any man shall take away from the words of the book of this prophecy, God shall take away his part out of the Book of Life."[1] This being so, all commentators, terrified by the fulmination of this decree, agree with remarkable unanimity in judging that in this passage those persons are particularly referred to, who, after first dying to themselves by living a life altogether in God's embrace, have afterwards the wonderful blessedness of dying in it also. See, therefore, how true it is that it is those who live in the Lord who will die in Him ! But what is dying to self ? It is to anticipate the work of death by detaching ourselves from everything which death will at last take from us : property, country, relations, vanities, pleasures, amusements, and above all from

---

[1] Apoc. xxii. 19.

an excessive love of ourselves; so as to live, if possible, without the body, even while in the body. It is to such persons that the Apostle was able to write: "You are dead, and your life is hid with Christ in God."[1] It is true that to be crowned it is not enough to begin well, we must go on perseveringly to the end. And, therefore, those are not here called "blessed" who merely die to themselves, but those who, after having died to themselves, die at last in the Lord: "Blessed are the dead who die in the Lord." Of what avail will it be that thou didst once die to thyself to live to the Lord, if afterwards thou risest again from that death, to live once more to thyself? Thou must be willing to continue dead to thyself till the time comes when thou shalt "die in the Lord."

IV. Consider fourthly, that if this first death fills thee with terror, the second should console thee, because it will at length give thee eternal rest from all thy labours. As so it is added: "From henceforth now, saith the Spirit, that they may rest from their labours." For who is the Spirit who now urges thee to suffer much for God, to work and strive and mortify thyself? Is it not the Spirit of God? Well, then, that same Spirit, Who now bids thee labour, will then bid thee cease from labour. "From henceforth" means for a time dating from a certain moment. Take notice, therefore, in the first place, that whatever spirit bids thee cease from labouring before that moment is certainly not the Spirit of God; it is either thy own spirit, or the spirit of the world, or the spirit of evil. The Spirit of God never says this to any one till that moment has come—"From henceforth now, saith the Spirit"—but not before. Oh, if thou didst but know how hateful it is to the Spirit of God for any one to live in sloth on earth! He would have every one work—work continually—work as long as he is able: "Labour as a good soldier of Christ."[2] And this is not to be wondered at, for just as doctors tell us that slothfulness produces two very pernicious consequences in the body, namely, weakness and humours; so too is it with the soul, it makes it weak in doing good, and inclined to evil. It is true that these effects are not immediately perceptible in the soul any more than in the body, but only when they have taken possession of it and become fixed habits; but this is only an additional cause for fear, since it is precisely the maladies which are secret and gradual in their development which are

---

[1] Coloss. iii. 3.   [2] 2 Timothy ii. 3.

the most incurable in the end. And such are the maladies arising from sloth. Next, observe that the Spirit of God tells these blessed dead that they are to rest because they have laboured enough: "That they may rest from their labours." This rest is of two kinds, negative and positive. The negative rest is the mere ceasing from labours, and the positive is the beatitude which adds to cessation of labour that perfect peace which is experienced by the soul in the secure possession of the desired object. Now, in the text, God intends to speak of both kinds of rest. He says that they rest from labours, which is negative rest; and He says that they rest because of those labours, which is positive rest. If He had only intended the former meaning, in the first place that would be saying little, for what great reward is the mere cessation of labour? And secondly, it would, in this case, have been enough to say "from labours," without adding "their;" for who ever rested from labours that were not his own? The pronoun is added to denote the second meaning also, to signify that they receive beatitude on account of labours, but that these labours are "not those alone which God endured for them, as some would have it who promise themselves Heaven solely on account of faith, even though unaccompanied by works. Is it not folly to lay claim to rest merely on account of the labours which others have borne for thee? If thou desirest the repose to be thy own, the labours must necessarily have been thine also.

V. Consider fifthly, that modern heretics triumphantly declare that this passage proves how ridiculous it is to believe in Purgatory, since every one who dies in a state of grace passes immediately into this rest: "From henceforth," that is to say, from that moment, "from henceforth now, saith the Spirit, that they may rest." How foolish they are! Have those persons who do no more than die in grace laboured in the manner here signified up to that moment? Surely not. How, then, can they claim to begin to rest from that moment, in this manner? Let those poor heretics take notice who the persons are of whom the Divine Spirit says "that they may rest." He says it to those who have laboured very hard, dying to themselves and living altogether to God: "From henceforth now, saith the Spirit, that they may rest from their labours," not "their labour," but "their labours." If they had chosen to labour but little to win Paradise, as those do who think to reach it, so to speak, in their carriages, they certainly would not gain admittance so quickly. They would first have to suffer the punishment of their slothfulness in the flames of Purgatory. But, because they laboured greatly, therefore are they so quickly called to the enjoyment of rest. Oh, if men did but understand all that is implied in those words, "from their labours"! But many do not understand their force, because they have never learnt it by experience. Besides, when it is said of these blessed dead, "From henceforth now, saith the Spirit, that they may rest," this is a different thing from saying, "The Spirit saith that they shall rest now." As soon as they have drawn their last breath, He pronounces the sentence of eternal rest in their favour : therefore it is said, "Now, saith the Spirit, that they may rest ;" but it is not said "that they may rest now," because, between the sentence and its execution there is generally some interval, longer or shorter, according to the debt which has to be paid. And therefore this text might with greater reason be turned against those persons who venture to use it against us. For if there were never any interval it would be said "that they may rest now ;" but because there is such an interval, it is said, "Now, saith the Spirit, that they may rest," that is, that they may rest when their time for rest has come. It is true, indeed, that this time comes quickly for those who have laboured much for God, and therefore it is here spoken of as though there were no interval at all, because these blessed dead are spoken of by the Apostle as those who "shall be saved, yet so as by fire,"[1] so brief will be their passage through these flames, even if they have to enter them.

VI. Consider sixthly, that because it is a sentence which is here spoken of, the expression chosen is of saying "that they may rest," and not rather of making them rest, although it is certain that the deed will correspond with the word. Neither will this be a sentence capriciously given ; for which reason it is added, "for their works follow them ;" because the works of those who have laboured so much for God will be faithful witnesses to their merit at His judgment-seat, according to the words, "Her works praise her in the gates."[2] And so these works of theirs are said to "follow them," because the works of the just are not corruptible, like the works of sinners which for that reason end with the life of those who do them : "Every work that is corruptible shall fail in the end."[3] The works of the just are solid, permanent, and lasting, therefore they follow him who does them. What

[1] 1 Cor. iii. 15.   [2] Prov. xxxi. 31.   [3] Ecclus. xiv. 20.

will be left to sinners in Hell of those roses with which they
loved to crown their brows? Only their thorns: that is, the
sharpness of remorse. Whereas the just will have reaped
the fruit of their labours—"The fruit of good labours is
glorious"—and so they will never cease to enjoy it, but will
find continual delight in remembering what they have suffered
for God. The works of the just are also said to "follow
them," because they will not take with them all the good
works which they did on earth, but will see many following
them one after another, according as they are gradually
perfected. For example, take all the glorious founders of
religious orders, how many ages have passed since their
deaths, and yet it may truly be said that at every moment
"their works follow them," because they are continually
reaping fresh fruits of their bygone labours: "Good things
continue with their seed."[1] Lastly, it is said that "their
works follow them," because, just as conquerors of old had no
more glorious retinue in their triumphs than that which was
composed of their works, that is, of kings in chains, of vanquished generals, subjugated rulers, representations of enslaved
cities, so likewise will it be with these blessed dead. They,
too, will have their heavenly triumph, in which they will
doubtless have the escort of a great multitude of angelic
legions, but yet there will be no retinue there comparable
to that of their works. This will be the most glorious of all,
and this, therefore, is the only part of it mentioned here:
"Their works follow them." It will not avail sinners to be
borne to the tomb with muffled drums and trumpets, and
mourning garments trailing in the dust for show: where are
the works that follow them? Bowed with shame, naked,
miserable, and alone, they must stand before the great
judgment-seat of Christ their Judge. Only the just will go
there with an honourable retinue, for they only will be followed
by their works: "For their works follow them."

[1] Ecclus. xliv. 11.

FIFTEENTH DAY.

Christ, therefore, having suffered in the flesh, be you also armed with the same thought (1 St. Peter iv. 1).

I. Consider first, that Christ, when He suffered so much in His Flesh, did so, not because this was needful for His Flesh, but for thine. He was most pure and most perfect in His Flesh : most pure, because He was never obliged to withdraw it from evil ; most perfect, because He was never obliged to incite it to good. Therefore, He never suffered anything from the need of His own Flesh, but from the very great need of thine, which is so slow to good and so prone to evil. It might seem, however, as though the Apostle ought properly to have said : "Christ, therefore, having suffered in the flesh, be you armed with the same suffering." For if Christ, in order to overcome thy flesh, which could not in any way harm Him, armed Himself with so many sufferings, with scourges, thorns, and sharp nails, how much more oughtest thou, who daily receivest from it so many injuries, to be thus armed in order to conquer it? Yet the Apostle, who knew thy weakness, did not say, "Be you armed with the same suffering, but "with the same thought." He would have thee arm thyself, if not with the suffering of Christ, at least with the thought of that suffering. What excuse will there be for thee if thou refusest to do so?

II. Consider secondly, that this armour must be of two kinds, defensive and offensive : defensive, to repel the assaults of thy rebellious flesh ; and offensive, to attack it, that is, to keep it humble and obedient, to compel it to pay to the spirit the homage which is its due. First, then, the thought of Christ's Passion will serve thee as a weapon wherewith manfully to repel the assaults of thy flesh, for it is universally taught that the most powerful remedy against sensual temptations is to think on what Christ suffered for us : "Thou shalt give them a buckler of heart, Thy labour."[1] How is it possible to set thyself to contemplate Christ on the Cross, to see Him, stripped of His garments, pouring forth all His Blood for thy sake, bruised and torn and tortured ; and at the same time to think of indulging thy body in forbidden pleasures? Rather wilt thou feel thyself inflamed with a holy indignation against thyself, rather wilt thou desire to punish thyself, to mortify

---

[1] Lament. iii. 65.

thyself, and to inflict on thyself due chastisement, which
consists not merely in defending thyself from thy flesh, but in
making war upon it. Take notice, therefore, that for this end
it is not enough to give a passing thought to the Passion of
Christ, it must be dwelt on attentively. And therefore the
Apostle does not here say, "Christ, therefore, having suffered
in the flesh, be you also armed with the same remembrance"
(*recordatione*) but "with the same thought" (*cogitatione*).
This is what is wanted, constant thought. Do not say that
men take up arms in time of need, and afterwards lay them
down ; for if the flesh is always either making war or preparing
to make war upon thee, can there ever be a time when it is
right to lay down weapons so efficacious against it ?

III. Consider thirdly, that in order really to derive great
help from this thought of the Passion, thou shouldst above
all things endeavour to have a clear apprehension Who it is
Who endured it for thee. This is why it is the Apostle says
absolutely, "Christ having suffered in the flesh ;" he does
not say "having suffered stripes, or wounds, or the Cross,"
but simply "having suffered ;" for this alone should be enough
for thee. If the True and Living Son of God had done no
more for thy salvation than taste that drop of gall which He
received for thee on the Cross, that alone ought to be enough
to make thee, a wretched worm of earth, ready to live steeped
in a sea of bitterness for love of Him. For here is the marvel,
not that He suffered so much for thee in His tender Body,
although this indeed was a very great marvel, since He had
to exercise even miraculous strength to be able to endure it ;
but that He condescended to suffer it. For just as Tobias,
considering the favours bestowed on him by the guide of his
young son, was ready to give him the half of his substance in
return, but on hearing that his benefactor was an angel, nay
an archangel who had come down from Heaven for his sake,
fell to the ground like one dead, and was incapable either
of looking on him, or answering him, or thanking him, but
thought that all he could do was to die at his feet, so, too,
doubtless oughtest thou to be greatly moved when considering
what Christ suffered for thee ; but when thou rememberest
that He Who so suffered was not an ordinary man, nor an
angel, nor an archangel, but the Son of God Himself Who
came down for thy sake from Heaven to earth, then surely
thou oughtest to be stupefied and utterly bewildered, and to
declare, if indeed the power of speech is left to thee, that thou

art ready, if He so pleases, to die there, prostrate at His feet. If thou hast not the heart of a tiger, thou canst not feel less than this at the remembrance of what "Christ suffered in the flesh" for thee, and therefore thou shouldst always keep it as vividly as possible before thy mind, in order to live as one dead to thyself, so that thy flesh may no longer be able to trouble thee : "I will be mindful, and remember, and my soul shall languish within me."[1]

SIXTEENTH DAY.

Exercise thyself unto godliness ; for bodily exercise is profitable to little, but godliness is profitable to all things, having promise of the life that now is, and of that which is to come (1 Timothy iv. 8).

I. Consider first, that piety is a virtue which inclines us to love that which is our beginning, for which reason it has been employed to signify the love of our parents and of our country. But since God is our First Beginning, therefore the first kind of piety is that which regards God with that particular inclination which is due to Him Who has so graciously given us our being. This, then, is in substance what the Apostle here means by the word "godliness" (*pietas*). He means the worship of God, but not that generic worship which we rather designate by the name of religion ; what he means is a more devout, loving, ardent worship, such as distinguishes those whom we are in the habit of calling pious persons. To this kind of piety very great rewards are promised, not only in the future but even in the present life. For as in the Decalogue a special reward even in this life was promised by God to that lesser piety which regards men—"Honour thy father and thy mother, that thou mayest be long-lived upon the land"—so much more, in the Gospel, was promised by Christ to the greater piety which regards God : "Seek ye first the Kingdom of God and His justice, and all these things shall be added unto you."[2] This is the virtue which has the blessing both of the right and the left hand from our Father, "of the dew of Heaven, and of the fatness of the earth,"[3] so dear is it to Him above all others. Dost thou not think, then, that the Apostle has good reason to exhort so earnestly to this virtue

[1] Lament. iii. 20.   [2] St. Matt. vi. 33.   [3] Genesis xxvii. 28.

one whom he loved as he did Timothy? How art thou inclined to works of piety? Dost thou practise them with a feeling of attraction or of repugnance? If the latter is the case, it is a sign that thou dost not as yet possess this beautiful virtue; for the mark by which a habit is known is a readiness to perform its acts.

II. Consider secondly, that in order to incline thee to this habit, the Apostle now says to thee too from Heaven, "Exercise thyself unto godliness," for it is in this way that the habit is formed, by the frequent practice of its acts. All the abstract science in the world is not enough to make thee pious, at least only after a very long time; it is practical science which does this quickly. Observe, therefore, that the Apostle does not here say, "Exercise thyself in," but "to godliness," because whenever thou findest no pressing occasion of exercising thyself in works of piety, thou art to act as soldiers do, who, when no battles are going on in which they may try their strength, have recourse to mock combats merely for purposes of training; and so the Greek word here used by the Apostle signifies the kind of exercise in use by athletes. In the *palæstra*, or lists where athletes practise fighting, running, riding, wrestling, these things are not done from any necessity of the moment, but simply as an exercise, so important is it considered to be ready in these matters. Thus, too, would the Apostle have works of piety done, if for no other reason, at least as an exercise—"Exercise thyself unto godliness"—for who can say how useful it is, when there is a necessity for these works, not to have to force oneself in any way, but to be able to practise them readily? And besides, in this world the prizes gained in a mock combat and in real battle are very different. In the former, they are of trifling value, a mantle, a collar, a girdle, a ring, things given merely as incentives to good practice, whereas in a real battle sometimes a whole kingdom is at stake. But with God it is not so. There is an equal reward for the man who fights in the arena as an exercise and for him who fights in the field from necessity. Who, then, would not desire to perform works of piety, since, in whatever way they are practised, the advantage is so great.

III. Consider thirdly, that it is in allusion to the subject I have been speaking of that the Apostle adds, "For bodily exercise is profitable to little, but godliness is profitable to all things." He compares the exercise of piety to the exercise of the body which athletes practised in the *palæstra*, which were then in the highest repute in Greece, and therefore he also makes use of their phrases. I say their phrases, because this one, "bodily exercise," corresponds to that which was in vogue in Greece, *gymnastica lucta*. This being the case, in order to show how much more desirous a Christian should be to be trained to piety than athletes to their feats of prowess, he says that the exercises of these latter, of whatever sort they were, are of little use, "profitable to little," whereas piety is "profitable to all things." The exercise of the athletes produces, at most, but two results besides the skilfulness which they attain in their practice: the first being health, which is fortified by continual motion of the body in such a manner as to prolong life; the second is the usual prize. But what are these things to the fruits which piety brings to those who practise it manfully? "Bodily exercise is profitable to little," because, though it prolongs temporal, it cannot bestow eternal life; it is able to give earthly, but it cannot give heavenly rewards. Piety, on the other hand, is "profitable to all things," because it not only lengthens temporal, but bestows eternal life; it not only gives earthly, but heavenly rewards, as is expressed in the words, "Having promise of the life that now is, and of that which is to come." There can be no doubt that piety brings with it eternal life and heavenly rewards, which are the promises relating to the life to come; but thou mayest be less sure about its bestowing temporal life and earthly rewards, which are the promises relating to the present life. Yet this too is most certain. For, with regard to life, the Wise Man says that "the fear of the Lord shall prolong days;"[1] and this is confirmed by reason, because the just live free from a countless multitude of disorders which shorten life far more than any sufferings endured for God; and as regards earthly rewards, it is certain that, speaking generally, piety is more prosperous than vice. It is true that when a pious person prospers it is not remarked, because it is a thing which appears in order, whereas, when a vicious person prospers, it is remarked at once and looked upon with distaste, as a monstrous thing; but the very fact of its being regarded as monstrous is a proof that it rarely occurs. It does, indeed, sometimes happen that God shortens the life of a just man and that He withholds earthly prosperity from him, but in such a case He does not fail in His promises,

[1] Prov. x. 27.

because, while withholding a lesser, He always makes up for it by bestowing a greater favour. If He shortens early life, He makes up for it by sending the just man more quickly to enjoy that which is eternal and far more glorious ; and if He withholds temporal prosperity, He makes up for it by the spiritual consolations with which He inundates the soul, by joy of heart and peace of conscience, that "hundred-fold" which is far more precious than all other gifts of God to the faithful soul in this world. Oh, how true is it, then, that "bodily exercise is profitable to little, but godliness to all things"! And yet—would it be believed?—there are many persons who go through excessive labour in exercising themselves in the arena of this world, and refuse to do so in that of God! This is the folly which is universal among men. What will thy lot be if it is thine also? Thou labourest greatly in the service of the world, that is, for the sake of gaining "a corruptible crown ; " why, then, dost thou not labour far more for the love of God, Who promises thee one which is incorruptible?

IV. Consider fourthly, that many persons have chosen to understand by the "bodily exercise" here spoken of, the penance which we call corporal, and so have gone very near to disparaging it on the authority of this passage. But, if we may believe St. John Chrysostom, who was so faithful an interpreter of the Apostle, such an explanation is entirely erroneous; and other eminent commentators have followed him in this: (1) Because corporal penance done from the love of God is a very real work of piety, which has been practised by all the saints, and consequently cannot be contrasted with it as an exercise which is opposed to it. (2) Because penance is not a bodily, but a spiritual exercise ; and it is manifest that it is the soul which is strengthened, and that the body is rather weakened by it. (3) Because it is not true that penance is only profitable "to little." It is profitable, even if this were all, for subduing the rebellion of the flesh ; and therefore it is true that though it does not contain all the benefits of holiness, because it is the prelude to it, yet it is profitable for obtaining it : "Profitable to all things." (4) Because it also has its promises, both as regards this life and that which is to come, like all other works of piety. (5) Because it cannot be shown that the Apostle has anywhere dissuaded from corporal penance ; on the contrary, he inculcates it, and that by giving us his own example : "I chastise my body."[1] And if in the .

[1] 1 Cor. ix. 27.

following chapter of this Epistle he bids Timothy relax his
bodily penance in some degree by making use of wine, his
manner of speaking cannot be much satisfaction to those who
are addicted to it, since he limits the quantity he is to take by
saying "use a little;" and the reason for which it is allowable
by saying "for thy frequent infirmities." So that it is evident
that when the infirmities ceased to exist, St. Paul would have
given him leave to drink no more wine. The Apostle could
never, therefore, have intended penance to be understood by
the expression "bodily exercise;" except, indeed, the penance
of those who perform it without any devotional feeling, in
which case it is a mere material act, and so there is no need
to wonder that it is of little value. Here, then, thou hast to
observe, for thy profit, that to all thy works of penance it is
necessary always to add the interior acts of compunction,
charity, and humility which are proper to them, in order that
they, too, may thus become works of piety. Otherwise it is
certain that no matter how severely thou mayest afflict thy
body with blows, sharp nails, or hair-shirts, thou wilt do
nothing more than exercise it, like the athletes in the *palæstra*,
with material acts. And so thou wouldst verify the interpretation of those persons who say that "bodily exercise
is profitable to little," meaning by "bodily exercise," penance;
because in that case thy works would be, if I may be allowed
the expression, rather those of a gladiator than of a penitent.

SEVENTEENTH DAY.

A hard heart shall fare ill at the last (Ecclus. iii. 27).

I. Consider first, that in order to understand what is
meant by this "hard heart," it is necessary to understand the
exact meaning of "hard." Material things may be regarded
as differing in three ways: some are hard, some soft, and
others liquid. Liquid substances are those which have no
form proper to themselves to distinguish them, but which
adapt themselves to every form; such, for instance, as water,
which immediately takes the form of the vessel in which it is
placed, allowing itself to be drawn off, raised or lowered to
another level at pleasure. Soft substances have a proper form,
but in such a way that they easily lose it to assume another;

just as water does when condensed into snow, which has, indeed, its proper form, but which, nevertheless, by simple manipulation, may be shaped into a ball, a pyramid, a statue, according to fancy. Hard substances not only have a proper form, but in such a way that they will not change it, do what thou wilt. They may be broken to pieces, but never made by mere manipulation to assume a form which is not their own ; take, as an instance, the same substance, water, when solidified into Alpine ice. Now, these three differences exist in the human heart to the eyes of God, though they are not so visible to ours. In some persons it is liquid like water in its purely natural state ; that is the case with those who, if we may say so, have no longer any will of their own at all, but to mould themselves after God's will that they let themselves be led by it as He pleases. This is the condition to which we are invited in these words : " Pour out thy heart like water before the face of the Lord."[1] In others, if not liquid, it is, at all events, soft, like water turned to snow ; and this is the state of those who are not, indeed, so easily conformed to the Divine will in all things as the first class, because they have still too much individual form, but who nevertheless would never altogether resist it, but, rather than displease God, will obey His will. Lastly, there are those whose heart not only is not soft, but hard as water in the state of ice ; and these are the persons who are so unwilling to submit themselves to the will of God that they are not afraid to offend Him grievously in order to live as they choose. This will at once show thee of what sort of heart the Wise Man here means to speak of when he says, " A hard heart shall fare ill at the last." A hard heart, to explain it properly, is a disobedient heart : "They make their heart as the adamant stone, lest they should hear the law."[2] Alas ! for thee, if thy heart is such a one as this ; it would be well to tear open thy breast and take it out with thy own hand. Pray to God, if thy heart is not liquid as water, at least to give thee one that is soft, a heart of flesh, which is human, not a heart of stone : " I will take away the stony heart . . . and will give them a heart of flesh."[3]

II. Consider secondly, why it is said that this hard heart "shall fare ill at the last," that is, at the hour of death. The reason is, that a heart of this kind will not be able to accept of death with proper resignation. Not having been accustomed

---

[1] Lament. ii. 19.   [2] Zach. vii. 12.   [3] Ezech. xi. 19.

to submit to the will of God in life, it will not be able to do so in death. For, if it found so much difficulty in doing this in things far less arduous, how much harder will it be to do it in the most difficult and painful of all things, namely, in dying? It will, indeed, see clearly that it has to submit, whether it will or no. But this very necessity will keep it in a state of agitation, anxiety, trouble, and dejection ; how, then, will it be possible for one in this disturbed condition to think rightly of his soul and to make the acts which are so necessary at that time? This is one reason why "a hard heart shall fare ill at the last." There is yet another, on the side of God : namely, that it is not enough for Him to make use of ordinary means to conquer this heart, because it is "a hard heart ;" the kind of grace He would have to employ is that which is described as "a hammer that breaketh the rock in pieces,"[1] that is, very powerful and even extraordinary aids. But how can it be expected that God will do this for one who has been so rebellious to Him? "With the perverse Thou wilt be perverted," said David to God, the meaning of which is what has just been said : "With the hard Thou wilt deal hardly."[2] And therefore such a heart will "fare ill at the last," because it will not have the grace necessary to win it. Take these words, then, in whichever of the two senses thou choosest, either in what is called the intransitive sense, as signifying "it will be in an evil state at the last" (*male se habebit*), and thou canst see that this is true, because it will be in evil dispositions ; and this is the reason we assigned as being on the part of man : or there is the transitive sense, "it will fare ill as to what it should have at the last," and thou seest that this also is true, for it will fare ill as regards the grace which it should then have in great abundance, which, as has been said, is the reason on the part of God. Such a heart will be in the case of a certain poor sinner, who, when his death was approaching, and the priest was assisting him by putting before him the usual motives which are most powerful in exciting to acts of compunction, confidence, or love of God, looked at him fixedly for a while, and then exclaimed, using a metaphor common to his class, "The loaf is hard and the knife does not cut," and so died. Poor, foolish man! he meant, it seems, by this expression, to divide the blame between two parties, the heart and grace, when he ought to have laid the whole on the former. If an ordinary knife will

[1] Jerem. xxiii. 29. [2] Bellarm. in Psalm xvii. 27.

not cut a loaf, what is to be done? Do people go and seek for a hatchet? No; it is not the knife, but the loaf that must be changed. We know very well that when it pleases God to do so, He can at once make use of means which no heart, however obstinate, can resist; but we also know very well that although He can, He is not bound to do so. See, then, how important it is not to harden the heart. But how do things become hard? By little and little. First water turns to snow, then to ice, then to impenetrable crystal: "The water is congealed into crystal, . . . and shall clothe the waters as a breastplate."[1]

III. Consider thirdly, what thou hast to do in order to remove from thy heart this sinful hardness, if unhappily this should be thy state. Powerful remedies must be used, and the chief among them is taught thee by the bride in the Canticles: "My soul melted when my Beloved spoke."[2] What dost thou suppose these words to imply? That she was dissolved in tears, fainting with the sweetness of which spiritual persons are often so desirous? Such a sense would ill-suit her high state of perfection. What she meant to convey in these words was that she was most perfectly willing to let herself be guided in all things by the will of her Beloved, without retaining any proper form of her own; just as we have said is the case in substances which are not only soft like snow, but liquid like water. But by what means had she attained so excellent a disposition? By hearing God's voice: "My soul melted when my Beloved spoke." This, then, is what is needful for thee for the purpose in question—to hear the word of God. Of this there are two kinds, the one living, the other dead. The former is heard in spiritual books, the latter in prayer. Devote thyself to both, to spiritual reading and to meditation, and thou wilt see that the heart, which now perhaps lies in thy breast more hard than the mountain ice, will gradually melt till it flows as readily as water. But if thou never listenest to that sweet voice of thy Lord, thou art lost; for then thou wilt never know how sweet He is, and so thou wilt not love Him. And if thou dost not love Him, how canst thou possibly run after Him with the ease of those substances which are liquid? It will be much if thou lettest thyself be moulded by Him, like those which are soft. But not even this will be thy lot; thou wilt go on growing harder and harder by preferring thy own caprice to

[1] Ecclus. xliii. 22.   [2] Cant. v. 6.

His law: "His heart shall be as hard as a stone."[1] In how evil a case wilt thou find thyself at the hour of death! "A hard heart shall fare ill at the last."

IV. Consider fourthly, that not only will the hard heart fare ill in death, but it fares ill in life too: not only *male habebit*, but *male habet*. Nevertheless, the Wise Man did not choose to say that it "fares ill" in life, but only that it "shall fare ill" in death, "at the last;" because he knew that such a heart, although its state is to be considered very evil even in life, does not know its own misery, and so is not concerned about it. Nay, there are none upon earth who think themselves happier than those who live altogether as they choose, and make light of every law: "Who is the Lord, that I should hear His voice?"[2] But when death comes, it is different. Then that heart, which in life did not know its misery, so hardened was it to the sharpest stings of conscience, will know it better than all besides, because it will see, better than all besides, that it is irretrievably lost. And therefore it will then, indeed, be so far softened as to be in the greatest agitation, but not far enough to feel compunction and confidence, and so to be saved: "When I consider Him," so will that wretched soul then say, speaking of God, "I am made pensive with fear. God hath softened my heart, and the Almighty hath troubled me,"[3] that is to say, the Lord Who, as God, shows me how much He deserved to be loved, "has softened my heart." But what then? Being Almighty, He at the same time shows me how He will punish me: "He hath troubled me." He has not touched me with compunction nor moved me to contrition, He has only "troubled me." And so, do thou draw the conclusion that "a hard heart shall fare ill at the last," more than in life. For if in life it fares ill, it does not know it; in death it not only will fare ill, but will also know it; and yet there will be no way of escape from it.

EIGHTEENTH DAY.

Be ye merciful, as your Father also is merciful (St. Luke vi. 36).

I. Consider first, that when it is said, "Be ye merciful, as your Father also is merciful," the word "as" does not enjoin equality, but similarity; for who is there who can possibly

Job xli. 15.   [2] Exodus v. 2.   [3] Job xxiii. 15, 16.

equal God in mercy, which is the virtue in which, above all others, He so much glories? It will be much to resemble Him in it, and it is to this that Christ invites thee in the text. And He does not say "have mercy" (*miseremini*), but "be merciful" (*estote misericordes*), that thou mayest aim, not merely at the act, but at the habit, which includes all perfection. Endeavour here to understand, as much as possible, each one of these perfections, so as to imitate them at least in some part, like one who proposes to improve himself by copying the works of an artist who stands alone in the world.

II. Consider secondly, that mercy is a willingness to assist and relieve others in their miseries. This willingness may spring from two causes, from charity or from compassion. When it springs from charity it is much more perfect than when it springs from compassion, because charity is a virtue, and compassion is not a virtue, but a natural affection of tenderness, which inclines us to grieve over the troubles of others even when we do not wish to do so. In God mercy has its source in charity, because it springs from a pure love, moving Him to relieve our necessities, not from a compassion which puts force upon Him to do so: "I will have mercy on whom I will."[1] God is not capable of such affections, for they imply weakness, as being given to supply the lack of the virtue. He who has true charity has no need of compassion to be induced to succour others in their misfortunes. All that he needs is to know them, and he is as much moved by hearing of them as by seeing them. And it is this which is required of thee in the words, "Be ye merciful, as your Father also is merciful." It is required of thee to be moved by charity, not by mere compassion, in succouring the miserable, that so thy action may be more meritorious.

III. Consider thirdly, that as the compassion which has been just spoken of, when it precedes the willingness to relieve, is not a virtue, but a natural affection which incites to the virtue, so, when it follows it, it is a virtue, and a very great one, because it is the result of a spontaneous act of the will, by which the succour is given with greater plenitude of charity. I say this, because it is evident that the man who exercises such an act not only desires to relieve the miseries of another with kindness, but to condole with him, that is, to feel for them as though they were his own: "Who is weak, and I am not weak?"[2] This is the great excess of charity which God

---

[1] Exodus xxiii. 19.   [2] 2 Cor. xi. 29.

showed when, not satisfied merely with that charity, immense and unexampled as it was, He was pleased to take upon Himself the tender compassion which made Him become Incarnate, herein doing far more than was necessary for our complete succour. And it is this to which thou too art exhorted in the words, "Be ye merciful, as your Father also is merciful," that is, thou art to be moved by charity to relieve others, but thou shouldst endeavour to superadd this affection of compassion, which will make thee feel the miseries of others as though they were thy own : "Put ye on, therefore, as the elect of God, bowels of mercy." [1] See how highly God esteems this act, when He has gone so far as to let Himself be spoken of as though, so to speak, He had not known what mercy was till He had done this : "It behoved Him in all things to be made like unto His brethren, that He might become merciful." [2]

IV. Consider fourthly, that this great compassion, even when it is a virtue from our having voluntarily chosen to feel it, although it certainly inclines us to relieve every one who is in trouble, yet it is chiefly those whose trouble has come upon them against their will ; for when a person has brought it upon himself by his own choice, we are disposed to say that it is his own affair : "Who will pity any that come near wild beasts ? " [3] Not so God. God has compassion on those who are the authors of their own misery, and therefore He has compassion on sinners. Nay, it is these He is bent on succouring beyond all others ; for although we are apt to think those most wretched who suffer evil against their will, yet in reality those persons are far the most so who bring it deliberately on themselves. This, too, our Lord requires of thee when He says, "Be ye merciful, as your Father also is merciful."

V. Consider fifthly, also, that this compassion makes us more inclined to grieve for the troubles of our friends than of our enemies ; nay, not only do we feel no sorrow whatever for the troubles of our enemies, but we are pleased at them. But God has compassion even on His enemies ; and not only in this world, where, in a certain manner, He defends from the anger of all creatures so many persons who are actually offending Him, and provides them with food and all things, but He has compassion upon them even in Hell, where, although He does indeed, because He is perfectly just, inflict punishment on them, yet, because He is merciful, He makes this punishment, severe as it is, less than they deserve. This, again, is

[1] Coloss. iii. 12.　[2] Hebrews ii. 17.　[3] Ecclus. xii. 13.

required of thee in the words, " Be ye merciful, as your Father
also is merciful," that thou shouldst be able to compassionate
the troubles, not only of friends, but of enemies, being equally
ready to succour both in their necessities, because this is the
example given thee by God, "Who raineth upon the just and
the unjust."[1]   Consider, also, that this compassion inclines us
to pity those, even amongst our friends, who are united to us
by relationship, or country, or some other tie, more than those
with whom we have no such bond : "Shall I take my bread
. . . and give to men whom I know not whence they are ?"[2]
It is otherwise with God.   No one is His neighbour, because
all men are at the same, that is an infinite, distance from
Him ; and yet He pours down His mercy upon all men from
this vast distance : "I will gather them from the ends of the
earth, and among them shall be the blind and the lame, . . .
and I will bring them back in mercy."[3]   This, also, is enjoined
thee in the words, "Be ye merciful, as your Father also is
merciful ;" not to confine thy mercy to those who belong to
thee, like a lake enclosed within its bounds, but to let it flow,
in due measure, over those at the greatest distance from thee,
like rivers which do not restrict their bounty to the particular
district in which they have their source.

VI. Consider sixthly, that the man who is very fortunate
in his condition, rich, powerful, prosperous, and strong, is, as
a rule, not very much inclined to compassion, from the fact of
his not being under any apprehension of calamity.   Now, no
calamity can possibly touch God, His happiness is perfect ;
nay, more, He is the Giver of every kind of happiness, and
yet He is more merciful than all those who are subject to
every kind of misery.   Therefore Christ said emphatically,
"Be ye merciful," not merely "as your Father," but "as your
Father also (*et*) is merciful," thus making the words more
forcible, as though saying, "He Whose happiness is so great,
is so great also in mercy."   Do thou remember this, and not
compassionate in others those evils only which thou knowest
by experience, but those also which thou dost not know :
"When I sat as a king with his army standing about him, yet
I was a comforter of them that mourned."[4]

VII. Consider seventhly, that, numerous as are the Divine
perfections, not one in particular is proposed for thy imitation
throughout the Gospel except mercy, because there is not one

---

[1] St. Matt. v. 45.
[2] 1 Kings xxv. 11.   [3] Jerem. xxxi. 8.   [4] Job xxix. 25.

which makes thee so like God. God's mercy is the highest
perfection, not in its nature (for in this all His perfections are
equal), but in its effects. For no attribute of God has caused
Him to do so much as mercy has; because, if mere charity,
goodness, kindness, and liberality made Him create mankind
and raise them to the state of grace, it was mercy which made
Him redeem them by His own Blood: "According to His
mercy He saved us."[1] There is, therefore, no virtue which
makes man so like God as mercy, because it makes him like
that which shines most strikingly in Him. Of all precious
stones, which most nearly resembles the diamond? That
which most resembles it in lustre. So, among all who set
themselves to imitate God, none will so much resemble Him
as those who are like Him in showing mercy. It is true,
indeed, that mercy is not absolutely the greatest virtue in man,
as it is in God, because God has no superior, and therefore
He can only do good to those under Him: "Pouring out
waters upon the thirsty land."[2] But man has God above him,
and the miserable beneath him; therefore he must first unite
himself to God by charity, and to the miserable by mercy:
"Put ye on, therefore, as the elect of God, bowels of mercy,
. . . but above all these things have charity." But yet it is
beyond a doubt that mercy is the greatest of the virtues which
bind man to his neighbour. It is the greatest in its nature,
because it is the kind of charity farthest removed from self-
interest, being exercised towards the miserable; and it is the
greatest in its effects, both because no other virtue gives so
much scope for noble actions as mercy does, and because it
is shown to all, even to the unworthy and ungrateful, and so
can be exercised without restriction. If, then, this is so, it is
no wonder that our Lord inculcates it above all others when
He says, "Be ye merciful, as your Father also is merciful."
Dost thou not feel inflamed with love for this virtue? If
not, thou art not only unlike thy great Father, but a most
degenerate child of His. And therefore Christ said, not "My
Father," as He might have said with equal propriety, but
"your Father," to remind thee of the obligation thou art
under to resemble Him.

VIII. Consider lastly, how Christ chose to call God in
this passage by the name of Father, because a true father has
in himself a perfect idea of the mercy whose qualities we have
here noted. A true father has no need to be endowed by

[1] Titus iii. 5.   [2] Isaias xliv. 3.

nature with especial tenderness in order to compassionate his own children; mere paternal affection is sufficient for this. Nevertheless, this is not enough to satisfy him; he can, when he pleases, feel for them the utmost possible tenderness of affection. He can compassionate them when they have brought trouble on themselves by their own disorders, he can excuse them, bear with them, love them more and more, even when he meets with no return of love, and, forgetful of himself, go in search of them however far they wander from him. He has no need of a personal experience or apprehension of their miseries to compassionate them most keenly, nay, he would deprive himself of happiness to bestow it upon them; and all this simply because he is a father. This, then, is the reason why Christ chose to call God "Father" when saying that He is merciful, in order to comprise in that name all the qualities distinctive of perfect mercy. "As a father hath compassion on his children, so hath the Lord compassion on them that fear Him."[1] Do thou in particular, who art in the position of a Superior, remember that the easiest method of practising mercy perfectly towards thy subjects is to act, in all things, as a father.

NINETEENTH DAY.

Enter not into judgment with Thy servant, O Lord, for in Thy sight no man living shall be justified (Psalm cxlii. 2).

I. Consider first, how all, even the greatest saints, have, without exception, feared the Divine justice. The mere thought of it filled them with confusion and dread, and all they did was to commend themselves to His mercy: "Although I should have any just thing I would not answer, but would make supplication to my Judge."[2] No wonder, therefore, that even David says to God, "Enter not into judgment with Thy servant, O Lord." See how genuine his fear is: he begs God not merely not to judge him, but not even to speak of judging him: "Enter not into judgment." And if thou dost not fear so terrible a Judge, how plain it is that thou hast all the greater need for fear, because it so evidently shows that thy conduct differs from that of all the saints: "If I would show myself innocent, He shall prove me wicked."[3]

[1] Psalm cii. 13. [2] Job ix. 15. [3] Job ix. 20.

II. Consider secondly, that this judgment is very terrible on the part of man, who is to be judged. For who is there that can say to God with confidence, Lord, I am clean? "Who can say, my heart is clean?"[1] Sometimes, indeed, a man may be able to say, "I am not conscious to myself of anything," but even then he is always obliged to add, "Yet am I not hereby justified."[2] Therefore the Psalmist here says: "Enter not into judgment with Thy servant, O Lord, for in Thy sight no man living shall be justified;" and he argues well in so saying, for if "no man living shall be justified," then he meant to say, How much less I, who am so miserable? Now, to return to our point, by saying "no man," he includes every one, and consequently thyself. See, therefore, how many reasons thou hast for fearing the Divine judgment, and how thou canst not venture to open thy mouth to justify thyself: (1) Because thou wert born a child of wrath, and therefore of a vile race, and so thou canst never venture to lift up thy head in God's presence: "Thy father," Adam, "was an Amorrhite," that is to say, a rebel, "and thy mother," Eve, "was a Cethite," that is, foolish. "Thou wast cast out upon the face of the earth in the abjection of thy soul, in the day that thou wast born."[3] (2) Because, although thou wert afterwards raised to a very high dignity by habitual grace in thy baptism, thou hast slighted it by sinning mortally, and so rendered thyself more guilty than before baptism. (3) Because, whereas thou art certain of having lost this habitual grace by, not one, but repeated sins, thou art not certain of having ever recovered it by the requisite penitence. (4) Because thou hast frequently neglected to make use of the right dispositions for acquiring the actual grace which God would otherwise have given thee in great abundance; and not only this, thou hast even put great obstacles in the way of receiving it. (5) Because again and again, notwithstanding these obstacles, God has lovingly given thee this grace, even superabundantly, and thou hast refused to correspond, and has slighted the lights and inspirations and invitations which He has vainly bestowed on thee. (6) Because, even when thou hast corresponded, it has been with very great coldness, and so a very great amount of grace has by thy fault produced no return at all: "Ten acres of vineyard shall yield one little measure."[4] (7) Because, not only art thou negligent in doing good, but thou daily committest much evil, at least by many venial sins of gluttony,

---

[1] Prov. xx. 9. [2] 1 Cor. iv. 4. [3] Ezech. xvi. 3, 5. [4] Isaias v. 10.

impatience, envy, or detraction, which are habitual to thee. (8) Because, even if thou doest more good than evil in the course of the day, yet it is a mere nothing compared with the countless favours which thou art daily receiving from God. (9) Because, insignificant as is the amount of good that thou doest, thou thinkest that it is a great deal, and so cherishest a vain esteem of thyself, at least, in comparison of others who are perhaps better in God's sight than thyself. (10) Because, in this small amount of good which thou doest, thou not only cherishest a vain esteem of thyself, but very often seekest, at all events secretly, the praise of men. (11) Because, in this good, thou, at the best, seekest thyself more than God, not loving Him disinterestedly as He loves thee, but rather serving Him faithfully from the hope of reward or the fear of punishment. (12) Because, lastly, even though thou shouldst be serving Him like a saint at the present time, thou art not sure that thou wilt persevere constantly to the end: "Behold, among His saints none is unchangeable."[1] Now, then, see whether thou hast not good reason to fear the Divine judgment. These twelve truths should be, as it were, twelve doors standing always open in thy soul to admit this chaste fear, so that it may every moment find free access to thy heart by whichever of them it may choose to enter.

III. Consider thirdly, that this Divine judgment is very terrible on the part of God, Who is the Judge; and this for two reasons: (1) because He has an infinite hatred of sin in others, (2) because He possesses infinite sanctity in Himself. In the first place, He has an infinite hatred of sin in others, which will cause Him to seek it out very carefully, and then to punish it very severely. Wouldst thou see how carefully He seeks for it? It is enough to tell thee that He seeks for it in the most secret recesses of the heart and reins: "All the Churches shall know that I am He that searcheth the reins and hearts."[2] And if He does this, what will become of us poor sinners who are so prone to evil? The motions of the concupiscent part are in "the reins," and of the irascible in "the heart." These motions are the least perceptible by us, for very often they are exceedingly strong, although they have arisen without our consent, and therefore it is very difficult to decide concerning them, whether they have gone so far as to be sins or not. Yet it is into these motions that God declares He will examine most seriously, searching at once both "the

---

[1] Job xv. 15.   [2] Apoc. ii. 23.

reins and hearts." In the next place, wouldst thou see whether when He has found the sin, He will punish it severely? He will not leave the very smallest unpunished : "Amen, I say to thee, thou shalt not go out from thence till thou repay the last farthing."[1] Then, not only does He hate iniquity in others, but He possesses perfect sanctity in Himself, and the consequence will be that any sanctity of ours, however bright it may be, will lose all its lustre as soon as it comes into His presence : "The heavens are not pure in His sight."[2] If, therefore, He will judge us according to the obligation we are under to be like Him in sanctity, who can be safe? And so it is that David here says to God, "In Thy sight no man living shall be justified," for even if a man might feel some degree of security when standing in the presence of his fellow-man, yet he needs must tremble when confronted with God : "Indeed, I know it is so, and that man cannot be justified compared with God."[3] Considering all these things, then dost thou not think that thou too hast great reason to say, "Enter not into judgment with Thy servant, O Lord, for in Thy sight no man living shall be justified"? Oh, how far better it is to keep at a distance from this judgment than to venture to provoke it !

IV. Consider fourthly, that perhaps thou mayest think this a vain prayer of David's, since, in spite of thy begging God not to enter into judgment with thee, He will not only do so, but will pursue and complete it, and, as has just been said, examine the inmost recesses : "The Triumpher in Israel will not spare."[4] And a triumpher is one who pursues after thee till the war is over. Yet, for all that, thou art mistaken. This is not only no vain prayer, but it is the very best and most useful that thou canst make. For, tell me, what it is that thou sayest to God when thou prayest Him not to enter into judgment with thee? Thou hast heard already. Thou sayest that thou ownest thyself to be a convicted criminal, covered with confusion, confessing thy guilt before the time. And if thou doest this, then God will not enter into judgment with thee, because thou hast already judged thyself : "If we would judge ourselves, we should not be judged."[5] This is the advantage of pleading guilty from the bottom of the heart before the Judge ; the man who does so is immediately absolved. I say from the bottom of the heart, because it is

[1] St. Matt. v. 26.　　[2] Job xv. 15.　　[3] Job ix. 1.　　[4] 1 Kings xv. 29.　　[5] 1 Cor. xi. 31.

necessary, first of all, really to consider thyself guilty, not merely to profess thyself so in words. Next, it is necessary to be, at the same time, firmly resolved to amend; otherwise what sort of confession would it be? Of what avail is it to confess thy sin at the time that thou art intending to continue to do the very thing which thou acknowledgest to be sin? Besides this, there is another advantage in this prayer which thou callest vain, its frequent use will greatly help to keep thee humble. See, then, how this is another way of escaping the Divine judgment, for, at all events, thou wilt thus escape its wrath: "Because they were humbled, the wrath of the Lord turned away from them."[1] It is the proud who will incur the most terrible judgment of God, because it is these who, instead of keeping it far from them, venture to provoke it. And how do they provoke it? In three ways: (1) By complaining that their prayers are not heard by God: "Why have we fasted, and Thou hast not regarded? have we humbled our souls, and Thou hast not taken notice?"[2] (2) By complaining that they are not rewarded for the service that they render Him: "Since we left off to offer sacrifice to the queen of heaven . . . we have wanted all things."[3] (3) By complaining that not only are they not rewarded, but afflicted with continual troubles when they do well, whilst others who do ill prosper: "Why doth the way of the wicked prosper?"[4] Men who act thus are those proud well-doers, who show that they fear God's judgment so little as actually to provoke it. Alas, wretched men! "Why will you contend with Me in judgment?" You shall see that I know well where the difficulty is, it is that "you have forsaken Me, saith the Lord."[5] Take every possible care not to be of this number. Keep thyself always in the present knowledge of thy misery, acknowledge it, confess it, declare it to God again and again. Say to Him continually with a contrite heart: "Enter not into judgment with Thy servant, O Lord, for in Thy sight no man living shall be justified," and thou wilt see how profitable this prayer, rightly used, will be to thee.

[1] 2 Paral. xii. 12.　　[2] Isaias lviii. 3.
[3] Jerem. xliv. 18.　　[4] Jerem. xii. 1.　　[5] Jerem. ii. 29.

TWENTIETH DAY.

Be thou faithful unto death, and I will give thee the crown of life
(Apoc. ii. 10).

I. Consider first, that the virtue which is valued in servants beyond all others is fidelity: so much so that the Wise Man says, "Let a faithful servant be dear to thee as thy own soul."[1] Thou art God's servant in the strictest sense, and therefore be not surprised that He incites thee with promises so great, to be always faithful to Him: "Be thou faithful unto death, and I will give thee the crown of life." Oh, how great is the fidelity required in a servant! In order to be a faithful friend, it is enough to esteem the interests of thy friend as thy own, and as such to promote and cherish them, because thy friend, however dear he may be to thee, cannot certainly be more than a second self. But this is not enough to make thee a faithful servant. As such, thou art bound to esteem thy master's interests much more than thine own, for he who is thy master is more than thyself, inasmuch as he is master of thyself. I willingly grant that thou art now, indeed, the friend of God, since He has done thee the wonderful honour of exalting thee to that rank; but thou art not the less His servant on that account. This is so essential to the nature of man, that Christ Himself, as Man, was called a servant in regard to God, although He was His Son by nature: "Behold My Servant, I will uphold Him: My Elect, My soul delighteth in Him."[2] Thou seest, then, what degree of fidelity He requires of thee when He says, "Be thou faithful unto death." It is the very highest possible degree. He would have thee esteem God's interests, not merely as thy own, but far more, because thou art not only His friend, but His servant; so that when it is a question of pleasing Him, everything must be sacrificed by thee, no matter what it may be—health, property, reputation, life itself. A faithful servant regards nothing of the sort; he esteems his lord more than himself. Dost thou consider that thou possessest such a fidelity as this? It is this which is necessary for the winning of the crown.

II. Consider secondly, that there are many servants who serve their masters for a little while with the fidelity we are speaking of, but very few who persevere in it to the end. And, therefore, our Lord says, "Be thou faithful unto death, and I

[1] Ecclus. vii. 23.   [2] Isaias xlii. 1.

will give thee the crown of life," for it is perseverance which chiefly proves fidelity. We do not call him a faithful servant who has kept inviolate the faith he owes his lord on one occasion only, but him who has been seen constant in keeping it through many trials. Thou art anxious only to die well, thou dost not concern thyself to live well. And why? Because thou art not a faithful servant. It is thine own interest that is thy object, for the salvation of thy soul is nothing else. Do not act thus: be a faithful servant to God; and so, set thyself to keep faith to Him not only in death, but "unto death;" say to Him from thy heart, that even if—which God forbid—thou wert to die ill, thou art still resolved to live well, because it is for His glory. And dost thou know what thy Lord implies in the words, "Be thou faithful unto death"? He implies, that if He were to send thee poverty to last till thy death, thou art to be faithful to Him "unto death" in that poverty; that if He were to send thee imprisonment to last till thy death, thou art to be faithful to Him "unto death" in that imprisonment; that if He were to send thee disgrace to last till thy death, thou art to be faithful to Him "unto death" in that disgrace: and so on. Fidelity is proved especially in adversity. "Was not Abraham found faithful in temptation?"[1] and, therefore, when true fidelity stands the trial, as his did, it is crowned: "Be thou faithful unto death, and I will give thee the crown of life."

III. Consider thirdly, that thou art alarmed at this expression, "unto death." But why is this? Because thou thinkest that thy life will be prolonged to the length of Adam's. How greatly thou art mistaken! It may be that death is waiting to strike thee as thou goest out of the house, when thou imaginest that it has long years to travel before it reaches thee. But take courage. Let us grant that thy life will be as long as is possible considering the age, whatever it may be, that thou hast already attained; shall I show thee how to make it immediately appear, not long, but exceedingly short? Think of eternity. Oh, then, even the sixty years to which thou mayest be looking forward, much more thirty, will seem but a moment! And see how our Lord, in order to remove the fear which thou mightest feel at hearing those words "till death," immediately adds, "and I will give thee the crown of life." This, therefore, is what He promises thee, an eternity; for this is the meaning of a "crown of life;" it means a life which

---

[1] 1 Mach. ii. 52.

will always describe a circle like a crown : "There shall be month after month, and sabbath after sabbath," and never come to an end. And is not the thought of so long a life, and one entirely spent in happiness, enough to take away all dread of the short sufferings thou wilt have on earth ? Rather shouldst thou complain to God that since the happiness is to be so long, He has made the time of suffering which precedes it far too short. Oh, what a wonderful crown is that "crown of life"!

IV. Consider fourthly, that the life which is in store for thee, if thou art a faithful servant "unto death," is called a "crown of life," not only because it will be an everlasting life, as we have seen, but because it will be a blessed, nay, a most blessed life : for it will be the crown of every blessed life that can be imagined, not of one life merely, but of every life. Whatever completes the perfection of a thing is said to be its crown—"Much experience is the crown of old men"—for it is certain that gray hairs do not of themselves alone make old men venerable. What makes them altogether so in the eyes of every one is the wisdom which they have acquired by long experience of the things of the world. The beatitude of Heaven, therefore, is so frequently called a "crown of life," because it completes the perfection of every life, and so crowns it. That life is esteemed happy which passes without any cloud of sorrow or weariness to darken its days, and the beatitude of Heaven will be the crown of this life, because it will make its days not only calm, but unchangeable ; and so it will be the crown of a peaceable life. Learned men esteem that life happy in which the mind is stored with many wonderful speculations, and the beatitude of Heaven will be the crown of this life, because it will bestow the science which comes not from streams, but from the fountain : and so it will be the crown of a learned life. Rich men who abound in wealth enabling them to satisfy their desires are reputed to have a happy life, and beatitude will be the crown of this life also, because it will give not a storehouse, but a mine of riches : and so it will be the crown of a wealthy life. In the same way go on passing in review every kind of desirable life that can be conceived, and know that the crown of them all will be that which is here promised thee by God, when He says, "I will give thee the crown of life." He has chosen to say "life" without qualifying the word by any adjective, in order that thou mayest add at thy choice whichever pleases thee : such

as "a calm life, a learned life, a rich life, a noble life, a
cheerful life, a safe life, a strong life," and so on without limit.
Dost thou suppose there is no greater good than mere life in
Paradise, when its glory is called "the crown of life"? This
would be a great error; if in that case we should be told of
"the blessing of life," not "the crown of life." But since it is
"the crown of life," there is more there than mere life. There
is the most perfect life, in every kind, there can be : there is
perfection. If there were only the blessing of such a life as
belongs to the young, and not that which belongs to men of
mature age, then the words would have been "the crown of
the life of youth ;" or if there were only such a life as belongs
to grown men, and not such as belongs to the young, then
they would have been "the crown of the life of man :" otherwise they would seem planned to deceive people by promising
more than awaits them. But the words being "the crown of
life," without limiting it to any one life in particular, this is a
sign that heavenly glory contains in itself the crown, in other
words, the perfection of every life, and so contains all good.
And is not such a crown sufficient to put into thy heart the
strongest desire of winning it? If thou desirest it, be faithful
till death : "Be thou faithful unto death, and I will give thee
the crown of life."

V. Consider fifthly, how marvellous it seems that God
should be pleased to give such a crown to a servant. And
yet this is most certain. And the word used is not *donabo*,
but *dabo*, as though the servant's fidelity were enough to earn
it. And, in truth, although this crown surpasses his merit,
yet still it is merited : it is a "crown of justice," because God
puts it before us as a reward. And why has He done so? To
show us how pleasing fidelity is to Him. And for this reason
He did not choose, in this passage, where He makes such rich
promises, to say "be strong, or courageous, or constant,"
but "be faithful," because the thing He most values in the
service we pay Him is not strength, or courage, or constancy,
but fidelity. This is a virtue which is greatly esteemed even
among men : "A faithful man shall be much praised." Every
one encourages and rewards it ; so that it is sufficient, of itself
alone, more than anything else, to raise not merely a friend
but even a servant to the highest possible position. Why was
Mardochai crowned, although his position was that of a
servant? On account of his fidelity to Assuerus. However
this may be, do thou delight in practising this fidelity to God,

which is so pleasing to Him, and tell Him, moreover, that thou wilt not practise it in future for the sake of the glorious crown which He has promised thee, but merely because He is what He is, to give Him pleasure, to do Him honour. So doing thou wilt attain the highest degree of a fidelity that can possibly be shown by a servant to his lord, which is to desire from him no wages except his favour.

TWENTY-FIRST DAY.

Give not place to the devil (Ephes. iv. 27).

I. Consider first, how strange a thing this is. If any one were to tell thee to take care not to admit into the house a poisonous dragon, a lion, a wolf, or even a man like thyself if he were a thief, who came for the purpose of robbing thee, thou wouldst laugh at him for giving thee advice which was officious rather than necessary, for thou art very well able to do this thyself without being urged to it by another. And yet thou needest urging not to give place in thy heart to the devil! Knowest thou not that he is a worse thief than any other, a thief who longs to rob thee of the most precious treasure on earth, namely, Divine grace? It is he who is the cunning wolf, the raging lion, the most venomous of dragons: "The great dragon, who is called the devil,"[1] whose very breath is enough to poison thee. How is it, then, that thou so readily admittest him into thy heart? If he were ever able to take forcible possession of it thou wouldst deserve excuse, but he can never enter it but by thy leave; and therefore it is said, "Give not place to the devil," because it rests with thee to give or to refuse him entrance. Think how little is required to overcome, no matter what great temptation assails thee: it requires no more than a determined "I will not." Who is there that would allow a dragon, a lion, a wolf, or a thief to enter his house if it were in his power to keep them away with a little trouble? No sooner are such intruders perceived, even at a great distance, than every one immediately gives the alarm. And yet thou sufferest one, who has it in his power to do thee more harm than all

[1] Apoc. xii. 9.

these together, to enter, not thy house only, but thy very heart : "Give not place to the devil."

II. Consider secondly, who they are who, strictly speaking, give place to the devil. Is it those who set wide open the door of their hearts to admit him? No ; for such persons not only give place to him in their hearts, but make him master of them. Those are properly said to give place to him who allow him a little chink to creep in by : they give him the means of access, they give him audience ; at all events, they give him the opportunity of tempting them, as Eve did in the terrestrial Paradise. This is what those do who indulge themselves in idleness, as she did, those who do not observe custody of eyes and ears, those who suffer themselves to be mastered by any disturbing passion, such as anger, over-anxiety, melancholy ; for this is the time of which the devil avails himself to approach : "Why hath God commanded you that you should not eat of every tree of Paradise?" Know, too, that the devil never asks for all thy heart at once. What he does ask is a "place" in it ; but alas for thee if thou givest it to him : "Give not place to the devil." And why not? Because he is never satisfied with the little that thou hast given. He soon wants to go on from little to much. First, he desires to know the prohibition which God has given thee ; then he finds fault with it, then he condemns it, and, last of all, he persuades thee to disregard it : "It goeth in pleasantly, but in the end it will bite like a snake."[1] Therefore, resist temptation, as thou oughtest to do, but resist it at the beginning, that is, when thou hast hardly begun to consider it as a temptation. For what says the Apostle? It is not enough not to give consent to the devil, we must not even "give place" to him. Examine carefully, and thou wilt see that, in the majority of cases, it is thy own fault when thou art tempted by the devil. It is by want of prudence and circumspection in thy manner of life that thou makest a way for him to approach to tempt thee.

III. Consider thirdly, the method taught by the Fathers of not giving place to the devil, no matter how importunate he may be in asking for it. It is to keep the mind employed in holy thoughts. It is true that he is a spirit, and that he enters by way of the eyes and ears, but, nevertheless, if he finds thy mind well guarded he must needs depart by the same road by which he came in. Therefore, whenever thou beginnest to feel a temptation which has passed through thy senses, and is

[1] Prov. xxiii. 31.

knocking at thy heart, do not answer, but think, instead of taking the trouble to argue with it, of the bier on which thou wilt be one day stretched, of the Judgment to come which awaits thee, of the reward and the punishment to be awarded, of the Blood which Jesus shed for thee on the tree of the Cross, and turning unto Him, say to Him with loving eagerness, "O Lord, let my heart and my body be pure, that I be not confounded."[1] Do this and thou art safe; there is no fear then that any bad spirit can go on to defile thy soul: "Henceforth the uncircumcised and unclean shall no more pass through thee."[2] Dost ·thou say that this method which I am teaching thee is troublesome to practise? It may be so; but if thou art to be saved it is absolutely certain that thou must undergo one of these two labours: either not to give place to the devil in thy heart, or, having given place to him, to drive him out again. Which of the two is easiest? If thou lackest courage to say at once to the enemy, "Thou shalt not enter," dost thou think that later on thou wilt have courage to bid him depart? What blindness to refuse to undergo some labour in order not to admit temptation into the heart, and so to have to undergo a far greater labour afterwards to drive it away. Therefore, "Give not place to the devil."

TWENTY-SECOND DAY.

ST. MARY MAGDALENE.

Thou wert naked, and full of confusion, and I passed by thee, and saw thee; and behold thy time was the time of lovers; and I spread My garment over thee, and covered thy ignominy; and I swore to thee, and I entered into a covenant with thee, said the Lord God, and thou becamest Mine (Ezech. xvi. 8).

I. Consider first, that in these beautiful words we have a description of God's wonderful dealings with a soul, when by an excess of mercy He attracts her to Himself, and of a great sinner makes a great saint. "Thou wast naked and full of confusion." Here we see this wretched soul "naked," because lacking every virtue; "full of confusion," because laden with every vice. God passes by the soul in this state, and beholds her: "I passed by thee, and saw thee." He passes by, like a king who, having gone out hunting, suddenly comes upon her.

"I passed by thee," and looks upon her; "I saw thee," that is, He looks upon her with one of those glances by which He is pleased to show His power in that soul. This is what our Lord meant when He said to Nathanael, "When thou wast under the fig-tree I saw thee,"[1] according to the interpretation of St. Gregory, for otherwise, what is there that God does not always see? And at what time is it that He watches this soul with so much love? At the very time when He sees her to be most given up to the things of the world, to amusements, pleasures, and vanity: "And behold thy time was the time of lovers." And yet it is at this very time (can it be believed?) that He determines to make her all His own, in order the more to enhance the triumph to be won by Divine mercy over human misery: "And I spread My garment over thee, and covered thy ignominy." Here we have the preventing grace with which God arrests the soul, so that she shall not escape from Him. And this is the force of the expression, "I spread My garment over thee," an action resembling that of the huntsman when he throws a net over the deer to stop her, although God did not choose to say "My net," but "My garment," because grace does not make it impossible for the soul to escape, only difficult, as would be the case if the huntsman threw his mantle over the deer instead of the net. Then comes justifying grace, which is to be distinguished from preventing grace, not in principle but in effect; for, when the soul has corresponded with God by conversion, the same spirit of charity, the force of which attracted her to God, next comes in to clothe her, as it were, by a glorious union with Him, in a garment of beauty. And this is what God means by adding, "and I covered thy ignominy," because no sooner does He draw a soul to Himself than He gives her feelings of compunction and contrition so keen as entirely to cover over the ignominy of the sins she has committed, far more than rich embroidery conceals the coarseness of the material on which it shines: "Charity covereth all sins."[2] It is in these dispositions, in which the sinful soul has become full of sorrow, that first the espousals and afterwards the bridal is celebrated. The espousals consist in the special rewards of love bestowed by God upon the soul in various gifts of devotion, sweetness, and tears, which in this state are but the "earnest of love;" but, above all, they consist in a strong confidence with which He inspires her, that He will be to her

[1] St. John i. 48.   [2] Prov. x. 12.

in place of all else, of which the soul is as certain as if she had heard it sworn in a sensible manner by the mouth of God, so that, all the more encouraged by this confidence, she resolves entirely to detach her heart from all creatures so as to belong to God alone : " Be Thou mindful of Thy word to Thy servant, in which Thou hast given me hope."[1] The bridal consists in the mutual union which soon after takes place between God and the soul, between the soul and God, in an entire union of will, so that at last she not only belongs to God, as every just soul does, but is all for God, that is, wholly devoted to His service : " My Beloved to me, and I to Him."[2] All this thou seest admirably expressed in the words which follow. " I swore to thee :" here are the espousals in which the soul does but receive an earnest from God. " I entered into a covenant with thee :" here is the bridal, which is the mutual covenant of wedded troth. " And thou becamest Mine :" the word is not *mea*, but *mihi*, to show that the soul is entirely devoted to His service, a state which belongs to those only whom God by a special grace chooses for Himself, in order either to people the stars with an elect race, or to enjoy with Him the delights of a sublime contemplation. Is there, in this picture, anything that thou canst recognize as God's gracious dealings with thy own soul? If so, oh, how greatly art thou bound to Him !

II. Consider secondly, that never was there a soul in which all this was wrought by God in a more sublime manner than that of Magdalene, and that therefore there is none to whom these words may be more justly applied. Consider her first at the time which was in truth " the time of lovers " with her. She was indeed " naked and full of confusion." But in whose eyes? In those of God. And here we have to admire the difference between the judgments of man and of God. In the eyes of men so far from being " naked " she was splendidly clad ; so far *from* being " full of confusion " she was courted, praised, and flattered. But of what value was all this when her condition was so shameful in the eyes of God? Alas for her, if God had not deigned to pass by her, and to cast on her a look of compassion when she was " a woman that was in the city, a sinner " ![3] This is the meaning of " I passed by thee and saw thee." First, He passed through her soul by the force of His Divine words, and so He does not say, " I passed before thee " (*ante te*), but " by (or through) thee " (*per te*),

[1] Psalm cxviii. 49.   [2] Cant. ii. 16.   [3] St. Luke vii. 37.

and so He illuminated her soul by the rays of His light, thus
seeing her long before He was seen by her; wherefore He
says "I saw thee," not "thou sawest Me," for that look was a
look of His mere love, such as might be that of some prince
who should look upon a mean peasant girl, and say boldly,
"This shall be my bride." Envy Magdalene this glorious
destiny, and do thou behold her with love, now that thou
seest her so richly clad and so full of glory in Heaven, because
God chose to regard her with love even in this world when He
saw her "naked and full of confusion," just as the prince we
spoke of loved the peasant girl, not for what she then was, but
for what he was able to make her by raising her to the rank of
queen.

III. Consider thirdly, that from the first state, which is
God's election of Magdalene, thou mayest go on to contemplate her in the second, which is that in which God first
prevented and then justified her with the abundance of His
holy grace. "When she knew:"[1] here was preventing grace
which stopped Magdalene in her career by a vivid perception
of the sins she had committed; and so when our Lord infused
this perception into her soul He spread His garment over her,
for then He was sure of her. "Standing behind . . . she
began to wash His feet with her tears:" here was justifying
grace enriching and adorning Magdalene by means of the
deepest penitence for those sins; it was then that He
altogether "covered her ignominy," for it was then that He
infused into her the grace which made her rich instead of
"naked," as at first, and decked with ornaments instead
of "full of confusion." It is said that the ignominy is
"covered" instead of "removed," not that this grace does
not in truth do away with the stain of sin, but because it also
imparts spendour, just as would be the case if, not satisfied
with cleansing a soiled material, thou wert also to cover it
with rich embroidery, which entirely remedied and did away
with the ignominy it had contracted by being soiled. Such is
the glorious covering here spoken of, when the sin itself acts
as a spur to the soul to rise to a greater height of sanctity, as
it is written: "Blessed are they whose iniquities are forgiven,
and whose sins are covered;"[2] that is, says St. Gregory,
"forgiven, inasmuch as they are blotted out, and covered,
inasmuch as an ornament is put over them." Note therefore
how, after this, our Lord "covered" Magdalene's ignominy,

[1] St. Luke vii. 37.   [2] Psalm xxxi. 1.

not merely defending her from the charges of the Pharisee, but praising her so far as to say "she loved much," a thing which can never be said in strict truth with regard to God. And yet these are His words: "Many sins are forgiven her because she hath loved much."[1] By saying "many sins are forgiven her," He at once took away the misery of her nakedness, because He enriched her with justifying grace, and by saying "she hath loved much," He covered her ignominy, because He showed that if she had greatly offended God, she had afterwards been able also to love Him greatly. How dost thou stand? Is thy ignominy to be seen covered henceforth in this glorious manner?

IV. Consider fourthly, how thou mayest go on from the second state of Magdalene penitent to the third of Magdalene exalted to great sanctity. And here we have, first, the espousals: "I swore to thee;" this was when, "sitting at the Lord's feet, she heard His word." Her part was only that of receiving consolation, joy, exceeding sweetness: God alone was the giver. Who can doubt that it was then that our Lord infused into her a superhuman confidence that He alone was able to be to her in place of all else, since in this state she did not even think of taking food? Then follows the bridal: "I entered into a covenant with thee;" this was when our Lord, taking her always with Him as His acknowledged spouse, not only gave her many marks of fidelity, but also received them from her, since this fidelity remained unshaken at the foot of the Cross, nor there only, but through arms and armed men even to the sepulchre, so faithful a spouse was she. Happy indeed is thy soul, if she, too, has contracted so glorious a bridal!

V. Consider fifthly, how from the third state, in which Magdalene was exalted to great sanctity, thou mayest go on to the last state, in which she became all for Christ: "Thou becamest Mine." This happened first when our Lord, after His glorious Resurrection, made use of her, as of an ardent huntress, to draw souls to Himself, installing her in that office by the words: "Go to My brethren, and say to them," &c.[2] Again, this was the case when He drew her from Judæa into the desert of Marseilles, and kept her there for forty years, no longer for herself or for others, but for Himself alone, in a state of perpetual contemplation. Art thou satisfied to be possessed by God, as is the case with every just soul, to whom

[1] St. Luke vii. 47.   [2] St. John xx. 17.

God can say, *Facta es mea?* Not so; but do thou strive that He may be able to say to thy soul also, *Facta es mihi*, either in labouring for others, or in contemplation; for although it is true that such gifts as these are gratuitous, yet what is there that in the end is not obtained from God by earnest prayer?

TWENTY-THIRD DAY.

I thought upon the days of old, and I had in my mind the eternal years (Psalm lxxvi. 6).

I. Consider first, that, speaking according to the grossness of our conceptions, there are three periods of time, past, present, and future. But there are really but two—past and future; since, if thou considerest closely, there is no present time. No sooner hast thou said that it is, than it is no longer; it has been. Imagine thyself seated on the bank of a rapid river; whenever thou choosest a fixed moment to say, "This water is here," thou sayest what is not true, for the water which thou saidst was here has already flowed far away. Time passes more rapidly than any river; thou canst not stop it. When thou wouldst do so by affirming it to be present, it escapes thee in the act of speaking, and is already past. The true present is with God only, with Whom there is no time: "With Whom there is no change."[1] Do not wonder, therefore, if the Psalmist, speaking in the text of time, mentions but two periods, the past and the future: "I thought upon the days of old"—that is, the past; "and I had in my mind the eternal years"—that is, the future. He did not give a thought to the present, either because it does not exist, or because, at all events, it is so brief as to be valueless. What is that which is present to us? If it exists at all, it is but a moment, that is, a point: "The joy of the hypocrite is but for a moment."[2] A single moment! this is all that the time comes to which thou possessest. All that preceded this moment is the past, all that succeeds it is the future. Therefore the Preacher said, "Whatsoever thy hand is able to do, do it earnestly."[3] The adverb is *instanter*, and it is to show that the past is no longer in thy power, to be employed for thy profit; and as for the future, thou knowest

---

[1] St. James i. 17.   [2] Job xx. 5.   [3] Eccles. ix. 10.

not that it ever will be. It is true indeed that, to speak still more correctly, the Psalmist was not so much here thinking of the past and the future, as of the past and the eternal. "I thought upon the days of old"—that is, "the days which have been"—he said, "and I had in my mind"—not the future, but—"the eternal years." All those years which will be eternal to us are future, doubtless; but all future years will not be eternal. The years which we have yet to live on earth are future, of course; but they cannot be called eternal, when after sixty, or at most seventy of them have passed, they will be at an end. The only eternal years are those which follow our death, for they will never end; and it is of these that David was thinking. Well is it for thee if thou, too, art in the habit of thinking of them, for the most salutary thought that our mind can harbour is this of our past days and of the eternal years; of the former, that we may see how swiftly they have flown, of the latter, to remember that they will never end: "I thought upon the days of old, and I had in my mind the eternal years."

II. Consider secondly, why it is that this thought will prove so salutary. It is because the thought of past days will make thee have a still greater value for the eternal years, which, as has been said, will never end; and, on the other hand, the thought of the eternal years will make thee think less and less of the past days, which have flown so quickly, as also of all those which may yet be in store for thee. Only I would bid thee observe, that in order to make this thought more efficacious, it is necessary to think neither merely of the past, nor merely of eternity, but of both together, as thou seest holy David did: "I thought upon the days of old, and I had in my mind the eternal years." The conjunction "and" shows the combination of the two. Wouldst thou know how valueless is all that passes away? Contrast it with that which never ends, and say to thyself, "Even though I should have to live, not merely my allotted number of years, which will very likely not reach eighty, but those of Noe, Nachor, or Mathusala, which were little short of a thousand, what would they be in comparison of those countless millions which are swallowed up in eternity? A mere nothing: 'As yesterday which is past.'[1] How then can I possibly prefer those years which will so soon be over to those which will have no end?" In the same way, in order to know how to form a just estimate

----

[1] Psalm lxxxix. 4.

of eternity, compare it with the past, and say to thyself: "When all these millions and millions of years are gone, where shall I be at the end of them? Nay, why do I say the end? I shall have to begin counting them again, as though they were but beginning. How then can I think less of a life which will never have a close than of one which will so soon be over? This is the right way to judge rightly of the two: to think of the past, to think of eternity, but to think of both together: "I thought upon the days of old, and I had in my mind the eternal years."

III. Consider thirdly, that the Psalmist says that he "thought" of the days of old, and that he "had in mind" the eternal years. For as to the past, thou canst turn it over all at once in thy mind, divide it, take it to pieces, at thy will; but not so with eternity. It will be a great thing if thou succeedest in having it in thy mind without examining it. Indeed, thy mind cannot contain it as a whole, only in portions, according to our feeble powers of conception. And so it is that it is possible for thee to have in mind "the eternal years," that is to say, those years which we have spoken of as endlessly succeeding one another; but not "eternity." That is too vast an idea to dwell in any human mind; it dwells in the mind of God only, which contains it in itself, and sees it in its entirety. It will be enough for thee if, like David, thou keepest in thy mind "the eternal years," often repeating to thyself, "When as many years of eternity have passed as there are leaves in spring, or sands on the shore, or atoms in the air, or stars in the heavens, will that portion of eternity which has been spoken of be really past in such a sense as that it will never return? Not at all. There will always be as much to come as is past." And, after all, who is there that can understand what eternity is? As it will be infinite, so will it be unknown. All that we can do here is to let our minds meditate on the "eternal years," which are things we can conceive of. Therefore, the conclusion of the subject is this: in this life there is no present, there is only past or future, as is the case in the waters of running streams which rapidly succeed each other, and all we mortals, thou shouldst think, are like these: "We all . . . like waters . . . fall down."[1] In eternity, on the other hand, there is neither past nor future, it is all present, as in the fountain from which the waters take their source. And such,

[1] 2 Kings xiv. 14.

thou shouldst think, is God : "Thou art always the self-same, and Thy years shall not fail."[1] What we speak of as past or future in eternity is not eternity itself, but only the time which flows forward in eternity. And this is what will belong to us, as it does now ; only that now it is for a short space, and then it will be for ever : "Their time shall be for ever."[2] And on this, as has been so often said, thou hast to think, so as to see whether it is for thy advantage to have a brief happiness and eternal suffering, or brief suffering and eternal happiness.

TWENTY-FOURTH DAY.

If we sin wilfully after having received the knowledge of the truth, there is now left no sacrifice for sins: but a certain dreadful expectation of Judgment, and the rage of a fire which shall consume the adversaries (Hebrews x. 26).

I. Consider first, who they are who are here said to sin "after having received the knowledge of the truth." They are apostates. For unbelievers sin only "after having heard of the knowledge of the truth;" apostates, "after having received" it. Now, if thou examinest closely, thou wilt see that these apostates are of two classes. There are those who rebel, not only against the precepts, but the dogmas of Christ, such are persons who from being Catholics, go over to heathenism, Judaism, or heresy. Others continue to hold the dogmas, but rebel against the precepts, since after knowing their beauty, loving, approving, and practising them for a certain time, they gradually became lax in their observance, and finally abandoned them. Now it is of both of these classes of rebels that the Apostle here intends to speak, and therefore it is of both that he says they "sin wilfully," or (as the original may be translated for the better specification of the sin) "fail, or degenerate ;" he says then of both, that for those who "sin," or "fail, or degenerate," "there is now left no sacrifice for sins." And how will it avail thee not to be in the first class of apostates if thou art in the second ?

II. Consider secondly, that both these classes of apostates are said to sin "wilfully," because both alike sin with full consent of will. We all know that every one who sins does

[1] Psalm ci. 27. [2] Psalm lxxx. 1

so because he chooses to sin; but still there are some who sin in hot, and others in cold blood. The first are overcome by passion, and do not rightly know what they are doing: "Fire hath fallen on them, and they shall not see the sun."[1] The second are superior to passion, and do know what they are doing, and yet they choose to do it because of the malice reigning in their hearts; nor do they know it merely, but very often they study it, they dwell and refine upon it, purposely turning their backs to the sun, that its rays may not strike too brightly on their eyes: "They have not known the light."[2] Therefore the first are said rather to sin "willingly" than wilfully; the second sin, not willingly only, but wilfully. And of this kind, if thou considerest rightly, are all the apostates above mentioned: "A man that is an apostate with a wicked heart deviseth evil."[3] What wonder then if it is said of all of them alike that for them "there is now left no sacrifice for sins"? There remains for them no propitiation of any kind. What is the greatest propitiation? It is our Lord Jesus Christ, the Victim prefigured by so many birds, and lambs, and rams, and last of all offered up for us on the lofty altar of the Cross. Now this precious and salutary Victim no longer actually sheds His Blood for us. It is beyond all doubt that Christ will never again be crucified for the redemption of man: "Christ, rising again from the dead, dieth now no more."[4] He has once done for us all that He could do: "What is there that I ought to do more to My vineyard, that I have not done to it?"[5] Therefore, He will do nothing again of the same sort, because He would thereby do no more than He has already done. But if this Great Victim remains for no one, in act—that is to say, that He cannot any more lay down His life—yet He does so remain in effect, which is giving that life to us. But for apostates He does not remain even in effect, therefore for them He "is now left" in no way whatever; for that which Christ once did when He died on the Cross will not avail these miserable men. What He said for all besides to His Eternal Father, "Father, forgive them, for they know not what they do," He cannot say for them; rather may He say for them, "Condemn them, for they know what they do." Even they indeed may, speaking absolutely, one day enter into themselves, be struck with compunction and converted, so as to profit by that perfect Sacrifice; but

---

[1] Psalm lvii. 9.   [2] Job xxiv. 16.   [3] Prov. vi. 14.
[4] Romans vi. 9.   [5] Isaias v. 4.

the case is so rare, that it may be left out of the question as if it did not exist : " A man that is an apostate shall suddenly be destroyed, and shall no longer have any remedy."[1] Hardly can there be found a single apostate of the first class who has returned to the true faith ; and so, whereas we see that among heresiarchs Berengarius, who was the first to deny the Real Presence of Christ in the Blessed Sacrament, was converted, yet Simon Magus, Arius, Montanus, Manes, Nestorius, Pelagius, Priscillian, Luther, Calvin, Carlostadt, Bucer, and many others, all died impenitent : they had "no longer any remedy." So too with regard to the second class of apostates ; there are but few who repent, and the reason is very plain, for what is the method to be followed to bring any sinner to amendment ? Is it not to put before him the enormity of his sin, the scandal caused to his neighbour, the offence given to God, the pleasure given to the devils, the imminent danger in which he lives of being lost ? But these persons know all this, and audaciously despise it. What hope can there be, therefore, of reclaiming them ? There is "no longer any remedy." Thou seest then how rightly the Apostle said that for those who "sin wilfully after having received the knowledge of the truth, there is now left no sacrifice for sins." For these sins of apostasy, being very difficult to retract, are also very difficult to be forgiven. Tremble at the spectacle of such a state, and be not over-confident because thou thinkest thyself very far from it now ; for it is by little and little that men fall into it.

III. Consider thirdly, that since these unhappy rebels do not choose to have Christ for their Mediator, they must needs expect to have Him for their Judge. Therefore, after having said that for those who "sin wilfully after having received the knowledge of truth, there is now left no sacrifice for sins," the Apostle immediately adds, "but a certain dreadful expectation of Judgment." He says "a certain" (*quædam*) expectation, because as yet these miserable men have not this expectation of Judgment in all its fulness ; if they had they would be "withered through fear ;" but they have it sufficiently to disturb their empty joy from time to time, and even in this measure it is said that that expectation will be "dreadful" to them, although its true terrors will only be known when it comes to them in its completeness. When will that be ? At the hour of their death. Think what will be the state of these wretched men when they hear that very shortly they

[1] Prov. vi. 15.

must appear before the judgment-seat of the Lord with Whom they have so shamefully broken faith : " I have heard, and my bowels were troubled." Why? Because they will not have the courage to say a word to excuse themselves: "My lips trembled at the voice."[1] A criminal who is guilty of a heinous offence, and yet has some excuse for it, fears greatly when he knows that he must soon stand before the judge; but far greater is the fear of one who has no excuse to allege. Such will be the case of these miserable men, who have sinned through malice by apostatizing from those truths, whatever they were, which they knew, sinning "wilfully after having received the knowledge of the truth." This expectation, of which we have hitherto spoken, is that of the Particular Judgment. There still remains that of the General Judgment. And when will this be? At the appointed time. Think, now, what will be their state when the trumpet arouses them from the graves in which their bodies have been long decaying, and they find themselves forced and driven by the devils to the valley of their destruction : "Nations, nations in the valley of destruction : for the day of the Lord is near in the valley of destruction."[2] Oh, what a dreadful expectation will that be ! Their terror at the expectation of Judgment will exceed that of all other sinners, because their shame in that Judgment will be greater than that of all others; for they knew how monstrous was the guilt which nevertheless they persisted in espousing like men mad and blind with love. And so, when those who have sinned through lack of knowledge will call on the rocks to hide them on that day, those who have sinned through malice will call on Hell itself to swallow them up, so overwhelmed with terror will they be at the expectation of that Judgment which they will see not, as now, afar off, but close at hand. None will be so terribly reproved by Christ, none so hateful, so execrated, so accursed as they. And why? Because no enemies are so detestable in the eyes of every prince as rebels. Think, therefore, within thyself, and say in thy heart, " If the mere expectation of their great shame will be so dreadful, what will be the anguish of the reality?"

IV. Consider fourthly, that not only will these sinners be more reproved by Christ on the Day of Judgment than all the rest of the reprobate who are condemned with them, but they will also be more severely punished. And so the Apostle adds, that there "is left" for them not only a "dreadful expectation

[1] Habacuc iii. 16.　　[2] Joel iii. 14.

of Judgment," but also a dreadful "rage of fire." No sooner will the final sentence of damnation be given, than this fire will seize upon the reprobate to drive them quickly into Hell ; and how far more eagerly will it cling to these miserable men than to all the rest, as being fuel that will burn most readily ! Remember, too, that the fire which God will then endow with supernatural power for the punishment of the reprobate, will not act as it now does. Now there is no difference in the degree of torment it inflicts on a martyr, or on a criminal, a thief, a murderer, an adulterer ; but then it will not be so. Then its action will be as though it were endued with reason, and the degrees of the torment it inflicts will be keen in proportion to the deserts of those who are tormented, for which reason some of the saints have called the fire of Hell, as it were, a "reasonable fire" (*rationalis ignis*). This is why the Apostle speaks of this fire as though it were eager to punish those wicked men : "A rage of fire." But, in truth, not the fire only, but all the elements will then be possessed by this fury of zeal, and will, so to speak, vie with each other in fighting to take vengeance for the insults offered on earth to their Lord. Then will be seen what the Wise Man so well described when he said that "the whole world shall fight with Him against the unwise,"[1] for all the elements will act as though filled not only with power, but with fury : "The shafts of lightning shall go directly." These are the fiery shafts which will strike, not blindly, as they now do, but with intention falling on those who deserve them : "They shall go directly from the clouds, and as from a bow well bent they shall be shot out, and shall fly to the mark "—not at random, as they now do. "And thick hail shall be cast upon them from the stone-casting wrath." Here we see the earth also, as it were, endowed with reason, and full of fury, making a tempest of stones like hail. "The water of the sea shall rage against them." Here is the water raging as though it, too, possessed reason. "And the rivers shall run together in a terrible manner "—as though the sea were not sufficient to accomplish the work of ruin and destruction, and they must hasten to its aid. "A mighty wind shall stand up against them." The air, too, acting as though by reason, here pauses (*stabit*), so to speak, to gather strength, and then "as a whirlwind shall divide them," by driving the wicked far from the good. But still, because in this battle of the elements

---

[1] Wisdom v. 21.

the fire will act, as it were, the part of leader—"a fire shall go before Him "[1]—therefore the Apostle says nothing here of the air, earth, or water, but only of the fire, and also because the ideal of zeal (*æmulatio*), which is an excess of ardour, is more appropriate to fire.

V. Consider fifthly, that this zeal is to consume the adversaries of thy Lord : "It shall consume the adversaries." These adversaries are most especially all the apostates here spoken of, for it is they who, more than all besides, make war on God by robbing Him of souls, by seducing, deluding, alluring people to sin. And on that day they will all be consumed, because they will be utterly destroyed. Therefore the word is not "enemies" (*inimicos*), but "adversaries" (*adversarios*), and there is much meaning in this. For observe, that these unhappy men will never cease throughout eternity to be the enemies of God, any more than all those will who are to burn with them in Hell, whether devils or lost souls ; but though they will all continue to be His enemies, they will not continue to be His adversaries, because they will no longer be able to set themselves to thwart His glory, as they once dared to do in the world. Therefore it is said that the "rage of fire" will "consume the adversaries," not "the enemies." How, indeed, could this be said, when we know so well that these miserable men will burn for ever in the terrible furnace of Hell, writhing, raging, but never consumed, because the fire which torments them will be of a kind which perpetually burns their vitals without destroying them, with so terrible a force of purpose will it be gifted. If the mere thought of this fire does not inspire thee with terror, thou mayest fear one day to be of the number of those apostates who are so bold as not only to rebel against the precepts of Christ which concern morals, but also against the dogmas of the faith.

[1] Psalm xcvi. 3.

TWENTY-FIFTH DAY.

ST. JAMES THE APOSTLE.

I beseech you, as strangers and pilgrims, to refrain yourselves from carnal desires which war against the soul, having your conversation good among the Gentiles (1 St. Peter ii. 11).

I. Consider first, that there are three ways in which men may be inhabitants of the earth, as citizens, as strangers, and as pilgrims. Of the first sort are those who own no country but this world: "They have set their eyes bowing down to the earth.[1] These are bad Christians, who although they are not on the earth as citizens by birth, for their origin is from Heaven, are yet citizens by choice, having pitched their tents here as though they would never have to strike them: "Their dwelling-places to all generations."[2] They are registered, enrolled here: "They have called their lands by their names," living like the heathen, who have "no hope."[3] Those live here as strangers who do, indeed, acknowledge that they have another country, that is, Paradise, to which they aspire, but who nevertheless live as though this earth were their home, because they are more occupied with earthly than with heavenly things: "How happeneth it, O Israel, that thou art in thy enemies' land? Thou art grown old in a strange country."[4] These are ordinary Christians. Lastly, those who are pilgrims on the earth are those who not only acknowledge Paradise to be their country, not only aspire to it, but cannot live content in this world: "Woe is me that my sojourning is prolonged;"[5] and, like men who are but spending a day here on their way, make use only of just so much food or other necessaries as they require from one day to another for their support: "Besides Thee, what do I desire upon earth?"[6] These are perfect Christians; and now, before going any further, pause and examine thyself, to see in which of these three classes thou now art, and in which thou wouldst desire to be at thy death.

II. Consider secondly, that the Apostle is not speaking in this passage to those who are living as citizens on earth, for in that case he would have had to urge them to refrain themselves not merely from "carnal desires," but from carnal

[1] Psalm xvi. 11.   [2] Psalm xlviii. 12.
[3] Ephes. ii. 12.   [4] Baruch iii. 10.   [5] Psalm cxix. 5.   [6] Psalm lxxiii. 25.

works. He is speaking only to those who are living here either as "strangers," or as "pilgrims;" and therefore he begs them, speaking in this way out of courtesy, to act in a manner worthy of what they are: "I beseech you, as strangers and pilgrims, to refrain yourselves from carnal desires which war against the soul, having your conversation good among the Gentiles," that is, those whom we have described as the citizens of this earth. But what are these "carnal desires" here spoken of? They are the three well-known desires which comprise everything pleasant to the flesh in wealth, reputation, and pleasure, chiefly sensual. It is true that, strictly speaking, the first are called avaricious, and the second ambitious desires. Carnal desires, properly speaking, are those of sensual pleasure. And it is especially from these that the Apostle would have thee refrain thyself when he says, "I beseech you, as strangers and pilgrims, to refrain yourselves from carnal desires," because it is these which, more than any others, attach us to earth. Therefore Daniel said of those two old men who were so possessed by carnal desires, that "they turned away their eyes that they might not look unto Heaven."[1] And among the fatal characteristics of lust may be enumerated darkening of the mind, imprudence, inconstancy, and rashness; but, above all, a horror of the world to come. Observe, therefore, that the Apostle is not contented with men's abstaining from the works of the flesh, the evil of which is openly manifested, but he would have them abstain also from carnal desires in which the evil is concealed; for if there is any vice which it is necessary to resist in its first beginnings, it is this of which we are speaking, and which is compared, more justly than any other, to a fire; because it may very often have its source in a mere spark, something read out of curiosity, a passing word or thought, a movement not instantly repressed: "Of one spark cometh a great fire."[2] It may be that thy own experience has taught thee the truth of this.

III. Consider thirdly, nevertheless, that the Apostle does not here say, "I beseech you to be free from carnal desires," but "to refrain yourselves from them;" because it is not given to every one to be free from those desires which arise involuntarily, but only when they are voluntary. And therefore this is what is necessary: to chase away, repel, reject, and keep them at a distance as soon as ever they arise

---

[1] Daniel xiii. 9.   [2] Ecclus. xi. 34.

in the soul. Nay, even this is not enough; for in that case
the Apostle would have said, "I beseech you to drive from
you carnal desires;" but he is not satisfied with so little. He
would have thee not only keep these desires away from thee,
but more than this, he would have thee keep away from them,
for this is the proper meaning of refraining oneself (*abstinere
se*), to keep at a distance: "He abstaineth from our ways."[1]
He would have thee, as long as possible, like a prudent
general, avoid engaging in battle. And therefore it is well
for thee to know that sometimes these carnal desires wage
active war against the soul, at other times their mode of
warfare is one described in the text by the word *militare*,
which is rather to stand ready prepared to fight. When they
are actually fighting, particularly when the battle is very hot,
then is the time for driving them off, repelling them, as has
been said; but in the other case, which is that presupposed
by the Apostle, when he says, *Militant adversus animam*, thy
course should be to avoid fighting (*abstinere te*), and to fly
rather than defend thyself: "Refrain from strife, and thou
shalt diminish thy sins."[2] Even when they do make war on
thee, thou shouldst, as much as possible, gain the victory in
the same way, by refraining from dwelling on them, by dis-
tracting and diverting thy mind. And the reason is that
although it is true that sometimes the act of thinking steadily
of the sin which is attacking thee is a means of lessening thy
propensity towards it, yet it is also true that it sometimes
increases it. When the former is the case, as in regard to
avarice and ambition, thou mayest fight by resisting the
thought, because the longer thou thinkest on the vanity of
the gains so much valued by avarice, or of the glory desired
by ambition, the more easily wilt thou come to despise them.
But when thinking increases the propensity, as is the case
with lust, which has the power of charming thee even when
thou art contemplating its ugliness, thou must fight, not by
striving with the thoughts it arouses in thee, but by flying
from them: "Turn away thy face from a woman dressed up."[3]
If, then, thou art to practise this art of conquering by flight
even when these carnal desires are actually making war upon
thee, how much more when they are only standing ready to
make it? *Militant adversus animam.* It is easy enough to fly
before the battle, but not so after it has begun. How wisely,
therefore, does the Apostle say, "I beseech you to refrain

Wisdom ii. 16.   [2] Ecclus. xxviii. 10.   [3] Ecclus. ix. 8.

yourselves from carnal desires, which war against the soul;" not to wait till they fight. And "whence," asks St. James, "are wars? . . . are they not hence, from your concupiscences which war in your members?"[1]

IV. Consider fourthly, what is the method of easily refraining from these desires. This is to mortify ourselves by shunning the occasions which have the power of arousing them, by custody of the eyes and ears, avoiding the reading of idle books, and the like. This, I say, is the only way. If this is neglected, not only wilt thou be unable to refrain from desires, but soon thou wilt not refrain from complacency, from consent, from act. It is necessary, then, to avoid incurring all these, to refrain from desires. And therefore, in conclusion, the Apostle would have thee be especially careful to have thy "conversation good." Whether thou art on this earth as "a stranger" or as "a pilgrim," thou must necessarily have frequent intercourse with those who unfortunately are there as citizens. These are here called "Gentiles" by the Apostle, either because, although they are Christians, they nevertheless are like the heathen in acknowledging no country but this world, or because, being much more numerous than either "pilgrims" or "strangers" on earth, they rightly come under this title of *gentes*, or the multitude. Since, then, thou art obliged to have intercourse with them, how necessary it is to conduct it with prudence and circumspection, so as not to be infected by their habits! Observe, too, that the Apostle, in speaking of this intercourse, says *inter gentes*, "among the Gentiles," not *cum gentibus*, "with the Gentiles;" to show that if it is necessary at times, for thy service or for theirs, there ought not to be any intimacy in it. But, even with regard to this obligatory intercourse, thou art to avoid every kind of it which is not evil only, but even suspicious; for it is in such intercourse more particularly that carnal desires are aroused. A look, a smile, an unguarded demeanour suffices to bring them out fully armed for the ruin of thy soul: "They war against" it. Examine thyself, and see whether thou art careful in refraining from these suspicious occasions, and refrain from them in future in order to refrain more easily from such desires.

[1] St. James iv. 1.

TWENTY-SIXTH DAY.

ST. ANNE.

The Lord . . . will be silent in His love, He will be joyful over thee in praise (Sophonias iii. 17).

I. Consider first, what is the act, so greatly desired, of being able to gain much at little cost: it is to exercise thyself in acts of the love of God, so that whatever thou doest thou mayest be always directed to Him by this express intention of doing it for love of Him. Then thou needest no longer complain if thy state does not allow of thy undertaking certain heroic actions for God, which others are able to perform in theirs; because God declares Himself to be satisfied with thee if thou, in thy state, dost not cease to love Him: "The Lord will be silent in His love." This is the most exact sense of these words, which are, therefore, calculated to give thee great consolation. Thou grievest, it may be, because thou art unable to practise the severe penances which many others do for God, all those disciplines and fastings, which, nevertheless, are justly demanded by thy sins. Make up for them by making frequent acts of the love of God, and then He will not insist on the rest. He "will be silent in His love." For it is quite evident that such acts, when made with all the heart, are sufficient to make thee capable even of escaping Purgatory. If thou canst do no more than serve Him in the pulpit: love Him, and He "will be silent in His love." If thou canst do no more than serve Him in the confessional: love Him, and He "will be silent in His love." Nay, if even thus much is out of thy power, and thou hast to devote thyself to domestic affairs, to bringing up children, to managing the household, or even to the performance of mere manual work; yet, if thou always dischargest these offices, as has been said, for love of Him, do not fear that God will not be as much pleased with thee in thy state, as with others in a more exalted one: "The Lord will be silent in His love," that is, not "in the work" done for Him, but "in the love" with which it is done. This is so consoling, that it should be a great incentive to thee to exercise thyself in these beautiful acts of love which are so agreeable to God. Hast thou any doubt about this? The saints attained to their wonderful holiness, not so much by means of the works they did, as of the love with which they

-did them. "Many sins are forgiven her," said our Lord of Magdalene, "because she hath loved much,"[1] not "because -she hath done much," but "loved much." At that time she had not shed one drop of blood for her sins, but her reward was the same, because she had shed so many heartfelt tears for them.

II. Consider secondly, that if to these acts of love, which we have spoken of, thou addest acts of praise, blessing God for everything which He orders for thee through the day, so that, far from complaining of anything that happens, however unfortunate it may be, thou even welcomest and approvest it, and sayest that whatever God does is well done, not only will He be silent concerning thee, as in the former acts of simple love, but He will "be joyful over thee in praise;" because in these acts thou addest to the love which is due to Him as a Father, the reverence, resignation, and respect which is due to Him as the Sovereign Lord. There is no praise which can be given to God so dear to Him as this which thou givest Him for His good providence. He is pleased with the praise which thou renderest Him for His infinity, His immensity, and all the rest of His sublime attributes; but most of all with the praise thou givest Him for His adorable providence. And why? Because this is that which, more than all besides, His enemies have disputed with Him. And hence, as we learn from the Apocalypse, this is the praise which above all is sung to Him even in Heaven, on those harps of the blessed: "True and just are Thy judgments."[2] "Faithful and true: and with justice doth He judge."[3] "Just and true are Thy ways, O King of ages."[4] It is as though Heaven would thus make reparation for all the accusations brought against God's inscrutable judgment, shaking from its lips the bit of which God spoke when He said: "For My praise I will bridle thee."[5] Take example, then, from Heaven, not earth, and praise God alike for everything that He orders in thy life: "His praise shall be always in my mouth."[6] Praise Him in prosperity, and in adversity, and thus thou wilt offer to God a sacrifice of praise so pleasing to Him that it will cause Him to "be joyful over thee in praise."

III. Consider thirdly, that in these few words of the Prophet is shown the shortest way of becoming a saint, not exteriorly but interiorly: to love and to praise. Love Him

---

[1] St. Luke vii. 47. [2] Apoc. xvi. 7. [3] Apoc. xix. 11.
[4] Apoc. xv. 3. [5] Isaias xlviii. 9. [6] Psalm xxxiii. 1.

always in thy work, praise Him always in His. Praise without love would be formal, love without praise false. Therefore the Prophet has here combined these two conditions of love and praise, because they can never be lawfully separated. The whole of human life is woven, so to speak, of two threads, the one, what we choose to do for God, the other what He chooses to do for us. In the actions which we do, God is above all things pleased with love; in His own, with praise. But observe how often thou doest the reverse of what thou art bound to do. In thy own works thou lovest thyself instead of God, seeking in them thy own interests, thy own ends rather than Him. In His works, far from praising Him, how often dost thou complain of Him, if thou dost not accuse Him? What wonder, therefore, that thou art so far from becoming a saint in thy state of life? The fault lies not with the state, but with thyself. Consider the Saint of this day, the glorious St. Anne, who attained so high a degree of sanctity that she merited to be the mother of her who was to be the Mother of God. And how did she attain it but by the practice of which we have been speaking? By loving God with her whole heart in the privacy of her life, and by praising Him during her long years of barrenness. So true is it, that if thou, too, art faithful in practising this exercise, "the Lord will be silent in His love, and will be joyful over thee in praise."

TWENTY-SEVENTH DAY.

Be at agreement with thy adversary betimes, whilst thou art in the way with him, lest perhaps the adversary deliver thee to the judge, and the judge deliver thee to the officer, and thou be cast into prison. Amen, I say to thee, thou shalt not go out from thence till thou repay the last farthing (St. Matt. v. 25).

I. Consider first, who, according to the most approved mystical sense, is "the adversary" here spoken of, as understood by the saints. It is the voice of conscience, and our Lord bids thee deal with it as thou wouldst do with a powerful adversary who has some just claim on thy person. In this case, art thou not anxious, when he is in the act of bringing thee before the judge, to give him, as far as thou art able, the satisfaction he requires, either by compounding with him, coming to a compromise, or paying something on account, so as to pacify him as much as possible? So, too, art thou to act with regard to conscience. And why? To avoid being condemned to pay in its full rigour the debt, which it is in thy power to discharge now at a far less inconvenience. If thou understandest this truth aright, thou wilt be less apt to go on slighting its just requirements.

II. Consider secondly, that this voice of conscience is called thy "adversary," not because it wishes thee harm (for in that case it would be called thy enemy), but because it opposes thy irregular desires, acting towards thee the same part as the Angel did to the obstinate Balaam, when he said to him, "I am come to withstand thee because thy way is perverse and contrary to me."[1] Sometimes this adversary seeks to check thee from running into evil, at others to urge thee to the good from which thou art holding back. In both cases he is alike withstanding thee, and so is always thy adversary (*adversatur tibi*). How then? Is this a reason for being angry with conscience? Nay, rather for loving it very greatly. Better is a friend who opposes than an enemy who flatters : "Better are the wounds of a friend than the deceitful kisses of an enemy."[2] The leaven of concupiscence is the enemy who flatters, and whom thou oughtest to hate ; the voice of conscience is the friend who opposes, and whom thou oughtest to love. And if thou desirest its opposition to cease, thou hast only to satisfy it : "Take away the adversary, and crush the enemy."[3] "Take away the adversary" by satisfying the dictates of conscience, and "crush the enemy" by quelling the leaven of concupiscence.

III. Consider thirdly, that our Lord bids thee come to an agreement with this adversary in the two things above mentioned : in abstaining from the evil from which it withholds thee, and in performing the good to which it incites thee. But He bids thee do this quickly : "Be at agreement with thy adversary betimes, whilst thou art in the way with him." He does not say "immediately," because sometimes it is well to take a little time for consideration ; but He says "betimes," that thou mayest lose no time, since it may easily happen that a man is at the end of the way when he fancies himself to be in the middle. This "way" is our mortal life : "Oh, that my ways may be directed,"[4] in which this friendly "adversary" continually accompanies thee. But how will it be if he has

---

[1] Numbers xxii. 32.
[2] Prov. xxvii. 6.　　　[3] Ecclus. xxxvi. 9.　　　[4] Psalm cxviii. 5.

not received due satisfaction in time? Then from being a friendly adversary he will become one who is very inimical, a legal accuser, for that is the word in the Greek text. Now, how does it stand with thee? Has this "adversary" of whom we are speaking any just claim on thee as to thy conduct? What does thy heart tell thee? Is there some good to which he is urging thee in vain? Oh, make haste to "be at agreement whilst thou art in the way with him," for at the end of the "way" thy regret at not having come to an agreement with him will be utterly unavailing. He will tell everything as it is, in strict justice : "All things that are reproved are made manifest by the light."[1]

IV. Consider fourthly, how ill it will fare with thee if thou hast not come to an agreement in time, as thou wert bound to do, with this adversary, who, as thou hast heard, will deliver thee into the hands of the judge. This Judge—thou knowest it well—is Jesus Christ. And it is into His hands that this adversary will deliver thee as thy accuser, for it is the voice of conscience which thou hast slighted which will arraign thee before Christ as a criminal, and not only arraign thee, but convict thee more fatally than all besides, so that on it more than on all besides will thy sentence depend. And so there is no doubt whatever that thy sentence will be in accordance with the testimony. Observe, therefore, that it is here written, " Lest, perhaps, the adversary deliver thee to the judge," but not, " Lest, perhaps, the judge deliver thee to the officer." It is said, " Lest, perhaps, the adversary deliver thee to the judge," so putting the matter doubtfully, because it is possible that, at least at the end of the way, thou mayest have made satisfaction to this "adversary" by so deep and intense a sorrow for the resistance thou hast made to him, that in consequence he has had to abandon all claim upon thee. But it is said absolutely, " Lest the judge deliver thee to the officer," because, as this adversary has become thy accuser, it is certain that he has gained the cause. It is also certain that the judge will have to give thee over to the officer, that is, to the angel charged with the execution of the sentence, and it is certain that this angel will have to take thee to the prison appointed for thee. Would it not, then, be a grievous error not to have come in time to an agreement with this adversary, to whom such deference will be paid in that great tribunal from which there is no appeal? " Be," then, "at agreement

---

[1] Ephes. v. 13.

with thy adversary betimes, whilst thou art in the way with him, lest, perhaps, the adversary deliver thee to the judge, and the judge deliver thee to the officer, and thou be cast into prison."

V. Consider fifthly, what this prison is which is here spoken of. It is two-fold—Purgatory and Hell, and one of these must certainly be thy lot, according to the nature of the sins thou hast committed. Whichever it may be, thou wilt have to give full satisfaction in it. It is our Lord Himself Who thus solemnly asseverates : "Amen, I say to thee, thou shalt not go out from thence till thou repay the last farthing." The word "till" sometimes implies a term, and signifies that which will come to pass in the future ; as in the passage of Job : "I expect until my change come."[1] Sometimes it does not imply such a term, and then it signifies what will never come to pass, as in that other passage of Job, where he says : "Till I die I will not depart from my innocence."[2] If, then, thou art to go to Purgatory, thou wilt certainly come out, but not till thou hast strictly satisfied every claim ; if thou art to go to Hell, thou wilt not come out again to all eternity. So that it is said in either case, "Thou shalt not go out from thence till thou repay the last farthing." Suppose that there are two debtors kept in prison at thy requisition, the one a man of large fortune, the other a bankrupt. If thou sayest to the former, "Thou shalt not go out from hence till thou repay the last farthing," thou tellest him that he is to come out ; but when? When he shall have given thee strict satisfaction. But if thou sayest the same thing to the bankrupt, thou tellest him that he is never to come out, because he is entirely unable to satisfy thee. The case we are considering is the same. In Purgatory souls are in a state in which they are able to make satisfaction, because they have a capital of grace ; in Hell they they are not in such a state, and so the former may be called rich, and the latter bankrupt. When, therefore, it is said to a soul in Purgatory, "Thou shalt not go out from thence till thou repay the last farthing," the soul is told that it shall indeed go out, but at its cost. But when this is said to a soul in Hell, the meaning is that it will have to remain in prison for ever. Whichever, then, may be the penalty here spoken of, whether temporal or eternal, how greatly will it surpass that which thou wouldst have suffered in coming to an agreement with thy adversary in the way ! If, therefore, thou

[1] Job xiv. 14.   [2] Job xxvii. 5.

art wise, make this agreement; do not delay, for time is passing : " Be at agreement with thy adversary betimes, whilst thou art in the way with him."

VI. Consider sixthly, that some persons would like to come to an agreement with this powerful adversary; but how? By bringing him over to their wishes; for they would fain induce the voice of conscience by plausible arguments gradually to approve of that which their appetite desires. But this can never be. And why not? Because it is thy part to be on his side, not his to be on thine. Observe how our Lord speaks : "Be at agreement with thy adversary betimes, whilst thou art in the way with him." He might have said equally well, "Whilst he is in the way with thee," for thou hast the voice of conscience in the depths of thy heart. But he did not choose to say so. He chose to say, "Whilst thou art in the way with him," to show that it is for thee to follow him, not for him to follow thee. Oh, how often dost thou seek to seduce, to deceive, or at least to silence him, that he may not cry aloud to remind thee of thy debt! What folly is this! "He that turneth away his ears from hearing the law, his prayer shall be an abomination."[1] And if it is in vain for that man to address himself to God who has purposely turned away his ears from hearing the voice of conscience, so abominable is his prayer, what will be the case of him who has gone so far as to try to silence it by perverting it? Remorse of conscience, which is, properly speaking, the voice which cries after the deed is done, may more allowably be neglected, especially by a person whose conscience is weak or timid, for fear of thus giving way to scruples; but the dictate of conscience, which is the voice that cries before the deed, must always be heard, at least in order to take time for deliberation ; and the louder the cry the more should it be listened to, because this shows more clearly that it is in the right.

---

[1] Prov. xxviii. 9.

TWENTY-EIGHTH DAY.

Thou shalt love the Lord thy God with thy whole heart, and with thy whole soul, and with thy whole mind, and with thy whole strength. This is the first commandment. And the second is like to it: Thou shalt love thy neighbour as thyself (St. Mark xii. 30).

I. Consider first, what it is that God requires of thee when He here says, "Thou shalt love the Lord thy God." In the first place, He requires thee to cling to Him "with thy whole heart," that is, with thy whole will. This is that part of man which is the ruling power, and therefore it is called by the noble name of the heart: "My son, give Me thy heart."[1] Next, in order to do this more perfectly, both in interior and exterior act, He would have thee call to thy aid first all the inferior appetites, which, rebellious as they are, seem more likely to interfere with this love; and then all the members of thy body, the tongue, the eyes, the ears, the hands, and so on. The appetites are here comprehended under the general name of the soul—"with thy whole soul"—and the members under that of strength—"with thy whole strength." But because thou art not able to attain this easily without the concurrence of the understanding, which is so important a part of man, He also tells thee to call in the assistance of the understanding with all its faculties. And this is here spoken of as the mind —"with thy whole mind"—so that one word may include them all. If, then, God lays this command upon thee, be of good cheer, for, by the very act of so doing, He binds Himself to give thee the strength necessary for fulfilling it. Do not, then, delay any longer to fulfil it. Let thy will be entirely given to God, let thy appetites obey no law but His. If thou desirest anything, let it be union with God; if thou rejoicest, thou shouldst rejoice in the honour paid to God; if thou grievest, thou shouldst grieve for the offences committed against God; if thou fearest, let it be the loss of His favour which is the subject of thy fear; and so of the rest. Let all thy members be employed in advancing the greater service of God, and let all thy mind be fixed on Him, so that in all thy studies and speculations thy object may be to discover how best to please Him. So shalt thou fulfil the precept here laid upon thee, "Thou shalt love the Lord thy God."

[1] Prov. xxiii. 26.

II. Consider secondly, that it is in Heaven, never on earth, that this commandment is perfectly fulfilled. But that should not dishearten thee; for it is to be observed that every one who issues a command has two things in view, to obtain the end of the command, and to obtain the operations which are the means contributing to that end. I will explain my meaning. When a general commands his soldiers under arms to carry by assault some strong position, what object has he in view? There is the taking of the post, which is the end of his command, and there are the operations which lead to that end, in accordance with good military laws, and which are the means of attaining the end. Now, he who accomplishes the end of the command, fulfils that command perfectly; and so, in time of war, the man who takes the post perfectly fulfils his general's command. The man who does not attain the end of the command, but who, nevertheless, so conducts himself that he follows the right rules for attaining it, to the extent of his power, does not, indeed, fulfil the command perfectly, but that does not matter; he fulfils it so far as not to be guilty; so far, indeed, as to be entirely praiseworthy as a soldier who, although he has not succeeded in taking the post, has not failed on his part in obedience to military laws. This being clear, thou seest that when God gives thee this commandment He has two things in view: one is, the end of the commandment, that is, thy perfect union with Him as thy last end; and the other, the operations which are the means contributing to it, that is, the exact observance of His law. It is certain that thou canst not attain this end perfectly on earth, for this is reserved for that state in which "God shall be all in all;"[1] but this is of little consequence. It is enough if thy actions are in accordance with the right rules which He has given for the attainment of this end. And if thou askest, why God thus chose to issue in clear and express terms a commandment to love Him with the whole heart, the whole soul, the whole mind, and the whole strength, when it is possible to do this perfectly in Heaven only, instead of suiting the injunction to our capacity, the reason is the same which causes the general to command the soldiers to take the post, a thing which certainly does not depend on them. God would have thee know in what direction to aim thy arrows, that is, thy operations; and how would this be possible without knowing what the target is? The target, in this case, is union with God by perfect

[1] I Cor. xv. 28.

love, which is that of the saints in Heaven. But since thou knowest what the glorious target is, look well whether all thy arrows are aimed at it, or whether they fly false: "The arrow of Jonathan never turned back."[1]

III. Consider thirdly, how justly God requires of thee to love Him as much as thou art able, in the manner here described, that is, with thy whole being. He is thy God, and therefore He is thy last end. And if so, is it not most just that thou shouldst devote every part of thyself to love Him? Look at the miser, who has made money his last end, and therefore his god: "Of their silver and their gold they have made idols to themselves, that they might perish."[2] Oh, how utterly does he give himself up to the love of that money! He loves it "with his whole heart," for his will desires nothing else; he is fully satisfied with it, so that he deprives himself of a thousand pleasures which he might have by spending it. He loves it "with his whole soul," for his appetites are employed in the service of little else. If he is angry, it is with some one who would dispute the possession of his money; if he is glad, it is when he has gained more money; if he is envious, it is of some one who possesses more money. He loves it "with his whole strength," for he spares it in nothing so little as in this; he fears neither fire nor water where money is concerned. And, above all, he loves it "with his whole mind," for it is in this that his mind is most faithful to him. How he broods and plans! he is incessantly finding out artful means of making greater gains. Now, if a man can go so far as thus utterly to devote his whole being, as thou hast seen, to the service of so false a god as money, why should he not be able to do as much for the true God? And if this is possible, it is just. Therefore when our Lord gave this chief commandment, He was not contented with saying, "Thou shalt love thy Lord," but He expressly chose to say, "The Lord thy God;" because if He is, as God, thy last end, is it not right that, as such, thou shouldst love Him with thy whole being? The reason why the miser so loves his money is because he thinks that in it he virtually possesses all good, although, in effect, he possesses none at all: "He that loveth riches shall reap no fruit from them."[3] How is it that thou canst not love God in the same way, nay, far more, since in Him there is, in effect, every good?

[1] 2 Kings i. 22.   [2] Osee viii. 4.   [3] Eccles. v. 9.

IV. Consider fourthly, that this same example gives thee the rule by which thou shouldst be guided in this love, and explains it to thee. What kind of love is due to our last end? That which puts it before everything. And this is what our Lord commands thee when He says, "Thou shalt love the Lord thy God." Thou art to do as the miser does, who follows his own inclinations and the inclinations of others in many things, but never to the detriment of his money; it is this which he must before all secure—it is his last end. And, besides rendering this honour to his money, that he prefers it to all besides, the miser also makes a formal act of love, which consists in loving money for money's sake, and not loving it, at least principally, for something it procures. And this, again, is the love thou art bound to give to God if thou wouldst love Him as thy last end. Thou art bound to love Him for Himself. It is not enough to love Him in order to escape the punishment which will be inflicted on those who do not love Him, nor merely for the sake of the reward which will be given to those who do love Him; for by so doing thou wouldst be guilty of not preferring Him to everything, thou wouldst put the punishment and the reward before Him. I do not deny that both of these may be incentives to make thee love Him more, but not to make thee love Him absolutely. So much so, that thou art bound to make often in thy life an explicit act of the love of God "above all things." I say explicit, for though it is certain that such an act is virtually contained in the keeping of the other commandments—and therefore Christ said, "He that hath My commandments, and keepeth them, he it is that loveth Me "[1]—yet it is not contained formally, and perhaps it was for this reason that Christ did not say "he loves Me," but "he it is that loves Me," as though to show that such keeping of the commandments, although it is certainly a mark of this love, nevertheless is not that formal love itself; for it is certain that those acts of implicit love which cannot be distinguished from the keeping of the commandments are rather acts of obedience and homage to God as a Master, than of love to God as our last end. Nevertheless it is most certain that we are bound to these also, for to teach the contrary is an opinion which is condemned. It is true that as the affirmative commandments are not binding at all times, but only in the circumstances requiring their performance, as is the case with fasting, confession, Communion, and almsgiving,

[1] St. John xiv. 21.

so, too, is it with this commandment by which we are bound to make these express acts of the love of God "above all." But what are those circumstances? Shall I say what I think? Make them as frequently as possible. Thou seest that our Lord fixes no time, as in the case of fasting, confession, Communion, almsgiving, and the like, because we are so strictly bound to love Him at all times. Those words, "All things have their season,"[1] apply to everything else, but not to this. Let this suffice for the present in explanation of this precept, which, being the chief of all contained in the Scriptures, rightly demands that more than one meditation should be devoted to it.

TWENTY-NINTH DAY.

Thou shalt love the Lord thy God with thy whole heart, and with thy whole soul, and with thy whole mind, and with thy whole strength. This is the first commandment (St. Mark xii. 30).

I. Consider first, that this commandment, which we explained in yesterday's meditation, is called "the first" for several reasons. (1) Because it is the first in the intention of Him Who gives the Law. It is the commandment to which all the rest are subordinate: "Charity is the end of the commandment."[2] And consequently it is first in intention, because it is the end of the other commandments. (2) Because it is the first in the obligation of him who receives the Law. For if he has to keep all the other commandments because they are subordinate to this, much more is he bound to keep this to which they are subordinate. (3) Because it is the first in dignity of all the commandments which make up the Law. And what commandment could there be more suited to the nobility of the soul of man? The noblest commandment is doubtless that which least offends thy liberty. And such a commandment is this: "Thou shalt love the Lord thy God," because it is the only exception to the list of those which are fulfilled by restraint of the will. The other commandments, "Thou shalt not steal, thou shalt not commit adultery, thou shalt not kill," and the rest, are more servile in their nature, because it is possible to fulfil them purely out of fear of the punishment attached to their transgression. Not so with this.

1 Eccles. iii. 1. 2 1 Timothy i. 5.

This is a precept of love, and so can only be fulfilled by loving.
If thou lovest because thou fearest, thou dost not love at all,
and so thou dost not fulfil it. There is no act more voluntary
than love, and therefore there is none nobler. This being so,
it is manifest that this is the first commandment in dignity:
"This is the first commandment." If there is no nobler act
than love, there can be no nobler commandment than that
which is concerned with that act. Do thou now consider
a while for thy profit, how greatly thou wrongest God by robbing
Him of this act to give it instead to mean creatures of earth.
Truly dost thou deserve that He should deal with thee as
He did with the serpent, and give thee, as a punishment, that
which thou art doing by choice, namely, never to raise thy
head from the earth: "He that is filthy, let him be filthy
still."[1] Even if God had not so expressly commanded thee
to love Him, thou shouldst most earnestly beg Him to allow
thee to do so, so noble an act is it. How, then, is it possible
not to love Him, when He has commanded it in the words of
the text?

II. Consider secondly, that as this commandment is first
in the dignity which it bears, so, too, is it first in the happiness
which it confers. For if it is love which sweetens the bitterness
of all the other commandments, how can it but be sweet itself?
The happiness enjoyed by the will in loving God is indescribable. It is happy in praising Him, in honouring Him,
in obeying Him, but incomparably happier in loving Him.
The reason is this: all pleasure comes from the harmony
existing between the faculty and the object; this is evident.
But this is not enough. There must also be the union between
them, so that the closer the union the greater will be the
pleasure. This is seen in the case of food, which is doubtless
always agreeable to the palate because of the harmony existing
between it and the food; but when is it most agreeable?
When the palate is most closely united with the food, that is,
by properly masticating it, not merely tasting and rejecting it.
Now, it is quite certain that there cannot possibly be any object
more in harmony with the will than God, Who is a Food which
satisfies and never cloys; and it is also quite certain that there
cannot possibly be a faculty more in harmony with God than
the will, which is a palate which feeds and is never satiated.
It follows, therefore, that the closest union between this faculty
and this Object must doubtless be the sweetest of all. And

[1] Apoc. xxii. 11.

this union is effected by love. Thou hast not experienced it? It may be so; but ask of all the saints who have done so; how eagerly will each one of them answer thee, "His fruit was sweet to my palate!"[1] If thou dost not experience it, the fault cannot lie either with the object or the faculty. With what, then? With the lack of the necessary union. Give thyself to the love of God, to contemplation, to compunction, and thou wilt see. But all that thou hast yet done is, at most, to taste the food, and then to reject it. It is said, "His fruit was sweet," not "to my lips," but "to my palate."

III. Consider thirdly, that as this commandment is the first in dignity and in happiness, so, too, is it first in utility. And the reason is, that it may be truly said that the payment given to him who keeps it is rather gained unfairly than earned. Properly speaking, it would be for us to pay God in order to induce Him to give us leave to love Him, not for God to pay us in order to make us consent to do so. See, then, how great a favour God has shown us by saying, "Thou shalt love the Lord thy God." He has made this love a commandment. And by so doing, He has declared that this love shall be meritorious of reward from Him, a thing which at first sight we might well doubt. We know that in the religious life even such things as going for a walk, taking food, recreation, sleep, are meritorious: and why? because they are performed out of obedience. So, too, with regard to this commandment, "Thou shalt love the Lord thy God," it is most certainly meritorious to love God, because we are obeying when we love Him. But what payment was ever due, in the nature of the case, to one who loves the Supreme Good? And yet it will be much if thou dost love Him, even with the incentive given by this commandment.

IV. Consider fourthly, how marvellous it is that not only thou, but so very many people, are negligent in the observance of a precept which is "the first" in every respect. It is true that it cannot be perfectly kept here, as was shown in the preceding meditation; but nevertheless they should strive to keep it, as far as they possibly can, by making use of those means which conduce to this end, and therefore they have no kind of excuse. But what are these means? The chief of them is this: to enter deeply into the knowledge of so great a Good as that which we are bound to love. The saints in Heaven know Him "face to face," and therefore it is that

---

[1] Cant. ii. 3.

they so greatly love Him. We ought to strive to know Him, at least afar off : " Being made eye-witnesses of His majesty."[1] Let this, then, be thy study, in thy measure : " Set thee up a watch-tower," saith the Prophet.[2] Know Him, and thou wilt love Him. And indeed, even when seen afar off, how worthy is He of love ! His creatures, even, cease not to bid thee love Him ; the heavens with all their stars, the air, the water, the earth, all bid thee love Him. All creatures with one voice repeat this commandment : " Thou shalt love the Lord thy God." If thou dost not hear them, the only reason is that thou dost not attend. If thou didst, thou wouldst act like a certain holy man, who as he went about kept striking with his stick the grass, the stones, the trees, and the flowers, bidding them not to cry so loudly that he must love God, for that he could not bear it. Thy mind must be distracted indeed not to hear these voices. And if thou hearest and dost not answer, what art thou ? I will tell thee what the devil replied when summoned to declare by the mouth of a possessed person who he was : " I am," he said with a terrible cry, " I am the creature who does not love ; " and that was the only account he gave of himself.

THIRTIETH DAY.

And the second is like to it : Thou shalt love thy neighbour as thyself
(St. Mark xii. 31).

I. Consider first, how sublime a commandment this of loving our neighbour is, since, although it is the second, it is nevertheless said to be like the first, namely, that of loving God, which we have considered in the last two meditations : " The second is like to it." And if thou wouldst know in what points it is like the first, I here put them down briefly : (1) Because, like the first, it is of obligation. To love God is not merely a thing that is useful, like voluntary poverty or virginal purity, but one that is necessary. It cannot therefore, be only recommended as a counsel, it must be a precept. So, too, of the love of our neighbour : " This is My commandment, that you love one another, as I have loved

[1] 2 St. Peter i. 16.   [2] Jerem. xxxi. 21.

you."[1] (2) Because it is, like the first, a Divine command-ment. To love God is not a precept given by men, but by God: "In His right hand a fiery law."[2] So, too, is the love of our neighbour: "This commandment we have from God, that he who loveth God, loves also his brother."[3] He says "from God," not "from man." And therefore whenever, directly or indirectly, men's dispositions, manners, habits, or traditions are in opposition to this command, they must give way: "We ought to obey God rather than men."[4] (3) Because, like the first, it is a moral law. To love God is not of the number of the ceremonial precepts which were abolished by Christ in His law, nor of the judicial precepts which were modified, but of the moral precepts which were confirmed. And the same is true of loving our neighbour, for which reason Christ devoted a large portion of His Sermon on the Mount, to shielding it from the false interpretations which had been given of it, to perfecting, to inculcating it; so much so, indeed, that on the last day of His life He was able to call it "a new commandment," because He had invested it with so much additional beauty, not only by His teach-ing but by His example: "A new commandment I give unto you, that you love one another as I have loved you."[5] (4) Because, like the first, it is natural. To love God is not, like Baptism, a positive Divine precept, it is natural; for nature itself teaches that every one should love his great Father. And so, too, as to the love of our neighbour; for nature teaches also that every one should love his brother: "Every beast loveth his like."[6] And therefore the love of our neighbour is not so much an action that is good because it is commanded, as one that is commanded because it is good. (5) Because, like the first, it is absolute. To love God is not a conditional precept, like that of penance, in enjoining which sin is presupposed; it is absolute. So, too, is the love of our neighbour, and therefore it is impossible for it to be attached to any condition; it is binding of its own nature, independently of any presupposition: "This is the declaration, which you have heard from the beginning, that you should love one another."[7] (6) Because, like the first, it is affirmative. It is not a negative precept, like that not to take His Name in vain, because it lays upon us what is good, and so is truly a commandment; it does not forbid what is

[1] St. John xv. 12.   [2] Deut. xxxiii. 2.   [3] 1 St. John iv. 21.   [4] Acts v. 29.
[5] St. John xiii. 34.   [6] Ecclus. xlii. 19.   [7] St. John iii. 11.

evil, which is merely a prohibition. And so with the love of our neighbour; therefore it is a nobler commandment than all the negative ones, since to do good is more than not to do evil. Besides, the negative does not include its affirmative, but the affirmative does include its negative. So that when thou art ordered not to hate—"Thou shalt not hate thy brother in thy heart"[1]—thou art not, at the same time, ordered to love; but when thou art ordered to love, thou art, at the same time, ordered not to hate: "The love of our neighbour worketh no evil."[2] (7) Because it is, like the first, universal. To love God is not a particular precept more binding on one sex, one state, one country than another; it is universal, and extends to all. The same is true of the love of our neighbour: "He that loveth not, abideth in death."[3] Nor is it universal only because it is given to all, but because it is given concerning all. As all are to love, so all are also to be loved, even our enemies: "Thy commandment is exceeding broad."[4] (8) Because it is, like the first, very clear. The commandment to love God, if thou considerest well, requires no explanation, and so it is that the simple often keep it better than great doctors. So, too, with the love of our neighbour: "The commandment of the Lord is lightsome, enlightening the eyes."[5] Every one knows that love is called "the commandment of the Lord" by antonomasia; and this commandment enlightens the eyes, because a man who does not love stands in need of many directions to teach him the laws of perfect friendship; but not so the man who loves, for love itself enlightens him: "He that loveth his brother, abideth in the light."[6] (9) Because, like the first, it is perpetual. To love God is not a temporary but an everlasting commandment, and therefore it does not end even with death: "Charity never falleth away."[7] And the same is the case with the love of our neighbour, by which we are so bound, while travelling here in the way, that we are not freed from the obligation even on our arrival at home: "He that is a friend loveth at all times."[8] Therefore, even if thou knowest no more than is here shown thee of this sublime commandment, ought not this alone to be sufficient to make thee in love with it? See what glorious privileges are attached to it. Dost not thou value them? The commandment to

---

[1] Levit. xix. 17.   [2] Romans xiii. 10.   [3] St. John iii. 14.
[4] Psalm cxviii. 96.   [5] Psalm xviii. 9.   [6] 1 St. John ii. 10.
[7] 1 Cor. xii. 8.   [8] Prov. xvii. 17.

love our neighbour is so "like" that to love God, that they are twin brethren, and the reason why to love God is called the first, and to love our neighbour the second commandment, is simply that we are to love our neighbour for the sake of God, not God for the sake of our neighbour. In other respects they are so closely united that they can never be separated. It is impossible for thee to love thy neighbour without loving God, or to love God without loving thy neighbour, and therefore the union is closer than that between twins, who are indeed born together, but who do not necessarily die together; whereas neither of these two commandments can possibly live without the other.

II. Consider secondly, what it is to love another. It is to wish him well. Therefore, thou lovest thy neighbour when thou wishest him well both as to what regards the body and the soul; and thou lovest him as thyself when thou wishest him all this as thou wishest it for thyself; and this is what our Lord means when He says, "Thou shalt love thy neighbour as thyself." From this thou shouldst draw three very useful consequences, which will help thee in the practice of this most important commandment, all of which are founded on these words. The first is, that it is impossible that the love of thy neighbour should make thee comply with him in anything which is either unreasonable or wrong: if thou doest this, far from loving, thou hatest him greatly, seeing that this is to wish him the same evil into which his chief enemies, the devils, are constantly endeavouring to draw him, namely, sin. This being so, thou not only art not fulfilling this commandment by such conduct, but thou art directly violating it; because our Lord bids thee wish well to thy neighbour (*diliges*), and thou not only dost not wish him well, but wishest harm to him, like a devil. The second is, that thou art bound to love thy neighbour for himself. Therefore, if thou lovest him because intercourse with him is pleasant to thee, or because dealings with him are profitable to thee, then, if thou art not actually violating this commandment, it is, at any rate, certain that thou art not fulfilling it, because what thou lovest is thy own will or profit, and consequently thou lovest not thy neighbour, but thyself, and what our Lord says is, "Thou shalt love thy neighbour." In such a case thou lovest him, not as a neighbour, but as a slave, because thou lovest him for thyself. And our Lord expressly made choice of this word "neighbour" to show thee that thou art to love him as such, and consequently to love him as an equal, not as a servant; since if he is thy neighbour he is equal in rank to thee, that is to say, he is capable of attaining the same everlasting beatitude as thou art. No matter whether he is high or low, a fellow-countryman or a foreigner, good or wicked, friendly or hostile, he is capable of being thy companion in Paradise: therefore he is thy neighbour. This is the teaching of the saints. The third is, that this good-will which thou owest thy neighbour must not be a cold, dull, unfruitful sentiment, which deserves no better name than that of velleity; for if this is so, thou dost not love him "as thyself," that is to say, not "in charity unfeigned."[1] Tell me, art thou contented, in thine own case, with a barren desire of this sort? On the contrary, how greatly dost thou labour to procure whatever thou thinkest will be profitable to thee! In the same way shouldst thou act with regard to thy neighbour: "Judge of the disposition of thy neighbour by thyself."[2] Otherwise, thou mayest persuade thyself that thou keepest this commandment by the good intention of thy heart; but this is not so, because the good intention does not pass into action: "They have spoken vain things, every one to his neighbour."[3] If, then, thou observest well, from the first to the last, thou wilt see plainly that very few persons in the world keep this commandment. For there are many who love their neighbour with an evil love, hating him when they imagine that they love him; many, again, love him with an interested love, and thus love, not their neighbour, but themselves; and more still love him with a love more dead than alive, because they will not labour, exert themselves, employ themselves, trouble themselves about him, and consequently they love him, not as themselves, that is, eagerly, warmly, practically, but like something which does not belong to them, that is, in an utterly lifeless way. And yet our Lord did not think it enough to say, "Thou shalt love thy neighbour as something belonging to thee." He said "as thyself." Is not this a very sorrowful thing? See to what the beautiful law of charity is reduced, it has countless multitudes of transgressors, and very few observers: "Blessed is he that findeth a true friend."[4] And yet this is the law which is as important as that which commands us to love God: "There is no other commandment greater than these."[5]

---

[1] 2 Cor. vi. 6.   [2] Ecclus. xxxi. 18.   [3] Psalm xl. 3.
[4] Ecclus. xxv. 12.   [5] St. Mark xii. 31.

THIRTY-FIRST DAY.

ST. IGNATIUS.

How can you believe, who receive glory one from another, and the glory which is from God alone you do not seek? (St. John v. 44).

I. Consider first, how injurious to thee is the love of human glory. It not only puts an impediment in the way of thy faith, but, as it were, makes it impossible for thee to have any: "How can you believe, who receive glory one from another, and the glory which is from God alone you do not seek?" Faith, to be right faith, must be true and living. The man who believes what the Church teaches, even though his works do not agree with what he believes, has nevertheless true faith, because false faith is found among heathens and heretics, but he has not a living faith, because it is not active: "Faith, if it have not works, is dead."[1] He whose faith is active has not only a true but a living faith, because a corpse cannot possibly do any works. Now, the love of human glory incapacitates thee in the greatest degree from both kinds of faith: "The beginning of the pride of man is to fall off from God."[2] When this love is excessively great it hinders thee from having even mere true faith, because faith requires a docile understanding, which is not merely easily won, but made captive in its service; and ambition, which is the same as the love of human glory, makes it proud, obstinate, and rebellious, so that if, even among Catholics, such an understanding were found secretly abiding in any of these insanely ambitious persons, they would be discovered really to possess no faith of any kind; and if they do not actually disbelieve certain troublesome articles, such as the immortality of the soul, and the like, they have, at all events, doubts concerning them. And if they doubt they have no faith at all, for it is not necessary expressly to disbelieve what it teaches in order to have none; it is enough to doubt about it. And even when this love is not so great it at least indisposes thee greatly for living faith, for, in order to act as a Christian ought, to forgive injuries, to restrain oneself, to give way, to humble oneself, it is often necessary to overcome much human respect, to despise applause and approbation, nay, to submit to very hard words. And how can this be done by a man

[1] St. James ii. 17.   [2] Ecclus. x. 14.

who has not entirely dethroned in his heart this idol of glory, but who, if he no longer worships, at any rate prizes it? "Many of the chief men believed in Him," says St. John, "but because of the Pharisees they did not confess Him." And what was the reason? "For they loved the glory of men more than the glory of God."[1] See, therefore, how necessary it is not only to dethrone this idol of ambition, but to break it to pieces, to grind it to dust, so that the very remembrance of it may not be left in thee: "To me it is a very small thing to be judged by you;"[2] not merely "a small thing," but the least possible, for the word is *pro minimo.* The ark containing the Law, which is the symbol of a faith which is not true only, but living, can never be on terms with this idol: either it throws it down or is driven away by it and departs.

II. Consider secondly, how great is the folly of those unhappy persons who love the glory which comes from man, since it prevents them, at least in a great measure, from obtaining the glory which comes from God. For the latter, being founded on merit, is solid; the former, being founded on opinion, is not merely not solid, but frivolous. The opinion entertained of them by men has three defects, which render it completely valueless. The first is that, generally speaking, it cannot form a just estimate of thee, and if it can, it is not willing to do so: "Chanaan, there is a deceitful balance in his hand, he hath loved oppression."[3] The second is that it is very uncertain of attainment, and therefore very often it is "filled with shame instead of glory;"[4] and the third is that it is exceedingly unstable after it is attained, according to the words which follow in the same text: "And shameful vomiting shall be on thy glory." Now the words of our Lord in the text are worthy of much attention. He says, not only that thou art not to seek the praise of men, but not even to accept it when they offer it to thee; and He says not only that thou art to accept it with great gladness from God, but also to seek for it: "How can you believe, who receive glory one from another, and the glory which is from God alone you do not seek?" Speaking of the praise which comes from men, He said "you receive," because even to accept it is very injurious; speaking of that which comes from God, He said "you do not seek," because it is very injurious not also to aim at obtaining it. And yet it will be well even if thou dost not very often do the reverse, namely,

[1] St. John xii. 42, 43. [2] 1 Cor. iv. 3. [3] Osee xii. 7. [4] Habacuc ii. 16.

seek after the praise of men and slight that which comes from God, that is, the testimony of a good conscience: "Our glory is this, the testimony of our conscience."[1]

III. Consider thirdly, that there are many persons who do desire to be praised by God, and seek after it also, but who, at the same time, wish for the praise of men too. This is displeasing to God, and therefore Christ said: "The glory which is from God alone you do not seek." He did not say "from God," but "from God alone," because, after all, it is in this that real virtue consists, in being satisfied to please God alone: "That we may glory in Thy praise."[2] Provided pleasing God do not involve displeasing men there are many who desire and endeavour to please Him, but if it does involve this, then they are not able even to desire to do so. How unspeakably dost thou, then, slight the glory which comes from God if thou art one of those who are not content with pleasing Him unless they please men also! When the general of the army praises thee in the presence of all those armed battalions, as his gallant soldier, art thou very anxious about what is said of thee at the same time by the common herd hanging about the baggage-waggons? Oh, if thou didst but know what it is to have glory in the sight of God! "Therefore shall a strong people praise thee."[3] Whenever He praises thee thou art praised also by countless hosts of angels, more in number than the atoms of the air and the sands on the sea-shore, by all the apostles, patriarchs, prophets, martyrs, and saints, by all the multitudes, in short, who always see His face. A "strong people" indeed! and not strong only, but prudent, wise, and so noble that they are to be called "people" only with reference to their numbers, for in other respects they are a "people" who are all kings. And yet thou art concerned about what may be said against thee by a set of persons from the dregs of society! For what better are the men of the world in the sight of God? Nay, they are of even less amount than this: "All nations are before Him as if they had no being at all."[4] The difference is that the opinion of men is manifest, and therefore it influences thee; that of God is hidden, and therefore it has no power to do so. But how is this, when that which is hidden is more certain than that which is manifest? That which is hidden is certain by faith, and that which is manifest is certain by appearance. Learn, then, not to esteem any glory save that which is recog-

[1] 2 Cor. i. 12.   [2] Psalm cv. 47.   [3] Isaias xxv. 3.   [4] Isaias xl. 17.

nized by this light, the light of faith ; for this is the true glory :
"That I may please in the sight of God, in the light of the
living,"[1] not "in the sight of men, in the light of the dead.
And this glory is that which comes from God alone. Dost thou
not see how great that praise is which is given thee by all the
inhabitants of Paradise, that "strong people"? And even their
praise would not be of any value of itself, if it were not simply
the echo of that which comes from God, so infallibly certain is
it that this alone is the true glory "which is from God alone."

IV. Consider fourthly, that thou oughtest so greatly to
value the esteem which God has of thee, that thou oughtest
not to value in an equal degree even beatitude itself ; because
beatitude presupposes thee to be estimable, and the esteem
which God has of thee makes thee so. Observe, therefore,
how our Lord here expresses himself : "The glory which is
from God alone you do not seek." He does not say "with
God," but "from God," in order to show what the glory is of
which He is speaking. Many seek after salvation, and therefore many seek the glory which is "with God ;" but few are
anxious to please God alone, without any interested thought
even of that glory, and therefore few seek "the glory which
is from God alone." Yet this it is to which Christ invites us
when He says "from God," because this, in the full strictness
of perfection, is to desire nothing but the glory which comes
from God alone, that is, to desire to please Him indeed, but
only for the sake of pleasing Him : "That he may please Him
to Whom he hath engaged himself."[2] I know that to seek the
glory which is "with God" is of no detriment whatever to a
living faith—indeed, it is rather an assistance to it, because it
helps it to act, but it is detrimental to the most intense degree
of that living faith, because it is a hindrance in the way of
acting from no motive but that of pure charity : "Charity
seeketh not her own."[3] Whoever desires greatly to please God
must entirely divest himself of all self-love : "Let no man seek
his own,"[4] that is, he must try to please God, but not even for
the sake of the advantage to himself ; he must do so purely in
order to do what God has commanded, that is, that we are to
try to please Him. This is to desire to please God alone, to
strive after that glory which God has, without considering the
glory which He will bestow in the Kingdom of Paradise :
"The just love Thee," not "the things that are Thine." It is,
if we may say so, a kind of strife of love with God— 'My

[1] Psalm lv. 13.   [2] 2 Timothy ii. 4.   [3] 1 Cor. xiii. 5.   [4] 1 Cor. x. 24.

Beloved to me, and I to Him "[1]—for it is to desire to love Him as He loves us for our sakes only. He loves me without self-interest, and so He is altogether "to me," not "to Himself;" and so too will I love Him, being altogether "to Him," not "to myself." "My beloved to me and I to Him." Nay, may we not say that in this beautiful strife of charity it is we who are, so to speak, conquerors, just as Jacob was, for without anything that we possess God is blessed in Himself, but what are we without what He gives us?

V. Consider fifthly, how marvellously all this was fulfilled by the great patriarch Ignatius, who went out from the house of his fathers, like a new Jacob, with only a staff in his hand, and saw, even in his lifetime, so noble a progeny given to him by God: "Thou shalt spread abroad to the west and to the east, and to the north and to the south."[2] He, too, sought, we know, to wed in his own person both those lives which are so praiseworthy, Lia and Rachel, the active and the contemplative; nevertheless, if there was any one thing on which he laid the foundation of his sanctity, it was not this, it was his utter contempt of the glory which comes from men: "Let not my glory be in their assembly."[3] These were the words of Jacob when dying, and these, too, were the words of Ignatius, who was dead in this life to himself, that he might live to God. And this was precisely the reason why he afterwards became so wonderful an instrument for advancing God's glory, because he despised, and that utterly, all glory of men. From this contempt proceeded in him, first that sublime faith which he possessed, a faith so strongly rooted in his understanding, and therefore so true, that he was wont to say that though all the world should rebelliously turn their back on Christ, he alone would remain faithful to Him for the sake of what he had learnt of Him at Manresa, when he, too, in his measure, might have said with Jacob, "I have seen God face to face, and my soul hath been saved"[4]—that soul which before was lost; a faith, too, so fervent in the will, and therefore so living, that he would fain have worked in all and everywhere for the glory of God, in the public streets, in churches, in prisons, in schools, in hospitals, in the country, with untiring energy through heat and cold: "Day and night was I parched with heat, and with frost, and sleep departed from my eyes."[5] And in all this he not only did not seek the glory which comes from men, but

---

[1] Cant. ii. 16. [2] Genesis xxviii. 14.
[3] Genesis xlix. 6. [4] Genesis xxxii. 30. [5] Genesis xxxi. 40.

would have none of it, he did not "receive" it ; nay, rather,
he shunned it with all possible care, as, among other instances,
when he avoided an honourable reception on returning to his
country, as though he feared it more even than Jacob did.
So much was this the case, that he was in the habit of saying
that he would gladly be thought mad by every one if it were
possible for him without sin to obtain this universal discredit.
In the next place, he sought the glory of God so earnestly that
he sought it alone, and for this reason chose a path of holiness
which apparently had in it little that was singular, austere, hard,
and therefore wonderful, merely because he considered that it
would do God greater service in helping the souls which were
dear to Him. He was never weary of again and again addressing
to Him those beautiful words of Jacob, which are much fitter
to be said to God with heartfelt emotion than to Esau with
cowardly flattery : " I want nothing else but only to find favour,
my Lord, in Thy sight."[1] What was there too great for him
to forsake in order to give God glory? He would have
renounced even the beatitude of Heaven. Surely this was to
seek nothing but the glory "which is from God alone"—nay,
not only *quæ a solo Deo est*, but *quæ solius Dei est*—not that
which is "with God," so bravely did he strive with God in
this magnificent conflict of charity. So that if he too was
"strong against God" in this matter, what wonder is it that he
"prevailed much more against men," and drew so many to
God's service. Do thou take this holy patriarch for thy patron
in this work of despising human glory, and do not suppose that
it is the partial affection which every son, however unworthy,
bears to his father which makes me thus propose him to thee,
since when Christ Himself desired to give His dear servant
Magdalen dei Pazzi a saint from Heaven to dictate to her very
sublime lessons of humility, He chose St. Ignatius from all the
rest to send to her, the Saint in whom was so marvellously pro-
minent that low opinion of himself manifested by Jacob when
he said to God, " I am not worthy of the least of all Thy mercies,
and of Thy truth which Thou hast fulfilled to Thy servant."[2]
Nay, it was still lower than even this ; for when he was about to
draw his last breath, this was the last favour which he asked of
his beloved sons, as he gave them his blessing, not that they
would bury him, as was the request of Jacob, in "a double cave,"
with the double honour paid to the corpses of the great, but
that they would throw him upon a dunghill like a dead dog.

[1] Genesis xxxiii. 15.   [2] Genesis xxxii. 10.

AUGUST.

FIRST DAY.

ST. PETER'S CHAINS.

A vain man is lifted up into pride, and thinketh himself born free like a wild ass's colt (Job xi. 12).

I. Consider first, that the man who is here called "vain" is the man who is devoid of wisdom, sense, and every other good quality; for this is the force of the word "Raca," from which it is derived, the same word which Christ made use of when He said, "Whosoever shall say to his brother, Raca, shall be in danger of the council."[1] And yet, incredible as it seems, this is the man who, as a rule, is more than any other given to pride—"A vain man is lifted up into pride"—and what is more, his pride reaches such a pitch that he believes himself to be his own master in the world; he will not submit as he ought to his Superiors, he does not reverence and obey them; indeed, what he claims is little short of exemption from every law, and all the time he does not perceive that in this his aspirations are exactly what an ass's colt in the woods vainly fancies of himself, believing with the greatest impudence that he was there born to liberty among the beasts. But how he deceives himself! For whereas the other wild animals are left in their independent state, he is sought for to be made a slave, and is easily compelled to bear the yoke, to toil and labour, and carry burdens, just like those of his species which are born in the stall: "A vain man is lifted up into pride, and thinketh himself born free like a wild ass's colt," which is so greatly deceived in the foolish opinion it has of itself. We must therefore understand that man is not born to live without a law, as he chooses, but that he must remain as quietly in his

[1] St. Matt. v. 22.

chains as St. Peter did in his. Dost thou not see that the
Apostle was actually able to enjoy a calm sleep in his chains?
"Peter was sleeping between two soldiers, bound with two
chains."[1] Such, too, should be thy behaviour, if thou wouldst
act as the servant of thy Lord, and not like a wild beast. Now,
there are three kinds of chains from which no one may hope
to be entirely free. First, there are the chains of the commandments, and these are the chains of all the just, and he who is
contented to bear these is entirely free from the other two,
which are the chains of sin and the chains of punishment.
But whoever is not contented to bear the chains of the commandments, falls at once into those of sin, which are the
chains of all sinners in the world ; and he who does not cast
off these in time, and submit to those of the commandments,
falls at last into those of punishment, which are the chains of
the damned in Hell. Choose, then, which thou wilt, the chains
of the just or those of sinners ; but consider well, for if thou
preferrest the latter to the former, thou must needs one day,
against thy will, bear those which thou wouldst fain escape,
namely, the chains of the damned of which we have spoken.

II. Consider secondly, in the first place, how desirable the
chains of the commandments are. It might at first sight seem
as though they bound thee very tightly, but it is not so ; on
the contrary, they enable thee, better than anything else, to act
with freedom, because they make thee act by reason, not by
impulse. No one is more truly a slave than he who is the slave
of his own concupiscence, for such a man is, as it were, compelled by himself to do, in his own despite, the things he would
not : "But I am carnal, sold under sin. . . . For I do not that
good which I will, but the evil which I hate, that I do."[2] The
only free man is he who is not the slave, but the master of his
own concupiscence. And to be so, he must obey the commandments. Dost thou think that there is any disgrace in
wearing such noble chains? So far from it, these chains of
the just man are like a necklace which do not in any way bind
his neck, but rather adorn and embellish it, and enable him to
lift up his head to heaven with more dignity. And even if we
have to acknowledge that these chains do, in a certain way,
bind the neck of the just man, by keeping him in subjection to
God, it is at any rate certain that they are no burden to him,
for as they are a great honour, so too they bring him great
pleasure and great gain. The pleasure is undeniable ; for the

[1] Acts xii. 6.　　[2] Romans vii. 14, 15.

truly just man, that is, the man who does right, not from an
exterior motive of fear, but only because he loves to do it, so
little feels the pressure of the law that it may sometimes be
said that he has none—"The law was not made for the just
man, but for the unjust"[1]—not that the just man is not subject
to the law as well as the unjust, but because whatever is laid
upon the latter is a burden, whereas the law is no burden, but
a delight to the just man, because it only obliges him to do
what is reasonable, that is to say, it obliges him to do what he
would do even if there were no law ; and so the law is given
to him, not imposed upon him. It is imposed only on the
unjust, who would fain shake it off as a heavy burden. As to
the gain, it is even more certain than the pleasure. For does
not the just man know the great advantage that he derives
from these chains which the law puts upon him ? It is enough
to say that they are the chains of salvation—"Her bands are a
healthful binding"[2]—salvation both temporal and eternal. For,
even as Joseph's bonds were the occasion of his being first
taken under God's special protection, and afterwards exchanging
chains for a throne, so, too, is it with the chains of the just.
First, they render God more gracious to him in the events of
this mortal life—He "left him not in bands"—and then they
cause God in the end to exalt him from these same chains to
the glory of Paradise : He "brought him the sceptre of the
kingdom."[3] For although it is true that this transition from
earthly bonds has but seldom happened in this world—"Out
of prison and chains sometimes a man cometh forth to a
kingdom"[4]—yet it is continually occurring from the chains of
which we are speaking. How is it possible, then, that thou art
not encouraged to remain wholly in them, if thou art already
bound by them, or if not, to put them on ? O blessed chains,
which make thee truly master of thyself, to thy exceeding
honour, keep thy heart filled with joy, and procure for thee, to
thy great gain, the Divine assistance in this life, and at thy
death, the Kingdom.

III. Consider thirdly, in the second place, how very different
from these chains of the commandments are the chains of sin.
It is impossible to decide which is the greater, the dishonour,
the pain, or the injury which these chains bring with them.
As to the dishonour, the same reason which makes the chains
of the just an honour to them, makes those of sinners their

---

[1] 1 Timothy i. 9.
[2] Ecclus. vi. 31.　　[3] Wisdom x. 14.　　[4] Eccles. iv. 14.

disgrace. And what shame can be greater than that of yielding,
like a brute, to the violence used to thee by lust, avarice, and
ambition, which are the three raging furies described by
St. John? " Immediately he followeth her as an ox led to be
a victim, . . . not knowing that he is drawn like a fool to
bonds."[1] And as to the gain, what content canst thou ever
feel in thy heart, when thy chains in the end bring thee into
extreme straits, and can but overwhelm thee with scruples,
anxieties, suffering, and disquietude?—burdens which cannot
be borne, but only dragged along painfully : " You draw sin as
the rope of a cart."[2] And as to the injury, not merely do they
deprive thee of the Divine protection, but they at once make
thee the slave of Satan, so that, if thou diest in them, thou art
lost. And it is useless to say that when that moment comes
thou wilt cast them off ; for I ask, What warrant hast thou for
this? " To whom will ye flee for help," says the Lord, " that
you be not bowed down " at that hour of your death—" under
the bond "—even more than you were before—" and fall with
the slain "[3]—if you go into final damnation? Thou must,
therefore, quickly shake off these chains which are so hurtful,
so hard, and so shameful, now when there is so much greater
certainty of the Divine assistance in the work : " Shake thyself
from the dust, arise, sit up, O Jerusalem ; loose the bonds
from off thy neck, O captive daughter of Sion."[4] And if thou
desirest to loose them, there are three ways of doing so—
contrition, confession, and satisfaction. Contrition will prevent
thee from ever again feeling ashamed of these chains, because
of that noble sorrow which has loosed them, or rather burnt
and reduced them to ashes by its ardour. " Behold, I see . . .
men loose . . . in the midst of the fire, and there is no hurt
in them,"[5] that is, nothing to mar the beauty of their aspect.
Confession will obtain for thee special relief from the burden
of all those scruples which weighed on thee so continually
(such is the power of the priest's hand in absolving thee from
all sin), so that thy chains, once so heavy, will not trouble
thee : " The bands of his arms were loosened by the hands of
the Mighty One of Jacob."[6] The special work of satisfaction
will be to prevent these chains from harming thee, by means
of the penance by which thou wilt have atoned for thy sins :
" Thus saith the Lord, . . . I have afflicted thee, and I will
afflict thee no more, . . . and I will burst thy bonds asunder,"[7]

[1] Prov. vii. 22. [2] Isaias v. 18. [3] Isaias x. 3. 4.
[4] Isaias lii. 2. [5] Daniel iii. 92. [6] Genesis xlix. 24. [7] Nahum i. 13.

so that they may no more drag thee down to Hell. Is it possible that thou wilt not even now avail thyself of means so profitable to thy salvation? Take good heed, for there is no alternative but to pass from the chains of sin to those of punishment; and this is why sinners are spoken of as "such as turn aside into bonds,"[1] because they turn aside from the commandments to sins which bring upon them the bonds of punishment.

IV. Consider fourthly, the number of these chains of punishment, which are the chains of the damned. The Sacred Scriptures reduce them to three, namely, darkness, torments, and God's immutable decree to keep these wretched beings to all eternity in their fatal prison. The first chains are those of darkness, which are enough of themselves to prevent all flight. In this darkness all the damned will alike be plunged: "Fettered with the bonds of darkness."[2] What will be their lot in this condition? In the horrible darkness which encompassed the Egyptians, the Sacred Scriptures say that none of them dared to stir a step from the place he was in, for fear of worse happening to him: "No man moved himself out of the place where he was."[3] No one hastened to his companion to lift him up or to help him: "For they were all bound together by one chain of darkness."[4] Think, then, what must be the state of the damned! In that place where those wretched ones will be, they will be all bound together by that dense night as "by one chain," like so many slaves who can, indeed, join in cursing their lot, but cannot help each other. The second chains are those of the torments, in which each will groan without hope of pardon, since it is written that the prince, when he is angry, "will not spare to do thee hurt, and to cast thee into prison."[5] And therefore, as God "will not spare to do hurt" in Hell, so neither will He spare any of the chains of that "prison" into which thou art cast. And who can describe these chains? How many kinds of them there are! Fire and sword, brimstone, serpents, scorpions, dragons, and all imaginable horrors: there is no need to enumerate them one by one, thou canst easily go through them thyself. And yet all these chains which torture the senses are as nothing in comparison of those which torture the soul: "Its bands are bands of brass,"[6] so far do they exceed the others. Lastly, the third chains are those which proceed from God's immutable

1 Psalm cxxiv. 5.   2 Wisdom xvii. 2.   3 Exodus x. 23.
4 Wisdom xvii. 17.   5 Ecclus. xiii. 15.   6 Ecclus. xxviii. 24.

decree, and therefore they are called "everlasting." The angels who kept not their principality . . . He hath reserved under darkness in everlasting chains, unto the Judgment of the Great Day."[1] And these are the chains which will drive the damned to utter despair. God said to His dear servant Ezechiel, "Behold, I have encompassed thee with bands, and thou shalt not turn thyself from one side to another;" but He instantly softened the severity of this decree by the consolation which follows: "Till thou hast ended the days of thy siege."[2] But there is no such consolation for the reprobate in Hell. Lastly, the days of Ezechiel's siege, which prefigured in his person the siege of Jerusalem, were but three hundred and ninety, and so they were soon over. But when will the days of that siege be ended which will surround the damned? A million of ages will pass, and "the days of the siege" will not be over; fifty, a hundred million, nay, more millions than there would be grains of sand to fill the vast space from earth to heaven, and still that siege will be beginning, and its days will not be ended. If, then, thou art of the number of the damned, what will become of thee who fearest to be bound for a few days in the chain of the commandments? There will be no help for thee through all eternity. The chains of the commandments end with this life, and there is the power of escaping from those of sin till the moment of death; but there is no end to the chains of punishment, and no escape from them to all eternity.

SECOND DAY.

Amen, amen, I say to thee: if any man keep My word, he shall not see death for ever (St. John viii. 51).

I. Consider first, what a great difference there is between an ignorant and uneducated peasant, who has never in his life learnt the virtue of plants, and an excellent herbalist, who knows how to distinguish each from each. If these two should be walking in summer time over a mountain rich in beautiful plants, the peasant will not cast a glance upon them, but will tread them all alike under foot as he goes along, while the herbalist will pause to admire their beauties, search for them, gather them, tie them together, and on returning home preserve

[1] St. Jude 6. [2] Ezech. iv. 8.

them with great care, so as to put them to most useful purposes. Now, consider that the very same thing occurs with regard to the words of Christ. Some persons are entirely ignorant of their virtue, and therefore make no more account of them than of ordinary sayings: "They have most wickedly transgressed My words."[1]   Others will understand their virtue, and therefore keep them with the utmost care. And it is this which Christ would urge on thee when He says, "Amen, amen, I say to thee: if any man keep My word he shall not see death for ever."   Suppose that there were an herb which had the power of warding off death from thee for ten centuries, wouldst thou not give it the best place in thy stores, displacing pearls and rubies, and even diamonds, for its sake?   How much more carefully, then, oughtest thou to keep the words of Christ, which possess a virtue so much greater?   The virtue which they possess will keep death from thee for ever.

II. Consider secondly, how true it is that our Lord's words possess this virtue.   Death is of two kinds, the death of the body and the death of the soul.   As to the former, our Lord says that whoever keeps His word "shall not see death for ever," not because he will not die (for that is common to all men, even to Christ), but because after dying he will one day have a more beautiful, happy, and perfect life than before—just as wheat decays for a short time in the ground to flourish anew—and so, although it is true that he will see death, yet he will not see it "for ever," like the damned, who will always have it before their eyes, so that the life which they will have will only be just so much as is needful to make those unhappy beings continually feel what the pain of dying is.   Then as to the death of the soul, which is sin, our Lord says that whoever keeps His word "shall not see death for ever," because he shall never sin mortally.   For what constitutes this grievous death but the not keeping His commandments?   The man who lives according to what our Lord teaches is certain never to lose grace, and so never to lose the life of which we are speaking: "Son, keep My commandments, and thou shalt live."[2]   Now, as the death of the body may come from three causes, namely, natural infirmity, accidental occurrences (such as falls, drowning, fire, and the like), and violent assaults, so too the death of the soul may easily happen from three causes. It may come from natural infirmity, by which I mean an interior condition excited in us by the disorder of the passions,.

[1] Jerem. v. 28.   [2] Prov. vii. 2.

and our Lord's words keep these in such subjection that they
do not suffer them to cause death. It may come from accidental occurrences, which are the dangers we meet, against our
will, in occasions of sin, and the words of our Lord preserve a
man from perishing in them. It may come from violent assaults,
which are the temptations of the devil, and the words of our
Lord have the power of repelling these so that they come to
nothing. See, therefore, how we ought to value these words,
which are full of so much power: "My son, hearken to My
words, and incline thy ears to My sayings, . . . for they are
life to those that find them."[1]

III. Consider thirdly, in what way thou art to keep these
words of thy Lord in order to derive from them this great
advantage. There are three ways of doing so: in heart, in
word, and in work. As to the heart, thou must keep them in
the understanding, by meditating on them at the proper times,
particularly in the morning, when the understanding is clearest;
in the will, by always loving them; and in the memory, by
frequently calling them to mind; above all, in the dangers of
sinning which occur to thee: "Thy words have I hidden in
my heart, that I may not sin against Thee."[2] As to the
tongue—"in word"—thou must keep them, not only by loving
to converse about them, but by showing that thou valuest
them; not like those who are ashamed of professing what they
teach in conversation: "With my lips I have pronounced all
the judgments of thy mouth."[3] As to the hands—"in work"—
thou must keep them by faithfully putting them in practice:
"I lifted up my hands to thy commandments, which I loved,
and I was exercised in thy justifications."[4] Now examine thyself carefully, and see what is thy diligence in keeping the
words of thy Lord in all these three ways. Thou thinkest,
perhaps, that this is very laborious. But if laborious, it is still
more profitable. Remember that these are "the words of
eternal life." What, then, will become of thee if thou slightest
them? If in the keeping of them thou wilt have life, what
must remain if thou dost not keep them? Everlasting death.

[1] Prov. iv. 20—22.
[2] Psalm cxviii. 11.   [3] Psalm cxviii. 13.   [4] Psalm cxviii. 48.

THIRD DAY.

The hour cometh wherein all that are in the graves shall hear the voice of the Son of God, and they that have done good things shall come forth unto the resurrection of life, but they that have done evil unto the resurrection of judgment (St. John v. 28, 29).

I. Consider first, how, when the hour of the Last Judgment approaches, the Archangel St. Michael, accompanied by a multitude of other angels, will awaken with a loud trumpet-blast all the dead, who will be, so to speak, sleeping in their graves : "Arise, ye dead, and come to Judgment." And this trumpet of which I speak is no figure of speech, as some have thought, but a reality—" For the trumpet shall sound "[1]—and it shall sound because it is an instrument very suitable for that office. For the Jews were in the habit of using the trumpet for four purposes : to call to confession, to announce war, to do honour to the great feasts, and to give the signal for striking the tents in their journeys ; and all these four reasons apply to the sounding of the trumpet on the Day of Judgment. For then will be made the fullest confession that was ever known in the world : "The Lord will enter into judgment with the ancients of His people."[2] Then will be declared a universal war on all the reprobate : "The whole world shall fight with Him against the unwise."[3] Then will be the greatest of all feasts for the elect : "Blow up the trumpet on the new moon, on the noted day of your solemnity, for it is a commandment in Israel, and a judgment to the God of Jacob."[4] It is "a commandment" in reference to men, who have to appear before God, and it is "a judgment" in reference to God, Who will pass sentence upon them. Then, too, as we may say, will the tents be struck for the last time, when the elect and the reprobate will journey on those different roads : "And they that have done good things shall come forth unto the resurrection of life, and they that have done evil unto the resurrection of judgment." And when thou hearest it here said that all the dead without exception, "all that are in the graves shall hear the voice of the Son of God " calling them to the General Judgment which is about to be held, do not suppose that the Son of God will have to call them with His own lips, for the dignity of the Judge requires that He should never use His own voice, but that of His emissaries in sum-

[1] 1 Cor. xv. 52.
[2] Isaias iii. 14.   [3] Wisdom v. 21.   [4] Psalm lxxx. 4, 5.

moning criminals to the bar. He will call them by the voice
of the trumpet. Nevertheless that voice is spoken of as "the
voice of the Son of God" (just as the priest's voice in adminis-
tering the sacraments is called both the voice of Christ and
the voice of His minister), because it will be both the voice
of His will and the voice of His power. Of His will, because
He will summon that great assize ; of His power, because He
will cause the dead to hear it and to live again. Therefore it
is written, that on that day God "will give to His voice the
voice of power,"[1] that is, He will cause "His voice," namely,
the voice of this trumpet, to be a "voice of power," because
He will endow that voice with such efficacy that at its first
sound all the bodies which have been for ages not only reduced
to dust, but scattered and dispersed, will instantly resume
their original form, and become reanimated by a miracle
which can only be effected by the Divine power : "The voice
of the Son of God." This is why Christ, Who, in speaking
of this same Judgment, was accustomed to call Himself the
"Son of Man" (as has been mentioned in a previous meditation), this time purposely calls Himself "the Son of God,"
because by the power which He possessed of instantaneously
restoring life to the dead He intended to prove on this occasion
the certainty of that Divinity which He claimed before the
rebellious Jews. Do thou think now for a while how wonderfully solemn will be that obedience which all the dead will
render to Him on that day ! This is the particular force of
the words "shall hear." Not that the dead will not also hear
the voice in a sensible manner with their ears (for being
aroused by the sound of those first words, "Arise, ye dead,"
they will easily hear those which follow, "come to Judgment"),
but because they will not only hear, but also have to obey
them. Such is the meaning of "shall hear ;" the words mean
both to hear and to obey : "My people heard not My voice."[2]
Oh, how many there are now who will not hear the voice of
Christ ! They will hear neither that which is directly addressed
to them in the inspirations which He sends them, nor that by
which He addresses them in the words of His ministers. But
these unhappy persons will not be able to act thus on that day.
It will be in vain then to stop their ears and to say with
Pharao : "Who is the Lord that I should hear His voice?"
Small and great, subjects and kings, poor and rich, ignorant
and learned, must all alike obey that voice : "All . . . shall

[1] Psalm lxvii. 34.   [2] Psalm lxxx. 12.

hear." Alas for thee if thou now slightest the voice of thy Lord, either direct or indirect. What will become of thee in that hour? Yet it is absolutely certain that that hour will come: "The hour cometh." He does not say "will come," but "cometh," because it is so certain that it will come, that it may be spoken of as if it were come already.

II. Consider secondly, how, in conformity with the prompt obedience which, as has just been said, will be rendered by all the dead to this voice, it is added that they will all "come forth" from their graves; but, oh, with what a difference! The elect will have their bodies restored to them, no longer frail and worn, wounded and weary with constant hardships, but exceedingly glorious. The reprobate, on the other hand, will find theirs perfect indeed as to the number of their limbs, but in other respects so disgusting, squalid, and loathsome, that the mere necessity of re-entering dwellings so foul will form a great part of their misery. And this is not to be wondered at, for the former "shall come forth unto the resurrection of life," and the latter "unto the resurrection of judgment." The elect will "come forth unto the resurrection of life," because they will rise not only to live a life which will know no death (for this will be the case with the reprobate also), but because they will rise to live that life which is true life, the life which is enjoyed in Paradise, the beatitude of which is often described by the word "life:" "For with Thee is the fountain of life "[1]— that is, the fountain of beatitude. And the reprobate will "come forth unto the resurrection of judgment," not only because they will rise to be judged (for this will be common to both them and the elect), but because they will rise to be condemned. For such is the force of this word "judgment." Sometimes it signifies discussion: "Judgment determineth causes,"[2] and sometimes it signifies condemnation: "They that were not amended by . . . reprehensions experienced the worthy judgment of God."[3] And this last is undoubtedly the sense of the word in this passage which we are explaining, for "judgment" is here opposed to "life." The bodies of men will then be different, just as the nature of their resurrection will be different. This being so, what misery for thee if it should be thy lot to have one of those loathsome bodies! Thou wilt not then flatter and consider and caress it; nay, rather thou wilt curse that excessive love which thou now bearest it without being conscious of doing so.

[1] Psalm xxxv. 10.   [2] Prov. xxvi. 10.   [3] Wisdom xii. 26.

III. Consider thirdly, that the word used to describe the manner in which all men will come forth from their graves when they rise again at the Last Day is not merely *exibunt*, but *procedent*, because they will go forth to meet Christ in order, not of time, but of rank. Not of time, since all, both good and bad, will rise together in an instant—"In the twinkling of an eye"[1]—in order to show more forcibly the power of that Divine voice which enables them to rise again, but of rank, because Christ will go first to meet the elect, who will come forward to receive Him "in the air,"[2] and afterwards the reprobate, who will have to wait for Him on the earth; and even among the elect those will be foremost who are most closely united to Christ by the abundance of their merits, afterwards will come the rest in duly ordered bands: "Every one in his own order."[3] Imagine, as thou contemplatest this spectacle, what a terrible separation that will be, when the good and the wicked, alike issuing from the grave, will take such different ways, for "they that have done good things shall come forth unto the resurrection of life, but they that have done evil unto the resurrection of judgment." At the division of these roads, as we may call it, the angels will be waiting who are appointed to make the great separation between the elect and the reprobate : "The angels shall go out and shall separate the wicked from among the just."[4] And oh, what weeping and wailing, what groans and cries will be heard among the wicked! "Comfort is hidden from my eyes, because He shall make a separation between brothers."[5] Not only will that separation bring exceeding shame on those especially who have been in the habit of ruling and lording it over others, and who will then see themselves sent to take their place among the very dregs of the universe; but it will also be an occasion of extreme suffering, being a clear token of the miserable fate to be allotted to each one of them in the final sentence to which they are summoned. And thus the same thing will take place which occurred in the famous division of Jordan which was made by Josue, who was a type of Christ : "The waters from above," that is, the elect, will at His bidding rise up with great glory; and "those that are beneath," that is, the reprobate, will fall down headlong till they are lost in the "Dead Sea."[6]

IV. Consider fourthly, that the only reason assigned for this widely different destiny of the elect and reprobate, when

[1] 1 Cor. xv. 52.   [2] 1 Thess. iv. 16.   [3] 1 Cor. xv. 23.
[4] St. Matt. xiii. 49.   [5] Osee xiii. 14, 15.   [6] Josue iii. 16.

the former "shall come forth unto the resurrection of life,"
and the latter "unto the resurrection of judgment," is this:
the difference of their works in the past. And thus take
notice, and let thy soul tremble in doing so, what are the
terms of which Christ, the Infallible Truth, makes use. He
says that those who "shall come forth unto the resurrection
of life" are not those who were noble, rich, or learned, nor
those who won for themselves on earth the applause of crowds,
but those only who employed themselves in doing well:
"They that have done good things." They "that have done
evil," be they who they may, even the most powerful monarchs,
shall never to all eternity "come forth" to that "resurrection
of life." To what, then? "To the resurrection of judgment."
What, then, dost thou say to this, thou who, it may be, now
estimatest everything more highly than good works? On that
day thou wilt see what it will be to have neglected their
performance in order to plunge deeper and deeper into worldly
concerns, to heap up wealth, to gain dignities, to take thy
pleasure. Blessed for ever will those be who "have done
good things," and for ever cursed those who "have done evil."
If there is not this treasure of good works, all else is worthless.
I know that these words are a clear condemnation of all those
who—slothful as they are—maintain that faith alone, even
without good works, is sufficient for salvation. But I will
not suppose thee to be one of these desperate madmen. On
the contrary, do thou derive this useful teaching from what
has been said: that that which ought to be valued beyond all
else in every one is good works. "Fear God," by refraining
from the evil which will be so terribly punished by Him on
the Day of Judgment, and "keep His commandments," by
doing the good which alone will be rewarded: "For this is all
man,"[1] that is, in this all is comprised.

[1] Eccles. xii. 13.

FOURTH DAY.

ST. DOMINIC.

The charity of Christ presseth us . . . that they also who live, may not now live to themselves, but unto Him Who died for them (2 Cor. v. 14, 15).

I. Consider first, what was the intention of Christ in going so far as to die for thee on the gibbet of the Cross. Was it only to redeem thee from the slavery of Hell? Surely not, for then it would have been enough to give one single drop of His Precious Blood. Therefore, by shedding it in abundant streams, by enduring so many tortures and insults, He desired so to possess Himself of thy heart, that even though thou shouldst desire to live to thyself, it might be impossible for thee to do so, being compelled to live to Him alone. Therefore did the Apostle, who was so well able to comprehend this truth, break forth into these beautiful words: "The charity of Christ presseth us." He does not say "invites us," nor even "impels us," but "presseth us," because it was a force which he was unable to resist. Even if he had wished to cease labouring in the service of his Lord, to cease journeying, preaching, and spending himself wholly for the salvation of the souls so dear to Him, it would have been impossible for him to do so. It was as though blazing torches were attached to his sides, which gave him no rest : "The lamps thereof are fire and flames"[1]—fire to inflame his own heart, and flames to make him enkindle the hearts of others. How is it as to thy experience of this blessed ardour of spirit? This it is which is the proof of being really a child of God: "Whosoever are led by the Spirit of God, they are the sons of God."[2]

II. Consider secondly, that it seems as though the Apostle might more truly have said, "The death of Christ presseth us, that they who live may not now live to themselves,"&c. Nevertheless, he says "the charity of Christ," because if thou oughtest to be greatly moved by what Christ has endured for thee, thou oughtest to be incomparably more moved by the love with which He endured it. See how much Christ deigned to suffer for thy salvation, and yet it was as nothing compared with what He would have suffered had such been the will of the Father: "Many waters cannot quench charity."[3] All those torrents of calumnies, reproaches, insults, treasons,

[1] Cant. viii. 7.   [2] Romans viii. 14.   [3] Cant. viii. 6.

scourges, buffets, piercings; of anguish and bitterness, of
torture and agony, were not enough to quench the thirst of
His burning love. Therefore, if what Christ has endured for
thee should move thee to resolve henceforth no longer to live
to thyself, but only to Him, the love with which He endured
it ought to compel us: "The charity of Christ presseth us."
Lastly, all His sufferings, excessive as they were, had the limits
prescribed to them by the providence of God; but His love
had no limit.

III. Consider thirdly, what it is to live to oneself. It is
to live to one's own will, one's own interests, one's own glory,
one's own pleasure. Now there ought necessarily to be an
end of all this in thee, since Christ has gone so far as to die
for thee with so much love. And the reason is very plain;
for if He died for thee, it is no more than right that thou
shouldst, at the least, die for Him. I say, at the least; for
if it were possible, it would be right that thou shouldst do
much more, seeing that there is no proportion of any kind
between thy life and the life of Christ. His was a life of
infinite value, and thine is a vile, soiled, and miserable life,
deserving of death. What great thing, then, would it be, if
thou wert even to die for Christ, since He deigned long ago
to die for thee? But if thou dost not even go so far as to
die for Him, then thou art bound at least to do what is so
much less a thing, namely, to live for Him; that is, to live
in order to love Him, and to endeavour to make all others
love Him, which was so wonderfully done by the great patriarch
Dominic and his illustrious sons: "To Him my soul shall
live, and my seed shall serve Him."[1]

FIFTH DAY.

OUR BLESSED LADY AD NIVES.

*Blessed is the man that heareth me, and that watcheth daily at my gates,
and waiteth at the posts of my doors. He that shall find me shall find life,
and shall have salvation from the Lord* (Prov. viii. 34, 35).

I. Consider first, that true devotion to the Blessed Virgin
has three steps which lead to the attainment of perfection in
it. The first is to forsake sin for love of her, for what honour

[1] Psalm xxi. 31.

that will be acceptable to her can be paid her by the man who refuses her this? The second is, to add to the first some special homage, as is rendered by those who fast on Saturdays in her honour, visit churches dedicated to her, recite her Rosary, or perform any other similar act of worship. The third is, to add to the second the imitation of her sublime virtues. And it is this which gives to this devotion its final perfection. Now, it is precisely these three steps which the Blessed Virgin here points out to us in the words which Holy Church has ascribed to her for many ages : "Blessed is the man that heareth me :" this is the first step ; "and that watcheth daily at my gates :" this is the second ; "and waiteth at the posts of my doors :" this is the third. If thou hast not yet begun ascending these steps, delay doing so no longer, that thou mayest speedily reach the summit.

II. Consider secondly, as to the first step, that our Lady says, "Blessed is the man that heareth me," because she desires of thee, beyond all things, that thou shouldst hear her whenever she bids thee forsake sin. If thou stoppest thy ears in order not to hear her in this matter, thou art lost. How canst thou expect her ever to receive or acknowledge thee as her loving client? There are in sin two detestable qualities which make it worthy of the utmost hatred—its deformity and its malice. The origin of the latter is its turning away from the Creator ; of the former, its turning to creatures. If, then, thou takest pleasure in the deformity of sin, how can she accept thy love when thou art like a devil in human form? and if thou takest pleasure in malice, how can she accept the love of one who is an actual traitor, renegade, and rebel to her Son? She is so pitiful that she will, indeed, graciously help thee to quit this state by obtaining pardon for thee ; but she will not help thee to continue in it by obtaining for thee the impunity which some persons desire. Hear her, therefore, by forsaking the sin which is so displeasing to her. If thou dost this, blessed art thou, for thus thou openest the way to her friendship : "Blessed is the man that heareth me."

III. Consider thirdly, that, as to the second step, our Lady says, "And that watcheth daily at my gates," for this is the practice of those who love, to watch at the gates of the persons they love, in order to show how greatly they love them. It is a characteristic of love that it banishes sleep. What, then, is the sleep which the love of our Lady should banish from thee? It is slothfulness. Thou oughtest to be zealous in the service thou payest her, therefore she says " watcheth ; " and thou oughtest to be persevering, therefore she says " daily." Suffer not a day to pass without honouring her in some special way. If thou doest this, blessed art thou, for thus she, on her part, will not let a day pass without returning thy service by some special aid : " Blessed is the man that . . . watcheth daily at my gates."

IV. Consider fourthly, that, as to the third step, our Lady says, "And waiteth at the posts of my doors," because he who loves greatly not only watches at the door of the beloved one, but tries unobserved to contemplate what she is doing ; he remarks her occupations and her demeanour, and when there is an opportunity, he imitates her to please her better. How beautiful are the examples which thou mayest find to copy in the Blessed Virgin if thou settest thyself to consider her attentively ! Imitate her, and then blessed indeed art thou, for thus thou not only solicitest but constrainest her to love thee. Service that is paid her causes her to love thee by election ; but imitation, by nature : " Blessed is the man that . . . waiteth at the posts of my doors."

V. Consider fifthly, that in the first step there are no gates of any kind mentioned, because he who is in this degree is preparing to be a more faithful client of Mary than he has yet become, and therefore he is still in the way. In the second degree, mention is made of gates (*fores*), but not of door-posts (*postes*), which are those wooden doors which shut close ; because he who is in this, although he is indeed a special client of the Blessed Virgin, yet he is, so to speak, on the threshold of this devotion, which is common to all ; he has not yet entered the private apartments. Lastly, in the third, there are not " gates " only, but " doors," because he who is in this degree is in the interior of the house, to which all are not admitted. But it is to this that thou oughtest to aspire with all thy mind ; if thou canst do no more, at least knock and beg, and it shall be opened to thee. Ask our Lady with all thy heart to make thee worthy to imitate her, and thou shalt do so.

VI. Consider sixthly, that our Lady adds that whoever finds her by means of practising this devotion, finds life : " He that shall find me shall find life." This life is Divine grace, which is the life of our soul, and he who finds our Lady will find Divine grace, because he will find her who has found that grace—found it both for herself and others ; for which cause the Archangel Gabriel said, "Thou hast found grace with God;" not only the grace "of God," which is the grace which makes her holy, but "grace with God," which is the grace which enables her to obtain sanctity for others also. How greatly should this urge thee to be devout to her! For when thou hast, at any time, been so unhappy as to lose Divine grace, what wilt thou do? Wilt thou go to God, to beg of Him another grace like that which thou hast lost? Alas! this is to declare thyself unworthy of obtaining it; for other jewels are, after all, lost against the will of the loser, but Divine grace is a jewel which, if lost, is lost wilfully. The first thing to be done, then, is to beg pardon for thy exceeding negligence in keeping it. And our Lady is especially appointed for the gaining this pardon for us, because she possesses a grace so eminent as to be able to merit for others also the grace which they have lost, and this is why she says, "He that shall find me shall find life," that is, "shall find grace." Therefore, as the other saints are our advocates to obtain for us, one strength in time of temptation, another obedience, another humility, another some other virtue, our Lady is our advocate to obtain for us Divine grace, because she obtains for us not merely habitual grace, which is the life from which all these virtues proceed, but likewise actual grace, which is the life by which they are maintained, increased, and perfected. This being the case, how important it is to take the utmost pains to find our Lady! If thou hast found her, thou hast found grace. And do not be alarmed, as though it would cost thee great labour to find her, for she desires no better than to be found. She is "easily found by them that seek her;" and the reason is that she "preventeth them that covet her, so that she first showeth herself unto them,"[1] such is the sweetness of her nature. Nevertheless, she says, "He that shall find me," because, if it is not necessary to spend labour in finding her, it is necessary to use diligence in offering her the acts of loving devotion which have been mentioned.

VII. Consider seventhly, that it would be but a small thing for the Blessed Virgin to obtain for thee the grace of God in this world, if she did not also obtain for thee glory in the next. Therefore, in the last place, she concludes in these words: "And shall have salvation from the Lord." This is salvation—final perseverance, by which thy salvation is secured. This is given to thee, as every one knows, by

[1] Wisdom vi. 14.

Christ. It is "from the Lord;" but it is given to thee through the Blessed Virgin, with this difference, that all the elect undoubtedly obtain their salvation by her means, but that her devout clients obtain it more easily. I say that all the elect obtain their salvation by her means, because no one is saved for whom she does not offer up special supplications for that intention. As the common advocate of all mankind, she has care of all alike. But her devout clients obtain it with greater ease, for she not only cares for them, but cares with the utmost solicitude, and therefore she obtains for them that Hell shall have less power to tempt them; and this is not all, for she assists them in a particular manner at their last hour, consoling and encouraging them, giving them confidence, and obtaining for them a very peaceful death. This is the meaning of these words "shall have salvation;" for the word here translated "have" is *haurire*, and it expresses that salvation is attained with little difficulty and expense. *Haurire* is a word with two meanings: the first is "to draw," as applied to water which flows from a fountain; and the second is "to drink." The first is without labour; the second is not only without labour, but with pleasure. Both these meanings are applicable to our subject; for the Blessed Virgin not only prevents her clients from experiencing much trouble in enduring what is necessary for their salvation, but she also causes them to find great pleasure in it, such is the fulness of consolation which she obtains for them. Hence we gather plainly how eminent a mark of predestination true devotion to our Lady is. The reason is, that her clients are saved with great ease because of the special patronage extended to them by this great Queen on all occasions, but most of all in the hour of death, which is the point on which their salvation depends.

SIXTH DAY.

THE TRANSFIGURATION.

This is My beloved Son, in Whom I am well pleased: hear Him (St. Matt. xvii. 5).

I. Consider first, how honourable is this testimony which is given by the Father to His beloved Son, when He says: "This is My Son." All the just are the sons of God, but in how different a way! Christ is the Son of God by nature, the just are His sons by adoption. Christ is a Son because He is a Son: "The Lord hath said to Me, Thou art My Son."[1] The just are sons because they are treated as such, being admitted to a close union with the Divine Nature, but not to the Hypostatic Union. Consequently, Christ is the Son of God by consubstantiality, the just are the sons of God by association: made "partakers of the Divine Nature."[2] And thus Christ is a Son equal to the Father, and the just are sons like the Father. How rightly, therefore, does the Father designate the Person of Christ by the pronoun "This," saying absolutely, "This is My Son," because none is so in the same way as He Who is so by nature. And so it is: Christ was Son by adoption neither as God nor Man; He was Son by nature; and therefore the word *meus* in this place does not signify dependence, as with us, but identity. What, then, canst thou expect who dost not truly rejoice with Him in His glory? "This"—He Whom the people of Capharnaum would fain have bound as a madman, He Whom so many accused of being in league with Beelzebub, Who was mocked as a fool, treated as one possessed with a devil, Whom the Nazarenes a short time before sought to cast down from a high rock; this is He of Whom the Father says: "This is My Son." What dost thou answer to this? Dost thou not rejoice that He Who was once subjected to shame so great, is now receiving so much glory? But the glory is given in secret, whereas the shame was allowed to be inflicted in public; and this is a sign that shame, not glory, is to be expected on earth.

II. Consider secondly, that not only is Christ called a "Son," but also a "beloved Son;" and He is called beloved in the same sense as He is called Son. For observe that there are two ways in which any one may be beloved by thee: either for his own sake, as thou lovest thy friend; or for the sake of another, as thou lovest thy friend's friends. All the just are beloved by God, but it is for the sake of another, that is, Jesus Christ, Who has gained that love for them: "He hath called us . . . according to His grace, which was given us in Christ Jesus."[3] But Christ is beloved for Himself, and therefore He is absolutely the Beloved: "My beloved Son." Therefore He is first Son, and then beloved, *Filius dilectus*, not first beloved and then Son. The just are sons by grace, and there-

[1] Psalm ii. 7.   [2] 2 St. Peter i. 4.   [3] 2 Timothy i. 9.

fore they are first beloved and then sons, because it is the love borne them by God which raises them to that lofty dignity. Christ is Son by nature, and therefore He is first Son and then beloved, because it is the dignity which He has in Himself which raises Him to this height of love. And this may be the reason why the Father did not here choose to say first *dilectus*, and then *Filius;* but first *Filius*, and then *dilectus*, in order by this means to distinguish Him from those who are first beloved and then sons ; because, although it is true that they are sons, yet it is merely by virtue of love that they are so. However this may be, it is by this beautiful title of "the Beloved," that Christ is so often called in the Scriptures : "I will sing to my Beloved a canticle."[1] "Come, my Beloved."[2] "My Beloved had a vineyard."[3] He is so called because this title belongs to Him essentially, and also by reason of the greater marks of love which He has received above all others who are sons of God : "The Father loveth the Son." Therefore what follows ? "And He hath given all things into His hand."[4] This is the great mark of love which He has received, that He has been appointed by the Father the Universal Ruler of all that is His ; so that it is not said, "He hath given Him all things," though that, indeed, would be very much ; but "He hath given all things into His hand," because Christ can do as He pleases with all things. How eagerly, then, shouldst thou endeavour to be united with that beloved Son, to Him from Whom, because of His Sonship, thou canst receive every good if only He will condescend to give it thee ! Love Him, follow Him, serve Him, and all shall be thine. Remember what He has said : "Whatsoever you shall ask the Father in My Name, that will I do."[5] It might seem a more correct way of speaking to say, "That will He do," because, the prayer being addressed to the Father, it might seem that it was His part to grant it. Yet our Lord did not say this, He said, "That will I do," because the request is made to the Father, and the Son, as His chief instrument, grants it : so greatly is He beloved.

III. Consider thirdly, that it was precisely in order to explain this, that the Father immediately added, "In Whom I am well pleased," because it was in His Incarnate Son that He was pleased to give all good things to men : "He hath blessed us with all spiritual blessings in heavenly places in Christ."[6] Thou mayest, therefore, take these words of the Father in two

---

[1] Isaias v. 1.   [2] Cant. vii. 11.   [3] Isaias v. 1.   [4] St. John iii. 35.
[5] St. John xiv. 13.   [6] Ephes. i. 3.

senses : either as signifying that the Father was "well pleased" in His beloved Son, as an artificer is pleased with the most beautiful work that ever came from his hands—which is indeed a true, but an incomplete sense ; or that in His beloved Son He was pleased to do all the good which He intended to do to the world, which is in the full sense, and one which leaves room to add what it is that is involved in that good pleasure, as though the Father intended to signify by these words, "This is My beloved Son in Whom I am well pleased," to redeem the unhappy race of man from the slavery of Hell, to give grace and glory and all My treasures to all men. This is the force of the word "well:" it is not used here to denote the goodness of pleasure, because every Divine pleasure is always equally good ; it is employed solely to denote plenitude, since it is impossible that there can be any pleasure greater than that which the Father had in this most beloved Son in Whom He determined to save the world : "He hath purposed . . . to re-establish all things in Him."[1] But is not this, on the other hand, a very marvellous thing ? It is easy to understand that the Father should be so well pleased to have such a Son— "A Father pleaseth Himself in the Son "[2]—but that He was well pleased that He should be the Saviour of us miserable men, this is indeed past comprehension, for what is God the better for our salvation ? In no way whatever. And yet how great is His pleasure in it ! "It hath pleased your Father to give you a Kingdom."[3] This is that love of God which is utterly beyond our comprehension, unless it may be said that God took pleasure in our salvation because such was His pleasure. It is impossible to assign any other reason, at least any antecedent reason, for the love of God, except the will of God : "Thou shalt no more be called forsaken, . . . but thou shalt be called My pleasure in her, . . . because the Lord hath been well pleased with thee."[4] If God loves us, it is because He chooses to love us, not because loving us will occasion Him any greater pleasure than He has in Himself without loving us. Observe, therefore, that He does not here say that the act which He performs by saving us in Christ has given Him pleasure ; He says only that He has pleased Himself in the act : "In Whom I am well pleased."

IV. Consider fourthly, how, having established the plenitude of purpose which the Father has formed of making every

---

[1] Ephes. i. 10.

[2] Prov. , 12   [3] St. Luke xii. 32.   [4] Isaias lxii. 4.

good thing which we receive pass through the hands of Christ, He adds as a consequence, " Hear Him." This is what a sovereign does when he has, in his exceeding love, placed in the hands of his first-born all the government of his great kingdom ; although it is still, as before, in his power to dispose of everything himself, if he chooses, yet he will reply to all who come to treat with him of important business : " Go and hear what is the opinion of my son, the prince, on the subject— 'hear him.'" And this is what the Heavenly Father here intends. There is no matter, small or great, of any sort, which does not altogether depend on Christ as absolute Ruler : " All power is given to Me in Heaven and in earth."[1] And although, at the same time, He makes Himself our Advocate, by praying for us to the Father, yet He does this as an act of supreme reverence, just such as would be performed by the prince we have spoken of, who, although his father had left him free to decide everything, nevertheless should never form any important resolution without first obtaining his father's sanction in express terms. What, then, should any one who desires something do? He should go to Him Who gives audience— " Hear Him "—and this is Jesus, Who is given to us by the Father for this reason, that, being a Man, like ourselves, we may have the greater confidence in dealing with Him : " The Lord thy God will raise up to thee a Prophet of thy brethren."[2] What excuse, then, will there be for those who will not have recourse to Him ? If a brother of thine were raised to the government of thy native country, so that it lay with him to decide according to his will who should be received at Court, to dispose of all offices, causes, and expeditions, what wouldst thou do? Is it possible to imagine pleasure greater than thou mightest every day enjoy in conversing with him? And yet thou makest so little account of having an audience of Christ ! He is thy Brother—" of thy brethren "—a Brother raised to a far higher rule than was given to Joseph. How is it, then, that thou dost not every day devoutly kneel before Him? If thou hast offended Him, He is ready notwithstanding to welcome thee with greater love than that which Joseph showed to those who were no longer his brethren but his betrayers. At least, do not disdain to approach Him, as though He were a Brother of Whom thou wert ashamed instead of proud. See how glorious He is to-day in His Transfiguration ! Yet that glory is but a very faint shadow of that which He has in

St. Matt. xxviii. 18.   [2] Deut. xviii. 15.

Heaven: His "lightnings enlightened the world."[1] How is it, then, that thou art sometimes ashamed to listen to Him, that thou slightest the teachings of the Gospel, hast no value for them, sometimes even hast the extreme audacity to find fault with them as though they were unbecoming a well-born man? Is this to hear Jesus Christ? It is rather to turn thy back on Him. If thou wouldst have Him hear thee when thou prayest to Him, thou must listen to His precepts. This also is enjoined on thee by the Father when He says, "Hear Him;" He means not only "hear," but "obey Him." "Hear and your soul shall live."[2] Know, therefore, that this is He Who was promised so many ages ago to the world, when God said to Moses: "I will raise them up a Prophet out of the midst of their brethren like to thee. . . . And he that will not hear His words, which He shall speak in My Name, I will be the avenger."[3] And yet it may be that thou art often more ready to hear Tacitus and Cicero than Jesus Christ! "Hear Him," rather than any of the many teachers who once stood so high and whose teaching has proved so fallacious.

SEVENTH DAY.

ST. CAJETAN.

Be you humbled, therefore, under the mighty hand of God, that He may exalt you in the time of visitation; casting all your care upon Him, for He hath care of you (1 St. Peter v. 6, 7).

I. Consider first, that perhaps the greatest evil that is in thee is thy unwillingness to be wholly guided by God. I will not suppose that thou art one of those who strive to exalt themselves against Him. But if thou art not of this number, thou mayest easily be one of those who proudly resent every misfortune that He sends them, and will not say with humility: "It is the Lord; let Him do what is good in His sight."[4] Therefore does the Apostle here plainly exhort thee to humble thyself "under the mighty hand of God," since, if thou wilt not humble thyself to Him, and so gain merit, He will surely humble thee to thy confusion: "The Eternal shall humble them."[5] Dost thou think that this will be hard for Him?

[1] Psalm lxxvi. 19. [2] Isaias lv. 3. [3] Deut. xviii. 18, 19.
[4] 1 Kings iii. 18. [5] Psalm liv. 20.

On the contrary, His hand is here said to be so mighty, because He can so easily do it. The hand which requires a lance, a sword, a scimitar to strike down a giant, is not mighty; that is a mighty hand which can strike him down with a sling, as the shepherd-boy David did. And such is the hand of God. It costs it no effort to humble thee: "As clay is in the hand of the potter, so are you in My hand, O house of Israel."[1] See how easily the potter can destroy his vessel. He has no need of a heavy hammer, such as would be required to break a vase of metal or of marble; he can shiver it into a thousand fragments merely by the blow of a stick. And so can God do with thee: "It shall be broken small, as the potter's vessel is broken all to pieces with mighty breaking, and there shall not a shard be found of the pieces thereof."[2] This being so, how is it that thou dost not yet humble thyself in the deepest reverence to the will of Him Who can so easily bring upon thee so much worse than what has already befallen thee? "Be you humbled under the mighty hand of God." This is the Apostle's meaning when he bids thee be humbled. He would have thee bow thy head, and confess humbly, in the midst of all that thou art suffering, that it is well.

II. Consider secondly, that just as the hand of God is mighty to humble thee if thou exalt thyself, so, too, is it mighty to exalt thee if thou humblest thyself. It is able to exalt thee, even in this world, by causing the misfortune which thou bearest patiently as coming from Him, to redound in the end to thy greater glory, as was the case with Joseph and his miserable slavery in Egypt: "You thought evil against me, but God turned it into good, that He might exalt me."[3] But even if He does not exalt thee in this world, He will exalt thee in the next, which is far better, when He Who is faithful will render to every one the reward of the submission they have paid to the Divine will: "He will exalt the meek unto salvation."[4] This it is which should be the sole object of thy desires; and therefore the Apostle says: "Be you humbled under the mighty hand of God, that He may exalt you in the time of visitation;" not "in this time," but "in the time of visitation," that is, in the great Day of Judgment. That will be the day of general visitation, appointed by God for the special purpose of revising and setting straight the accounts of the whole human race, so that no one will be able to

---

[1] Jerem. xviii. 6.
[2] Isaias xxx. 14.　　[3] Genesis l. 20.　　[4] Psalm cxlix. 4.

complain of being wronged : " Behold the day of the Lord shall come . . . and I will visit the evils of the world."[1] How great an honour will it be on that day of visitation to be acknowledged in the sight of all men as God's faithful servant, that is, a servant who would never defraud Him of the least atom of His glory, but was satisfied to endure any suffering, any injury, provided only that God was glorified ! How will God then hold Himself bound to exalt so generous a servant ! He will embrace him with the utmost tenderness, lavish caresses upon him, speak to him words of praise, and give him a far more glorious crown than Assuerus placed on the head of the despised Mardochai : " He hath lifted him up from his low estate, and hath exalted his head."[2] Be content, then, for a little while, to bow thy head with humility in the misfortunes which may very likely befall thee, for at last there will come a day when thou shalt lift it up : " Be you humbled under the mighty hand of God, that He may exalt you in the time of visitation."

III. Consider thirdly, that the reason why thou dost not let thyself be guided as God sees fit is, more than all besides, because thou dost not trust Him. Thou thinkest in some sort of way, that He is occupied in thinking of the good of so many persons, that He cannot give a thought to what concerns thee, and so almost allows things to happen to thee by chance : " And thou sayest . . . He judgeth, as it were, through a mist."[3] Alas, in what a delusion thou art living ! Be very sure that He has a special care of thee, as He has of all : " For He hath care of you." Take courage, therefore. What do those words, " He hath care of you," mean ? They mean not only that He thinks of thee, but He thinks of thee in such a way that no trouble or sorrow can ever come to thee that is not sent by Him for thy greater good. What is it to take care of a sick person ?—" Take care of him."[4] Is it to watch by his bed and give him continually the most injurious things if he asks for them ? Surely not. It is to watch by him so as to give him even bitter medicines when it is for his good. And this is the way God deals with thee also. Thou too art sick : " An unactive man that wanteth help."[5] He knows thy needs ; and therefore it is said that " He hath care of thee," because He gives thee, not what is pleasant, but what is profitable to thee. Imagine, then, that thou seest

---

[1] Isaias xiii. 9.   [2] Ecclus. xl. 13.
[3] Job xxii. 13.   [4] St. Luke x. 35.   [5] Ecclus. xi. 12.

Him assisting thee in person with the love of a true Father ; now crossing a purpose which He knows to be injurious to thee, now preparing for thee some humiliation, or sending thee some conflict, in short, ordering most perfectly all that befalls thee from day to day. Never wilt thou be able to bring fully before thy mind all that God does for thee in this respect : "He hath care of you," He Himself, in His own Person, not in that of His ministers.

IV. Consider fourthly, what great fruit thou wilt gain from this conviction if thou dost always keep it vividly before thy mind. This fruit will be to cast thyself without solicitude into the arms of God, in such a way that thou wilt desire to know nothing of thyself, even as a wise son who thinks nothing about himself, because he knows that he has so good a father. And this it is to which the Apostle would have thee attain. Wherefore he says : "Be you humbled under the mighty hand of God, . . . casting all your care upon Him, for He hath care of you." He does not say "laying," but "casting," so injurious does he consider this care for self. Oh, if thou didst but understand how hurtful to thy spiritual life is this excessive, anxious, painful care of thyself. It is this which, more than all else, holds thee back from giving thy heart wholly to God. Therefore thou shouldst not only shake it off as quickly as possible, but fling it from thee as a man would a serpent which he found in his bosom. And is not excessive prudence a serpent? Nay, is it not a worse serpent than any other, since it is the one which made our first parents distrust God in the earthly paradise? Cast it away, then ; cast it from thine own bosom into that of God—"upon Him." He will accept even this serpent from thee as a more acceptable gift than He did the doves of old, as a gift which will move Him every day to have a greater care of thee in proportion as thy reliance on Him is greater: "Cast thy care upon the Lord," that is to say, this painful anxiety, "and He shall sustain thee."[1] And the word is not merely *nutriet*, but *enutriet*, to show that He will do this with a very special affection. This is the great advantage gained by the man who trusts in God, that by very little on his part he claims of God the greatest favours : "Thy life shall be saved for thee, because thou hast put trust in Me."[2]

V. Consider fifthly, that there never was one on earth who better understood this truth than the glorious St. Cajetan,

[1] Psalm liv. 23.   [2] Jerem. xxix. 18.

whom we commemorate to-day, for he laid all his sons under
a particular obligation to depend on Divine Providence, not
only for those things that are less essential, but even for
those which are absolutely necessary, such as food and
clothing ; so much so, that he would never allow them to ask
alms for God (which is freely done by others), but to wait for
it : " In the Lord I put my trust : how then do you say to
my soul, get thee away from hence to the mountain like a
sparrow ?"[1] It is said that when the sparrow leaves the valley
to fly to the mountains, she carries a grain of corn in her bill,
as though she were doubtful of being able to find food there
easily. But David says, I will not act thus. If the persecution
of Saul drives me into the mountains, I will not be in the
slightest anxiety as to finding needful provision there : everywhere I shall have God ; in Him I put my trust, I shall never
lack necessary support.   It may be that the foolish and
uncourteous Nabal may rudely refuse me a slight refreshment,
but in that case Abigail will make up for Nabal's shortcomings.
And so, too, did this great Saint say ; nay, he even surpassed
David, for whereas David asked food of Nabal, St. Cajetan
would not ask of any one, but waited till it came.   Do thou,
if thou canst not attain to this degree of confidence, at least be
ready to believe that God will not fail to provide thee in due
time with what is profitable for thy state, and so abstain from
procuring it for thyself by means which, if not unlawful, at
least savour of imperfection : " Am I become a wilderness to
Israel, or a lateward springing land ?"[2]   Not only is God not
like a barren soil which does not give fruit to those who trust
Him, but He is not even like a backward soil, which does not
give it in due time.

EIGHTH DAY.

He that contemneth small things shall fall by little and little
(Ecclus. xix. 1).

I. Consider first, that God does not here say that he who
commits venial will gradually fall into mortal sins ; but he
who despises them—" He that contemneth :" for who is there
that does not commit venial sins every day? "There is no
just man upon earth that doth good and sinneth not."[3] But

[1] Psalm x. 1.   [2] Jerem. ii. 31.   [3] Ecclea. vii. 21.

it is one thing to commit them, and another to despise them. He despises them who takes no trouble about them, as though there were no necessity to abstain from them in order to be saved. Art thou one of these unhappy persons? If so, in what great peril art thou of being lost eternally, seeing that these words of God are infallible; that the man who despises small faults will fall by degrees into great ones: "He that contemneth small things shall fall by little and little," that is to say, he shall fall from perfection, from piety, from goodness; in a word, from the state of grace into that of sin. This is the interpretation of holy writers. For what matter is it that the cracks in a vessel out at sea are small, if through neglect they cause as much damage as great ones? Small though they may be, they place the vessel in a state of ruin, not indeed imminent ruin, such as would be the case if they were large, but at any rate remote, since by little and little they allow the water to leak through and sink the vessel: "By slothfulness a building shall be brought down."[1]

II. Consider secondly, that there are three reasons why God says that "he that contemneth small things shall fall by little and little;" one on the side of man, another on that of the devil, and another on that of God: and all three are equally terrible when we reflect upon them. The first is on the side of man, since he who despiseth a small sin does two very grievous injuries to himself. One of these is that he gradually loses the fear which keeps him from a great one, and the other is that he increases the inclination which impels him towards it. He loses fear, because, as small faults do not produce their sad results as speedily as great ones, but more, as it were, after the manner of wounds inflicted by a blunt knife, it happens that after a while a man begins to persuade himself that such faults are really of a kind that do no harm; and thus, becoming bold, he not only does not hesitate to continue in them with the greatest freedom, but goes on at length to increase them till they cause death. Poison at once shows the mischief that it brings with it, and therefore every one avoids it; sour fruits only show it very slowly, and therefore some persons get in the habit of eating them very eagerly. And yet, in the long run, it is possible for sour fruit to be as fatal as poison, with this difference, that the poison causes death by the dangerous properties inherent in it, and the fruit by those which it produces in course of

[1] Eccles. x. 18.

time. So it is in the case we are considering. In the next place, just as a man, by despising small faults, loses the fear which restrains him from sin, so, too, does he in an equal degree increase the inclination which impels him towards it. For this inclination is nothing else, in all cases, but irregular concupiscence ; and we all know that the more that is granted to it, the more greedy does it grow in its demands. It resembles fire : "Lust is enkindled as a fire."[1] And therefore, just as, even in a stubble-field, there must be, in the first instance, some one to kindle a fire before it can spread ; but when once this first fuel has been given to it, it gains strength, and becomes so insatiable, that it will even devour what is refused to it ; so is it with concupiscence, which at first is so moderate that it requires something to excite it, but which, no sooner does it see itself supplied with what it desires, than it becomes insatiable : "It never saith, It is enough."[2] It is always craving, always seeking, always increasing in violence, and so long as it has anything to hope for, is never pacified : "A hot soul is a burning fire, it will never be quenched till it devour something."[3] Added to this, in course of time the pleasure which it has in small faults becomes so habitual, that it is not very keen, and so nothing remains but to seek for a greater in more grievous faults. Judge, therefore, thyself, whether it is in the power of any man to refrain, of himself, for any length of time from grievous sins when once he has gone so far as to make nothing of small ones. To expect this is like giving the rein to a young horse and yet desiring him to keep in the straight road.

III. Consider thirdly, the second reason, which is that on the devil's side, because in the conduct we are describing the devil has just what he wants. Every one knows that this is always the devil's way, to ask for the greatest possible evil, but to do so gradually. If, however, at the very first, he were to ask for adultery, theft, rape, and murder, every one would drive him away as an open enemy. And, therefore, to begin with, he only asks for some friendship a little more close than duty permits, some attachment to worldly goods which goes too far, a somewhat too anxious care for reputation, an act of bad faith which seems rather more politic than malicious ; and by these means having made a breach in the unguarded heart, he has no fear of not being able to conquer it at the first onset. What art thou about then, when thou accustomest

---

[1] Ecclus. ix. 9.   [2] Prov. xxx. 16.   [3] Ecclus. xxiii. 22.

thyself to commit many faults fearlessly, because they seem to thee but small? Thou art sparing the devil all the preparatory labour, which is the hardest. And so, nothing remains for him but boldly to follow up the victory which thou thyself givest him by abandoning those trenches on which he would otherwise have had to concentrate all his first efforts : "Israel hath cast off the thing that is good"—that is, by forsaking the more pious, strict, and religious life which he led formerly— "the enemy shall pursue him,"[1] till he has brought him to lead one which is scandalous.

IV. Consider fourthly, the third reason, which is that on the side of God; for all the holy doctors without exception teach that God punishes one who indulges in lesser sins by allowing him to fall into greater. He does not, indeed, inflict so terrible a punishment without first sending the sinner many salutary warnings, just as the gardener only leaves the tree without check or training after having spent much patient care on it in vain. But when He sees that a man does not listen to Him, then, at last, He leaves him to gratify all his desires, even those which are the most unlawful : "My people hear not My voice, and Israel hearkened not to Me." Therefore, what follows? "So I let them go according to the desires of their heart," so that these unhappy persons "shall walk in their own inventions,"[2] till they reach the goal to which the free and wilful road which they have followed leads, that is, final impenitence. Do not, therefore, abuse the goodness of God by saying to thyself, He will bear with my faults patiently because they are small ; for these small faults will, in time, become intolerable by the increasing frequency with which they are committed. This it is to which God would seem to allude when He says : "Behold, I will screak under you, as a wain screaketh that is laden with hay."[3] Observe how labourers proceed in loading their waggons. When they are to be laden with timber, earth, or heavy stones, they are exceedingly careful not to make the weight too heavy, but when hay is to be carried in the field, they heap it up to such a height that it is surprising ; so that often the waggons creak under a load of hay more than under one of stones. Do not say, then, All my faults are like hay, they are so light ; for if they are light they are very numerous, so that it may be said that they make God groan under thee, as though He complained that the load was too heavy ; that thou art wearying

[1] Osee viii. 3.   [2] Psalm lxxx. 12.   [3] Amos ii. 13.

Him, and abusing His gentleness in bearing with thee ; and if
He does not at once deprive thee of His grace in consequence,
as He does in the case of mortal sins, He will at least deprive
thee of His protection by justly taking from thee those special
and superabundant aids without which thou wilt speedily come
to lose even grace itself.   These are the three reasons why it.
happens that "he that contemneth small things shall fall by
little and little ;" not "at once," but "by little and little," and
all others which may occur to thee may be reduced to these
three heads.

NINTH DAY.

Be astonished, O ye heavens, at this, and ye gates thereof be very deso-
late, saith the Lord : for My people have done two evils. They have forsaken
Me, the fountain of living water, and have digged to themselves cisterns,
broken cisterns, that can hold no water (Jerem. ii. 12, 13).

I. Consider first, that there are in sin two very terrible evils
which act mutually the one on the other, so that each is
increased and aggravated. These are, the turning from the
Creator, and the turning to things created by Him.   Suppose
that there were no other harm in sin than that of turning the
back upon God, dost thou not think that this would be a very
great enormity?   What, then, must it be, when, in addition,
men turn their backs on Him in order to go after contemptible
creatures, which, at best, are but the work of His hands? And,
on the other hand, if nothing unreasonable were done in
sinning except the going after these creatures to pay them a
homage which they do not deserve, would not this alone be
sufficient ground for abhorrence? What, then, must it be,
when, in order to pay them this homage, they also turn their
backs on God?   These two evils combined God declares to
have been committed by His people : "My people have done
two evils."   And, therefore, as though He were Himself filled
with horror at so strange an audacity, He not only bids the
heavens be astonished, but also bids the floodgates of the
heavens be broken open and pour down furiously on this
guilty people storms, whirlwinds, tempests, darts, and every
other terrible destruction due to them : "Be astonished, O ye
heavens, at this, and ye gates thereof be very desolate, saith
the Lord."   But how, if He were able to say that thou too art

equally guilty of both these two evils combined? I know, indeed, that just as God, in this great complaint of His, intended to signify Himself by this "fountain" of which He speaks, so, too, He intended to signify idols by the "cisterns;" but this is only the first sense, for it is also very certain that in the second sense He also meant by cisterns those men from whose sinful friendship His people could not separate themselves, such as Egyptians, Assyrians, and the like, who were only fit to teach them evil. If, then, thou art one of those who value the friendship of men more than that of God, take to thyself these words, for they apply to thee. And how likely is it that thou hast been, it may be for a long time, of this number!

II. Consider secondly, how great the difference is between a fountain and cisterns. The water of a fountain springs directly from it, and is fresh, unlimited, and unfailing, and no matter how abundantly it is given to every one, the supply is never diminished. Cisterns only have just so much water as their narrow limits can contain, neither is the source in themselves, so that they only have just so much, and no more, as they receive from the pipes which feed them. And this is precisely the difference which there is between thy God, and those persons whom thou lovest, and dost not hesitate sometimes to prefer to Him. He is that full fountain of all good which depends on no one: "With Thee is the fountain of life."[1] On the other hand, what is there of any value in all the persons we are speaking of? Nothing whatever. They have nothing but what God has graciously given them, and even this they have by measure, and that a scant and limited measure: "Behold the Gentiles are as a drop of a bucket."[2] And yet dost thou forsake God for them? Oh, what an unutterable injury art thou doing Him! What is the motive which induces thee to prefer the friendship of men to that of God? It must certainly be either a motive of honour, profit, or pleasure; there is none other possible. As to the first, dost thou not esteem it a greater honour to have in thy garden a beautiful fountain rather than a cistern of mere rain-water, which is never thoroughly clear? As to the second, which wouldst thou choose to possess as a greater source of revenue, a common cistern of water which hardly suffices to quench the thirst of thy poor labourers, or a living fountain, capable also of watering thy herds of cattle, and all thy plants and meadows?

[1] Psalm xxxv. 10.　　[2] Isaias xl. 15.

Then, as to the third point, what dost thou do when thy throat is parched in a long journey? Dost thou not hasten to the fountain? It is only in case of necessity that thou wouldst go to the cistern: for there is no enjoyment in drinking full draughts of these borrowed waters, as there is in drinking from a fountain? How, then, is it possible that none of these reasons is sufficient to make thee choose to love God more than men? God is the fountain; men, as we have said, are the cisterns. And yet thou considerest men far more than God: "They have forsaken the fountain of living water, and have digged to themselves cisterns." Well indeed may God say, "They have digged to themselves." He does not say that His people have found the cisterns ready made; He says that they have been so deluded as to make them, by their own choice, for themselves. So it always is. Every one makes his own cistern, according as his inclination leads him. He does not consider the creature, as it is in itself, devoid of all good, but as he pictures it in his own mind (just as idolaters do with the idols they worship), and so, if he does not worship it, at all events he loves it more than is right. Do thou, as I bid thee, act otherwise. Keep always vividly before thy mind this maxim, that men never have any good thing of themselves, but that all they have is from God; and then it can never be possible for thee not to love God more than men.

III. Consider thirdly, that this conduct which is blamed in the text would be more tolerable if men, being as it were, so many cisterns, were, at any rate, sound and whole cisterns, which were able to hold the small quantity of water for which they are sought. But the worst part is that they are all broken cisterns, leaking on all sides, so that they speedily become dry. This is the further point which God intends to express when, after saying of those who run after human friendships, "They have digged to themselves cisterns," He adds with very strong emphasis, "broken cisterns, that can hold no water." For, if the persons who were so dear to thee on earth were immortal, there would then be some excuse for thy valuing them so highly. But dost thou not see how very shortly they must all die? Alas, they are all "broken," that is, full of disease and misery, which gradually rob them of all value, and therefore "they can hold no water." They may strive as they will to live a very long life, they cannot obtain it. The water which they contained is rapidly exhausted. Beauty, wisdom, prudence, grace, all their gifts suddenly vanish, and nothing remains but

L

corruption : "They shall sleep together in the dust," with the meanest persons in the world, "and worms shall cover them."[1] If, then, thou wouldst detach thy heart from all creatures to give it to God, as is right, imagine that thou seest them in the grave, decayed and fleshless and turned to dust : then, indeed, thou wilt see them to be "broken cisterns," which "can hold no water," even though they were able to receive a whole river. How, then, when thou seest them to be what they are, canst thou forsake God, Who cannot die, for their sakes?

TENTH DAY.

ST. LAURENCE, MARTYR.

I suffer, but I am not ashamed: for I know Whom I have believed, and I am certain that He is able to keep that which I have committed to Him against that day (2 Timothy i. 12).

I. Consider first, that perhaps the greatest temptations which assail thee in the spiritual life are temptations to distrust. Thou thinkest at times that all that thou doest in it for God is lost, and that in spite of it thou wilt be damned. Now, as a weapon against them, take this beautiful passage of the Apostle, which I here propose for thy contemplation. Listen to the first word which pain seems, as it were, to force from his lips : "I suffer." He confesses openly that his suffering is great, but immediately after he adds, "but I am not ashamed," or, as it is literally, "confounded." Thou often thinkest that because the saints were so greatly inflamed with the love of God they were insensible in the midst of their sufferings, as certain of the martyrs were on the cross or at the stake. Not so ; they were very sensible both of the injuries inflicted on them, and of troubles, trials, and sickness, but they felt these things without losing courage. They said boldly with the Apostle : "I suffer, but I am not ashamed." And why did they say so? Because they knew to what a Master they had committed themselves : "I know Whom I have believed." Wonder not, therefore, if thou, who art so feeble of mind, feelest suffering very keenly ; if thou didst not feel it, it would not be suffering. It is enough if when thou sufferest thou art not confounded, that is, if thou dost not abandon the strong

[1] Job xxi. 26.

faith and confidence which thou shouldst have in God: "Thou shalt know that I am the Lord, for they shall not be confounded that wait for Him."[1] Oh, how much reason hast thou to say in this matter with the Apostle: "I know Whom I have believed!" If thou hast a master whom thou knowest well, thou dost not suffer thyself to be turned against him by those who would fain discredit him in thy eyes, as though he took no thought for thee; on the contrary, thou treatest their words with contempt, saying to thyself, I know who it is I am trusting. So, too, oughtest thou to say in the case we are considering. What does it matter that thy thoughts are overcast by a thousand fanciful shadows and clouds, representing to thee that thou art serving a Master Who will in the end abandon thee on account of thy sins? Do not think of combating them, but simply say to thyself: "I know in Whom I have believed." This is the easiest way of putting them to flight.

II. Consider secondly, what it is that the Apostle here understands more particularly by this, "I know Whom I have believed." He understands two things, which, after all, are but one. First, I know Who it is in Whom I have trusted—"Whom I have believed;" and secondly, I know to Whom I have trusted all the good that I do. He says, "I know Whom I have believed," not "what I have believed," to show that it ought to be enough for thee to know certainly how faithful a Master He is Whom thou servest, how good, how gracious, how prone to mercy, because He is God. And for the rest, if thou art unable to solve the difficulties which are suggested by thy thoughts to perplex thee concerning the grace which He gives to others and not to thee, concerning predestination, perseverance, and the like, which are obscure even to the learned, be not troubled at this; enough for thee that thou knowest on Whom thou dependest: "I know Whom I have believed." Is not faith a stronger ground of assurance than any amount of revelations which could possibly be made to thee in these matters? They are liable to delusion, faith is not. And therefore it is not necessary to understand all about such matters in order to do right; it is enough to believe them by making an act of faith. And so, it is not even necessary to be able to say, "I know Whom I believe," so long as thou canst say, "I know Whom I have believed;" and if at times thou art in such a state of mental darkness, dryness, and

---

[1] Isaias xlix. 23.

distress as to be incapable of eliciting that faith from thy heart, then habitual faith is sufficient. Call to mind those acts of confidence which thou hast formerly made, and hold fast by them, and though past, they have the power of securing thee for the present: "I know Whom I have believed, and I am certain." Observe, it is not said, "I was certain," but "I am certain."

III. Consider thirdly, what it is that is committed to God, of which the Apostle speaks when he says, "I am certain that He is able to keep that which I have committed to Him against that day." It is the sufferings which he endured for God, the journeyings, the preachings, the imprisonments, the scourgings, and the rest, which thou mayest go through in thy mind. All these he includes in the word *depositum*, because he had once committed them into God's hands, and determined never again to think of himself, not even in matters concerning his salvation, but only of God. Oh, how beautiful an act was this! Why, then, dost thou not endeavour to imitate it, so far at least as thy misery allows? Do thou, too, give up everything into the hands of thy God, even the concern of thy everlasting salvation, which sometimes makes thee so anxious; and, instead of occupying thy thoughts with troublesome fancies and arguing with thyself whether or not thou wilt be saved, betake thyself rather to making acts of the love of God, labour for Him, study for Him, perform thy devotions for Him, protest that all thy desire is to depend on Him alone: "My lots are in Thy hands,"[1] and by this means thou wilt gain the time which thou art now losing in thoughts which are either useless or disquieting.

IV. Consider fourthly, why it is that the Apostle does not say, "I am certain that He will keep that which I have committed to Him," but only "that He is able to keep" it. He says this to make the expression stronger, saying less but meaning more: Dost thou not believe that God is very well able to keep safely all that thou hast suffered for His sake? If, then, he can consider it absolutely certain that He will do so, because, in our opinion, thou wouldst do God a more grievous wrong in questioning His fidelity than His power—"He is able to keep"—if so, what is there to fear? If He is able to keep He will keep. "God is not unjust," said the Apostle to the afflicted believers, "that He should forget your work, and the love which you have shown in His name."[2]

[1] Psalm xxx. 16.   [2] Hebrews vi. 10.

But is not this a strange manner of speaking? Should he not rather have said, "God is not unmindful" than "God is not unjust"? Nevertheless, this is what he did say, to show what that "deposit" was of which we are speaking. We men may now and then forget some small matter which has been committed in trust to us without blame, but not so God. He "is able to keep" in the deep store-house of His Divine mind the very smallest straw picked up from the ground for His sake. And therefore if He is able to do so, He is bound to do so; and if He is so bound, it would be impossible for Him without injustice to be unmindful of the least item of anything done for Him. Hence it is that where men are concerned the prudent warning of the Preacher is very applicable: "Deliver all things in number and weight, and put all in writing that thou givest out or receivest in;"[1] but with God it would be superfluous, and therefore a wrong to Him to do so. Leave, then, the care of everything to Him, and let it suffice thee to know that He is very well able to keep all that thou hast committed to Him, in order also to know that He will so keep it. Art thou afraid that if He keeps it for thee He will not one day faithfully restore it to thee? This is the way with men, but never with God.

V. Consider fifthly, why the Apostle said, "I am certain that He is able to keep that which I have committed to Him against that day," that is, the Last Day. Could he not, if we may say so, have caused God to restore this "deposit" to him before that day, by receiving from Him, even on earth, a great part at least of the recompense which had been merited by the successive trials borne for God? But it was enough for him that what was due to him should be kept till the day we are speaking of. Men who are wanting in prudence, when they are engaged on some work involving great labour and outlay, insist on being paid day by day, and so they never become rich, but wiser persons prefer, on the contrary, to receive the whole in one sum, when the work is completed. How is it, then, that thou complainest in thy heart, as though God had wholly forgotten thee? Thou wouldst fain have Him pay thee from time to time? Not so; be satisfied to wait till the Last Day—"that day"—so shalt thou be far richer. And what is this Last Day? It is the day both of the Particular and the General Judgment. On the former God will reward thee most exactly for all that thou hast borne for

[1] Ecclus. xlii. 7.

Him ; and on the latter He will, in addition, restore to thee the body in which thou hast borne it. This, then, is another "deposit," of which the Apostle may have intended to speak when he says, "He is able to keep that which I have committed to Him," that body of his which has been so wearied and worn, so mortified and afflicted. The first "deposit" appertains to the former of these two days, and the second to the second, and the latter is called "that day," without any adjective, because there is no other like it either to the good for blessedness, or to the wicked for misery. This is the day which thou shouldst have always vividly in mind for thine encouragement, saying to thyself: "I suffer, but I am not ashamed: for I know Whom I have believed, and I am certain that He is able to keep that which I have committed to Him against that day," not "for that day," because then there will no longer be a question of keeping, but of restoring, but "against that day," because it is only up to that day that He will have to keep it: "Behold, I come quickly, and My reward is with Me, to render to every man according to his works."[1]

VI. Consider sixthly, that it may be gathered from this passage that even the greatest saints are permitted, especially in times of affliction, suffering, and adversity, to encourage themselves with the hope of their certain reward; indeed, it has always been their custom to do so, as thou wilt see by consulting the Sacred Scriptures. It is true that sometimes, in order to drive away the devil with greater ignominy, and prevent him from ever again returning to disturb thee by temptations to mistrust, thou mayest have to say to him, "I know Whom I have believed, and I am certain that He is able to keep that which I have committed to Him against that day;" but even if He refused to keep it, and were to forget it, so as to allow me to be lost, as He might do, I would nevertheless, in spite of thee, go on serving Him to the utmost, because He is so glorious a Lord that He deserves to be loved for Himself alone, even by those whom He might hold in abhorrence. These were the words of those three courageous children to King Nabuchodonosor, when he tempted them to idolatry, under the pretence that their God would never deliver them from his hands: "Who is the God that shall deliver you out of my hand?" But they answered: "We have no occasion to answer thee concerning this matter,"

[1] Apoc. xxii. 12.

because to do so would be waste of time; "for behold our God, Whom we worship, is able to save us from the furnace of burning fire, and to deliver us out of thy hands, O King. But if He will not, be it known to thee, O King, that we will not worship thy gods, nor adore the golden statue which thou hast set up."[1] O Divine answer! Such a one shouldst thou give to the devil whenever he would tempt thee to worship his idols, which are sin and vanity, under pretence that after all thou wilt be lost; say, "I have no occasion to answer thee concerning this matter. I will not stop to argue with thee, O King of darkness. I know that my God is able to do far more good to me than I deserve: 'Behold my God, Whom I worship, is able to save me from the furnace of burning fire,' in which thou hast been tormented for so many ages, 'and to deliver me out of thy hands.' But even if He will not do so, because of the grievous injuries He has received at my hands, 'if He will not,' still I say to thee, 'Be it known to thee,' that even then I will strive to serve Him with all possible fidelity till death. I will love and adore Him, and never consent to bend the knee to any save to Him alone: 'Be it known to thee, O King,' but King of darkness, 'that I will not worship thy gods, nor adore the golden statue which thou hast set up;' that is, the happiness which thou falsely promisest." By this means the devil will be forced to cease from tempting thee to distrust in this matter of thy salvation, perhaps the most cruel of all temptations.

And now, if thou preferrest on this day to apply this glorious passage of the Apostle to the heroic martyr St. Laurence, whom it so wells suits, thou wilt be able to do so very easily for thyself. How well might he have said in his heart, when he was stretched upon that terrible gridiron: "I suffer, but I am not ashamed, for I know Whom I have believed, and I am certain that He is able to keep that which I have committed to Him against that day."

---

[1] Daniel iii. 15—18.

ELEVENTH DAY.

If any man think himself to be something, whereas he is nothing, he deceiveth himself (Galat. vi. 3).

I. Consider first, that if the injunction here set before thee for contemplation by the Apostle were well understood, there would be an end of vainglory in the world. How is it that so many persons become prouder every day? " The pride of them that hate Thee ascendeth continually."[1] Because they become more blind every day to the knowledge of themselves. They think in their hearts that they are something in themselves, whereas they are in reality simply nothing. Listen, therefore, to this general declaration, which excludes no one : "If any man," be he who he may, "think himself to be something"—the Apostle does not say "something great," he says "something"—"whereas he is nothing, he deceiveth himself." This, then, is the important truth of which thou shouldst one day be firmly persuaded, that of thyself thou art nothing. And why? Because of thyself thou hast nothing but sin, which is utter nothingness. Everything which is thine, with the exception of sin, is from God. This is the way to attain true humility, to enter deeply into this knowledge. For although the essence of humility resides in the will, which modestly submits itself, yet when it does so submit itself, either to a greater or less degree, it is always the understanding which prescribes the rule.

II. Consider secondly, that thou mayest, in the first place, consider thyself merely in thy natural state : and here, if thou "thinkest thyself to be something," thou deceivest thyself, since in thyself thou art nothing. Thou art "nothing" as to thy being; "nothing" as to the operations which proceed from that being, as belonging to it : "Where is then thy boasting?"[2] If thou considerest thy being, what art thou at this moment? Just what thou wert ages before thy birth. Contemplate thyself in that abyss : how deep it is ! The longer thou seekest thyself amongst those shadows, the more impossible it becomes to find thyself. Well, then, thou art at the present moment what thou wert in those long ages before thou wert born : thou art nothing, because thou camest out of nothing. If thou hast an existence, it is only because God gave it thee, and preserves thee in it. And if this is so,

[1] Psalm lxxiii. 23.　　[2] Romans iii. 27.

thou hast no existence of thyself. Wouldst thou say that the reflection in a mirror has any existence of itself, vividly as it represents thy person? Certainly not: and why? Because it is entirely dependent on thee. Just so art thou with regard to God, Whose image thou bearest—an image, indeed, which is real, not apparent: "Man was made to the image of God."[1] If He does but turn His face away from thee for a moment, thou returnest at once to thy original nothingness: "I will bring thee to nothing, and thou shalt not be, and if thou be sought for, thou shalt not be found any more for ever, saith the Lord God."[2] If thou considerest the operations which proceed from that being as belonging to it, whose are they? They are His Who gave thee that being, and Who preserves thee in it. Whence, thinkest thou, does the fruit of a fine tree proceed? From the branch which immediately produces it, or from the root which gives its being to that branch? If there is nothing of thyself in thy being, neither is there in its operations: "Behold you are of nothing, and your work of that which hath no being,"[3] that is, "none which is yours." What more beautiful operation is there than is performed by the shadow of a well-regulated dial which accurately marks the hours? And yet no one attributes it to the shadow, but to the sun, on which the shadow depends. Even so dost thou depend on God. The only difference between thee and the shadow is that its operations are involuntary, and thine are voluntary. But then that very will of thine comes from God, Who gave thee free-will at the beginning, and ever since concurs in each successive voluntary action which thou performest, although that concurrence is always in harmony with this faculty of free-will; that is, it gives thee the power of action, but does not force thee to act. If this is so, "Where then is thy boasting?" Whosoever depends entirely on another in his natural state has no existence of himself. And therefore the Apostle said, "If any man think himself to be something" (he means "of himself"), "whereas he is nothing, he deceiveth himself," because, in strict truth, only He Whose being is of Himself can be said to be: "I am Who am,"[4] that is, "Who am of Myself." And this is the beautiful instruction which God gave to His dear servant, St. Catharine of Siena, saying, "Dost thou know what is the difference between Me and thee? I am Who am: thou art who art not," that is, "who art not of thyself;" and, therefore, thou "art not."

[1] Genesis ix. 6.   [2] Ezech. xxvi. 21.   [3] Isaias xli. 24.   [4] Exodus iii. 14.

III. Consider thirdly, that in the second place, thou mayest contemplate thyself in the state of grace. Thou thinkest it perhaps easier to conceive some good opinion of thyself in this state, and that thou, too, mayest say, "I am not as the rest of men."[1] But the very reverse is the case. If thou "thinkest thyself to be something" in this state, it is a worse error than the former, because it is still clearer here that of thyself thou art "nothing." By calling it a state of grace, thou art taught by that very word that there is indeed matter here for thankfulness, but none for boasting. The reason is this. With all thy gifts of nature is it possible for thee to attain to making an act which is meritorious of eternal life? Certainly not. Every such act requires a two-fold grace, habitual and actual; habitual grace is that which makes thee just and so gives thee the power of performing good actions, and actual grace is that which makes thee act in that capacity as a just man, and so gives thee the action. It is not enough, in order to see well, that the pupils of the eyes be perfectly sound; the ready concurrence of light is also necessary in the case of every object which has to be looked at. So, too, in our case, it is not enough for the soul to be in a healthy state by the habitual grace which it possesses, for that only makes it capable of operating; the concurrence of actual grace is also required for every individual operation proper to that state: "Where, then, is thy boasting?" Thou wouldst fain attribute it to thyself, it may be, the co-operation which thou affordest to this grace? But how so, when this very co-operation is effected by God's grace concurring with thee in order to enable thee to co-operate? "Without Me you can do nothing?"[2] Christ said not only "you can do nothing easily," as the Pelagians would have it, but "you can do nothing" in any way. Light not only causes the eye to see clearly, but to see. And so it is not only at the beginning of the spiritual life that thou needest this grace, but successively, consecutively, and continually, till thy last breath. No habit of acting virtuously, however long, can possibly suffice in the stead of grace. To keep to the illustration of the eyes, which is the most forcible: no matter how long they have been trained from the early morning in seeing perfectly, they are as much in need of light at the last as at the first hour of the day, if they would continue to see. As light fails sight fails. So thou, too, if thou desirest to go on doing well,

[1] St. Luke xviii. 11.   [2] St. John xv. 5.

needest grace to the last moment. And why? Because of thyself thou canst do nothing : "Our sufficiency is from God."[1] Consequently, what art thou, of thyself, in the state of grace? Simply nothing : "If any man thinketh himself to be something" in that state, "whereas he is nothing, he deceiveth himself."

IV. Consider fourthly, that thou mayest, in the third place, contemplate thyself in the miserable state of sin ; and in this state, "if thou thinkest thyself to be something," thou art mad, for thou art not merely nothing, but less than nothing. And the reason is, that thou art reduced to a condition which is worse than nothing : "It were better for him if that man had not been born."[2] This is a state which comes entirely from thyself, and therefore is worse than nothing, because, of thyself, thou canst do nothing but evil. Of what avail, then, is it to thee to be, if this being comes from thyself? It would be far better not to be. "Where then," in this state, "is thy boasting?" Thou boastest, perhaps, of the talents, the sagacity, the acuteness, which thou displayest in sinning, as all do who "are wise to do evils."[3] But all these qualities come from God ; what thou doest is merely to abuse them. There is never anything of thine in a sinful action but the pure malice of it. And is it for this that thou wouldst think much of thyself? On the contrary, it is the only thing in the world of which thou oughtest to be ashamed. Poverty, a mean condition, a poor capacity, are no grounds for shame to thee, because they do not come from thee. The only ground for shame, rightly considered, is the malice which proceeds from thyself : "Be ashamed at your own ways, O house of Israel."[4] Who, then, can express the confusion which should overwhelm thee when, setting before thy eyes the mountain of sins which thou hast committed, thou mayest so well say, "My iniquities are gone over my head."[5] Think how many there are of commission, and still more of omission ! It may be that thy whole life has hitherto been one continuous sin. Is there, then, any reason why in such a state thou shouldst not desire altogether to cease to exist ? There is one only, namely, that thou mayest be able, by repentance, to quit this state. Were there not this reason, thou mightest undoubtedly with good reason wish not to be. The existence of the damned is given to them as a punishment : "He shall

---

[1] 2 Cor. iii. 5.   [2] St. Matt. xxvi. 24.
[3] Jerem. iv. 22.   [4] Ezech. xxxvi. 32.   [5] Psalm xxxvii. 5.

be punished for all that he did, and yet shall not be consumed."[1] Rightly, then, may they say that it is worse than not to be. That is my firm belief; but this happens also in the case we are considering. God may also give existence as a punishment to a sinner in this world, if He foresees that he will not employ it to repent, but to continue in sin. Such a sinner, therefore, who persists in sinning in this world and in remaining impenitent, is in a state which is worse than nothingness, being worse than existing, for "it is better not to be than to be in a state of sin."[2]

V. Consider fifthly, that hitherto thou hast been contemplating the absolute nothingness which is in thee. It remains for thee to see thy comparative nothingness, that is to say, that which is more evident because it is seen by comparison. Confront thyself with those great saints who are reigning in Paradise, the apostles, patriarchs, prophets, martyrs, and all the other glorious spirits, who once lived in the world as thou dost, but whose lives were so much better than thine, and what dost thou think that thou art before them? Dost thou see thyself? dost thou recognize thyself, dost thou "think thyself to be something"? It is impossible that thou shouldst not begin at least to grow less in thy own estimation, more than a dwarf would in the presence of an army of giants: "He shall look upon men, and shall say, I have sinned, and indeed I have offended, and I have not received what I have deserved."[3] Then go further still, and after passing through all the heavenly ranks, pause before the throne of the Blessed Virgin, who surpasses all the saints we have spoken of as much as they surpass thee: "The mountain . . . on the top of the mountains."[4] What is there left of thee now? How dost thou find thyself vanish almost into nothing, like a grain of sand before Olympus! But do not stop even here: rise still higher, unto the sublime presence of God Himself, and after one glance on Him, cast down thine eyes, and see what thou art. Oh, here indeed thou art nothing, even as a tiny spark before the sun. If, in His presence, all the apostles, patriarchs, prophets, martyrs, with all the rest of the saints, including His most holy Mother, appear as nothing—"All nations are before Him as if they had no being at all"[5]—how will it be with thee, miserable sinner? Does it not seem as if thou hadst returned to the

[1] Job xx. 18. [2] St. Jerome on Jerem. xx.
[3] Job xxxiii. 27. [4] Isaias ii. 2. [5] Isaias xl. 17.

original nothingness in which thou wert buried for a whole eternity? How, then, can it ever enter thy mind to be so lifted up with pride in the sight of God, as to make more account of thyself than of His law? This, therefore, is what thou hast to do: to keep vividly before thy mind, first thine absolute, and then, if that is not enough, thy comparative, nothingness. This will make it impossible for thee ever to think thyself "to be something," for to do so would be like determining to see dimly in the noonday light; for this is why the Apostle says, "If any man thinketh himself to be something . . . he deceiveth himself," because the man who has a high opinion of himself is deceived only because he chooses to be deceived: it is not said "he is deceived," but "he deceiveth himself," so evident is the self-deception, and yet he chooses it.

TWELFTH DAY.

ST. CLARE, VIRGIN.

Be ye like the dove that maketh her nest in the mouth of the hole in the highest place (Jerem. xlviii. 28).

I. Consider first, that when God was about to bring destruction on the land of the Moabites, He not only showed them the mercy of causing His servant Jeremias to predict it beforehand to them, but He further vouchsafed to favour them with this remarkable admonition, that they should all imitate doves, which do not make their nests in the heart of the little hollow to which they repair, but just at the entrance, so that they may the more readily escape when their abodes are threatened with ruin: "Be ye like the dove that maketh her nest in the mouth of the hole in the highest place." And this warning should be received by every one in the world as given to him by God in a spiritual sense. Oh, how terrible a destruction it is which threatens this miserable world! What, then, must we do? We must always keep vividly in mind that our dwelling is falling to decay, and therefore that although we must, indeed, remain in it as long as God pleases, yet we should be always ready to leave it, or rather to flee from it, and so we should dwell "in the mouth of the hole in the highest place." Never should we occupy ourselves here so intently as though our dwelling-place were permanent; on the contrary, we should rid ourselves of every obstacle, every cause of delay, so as to be at every moment ready to take flight, "like the doves," as Isaias says in another passage of Scripture, "to their windows." Blessed are they who perfectly accomplish this injunction : they are truly exiles upon earth.

II. Consider secondly, that if ever it was rightly accomplished by any one, it was by the great St. Clare and her numerous band of religious daughters, who to this day observe her Rule in its primitive strictness. They are in very truth doves in the world, who desire nothing of it. They are indeed doves, as all know, for many other reasons : on account of the great purity in which they live, of their solitude, their simplicity, and the sublime heights to which they soar upwards in their secret contemplations : they are doves, too, in the burning charity which consumes them, in the continual compunction which makes them weep, and in the chaste fear of God which makes them tremble at the danger of even a light sin. But then there may be others among the brides of Christ who are their equals in these points : that in which they undoubtedly surpass all the rest is, that they are those doves here spoken of by Jeremias, that is, those who ask the very least that is possible of this miserable world in which they are compelled to live. See how truly they have made their nests "in the mouth of the hole in the highest place." They possess nothing : their dwellings are the meanest, their food the coarsest, their habit the poorest, their bed the hardest, if, indeed, that can be called a bed which rather induces watching than sleep. How is it possible to have less of the world than they have ? And what wonder is it that they are so ready to leave it at the last ? They are unfettered, disengaged from all things : they dwell "in the mouth of the hole in the highest place." And so, no sooner do they hear the Bridegroom's voice saying, "Arise, make haste, my dove, and come,"[1] than they are ready to take the long flight from this world to the next. How, then, will it fare with those who, unlike these chosen souls, are so deeply sunk in this world that they live, as it were, in its very centre ? Can they be said to live "in the mouth of the hole in the highest place"? Alas! how many are there who are always striving to dwell more and more deeply within it !

Cant. ii. 10.

III. Consider thirdly, that there can be no doubt of thy folly, if thou art not afraid of living like these imprudent persons. Dost thou not see that very shortly thou wilt be obliged, in spite of thyself, to leave this world? Why, then, dost thou cling to it as though thou wert sure of having a lasting habitation here? Those are the wise doves which make their nests "in the mouth of the hole in the highest place;" those who make them far in the hole are deluded: "Ephraim is become as a dove that is decoyed, not having a heart."[1] And why so? Because they have suffered themselves to be allured by the small amount of comfort which they enjoy in their prison, and so they have ceased to desire their liberty. They have not, alas for them! the heart for it: "Not having a heart." They look upon the loveliness of plain and valley, of rivers and streams and grassy meadows; and, although they raise their eyes to heaven which invites them, they have not the heart to abandon for it the miserable dwelling, the love of which has "decoyed" them, and yet every day they suffer countless tortures from the gaoler who feeds them indeed, but only to kill them. Art thou not ashamed of thy folly in following the example of such persons? Copy those doves whom God praises, not those whom He blames. Set thyself now to examine what are the ties which bind thee most closely to this world, and then shake them off, break them, for thou hearest that God threatens to bring destruction on thy dwelling. Every day death draws nearer: and what will become of thee if, instead of being "in the mouth of the hole in the highest place," as is right, thou art so far within?

THIRTEENTH DAY.

This is My commandment, that you love one another, as I have loved you (St. John xv. 12).

I. Consider first, how gladly this commandment of fraternal charity ought to be fulfilled which our Lord has called His own: "This is My commandment." By this He was pleased to honour it beyond all the others which He gave us, either because it is the most eminent considered in particular, or because all the rest may be reduced to this, if it is taken in general:

[1] Osee vii. 11.

"For he that loveth his neighbour hath fulfilled the law."[1] Neither can it be said that the same commandment had been already given by God on Mount Sinai under the Old Law, for it had never been given in these lofty and sublime terms in which it was promulgated by Christ when He said, "This is My commandment, that you love one another, as I have loved you." Rightly, therefore, might He call it His, since, if not His as to substance, it certainly was His as to manner. And it is precisely to the consideration of this manner that He invites us by the word "as." It is beyond a doubt that He does not hereby intend to determine the amount of the love, seeing that His was infinite and immense. Who, then, can attain to one that is equal to it? It is merely the quality which He denotes, and to this He obliges us by a very precise law, so that if it is beyond our power to equal His love, we should at any rate attain so far as to resemble it. Do thou beg of our Lord Himself a clear light to understand rightly what was the rule that He observed in loving us, so as to be able, at least, to conform thyself exactly to it, as he does who copies a pattern perfectly free from any flaw.

II. Consider secondly, in the first place, that Christ loved us with perfect rectitude. Rectitude requires three things in the love we bear to any one who is our neighbour. (1) That we should be able to distinguish between substance and substance in him, that is, between soul and body, in such a way as to love the soul for God, and the body for the soul; and consequently the soul more than the body: "He set in order charity in me."[2] This was what Christ did, Who, therefore, in the Apostles He held so dear, only loved the body for the sake of the soul, commanding them to expose it, for the good of the soul, to the severest labours, to poverty, sufferings, and slaughter: "Be not afraid of them who kill the body."[3] And He loved the soul for the sake of God, calling them to Himself neither in order to converse with them nor that they might follow Him, but solely to make them all saints: "He chose us before the foundation of the world that we should be holy."[4] How dost thou keep this rule, who readily givest corporal alms to thy neighbour if thou seest him in want, but not so spiritual alms if thou seest him in error, nay, who art not afraid sometimes even to give him advice prejudicial to his everlasting salvation, if it seems to be

[1] Romans xiii. 8.
[2] Cant. ii. 4.   [3] St. Luke xii. 4.   [4] Ephes. i. 4.

expedient for his temporal interests—"Charity dealeth not perversely"[1]—as is the case when order is not observed in love? (2) Rectitude in the love of our neighbour requires us to be able to distinguish in him between substance and accident, so that whilst hating the sin which he has from himself, we may still love the nature which he has from God: "Every one that loveth him who begot, loveth him also who is born of him."[2] This is what Christ did, Who, although He utterly hated the malice in Judas, nevertheless ceased not to assist him by all kinds of acts calculated to attract him to do well; He even knelt at his feet like a servant, washing and wiping them, embracing and kissing them, with an excess of tenderness unheard of till then, not hesitating to do him honour by calling him friend, even when He saw him in the very act of sacrilegious treason: "Friend, whereto art thou come?"[3] How dost thou keep this rule when thou art continually confounding the sin with the sinner in regard to thy neighbour, and when he has done thee an injury art instantly ready to call down fire from heaven to consume him? "Charity is not provoked to anger" against the sinner, but against the sin. (3) Rectitude in the love of our neighbour requires us also to distinguish between one sort of accident and another, because they are not all of the same kind. Some accidents are good, such as virtues; others are bad, such as vices; some are indifferent, such as noble birth, looks, talents, fortune, and the like, whether they be natural or adventitious gifts. It is very easy for an unwise love to confound these different accidents when they meet together in the same person, so that it may be possible to think that Susanna is loved because she "fears God," and not to discover that the real reason is because she is "very beautiful." Not so Christ: He loved all men from what He saw from time to time to be worthy of love in them. And therefore, as on one occasion He called Peter blessed because he spoke according to the spirit—"Blessed art thou, Simon Bar-Jona, because flesh and blood hath not revealed it to thee"[4]—so on another He called him Satan, when he spoke according to the flesh: "Go behind Me, Satan, because thou savourest not the things that are of God, but that are of men."[5] How dost thou keep this rule when thou often lovest thy neighbour on account of every quality except that which really

---

[1] 1 Cor. xiii. 4.   [2] 1 St. John v. 1.
[3] St. Matt. xxvi. 50.   [4] St. Matt. xvi. 17.   [5] St. Mark viii. 33.

deserves love, namely, his virtuous life? "Charity rejoiceth not in iniquity, but rejoiceth with the truth."[1]

III. Consider thirdly, that Christ loved us not rightly only, but efficaciously, because He loved us not with the heart only, but with acts. Go through His life for thyself, and thou wilt see how much He did for our good. There was not a moment when He was not thus acting. And what more could He do than He did when at last He died for us on the Cross, mocked, naked, and abandoned, between two wretched thieves? "Greater love than this no man hath, that a man lay down his life for his friends."[2] Nay, He did far more than it was necessary to do. For He might have obtained the same salvation for us by one sigh or tear, and yet He chose to purchase it for us by His Blood: He "loved us, and washed us from our sins in His own Blood."[3] How canst thou flatter thyself that thou lovest thy neighbour with similar efficacy so long as thy love is barren, and instead of bearing fruits expends itself entirely in the leaves of fair words? "Where there are many words there is oftentimes want."[4]

IV. Consider fourthly, that Christ loved us, not only efficaciously, but truly, because He loved us as much as His acts showed us that He did. Nay, He loved us far more than this, because, although He did so much for our sakes, yet all His acts fell short of the great love He bore us, because that love was infinite. But how often is thine a feigned love! "The holy spirit of discipline will flee from the deceitful."[5] When does this happen? In three cases: (1) When thou professest to love thy neighbour more than thou dost in reality love him; and this is a dissimulation which is bad, because it is one of exaggeration: "He will give a few things, and upbraid much."[6] (2) When thou professest to love him, and dost not love him at all; and this is a worse dissimulation, for it is one of flattery: "The dissembler with his mouth deceiveth his friend."[7] (3) When thou professest to love him, and not only dost not love, but hatest him; and this is the worst kind of dissimulation, being one of treachery: "Woe to him that giveth drink to his friend, and presenteth his gall."[8] It may be lawful sometimes for a right motive to seem to love another less than we do love him, as our Lord once did, for instance, in the case of His Blessed Mother—"Who is My Mother?"[9]

---

[1] 1 Cor. xiii. 6. [2] St. John xv. 13. [3] Apoc. i. 5.
[4] Prov. xiv. 23. [5] Wisdom i. 5. [6] Ecclus. xx. 15.
[7] Prov. xi. 9. [8] Habacuc ii. 15. [9] St. Mark iii. 33.

—but not to love another less than we profess to love him: "Let love be without dissimulation;"[1] and the word is not "dissimulation" here, but *simulatio*.

V. Consider fifthly, that Christ loved us not only truly, but gratuitously, because He loved us without any interest whatever of His own. All that glory which He received from His Father as Redeemer He might, had He so pleased, have received equally by the mere right of being His Son by nature, holy, innocent, unspotted, and separate from the rest of sinful men; and therefore, if He loved us, it was because it pleased Him to love us: "I will heal their breaches, I will love them freely."[2] He did not love us because He had received any benefit from us, for, on the contrary, He had received from us innumerable injuries; neither did He love us because He expected this, for He saw that those whom He loved were either ungrateful or helpless, and so His love for us was the purest possible love of benevolence, for He not only sought our good as His own, but He sought our good alone: "Christ did not please Himself."[3] How far art thou in reality from so noble a love? "Charity seeketh not her own."[4]

VI. Consider sixthly, that Christ loved us not only gratuitously, but with all His strength, to the very end: "Having loved His own who were in the world, He loved them to the end."[5] His love was not, as that of most men is, a fickle and unstable love, but one that was very strong at all times, even on the Cross, for it was on the Cross that He prayed to the Father for those very murderers, those barbarians who so cruelly nailed Him to it: "Father, forgive them, for they know not what they do."[6] And therefore His love was not only strong "unto death," but strong as death, nay, stronger than death. It was strong as death, because He was not forced by death to cease loving those who inflicted it; and it was stronger than death, because He conquered it by dying even for them. What is the strength of thy love for thy neighbour? "He that is a friend loveth at all times,"[7] and therefore he who can love at certain times only is not a friend, not even at the time when he does love. These are the five characteristics of the love of Christ, which it is possible for each one of us to imitate. When, therefore, thou hearest Him say in future, "This is My commandment, that you love one another as I have loved you," thou wilt know at once what He meant by

[1] Romans xii. 9.   [2] Osee xiv. 5.   [3] Romans xv. 3.   [4] 1 Cor. xiii. 5.
[5] St. John xiii. 1.   [6] St. Luke xxiii. 34.   [7] Prov. xvii. 17.

this word "as." He meant rightly, efficaciously, truly, gratuitously, and unchangeably ; and all other ways may be reduced
to these heads. Only here it may be observed that we ought,
if it were possible, to love each other even more than Christ
loved us, because union among ourselves is of very great help
even to our attainment of everlasting beatitude—"A brother
that is helped by his brother is like a strong city "[1]—seeing that
each one of us is, by himself, exceedingly weak, whereas Christ
was able, of Himself, to do as much as when associated with all
men ; so that, properly speaking, His love for us is that of a
father, and our love for each other is that of brethren. Is it
not, then, very wonderful that Christ should set before us His
own example to urge us to mutual love? For even if this
mutual love were not a matter of obligation, we ought to beg
Him to make it so, seeing how advantageous it is to us.

FOURTEENTH DAY.

*The souls of the just are in the hand of God, and the torment of death
shall not touch them. In the sight of the unwise they seemed to die, and
their departure was taken for misery, and their going away from us for
utter destruction, but they are in peace* (Wisdom iii. 1—3).

I. Consider first, that, as long as they live, the just are
incessantly offering their souls to God. Therefore, just as the
priest holds the Sacred Host in his hands when he offers it to
God from the altar in these words, "Accept, O Holy Father,
this immaculate Host," so, too, it is said of the just, that they
hold their souls in their hands : "My soul is continually in
my hands."[2] Then when, at the close of life, this act of
offering is over, the soul passes from the hands of the just into
those of God, just as the Host does when it has been offered
by the hands of the priest. And this is the precise reason why
it is here said, "The souls of the just are in the hand of God,"
because, as the context shows, those who are spoken of are the
just who have ceased to live, and who consequently have completed their offering which is so dear to God. So long as they
live it may be said more correctly that God keeps His hand
over their souls—"Thou hast laid Thy hand upon me "[3]—for
then is the time for defending them ; after death it is more

[1] Prov. xviii. 19. [2] Psalm cxviii. 109. [3] Psalm cxxxviii. 5.

properly said that He holds their souls in His hands, for then it is not the time for defending them, but for accepting them. And for what object? To embrace, enrich, and reward them, that is to say, to crown them as triumphal victims. Oh, happy wilt thou be if then thou art one of those just ones who continually make to God this acceptable oblation of their souls! See how glorious a reward there will be! Thou, too, wilt place thyself in His hands: "The souls of the just are in the hand of God."

II. Consider secondly, that the just who are particularly spoken of here are such as have suffered like the martyrs, or such as have brought themselves to live a hard, poor, penitential, and mortified life in this world for the love of God. These have indeed made a solemn oblation of themselves, and therefore there is all the more reason why He should receive their souls in His hands, and bear them as precious victims to the glory of Paradise when He sees them come forth from bodies so wounded or macerated for the love of Him. This is why it is here said that these just ones die so gladly that, so to speak, they do not even know what the torment of death is: "The torment of death shall not touch them." Then it is that they see their triumph to be so near that there is reason rather to exult and rejoice than to grieve. If after death they had to fall into the hands of Satan in company with those who have chosen to have their Paradise in this world, like the rich glutton, who "received good things in his life,"[1] then, no doubt, they would be very loth to die. But when they know that, like Lazarus, who received "evil things" here, they will be borne by the angels to rest in the hands, not even of the great Patriarch Abraham, but of God Himself, oh, how joyfully do they die! "Rejoice, O Zabulon," thou tribe who wert so abject, so toil-worn in Egypt, "rejoice in thy going out," for thou shalt possess the richest cities of the coast, "and the hidden treasures of the sands."[2] Now, if it is asked what is the precise torment which is here called "the torment of death," it is enough to see what sinners feel in that sore strait. It is a torment composed of three bonds, each more painful than the other, which all combine at that hour to torture the hearts of the wicked: and these are the past, the present, and the future. The past will afflict them with the bitter memory of all their sins of gluttony, sensuality, and revenge, and of all the good which they have neglected to do. The present will

[1] St. Luke xvi. 25.   [2] Deut. xxxiii. 18, 19.

torment them with the sight of all the beloved objects which they must part from, such as riches, rank, pleasure, friends, and, above all, of their own body, the parting from which will cause the keenest anguish to the soul. And, lastly, the future will terrify them with the anticipation of that terrible Judgment at which they must appear laden with the burden of so many sins. This cruel torment "shall not touch" the just, least of all when they have offered to God that solemn sacrifice of themselves which we have spoken of. For, as to the past, if they have committed sins, they have grieved over them, and made what satisfaction was in their power for them. As to the present, their hearts have been long since detached from everything they are leaving. And as to the future, if in themselves they are afraid, being conscious of their own misery, yet they have confidence in their certainty of the mercy of the God Who is then calling them to Himself so lovingly. This being so, it is evident that the torment of death is not for them— " The torment of death shall not touch them "—because none of the three bonds constituting that torment has any power over them. What, then, must thou do if thou wouldst fain resemble them in this blessed death? Thou must first resemble them in life by making to God that perfect oblation of thyself which He rewards so richly : " Why do I tear my flesh with my teeth," except for this reason (so said Job, who was continually adding suffering to suffering), "and carry my soul in my hands?"[1]

III. Consider thirdly, that, from what has been said, it is evident how greatly all those foolish worldlings are mistaken in their judgment of these just ones when they are at the point of death. They think that at the moment of death they experience terrible bitterness, and after death complete annihilation. But the very reverse is the case. And therefore the next thing said of these just persons is this : " In the sight of the unwise they seemed to die," that is, "they were seen to die by the eyes of the unwise." They were seen to die, as indeed they did die, by the clouded eyes of all those who are without faith ; and immediately their departure was, in the sight of those miserable men, esteemed affliction, and their journey to the next world annihilation : "Their departure was taken for misery, and their going away from us for utter destruction." This departure is that passage which doubtless brings with it very bitter misery to the wicked, for the three reasons which have been mentioned,

---

[1] Job xiii. 14.

all of which combine to torment them—the past, present, and future. But they cannot bring misery to the just, because of what we have already seen; and thus it is that so many of them actually rejoice at that moment far more than the Jews did when they were released from their sad captivity in Babylon: "When the Lord brought back the captivity of Sion, we became like men comforted."[1] It is not absolutely said that they were "comforted" (because there can be no perfect consolation till the dear Jerusalem is reached), but "like men comforted," because that dear Jerusalem begins to be seen very near. And the journey, the "going away," how can that be called "destruction"? It is the journey which is taken by the just when they go from earth to Heaven; it is a "going away from us" to God. But it is a journey which is not believed in by those who only judge by the senses. And therefore that which is in reality merely the passage from one world to another, is deemed by them to be "destruction," because they think that when the body dies the soul dies with it. But what error can be more wicked or more irrational? For not only do the just really take, after death, the journey which has been spoken of, but it is a journey the like of which was never made by any Roman conqueror of bygone days, when returning in triumph from the provinces he had subdued and devastated, to be crowned in the Capitol. What, then, is required in order to believe in this journey? To judge not merely by sight, as so many foolish persons do, but to judge by reason, or rather by those principles of faith which are the only ones in the world which are not liable to deceive us. Close thine eyes, and thou wilt see how glorious is this journey of the just, which nevertheless is deemed by so many persons to be destruction: "There is the way by which I will show him My salvation," says God, that is, "by which I will show him Myself;" not "I will give," seeing that this is reserved for the end of the journey, but only "I will show," which is the most that can be granted during the journey.

IV. Consider fourthly, that in order still further to enhance the folly of the judgment formed by bad Christians concerning the death of the just, the Wise Man says, in conclusion, that not only do they not go away, as so many think, to destruction, but that they are in the enjoyment of the greatest peace: "But they are in peace." Peace, in the Sacred Scriptures, when thus generically described, has a two-fold meaning, negative and

[1] Psalm cxxv. i.

positive. In the first sense, it expresses the cessation of all evil. This is the meaning in the following passage: "Jerusalem, . . . blessed are all they that love thee, and that rejoice in thy peace;"[1] for it is immediately added, as though to explain what this peace is: "My soul, bless thou the Lord, because the Lord our God hath delivered Jerusalem His city from all her troubles."[2] In the second sense, it expresses, in addition, the fulness of all good, as in the following passage: "Rejoice with Jerusalem, and be glad with her, all you that love her, . . . for thus saith the Lord, Behold I will bring upon her a river of peace."[3] Now, both these kinds of peace will be enjoyed after their death by those just ones of whom we are speaking. They will enjoy the cessation of all evil, for suffering will then be at an end: "He hath delivered Jerusalem from all her troubles." And they will also enjoy the fulness of all good, because then will begin for them everlasting life, beauty, health, wisdom, riches; in a word, everlasting happiness: "God will bring upon her a river of peace." It may seem that the Wise Man might have said, in much more definite terms, "But they are in the Kingdom of Heaven," rather than "they are in peace," seeing that the Kingdom of Heaven comprises equally both kinds of peace. Nevertheless he did not say so, and this for two reasons: first, because at the time when he lived, the just, after death, did indeed cease from all evil when they passed into the peace of Limbus, where all the good then rested, but they did not enjoy the fulness of all good, for that comes only from the clear vision of God; and therefore, as they had not then both kinds of peace, the negative and the positive, but only the negative, they could not at that time be said to be in the Heavenly Kingdom, which alone can give them both, but only in expectation of it: "Thou wilt keep peace; peace, because we have hoped in Thee."[4] The other reason is that this expression, the Kingdom of Heaven, was never used in any of the Old Testament writings. It was first used by St. John, the Forerunner of Christ, when he lifted up his voice on the banks of Jordan and began to say, "Repent, . . . the Kingdom of God is at hand."[5] This Kingdom of Heaven was, indeed, spoken of before his time, but in less sublime terms, as a land of promise, a city, a house; as chosen, but yet earthly tabernacles, as riches, rest, and life; and so it is here spoken of by the name of peace,

---

[1] Tobias xiii. 11, 18.   [2] Tobias xiii. 19.   [3] Isaias lxvi. 10, 12.
[4] Isaias xxvi. 3.   [5] St. Matt. iii. 2.

although without limitation, because although at that time all the just who were then dwelling in that much-desired Limbus had in possession (*in re*) only the first kind of peace, that is, the cessation of all evil, yet, as has been already said, they did possess in hope (*in spe*), and that not an uncertain hope as ours is on earth, but a firm and solid hope, the second kind also, namely, the fulness of all good : " Peace, peace." If, then, thou wouldst attain to this two-fold peace which is so greatly to be desired, thou shouldst now make an oblation of thyself to God, by continually offering Him thy soul, as a victim more acceptable to Him than many flocks and herds : " As in thousands of fat lambs, so let our sacrifice be made in Thy sight this day, that it may please Thee."[1] If thou doest this, then, at thy death He will take this sacrifice into His own hands, and keep it in peace with Him, "and there shall be no end of peace."[2]

FIFTEENTH DAY.

THE ASSUMPTION OF OUR LADY.

Humility goeth before glory (Prov. xv. 33).

I. Consider first, the greatness of the glory bestowed on the Blessed Virgin on this day when she was exalted above all the angelic hierarchies, the martyrs, prophets, patriarchs, and all those holy apostles who were so dear to God, being seated in Heaven on so magnificent a throne, as Sovereign Empress of the Universe. Now, incredible as it seems, the humility by which Mary was prepared to receive all this sublimity of glory is of greater value than that glory itself. And therefore it is that the Wise Man here declares that "humility goeth before glory;" and this it does in three respects, by merit, by origin, and by order. And it is on these points that thou wilt make this meditation, so that it may be both for our Lady's honour and thy profit.

II. Consider secondly, that humility goes before glory, because it precedes it by merit. Hence it is that if the Blessed Virgin had to part either with the glory which she won by humility, or with the humility by which she won that glory, she would most certainly prefer to lose the whole of the glory

[1] Daniel iii. 40.   [2] Isaias ix. 7.

rather than the least portion of humility. How is it, then, that thou art so proud in thy sentiments as not to hesitate to prefer to it even so worthless a thing as earthly glory, when even that which is heavenly cannot be preferred to it? There might, however, be some little excuse for thee if it were in Heaven only that humility is more valued than glory; but this is not the case: it is more valued even on earth. In proof of this, who are those who are, in the end, the most honoured, loved, and admired of men? Those who, yielding to ambitious impulse, eagerly followed after glory? Not so; but those who fled from it even when courted by it. A Francis of Assisi, a Francis of Paula, a Romuald, an Arsenius, an Antony, a Giles, and others like them, who retired even into caves there to bury the very knowledge of their names, these are the men who are exalted in the end: "The humble were exalted."[1] Consider the subject, and thou wilt see the truth of this. It is a proof, therefore, that even upon earth glory is obliged to give place to humility, since even upon earth the man who has modestly espoused humility is more highly esteemed than the man who has rejected her in order to pay continual court to glory as her foolish admirer. Here, then, thou hast the first reason why it is said that "humility goeth before glory," because it does so as to merit. Wilt thou, then, disdain it?

III. Consider thirdly, that humility goes before glory, because it does so in its origin. If the Blessed Virgin was on this day exalted to so great a height of glory as has been said, why was she so exalted? Because she humbled herself. And so of her, too, may be said on this day what was said of Christ: "Now that He ascended, what is it but because He descended first into the lower parts of the earth?"[2] Wherefore, although it is true that her devotion, obedience, virginity, faith, and the rest of her virtues render her dearer to Him than all besides: so much so, that the saints tell us that it was this which prevailed with Him at all events to hasten the moment of His Incarnation, in order that it might be seen that as it was the pride of a woman which had moved Him to such grievous anger against mankind, so, too, it was the humility of a woman which inclined Him to be appeased. So, too, the Blessed Virgin herself plainly said that it was especially her humility which God regarded in her—"He hath regarded the humility of His handmaid"—not that He did not also regard all the other virtues which, as it were, vied with each other in

---

[1] Esther xi. 11.　　[2] Ephes. iv. 9.

making her so perfect, but that it was especially on account of her humility that He raised her to the sublime dignity of Mother of God; and it is to this also that she appears to allude, although more obscurely, in these words: "While the King was at His repose, my spikenard sent forth the odour thereof."[1] "The King at His repose" was the King of glory in the bosom of the Father, as is evident. And yet it was in the power of a poor maiden to bring Him down from the bosom of God into her own, so great was the fragrance which rose to Heaven, not from the balsam, cedar, cypress, cinnamon, or any other of the sweet-scented plants which symbolize her, but from the spikenard, that is the aspic, which is the humblest and lowliest of all those which are used as types of her. If, then, it was especially by humility that the Blessed Virgin attained the dignity of Mother of God, what wonder is it that by it also she attained the glory which as such she now enjoys above the stars, where, in herself alone, she constitutes a separate choir, far surpassing that which the blessed form, and only second to that which is formed by the King, her Son? This, then, is the reason why, in the second place, it is said that "humility goeth before glory," because it precedes it as its cause: "He that hath been humbled shall be in glory."[2]

IV. Consider fourthly, that humility goes before glory, because it does so in order. For if it is by humility that glory is given, it follows that humility comes first, and then glory; not first glory and then humility. And here observe attentively the way in which the Blessed Virgin humbled herself before attaining her glory, in order that so beautiful an example may be the more profitable to thee in to-day's meditation. She humbled herself by the low estimation in which she held herself, by despising herself, and by loving to be despised. All degrees of humility may, if thou examinest carefully, be reduced to these three, which, therefore, are sufficient for thee to consider at present. The Blessed Virgin humbled herself, then, by the low estimation she had of herself, which is the first of these degrees of humility—"I will be little in mine own eyes"[3]—not that she did not perfectly know what sublime gifts she had received from God, but because she thoroughly understood them to be gifts, and, as such, did not attribute them to herself, but to the graciousness, kindness, and liberality of the Giver, and so, no sooner did she hear Elizabeth praising

[1] Cant. i. 11.   [2] Job xxii. 29.   [3] 2 Kings vi. 22.

and almost envying her—"Blessed art thou that hast believed"
—than she instantly replied, "From henceforth all generations
shall call me blessed." I grant it, but why? "Because He
that is mighty hath done great things to me." "He," not "I."
And if, as St. Gregory observes,[1] there is this difference between
the humble and the proud, that whereas, when the latter have
anything in them worthy of respect, they keep their minds
constantly fixed upon it and averted from what is contemptible
in them; the humble do the very opposite. Do not suppose
that she was continually going over those gifts in her mind.
Rather did she choose to dwell in thought on her own vileness,
so that even at the moment when she was chosen for the
dignity of Mother of God, she was not able to forget it; and
instead of thinking that it was her lot to conceive and bring
forth and govern her God, she thought only that she had to
minister to Him in mortal flesh: "Behold the handmaid of
the Lord." And, lastly, if she did not dwell in thought, still
less did she dwell in words on these gifts. Hence, it was
characteristic of her to dislike to hear herself praised, and, if
she could do nothing else, to be greatly troubled by it, as was
the case when she heard the Archangel salute her with a title
never before heard of, of a person filled with grace: "Full of
grace." Thus much as to the first degree of humility which
has been noticed, and which consists in despising oneself:
"I will play, and make myself meaner than I have done."[2]
It was accomplished most perfectly by our Lady in three
ways: (1) By carefully concealing all that was great in her
(so that she was compared to "a garden shut up," because she
never made a display of the fruits which adorned her), her
knowledge, wisdom, and sanctity, and the power she certainly
possessed of working miracles. (2) By submitting to laws by
which she was not bound, even though it involved serious
prejudice to her reputation; this she did on many occasions,
especially when she presented herself, like a sinful woman, for
purification after childbirth. (3) By subjecting herself also to
persons who were so greatly her inferiors, as Joseph and John,
and each of our Lord's disciples, amongst whom, as it is
evident, she did indeed take her place in the Cenacle, but
in the lowest place. Lastly, as to the third degree, which
consists not merely in despising oneself, but in loving to be
despised, this also was even deeply practised by our Lady in
three ways: First, by willingly encountering the contempt which

[1] *Moralium*, cap. vii.   [2] 2 Kings vi. 22.

might be shown her on account of her humble condition, as when she was repulsed by all the inhabitants of Bethlehem, and did not shrink from betaking herself to a stable there to await the birth of her Child. Secondly, by willingly enduring the reproaches which might be addressed to her on account of the disgraceful position in which those nearest to her might be placed, so that although she left Jerusalem when her Son entered it in triumph, she hastened thither with all speed when He was going thence mocked, buffeted, scourged, and dragged by the people, like a highway robber, between two thieves, to Calvary. Thirdly, by most willingly submitting to the blame which she might meet with for faults from which she was entirely free, so that she courageously put herself in the way of rebukes and reproaches, enduring with the utmost tranquillity the certain answers which her Son Himself judged well in His secret providence to give her on several occasions, especially when He appeared to pay no attention to her entreaties, modest as they were: "Woman, what is to Me and to thee?"[1] It was of these exercises of humiliation, therefore, as of so many firm and solid steps, that the Blessed Virgin made a ladder to that glory which she now possesses in Paradise. And therefore it is declared, lastly, that "humility goeth before glory," because, just as it precedes it as its cause, so, necessarily, does it also precede it in order. We must go up to the mountain from the valley. Do thou, also, if thou wouldst attain to the glory which God has prepared for thee in Paradise, humble thyself as much as possible on earth, for this is the universal rule for every one, be he who he may, that he must begin low in order to rise high : " Before he be glorified, the heart of a man is humbled,"[2] that is, he must be humbled in this life before he can be glorified in the next. And if thou wouldst see how true it is that this rule is universal, as I say, consider that our Lady submitted to it. Nay, why do I speak of our Lady? Her Divine Son Himself had to submit to it, for it is written of Him : " He shall drink of the torrent in the way, therefore shall He lift up the head."[3]

[1] St. John ii. 4.   [2] Prov. xviii. 12.   [3] Psalm cix. 7.

SIXTEENTH DAY.

Come to Me all you that labour and are burdened, and I will refresh you (St. Matt. xi. 28).

I. Consider first, who they are who labour, and instead of receiving a recompense, receive a burden: "They labour and are burdened." Speaking broadly, but truly, they are all those persons who seek for consolation in the goods which are called worldly, such as bodily pleasures, greatness, glory, riches, and other similar things, if, indeed, there are any which cannot be reduced to these heads. It is certain that all these persons endure very great labour in the search after such consolation, because they seek it where it cannot be found: for the goods which we have enumerated, be they what they may, are like salt streams, which have the effect of increasing, instead of quenching thirst: "Whosoever drinketh of this water shall thirst again."[1] And so thou seest that the more they have indulged their body, the more are they obliged to seek for new ways of indulging it, because they get to despise those to which they are accustomed; the higher their position, the more they strive to rise; the greater their reputation, the more they desire to shine; the richer they are, the more they study to heap up wealth; and in this way they go through excessive toil: "Thou hast been wearied in the multitude of thy ways: thou saidst not, I will rest."[2] All the more, because these goods can never be acquired by any one without serious loss not only of the health which is worn out in the pursuit, but sometimes of life itself. And yet, incredible as it may seem, these very persons of whom we are speaking, after enduring all this labour, instead of receiving the consolation which was its end, receive a burden by loading themselves with grievous sins, and so filling up the measure of their evil: They "labour and are burdened." And dost thou suppose that sins are a light burden? On the contrary, they are the heaviest of all: "My iniquities, as a heavy burden, are become heavy upon me."[3] Now, every burden that is really heavy has three qualities: it distresses, bows down, and sometimes makes a person fall, and fall headlong. So is it with sins. In the first place, it is certain that they distress thee more than other burden, for every other afflicts only the body of him who bears it, but this burden afflicts the heart also by exciting in it

[1] St. John iv. 13.   [2] Isaias lvii. 10.   [3] Psalm xxxvii. 5.

that deep anguish which proceeds from a bad conscience: "I roared with the groaning of my heart."[1] Next, they bow thee down more than any other, by weakening the strength which is the most valuable. I mean spiritual strength, thus making thee utterly incapable of doing good: "Their strength hath failed, and they are become as women."[2] And lastly, they make thee fall down the most terrible of all precipices, the abyss of Hell, from which whosoever falls can never hope to rise again for all eternity: "The iniquity thereof shall be heavy upon it, and it shall fall, and not rise again."[3] Dost thou not think, then, that it is but too true that those who seek consolation in the goods of this world do indeed labour most grievously, and instead of afterwards receiving a recompense, receive a burden? They "labour and are burdened." What is to be done if, unhappily, thou shouldst be one of these? Strive earnestly to understand the misery of this condition in order to prepare thyself to come out of it.

II. Consider secondly, that if thou sincerely desirest to come out of this condition, it is in thy power to do so very easily; and how? Because thou canst speedily have recourse to Christ, Who will give thee what thou vainly seekest elsewhere. Listen to this loving invitation from His own lips: "Come to Me, all you that labour and are burdened, and I will refresh you." Oh, words which should well-nigh break thy heart with tenderness! But before going any further, pause to consider these four: "Come to Me, all," and think seriously in thy heart Who it is Who calls thee. It is thy God, Who has no need of thee whatever; and yet it is this very God Who deigns to say "come;" nor is this all, but "come to Me," and further still, "come to Me, all." In all justice it would be for thee, a vile worm of earth, to beseech thy Lord with pressing entreaties to condescend to receive thee among His servants, and yet it is He Who makes the first advances by saying "come." In the next place, He might, when He calls thee, do so in order to communicate to thee those gifts only, whether of grace or of glory, which are distinct from Himself; but this does not satisfy Him. He calls thee in order to give thee Himself, Who contains in Himself all good, that is, in order to give thee a good which is infinite, and therefore He says not "come" only, but "come to Me." And lastly, even while calling thee to receive such a gift as this, He might do so when He perceived in thee some

---

[1] Psalm xxxvii. 9.   [2] Jerem. li. 30.   [3] Isaias xxiv. 20.

disposition which was requisite in order to merit such an honour : but He so loves to be beforehand with thee, that He calls thee with all thy want of disposition ; and therefore He says not only "come to Me," but He adds "all," without making any exception. Is it possible, that, when He so graciously invites thee, thou on thy part art not ready to accept His invitation? Would He not have good reason to complain of thee by saying : "I called My servant and he gave Me no answer, I entreated him with My own mouth"?[1]

III. Consider thirdly, with what good reason, when Christ calls thee to Himself, He promises to refresh thee. And so, having pondered the words "come to Me, all," pass on, in the last place, to meditate on those which remain : "And I will refresh you." This refreshment is of two kinds, negative and positive, both of which are promised in these words of our Lord. The negative refreshment is the deliverance from the labour and burden. For, when thou acceptest Christ's invitation, the first thing He does will be to unburden thy soul of the sins with which it is now overwhelmed : "And it shall come to pass in that day that his burden shall be taken away from off thy shoulder."[2] And next He will relieve thee from all the labour which thou art now enduring, while vainly seeking in earthly goods the consolation which can only be found in God : "And it shall come to pass in that day, that God shall give thee rest from thy labour."[3] The other kind of refreshment is, as has been said, positive, and will consist in filling thy heart with that consolation which thou hast sought fruitlessly elsewhere, as it is written, "He satisfieth thy desire with good things,"[4] by producing in thee three effects contrary to those which were occasioned by the burden of thy sins. For whereas they kept thee in a state of torment by the anguish caused by an evil conscience, He will keep thee in gladness by the peace proceeding from a good one : "My thoughts are dissipated, tormenting my heart."[5] And whereas they deprived thee of all power to do right, He will instantly make thee vigorous by the interior force of the grace infused so specially in the sacraments, which are that marvellous water known as the "water of refreshment," which restores strength immediately : "He hath brought me up on the water of the refreshment : He hath converted my soul,"[6] that is, He has

<hr/>

[1] Job xix. 16.   [2] Isaias x. 27.   [3] Isaias xiv. 3.
[4] Psalm cii. 5.   [5] Job xvii. 11.   [6] Psalm xxii. 2, 3.

made it strong instead of weak. And lastly, whereas they would have cast thee headlong to perdition, He, on the contrary, will lift thee up to a sure hope of that glory which He keeps in store for thee in Heaven, where at length there will be perfect refreshment : "I rejoiced at the things that were said to me : We shall go into the house of the Lord."[1] If, then, this is so, thinkest thou not that it was with good reason that Christ said, "Come to Me all you that labour and are burdened, and I will refresh you"? It is true, indeed, if we consider aright, that "man is born to labour."[2] And so there is a certain degree of labour which must necessarily be endured in the service of God, because of the strictness required by Him in the keeping of His commandments. But how light is the labour borne in His service compared with that which is borne in the service of the world! "Being made free from sin," which is a heavy burden, "you are become servants to God,"[3] which is a most sweet one. It is necessary, therefore, in the first place, to hear the expressions used by Christ in this same passage when He invites us to keep those commandments, and which will suggest the matter for to-morrow's meditation.

SEVENTEENTH DAY.

Take up My yoke upon you, and learn of Me, because I am meek and humble of heart, and you shall find rest to your souls (St. Matt. xi. 29).

I. Consider first, that Christ called His most holy law a yoke, because of the resemblance that there is between the two. For a yoke, if thou observest rightly, has two qualities : it obliges two animals to go together, which would not keep company when loosed from the yoke, and it obliges those who bear it to keep the straight road, according to the choice of the driver, and this is what the Gospel law does. In the first place, it has united under it the Jewish and Gentile nations, which were once so widely separated ; and secondly, it causes men to live, not after their own pleasure, but according to the commandment given by God to walk in the straight road which leads to Heaven : "Thy ears shall hear the word of one admonishing thee behind thy back : This is the way, walk ye in it, and go not aside, neither to the right hand nor to the

[1] Psalm cxxi. 1.   [2] Job v. 7.   [3] Romans vi. 22.

left."[1] This is a yoke, therefore, not servile, like others, but very honourable, so that Christ dignified it by calling it His: " My yoke." It is His because it was ordained for us by Him as God, and His also because it was borne by Him, as Man, with invincible constancy for thirty-three years, and so borne, that none have ever begun to do so from a tenderer age : " It is good for a man when he hath borne the yoke from his youth."[2] So, too, in regard of His yoke, Christ has displayed both meekness and humility: meekness, in ordaining it as God, that is, as a Prince, not harsh and austere as tyrants are, but full of gentleness ; and humility, by Himself bearing it as a man and not accepting even the least exemption from it. This is why, after saying, "Take up My yoke upon you," He immediately added, "and learn of Me, because I am meek and humble of heart," as though He had said : Only begin to bear My yoke a while, and you will know by experience that I am not a cruel, but a gentle Master, and therefore, that the law imposed by Me is not severe, like that of the world, but easy ; and that I am not a proud but a humble Master, so that unlike the world, I do not disdain to submit Myself to the law which I enjoin on others. How is it that thou dost not feel thyself greatly encouraged by these words to choose such a yoke as this? It is ordained by God : what more canst thou desire ? It is impossible that He should ever lay upon thee a yoke beyond thy strength. And it was borne by God made Man long before thee. How, then, canst thou refuse to bear it after Him? Ponder well these two points, and they will suffice to make thee patient under this yoke : " And you shall find rest to your souls."

II. Consider secondly, that this yoke is made for men, not for animals ; and therefore, it is necessary for thee to bring thyself to bear it voluntarily. Hence it is that Christ says so expressly, "Take up My yoke upon you." He does not merely say "bear," but "take up," because it is not His will to violate thy liberty in the least degree : "God made man from the beginning, and left him in the hand of his own counsel." It is said only that " He added His commandments and precepts :"[3] His commandments, as regards the natural, and His precepts as regards the written law. " If thou wilt keep them, they shall preserve thee ;" not "if thou keepest," but "if thou wilt keep" (*volueris servare*). For after all it is in this, in thy will, that the whole of the merit lies. Why, then,

---

[1] Isaias xxx. 21. [2] Lament. iii. 27. [3] Ecclus. xv. 15.

dost thou not will? Now, if thou wouldst know more precisely what part of thee it is that thou hast to subject with the most docile submission to this yoke, here thou hast a plain answer. Thou hast to subject to it not the lowest part of thee, as oxen do, which place their bodies only under the yoke, and do even that reluctantly, but the noblest part; and so it is not so much the body as the soul which must be put under it: "Your reasonable service."[1] And thus Christ was not satisfied with saying, "Take up My yoke," but He chose to add "upon you," to teach thee in very plain terms that it is that which is, properly speaking, thyself that must be put under this yoke. It is not very difficult to thee sometimes to put thy flesh under the yoke of Christ by chastising, afflicting, and mortifying it; but oh, how difficult dost thou find it to put thy soul under it continually! And it is this which more than anything else must be subjected, thy pride, thy obstinacy, thy ambition, thy love of rule. If thou doest this, then, indeed, thou wilt truly bear the yoke of Christ upon thyself, which is the part of a man, and not only upon thy members, which is common also to beasts: "Submit your neck to the yoke"—this shows that the subjection to the yoke must be voluntary—"and let your soul receive discipline," which shows that it is especially the soul which must be subject.

III. Consider thirdly, that there are two vices which, more than all besides, make people rebel against bearing the yoke of Christ: impatience and pride. Impatience makes them shake it off as heavy—"Let us cast away their yoke from us"[2]—and pride makes them disdain it as disgraceful: "Of old time thou hast broken My yoke . . . and thou saidst, I will not serve."[3] This, then, is the other reason why Christ here bids us learn from Him to be meek and humble, because these two virtues will make thee in future remain perfectly quiet under the yoke: "Learn of Me, because I am meek and humble of heart, and you shall find rest to your souls." Therefore, although the most literal sense of these beautiful words is that which we have already pointed out, namely, that by submitting to Christ's yoke thou wilt see clearly, to the great peace of thy soul, how good a Master He is Whom thou servest, not a cruel one who lays on thee an intolerable yoke as the world does, nor a proud one who will not help thee to bear it: yet that does not prevent this other sense, which is

[1] Romans xii. 1.   [2] Psalm ii. 3.   [3] Jerem. ii. 20.

given by many saints, from being a very proper, although not so exact a one, namely, that it teaches thee, by Christ's example, to be meek and to be humble, as He was throughout His whole life ; because in these two virtues resides that deep peace which thou hast vainly sought for in the pursuit of earthly goods. Dost thou think thou possessest in any degree these two virtues so befitting a Christian ? Art thou meek, or passionate, in thy conduct ; humble, or proud ? But since this is a subject which merits the fullest consideration, it will be well, in order to study it aright, if I propose it as constituting in itself the subject-matter of the following meditation.

EIGHTEENTH DAY.

Learn of Me, because I am meek and humble of heart, and you shall find rest to your souls (St. Matt. xi. 29).

I. Consider first, that man naturally desires to find rest, but he does not attain it : "My inner parts have boiled without any rest."[1] And the reason is that he takes a road directly contrary to it. What does a man do who seeks rest in a natural way ? He tries to avoid whatever may disturb him, he is angry with every one who causes such disturbance, he resents it, he takes revenge for it ; and what is this but to desire the waves of the sea not to buffet him ? What is required, therefore, is not so much to fly from disturbing influences (a thing which is impossible to a man obliged to live in the midst of those waves) as to know how to dwell amongst disturbing influences without being disturbed, so as to be like a rock in the midst of the waves : "I will not fear thousands of the people surrounding me."[2] There have been some among the philosophers who professed to teach a similar doctrine, but they did so rather brilliantly than solidly. None ever taught it on a true foundation till Christ brought it down from Heaven to earth. Therefore it is that He here says, "Learn of Me," and by so saying He shows that the doctrine is certainly worthy of the Teacher. He might have bade thee learn from Him to foretell future things, to heal the sick, to raise the dead, to walk dryshod upon the waters. But what would it be to say such things as these ? Christ was not so

[1] Job xxx. 27.   [2] Psalm iii. 7.

worthy of admiration for the infinite miracles which He wrought upon the earth, as for the infinite examples which He gave of meekness and humility, such as were never seen or heard of in all the ages before Him. With good reason, then, does He say, " Learn of Me, because I am meek and humble of heart." If thou knowest well how to practise these two virtues taught thee by Christ, thou hast found the rest which thou desirest. Prepare thyself, then, like a diligent scholar to hear His doctrine, since it is upon it that thy rest must be founded if it is to be lasting : upon "everlasting foundations," which can never be shaken, built "upon a solid rock."[1]

II. Consider secondly, that all things which have the power of disquieting the soul come either from without or from within. Slights, misfortunes, and such like evils come from without. Those which come from within are all thy defects, whether physical or moral, which sometimes trouble thee more than all the evils which come from without. Arm thyself with meekness against the former, and with humility against the latter. Meekness makes thee restrain thy anger, and so especially prevents thy being disturbed by those evils which come to thee unexpectedly from without. Humility makes thee restrain thy pride of spirit, and so prevents thy being disturbed by those evils which proceed from within, such as thy own defects, because it shows thee that they are what might be expected. What can the soil of thy heart produce but vile weeds ? If, then, by means of these virtues thou succeedest at length in remaining undisturbed, thou hast attained rest. Therefore Christ said, " Learn of Me, because I am meek and humble of heart." He did not merely say, "Because I am humble of heart," seeing that interior humility without the exercise of daily forbearance is not sufficient of itself to restrain the irritation produced by external causes. Neither does He merely say, " Because I am meek," seeing that the exercise of daily forbearance without interior humility is not lasting. Besides this, meekness alone makes thee endure the slights and misfortunes which befall thee ; but humility combined with meekness makes thee love them. Humility alone makes thee endure the defects which thou seest in thyself, as being what are to be expected from thee (to love them in the case of moral defects is never lawful), but meekness, combined with humility, makes thee not merely

[1] Ecclus. xxvi. 24.

endure, but overcome them, at least to a great extent, by helping thee to conquer at all events those defects which arise from the irascible part, which are generally the easiest to vanquish and the most frequent. And when once thou hast attained to this state, see how sweet a rest is thine! "I have laboured a little, and have found much rest."[1] Thou wilt be not only like a rock amidst tempests, which does not heed them because they are outside it; but thou will be like a lofty mountain, which does not feel them, because they are below it. But observe that both the meekness and the humility must be really from the heart—"Meek of heart, humble of heart"—and therefore Christ says so plainly, "Learn of Me," because all who came before Him did not so much teach men to possess these two virtues as to affect them. Dost thou affect, or really possess them?

III. Consider thirdly, that it is very easy to understand theoretically this beautiful doctrine taught by Christ, but that the difficulty is to reduce it to practice. For this reason also He says to thee, "Learn of Me." Go to the school in which it is the heart more than the mind that studies, and thou wilt learn this. Go to the school of prayer, converse there frequently with Christ, ask Him to teach thee how He acted in circumstances so far more grievous than thine, and thou wilt see how quickly thou wilt become learned: "They that approach to His feet shall receive of His doctrine."[2] How great was His meekness amidst exterior trials of every kind: "Like a lamb without voice before His shearer so opened He not His mouth."[3] And how great was His humility in the interior weaknesses of nature, nay, amidst the sins which were not indeed His, but which He had to take upon Himself as though they had been His: "Far from my salvation are the words of my sins."[4] It is impossible to decide whether He was more humble in His meekness, or more meek in His humility; the one always kept pace with the other: "I am meek and humble of heart." It is profitable, in the next place, to observe that the utmost that other teachers in this school could do would be to infuse this doctrine in thee by teaching it, but Christ teaches it by infusing it into thee. O best of Teachers! He first enables thee to practise the doctrine, and then to know it. Such is the force of His most holy grace. And therefore He says so expressly, "Learn of Me," not.

---

[1] Ecclus. li. 35.
[2] Deut. xxxiii. 3.   [3] Acts viii. 32.   [4] Psalm xxi. 2.

from My angels even, nor from My prophets, preachers, and books, but from Me. We must go and converse directly with Christ in prayer : " Because the Lord giveth wisdom."[1] Others teach it, but He gives it. There cannot be found a man in the world who has ever learnt this practice in any school but that of which we have been speaking, the school of prayer. What wonder is it then if thou dost not learn it. Thou leavest school too soon.

IV. Consider fourthly, that Christ selected these two virtues of meekness and humility as in a special manner His own, out of so many others which He might have commended, because it was these especially which He came to bring down from Heaven. And so He was like a merchant who, although he is rich in all kinds of wares, especially delights in displaying those which come from the most distant countries. Oh, how destitute was the world of these treasures before Christ ! " Seek the just," said Sophonias of old, " seek the meek,"[2] as though it were a great marvel of which he was speaking. And what did he mean by "just"? He meant humble, as is shown by another text : "The just is first accuser of himself."[3] Nevertheless, He did not use the word humble, because in those days he would scarcely have been understood, so rare was it to find any one who practised thoroughly the thing signified. Besides, it is evident that these are the two virtues which most frequently have to be practised by a Christian, who is born to suffering. There is not always an opportunity of practising liberality, compassion, charity, or obedience, but there is always an opportunity of practising both meekness and humility, who, like sisters, always go hand in hand, especially in difficult circumstances. Lastly, Christ came to bring into the world the blessing which it had never been possible to find but in Him, that is, peace of heart, and therefore no sooner was He born, than the angels descended joyfully to declare it : "Glory to God in the highest, and on earth peace to men of good-will." And these are the two virtues which, as we have seen, contribute more than all the rest to the securing it, therefore they are those also which beyond all the rest, Christ set Himself to teach : " Learn of Me."

[1] Prov. ii. 6.　　[2] Sophonias ii. 3.　　[3] Prov. xviii. 17.

NINETEENTH DAY.

For My yoke is sweet and My burden light (St. Matt. xi. 30).

I. Consider first, that when thou hast thoroughly learnt meekness and humility, chiefly from the example of Christ, thou wilt not only have found that profound rest of which we treated more generally in the last meditation, but thou wilt see plainly (as was set down in the last but one, which more particularly introduces our present subject) that the yoke to which Christ invites us as our most meek and humble Lord, is incomparably easier to bear than that which the world imposes as a proud and arrogant tyrant ; and this is a powerful consideration for calming the perplexity of any one who doubts which of the two yokes to accept.   Therefore, after saying, "Take up My yoke upon you, and learn of Me, because I am meek and humble of heart, and you shall find rest to your souls," He goes on to say, "for My yoke is sweet and My burden light."   It is certain that this "yoke" signifies His evangelical precepts, which are not intolerable, but sweet ; and by "burden" we may very well understand His counsels, which, in a certain sense, are superadded to the yoke, and, nevertheless, instead of increasing, alleviate its weight ; and this is all that remains for us to consider in order to complete this exhortation of Christ, which has already been divided into several meditations.   But when He says that His yoke is sweet and His burden light, was He speaking of this yoke and this burden absolutely or only respectively to those of the world ?   He spoke in both these senses, but more particularly in the latter, for His object was to induce those unhappy slaves of the world, who "labour" in bearing its yoke, and who "are burdened" by lying under the weight of the sins with which they are laden, to make an exchange, once for all, both of yoke and burden, and so to see by experience how wise they have been in making it.   This is the connection between these words and the preceding ones.   And if thou thus attainest to a clear understanding of the difference that there is between the service rendered by the wicked to the world and that rendered by the good to Christ, dost thou not think that thou wilt have gained an immense advantage ?

II. Consider secondly, that the law of the world, which is the yoke it imposes, seems at first sight much sweeter than the law of Christ, because all that the world requires of thee in order to be one of its followers is to study to satisfy thy concupiscences as much as possible : the concupiscence of the flesh, by indulging in every sensible or sensual pleasure ; the concupiscence of the eyes, by striving continually to rise, to increase, and heap up riches ; and the concupiscence of the mind, if we may use the expression, which St. John calls "the pride of life," by acquiring all kinds of greatness and glory. Whereas, what Christ requires of thee is the very contrary, namely, to mortify these concupiscences to the utmost. And yet it is true that herein the law of Christ is beyond comparison sweeter than that of the world. For there is no one, be he who he may, who cannot by degrees so accustom himself to mortify his concupiscences, that in the end he can attain to doing it with ease. But who can ever succeed in satisfying them? On the contrary, the more any one indulges them the more insatiable he makes them, for they are just like the flames of a furnace, the hunger of which, far from being appeased, is only stimulated by feeding them. And what a law is that which forces thee to strive after a thing which it is impossible to attain? Instead of bringing thee happiness it makes thee restless. See, then, how much sweeter Christ's law is as regards the end : "My yoke is sweet." And how much sweeter also is it as regards the means. For if Christ requires of thee something which is repugnant to human nature, yet, after all, He furnishes thee with such helps of grace as enable thee to act supernaturally, strengthening, assisting, and giving thee power to bear the greatest burdens : "The Spirit helpeth our infirmity."[1] But not so the world. The world abandons thee to thy own nature, and whilst commanding thee to aim at living in pleasure, dressing splendidly, spending lavishly, and pushing thy fortunes, it nevertheless furnishes thee neither with capital, talents, prudence, courage, nor strength for the purpose ; nay, rather does it deal with thee as Pharao did with the wretched Jews, when he commanded them to make bricks without giving them either stones or straw, still less the money necessary for providing them : "Go and gather it where you can find it ; neither shall anything of your work be diminished."[2] What doubt is there, then, that it is much more advantageous to serve Christ, Who is a considerate Master, than the world, which behaves like a tyrant? How much sweeter, then, is the yoke of Christ : " His command-ments are not heavy."[3] How is it possible that, although

[1] Romans viii. 26. [2] Exodus v. 11. [3] 1 St. John v. 3.

thou knowest, even by experience, the certainty of these truths, thou art nevertheless unable to detach thyself from the world to give thyself to Christ? Oh, how greatly art thou deceived in believing these laws to be not what they are in reality, but what thou imaginest them to be. What can be more iniquitous than such a judgment? "Doth the seat of iniquity stick to thee who framest labour in commandment?"[1] so as to be guided by thy own caprice, and to esteem that light which is heavy, and that heavy which is light.

III. Consider thirdly, that as the yoke of Christ is sweeter than that of the world, so, too, is His burden lighter: "And My burden is light." This burden, as we have said, is that of the evangelical counsels, which, when added to the precepts, do but add to them a greater perfection in the keeping of them. And this burden also is contrasted with that which the world lays upon thy shoulders, which is that of the sins with which thou art laden in its service, sins which are neither few in number or small in kind. How evident is it, therefore, that Christ's burden is lighter than that of the world! Shall I show thee how much lighter? See how much more delightful the life of the perfect is in itself than that of sinners. In the first place, if thou fulfillest Christ's precepts with a perfection greater even than that to which thou art bound, thou wilt attain that perfect tranquillity of conscience, to which no earthly pleasure can be compared: "The grace of God, which surpasseth all understanding."[2] And this it is which is contrasted with the unspeakable distress which the burden of thy sins causes thy heart, as we saw in the first of this series of meditations. Secondly, if thou fulfillest Christ's precepts perfectly thou wilt find them easier and easier to fulfil, for this is peculiar to the service of God, that the more any one mortifies himself in it, the more his life is renewed: "When I am weak, then I am powerful."[3] And this is contrasted with the excessive feebleness produced in thee by the burden of thy sins, which gradually rob thee of thy strength, so that at length they weaken, or rather utterly destroy thy power of doing right. And in the third place, if thou fulfillest Christ's precepts perfectly, thou hast an all but infallible certainty of salvation: "I have fought a good fight. . . . As to the rest" (that is, in the future) "there is laid up for me a crown of justice, which the Lord, the just Judge, will render to me in that day."[4] And this is opposed to the great fear which thou

[1] Psalm xciii. 20. [2] Philipp. iv. 7. [3] 2 Cor. xii. 10. [4] 2 Timothy iv. 7, 8.

oughtest to feel, as long as thou art in sin, of falling headlong
into Hell, to which that weight impels thee. Even though
we should be willing to grant that the burden imposed by
Christ was really very heavy in itself, thou seest that it at once
ceases to be so by means of the numerous good effects
produced by it which lighten by counterbalancing it. But
how can the evangelical counsels be called heavy in themselves, if they lessen the weight of the precepts? And therefore their weight is compared by the saints to that of the
wings of eagles and herons, which seem at first sight as though
they must weigh down those birds by their great size, and yet
which are so far from doing this that they enable them to lift
their bodies with great agility not only above the mountain-
tops, but above the clouds. How, then, canst thou excuse
thy timidity in taking such a burden on thyself? I grant that
thou art not obliged to bear it, for which reason Christ said,
"Take up My yoke upon you," but did not say, "Take up
My burden," because He enjoined His precepts, which are
the "yoke," on all men ; but His counsels, which are the
"burden" added to this yoke, He enjoined on no one. But,
notwithstanding, what does it signify that thou art not obliged?
If a very great gain is attached to a certain action, who would
wait for an obligation to be laid on him before performing it?

IV. Consider fourthly, that, without a doubt, that which
more than all besides makes Christ's yoke sweet and His
burden light to His followers is the love which they bear Him.
For this is the effect of a very great love, to make the lover
not feel anything that he bears for the beloved one : "Jacob
served seven years for Rachel : and they seemed but a few
days because of the greatness of his love."[1] But how can
any one reasonably have such a love for the world, which
proves, in the end, not only a hard, but a bold, faithless,
deceitful, and treacherous master, although, at the beginning,
it flatters credulous persons with fair words? It may be that
thou clingest to the world rather by the innate desire which
inclines thee to indulge thy irregular concupiscences as it
teaches thee to do, that is, by self-love. But I assure thee
that self-love never causes any one so much pleasure as the
love of Christ does. Thou canst not, perhaps, understand
this truth in thy present state. But at least believe the many
saints who know it by experience. Which of them would
consent to exchange for a single day his purity, his poverty,

[1] Genesis xxix. 20.

fasts, and disciplines, nay, the very scorn with which he is treated, which is the hardest of all, for whatever could be promised him by the concupiscence of the flesh, the concupiscence of the eyes, or the haughty pride of life? "For which cause I please myself in my infirmities, in reproaches, in necessities, in persecutions, in distresses, for Christ."[1] Observe in what the Apostle says, "I please myself:" not in miracles, nor in praise, nor in applause, nor in the triumphs of his divine eloquence, but in the multiplicity of the sufferings which he endured for Christ. It is true, indeed, that such feelings are only for adepts in sanctity. But what is it that thou, in thy state, oughtest to do? Strive to love Christ as much as possible, and then thou wilt see if there was any exaggeration in His words when He said that His yoke should be sweet and His burden light to the shoulders of His followers: "My yoke is sweet and My burden light."

TWENTIETH DAY.

Thy arrows pass: the voice of Thy thunder in a wheel (Psalm lxxvi. 19).

I. Consider first, what all those evils, trials, and misfortunes are which God sends us on earth. They are, rightly considered, so many arrows which He aims at us from Heaven, either to punish us, to try us, or to stop us when we are fleeing from Him: arrows which, it cannot be denied, are very terrible, sharp, and keen, and which sometimes pierce so deep as to draw from us not only our heart's blood, but the very spirit within us: "The arrows of the Lord are in me, the rage whereof drinketh up my spirit;"[2] but after all, arrows which pass: "Thy arrows pass." Thou losest a child, it is a stroke that passes; thy character is taken away, it is a blow that passes; thy property is taken away, it is a blow that passes; judgment is given against thee in a court of law, it is a blow that passes: all these things are "arrows that pass." What is that which will never pass? It is the terrible voice with which Christ will thunder in the ears of sinners, when on the Last Day He will drive them away from Him, and say in His anger: "Depart from Me, ye cursed, into everlasting fire." That will be a voice sounding eternally in

[1] 2 Cor. xii. 10.		[2] Job vi. 4.

the ears of the reprobate, eternally torturing, eternally accusing them, so that they will never be able to exclude it from their minds, but will hear it every moment through the course of ages, as plainly as though it were then in the act of being uttered by Christ, their Judge. It will not, then, be a voice which passes quickly away like our voices, but a strong and lasting voice, like the voice of God ; and if there is a sense in which it passes on, in its effects, from time to time, yet it passes on without ever ceasing to do so, seeing that it will always be in perpetual motion on the great wheel of eternity, producing the same effects in the hearts of the reprobate as when first it was uttered : "The voice of Thy thunder in a wheel." How is it possible, then, that thou art so disturbed by temporal evils which press like arrows, and consequently have no power ever to come back, and so little moved by those which are eternal, which do indeed pass, but pass, as it were, in a circle, so that they are always passing, and yet never passing away ?

II. Consider secondly, why it is that the voice in which Christ will utter the final sentence of damnation on the reprobate is called a "voice of thunder." There are three reasons for this : its origin, its character, and its effect. (1) It is called so on account of its origin. For dost thou not well know from what the voice of the thunder proceeds ? It proceeds from the victory gained at last by the vapour when, bursting the clouds in which it was condensed and confined, it comes forth, no longer a prisoner, but free, to course through the air with furious force. And this terrible voice of Christ, the Judge, will have a similar origin. It will proceed from the victory which His most just anger, so long restrained and kept back by His patience, will at length gain on that day, a day which, for that very reason, is called "the day of wrath," because that wrath will no longer be shut up, as it now is, within the Heart of Christ, but will break forth to vent itself the more vehemently on those insolent offenders, because of the length of time that it has delayed to do so : "I have always held My peace, I have kept silence, I have been patient; I will speak now as a woman in labour."[1] And yet thou venturest to provoke him so greatly to anger now, because He is silent. (2) This voice of Christ is called a "voice of thunder," because of the quality it possesses of sounding very loudly : "The noise of His thunder shall strike the earth."[2]

[1] Isaias xlii. 14.   [2] Ecclus. xliii. 18.

The angels, too, will be heard on that day, but what will their voice be? As the voice of a trumpet only: because the one of their number who is heard in one of the four quarters of the world, will not be heard in the others. If it were not so, what would it avail to send many of them? But Christ will speak with a voice of thunder, and very terrible thunder: "He shall thunder with the voice of His majesty,"[1] because He will be heard at the same moment by all the four parts of the earth, which is also the reason why it is said that His voice will sound "in a wheel," that is, "in the globe." "The voice of Thy thunder in a wheel." (3) This voice of Christ is called a "voice of thunder," because of its effect, which will be the unutterable terror produced by it: "At the voice of Thy thunder they shall fear."[2] It will be so great that the damned will not only begin in their anguish to beseech the mountains to fall upon them, the rocks to crush them, the stones to break them to pieces, but even the earth to open and swallow them up at once in its depths. Pause a while, and bring vividly before thee those words, "Depart from Me, ye cursed, into everlasting fire," which are the very strongest possible expression of the Divine wrath, and thou wilt see how true it is that they will fill thee with exceeding terror. It may be said that God never now speaks in this voice of thunder, because He is never vehemently angry—"He doth not now bring on His fury"[3]—the day of His wrath not being yet come, and yet thou seest what terror He causes whenever it happens that He raises either a storm in the air or an earthquake among the habitations of men! What will it be, then, when He speaks in His "voice of thunder"? "Seeing we have heard scarce a little drop of His word, who shall be able to behold the thunder of His greatness?"[4]

III. Consider thirdly, that the voice of this thunder is said to be "in a wheel," not only because it will fill the whole circuit of the earth with its strength, as has been shown, but also because it will fill the whole circuit of the never-ending eternity with its wrath. And here meditate a while within thyself on the vast space it has to fill. If the task were given thee to find the sum total of minutes necessary to fill the circuit of eternity, couldst thou do it, however skilful an arithmetician thou mayest be, even with as much ease as thou couldst find the number of grains which it would take to fill

---

[1] Job xxxvii. 4.
[2] Psalm ciii. 7.   [3] Job xxxv. 15.   [4] Job xxvi. 14.

the whole circuit of the earth up to the highest pitch of the firmament? Let it be supposed that ten thousand millions of grains, as small as the tiniest poppy seeds, multiplied again, and again, and again, by the million, would be sufficient to do this, yet what would all this be in comparison with eternity? After having placed all these millions and millions, not of minutes, but of centuries, in its gigantic circuit, thou hast still done simply nothing; and if thou repeatest the operation any conceivable number of times, it is still nothing : literally and absolutely nothing. There will remain as much empty space to fill as there was at the beginning. Eternity swallows up everything. What a devouring grief it is! And what will become of thee if thou fallest into the abyss where there is nothing ever done but to suffer incessantly? Strive, then, with all thy might to ascend, otherwise thou art lost : "Thou shalt be destroyed for ever, saith the Lord."[1] For the wheel of eternity, both of the good and the wicked, does not turn, as foolish men say that of fortune does : it is always stationary. He who once is above is always above, he who is below remains below. Time will go on, but there will never be any turn in the fate of the man who has no longer the time for doing well. Do it, therefore, now, while thou hast time.

TWENTY-FIRST DAY.

What things a man shall sow, those also shall he reap. For he that soweth in his flesh, of the flesh also shall reap corruption. But he that soweth in the spirit, of the spirit shall reap life everlasting (Galat. vi. 8).

I. Consider first, that this life is our seed-time, and that the next will be harvest-time. The harvest, therefore, which will be reaped in the next world will correspond with what is sown in this : "What things a man shall sow, those also shall he reap." The seed is our works, and the harvest is, as we all know, the retribution answering to those works, either of reward or of punishment. The man who has sown grain will have grain, and the man who has sown cockle will have cockle, which is the same as saying that it will be well with him who has done well, and ill with him who has done ill : "The Son of Man shall come in the glory of His Father with

[1] Jerem. li. 26.

His angels, and then will He render to every man according to his works."[1] And there is no law, in my opinion, so terrible as this, because it admits of no exception whatever; it concerns all, it binds all, it strikes all; and therefore the Apostle enunciates it in these universal terms: "What things a man shall sow, those also shall he reap." Be he who he may, prince or beggar, there will be no regard paid to any one's position; he is a man, and, as such, he is subject to this law in such a way that for all eternity he can hope for no exemption from it. What art thou thinking of, then, when thou considerest so little, sometimes, about the works that thou doest? Take good heed, for all these works are so much seed which cannot remain unfruitful; do not, then, allow them, as it were, to slip by chance out of thy hands. Take example by the sower, and see first whether they are good or bad. If they are good, scatter them gladly, for they will bring thee in a good harvest: "Say to the just man that it is well, for he shall eat the fruit of his doings."[2] But if they are bad, alas for thee! fear and tremble, for thy prospects are bad indeed: "Woe to the wicked unto evil, for the reward of his hands shall be given him."[3]

II. Consider secondly, that the prudent husbandman, in order to secure a good harvest, not only does his best to sow good seed, but also to sow it in a good soil; otherwise, the result would be the same as though the seed were bad: "What he shall sow, those also shall he reap." For although it is true that a good soil cannot make bad seed good, yet, on the other hand, a bad soil can make good seed bad by corrupting it, so that in the end the harvest will be bad: "They have sown wheat and reaped thorns."[4] Thou, too, oughtest to practise the like care in thy actions. Thou hast in thyself two very different kinds of soil, the flesh and the spirit. The spirit is a pure, rich, fertile soil, but the flesh is so foul a soil that it corrupts even the good seed which is put into it, and makes it so bad, that from being worthy of recompense it degenerates into being worthy of punishment. And therefore the Apostle here says that "he that soweth in his flesh, of the flesh also shall reap corruption; but he that soweth in the spirit, of the spirit shall reap life everlasting." That man is said to sow in the flesh who does his works for the sake of the flesh; and he is said to sow in the spirit who

---

[1] St. Matt. xvi. 27.
[2] Isaias iii. 10.   [3] Isaias iii. 11.   [4] Jerem. xii. 13.

does them for the sake of the spirit. Take heed, therefore, for it is not enough that thy works are good in themselves, they must also be sown in the spirit, that is to say, thy intention in doing them must be for the sake of the spirit. If thy intention is for the sake of the flesh, then thou art really sowing in the flesh, and consequently thou art lost; it is impossible for thee to reap anything but corruption. I will explain my meaning. If thou art liberal in spending, thou art sowing good seed; but if thou spendest in this way for thy pleasure, in feasting, luxury, and debauchery, then thou art sowing in the flesh, because thou art spending for the sake of its sensuality. If thou preachest, thou art sowing good seed; but if thou preachest for gain, thou art sowing in the flesh, because thou art preaching for the sake of its avarice. If thou sufferest, thou art sowing good seed; but if thou sufferest to gain praise, thou art sowing in the flesh, because thou art suffering for its ambition. And this being so, thou canst expect nothing but a harvest of rottenness, like the soil in which thou art sowing, because a corrupt soil makes the seed corrupt; and this is why the Apostle says that such a one "shall reap corruption," because every corruptible action is lost: "Every work that is corruptible shall fail in the end."[1] If thou wouldst reap a profitable harvest, sow not only good seed, but sow it in the spirit—that is, do not act for the sake of gratifying any of the three irregular appetites which reign in the flesh, but do the good that thou doest from motives of eternal life, so shalt thou not merely avoid reaping corruption, but thou shalt reap life, and life everlasting. The spirit is the giver of life: "It is the spirit that quickeneth."[2] And so it is the spirit which gives the germ of life. The spirit is ever-lasting, because it never dies, and so the spirit gives the germ of life, and of life everlasting. In which of these two fields, therefore, thinkest thou that it is right to sow thy seed, in that of the flesh or in that of the spirit? Most certainly, in the case of a thing that thou possessest, thou wouldst not choose that which would make all thy toil fruitless; how, then, canst thou choose it, when it is thy very self which is concerned? And observe, that when the Apostle is here speaking of the man who resolves to spend his labour on behalf of the flesh, he says "in his flesh;" but when immediately afterwards he is speaking of him who resolves to spend it on behalf of the spirit, he does not say

[1] Ecclus. xiv. 20.   [2] St. John vi. 64.

"in his spirit," but "in the spirit," because the flesh belongs
to us, in a certain sense, of ourselves, and therefore is with
more reason attributed to us, whereas the spirit is entirely
given to us by God. And if so, what wonder is it that the
flesh and the spirit produce germs so different? We cannot
expect any from ourselves, being what we are, but such as are
altogether evil: "My son, sow not evils in the furrows of
injustice," that is, of the flesh, "and thou shalt not reap them
seven-fold."[1]

III. Consider thirdly, that if, as soon as a man did well in
this world, it was well with him, and that as soon as he did ill
it was ill with him, he would certainly be much more careful
in all his actions. But do not suppose that this alters the
case; for this is precisely another point in which the works of
man resemble sowing, that the reward and punishment do not
follow them immediately—a certain time has to elapse: "What
things a man soweth, those shall he reap," not "he reaps," but
"he shall reap." Sometimes, indeed, in this just judgment of
some particular person, God does either punish or reward
immediately. But such cases are exceptions to all rule, as
when we are told that in the same year in which Isaac sowed
the land of Gerara, he reaped it, and even reaped a hundred-
fold, because God blessed him in a special manner: "And
Isaac sowed in that land, and he found that same year a
hundred-fold, and the Lord blessed him."[2] The ordinary rule
is that the time pre-ordained by God for the harvest must be
waited for: "The time of repaying,"[3] which is not this life—
for this is the seed-time—but the life to come. If, therefore,
thou hast done evil, do not say, "I have done it and yet I
have not suffered for it in any way: "Say not, I have sinned,
and what harm hath befallen me."[4] For if thou hast sinned,
thou hast sown; let that suffice thee. Thou wilt reap, all too
surely and abundantly, in due time the evil that thou hast
done: "He that soweth iniquity shall reap evils."[5] And if
thou hast done well, do not say, I have gone on doing well
for so long, and yet I have not so much as begun to reap the
fruit of it: "Why have we fasted, and Thou hast not regarded:
have we humbled our souls, and Thou hast not taken notice?"[6]
Have patience, for thou shalt reap a fuller harvest than thou
thinkest: "To him that soweth justice there is a faithful
reward,"[7] not "a speedy," but "a faithful reward," faithful as

[1] Ecclus. vii. 3. [2] Genesis xxvi. 12. [3] Ecclus. xviii. 24.
[4] Ecclus. v. 4. [5] Prov. xxii. 8. [6] Isaias lviii. 3. [7] Prov. xi. 18.

to its certainty, its superabundance, and its permanence. See how patiently the husbandman waits for his harvest, even though for a little while he should suffer hunger : " Behold the husbandman waiteth for the precious fruit of the earth, patiently bearing till he receive the early and the latter rain,"[1] that is, the morning showers and those which fall late in the day. Do not, then, long for thy harvest while the corn is green, by desiring that God should reward thee in this life, seeing that if He did so, it would certainly not be to thy advantage. Wait till the next life, which will not fail to come at last : " Be you therefore also patient, and strengthen your hearts, for the coming of the Lord is at hand."[2]

TWENTY-SECOND DAY.

And in doing good let us not fail.   For in due time we shall reap, not failing (Galat. vi. 9).

I. Consider first, how, in conformity with all that we have been considering, especially at the end of the last meditation, the Apostle, after saying, " He that soweth in the spirit, of the spirit shall reap life everlasting," immediately added these words, which will form the subject of the present one, " And in doing good let us not fail.   For in due time we shall reap, not failing," to teach us that since there is so much to be gained by every one who sows in the spirit, he should lose no time if he is wise : " In the morning sow thy seed," by beginning to act virtuously in thy youth, "and in the evening let not thy hand cease," by continuing to do so in old age : "for thou knowest not which may rather spring up, this or that," that is, thou canst not tell whether the morning or the evening sowing will be the most fruitful : "and if both together it shall be the better,"[3] for it will indeed be better if they both bear fruit at the same time.   There are three things which may cause a sower at last to abandon an undertaking of so much labour as his : namely, weariness, fear, and sadness. And the same causes may also make thee abandon the work of well-doing, if thou dost not overcome them.   The first is weariness, because, in course of time, sowing becomes tedious, being an occupation in which there is no admixture of

[1] St. James v. 7.   [2] St. James v. 8.   [3] Eccles. xi. 6.

amusement, and so it may easily happen that the work is abandoned in the very midst through sloth. The same thing happens in doing good, especially in seasons of supineness. In such circumstances, try to shake off this wretched weariness by reminding thyself that he who sows little reaps little : " He that soweth sparingly shall also reap sparingly."[1] If we would have a full harvest we must sow abundant seed, and to do this requires diligence. The second cause is fear, because a sower is exposed to the inclemencies of the weather, and for fear of them often returns home sooner than he ought to do. So, too, in the case of a man who is doing good : he sometimes leaves off, and why? Because if some troublesome wind arises, either of temptation, or trial, or evil counsel, which has been given him by careless livers, in this case, he should recall to mind that "he that observeth the wind shall not sow."[2] The man who would sow much seed must not heed the roughest blasts, and so, in the second place, courage is needed. The third cause is sadness, because the sower deprives himself of the grain which he possesses, and therefore, although he knows that he is not throwing it at random, but, so to speak, putting it out at interest, yet he may come, in the end, hardly to believe himself, and so he does not work with the alacrity of him who reaps : "Going they went and wept, casting their seeds."[3] And the same thing happens in the case we are considering. Men have so little faith that it almost seems to them that they are throwing away their labour, when they are engaged in sowing in the spirit, which in the end will bring them in a hundred-fold. Stand firm, therefore, on the promises of Christ, which will enable thee to labour not only with diligence and courage. but also with gladness : "That both he that soweth and he that reapeth may rejoice together."[4] For there is this difference between material and spiritual sowing, that the former sometimes comes to nothing, and therefore it is no wonder that the sower cannot be so joyful as the reaper. But spiritual sowing is always certain, and therefore he who is employed in doing good ought to be as glad as if he had already received his reward : "The fruit of justice is sown in peace,"[5] because, in this case, there is no fear of its being scattered by any storm.

II. Consider secondly, that nothing is more calculated to lighten the poor husbandman's labour in sowing, and to give

---

[1] 2 Cor. ix. 6.   [3] Eccles. xi. 4.
[2] Psalm cxxv. 6.   [4] St. John iv. 36.   [5] St. James iii. 18.

him both courage and gladness, than the thought of the
harvest : "He that plougheth should plough in hope."[1] And
so the Apostle says, "In doing good let us not fail," and
immediately adds, "For in time we shall reap, not failing."
But what is the meaning of the words "we shall reap, not
failing"? It is this : "We shall reap if only we do not fail."
For this is a very necessary condition for every one who
desires to reap the precious fruit of the beatitude which God
is preparing for us in the next, not to cease sowing in the
spirit, in this life, no matter what obstacles are in the way :
"He that shall persevere unto the end, he shall be saved."[2]
No sooner do we cease sowing in the spirit, and begin sowing
in the flesh, than all is lost : "Judgment shall spring up as
bitterness in the furrows of the field."[3] Some of the saints
do, indeed, give two different explanations of these words.
The first is that "we shall reap, not failing," because the
harvest of consolation, happiness, and contentment which we
shall reap in the glory of Heaven will be a never-ending
harvest : "He that soweth in the spirit, of the spirit shall reap
life everlasting." If, then, the harvest is to be perpetual, is it
not just that during the short days of life which God has
appointed for us to spend in sowing we should not draw back
our hands from the work on account of a little weariness?
"Linger not in the time of distress"[4] (for the word *angustia*,
which is here translated "distress," means also a short,
restricted time, such as the present), seeing that the reward
set before us by God will be continually beginning : "If man
works without ceasing, God will also reward him without
ceasing."[5] The second interpretation is that "we shall reap,
not failing," because there is similarity between the reaping of
Paradise and the reaping of earth. Here it is certainly a
joyous, but still a laborious work, which soon exhausts the
strength even of robust persons. But in Heaven it is an
occupation of unmingled pleasure, in which there will be no
fear of one growing weary though we employ all our faculties
in it : "We shall reap, not failing." But what does this prove
but the deep delight which is felt in it? All pleasures of this
world pall in the end ; this will find us always fresh and eager
as when an excellent dramatic representation is just beginning :
"Who shall be filled with beholding His glory."[6] What is to
be considered, then, from both these allowable interpretations,

[1] 1 Cor. ix. 10.   [2] St. Matt. x. 22.   [3] Osee x. 4.
[4] Ecclus. x. 29.   [5] St. Augustine.   [6] Ecclus. xlii. 26.

which I have cited, but that we must sow incessantly, in this world, for the sake of the spirit, even though to do so should occasionally cause us some trouble? "Be not weary in well-doing,"[1] because the harvest will be rich beyond belief: "They that sow in tears shall reap in joy."[2]

III. Consider thirdly, that in order to obtain this blessed harvest, which, as we have said, is not only certain but most abundant, it is not enough either to sow good seed, or to sow it in good soil, or to do all the rest that has been laid down in the two meditations which have been given on the subject. The seed which has been sown must also be protected from the birds which are watching to carry it off, for this also is one of the duties of a good sower, although it is not here particularized. And how is the seed protected? By covering it. This is what is done by the humble, they are very careful to hide all the good that they are continually doing in the spirit, and therefore it is that they gain so great an advantage in the end; while vainglorious persons, on the contrary, are very ready to let it be seen by others, and thus, if they do not lose it entirely, they lose, at least, a great part of it: "You have sown much, and brought in little."[3] Why is it then, that thou gainest so little from the good that thou doest? Because thou dost not cover it, as it is thy duty to do: "The fowls of the air," that is, thy frequent vainglorious thoughts, "have devoured it."[4]

TWENTY-THIRD DAY.

Dreams have deceived many, and they have failed that put their trust in them (Ecclus. xxxiv. 7).

I. Consider first, that there is so much similarity between temporal goods and dreams, that commentators on the Scriptures take dreams in this passage, speaking broadly, to mean possessions of this kind. And it is certain, that just as dreams are only regarded by persons who are asleep, and derided by those who are awake, so, too, is it with the goods of this miserable world. Who are those who so greatly value them? Those who sleep, that is, those who, having their understandings clouded by the vapours of sin, judge of things, not

[1] 2 Thess. iii. 13.
[2] Psalm cxxv. 5. [3] Aggeus i. 6. [4] St. Luke viii. 5.

as they really are, but as the imagination represents them. The saints, on the other hand, who, in obedience to Christ's frequent injunctions, are always watching, that is, who never allow their understandings to be dulled. Oh, how cheap do they hold them! What art thou to do, in order to despise them as they ought to be despised? To watch: "Therefore, let us not sleep as others do, but let us watch."[1] The devil does his utmost to produce in thy soul this mischievous sleep. For this purpose he induces thee to abandon mental prayer, which is more efficacious than anything else in driving it away; for this he inspires thee with a hatred of penance; for this he invites thee to banquets, amusements, and gaieties, so as to increase those vapours which mount to the head, and in the end force thee to close thy eyes against thy will. Do not yield to him: do the very opposite of all these things to which the devil would persuade thee, and thus, by keeping always wakeful, thou wilt despise that which there will be so much danger of thy valuing if thou art overtaken by slumber. For thou hast been told who they are who so greatly love dreams—those who sleep—according to the words of Isaias, "Sleeping and loving dreams,"[2] so close is the connection between the two things.

II. Consider secondly, that it is here said that "dreams have deceived many." So, too, the goods of this world have deceived many, nay, how many are there whom they still deceive, and with precisely the same kind of delusion which takes place in dreams. There are two ways in which pleasant dreams deceive thee: either by persuading thee that thou art happy when thou art miserable, as though a beggar were to dream that all the ships which came into port were his; or by promising that thou shalt be happy in the future, as in the case of persons who go on dreaming that they will be made prelates or popes, but who, on arriving at Rome, cannot even get an introduction to Court. It is the same with the goods of this world, riches, patronage, applause, dignities: these things make thee at once imagine thyself to be happy. But it is not true, nay, thou art really more miserable than ever, because thou art in imminent danger of being lost without knowing it: "Dreams lift up fools,"[3] that is, they lift them out of themselves, so suddenly do they even make them lose their senses. And if they cannot always succeed in making thee imagine thyself happy, so great are the uneasiness, the bitterness, and the mental anguish which thou feelest even while possessing

---

[1] 1 Thess. v. 6.   [2] Isaias lvi. 10.   [3] Ecclus. xxxiv. 1.

them, yet they persuade thee that in future thou wilt be happy.
But do not believe them; for if it were possible for them to
make thee happy, they would already have done so. What is
the man doing who "trusteth to lies"? "He feedeth the
winds," that is, his own ambitious thoughts; but he himself
"runneth after birds that fly away,"[1] because it is impossible
for him ever to reach the goal he is striving after. Besides,
thou knowest the common saying about dreams, that they
predict the reverse of what is about to happen; and this, too,
is what occurs in the case of worldly goods. They promise
thee happiness, and give thee misery. Therefore it is said,
"They have failed that put their trust in them," that is, they
have failed of the hope which they were deceived in forming.
This was the case with that man spoken of in St. Gregory's
Dialogues, who dreamed that he had many years to live, and
so spent his time in collecting, scraping together, and storing
up the means which he was afraid would one day fall short for
the numerous journeys he expected to take. This poor man
so wore himself out with these labours, that in a very short
time he died, finding, to his confusion, that it was not the
means that were wanting for the journey, but that the journey
itself would never be taken. Take care that this is not thy
case; thou dreamest that thou hast a long time before thee:
"Soul, thou hast much goods laid up for many years." And
so thou livest just as though thou wert perfectly certain of
attaining the object which thou hast placed before thy imagination. Now, then, take heed, lest this very night a terrible
voice should sound in thy ears, crying: "Thou fool, this night
do they require thy soul of thee, and whose shall those things
be which thou hast provided?"[2]

III. Consider thirdly, that although the Wise Man here
says that "dreams have deceived many," yet he does not add
that "they have failed that had them," but "that put their trust
in them." For it is not in having even the most seductive
dreams that the danger lies, but in giving credit to them. So,
too, shouldst thou believe to be the case with regard to
earthly goods. It is true, indeed, that, as a rule, "Where
there are many dreams there are many vanities,"[3] because it is
difficult not to set at least some slight value on them. Still,
after all, the harm does not consist in possessing these things,
even in great abundance. They were possessed by the Emperor
Henry, who was so renowned for sanctity, and by such Saints

---

[1] Prov. x. 4. [2] St. Luke xii. 20. [3] Eccles. v. 6.

as Gregory, Charles, Casimir, and Louis of France ; and yet none of them were the worse but the better for them, because they knew how to employ them profitably for God. The harm is in putting trust in these goods, as though they had the power of making the man who has them in abundance happy. It is against this that thou shouldst be especially on thy guard. This is why the Wise Man here says, speaking of dreams, " For dreams have deceived many." And, in further proof of this, observe that the word translated "deceived" is not *deceperunt,* but *errare fecerunt,* because, strictly speaking, dreams do not really deceive any one, as we said just now, speaking more broadly and carelessly, but rather give people an occasion of deceiving themselves. And so God said : " Let there not be found among you any one that observeth dreams."[1] It is not earthly goods that deceive thee, for they leave thee free to believe what thou pleasest about them, as we see how differently foolish and sensible persons think about them : it is thou who deceivest thyself concerning them, because in thy sleep thou esteemest them to be much more worth than they are. Do but open thy eyes, and thou wilt disdain them : "As the dream of them that awake, O Lord, so in Thy city Thou shalt bring their image to nothing."[2] What is the "image" here spoken of ? It is the happiness of worldlings, a happiness which is not real, but imaginary. Now, if thou desirest it, this happiness, which seems to many persons so solid a substance, shall vanish instantaneously like a dream as soon as thou awakest. But where art thou to awake ? In prayer. This is that city of Jerusalem in which God manifests Himself on earth in so vivid a light, that the sleeper awakes, and in the moment of waking despises that which he so greatly esteemed while sleeping, and he too acknowledges that the happiness of worldlings all comes to nothing : "The joy of the hypocrite is but for a moment . . . as a dream that fleeth away, he shall not be found, he shall pass as a vision of the night."[3]

[1] Deut. xviii. 10.　　[2] Psalm lxxii. 20.　　[3] Job xx. 4, 8.

TWENTY-FOURTH DAY.

ST. BARTHOLOMEW THE APOSTLE.

Strive for justice for thy soul, and even unto death fight for justice, and God will overthrow thy enemies for thee (Ecclus. iv. 33).

I. Consider first, that our justice is the grace of God, for it is this that makes us just. Now, whenever this grace is spoken of, thou shouldst not think that the whole of man is spoken of: "For this is all man."[1] What is any man without the grace of God? He does not so much as deserve the name of man, being far more wretched than any brute, stock, or stone, which, in a certain sense in the state in which they are, are loved by God, whereas he is in a state in which he is rather an object of hatred to Him. Therefore, whenever there is question of suffering for the sake of preserving the grace of God—"for justice"—it is really "for thy soul," that is, "for thy virtue," that thou art suffering, by exerting every part of thyself. Yes, thou must "strive:" thou must, if needs be, strive in a very agony, so that there may be no single thing, either in thyself or in those belonging to thee, from which thou art not ready to be detached. Such is the great value of the grace of God, that for its sake thou must bring thyself to part from even such things as it is agony to cut off: "Strive for justice for thy soul."

II. Consider secondly, that not only oughtest thou to bring thyself, for the sake of the grace of God, to the state of one who thus strives, that is, who wrestles with death to the utmost, in order that he may not be robbed of his life by it, but thou oughtest also to bring thyself to the state of one who dies, letting thy life be taken from thee rather than be induced to part with this grace. This is the force of the words: "Even unto death fight for justice." They do not merely mean that thou art to fight faithfully up to the moment of death, never laying down thy arms; but they mean that thou art to fight even to the point of bearing death with great fortitude. There are two kinds of death: a real and a metaphorical death. Those who die a real death "for justice" are all those who choose rather to be consumed by flames, cut to pieces by the sword, torn by wild beasts, than ever to consent to what is wrong. All that is required of thee with regard to this death

[1] Eccles. xii. 13.

is to live in a state of preparation, by recalling to thy soul that, in such a case, this is thy duty. Rather die by being burnt, stabbed, or bled to death than commit one sin : "Even unto death fight for justice." As to the metaphorical death, those persons die by it every day who are able also to say with the Apostle, "I die daily,"[1] that is, by that entire abnegation of self by which they, so to speak, renounce their life. To this death (which is so necessary for thee in order to preserve the grace of God) thou too shouldst be ready to subject thyself, and not shrink even from entering religion in order to attain to this. I say this, because it is only in religion that this blessed death really is found, a death which is considered equal to martyrdom, because of the solemn vow of perpetual obedience made in it. Nor is this surprising : for while exercising self-denial by the practice of the other two evangelical counsels, poverty and chastity, it is possible for thee to live as thou pleasest : thou mayest be poor, and yet live to thyself; chaste, and yet live to thyself; because in all other respects thou mayest follow thine own will. But when thou deniest thyself by practising perpetual obedience, it is no longer possible for thee to live to thyself, thou art like one already dead ; for obedience is the sword of which St. Gregory says that it cuts thy head from thy body, so that it is impossible for thee to be governed any more by thy head : "Having beaten down the pride of thy will, He slays thee with the sword of His commandment."[2] And, as I said, justice obliges thee to submit willingly to this death. What do I mean by this ? That if thou knowest that the true, perhaps the only way, for thee to continue in the grace of God is to enter some religious order, and so thou dost enter one, then thou art bound to "fight even unto death," that is, including death, "for justice." Is not Divine grace a life which is worth much more than any life in the world? "The grace of God is life everlasting."[3] Is it a great matter, then, to stake temporal for eternal life ?

III. Consider thirdly, that the things of which we have been speaking cannot be thought of by the inferior part of the soul without great dread, because it has naturally a strong repugnance to every kind of struggle, and still more to every kind of death. And this is the very reason why thou are told to "strive for justice." The superior part must struggle manfully with the inferior till the victory is won. Does not

---

[1] 1 Cor. xv. 31.
[2] *Moralium*, lib. xxx. cap. x.   [3] Romans vi. 23.

the inferior part shrink from facing the fire of musketry? And yet how many do this in war, some to please an earthly sovereign, some for love of gain, some from the desire of glory, or other similar worthless motives. The superior part, therefore, should say to the inferior in the case we are concerned with: Have patience, and be willing that I should do for God what multitudes fearlessly do every day in the service of the world. The crown is only awarded to the man who has gained it sword in hand: "He is not crowned unless he strive lawfully."[1] Justice, too, has her crown, the brightest of all—"It triumpheth crowned for ever"—because justice, that is Divine grace, can only be preserved by a long and persistent conflict, both exterior and interior: "Winning the reward of undefiled conflicts."[2]

IV. Consider fourthly, that men fight not only to keep what they have gained, but to gain it in the first instance. Therefore, when it is here said, "Strive for justice," and afterwards, "even unto death fight for justice," the meaning is not only "to preserve justice," but "to win justice;" because every additional degree of Divine grace is well worth any painful separation, any struggle, any trial, nay, even the most cruel death that can be endured for its sake. And if this is so, even though thou mayest save thy soul, yet how much more grace is it in thy power to gain in religion! Why, then, for the sake of gaining it, dost thou not die courageously to thyself, by presenting thy neck to the sword which will make thee equal to the martyrs in Paradise? I grant that the many fervent practices of piety, customary in religion, may perhaps hasten thy natural death if thou livest in that state; but what a store of grace thou canst acquire by means of them! Is it not better, then, to live ten years less, and to acquire it? "The purchasing thereof is better than the merchandise of silver . . . and gold."[3] And how many there are who shorten their lives in the world for the sake of silver and gold! Thou wilt have to endure the furious attacks of the wolves, from which thou seekest to snatch those wretched sheep which they have torn away from Christ. But do not thou fear their attacks. Oh, how much wilt thou have gained by winning a soul from evil ways! If thou shouldst lose thy life in so worthy a cause, say with the Apostle: "I fear none of these things, neither do I count my life more precious than myself."[4] This is,

---

[1] 2 Timothy ii. 5.   [2] Wisdom iv. 2.
[3] Prov. iii. 14.   [4] Acts xx. 24.

indeed, not only to fight for the sake of preserving justice, as was shown in the last point, but also of gaining it.

V. Consider fifthly, that thou mayest perhaps allege thy lack of strength as a reason for holding back from these great undertakings. But, in answer to this, the Wise Man adds in conclusion, that God will be thy helper: "God will overthrow thy enemies for thee." What, then, dost thou fear? It is not said, indeed, that He will "fight" (*pugnabit*) for thee, for that is thy part; but it is said that He will "overthrow" (*expugnabit*) "thy enemies for thee," because it is He Who conquers in thy stead. Thou canst do nothing of thyself, that is certain. But do the little that is in thy power by means of the help which God imparts to thee from time to time, and meanwhile pray to Him, pray with all thy heart, pray without ceasing, that He will vouchsafe to overthrow the rebels within thee. These are thy disorderly passions, thy love of strife, thy love of worldly society, of thy own comforts, of human applause. If once these are reduced to submission, with what enemies canst thou fear to engage? "The men shall perish that strive against thee, thou shalt seek them and shalt not find the men that resist thee."[1] Thou mightest justly doubt, if both fighting and conquering were allotted to thee. But it is not so. Thy part is only that of fighting: "Even unto death fight;" the conquering rests with God: "He will overthrow for thee." Nay, it is God also Who gives thee strength to fight: "I am the Lord thy God, Who take thee by the hand, and say to thee: Fear not, I have helped thee."[2] There remains, therefore, but one thing, never to be weary of co-operating with His most holy grace, never to throw down thy arms; for then, indeed, thy enemies will greatly triumph over thee. As to the rest, if God has not as yet overthrown them, do not be troubled, for most surely He "will overthrow them;" it is a promise, as thou seest, which is faithful: "The lip of truth shall be steadfast for ever."[3]

---

[1] Isaias xli. 11, 12. [2] Isaias xli. 12. [3] Prov. xii. 19.

TWENTY-FIFTH DAY.

ST. LOUIS, KING OF FRANCE.

Fear ye not the reproach of men, and be not afraid of their blasphemies: for the worm shall eat them up as a garment, and the moth shall consume them as wool, but My salvation shall be for ever (Isaias li. 7, 8).

I. Consider first, how injurious it is to thy spiritual life to be afraid of the mockery which thou hast to bear from those whose conduct is different. Therefore God encourages thee in these words to give no heed to it. Such mockery must be either in deeds or in words : is it not so ? Now, imagine the very worst that can happen to thee when this mockery is in deeds : it may oblige thee to suffer not only dishonour and contempt, but actual disgrace. Imagine also the worst that can happen to thee if the mockery is in words : it may oblige thee to suffer not only sneers but curses, nay, even execrations, such as persons are mad enough to utter against Heaven. Yet even of these God bids thee take no heed : " Fear ye not the reproach of men, and be not afraid of their blasphemies." And why? Because the evil that thou receivest from this mockery on earth comes to thee from men who will soon fall to decay ; and the good that will afterwards result to thee from it in Heaven will come from Him, and therefore will last for ever. There are a thousand other motives which God might assign for thy not fearing this mockery. But for the present He is content with this one, because it is the most powerful in banishing fear. Fear arises from the apprehension of an evil which is difficult to bear ; and the motive which is here alleged shows thee that this is very easy to bear, and so at once banishes fear. But how does it show thee this ? Because it points out that, on the one hand, this evil, be it what it may, will quickly come to an end ; and on the other, it makes thee merit an everlasting reward. Meditate on these two points, and not only wilt thou cease to dread this evil, but wilt desire it : " Blessed are they that suffer persecution for justice' sake."[1]

II. Consider secondly, that the evil we have spoken of, if we observe closely, consists, after all, in opinion ; because it consists in the low esteem of thee which is evinced by men. If, then, thou desirest not to fear that opinion, set thyself to consider who these men are. They are men subject to corruption, not saints reigning in Heaven, who, on the

[1] St. Matt. v. 10.

contrary, if thou actest virtuously, hold thee in high esteem. They are mortal men who, as such, are fallible, unjust, and fickle in their judgments, and who, if they had no other demerit, will very soon all cease to be. See, therefore, in what strong colours God represents their mortality, saying that "the worm shall eat them up as a garment, and the moth shall consume them as wool." The worm comes from without, from dirt which adheres to cloth ; the moth is bred within it. The articles which are most liable to the worm are those which are used as garments, as coverings for chests and cupboards, or as hangings for walls, and the like ; because they are the most exposed to constant soils. The stuffs which are most liable to the moth are those which are not in use, but stored in trunks. And so, in this passage, the worm is ascribed to the articles in use which are described as "a garment," and the moth to stuff in the piece, which is merely called "wool." What is it that God would have thee under- stand by this distinction? That every one, whosoever he may be, whether he thinks of it or not, will at length have to die. The "worm" signifies injuries which attack men from without ; the "moth" those which attack him from within. So that even if there were no worm, which is the first thing mentioned, to occasion his speedy death, there will yet be the moth to come, that is, his natural frailty : "They shall be consumed as with the moth."[1] This frailty is inherent in the vitals of man ; and precisely for this reason it is like the moth, which not only eats him like the worm, but consumes him, not sparing his very bones. See here the state of mortal man briefly described by God. Now, therefore, see whether he is to be more esteemed than God Himself : "Who art thou, that thou shouldst be afraid of a mortal man, and of the son of man, who shall wither away like grass? and thou hast forgotten the Lord thy Maker."[2]

III. Consider thirdly, that having contemplated on earth these men who mock at the good thou doest, thou mayest very well now contemplate them in Hell in order to encourage thyself still more to disregard them : since there are none more sure to go there than those who not only will not do good themselves, but cannot endure that others should do it. If, then, thou contemplatest them in Hell, thou wilt be able very rightly to understand in a moral sense that worm and that moth, the literal sense of which thou hast just now seen by

[1] Job iv. 19.   [2] Isaias li. 12, 13.

contemplating them on earth. Be very sure that when these unhappy men are in the depths of Hell: "The worm shall eat them up as a garment, and the moth shall consume them as wool." The "worm" there will be remorse for the exceeding evil which they did on earth in mocking thee, and the "moth" will be their envy of the exceeding good which their insults have won for thee in Heaven. And who can express the torments which such a worm and moth as these will vie with each other in eternally causing them? The remorse which they will suffer will certainly be in their hearts, but still, it will not go so deep as the envy which will penetrate into their very bones: "Envy is the rottenness of the bones."[1] And therefore the former is represented by the worm, and the latter by the moth. Remorse will indeed gnaw them: "The worm shall eat them up as a garment;" but envy will utterly devour them: "The moth shall consume them as wool." For it cannot be doubted that although remorse will be a terrible anguish to the damned, yet envy will be incomparably worse; because these wretched men will not be so much grieved for the evil they have committed as for the good they have lost, all the more when they see it enjoyed by those whom they so greatly despised: "These are they whom we had sometime in derision."[2] Envy which is felt towards one who has gained some good always causes torment, but this torment is never so great as when the good is exceedingly great, and he who has gained it is an enemy. And these two things never are combined in any envy as they are in that which is felt in Hell. What is it then to thee if wicked men now insult thee either in word or deed? Be very sure that if they insult thee now for a few years, they will envy thee hereafter to all eternity.

IV. Consider fourthly, that this is why God immediately adds: "But My salvation shall be for ever." It might have seemed as though He should have contrasted with the mockery thou receivest on earth the honour which this mockery will result in for thee in Heaven; but this was too little to satisfy Him, so He contrasted with it His "salvation," which embraces everything, in order to show thee how much fruit this mockery will produce for thee if thou bearest it patiently, and this for two reasons. First, because it will detach thee from the love of creatures to whom thou wouldst cling if, instead of mocking the good that thou doest, they were to unite in paying thee honour; and next, because they will win for thee the love of

[1] Prov. xiv. 30.   [2] Wisdom v. 3.

the Creator. Dost thou not know how much greater His love for thee will be if thou art despised on account of the service thou renderest Him? If thou wert honoured, thou wouldst be bound to Him for it; if thou art despised, it is He, on the contrary, Who is bound to thee. This, then, is that happy state in which God becomes wholly thine, when for love of Him thou doest that which is good, and receivest that which is evil: "If you be reproached for the name of Christ, you shall be blessed."[1]

V. Consider fifthly, that God calls this salvation which will be bestowed on thee "His." It might seem that He should rather have said "yours." But He says "My salvation," to teach thee that although thou hast to concur in this salvation by thy co-operation, yet He concurs incomparably more by His most holy grace: so much more, that it may be absolutely said that He does all. If, therefore, this salvation is called thine by reason of acquisition, much more may it be called His by reason of gift. But if this is so, is it not clear how important it is for thee, on this very account, to gain His love by enduring contempt for His sake, since by this means thou layest under obligation to thyself the God on Whom thy salvation depends? "Destruction is thine own, O Israel: thy help is only in Me."[2]

VI. Consider sixthly, that if ever any one understood this doctrine, it was the saintly King whose memory we venerate to-day. He chose, in His royal dignity, to do what all great men should, to espouse a virtue of no mean degree, but the most sublime and magnificent possible, I mean sanctity; for which reason he was depised by foolish worldly politicians, because in his way of governing, living, dressing, and conversing with others, he followed maxims wholly opposed to theirs, preferring humility to pride, simplicity to deceit, plainness to pomp, charity to the poor to the love of money. Far wiser than they, he utterly despised the being despised, and thou seest to what a height of greatness he has been raised in consequence. The men who mocked him are now envying him in the abyss of Hell, while he is not only in the glory of Heaven, but even upon earth he is regarded as the greatest King that France can boast of. And therefore with good reason are these words applied to him on this day: "Wisdom showed them to be liars who had accused him, and gave him everlasting glory."[3]

---

[1] St. Peter iv. 14.   [2] Osee xiii. 9.   [3] Wisdom x. 14.

TWENTY-SIXTH DAY.

Doing the truth in charity, that we may in all things grow up in Him Who is the Head, even Christ (Ephes. iv. 15).

I. Consider first, what it is that the Apostle desires of thee when he says, "That we may grow up in Him." He would have thee try to grow, not, as so many do, in esteem, in wealth, in dignities, or other similar goods: He would have thee try to grow in Jesus Christ: "Grow up in Him Who is the Head, even Christ." And what is to grow in Christ but to become more and more deeply rooted in Him? This is the case when thou hast so laid up thy heart in Him as no longer to care for anything beside Him, finding all things in Him, honour, wealth, dignities, everything which others seek for out of Him. See, therefore, how well the Apostle here spoke when he said, "That we may grow up in Him." "To grow up to Him" is a different thing from growing up "with Him," and this, again, from growing up "in Him." Those persons grow up "to Him" who have forsaken evil and given themselves to good by an ever-strengthening determination to follow Christ. And these are they who are called beginners in the spiritual life. They grow up "with Him" who, having already given themselves to follow Him, bear Him company whithersoever He goes, even to Calvary, imitating His Divine virtues more and more closely every day. These are those who are advancing in the spiritual life. And they who grow up "in Him" are those who, having practised the imitation of Christ so far at least as is possible to human weakness, endeavour more and more to find their rest in Him, desiring no other good. These are the perfect. And here thou shouldst consent, as it were, to go in search of thyself, to see whether thou art in any of these states, to humble thyself if it should prove that thou art not even in the second, but barely amongst the first, and desire earnestly the blessed lot of those who are in the third.

II. Consider secondly, that it is easy to understand how persons in the first and second states must all grow; but that this is not so easily seen in the case of the third. And yet how well does the Apostle here teach that this is so, when he says, "Doing the truth in charity, that we may . . . grow up in Him." Thou knowest that "truth" refers equally to thoughts, words, and works; and hence, in the Sacred

Scriptures, it often briefly signifies all the good which a just person can possibly do: "Open ye the gates, and let the just nation that keepeth the truth enter in."[1] When thou hast attained to such a state, that whenever thou thinkest, speakest, or actest, thou doest what is right, "doing the truth," and still further doest it in the right way, "in charity," that is purely for the love of God, and not "in cupidity," that is from the desire either of pleasure, or gain, or glory; when, I say, thou hast attained even to so sublime a state as this, still thou oughtest always to try to increase in growth: "Doing the truth in charity, that we may grow up in Him." Dost thou suppose that growth concerns beginners only, or at most those who are making progress? Thou art greatly mistaken, it belongs also to those who are perfect. And therefore God was not satisfied with saying, "He that is just, let him be justified still," but immediately adds, "and he that is holy, let him be sanctified still."[2] Well is it for thee, if, in whatever state thou art, thy heart is inflamed with this desire of growing as is right to do. God grant that the very beginning may not be as yet unmade.

III. Consider thirdly, how, to take away every excuse from thee, the Apostle not only says "that we may grow up in Him." For, even if thou thinkest that thy whole heart is already fixed in Christ, thou hast still to diffuse thyself on all sides, so as always to do as much as possible for Him. Thy mind must always strive to think of Him, thy tongue to praise Him or preach of Him, thy hands to labour for Him, and so of all the rest; for just as bodily growth must be universal, although with due proportion in the different parts, so, too, must it be with spiritual growth, "that we may in all things grow up in Him." Oh, that thou knewest the importance of thus growing! In the body there is a certain point beyond which it is of no consequence to consider about growth; but in the soul there is no such point. Nay, no sooner dost thou cease striving to grow, than thou at once beginnest to deteriorate: "Not to go forward is to go back." And the reason is, that if thou ceasest striving to grow, it is a clear proof that thou supposest thyself to have grown enough; and this, in itself, is to retrograde. This is seen by the Pharisee, who no sooner thought that he had reached such a pitch of perfection as not only to be equal to other men but to surpass them—"I am not as the rest of men"[3]—than he was at once

[1] Isaias xxvi. 2.   [2] Apoc. xxii. 11.   [3] St. Luke xviii. 11.

less than the very publican above whom this proud man
exalted himself. If, instead of vainly dwelling on the good
that thou hast done, thou wouldst rather consider that which
thou lackest, oh, how greatly wouldst thou find thyself in
need of growth! The reason why thou thinkest thy growth
sufficient is that thou dost not raise thy eyes high enough.
Thou comparest thyself with thyself; thyself as thou now art
with thyself as thou wast in the past. Rather, compare thyself
with the saints, who did so much more for God than thou
hast done, and thou wilt see how far short thou art of reaching
their height, or even of approaching it. Imitate the glorious
St. Charles, who often entered into himself to think seriously
what more he could do, in the state he was in, for the love of
God, in order really to grow "in all things." And even if
thou dost not see what more thou canst do, if thou canst not
grow any more "in all things," then grow "in Him" so as to
fix thy heart more and more deeply in Christ: "My glory
shall always be renewed."[1] Neither do thou take fright, as
though I desired to lift thee too high, for, as thou hearest, if
thou art obliged to grow, it is "in Him, Who is the Head."
And so it is from Him that all thy strength must come; it
must come from the Head. All that is necessary for thee is
not to be separated from it: "That we may grow up in Him
Who is the Head, even Christ." Only acknowledge Him as
such, and beg this strength of Him.

IV. Consider fourthly, the signs by which thou mayest
perceive, so as to know how thou art to grow, in which of the
three states that have been mentioned thou art; whether in
that of the beginners, of those making progress, or of the
perfect. Thou mayest perceive this by examining thyself,
and seeing what it is to which thy chief attention should be
directed. If it is to the escaping from vices, thou art in the
state of a beginner; if it is not so much the escaping from
vices as the advancing in virtues that claims thy care, then
thou art in a state of progress; and if it is not even this so
much as to be very closely united to thy God, it may be said
that, by His grace, thou hast attained to the state of perfection.
Do not, however, imagine, that those who are making progress
do not also require to keep themselves from vices, or that
beginners are not also obliged to cultivate virtues, for all this
is common also to the perfect, even as it is common to the
two first classes to endeavour after a measure of union. But

[1] Job xxix. 20.

this is not their main study. Therefore it is from this that may be gathered the nature of a man's state : from seeing with regard to what particular point he has to live most carefully day by day. I know that it often happens that a person aims at attaining that state which belongs only to the perfect without having first duly passed through the other two. But this cannot be done. Take example by what happens to the body after birth. First, it is nourished merely that it may not die ; then, being nourished, it is strengthened ; and having become strong, it comes to perfection. So, too, must it be with regard to the soul. How can it attain such a degree of perfection as to have God for its only good, and to rest in Him, if it has not first been strengthened by the practice of virtues? And how can it be strengthened by the practice of virtues, if it has not been previously nourished with that first food which keeps it from dying? There are no sudden starts of bodily growth, and the same is true of that which is spiritual : "They shall go from virtue to virtue."[1] Observe, "they shall go," not "they shall leap." Therefore, also, it is that the Apostle here says, "Doing the truth in charity, that we may grow up in Him." First, do all that is right in every way—"do the truth"—and still further, do it in the right way, that is, "in charity;" and then thou wilt go on with ease to grow likewise "in Him Who is the Head, even Christ."

TWENTY-SEVENTH DAY.

They shall go into the lower parts of the earth : they shall be delivered into the hands of the sword, they shall be the portions of foxes (Psalm lxii. 11).

I. Consider first, that the cause of the sinful prevarications of so many persons lies in the three affections which have been again and again mentioned, but which have never been sufficiently detested : the love of glory, of pleasure, and of worldly gain. Now, that these affections may not spring up, at least too thickly, in thy heart, accustom thyself to cut them down frequently, since it is not possible to root them up entirely. With this object meditate on these words of the Psalmist, which will speedily show thee what will be the last

[1] Psalm lxxiii. 8.

end of those unhappy persons who allow themselves to be unduly influenced by these affections. They have been too much in love with glory, and therefore it is said that "they shall go into the lower parts of the earth." They have been too eager after pleasure, and therefore it is said that "they shall be delivered into the hands of the sword." They have been too much attached to money, seeking, for love of it, the blood of the poor, and practising a thousand falsehoods and deceits; and therefore it is said, in the last place, that "they shall be the portions of foxes." Beg of God to give thee light to understand more clearly the nature of all these three punishments which will be inflicted on the damned, in order that thou mayest escape them.

II. Consider secondly, in the first place, that the damned "shall go into the lower parts of the earth," that is, into the very centre of it, where it is with most reason believed that Hell is situated, in order that it may be, in all its parts, equally distant from the empyrean heaven: "They are all delivered unto death to the lowest parts of the earth."[1] Therefore, even if these wretched beings had no other punishment to undergo than that of being shut up for ever in a pit so deep, and consequently so fetid, foul, and dark, how great a one would it be! Imprisonment for life, even in this world, is esteemed a penalty equivalent to death, even though sometimes a comfortable house or a convenient room is allowed to the prisoner. What, then, will it be to dwell in the most horrible dungeon that the mind can conceive? For, if Hell is situated in the centre of the earth, it must be the great sewer of the human race, into which, therefore, flow all the offscourings of the world, which, great as they now are, will be incomparably greater after the Day of Judgment; for in the universal purification of the elements which will then take place, rendering the surface of the earth clear as glass, the water like crystal, the air like the ether, the fire like the stars, all the impurities of every kind which had till then obscured them will, after this process of renovation, be cast down like rain upon the damned to intensify their torment. This is why Hell is so frequently called a pit in Scriptures: "Thou art gone down to the bottom of the pit."[2] "They shall be gathered together, as in the gathering of one bundle into the pit,"[3] because all the filth which flows down thither will not be able to pass off, as it does on the earth, but must needs

[1] Ezech. xxxi. 14. [2] Isaias xiv. 19. [3] Isaias xxiv. 22.

remain there. Judge, then, whether it will be possible to dwell in this sink of Hell! Yet so it is; in this foul and fetid pit the damned will have to dwell for endless ages, as in a dungeon, hewn not out of any rock, but out of the very depths of the earth, for which reason, after it is said that "they shall be gathered together in the gathering of one bundle into the pit," it is immediately added, "and they shall be shut up there in prison." Alas for thee, if ever thou art taken to that fatal prison! But there is one thing from which thou mayest now take comfort, namely, that no one is ever taken there by force. Whoever goes there, goes by his own choice. Observe, therefore, that whereas it is said of the reprobate that "they shall be delivered into the hands of the sword," and that "they shall be the portions of foxes," it is not said that "they shall be dragged down into the lower parts of the earth," but only that "they shall go" there; because, granted that when they are once in Hell, they must, whether they will or no, endure all the punishments which are there in store for such as they, yet, after all, they need not be there, since it is a matter of their own free choice to enter or not to enter it. All that is required is to keep from sinning, or having sinned, to repent and amend their lives. If thou committest an offence against thy sovereign, he puts thee in prison whether thou repentest or not. But God does not deal thus with thee. He only puts thee in prison when thou sinnest against Him and refusest to repent. Is it not plain, therefore, that if thou art damned, it is only because thou choosest to be damned: "They shall go into everlasting punishment."[1]

III. Consider thirdly, in the second place, that the damned will be "delivered into the hands of the sword," which means that they will all be given up to the judgment of God, which will be, as it were, an unsparing sword eternally slaughtering them: "Flee from the face of the sword, for the sword is the revenger of iniquities: and know ye that there is a judgment,"[2] that is, "know that this sword is the judgment of God." Who, then, can possibly describe in any degree such a sword as this? It is a sword which will pierce, and rend, and inflict all dreadful wounds upon the damned: "This is the sword of a great slaughter, that maketh them stand amazed," that is, in bewilderment at miseries which they had never thought possible, "and languish in heart,"[3] that is, with sorrow. Besides, this will be a two-edged sword, in order to

[1] St. Matt. xxv. 46.   [2] Job xix. 29.   [3] Ezech. xxi. 15.

wound more cruelly—"Let the sword be doubled"[1]—one edge will smite the body with the pain of sense, and the other the soul with the pain of loss. And if a sword is to be dreaded in proportion to the strength of the arm that wields it, think what blows will be dealt by this sword which is wielded by an omnipotent King! Who can withstand such a sword as this? And therefore it is said that the damned "shall be delivered into the hands of the sword." It will, then, be free to execute the necessary slaughter according to the ample powers with which God invested it when He said, "Be thou sharpened; go to the right or to the left, which way soever thou hast a mind to set thy face."[2] Only two ways of deliverance might be possible from this sword: the one, that God should one day put it into the scabbard; the other, that the damned might escape from it by flight. But the first of these cannot be hoped for in any way, since God has spoken expressly on this point, saying, "That all flesh may know," that is, all who, for their sins, are in the prison of Hell, "that I, the Lord, have drawn My sword out of its sheath, not to be turned back."[3] Neither is the second way possible, for whithersoever the damned may turn to flee from their dungeon, they will everywhere see the same sword before them; and, therefore, there cannot be one among them who "believeth that he may return from darkness to light, looking round about for the sword on every side."[4] The only help is in fleeing now, when this sword does not yet strike, but only gleams to terrify men into flight. "Thus saith the Lord God: say, The sword, the sword is sharpened and furbished. It is sharpened to kill victims; it is furbished that it may glitter."[5] If, indeed, God purposely now hid the sword from thy eyes, thou mightest think that He desired to use it for thy hurt; but since He makes it glitter by as many blades as are the tongues of His ministers, for this very reason that thou mayest see it from afar, it is thy own fault if thou dost not escape from it betimes. Hear how repeatedly preachers busy themselves in continually crying from the pulpits: "The sword, the sword!" If, then, thou believest these things, what more is necessary to make thee amend thy life?

IV. Consider fourthly, that it is said in the third place, that the damned "shall be the portions of foxes." All commentators understand by these "foxes" the devils whom we

---

[1] Ezech. xxi. 14.   [2] Ezech. xxi. 16.
[3] Ezech. xxi. 5.   [4] Job xv. 22.   [5] Ezech. xxi. 9, 10.

in this world too much honour by esteeming them lions and wolves, but who will be clearly seen in Hell rather to be foxes who have conquered us, not by strength, but by cunning : "And there came forth a spirit, and stood before the Lord, and said, I will deceive him."[1] The damned, then, will be the portions of foxes, which are the worst of all, for they will be given up as a prey to these executioners, the devils, who are the more hateful from having themselves occasioned the sin which they afterwards punish. Pause now for a while, and think what it will be if ever (which God forbid) it should be thy lot to see in Hell how those same devils who were in thy lifetime such alluring tempters, continually flattering and enticing thee, and inviting thee to sin by a thousand fair promises, are changed there into such cruel, fierce, and inexorable tormentors. "Ah, accursed, wicked, and malignant foxes!" wouldst thou say to them with implacable fury ; "these, then, are the fair promises of happiness which you made me under the guise of faithful friendship !" "I called for my friends, but they deceived me."[2] But since all such reproaches will then be in vain, open thy eyes now, and be not deceived ; for those same devils who are now thy special tempters will, if thou sufferest thyself to be seduced by them, be assigned to thee in Hell as thy special tormentors ; for God knows perfectly both them and thee : " He knoweth both the deceiver and him that is deceived."[3] And so, for this reason also, it is said that the damned are "the portions of foxes" in Hell, because the devils will take each his share of that wretched crowd, just as pirates do with those whom they have made captives. What an infernal festival will that be ! "They shall rejoice . . . as conquerors rejoice after taking a prey, when they divide the spoils."[4] And it will be a truly infernal festival, because it will be impossible to say whether joy or grief predominates. On the one hand, it might be thought that it will be joyful, because of the hatred borne by the devils to the damned souls which they have to torture ; and on the other, it will be full of sorrow, because of the still greater hatred which they bear to the Divine justice whose executioners they are compelled to be, in order to render it more glorious. So that it will be a festival of pure fury which will burst forth in terrible rage upon the damned, and most of all upon those who were the most caressed by the devils who tempted them on earth.

[1] 3 Kings xxii. 21.　　[2] Lament. i. 19.　　[3] Job xii. 16.　　[4] Isaias ix. 3.

V. Consider fifthly, how, in these words on which thou hast been meditating, are seen combined the three things which concur in making Hell so terrible : the depth of the abyss—"They shall go into the lower parts of the earth ;" the severity of the puhishment—"they shall be delivered into the hands of the sword ;" and the company of the devils—"they shall be the portions of foxes." Doubtless these three miseries will be alike common to all the damned, yet will they torment each one in a measure proportionate to his sins. We must, therefore, assume as certain that the damned will take with them into Hell the irregular affections which they had on earth : "They went down into Hell with their weapons."[1] And if this is so, what a torment will it be to those who in their lifetime were always seeking high places, desiring to rise, to command, when they find themselves cast down into those abysses of darkness "in the lower parts of the earth." No longer, then, will these miserable men be able to exalt themselves when God will say to them as He did to Lucifer, "Thy pride is brought down into Hell."[2] And what a torment will it be to those who have too greatly loved their pleasures to see themselves condemned to so cruel a slaughter of body and soul as that will be which is wreaked on them by the sword of Divine justice when "they shall be delivered into the hands of the sword." It is a sword which expresses in one word every kind of torture, because whatever orders it may receive from the will of God, those will it be most ready to execute. And therefore, whereas men hold the sword in their hand whenever they use it, God is said to hold it in His mouth—"From His mouth came out a sword."[3] "I will fight against him with the sword of My mouth"[4]—because it is certain that men have to endure great labour if they would wield the sword valiantly ; but God has only to speak. What, then, will it be to be continually driven at the point of such a sword ? And what will it be, above all, to those who in the world were continually sucking the blood of the poor, and overreaching them by numberless frauds and artifices, to find themselves at last given over as a prey to the devils, who are, as it were, foxes—nay, far worse ? "Woe to thee that spoilest : shalt thou not thyself also be spoiled ?"[5] From what has been said, thou canst perceive how true it is that the nature of the punishment corresponds to that of the

---

[1] Ezech. xxxii. 27.   [2] Isaias xiv. 11.
[3] Apoc. i. 16.   [4] Apoc. ii. 16.   [5] Isaias xxxiii. 1.

sin: "I will visit upon you according to the fruit of your doings,"[1] in order that not only the severity but the wisdom of God's justice may be the more manifest. "The Lord of hosts is thy name, great in counsel, and incomprehensible in thought: Whose eyes are open upon all the ways of the children of Adam, to render unto every one according to his ways."[2]

———

TWENTY-EIGHTH DAY.

ST. AUGUSTINE.

The fire of the Lord is in Sion, and His furnace in Jerusalem
(Isaias xxxi. 9).

I. Consider first, that the fire here spoken of is Divine love, which is, in truth, here in our Sion, that is, the Church Militant, but the furnace of which is not here; for that furnace, glowing with fervent heat, is above, in the Heavenly Jerusalem, the Church Triumphant: for it is there that God is truly loved. We on the earth can hardly boast of loving Him in any degree. Nevertheless, observe that these words of the Prophet show that our fire is not different in kind from that of Paradise; for otherwise, as St. Thomas remarks, it could not be called the fire of that furnace: "The fire of the Lord is in Sion, and His furnace in Jerusalem." But if it is not different in kind, it is far inferior in perfection, even as a fire which is out of the furnace, that is, out of its proper place. What, then, should he do who desires to love God with a right love? To strive, as far as possible, to conform it to the blessed in Heaven. This is what St. Augustine did: and thus it was that his love was so perfect. There are five qualities in which, if we consider aright, the love of the blessed to God surpasses ours: it is pure, recognizable, great, inextinguishable, and unchangeable. He, therefore, who endeavours to imitate it, will also have a fire resembling if not equal to that which is in the furnace of which I have been speaking, such as St. Augustine had. Beg of God, if, of thyself, thou art not able to rise to the understanding of the qualities belonging to this heavenly fire, to vouchsafe to bestow on thee at least some spark of it from above. So shall experience alone supply the

---

[1] Jerem. xxi. 14. [2] Jerem. xxxii. 18, 19.

place of all explanation : "From above He hath sent fire into my bones and hath instructed (*erudivit*) me."[1]

II. Consider the first quality of the fire of Paradise, which is that it is pure in its material, which is God only. The blessed love God, and we love Him ; but the blessed love nothing but God, and we love ourselves as well as God, loving ourselves, if not more than God, at least together with Him, and, in consequence of this mixture, our love becomes less noble, and as fire would which was made of cinnamon and chips of wood. The heat of a fire corresponds to the material of which it is made : "As the wood of the forest is, so the fire burneth."[2] And therefore the fire of Paradise could not possibly be nobler than it is, because, as the blessed love nothing whatever but God, so neither do they even love themselves out of God, although they love God in themselves. And this, because the love of God is so strong in them that, in the end, it destroys every other love : "He is a consuming fire."[3] See, therefore, how thou hast to purify the love that thou bearest to God, by loving Him alone, in such a way as to love no creature directly out of Him, but Him in all creatures. He does well who loves creatures in God, for this is to love the fruit in the plant ; but he does better who only loves God in these creatures, for this is to love the plant in all its fruits. He who does this has a pure fire, because he feeds it with the purest possible material, which is God, and God alone. This was done by St. Augustine, who, from the time when he gave himself up to the true love of God, was able to love nothing in any creature but Him Who created it : "Whenever we leave God," he was wont to say, "to love anything which comes from God, it is wrong love."[4]

III. Consider the second quality of the fire of Paradise, which is that it is very easily recognizable. For in Heaven those who love God know that they love Him, and show all others that they love Him, so that not only the hearts, but even the faces of the blessed were compared by Ezechiel to "burning coals."[5] Here, on the contrary, not only is our fire hidden from those who see us, but even from ourselves, who yet have it always in our breasts ; because we love God, but are not certain, at the same time, that we do love Him, so thickly is that fire covered, or, we might rather say, choked with ashes. It is true, indeed, that if our fire is not recogniz-

---

[1] Lament. i. 13.   [2] Ecclus. xxviii. 12.   [3] Deut. iv. 24.
[4] *Conf.* lib. iv. cap. 12.   [5] Ezech. i. 13.

able in itself, like that which is in the furnace of Jerusalem, it does not fail to manifest itself sufficiently in its effects. So that, in the long run, it is very possible to distinguish who the just are who really love God on earth. See, then, what thou hast to do in order that thy fire may, as much as possible, resemble in brightness, also, that of the blessed. Thou art not to keep it purposely hidden in thy heart, as though it were something to be ashamed of to be recognized amongst thy fellows as one of those who profess to love God. Nay, if thou canst not make it evident that thou lovest Him, at least make it evident that thou professest to love Him by overcoming all the vain human respect which keeps thee back from doing so. No sooner had St. Augustine given his heart to God, than he began to make open war on all his enemies by rooting them out: so he fought against Manichæans, Pelagians, Priscillianists, Arians, and countless others; neither could he be satisfied with keeping this fire confined to his city of Hippo, as to one little corner of the world, but made the flames of it spread throughout all Africa.

IV. Consider the third quality of the fire of Paradise, namely, that it is very great in quantity, whereas ours is so scanty that, contrasted with it, it is like the fire from a handful of coals compared with that of Etna. Neither is this wonderful, for the love of God is in proportion to the knowledge of Him. Our knowledge of God here is only partial: "We know in part," and so our love too is only partial. In Heaven we shall know Him fully, and therefore also we shall love Him fully: "When that which is perfect is come, that which is in part shall be done away."[1] What, then, shouldst thou do, in order to love thy God as much as possible on earth? Strive, like St. Augustine, to know Him also as much as possible on earth. Think often of His sublime attributes, contemplate them, dwell on them, read about them: beg of God to vouchsafe to manifest Himself to thine eyes, as He often does to the eyes of those who serve Him faithfully. But if thou takest no pains to know Him, what wonder is it that thou lovest Him so little? "In my meditation a fire shall flame out."[2] This may have been the chief reason why St. Augustine loved God with his will more than so very many other saints, who did indeed love Him, but not with the same intensity: because he was always striving to know Him better by his understanding.

[1] 1 Cor. xiii. 9.   [2] Psalm xxxviii. 4.

V. Consider the fourth quality of the fire of Paradise, which is that its heat is inextinguishable, and this, because it is a fire which is in its furnace. Not so ours, which is continually going out, and for two reasons : sometimes because it lacks air, sometimes because it is damped by water. By air are signified those spiritual aids which we are obliged to make use of from time to time to revive it ; and by water, the carnal concupiscences which are always tending to stifle this fire, as being the greatest enemy that they have upon earth. And would that they did not so often succeed ! But in Heaven these two evils find no place, and therefore the fire is sure to burn for ever there. There are no floods of water there, for the flesh will be not only subject to the spirit, but conformed to it ; neither is there any need there of spiritual aids, because the fire being in its proper sphere will need no fanning, as it does on earth. Knowing, then, how apt thy fire is to go out, what is to be done but strive, at any cost, to keep it alive? So will thy fire be like that of the blessed in Heaven, because it will be an undying fire, as it may well be said was that of St. Augustine on earth, who, from the moment it was enkindled in his heart, never suffered it to die out, ever striving both to repress the carnal concupiscences by which he was once governed, and to avail himself more and more of the highest spiritual aids : "The fire on the altar shall always burn."[1]

VI. Consider the fifth quality of the fire of Paradise, which is that it is a fire, the flame of which is not only inextinguishable, but steady, fixed, and lasting ; not liable, like ours, to increase and diminish, because, being a fire in its sphere, it is perfectly tranquil. Ours is restless, because it is striving after its sphere, and so is always in motion, both because it moves, and because it is moved. Here the love of God should be a matter of merit, and not of reward, and consequently it is necessary for it to be always in movement in order to act, and never to be at rest: "Fire," so long as it is here, "never saith, It is enough."[2] In Heaven, on the contrary, it is a matter of reward not of merit, and so there it does not act, but rests from action, and is altogether occupied in the enjoyment of its beloved object. Besides, here on earth, it is subject to change of state from many persons, who therefore may be said to move it. In Heaven it can be disturbed by no one. Now, since thou neither canst nor

[1] Levit. vi. 12. [2] Prov. xxx. 16.

oughtest to expect such a love as this on earth, do thou at least endeavour, like St. Augustine, to prevent it from having any motion but that which is natural to fire, namely, that of soaring upward by continual aspirations.

VII. Consider seventhly, that there is, in contrast with this furnace of Jerusalem, the furnace of Babylon, in which the three children, who were types of the just, remained unharmed; whilst the Chaldeans, who were types of the wicked, were consumed like straw. This is the furnace of self-love, which is opposed to Divine love, and it is from this that each of these miserable men kindles his fire, although in each case it is different, according to the variety of the false goods which they love more than God. If, then, thou considerest all these fires well, thou wilt find that they are kindled with three kinds of fuel : the first being fed with ordure, the second with vine-shoots, and the last with dry wood. The first is that of the luxurious, the second of the ambitious, the third of the avaricious. The luxurious love their unclean sensuality more than God, and therefore theirs is a fire which is fed with dung, and which imparts as much infection as warmth to him who bears it in his bosom, a fire which has no brightness, and only disgusts all who approach it with its stench. Ambitious men love their vainglory more than God, and so their fire may be compared to one made of vine-shoots, which is bright enough to look at, but of brief duration : "I passed by, and lo, he was not."[1] Avaricious men love the money which they keep in a chest more than God, and therefore their fire may be described as one made of dry wood, which lasts for a while, but is of no use. Dost thou think, then, that the furnace of Babylon is better than the furnace of Jerusalem? Alas ! there is nothing to be done, as regards the furnace of Babylon, but to pass from it to that of Hell, in which whosoever burns, burns with a fire, not of love, but of fury : fury against God, against the devils, against the damned, against himself. Whosoever loves God more than himself in this world will never change his love to all eternity, because its fire is the same as that in the furnace of Jerusalem ; he will only bring it to perfection, so that all that is painful will vanish, leaving only beatitude. Whoever loves himself more than God will change his love into fury ; so that he will curse his fate as often as he remembers his existence.

[1] Psalm xxxvi. 36.

TWENTY-NINTH DAY.

DECOLLATION OF ST. JOHN THE BAPTIST.

If thou wilt separate the precious from the vile, thou shalt be as My mouth (Jerem. xv. 19).

I. Consider first, the first meaning of these words, which is, that if thou duly separatest the precious from the vile in thyself, by ascribing to God what thou hast from Him, which is all that is precious in thee, and to thyself what thou hast from thyself, which is all that is vile in thee, thou shalt be as the mouth of God Himself, because by so doing thou wilt always speak the truth: "If thou wilt separate the precious from the vile, thou shalt be as My mouth." What is meant when every man is said to be false?—"Every man is a liar."[1] It is that he does not make this separation which is so entirely reasonable. He ascribes to himself that which is not his: "Their mouth hath spoken proudly."[2] What hast thou of thyself that is precious? Noble birth, intellect, talent, riches, wisdom, health, beauty? All these things are from God; of thyself thou hast nothing. And as all gifts of nature are from God, much more are all gifts of grace from Him, for which reason, indeed, they are called gifts. Of thyself thou hast nothing but sin. This is a truth which thou dost not rightly understand, and therefore thou so often speakest lies by boasting. Make the right separation by always ascribing to God what belongs to Him. This has always been the language of the saints: "The Lord our God slew him by the hand of a woman. . . . The Lord hath killed the enemy of His people by my hand this night."[3] In this way did the valiant Judith always speak when she had to say that she had slain wicked Holofernes. Never did she presume to say, "I slew him," for she saw that it would be greatly wronging God if she had in any way ascribed to herself the issue of her noble deed. But thy conduct is the opposite of this: thou ascribest to thyself all the good that thou doest, and what dost thou reserve to God? Very often the blame of the sin thou hast committed. For when thou sinnest, instead of ascribing it to the malice of thy will, thou ascribest it to the evil nature which God has given thee, to thy weakness, to thy concupiscence, to the difficulty of the Gospel precepts, which thou thinkest made on

[1] Psalm cxv. 2. [2] Psalm xvi. 10. [3] Judith xiii. 18, 19.

purpose to render the glory of Paradise difficult of attainment : "Behold Thou dost cast me out this day from the face of the earth."[1] Do not speak thus ; acknowledge from thy heart that if there is any particle of good in thee, it is not thy work, but God's, that of thyself thou doest only evil. Thus, making the separation in the right way, thou shalt be like the mouth of God Himself, because thou wilt always declare the infallible truth that all the good is from God, and all the evil from thyself : "If thou wilt separate the precious from the vile, thou shalt be as My mouth."

II. Consider secondly, the second sense of these words, which is, that if thou exercisest a wise judgment by separating what is worthy of esteem on the earth from what is worthy of contempt, thou shalt be as the mouth of God, because thou wilt always speak the language of God, which is a right language : "All thy words are just," Wisdom cries aloud.[2] Thou wilt never speak the language of men, which is most perverse. What is the language of men ? They call that man happy who possesses great wealth, who has the government and arrangement of affairs, who takes his pleasure : "They have called the people happy, that hath these things."[3] What is the language of God ? He calls that man happy who has laid up all his bliss in Him : "Blessed is the nation whose God is the Lord."[4] Is this the language of the present time ? Oh, how likely is it that it is rather like the low speech of men : "Thou shalt be brought down, thou shalt speak out of the earth, and thy speech shall be heard out of the ground."[5] Thou must learn how to make the due separation in thy mind between what is precious and what is vile. There is only one good upon earth worthy of esteem, and that is the grace of God. All others in themselves are utterly valueless, and he who possesses them "is as it were rich, when he hath nothing."[6] If they have any value, it is only because they may be despised in order to acquire this grace. How is it, then, that thou art so slow to understand a truth which is so certain, and art not ashamed so often to take gold for dross and dross for gold ? How base is the language of those who so extol the goods of this world, admiring the persons who possess them, and praising those who pursue after them, and utterly disdaining the man who hides the rich treasure of Divine grace beneath the rags of poverty ! And all the while the

---

[1] Genesis iv. 14. [2] Prov. viii. 8. [3] Psalm cxliii. 15.
[4] Psalm xxxiii. 12. [5] Isaias xxix. 4. [6] Prov. xiii. 7.

former appears to be rich and is really poor, and the latter, though he appears to be poor, is exceedingly rich : "One is as it were rich, when he hath nothing, and another is as it were poor, when he hath great riches."[1] Make this separation, which is all important ; if once thou makest it in thy mind thou wilt very soon make it also in thy words, and so thou wilt become like the mouth of God, which speaks of things as they are in reality, not in appearance : "If thou wilt separate the precious from the vile, thou shalt be as My mouth."

III. Consider thirdly, the third sense of these words, which is, that if thou art occupied in bringing souls out of sin, so separating the precious from the vile, thou shalt be as the mouth of God Himself, because He will speak by thy mouth, making use of thee as a mediator to recall to Himself those who have turned their backs on Him. This is the office which is discharged by all those who are employed in withdrawing others from evil, the office of God's ambassadors. How pleasing, then, must it be to God if it is faithfully discharged ! It is the office which Jesus performed on earth ; He acted as His Father's mouth : "As the Father hath taught Me, these things I speak."[2] It is the office also which was discharged by the Apostles, and all their lawful followers ; they, too, acted as the mouth of Jesus : "For Christ, therefore, we are ambassadors, God, as it were, exhorting by us."[3] Not that this is the only reason why God says that the man who so does shall be "as His mouth." There is another and more sublime reason, and it is this, that the man who discharges this office imitates by his words the sovereign power which belongs to the mouth of God. The mouths of men have the power of speech, but not of action ; the mouth of God both speaks and does : "He spoke, and they were made."[4] Thou knowest that God, merely by the power of His word, created all things out of nothing. Now, there is only one work of which it can be asked whether it is greater or less than the creation of the world. What is this work ? It is the justification of the wicked. St. Augustine teaches that it is greater, not by reason of the mode of operation, but by reason of the excellence of the work, because the creation of the world was in order to a natural, and justification to a supernatural good. If, therefore, thou separatest the precious from the vile, by taking souls out of the sin in which they lie buried more deeply than the universe lay in its nothingness

[1] Prov. xiii. 7.　[2] St. John viii. 28.　[3] 2 Cor. v. 20.　[4] Psalm clxviii. 5.

before the creation, thy mouth will become like the omnipotent mouth of God ; for if His work then surpassed thine in
the mode of operation, when He drew all things out of nothing
without their co-operating in the smallest degree in issuing
from that state, thine now surpasses His in the excellence of
the work.   And it is to this, according to the judgment of the
same Saint, that Christ intended to allude when He said :
"Amen, amen, I say to you, He that believeth in Me the
works that I do, he also shall do, and greater than these shall
he do."[1]   Art thou not even yet inflamed with an earnest
desire of being able thyself, according to thy state, to separate
a beautiful pearl from corruption, which is, in other words, to
save a soul from sin ?   What greater inducement could God
give thee to do this than by actually saying, "If thou wilt
separate the precious from the vile, thou shalt be as My
mouth?"   Thou canst now easily see, if thou wilt, how rightly
these words apply to the great Forerunner, St. John, whose
decollation we commemorate this day, who did indeed
separate the precious from the vile in all the three ways above
mentioned : for he always spoke in the loftiest terms of Christ,
and in the lowest of himself; he boldly despised worldly
pomp, even in the palace of the King, and not only in woods
and caves, where he announced the Kingdom of Heaven to all
men as the only thing to be valued ; and he employed his
whole life in bringing both unbelievers and hardened sinners
to repentance.   Well, then, may he be said to have been as
the mouth of Christ on earth—nay, he was His very voice :
"I am the voice."[2]   And if that voice was at last silenced on
this day, the reason was because he had raised it so power-
fully in his efforts to bring the lovers of sensuality more out
of the mire in which they were lying.

THIRTIETH DAY.

*Jesus, that He might sanctify the people by His own Blood, suffered
without the gate. Let us go out therefore to Him without the camp, bearing
His reproach* (Hebrews xiii. 12).

I. Consider first, that Christ our Lord did not die within
the city of Jerusalem, but outside, on a hill destined for the
public execution of malefactors: "Jesus, that He might

[1] St. John xiv. 12.　　[2] St. John i. 23.

sanctify the people by His own Blood, suffered without the gate." He ordained it thus for three reasons. First, for the greater encouragement of those who should be willing to avail themselves of His death, since He hereby showed that He did not die merely for the private benefit of those who lived within the circuit, ample though it was, of those walls, but for the public benefit of the whole human race : for which reason it was commanded in the Old Law that the victim whose blood had been offered in expiation for all the people, should never be burnt within the enclosure of the tents (as those constantly were, the blood of which had been offered in expiation for some individual), but beyond its limits—"without the camp."[1] Secondly, for the greater terror of those who were not to profit by it. For it is beyond doubt, that a public execution, especially when it is also very terrible in its nature, strikes far more terror than a private one. But what execution was ever so terrible as this : when not a common man, not a citizen, not a consul, not an earthly king, but—O horror of horrors !—the King of Heaven, was stripped naked, and nailed with cruel nails upon a cross for crimes with which He had never been soiled, but only seemed to be so? So horrible an execution as this, even had it taken place, not in a public place in Jerusalem, but in a well-guarded and secret tower, was enough to astound the world, even those who had not witnessed it, by the mere report. Far more, then, must this be so, when it took place, not merely in the square of a city, but on a mountain accessible to a host of people who were able to come from all sides to behold it. Ought not every sinner to infer from this the fearfulness of the punishment which the Divine wrath would have wreaked on him ? "If in the green wood they do these things, what shall be done in the dry ?"[2] Thirdly, for the greater shame of Christ Himself, Who chose, not only to be fed, but to be satiated with insults : "He shall be filled with reproaches."[3] Was it not, then, disgrace enough to die within the walls of a city so famous and populous, and especially at that time of the Paschal festival, so crowded ? Surely it was. But this was not enough for Christ ? And, therefore, even as he preferred Bethlehem to Jerusalem for His birth-place, and Jerusalem to Bethlehem for the scene of His death, so out of all the parts about Jeru-salem, He chose for Himself the most disgraceful and infamous, namely, Calvary : a place very near the city, which

[1] Levit. xvi. 27.   [2] St. Luke xxiii. 31.   [3] Lament. iii. 30.

was reached through that gate the very name of which showed
its vileness, being called the Dung Gate. Out of this gate,
therefore, behold thy Jesus coming forth between two thieves,
bearing a heavy Cross on His shoulders, whilst the immense
crowd assembled there in broad noonday, greeted Him with
jeers instead of drums and trumpets. Go, if thou canst,
after such a spectacle, take thy fill of that worldly glory which
thou so greatly covetest.

II. Consider secondly, that such was not the desire of the
Apostle when he said, "Let us go out, therefore, to Him
without the camp, bearing His reproach." This is the true
consequence to be drawn from the generosity of our Lord, not
that of studying to be in high esteem. But what is this
"reproach" of which the Apostle here speaks? Literally, it is
the name of Christian, which was, in his time, a name of
derision, as signifying the follower of one Who had been seen
so short a time before dying the shameful death of the Cross
on Calvary, and Whose victories were not yet known. It was
to the bearing this name courageously "without the camp" of
all human respect, even before tribunals, synagogues, senates,
and royal courts, that the Apostle exhorted the converted Jews,
who hesitated to relinquish accredited legal observances from
shame at openly showing their allegiance to a despised Legislator: "I am not ashamed of the Gospel."[1] In our days there
is no longer any reproach in the mere name of Christian, but
there is in the name of practical Christian, a Christian who
is poor, chaste, patient, and mortified; for all are ready to
ridicule such a Christian as this: "The simplicity of the just
man is laughed to scorn."[2] This is the reproach that thou
hast to bear. Therefore observe that the Apostle does not
here say, "Let us go out to Him without the camp, bearing
His shame," but "His reproach," for this is the most difficult
thing of all, to have to hear with thy own ears the mockeries
of those who jeer at thy manner of life, to endure, and even
glory in them. Yet it is to this that thou hast to encourage
thyself if thou wouldst return with thy love all that Christ
vouchsafed to suffer for love of thee. Consider a while what
a reproach it was that He bore when, as He staggered beneath
His Cross, He heard so many pouring forth all kinds of bitter
words against Him, and there was not even one in all the
crowd who ventured to defend Him! One man would revile
Him as a false prophet, another as a hypocrite, another as one

[1] Romans i. 16.   [2] Job xii. 4.

possessed by a devil, another as a disturber of the people ; and
yet He never shrank from bearing this public disgrace to the
very end, although He was able to cover with shame and give
the lie to all those ribalds by immediately putting forth His
power and silencing them by a miracle. Why, then, dost thou
hesitate any longer to "go out to Him, without the camp" of
thy shameful reservations? It is not enough to bear thyself as
a true Christian within the private walls of thy chamber or thy
house: thou must go out openly "beyond the camp." And if
men deride thee for so doing, let them deride thee : thou wilt
be derided with Christ.

III. Consider thirdly, that this is precisely why the Apostle
was not content to say, "Let us go out beyond the camp,
bearing His reproach," but chose to add "to Him," because
in these words lies the whole encouragement. Imagine that
thou beholdest thy Lord coming out of the gate of Jerusalem,
bearing the reproach which has just been described. What
must thou do to return such an excess of charity? Shouldst
thou wait for Him to send to summon thee from the enclosure
of those walls into which thou hast retired, as though thou wert
ashamed of the name of Christian which belongs to thee?
Thou oughtest, on the contrary, to hasten of thy free choice
to meet him with all possible alacrity and gladness, unheeding
all that may be said of thee by idle babblers. And the Apostle
did not choose to say more than this, "to Him," in order to
leave free scope to affection. By saying "to Him," he says
everything : to follow Him, to honour Him, to glorify Him in
every possible way. He says "to follow Him," as they do
who choose to forsake the world in order to give themselves
up to that perfect following of Him which consists in the
keeping of the three evangelical counsels : "Leaving all things,
they followed Him."[1] He says "to bear Him company," as
those do who, while thus following Him, keep closest to Him
by the denial of all their appetites, great and small, being
firmly resolved to die with Him on His Cross: "Let us also
go that we may die with Him."[2] He says "to preach Him,"
as those do who carry His name to others who either do not
know Him, or do not care about Him, and who are not
ashamed everywhere to preach Jesus, and Jesus Crucified :
"The Jews require signs, and the Greeks seek after wisdom ;
but we preach Christ crucified, unto the Jews indeed a
stumbling-block, and unto the Gentiles foolishness."[3] He

[1] St. Luke v. 11.   [2] St. John xi. 16.   [3] 1 Cor. i. 22, 23.

says "to confess to Him," as those do who not only uphold His name in the pulpit, but even before tribunals, and in prison, and to death, and that by the most fearful tortures : " I am set for the defence of the Gospel."[1] He says "to honour Him," as those do who pay Him due honour, at least among the rest of the faithful, frequenting the churches with especial reverence, often receiving the holy sacraments, praying, reciting psalms, offering sacrifice, and giving to His worship the high homage, which it merits. He says, lastly, "to glorify Him" in every possible way, as they do who are never weary of promoting His glory, in whatever manner they are able to do so, of advancing it in their own souls, of conveying it to the souls of others, either by life or death, without any other consideration but what will most redound to the honour of Christ : "In nothing I shall be confounded, but with all confidence as always, so now also shall Christ be magnified in my body, whether it be by life or by death."[2] Now, it is quite certain that in all these ways of going to Him, we must go with a mind prepared for every sort of contempt, otherwise we have done nothing. And therefore the Apostle says, " Let us go out to Him," but it must always be " bearing His reproach." For if thou desirest to go out "to follow Him," thou must bear the reproach which will fall on thee from thy closest friends, who will tell thee that thou art mad to forsake the world in the bloom of thy years, and at the height of the friendships thou hast formed, and that thy doing so will seriously injure thy family. If thou desirest to go out "to bear Him company," thou must bear the reproach which will fall on thee from those who live with thee, and who will tell thee that thou wouldst fain do more than others, whereas thou hast less strength than they. If thou desirest to go out "to preach Him," thou must bear the reproach which will fall on thee from those who will laugh at thy style as lacking grace, point, sublimity, and depth of doctrine, and who will leave thee to hear others who speak more to the ear than the heart. If thou desirest to go out "to confess Him," thou must bear the reproach of those who will mock thee for meditating a journey to the Indies when thy stock of virtues is so small, and for venturing to cross the seas and brave the spears of the Japanese when thou art not able to endure the stings of mosquitoes in thy cell. If thou desirest to go out " to honour Him," thou must hear the reproach which will fall on thee

[1] Philipp. i. 16.   [2] Philipp. i. 20.

from those who, when they see thee behaving thyself with
unusual devotion in church, confessing, communicating, and
performing other acts of piety, will say that thy object in all
this is to obtain by means of sanctity that credit which thou
art not able to attain by means of thy talents. Lastly, if thou
desirest to go out "to glorify Him" in all ways that are possible
to thee, it is in this matter more than all besides that thou hast
to be on thy guard. Thou wilt have to bear reproaches of
every sort, and to hear even the most forbearing say that thy
zeal exceeds thy prudence. Think, at such times, of thy Jesus
coming out of the meanest gate of Jerusalem arrayed in so
glorious a robe of scorn, and say to thyself, "Jesus, that He
might sanctify the people by His own Blood, suffered without
the camp;" and shall I be coward enough to remain within
my entrenchments? Far be this from thee: "Let us go therefore to Him without the camp, bearing His reproach." But
when, after all, wilt thou have to bear "His reproach"? It
will be much, if thou ever hast any reproach to bear for Him
which deserves so much as to be called like His.

THIRTY-FIRST DAY.

'I know thy works, that thou art neither cold nor hot. I would that thou
wert cold or hot. But because thou art lukewarm, and neither cold nor hot,
I will begin to vomit thee out of My mouth (Apoc. iii. 15).

I. ¹Consider first, that this tepidity which is so hateful to
God is without doubt that which is found in the service of
God. And if this is so, it is plain who, according to the best
interpretation, are the "lukewarm" here mentioned, and who
are the cold and the hot. Those are cold in the service of
God who, never having received light to know its hidden
treasures, have never felt inflamed to embrace it; those who
are hot are those who, having embraced it, are, as they ought
to be, very fervent in paying it. The tepid, or lukewarm, are
those who do indeed render God service, but a service that is
very lax. It is necessary, therefore, in order to attain to a
right understanding of this passage, which is not very easy,
to observe closely that there are two causes of tepidity: one
when things pass from cold to hot, the other when they change
from hot to cold. Now, it is very certain that God is not here

speaking of those who, having left their state of coldness, are indeed as yet tepid in His service, but only because they are in a state of transition from cold to hot. Such persons, in spite of their tepidity, are progressing towards a good state, and consequently cannot be displeasing to God. He is speaking of those who have degenerated from their first fervour, and are in a state of transition from hot to cold. And such persons are not only displeasing, but hateful to God, by reason of the foolish resolution they have made. Of which number art thou? If of those who are passing from cold to hot, take courage to effect quickly so desirable a change : "Take courage and do it."[1] But if thou art one of those who are passing from hot to cold, alas for thee! Tremble and fear, for thou art in the fatal number of those tepid persons in detestation of whom God here exclaims, "I would thou wert cold or hot," so greatly does He abhor thy state.

II. Consider secondly, that this expression, "I would," appears to present much difficulty. For if to be "cold" means not to have known and embraced the service of God, how is it possible that a state of coldness can be preferred by God to one of tepidity, which means having once embraced it with great fervour, and afterwards becoming negligent in it? But it is not said that coldness is preferable to tepidity in God's sight, but only that it is less vexatious. For observe that by the words, "I would," God does not here express the desire of a positive, but of a negative good, that is, in clearer words, of a lesser evil, such as was the desire of those who said of old, "Would God we were sold for bondmen and bondwomen."[2] "Oh, that I had been consumed that ye might not see me!"[3] And it is a less evil never to have known and embraced God's service, than to have embraced it with great fervour and then grown negligent in it. I say this, because if it is not a less evil in itself, it is less in the consequences it involves. For if thou supposest tepidity ever to be an abiding state, thou art greatly deceived. It is a state in which no one can possibly continue, even if he wishes to do so, but must necessarily become worse and worse till he perishes. A vessel of water which is removed from the fire is not only incapable of retaining the great heat which it had acquired while close to it, but it cannot even retain that small degree to which it has cooled down by being removed from it ; it must gradually become completely cold. So, too, is it in

<hr>

[1] 1 Paral. xxviii. 10.   [2] Esther vii. 4.   [3] Job x. 18.

the case of the tepid man whom we are considering. He has gone away from the fire, he has begun to neglect prayer, he takes no more pleasure in spiritual books, he no longer mortifies or restrains himself, he is given up to pleasures, which are useless, if no worse. And he imagines that it is possible for him to continue in this state for a long time. How is he deceived! He will have to go on falling from bad to worse— and to what point! till he has become entirely cold. And therefore God, seeing the evil dispositions he is in, so greatly abhors him in his tepidity, that He goes so far as to utter this exclamation, which at first sight appears exaggerated: "I would thou wert cold or hot." And who can say that thou mayest not be the unhappy man here spoken of?

III. Consider thirdly, that even yet thou mayest not be satisfied. For if tepidity is so great an evil for this reason that it gradually leads to coldness, then coldness must be a much greater evil than tepidity. And if so, how can God possibly desire that thou shouldst be cold rather than lukewarm? But remember the distinction which I pointed out at the beginning as necessary to be laid down for the right understanding of the passage. There is a difference between the coldness which precedes the fervour so becoming in the service of God, and that which follows it. The first is excusable in His sight, because, as has already been said, it arises from want of proper knowledge; but not so the second, which implies this knowledge, and therefore is without excuse. When God here says, therefore, "I would thou wert cold or hot," of which coldness did He mean to speak? Certainly not of that which is subsequent to heat; for this is that worst of evils in which issues the tepidity of a man who relaxes in the good which he once practised, and it is therefore certainly worse than tepidity. He meant to speak of the coldness which is antecedent to fervour, and therefore observe that He does not say, "I would thou wert hot or cold," but "I would thou wert cold or hot." And as often as the words are repeated the same order is followed, first "cold," then "hot" is spoken of, to show thee that the coldness in question is not that of a man who has gone back from hot to cold, but of one who has not yet passed from cold to hot. Neither is there anything to wonder at in this. It is not surprising that God should show mercy to a man who is in the state of not having yet embraced good because he has not known it (which is the coldness we have called antecedent), by bringing him at length to one of very

great spiritual fervour, because He sees that if this unhappy
man has sinned, he did so purely in ignorance, and it is for
this reason the Apostle has left on record that he himself
obtained pardon for his persecuting fury: "I obtained the
mercy of God, because I did it ignorantly in unbelief."[1] But on
what grounds could God show the same pity to a man in the
other state, that of having embraced and then forsaken what is
good (which is the coldness we have called subsequent)?   He
must needs leave him in the coldness he has chosen.   And so
we read of many who from sinners have become saints, and
very great saints (because it is possible to pass from one
opposite to another), but of exceedingly few who after being
perverted have retraced their steps and become saints ; because,
according to the doctrine of the Philosopher, there is no going
back from privation to habit, at least ordinarily speaking, and
it is precisely this which is confirmed by those words of the
Psalmist, which say that man is "a spirit that goeth and
returneth not,"[2] because it is easy to pass from good to bad,
but not from bad to good.   That requires an evident miracle
of grace : "It is impossible," that is, exceedingly difficult, "for
those who were once illuminated . . . and are fallen away, to
be renewed again to penance."[3]   This then is the reason why
God would rather have thee cold, as thou wert before conversion, than lukewarm, as thou art when beginning to be
perverted, because this tepidity tends to a far more deplorable
state than that of thy first coldness : "I would thou wert cold
or hot."   This too is the reason why He adds, "But because
thou art lukewarm, I will begin to vomit thee out of My
mouth," for if thou art preparing by thy tepidity to come out
from the bosom of God, what wonder is it if God does not
wait for thee to come out of thy own will, but vomits thee
forth because He cannot endure the disgust thou causest
Him?

IV. Consider fourthly, what this terrible vomiting forth is
with which God threatens thee.   Is it thy damnation?   Nay,
God cannot damn thee for mere tepidity in His service as He
can for coldness, whether antecedent or subsequent.   And the
reason is that coldness implies grievous sin, whereas tepidity
implies only that which is venial though voluntary.   This
"vomiting" then is not strictly speaking damnation, but
rather a disposition for damnation.   God may be said to
"vomit" thee forth when He begins to cease from the loving

[1] 1 Timothy i. 13.   [2] Psalm lxxvii. 39.   [3] Hebrews vi. 4.

care of thee which He had at first. He no longer caresses
thee with spiritual sweetness, which, according to some, is the
first degree of this "vomiting." He suffers thee to be overwhelmed with great aversion to all that concerns His service,
with sadness, weariness, temptations, which is the second
degree; and at last He lets thee fall into the reprobate sense,
which is the third degree, to be followed in the end by
irremediable damnation. Observe, therefore, that God says,
"I will begin." He does not vomit thee forth all at once, but
gradually. If, therefore, He has not done so entirely, make
haste and repent whilst yet thou hast time to continue in His
bosom, although still *that* time presses. Renew thy resolutions to serve Him well, amend thy life, rekindle thy fervour,
for this is why He says, "I will begin," in order to afford thee
time to give Him such consolations that He may no longer
feel disgust at thee.

V. Consider fifthly, why it is that God is not content to
say, "But because thou art lukewarm, I will begin to vomit
thee out of My mouth," but after the word "lukewarm,"
further adds, "and neither cold nor hot." Would it not have
been enough to say "lukewarm"? Enough, doubtless; nevertheless, as so important a matter was in question, God chose
to say more than was necessary, rather than fail in clearness of
expression; and so to explain Himself, that no one should
wrongly understand by "lukewarm" one who is rather cold or
rather hot. The degree of cold and heat may be small, but
it remains cold and heat still. A man who is "lukewarm" is
neither cold nor hot. Therefore, supposing thee to be cold in
the sense of being ignorant of the blessings attached to God's
service, and so not having hitherto embraced it, then God has
not yet received thee into His bosom, and so could not
"vomit" thee from it. And if thou wert hot, He would
gladly keep thee in it. But He says He will "begin to vomit
thee out," because thou art neither cold nor hot. It is in this
then that tepidity consists: in knowing the obligations that bind
thee to a God Who has been so good in granting thee all the
graces which He has bestowed on thee ever since thou didst
first pledge thyself to His service, and yet in neglecting that
obligation. Oh, what great reason hast thou to fear if the
guilt of so unworthy a negligence attaches to thee! Does
it not excite thy disgust when thou seest one whom thou
hast specially distinguished, caressed, and favoured, begin to
meditate forsaking thee, when thou wert justified in believing

him devoted to thee? And yet this is what thou too art
doing when thou art negligent in God's service. Thou art
meditating forsaking Him, because, as has been already said,
tepidity is not a state in which thou canst remain long. It is a
transition state from which thou wilt quickly pass to one of
coldness, and that a far more obstinate coldness than that
which was thy condition before passing from cold to hot, so
that it will one day be said of thee as it was of faithless Jerusalem, "As a cistern maketh its water cold, so hath she made
her wickedness cold."[1] For a cistern imparts to the water a
far sharper cold than originally belonged to it, and yet it does
not do this suddenly, but by degrees. So, too, is it with the
soul which, faithless as Jerusalem, has in the end withdrawn
its love from God.

[1] Jerem. vi. 7.

SEPTEMBER.

FIRST DAY.

Blessed is the man whom Thou shalt instruct, O Lord, and shalt teach him out of Thy law (Psalm xciii. 12).

I. Consider first, that nothing so much incites pupils to study as the kindness of their master. But what master in the world was ever so kind as Jesus Christ? He was the Master promised to us so long ago by Isaias in the words: "Thy eyes shall see thy Teacher."[1] One would have thought then, that all men would hasten in crowds to His school. And yet, far from hastening, there are very few who go thither at all. Do not wonder, therefore, that the Psalmist exclaims in this passage, "Blessed is the man whom Thou shalt instruct, O Lord," for let every one observe that he says "the man," not "the men," because it is rare to find one who can be induced to study in earnest under this Master, excellent as He is. Rather do men continually flock to those who have "itching ears," that is to say, deceitful teachers, who, if thou listenest to them, promise to make thee happy at once by the maxims which they propose; such as to revenge oneself, to desire high places, to heap up riches, to indulge the senses, and afterwards betray thee by condemning thee to everlasting misery: "O My people, they that call thee blessed, the same deceive thee."[2] That man will be truly happy, who makes himself a really faithful scholar of Jesus Christ; for he only will attain the chief good. Give God thanks, that thou mayest so easily, if thou wilt, now profit by this great Teacher, especially in His choicest school, which is that of prayer; and humble thyself, if thou hast not yet listened to Him.

II. Consider secondly, that sovereign legislators do indeed issue laws to their subjects, but do not explain them. They

[1] Isaias xxx. 20.   [2] Isaias iii. 12.

leave it to other teachers to do this from the professor's chair. Not so this Divine Master. He, after having promulgated the Law with His own mouth on Mount Sinai, came down in Person to seat Himself in the chair of the teacher, that He might explain the law which He had already given, and show all men what was the way of practising it with the greatest possible perfection, in order that there might be no excuse for those who do not practise it: "I Myself, that spoke, behold I am here."[1] Therefore does David here say, "Blessed is the man whom Thou shalt instruct," not "thy servants," but "Thou." He did not envy those whose lot it had been to have Moses, nor those who should have Isaias or Jeremias, but those who should one day have Jesus Christ, the Son of God, for a Master. And we are these very persons. How is it possible for us, then, not to have our hearts enkindled with the desire to learn of Him. All the more ought this to be so, because other masters, whoever they may be, can, at most, give their laws to the ear, but not to the heart. It is He only Who "gave a way for the sounding storms,"[2] and enables the voices of His ministers, preachers, and prophets, to awaken and enlighten the mind. And therefore not only may it be truly said that He is the best Master in the world, but the only one—"Neither be ye called masters, for one is your Master, Christ"[3]—because all others who pass for masters are so only exteriorly, not interiorly. There is none besides who makes thee both understand and put in practice what thou art taught.

III. Consider thirdly, that even when sovereign legislators do consent to teach their laws (which is a very rare case), they will not go through the labour of instructing their scholars in the elementary beginnings, but leave to others the care of forming them, step by step, in the lower schools, so that when they receive them they are always prepared for higher teaching. Not so our Lawgiver: "I am the Teacher of them all."[4] It is He Who instructs those who are already formed, and it is He Who forms them: "Blessed is the man whom thou shalt instruct, O Lord, and shalt teach him out of Thy law." And who can describe the patience with which He performs this work of forming scholars? "As a man traineth up his son," and that when the son is most incapable and dull, "so the Lord thy God hath trained thee up."[5] Thou canst easily

---

[1] Isaias lii. 6.   [2] Job xxviii. 26.
[3] St. Matt. xxiii. 10.   [4] Osee v. 2.   [5] Deut. vi 5.

understand, for thyself, what trouble thy Lord had to go through with thee in the first lessons He gave thee, in order to form thy mind, that is, to take from thee those bad inclinations which prevented thee from being able to learn His law properly, to do away with thy pride, thy ambition, and thy excessive self-love: "I was instructed," so, perhaps, mayest thou too say with truth, "as a young bullock unaccustomed to the yoke."[1] And well will it be, if after all the years He has kept thee at school, thou art even yet sufficiently broken in. Yet thou wonderest that He does not as yet give thee those sublime lessons which are peculiarly His! The reason is that He finds thee still too unformed. Thou wouldst fain receive at once those lights in prayer which He bestows on the saints. I understand the desire. But this is thy mistake: that thou wouldst be taught the doctrines of so great a Master without being first formed. Let Him begin by ridding thee of the excessive attachment which thou still retainest to thy own comforts and fancies; and then do not be afraid that He will not give thee the lofty lessons which thou longest for. But unless thou first submittest to be thoroughly formed by Him, there is no chance of His ever so teaching thee. Listen to the words of the Psalmist: "Blessed is the man whom Thou shalt instruct, O Lord, and shalt teach him out of Thy law." He first "instructs," then "teaches," not *vice versâ.*

IV. Consider fourthly, that it belongs to such a Master as this to teach, not only what concerns the fulfilment of His law, but also the sublime mysteries of the faith, such as those of the Three Divine Persons, Predestination, Providence, Grace, and others never before heard of: "I will utter things hidden from the foundation of the world."[2] Nevertheless, notice what is very wonderful; the Psalmist does not here call any of those blessed who are taught of God in such mysteries as these, but in His law: "Blessed is the man whom Thou shalt instruct, O Lord, and shalt teach him," not "Thy secrets," not "Thy judgments," not "Thy incomprehensible wonders," but "out of Thy law." Because thou mayest be saved without the speculative knowledge of these sublime mysteries, but not without the practical knowledge of the law, which we have spoken of. Think, then, for a while, how important it is for thee to aim at making progress in this knowledge above all others. This is the knowledge which will make thee blessed. Now beatitude is two-fold: one is perfect, and is that which

[1] Jerem. xxxi. 18.   [2] St. Matt. xiii. 35.

belongs to our country; the other imperfect, which belongs to the journey thither. The beatitude of our country is the vision of God. The beatitude of the journey is to walk in the straight way leading to that country. But this way is not even the most perfect knowledge of mysteries, but the knowledge of the law: "Blessed are the undefiled in the way, who walk in the law of the Lord."[1] Observe who they are who are called "blessed in the way"—those who, keeping themselves undefiled by the mire of which there is so much to pass through in this way which leads to Heaven, walk on with steadfast pace in the Divine law: they "walk in the law of the Lord." And if this is so, how plain it is that thou shouldst apply thyself far more to the knowledge of God's law, than to that of all worldly science? What will thy knowledge of poetry, civil or canon law, moral doctrine, or the sublimest theology avail thee, if thou art damned? And yet it is possible to be damned with these sciences and all the rest, almost infinite as they are; but not so with that of the Divine law, if the knowledge which thou hast of it is that practical knowledge which, as has been said, is taught directly by God. The speculative knowledge of His law may be learnt also from spiritual books which treat of it, the practical knowledge from God only. And therefore, too, the Psalmist here says, "Blessed is the man whom Thou shalt instruct, O Lord, and shalt teach him out of Thy law," because, not even in so important a matter as this of the Divine law, did he so much regard speculative knowledge. Which of the two canst thou boast of possessing. Perhaps neither the one nor the other, but only the profane sciences which the world values.

SECOND DAY.

The life of man upon earth is a warfare (Job vii. 1).

I. Consider first, that these words are well known to every one, but not so the very useful consequences to be deduced from them. Do thou, therefore, set thy mind to extract these consequences from their hidden depths, and be not like so many who are satisfied with those considerations in the Sacred Scriptures, which are like gold melted into coin. Thou

[1] Psalm cxviii. 1.

oughtest also to seek for those which are like gold buried in caverns : "If thou shalt seek wisdom as money," which is the most that is done by the former, "and shall dig for her as for a treasure," which is that which the latter do, in addition, "then shalt thou understand the fear of the Lord," which is that which suffices to keep the will in its due exercise, "and shalt find the knowledge of God,"[1] which is that which likewise enriches the understanding, to the great advantage of the will which depends on it. To return, then, to our subject. The life of man is a warfare in which God is the general, and those who stand in His place on earth are the subordinate officers, whilst the soldiers are men who are obliged to fight during the whole of their lives ; for which reason it is not said that "there is a warfare in the life of man," but that "the life of man is a warfare." The battlefield is this earth, on which are ranged in various ranks all men according to their state ; the uniform is the glorious name of Christian ; the arms are the prayers which they use in the battle, the Scriptures, sacraments, penances, and other like spiritual aids ; the enemies are their irregular passions, strengthened by the infernal spirits in league with them ; the flag is the encouragement bestowed by grace ; the losses are falls into sin ; the conquests are noble acts of virtue ; defeat is damnation, and the triumphs the glory of Paradise, which in the end crowns him who has ended his warfare victoriously. But these are things universally known ; do thou reflect now on those useful consequences, which are to be deduced by thee from them for thy greater profit.

II. Consider secondly, that if the life of man is a warfare, it follows that it must be a time of continual labour, not of rest ; and so it is that the laws teach us that there is no vacation-time in warfare (*in militia nullas ferias admitti*), because even if there is a cessation from fighting (and in the spiritual warfare with which we are concerned this very rarely happens), there is nevertheless no cessation of labour. Even when not actually fighting, because the enemy is not attacking us, every one should be in a state of readiness to fight : "Stand, therefore, having your loins girt."[2] He should polish and sharpen his weapons ; he cannot go wandering about hither and thither as unoccupied persons do, he must keep in his quarters, at his post, at his drill—"I will stand upon my watch"[3]—even though he has been exposed in consequence to the inclemencies of every season, to shiver

Prov. ii. 4.   [2] Ephes. vi. 14.   [3] Habacuc ii. 1.

with cold, to faint with hunger, and to endure grievous suffering. How, then, canst thou desire to take thy ease all through thy life? Is this the way to behave during a warfare? "The life of man upon earth is a warfare," not all amusement.

III. Consider thirdly, that if the life of man is a warfare, it follows also that it is a time not of reward, as some would have it, but of merit. What wonder is it, then, that it fares ill in this world with so many just persons? It is not the practice of a wise general to place his best soldiers beyond musket range; on the contrary, it is these whom he selects from all the rest to put them in the foremost ranks, and by so doing he shows that he knows them, and loves and values them beyond others. It is enough that when the victory is won he rewards them more than the rest. What, then, can be said of thee, who art so ready to blame Providence, because in this world so much has often to be suffered by just men? "In the world you shall have distress."[1] It is the treatment valiant men meet with: "The life of man upon earth is a warfare." Wait for the end, and thou shalt see that those who have laboured more than others will also be more rewarded by God than others. All that can be demanded is that He should give them wages in proportion to the labour He requires of them; for "who serveth as a soldier at any time at his own charges?"[2] and that, therefore, He should bestow on them greater supplies of grace than on others.

IV. Consider fourthly, that if the life of man is a warfare, it is also a time for obeying humbly, and not for acting as he pleases. Every one knows what exact obedience is required, and always has been required, in warfare. There is no greater obedience in the world. A soldier, therefore, must not discuss the orders of his captain, all he has to do is to execute them blindly: "I have under me soldiers, and I say to this, Go, and he goeth, and to another, Come, and he cometh."[3] And this obedience is exacted, not only in easy things like going and coming, but in those which are very hard. And therefore those soldiers are most severely punished who dare to turn upon their captain even when he lifts his baton of command to strike them. How, then, canst thou desire to have no law but thy own will in this world? "The life of man upon earth is a warfare." If, then, life is the time for fighting, it is also the time for obeying perfectly, and not for complaining, not

[1] St. John xvi. 33.   [2] 1 Cor. ix. 7.   [3] St. Matt. viii. 9.

even of the stripes which are given by the hand of the general
or of those who are acting for him.

V. Consider fifthly, that if the life of man is a warfare,
it further follows that it is a time of great danger, and not of
safety. Who can doubt it? "Know it to be a communication
with death." This is the declaration made by the Wise Man
to him who finds himself enrolled at his birth, whether he
will or no, amongst the soldiers in this great warfare of which
we are speaking. "Know it to be a communication with
death," that is, let every one understand that so long as he
lives he is in danger of damnation, like all other men. And
why? "For thou art going in the midst of snares, and walking
upon the arms of those that are grieved."[1] The reason is,
that thou art continually threatened by thousands of ambushes
and by thousands of attacks. These ambushes are the
occasions of sins which are unexpected, and the attacks are
those which thou dost indeed expect, but which thou art not
prepared to resist manfully. The first are formidable from
their number, the last from their violence; and this is why,
in the former case, it is said, "Thou art going in the midst
of snares," and in the latter, "Thou art walking upon the
arms of those that are grieved." Oh, if thou couldst but look
down from above on the earth, which is the vast battlefield
thou art in, thou wouldst see it to be, as it were, strewn with
the arms which were basely dropped by the hands of those
unhappy men who are now vainly bewailing their losses in
Hell. And what are these losses but tokens of the defeats
which are daily sustained in these attacks: "The arms of
those that are grieved." And yet thou makest as sure as
though thy salvation were already within thy grasp! Thou
art greatly deceived: "The life of man upon earth is a
warfare." Stand, therefore, upon thy guard, for thou, too,
mayest be overcome: "Various is the event of war, and sometimes one, sometimes another, is consumed by the sword."[2]

VI. Consider sixthly, that if the life of man is a warfare,
it follows also that it is a time of trial, and not a time for
presumption. Oh, how many virtues dost thou think, perhaps,
are dwelling in thy heart! But, if so, thou must needs come
to the proof. And this, too, is one of the special ends of the
warfare which in this passage is called in the Septuagint by
the name of "trial." "The life of man upon earth is a trial."
This trial is ordained in order to prove the constancy of

[1] Ecclus. ix. 20.   [2] 2 Kings xi. 25.

one and the cowardice of another, for nowhere are both of these better tested than in a field of battle. And thus, it is said in the Fourth Book of Kings, that "Sophar, the captain of the army, exercised the young soldiers of the people of the land."[1] The Hebrew version has "proved" for "exercised." Only, there is this difference: that in that warfare none but beginners—"the young soldiers of the people"—were proved, whereas in the warfare of which we are speaking even veterans are subjected to the same trial: "God tempted (or proved) Abraham."[2] And this, because the proofs to which God puts men, who are His soldiers, last to the very end. How is it, then, that thou so readily believest thy own pride whenever it tells thee that thou hast already attained sanctity? It is false —it is false! The trials are not over yet: "The life of man upon earth is a warfare" (that is, a "trial"); when it is ended it shall be seen what thou art.

VII. Consider seventhly, that if the life of man is a warfare, it follows, lastly, that it is not a free but an appointed time. What do I mean by this? There were certain philosophers who, in order to dignify the height of despair with the name of courage, maintained that it was a praiseworthy action for a man to kill himself in order to escape from some misfortune, disgrace, malady, or any other evil which was too difficult for him to bear. But what can be a greater error? "The life of man upon earth is a warfare." And how can it be lawful for a soldier to fly from it without leave from his general? Such conduct has always been regarded as exceedingly wrong and insolent, and as such is still severely punished by all nations. And, if so, how can it be praiseworthy? A soldier, especially when wearied by the burden of long labour, may indeed earnestly beg of his general to be discharged from the service, but he is never free to leave it by his own will. The same thing may be asked by man of God: "It is enough for me, Lord; take away my soul, for I am no better than my fathers."[3] So it was that Job, when he saw that his friends were greatly scandalized, by hearing that he earnestly desired death, as though out of impatience in bearing his heavy trials, broke out at length into these words: "The life of man upon earth is a warfare;" by which he meant to show them that he well knew his duty upon earth was to fight, and consequently to suffer much, but that this did not prevent him from desiring a speedy death, since it never was forbidden

[1] 4 Kings xxv. 19.   [2] Genesis xxii. 1.   [3] 3 Kings xix. 4.

to a soldier to long for the close of his warfare, and to ask for it, which is what he says in another place: "All the days in which I am now in warfare I expect until my change come."[1] And so, when any one loves a long life upon earth, as worldlings do, what does this show? It shows that such a man is a soldier who is but little wearied with his labours, because he has made it his object to avoid them.

THIRD DAY.

There hath stood one in the midst of you Whom you know not (St. John i. 26).

I. Consider first, how great is the mistake of some persons, who seek after God as though He were very far from them, and who would fain bring Him down to them with sighs, tears, and lamentations, while all the while He is within them. To such men it may, indeed, be truly said: "There hath stood one in the midst of you Whom you know not." Their conduct is as foolish as that of persons who have a fountain in their dwellings, and yet go out to draw water. If, then, they have gone out, they must come back again, reflect and go into themselves, where they will at once find what they sought for vainly outside, while wandering about the public roads. This is the right rule: how is it, then, that thou dost not at once follow it? If thou wouldst find thy Lord, so as to be easily united to Him, do not suffer thy imagination to go rambling out of thyself. For although it is true that creatures may show Him to thee, yet very often they may divert and lead thee away from Him; and after all, the most that they can do is to convince thee that He is within thee. Rather penetrate into the inmost recesses of thy heart, and try to comprehend, as a thing which is most certain, that there thou possessest thy true and living God, and that there is no need to seek for Him elsewhere: so will it always be easy for thee to remain in His presence, as those just persons do who strive to imitate the blessed by never losing sight of Him. Is it not a crying shame that God has been within thee so long, and that thou hast hardly yet begun to know Him? "So long a time have I been with you, and have you not known Me?"[2]

[1] Job xiv. 14.   [2] St. John xiv. 9.

II. Consider secondly, that this word "not to know" (*nescire*) has two meanings in the Sacred Scriptures, one of which concerns the understanding, and means strictly not to know—"They know not Him that sent Me"—and the other concerns the will, and means not to value: "I know you not."[1] Both senses are applicable to the passage thou art considering. For Christ was dwelling in the midst of Judæa, and yet most of the inhabitants did not know Him; and those who knew Him did not value Him, but regarded Him as mere man like the rest of the world. And this, it would seem, is the case with thee, too. Thou hast thy Lord always within thy heart, and yet *nescis illum : nescis*, because thou dost not know Him; and *nescis*, because thou dost not value Him. What wonder is it, therefore, that thy progress in virtue is so small? "God is not before his eyes: his ways are filthy at all times."[2] How, indeed, could it be possible for thee to be persuaded to do anything to displease God, if thou wert always present to Him as He is to thee? What man would venture to be guilty of an act, I will not say of rebellion, but even of disrespect or rudeness, in the presence of his sovereign? If, then, thou wouldst take a short road to perfection, this is what thou hast to do : adopt this practice of the Divine presence, so strongly recommended by all the saints as not important only, but necessary: "Walk before Me, and be perfect."[3] And if thou askest more particularly in what this practice consists, thou hast not far to seek. It consists in doing the reverse of what is signified by the word *nescire :* it consists in knowing God, and in valuing Him. It consists in earnestly applying the understanding to know how truly He dwells within us, and also applying the will to make Him a return by those devout affections which one who really values Him will at once pour out to Him from the depths of his heart.

III. Consider thirdly, in what manner thou art to apply thy understanding to know that God dwells within thee. Thou art to do this principally by the way of faith, believing firmly that it is so because faith teaches it : "He is not far from every one of us."[4] This is a much easier, and at the same time a more fruitful method than that of exercising the imagination. But in order the better to prepare thyself to lay hold by the understanding on that which thou believest, thou shouldst, in the first place, consider that God dwells within

[1] St. Matt. xxv. 12. [2] Psalm ix. 25. [3] Genesis xvii. 1. [4] Acts xvii. 27.

thee precisely as a king does in his kingdom. A king dwells in his kingdom by his actual presence, by the knowledge which he has of everything, and by the power which he exercises in it. So, also, does God dwell within thee. The first way is that of essence, because God is within thee, as the king is in his kingdom when he resides there in person, and not by means of some viceroy, with this difference, that the king does not reside in person in every part of his kingdom, but only in one, whereas God dwells in every part of thee. The second way is that of knowledge, for as the king knows everything that is done in his kingdom, and so is said to be present at everything, so does God know everything that takes place within thee; only that if the king does really know everything, he knows it because it is told him by others, whereas God knows it because He sees it with His own eyes. The third way is that of power, for as the king can do as he pleases with everything in his kingdom, so, too, can God with everything in thee, only that there is but little that the king can do personally, and God does everything. It is this that he who spoke these words would have thee clearly understand: "There hath stood one in the midst of you Whom you know not." He says "in the midst," not merely to describe the local site which God occupies, for that is unlimited, but His sovereignty. He is said to be "in the midst," because He can reach to every part with equal ease, as the king can who resides in the centre of his kingdom. And if so, how is it possible for thee ever to lose sight of Him? See how gracious a King thy God is! That thou mayest have no excuse for saying that thou canst not seek Him in His Kingdom beyond the stars, He has set up His Kingdom within thee: "Behold the Kingdom of God is within you."[1]

IV. Consider fourthly, in what manner thou art also to apply thy will to show that thou truly valuest thy God, and wilt not leave Him solitary within thee, like a king forsaken in his kingdom. Thou art to do this by the frequency of the devout affections which thou oughtest to pour forth to Him during the day, such as adoration, love, oblation, glorification, joy, thanksgiving, confusion, contrition, and the like; but above all by frequent invocation. This is to treat Him really as a King, for thus dost thou show thy entire dependence on Him. Call upon Him, to direct thee in thy ways, to strengthen thee in temptation, to comfort thee in trouble, to enrich thy

[1] St. Luke xvii. 21.

poverty; and in a very special manner, to vouchsafe to assist thee at the hour of death, which may be any hour as it comes. God being so good a God, only desires to bestow favours on thee; but, being a King, He will also be entreated. This is that practice of the Divine presence which may so easily be performed by any one. And there are two motives which should especially incite thee to it: gratitude and necessity. Gratitude, because He is always within thy heart to lavish favours on thee. Is it not right, therefore, that if He is incessantly thinking of thee, thou shouldst think of Him, I do no say incessantly, for this is beyond thy power; but at least that thou shouldst always be striving to do so more and more? Necessity, because when thou losest sight of thy God, thou art like a piece of ground from which the sun is hidden by some high wall—no longer capable of producing flowers or fruits, but only weeds: "Thy land is like a garden of pleasure before Him, and behind Him a desolate wilderness."[1]

FOURTH DAY.

You are the temple of the living God, as God saith: I will dwell in them, and walk among them, and I will be their God, and they shall be My people (2 Cor. vi. 16).

I. Consider first, that if God dwells in a general way in the hearts of all men by essence, knowledge, and power, as has been explained in the preceding meditation, He dwells in a more intimate way in the hearts of the just, because in them He also dwells by grace; and therefore, if He is in all men like a king in his kingdom, He is, in the just, like the same king in the palace which is his residence. And this is what the Apostle here means when he says, "You are the temple of the living God," for temples are the palaces of God on earth, and therefore are becomingly adorned and beautified as being destined for the splendid habitations of the King of kings: "I have chosen this place to Myself for a house."[2] And so the just are called temples, and temples "of the living God." They are temples, because they are dwellings consecrated to God; and they are temples of the living God, because they are not dedicated to a false god like the heathen

[1] Joel ii. 3.   [2] 2 Paral. vii. 12.

temples, but to the true God. And do not suppose that these temples are bare. Could any one but enter, and see the richness of their furniture and the magnificence of their decorations, he would acknowledge that there is all the difference between them and the great Temple of Solomon which there is between the type and the thing typified. " Rich men in virtue:"[1] such is the description of the just. It is not said "in substance," for very often they have nothing in this respect; but so much the more are they rich in all the treasures of virtue. Therefore, if there were no other inducement for thee to live justly than this, to know by faith that in such a state thou art the temple of God, ought not this to be sufficient? " The Lord is in His holy temple."[2] And this is the temple which is truly holy, that is, the spiritual temple; because it is holy, not by exterior sanctity, like the material temple, but by that which is interior.

II. Consider secondly, that there are four operations of God in His material temples on earth. The first is His dwelling there; the second, His favouring us there by visiting us in a more intimate manner; the third, His hearing our petitions there more particularly; and fourth, His receiving there from us, in a more special manner, the worship which is nevertheless equally His due in other places. And it is from these same four operations that the Apostle proves the just to be God's temples: "You are the temples of the living God, as God saith." And what reasons does he give? First, because "I will dwell in them;" secondly, because "I will walk among them;" thirdly, "I will be their God;" and fourthly, "They shall be My people." First, then, the just are the temples of God, because God dwells in them by His sanctifying grace: "I will set My sanctuary in the midst of them for ever, and My tabernacle shall be with them."[3] And so it is indeed said of the rest of men, that He is within them, as He is everywhere—"All the earth is full of His glory"[4]— but it is never said that He dwells there. This expression is confined to the just in the Sacred Scriptures: "Sing ye to the Lord, Who dwelleth in Sion."[5] "The Spirit of God dwelleth in you."[6] "That Christ may dwell by faith in your hearts."[7] And the reason is, that He is in others only by that action of His own by which He unites Himself to them, preserving their existence, ruling them, beholding

[1] Ecclus. xliv. 6.   [2] Psalm x. 5.   [3] Ezech. xxxvii. 26.
[4] Isaias vi. 3.   [5] Psalm ix. 12.   [6] I Cor. iii. 16.   [7] Ephes. iii. 17.

them, without any correspondence on their part ; whereas He is in the just also by that mutual action in which they are united to Him, loving, obeying, and reverencing Him, and so giving Him an abode within them. So that, even though God were not in the just, as He is everywhere by essence, knowledge, and power, He would be bound to be in them by love, which is a more powerful motive. And this is what He means by saying "I will dwell in them ;" not "I will be in them," but "I will dwell," just as a king says that he is in his kingdom, but that he dwells in his palace. Secondly, the just are the temples of God, because He visits their souls in a more special way, continually imparting to them new lights, inspirations, or spiritual consolations, whereby He incites them to well-doing. These are never permanent, but come and go at different times, so much so that God is said sometimes to draw near to the souls He loves, and sometimes to withdraw from them : "If He come to me I shall not see Him ; if He depart I shall not understand."[1] Not that He forsakes His dwelling, but because His motions in it are various, passing from the understanding to the will, and from the will to the understanding, by means of incitements answering to these respective faculties. This is signified when God says, "And I will walk among them," for the word is not simply *ambulabo*, but *inambulabo*, because He is always alike within the soul ; and though He moves from one room to another of His royal palace, He does not leave it. Thirdly, the just are the temples of God, because He listens to their petitions in a special way, and grants them, showing Himself in different circumstances their Friend, their Father, their Protector, their Deliverer, their All. And this is signified by the words "I will be their God," not "I will be God among them," but "their God," because He makes Himself so entirely their own that they are able to dispose of Him at their will, as of something that belongs to them, just as the palace can, on occasion, dispose of its sovereign in a way that the kingdom at large cannot. Fourthly, the just are the temples of God, because He receives from them in a special manner the worship which is His due, whereas others withhold it from Him, or only render it in a material way, not uniting with it the reverence and obedience which the just, as His own people, always pay to Him : "The Lord thy God hath chosen thee to be His peculiar people of all peoples that are upon the earth."[2] And this, in the last

[1] Job ix. 11.   [2] Deut. vii. 6.

place, is what God signifies by the words, "And they shall be My people," not merely *meus populus*, but *populus mihi*, because in them He has, as it were, a people consecrated to His service, like those select and noble people forming the royal court in the palace. Such are the grounds on which all the just are called "the temple of the living God." Thou shouldst now examine whether thou thinkest to recognize them in thyself, in order thence to conclude whether God dwells in thee in so much more excellent a way than He does in all men generally.

III. Consider thirdly, that if there are probable reasons for hoping that thou art in the blessed number of those who are the temples of God, thou art all the more bound to watch thyself with extreme care and circumspection, so as to allow nothing within thee which is the very least profane: "What agreement hath the temple of God with idols?"[1] For if so much reverence is due even to material temples which are holy on grounds purely exterior, how much more is this so in the case of spiritual temples which are holy by reason of that true, effective, essential holiness which grace produces in them: "Unto the sanctification of the spirit."[2] But if this is so, how is it that thou allowest thoughts to lurk in thy mind which are either useless, idle, or vicious, or, at the best, more earthly than heavenly? "The Lord is in His holy temple," and therefore, what follows? "Let all the earth keep silence before Him."[3] Thou art all the more bound to give thyself to the practice of the presence of God, because of His dwelling in thee in a more intimate manner, that is, like a king in his palace: "In His presence all shall speak His glory."[4] The whole kingdom has relations with the sovereign, but they are more distant; the inmates of the palace have free access and intercourse.

FIFTH DAY.

Be sober and watch: because your adversary the devil, as a roaring lion, goeth about seeking whom he may devour. Whom resist ye, strong in faith (1 St. Peter v. 8).

I. Consider first, that in order not to be overcome by attacks so violent as those of the devil, it is not enough to wait till they are made, and then repel them vigorously: they must

[1] 2 Cor. vi. 16.　[2] 1 St. Peter i. 2.　[3] Habacuc ii. 20.　[4] Psalm xxviii. 9.

be prevented by prudence. Therefore the first thing which St. Peter here says is, "Be sober and watch," because, since we have to do with so formidable an enemy, the first thing necessary is not to let him take us by surprise. This is the meaning of the watching spoken of. It means that supervision which is practised by the man who fears ambushes and snares that may be laid for him when he least expects it. This watching, then, belongs to the soul; but it cannot be accomplished without the concurrence of the body. And, therefore, not only does the Apostle say "watch," but "be sober." Nay, he first says "be sober," and afterwards "watch," because it is sobriety which so especially avails to keep the mind wakeful. When "meat was not set before" the king, what followed? "Sleep departed from him."[1] Excess, on the contrary, so oppresses it, that before long it is forced to close its eyes in deep sleep: "Holofernes lay on his bed fast asleep, being exceedingly drunk."[2] How stands the matter, thinkest thou, with thyself? Dost thou watch sufficiently? If not, that is, if thou dost not keep good guard over thyself, examine into thy state, and thou wilt surely see that the principal cause of this is the indulgence of thy appetite. Set thyself to mortify it, as the saints did with such unspeakable anxiety: "Hold back thy bread, and give it not,"[3] and thou wilt see how much easier it will be to keep wakeful. Otherwise, the more indulgent thou art to thy appetite, the worse will it fare with thee both as to body and soul: "For thou shalt receive twice as much evil for all the good thou shalt have done to it."[4]

II. Consider secondly, that although this watchfulness may be somewhat troublesome, it is not enjoined on thee without good reason, for indeed it is very necessary. How terrible an enemy is it with whom thou hast to do! It is no other than Lucifer. He it is who is described in these tremendous words: "Your adversary the devil, as a roaring lion, goeth about, seeking whom he may devour." For, although it is true that, being cast down from his lofty station into the depths of Hell, he does not go about in person on the earth, yet he does go about in the person of those innumerable ministers of his whom he keeps here, and this is even worse. And observe the reason which makes him especially formidable: it is his desire of harming, which is greater than can possibly be described. Therefore the Apostle says of him, first, that he is "your adversary the devil;" then,

[1] Daniel vi. 18. [2] Judith xiii. 4. [3] Ecclus. xii. 6. [4] Ecclus. xii. 6.

that he is "as a roaring lion;" and lastly, that he "goeth
about, seeking whom he may devour." First, then, he says
"your adversary the devil," that thou mayest know that the
devil is not an enemy who contents himself with hating thee.
If he were, the Apostle would have been satisfied with calling
him simply an enemy, not an adversary. He is an enemy
who is always attacking thee, plotting against thee, pursuing
thee, and doing all in his power to ruin thee : "All the day
long he has afflicted me."[1] And therefore the Apostle says
"your adversary the devil," not "your enemy." For the same
reason the Apostle goes on to say "as a roaring lion," not
merely "as a lion," which would be bad enough, but "as a
roaring lion," to show that not only is the devil fierce, powerful,
determined, and proud, as all lions are, but that he is, in
addition, a hungry lion. A lion roars when, being inwardly
tormented by hunger, he sees the coveted prey, and already
devours it with the hope of making it his own. Thus it is
that the devil is a lion which is always roaring, because he is
always devoured by a raging hunger after souls, and always
hopes to capture them however much they flee from him.
Indeed, he roars to hinder them from fleeing, for this is the
lion's object in roaring when he sees his prey, the very time
when one might expect him to keep quiet in order not to
discover himself. He does this in order to strike such terror
into his victim that it may lose all power of breathing or taking
to flight ; and this is just what takes place, for we are told that
at the mere sound of his voice almost all animals are stupefied
with terror. Such, too, is the devil's object in roaring so
terribly. Well does he know how to terrify souls, especially
those of spiritual persons, which are those he hunts most
eagerly, by his temptations to mistrust, by the fears and
anguish with which he fills the heart : "The lion shall roar,
who will not fear?"[2] And therefore the first thing he does,
as a rule, is to strike terror by roaring, and then he goes on to
make his attack by saying that the best way is to take one's
pleasure as long as possible, to indulge one's fancies and
desires, since it is mere waste of time to think of acting
virtuously. And this is why the Apostle adds, in the third
place, "goeth about, seeking whom he may devour," in order
to teach us that the devil is never satisfied with any harm that
he does to us, but is always longing to do as much as he
possibly can; he "goeth about, seeking," not "whom he

[1] Psalm lv. 1.   [2] Amos iii. 8.

may bite," nor "whom he may slay," but "whom he may devour," an expression the force of which, as applied to the devil, is to show his exceeding fury in destroying souls, so that, if he could, he would swallow them all at once. Thus, whereas the lion, after having eaten abundantly, is at length satisfied, the devil's hunger, on the contrary, is only increased by eating. It is an insatiable hunger, and therefore it is in vain to hope that it will ever be diminished, or that he will ever become less savage, as the lion does when he has had enough. And, if this is so, is not incessant watchfulness required against an enemy so eager to injure us?

III. Consider thirdly, that if all we have to dread in the devil were his desire of harming us, it would be more endurable. But the worst is that, in addition to this desire, he possesses the sagacity, prudence, and cunning which enable him to do it. And so it is most justly that the Apostle says, not only that he "seeketh whom he may devour," but that he "goeth about seeking." This lion, furious as he is, does not set himself to attack his prey directly, as his audacity might suggest, but he does so artfully, which is represented under the figure of tortuous and indirect ways: "I have gone round about the earth, and walked through it."[1] There are three principal senses in which holy writers understand this expression. The first is, that the devil acts like a hunter, who, in order the better to deceive his game, does not always set his snare in the same place, but changes it, sometimes going from an open to a concealed spot, sometimes from a high to a low one. And so the Apostle says that he "goeth about," to show thee that he will seek thee out in all thy haunts, at home and in church, in the street, the court, the cloister, and the most retired paths, but in different ways, so that it is never easy to guess where he will most frequently lie in wait; but this very fact ought to prove to thee how necessary it is to let thy watchfulness extend to every place, because he "goeth about, seeking whom he may devour." The second sense is, that the devil acts like a captain who, before attacking the fortress which he intends to storm, makes the circuit of it, observing every part in detail, in order to make his assault on the weakest side. And therefore the Apostle says that he "goeth about," to show thee how well he understands going round thee on all sides to make his observations. Be very sure that he is always watching thy mind, eyes, ears, tongue, and every

[1] Job i. 7.

part of thee, and that he will make his attack on the side
where thou art weakest : "The sinner shall watch the just
man, and shall gnash upon him with his teeth."[1] So that,
although to defend thyself from him thy watchfulness ought
to be universal over every part of thee, yet it should be more
especially directed to that part of thee which most requires it,
on account of thy weakness.   The third sense is that the devil
acts like an assassin, who would fain, if he could, enter thy
house by night to kill thee, and so gain the reward offered him
by the man who desires thy death.   But because thou art well
protected, he waits for thee outside, going round about in thy
neighbourhood, in readiness to strike thee dead as soon as
thou art a step from thy house.   And so the Apostle says that
he "goeth about," because, if thou keepest closely shut up
within thy walls God does not, as a rule, give him leave to
effect an entrance ; but then, if the traitor cannot go *in*, he can
go *round*, for he is continually walking about, on the watch to
see whether thou wilt step out of the house on any side, and
then to rush upon thee.   These walls are the particular commandments of thy state by which thou art hedged in : thy
rule, the direction of thy spiritual father, the frequentation of
the sacraments, examination of conscience, retirement, silence,
penance, and other things of the like sort, which keep the
devil from approaching thee.   And thy watchfulness should
be directed to this point : never to leave the shelter of these
walls carelessly, by negligence in these exercises.   If thou doest
this thou art lost, for the devil is waiting for thee "as a lion
prepared for the prey."[2]   What will become of thee?   How
wilt thou be able to escape from his power ?

IV. Consider fourthly that even if it should happen, either
with or without thy fault, that the devil, that malignant lion,
rushes upon thee savagely to destroy thee, still, in spite of
what has been said, thou must not give thyself up as conquered,
because, although it is far better, as was said at the beginning,
to prevent his assaults than to be driven to the necessity
of repelling them, nevertheless they must most certainly be
repelled when they are made, because he is, after all, a lion
whose power extends just so far as we allow it to extend.
Therefore St. Peter concludes by the words, "Whom resist ye,
strong in faith," for he knows well that if we choose we can
resist him.   And how are we to do so?   Thou hast heard : by
strong faith, or, more properly speaking, by being strong in

---

[1] Psalm xxxvi. 12.   [2] Psalm xvi. 12.

faith. For faith is always equally strong in itself, but we are not always equally strong in it. From which it is easy to see why the Apostle did not here say "resist in strong faith," but "resist, strong in faith." When, therefore, thou feelest that the devil is assaulting thee, immediately thou shouldst revive in thy mind those great maxims which are called the maxims of faith: "This is the victory which overcometh the world, our faith."[1] Such are the following: That contempt is true glory, suffering true pleasure, poverty true riches, that to please God is true wisdom, that there is but one important thing in the world—to save our souls. And in the same way do thou go through the rest of the maxims which are the most especially opposed to the temptation which most especially assails thee. And then thou hast to be strong in this faith, not giving ear to what the devil suggests in contradiction, to delude thee, but to what Christ tells thee, Who is standing by, looking on at thy conflict, to reward or punish thee when it is over, to all eternity, according to thy deserts. But since faith also means confidence, thou shouldst, at the same time, have recourse to this same Lord Who beholds thee, that He may give thee help. And in this faith also thou must be strong, holding it for certain that the devil may roar like a lion as much as he will, that he may storm and rage and make a tumult, but that he can do nothing at all to harm thee if only thou resistest him by this two-fold faith of which we have been speaking, saying to him boldly, Away with thee! "Resist the devil, and he will flee from you."[2]

SIXTH DAY.

The sensual man perceiveth not these things that are of the Spirit of God: for it is foolishness to him, and he cannot understand (1 Cor. ii. 14).

I. Consider first, that there are two beatitudes for man: one in Heaven, and one on earth; to enjoy God in Heaven, to suffer for Him on earth. In Heaven it is to enjoy God, because the enjoyment of God is the end for which man was made; and, therefore, as soon as he comes to the enjoyment of Him, he becomes blessed, because he has attained his end, and his last end, which is that in which alone he can rest with that profound peace which all things find on reaching their

[1] 1 St. John v. 4.   [2] St. James iv. 7.

centre. On earth it is to suffer for God, because it is this
which above all things gives us the assurance of one day
coming to the enjoyment of Him in Heaven. And so, just as
the first beatitude is the attainment of our end, so the second
is a well-grounded hope of attaining it. And who can have
such good grounds for hoping this, as the man who suffers for
God on earth? "If we suffer, we shall also reign with Him."[1]
And therefore Christ called the poor, the persecuted, and the
mourners blessed. And He called them so because they have
so sure an earnest of salvation: "Blessed are ye that weep
now, for you shall laugh."[2] And so, if we examine the matter
very closely, we shall perceive that to suffer for God is a
greater beatitude on earth than to enjoy Him by receiving
from Him visitations, converse, lights, or sweetest raptures in
prayer, because all these are gratuitous gifts. But we can
never have so great an assurance of Paradise from that
which is a gift as from that which is a merit, which suffering for God is. Now, all such language as this, plain as
it is, is like some barbarous tongue to every one who is
living conformably to that part of him which he has in
common with brutes, and therefore does the Apostle here
say, "The sensual man perceiveth not these things that
are of the Spirit of God." He perceives neither those which
are in Heaven, nor those which are on earth. For we all
know that there are two qualities by which a brute differs from
a man. The first is, that his desires are guided by appetite,
not by duty. The second is, that his judgment is governed by
his perceptions, not by reason. This being premised, "the
sensual man," who lives like a brute (for the word translated
"sensual" is *animalis*), "perceiveth not the things that are of
the Spirit of God" in Heaven, because he too is guided by his
appetites, and so can imagine none but a Mahometan Paradise.
Now, there is no such Paradise in Heaven, because there all
pleasures are "of the Spirit," so much so, that even the bodily
pleasures which will be found there will be spiritualized, that
is, they will be like those of the spirit: "It is sown a natural
body, it shall rise a spiritual body."[3] Neither does such a one
perceive "the things that are of the Spirit of God" on earth,
because he too is governed by his perceptions, and so can
judge only by appearance. And thus, no matter how often he
hears it, he will never come to understand how those who
weep can be blessed: "Blessed are they that mourn." He

[1] 2 Timothy ii. 12.	[2] St. Luke vi. 21.	[3] 1 Cor. xv. 44.

considers the mourners, the poor, the persecuted miserable, because they appear miserable. Do thou bewail the unhappiness of such a state as this, if thou canst understand how unhappy it is. If thou dost not understand it, then not only shouldst thou bewail this state, but weep for thyself, because it is an evident proof that thou art one of those who are living in this state: "The sensual man perceiveth not these things that are of the Spirit of God."

II. Consider secondly, the reason why the Apostle says that the man who lives like the brutes "perceiveth not these things that are of the Spirit." The reason is, that he is a fool: "It is foolishness to him." And, being a fool, not only does he not understand these things, in the same way as this may be said of an inexperienced person, but he is unable to understand them: "He cannot understand." A man who has an unimpaired palate, but who has never in his life tasted sugar, certainly does not know what is meant by the flavour of sugar. But although he does not know it, he is capable of knowing it; all that is necessary is for him to taste it. But the man whose palate is dull, neither knows nor is capable of knowing it. Now, just such is the misfortune of the man who has given himself up to live like the brutes: "It is foolishness to him." His understanding, which is the palate of the soul, is dulled; if, indeed, it is not altogether dead, because it is never applied to any but sensible or sensual things, and is consequently incapable of conceiving those that are Divine. "He cannot understand," because he is incapable of relishing them; they are too much above him: "Many things are shown to thee above the understanding of men."[1] It is evident that Divine things can be understood in no other way but by tasting their recondite flavour. And therefore Moses said of his foolish countrymen: "Oh, that they would be wise, and would understand, and would provide for their last end."[2] It might seem that he should rather have said: "Oh, that they would understand and be wise," because understanding precedes being wise. Yet he did not say this: he said purposely, "Oh, that they would be wise, and understand," because although it is true of natural things that they are first understood and then known, yet supernatural things, being those which concern a future life, are first known, and then understood: "Taste and see."[3] And how can this be when the understanding is already blunted by a sensual life?

[1] Ecclus. iii. 25.   [2] Deut. xxxii. 29.   [3] Psalm xxxiii. 9.

But it must be remarked that it is not merely want of ability which prevents such persons from understanding these things; it has, perhaps, a still deeper cause in the withdrawal, if I may so express myself, of the principle of understanding. For to none is the Spirit of God so unwilling to impart Himself as to those who live like brutes. How does He turn away from them! How does He abhor them! God will only allow Himself to be enjoyed by the man who, in Heaven, is altogether dead to his senses, and by the man who, on earth, is mortified in them. "Man shall not see Me and live:"[1] these are the plain words He has spoken of Himself. But why did He so speak, except because He absolutely requires one of two things from every one who has any desire to enjoy Him: he must either be entirely dead to himself, or he must mortify himself. See, therefore, how important it is to abandon such a manner of life, which is, moreover, an animal life, that is to say, a life which disposes thee so greatly to love and prize the pleasures of sense. If thou dost not abandon it, thou art making thyself incapable of all Divine pleasures, for these are all of the spirit: "God is a spirit, and they that adore Him must adore Him in spirit and in truth."[2]

III. Consider thirdly, that if it is so necessary to abandon that life even which is so far like that of the brutes that it makes us indulge more than is right in sensible pleasures, far more necessary is it to abandon that which makes us indulge in those which are sensual. This, certainly, is that kind of "sensual" life which is beyond all others here condemned by the Apostle: "The sensual man perceiveth not these things that are of the Spirit of God: for it is foolishness to him, and he cannot understand." For if it is impossible for the man who is too much given to sensible pleasures ever to understand Divine things, the man who is given up to sensual pleasures is scarcely able even to believe them. But so it is, that little by little lust deprives thy heart of faith, even though thou mayest sometimes deceive thyself by thinking that thou still possessest it. Consider the heresiarchs of our times, at least the most famous among them. They all began by a life which was first impure, and then sacrilegious. Indeed, the same Apostle, when writing to the Colossians, signified luxurious persons by those who are unbelieving in that passage which says, "For which things the wrath of God, that is, the Deluge, cometh upon the children of unbelief."[3] And this is

[1] Exodus xxxiii. 20. [2] St. John iv. 24. [3] Coloss. iii. 6.

not to be wondered at, for in time lust has the effect of making thee despair of obtaining the blessings of a future life, because thou art told that, in order to do so, it is necessary to forsake those pleasures and habits to which thou clingest, like a vulture to the carrion he feeds on. And because thou thus despairest, thou wouldst fain have it believed, for the sake of sparing thyself trouble, that the goods promised thee in the next world are poor, nay, even false, and so thou art, at least secretly in thy heart, a traitor to the faith, without being thyself aware of it. Ask of this same Apostle who those are who are "alienated from the life of God, because of the blindness of their hearts." He will at once answer that they are those "who, despairing, have given themselves up to lasciviousness."[1] Wretched men who have come to such a state as this! Yet how many are there, even amongst Christians, who are continually falling into it! What, then, shouldst thou do, to keep really at a distance from it? Abstain, as far as possible, even from sensible pleasures, for if thou hast an excessive love for these, the mournful result will be that thou wilt gradually pass from them to such as are sensual.

SEVENTH DAY.

Christ suffered for us, leaving you an example, that you should follow His steps (1 St. Peter ii. 21).

I. Consider first, that it was for three most sublime ends that Christ our Lord came into the world, and, not regarding the joys and the glory which He might have justly taken for His own in the world, submitted to a life so full of sorrow. The first was that He might redeem us by His Blood, the second that He might enlighten us by His doctrine, the third that He might both direct and encourage us by His own most holy example. And it seems that He might well have applied to these ends those three glorious titles which He gave Himself when He said, "I am the Way, the Truth, and the Life,"[2] for He was to us the Way by example, the Truth by doctrine, and the Life by redeeming us with His Death. But in this passage, the Apostle passed by the other two ends, glorious as they are, and judged well only to speak of that of example,

[1] Ephes. iv. 19. [2] St. John xiv. 6.

which is the most needful for us in this world ; since, though
we are indeed redeemed and enlightened by Christ, we cannot
be saved unless we firmly resolve to follow Him in the way
of suffering which He trod. Keeping his eyes fixed on this,
therefore, the Apostle spoke in this manner: "Christ suffered
for us, leaving you an example that you should follow His
steps." Since, then, he said that He "suffered for us," we
might have thought that he would say also "leaving us an
example," but he did not say so; he said "you," because
although it is true that Christ gave His Apostles an example
of suffering greatly, yet He did not leave it: "I have given
you an example, that as I have done to you, so you do also."[1]
He left it to those who were to come after, and so St. Peter
said "leaving you," that is, "leaving behind Him." It was for
us, then, that our Lord arranged with provident care, that all
the examples He had given, but especially those in the matter
of suffering, should be so fully set down by four most accurate
Evangelists, so that as it was not possible for us to witness them,
as the Apostles did, with our eyes, we might at least learn
them by diligently meditating on those most holy books. But
of what avail is this care, if, instead of studying, thou turnest
away from them with distaste? Oh, what injury art thou
doing thyself by continually reading useless and foolish books,
which, by flattering thy corrupt senses, gradually alienate thee
from suffering instead of inciting thee to it! If, then, thou
art not ready to follow Christ courageously, the fault is thine.
He has left thee an example : if thou dost not take it, it must
be attributed to thyself that thou voluntarily rejectest, as it
were, thy inheritance, as something more burdensome than
it is lucrative. Oh, in what a state of deception thou art
living !

II. Consider secondly, that in order to remove the alarm
which thou mightest feel at hearing that thou art bound to
take example by Christ, Who suffered so greatly, the Apostle,
with great wisdom, adds that Christ left thee that example
that thou mightest follow it, not that thou mightest attain
it : "That you should follow His steps." He says, not *ut ad-
sequamini*, as Tertullian wished to understand it,[2] but *ut
sequamini :* for who amongst us could ever equal the example
of Christ? It is enough if we follow it. But how can that man
be said to follow it, who walks in a path entirely opposed to
it ? Thou complainest of thy weakness, but without reason.

[1] St. John xiii. 15. [2] Scorpiac. xii.

For thy weakness does indeed prove thy inability to keep pace
in the way of suffering with thy Lord, Who ran in it like
a giant—" He rejoiced as a giant to run the way "[1]—but it
does not prove thy inability to walk in it, if only thou co-
operatest with those aids which are imparted to thee by grace
for this end.   But thou art unwilling to follow Christ, even as
St. Peter did on the night of the Passion, who followed Him
indeed, but afar off, " because he was afraid."   Thou wouldst
even turn thy back upon Him boldly, seeking thy own
advantage to the utmost, seeking after pleasure and luxury
and all kinds of excessive self-indulgence.   It is not thy
weakness, therefore, but thy want of good-will which prevents
thy following Him.   If thou art not able to suffer as much
as Christ, at least be willing to suffer with Christ : " May it
please my lord to go before his servant, and I will follow
softly after him."[2]

III. Consider thirdly, that there are some who do indeed
walk in the same way as Christ, that is, the way of suffering,
and yet who cannot be truly said to follow Him, because
although they walk in that way, it is by compulsion ; they
suffer, because they cannot avoid it on account of the misery
of the state they are in, whether it be one of poverty, sickness,
disgrace, or any other misfortune which has overtaken them.
And, oh, how unwillingly do they suffer !   Such persons
certainly are not following their Lord, although they are
walking in His way, namely, that of great suffering.   And
therefore the Apostle was not satisfied with saying, " Christ
suffered for us, leaving you an example that you should follow
His way," but " His steps."   These are his words, and in
speaking thus, he spoke rightly.   It is one thing merely to
walk in the same road with a person, and another to tread
in his footsteps.   And, therefore, it is not enough for thee to
walk in the way Christ walked in, which is the way of suffering ;
but it is also necessary for thee to walk in it as Christ walked,
with the same submission of will, the same peace, the same
patience, the same perseverance, and, if possible, the same
gladness.   This, if we consider aright, is following His steps :
" My foot hath followed His steps."[3]   Never, indeed, canst
thou tread in them as deeply as He did, but still thou must
keep in them.   Of what avail is it to suffer greatly, if all the
while thou art inwardly murmuring against the troubles which
God sends thee, or if thou dost indeed suffer, but as thou

[1] Psalm xviii. 6.   [2] Genesis xxxiii. 14.   [3] Job xxiii. 11.

choosest, performing what penances, fasts, disciplines thou pleasest, but unable to endure one which is imposed on thee to correct thy faults? If, then, thou wouldst incite thyself to do so, as is right, remember how just it is that thou shouldst suffer, and say within thyself: "Christ suffered for us." Oh, what a disparity—"Christ for us!" A Lord of such majesty for a miserable worm of earth! A Master for the servant! The Prince for the subject! God for man! "Christ for us!" If, then, Christ suffered for me, how (so shouldst thou go on to say) is it possible that I should not suffer, and suffer gladly for Him? So wilt thou not only follow the same way with Him, but follow it by walking in the same steps: "He stuck to the Lord, and departed not from His steps."[1]

EIGHTH DAY.

THE NATIVITY OF OUR LADY.

The Lord possessed me in the beginning of His ways, before He made anything from the beginning (Prov. viii. 22).

I. Consider first, that these words, which, according to the most ancient interpretation of all the Greek and Latin Fathers, the Wise Man placed, in the first instance, in the mouth of Christ, the Incarnate Wisdom, have been also, from the earliest times, placed by the Church in the mouth of the Blessed Virgin Mary, by virtue of the privilege she enjoys of sharing all those glorious titles of Redeemer, Life, Way, Light, Hope, Salvation, and Harbour which properly belong on earth to Christ only. On this day, therefore, thou hast to receive these words from the mouth of Mary, so as to incite thee to a love for her corresponding to the love which God bore her from all eternity, a love which is inexplicable. Suffice it to know that from all eternity He predestined her to be the Mother of His Blessed Son, and thus He also elected her with Him from all eternity to one and the same order, an order composed of these two only, and superior to that of all the rest of the predestined—with this difference, that Christ was chosen for Himself, whereas Mary was chosen because Christ was chosen. And it is this which

[1] 4 Kings xviii. 6.

the Blessed Virgin here points out to thee in a few words, saying, "The Lord possessed me in the beginning of His ways, before He made anything from the beginning." She teaches thee that she was chosen by God "in the beginning," not of time, seeing that she was chosen by Him, "before He made anything from the beginning," but of His Divine decrees, "of His ways," that is, the very same "beginning" in which Christ was chosen, absolutely independently of all others. Now, this is enough of itself to prove to thee how great is the love borne to the Blessed Virgin by God. Out of an infinite multitude of creatures He elected her to that sublime height where she stands so far above all the predestined, that all of them, all their orders of apostles, prophets, pastors, doctors, and the rest, are at as great a distance from her as they are from that most sublime order in which Christ, their Supreme Head, stands. Let this thought move thee to love her as thou art bound to do. It is not enough to love her for the benefits which thou art continually receiving from her, for this is rather to love thyself than her; thou oughtest to love her for what she is in herself, for her gifts, for her greatness, for this is to love her truly.

II. Consider secondly, in the first place, that Mary says "the Lord." She does not say "God," a word which has a somewhat severe sound, which seems to say that He is just, that He is a Judge. She says "the Lord," that is, the Sovereign Master, to show that the special attribute exercised by God in exalting the Blessed Virgin is that of sovereignty; because He would not, in her instance, be bound by any of those laws which, as God, He has laid down for all others. See, therefore, how many are the prerogatives He has bestowed on her. She was a woman, formed like other women of human flesh, but without any leaven of concupiscence; an infant, but with a free-will operating in acts, with reason, and with wisdom; impeccable, but with merit; inviolate, but without sterility; fruitful, but virgin in her conception, and without suffering before or in childbearing; beautiful, but with a beauty which inspired those who looked on her with purity; . dying without pain, and incorrupt after death; she was a pilgrim on earth for more than seventy years, yet she never knew weariness or langour, but all her actions were always done with that perfect virtue which belongs to the blessed in Heaven. And so, even as the Tables of the Law written by God were broken at the foot of Mount Sinai, so too may it be

said that all the laws common to others were abrogated at the
feet of Mary, who was the mountain typified by Sinai—"A
mountain in which God is well pleased to dwell"[1]—so truly did
God choose to act as a Sovereign in her.   And what should be
thy part, but to exult and rejoice exceedingly in all these prerogatives with which thou seest thy Queen adorned above all
others?   Oh, well will it be with thee if by the homage that
thou renderest her, thou at last winnest her favour ; then thou
wilt be safe.   For even as God did not choose to be bound by
any laws in exalting the Blessed Virgin, so neither will He be
bound by any in granting her requests.   *Dominus*, He is "the
Lord."

III. Consider thirdly, next, that Mary says, "He possessed
me," to show that she always belonged to God, not by right
only, but by possession, which, with the exception of Christ,
is true of no other human being who is predestined to glory.
All other human beings predestined to glory are either our
first parents, or such of their posterity as are saved.   But,
with the exception of the Blessed Virgin, every one of their
posterity has been possessed by the devils before God, because
the devil robbed Him of all of them before their birth.   It is
true indeed that in the case of our first parents God was their
possessor before the devil ; but they were, too, soon torn from
Him.   Of Mary alone is it true that God was always both her
Possessor and Master ; because the devil was never able to
seize upon her, neither before God possessed her in mortal
flesh, nor after.   Not before, for God by His mighty arm
preserved her from that original sin from which He, as
Sovereign Lord, decreed that she should be exempt.   Not
after, because, by the aid of that same arm, she always kept
herself from actual sin.   Rejoice with all thy heart with the
Blessed Virgin in this sublime distinction, granted her by God
of being always all His own, both by right and by possession ;
and at the same time be confounded on thine own account,
because, whereas God is thy Lord by so many titles, thou yet
sufferest Him to possess thee in so small a degree.   It was thy
misfortune that thou wert possessed, before He possessed thee,
by that universal robber, the devil ; but it is by a strange
perversity on thy part that this robber continues to possess
thee after God has redeemed thee from him.

IV. Consider fourthly, that in the next place Mary says,
"In the beginning of His ways."   These "ways" are the

[1] Psalm lxvii. 17.

Divine decrees, as has been said. But these Divine decrees
have reference to two kinds of works, works of mercy and
works of justice. They can all be reduced to these: "All the
ways of the Lord are mercy and truth." But they all follow
this perpetual law, that the works of mercy always go before
those of justice; because as God is, by His nature, prone to
pity and slow to anger, so, when He begins to act, He always
begins by those works which are most spontaneous in Him,
namely, works of mercy. According to this law, then, what is
it that Mary says in these words, "The Lord possessed me in
the beginning of His ways"? She says, that if the Lord has
so highly exalted her, thou art not to marvel that when from
all eternity He deigned to take possession of her by decreeing
her birth in the world, He did so by this first way, the way of
mercy, not that of justice. He did not regard what justice
required in a woman who was to be of the race of Adam, a
mean, poor, and sinful race, He only regarded the exercise
of mercy; and so, adding to His freedom to act as Sovereign
Lord His inclination to do good, think what treasures of grace
He must have poured into her heart! And yet there is more
than this, for not only did the Lord begin by those works
which are works of mercy—"in the beginning"—(which is His
way of acting in all cases), but He then first began to do these
works—"in the beginning"—because the very first work of
mercy decreed by God was the election of Christ, and in the
same instant the election of Mary as His Mother, and to these
He added all the other works of the same kind which He
afterwards decreed in so vast a multitude, which were works
of mercy indeed, but secondary works. What wonder, then,
if the first of all those works with which God always begins,
was so perfect in its nature, that is to say, so perfect in the
way of mercy? Entreat Mary, that as she has the mercy of
God shown her in such rich abundance, so she may deign, on
this day of her blessed Nativity, to obtain a small share of it
for thee too; remembering, however, that the mercy which
thou needest is very different from that which she saw to be
exercised towards her. The mercy of which thou art in need
is the mercy which pardons, that which was shown to her was
the mercy which saved her from so sorrowful a necessity.

NINTH DAY.

Where envying and contention is, there is inconstancy and every evil work
(St. James iii. 16).

I. Consider first, that the zeal here spoken of—for the word translated "envying" is *zelus*—is the same thing which the Apostle has just before described as "bitter zeal."[1] And that, therefore, it is envy which is meant, for it is frequently called "zeal," because that from which it springs is jealousy of our own reputation; there being always this difference between envying another and hating him, that both do indeed make a man grieve at the good of another, but whereas he who hates his enemy is grieved directly because of evil will against him, he who envies another grieves on account of the love which he bears himself, because he thinks that the exaltation of his adversary must be his own abasement! "Saul was exceeding angry . . . and he said, They have given David ten thousands, and to me they have given but a thousand, what can he have more but the kingdom?"[2] Hence, as St. Augustine says, an equal envies his equal, because he sees that he is on a level with him; the inferior envies his superior, because he is not on a level with him; and the superior envies his inferior because, although he sees that he is not on a level with him, he fears that he may become so. Now, this envy is sometimes shut up in the heart, and then it is simple envy; and sometimes it breaks out into outward act, and then it becomes "contention," which is odious, in proportion to its ambitious character; for contention, according to the same Saint, is nothing else than a quarrelsome persistency in having the precedence in all possible ways, lawful or unlawful, since its object is not the advancement of merit, but of self. Wherever, therefore, an audacious envy of this kind lurks, there, the Apostle declares by Divine inspiration, "is inconstancy and every evil work:" "inconstancy" in the understanding, and "evil work" in the will. Oh, how necessary it is for thee to abhor this miserable condition! Beg of God, therefore, to make thee know the evil of it, so that thou mayest speedily come out of it.

II. Consider secondly, that wherever there is the envy we have spoken of, there is, in the first place, "inconstancy" in the understanding. And this is so, because envy not only

---

[1] St. James iii. 14.　　[2] 1 Kings xviii. 8.

darkens it, as every other passion does, but distorts it—"Saul did not look on David with a good eye from that day forward"[1]—so that the man who once appeared to thee deserving of all good, will seem utterly different from what he was, so soon as thou hast begun to envy him. What once appeared to thee to be devotion in him is at once changed into hypocrisy, generosity into effrontery, affability into affectation, and so on with all the qualities which once adorned him in thy eyes. And the reason is, not that the man himself is changed, but that thou art changed towards him : thou dost not "look on him with a good eye." It is that accursed spell, of which we are speaking, that has changed thee : it is envy. Thy understanding, no longer constant, but wavering, has been led by envy to change its judgment, or rather, it is never fixed in one opinion. For envy will sometimes make thee think that thy enemy is really deserving of the honours which he has gained, and at other times that he is undeserving of them. And it is impossible to say whether it disturbs thee most when it represents him as justly or unjustly honoured. Hence it is that this wretched inconstancy of judgment must necessarily show itself in thy manner of speaking of him, sometimes making thee appear reluctant to believe in the greatness of the credit which thou hearest attributed to him, so that thou persuadest thyself that it is less than is represented ; and again, making thee believe it to be even greater than it is : and thus thou art never at rest. And although thou wouldst fain hide the poison hidden within thy heart, thou art unable to do so, but despite thyself it will in the end betray itself in thy words, so far will thy agitation carry thee : "An evil spirit troubled him."[2] Hence it comes to pass that, in speaking of thy rival, it is impossible for thee to do so consistently, but after joining in praising him to some extent with those who speak well of him, for fear of openly showing thy envious disposition, by and bye in talking with others who blame him, thou wilt even go beyond them, in order to embrace an opportunity of depreciating him. And thus it is very easy for those who observe thee to perceive thy inconstancy: "And the servants of Saul said to him, An evil spirit troubleth thee." It must be remarked, however, that there is a master-stroke of art which always is practised by thee in blaming another, and which is the exact reverse of that which flatterers generally adopt, availing themselves of that affinity which exists between a

[1] 1 Kings xviii. 9.   [2] 1 Kings xvi. 14.

man's virtues and vices; so that whereas they, for instance, call the cunning of some prince prudence, avarice thrift, insolence courage, ferocity justice; thou, on the contrary, art wont to call the justice of thy enemy ferocity, his courage insolence, his thrift avarice, his prudence cunning; and in this way thou makest a bad use of this affinity between vices and virtues in order to give a colour to the malignity of the passion that troubles thee: "Why was not this ointment sold for three hundred pence, and given to the poor?"[1] What has been here said is enough to show thee whether envy has taken root in thy heart, for the signs which have been mentioned are considered by many the very clearest which betray its existence.

III. Consider thirdly, that just as there is inconstancy in the understanding wherever there is this envy of which we have spoken so much—"there is inconstancy"—so also is there "every evil work" in the will. To explain this, it is commonly said that envy transports a man to the greatest excesses of wickedness, so that, when he perceives that he can do no more to injure his adversary's reputation by words, he accomplishes it by deeds, and in this way goes on to deceit, plots, treachery, violence, and even to the most terrible murders that have been known in the world: "And Saul became David's enemy continually."[2] But there is another reason why it may be said that where envy is, there is "every evil work," not only "there will be," but "there is;" because envy is in itself an epitome of all wickedness: "An evil beast."[3] And so, if thou examinest other vices, thou wilt see that each one of them is opposed to some virtue, but only to its opposite, and that therefore that vice which is opposed to one particular virtue is not opposed to another. Gluttony is opposed to temperance, but not to liberality; cruelty to mercy, but not to chastity; anger to meekness, but not to frugality; deceit to loyalty, but not to long-suffering; and so on with all other vices: but it is not so with envy. Envy is the only vice which is opposed to all the virtues: because it is tormented by the sight of them all, as though they were all its opposites, and so it would fain either lessen, or uproot, or turn them all into vices: "Wherefore the Philistines, envying him, stopped up at that time all the wells that the servants of his father Abraham had digged, filling them up with earth."[4] In the same way,

---

[1] St. John xii. 5.
[2] 1 Kings xviii. 29.
[3] Genesis xxxvii. 20.
[4] Genesis xxvi. 15.

when any other vice occasions an evil of one sort, it necessarily
hinders another: if it makes a man avaricious, it prevents his
being prodigal; if it makes him audacious, it prevents his
being pusillanimous, and so on. But not so envy: envy, so
far from preventing any evil, encourages all. Therefore, as
thou seest, it was envy which brought them all into the world:
" By the envy of the devil death came into the world."[1] And
so there is something devilish in the conduct of the envious,
for they, like the devil, are grieved at the good, and rejoice at
the harm which happens to man. This is the reason of the
word used by the Apostle, which is translated "evil;" it is not
opus malum, but *opus pravum,* because the evil of the works
wrought by the envious is not an evil done by chance, but of
purpose; it is sharpened by malice and poisoned by malignity,
and so it is an evil which springs from an utterly perverted
will, like the evil wrought by the devil. And canst thou give
admittance to such an evil as this within thy heart?

IV. Consider fourthly, that although envy is really so very
hard to cure, that it is on that account compared to a corruption lurking in the bones—" Envy is the rottenness of the
bones "[2]—yet even it can be cured by the grace of God. But
the remedies must be applied in time, otherwise the gangrene
will become a grievous malady, only to be healed by a miracle;
and this is why envy, when it has reached the highest pitch of
iniquity, is counted among those sins which are said to be sins
against the Holy Ghost, for it is not just that He should do
good to one who grieves at the good which He does to others.
These remedies are of two kinds: one speculative, the other
practical. The first is to attain to a clear knowledge of the
injury thou art doing to thyself by envy. For whereas, if thou
accustomest thyself to rejoice in the good of another, all his
good becomes thine by means of that beautiful act of charity,
so that thou, too, mayest say to God with exceeding joy, " I
am a partaker with all them that fear Thee,"[3] so, on the
contrary, if thou art displeased at it, all the good of another
is changed instantly into thy evil, and a very grievous evil:
bodily evil, which afflicts, agitates, and torments thee in vain;
and mental evil, which makes thee as hateful to God as a
devil, who persecutes with enmity the good which God does
to the world. What a mad bargain is this! To change all
the good of others into evil for thyself, when thou mightest
so easily turn it to thy own good: "With a good eye do

according to the ability of thy hands; for the Lord maketh recompense, and will give thee seven times as much."[1] The second remedy is to be prompt in repressing the first motions of this detestable vice, so that if the devil, like a serpent, as a rule lurks at thy heel, that is, at the end of every good work, to prevent its proceeding to a good issue—"Thou shalt lie in wait for her heel"[2]—thou, on the contrary, shouldst at once endeavour to crush his head by attacking the temptation which he arouses in thee at its first beginnings: "She shall crush thy head." And this is to be done, in the case we are considering, in three ways: in the heart, in words, and in works. In the heart, by praying immediately to God for the person against whom the devil is striving to excite thy envy, and by wishing him all prosperity, grace, honour, and happiness. In words, by purposely speaking well of him when thou hast an opportunity, and still more by not saying anything to the contrary when thou feelest annoyance at hearing him spoken well of by others. In works, by endeavouring, if possible, to assist in advancing his interests in any matter, so far as justice allows. Do this, and the gangrene will be cured as by the application of steel and fire. The steel is the first remedy, coming from the understanding, and probing the hideous wound so as to discover all the corrupt matter contained in it. The fire is the second remedy, which arises from the will, and which absorbs this matter by acts of charity, which are effectual in proportion to their fervour.

TENTH DAY.

I therefore so run, not as at an uncertainty: I so fight, not as one beating the air: but I chastise my body, and bring it into subjection, lest perhaps, when I have preached to others, I myself should become a castaway (1 Cor. ix. 26).

I. Consider first, that, rightly regarded, the life of a Christian is nothing but a perpetual race, a perpetual conflict: a race for the prize, a conflict against those enemies who would fain impede us in our race. The prize is that perfection to which God calls us in our state: "The prize of the supernal vocation of God."[3] The enemies are those irregular appetites which we have within us: "A man's

---

[1] Ecclus. xxxv. 12, 13.   [2] Genesis iii. 15.   [3] Philipp. iii. 14.

enemies shall be they of his own household."[1] It is necessary, then, to set thyself with manly courage to do both these
things, to run and to fight.   But observe well the manner of
doing so which the Apostle teaches us.   It is not to act at
hazard, but to place before thee very particularly the goal
which is the end of the race, and the enemies whom thou
desirest to overcome by fighting.   That man runs "at an
uncertainty" who, while desiring to attain that perfection
which is the final goal, does not set before him, one after
another, the virtues which he specially wishes to acquire ; and
he fights "as one beating the air" who, while desiring to
subjugate his passions, does not fix on any one in particular.
How stands it, thinkest thou, with thyself in this matter?   If
thou wouldst do right, examine which is the virtue in which
thou art most wanting, and let that be the object of thy race ;
examine which is the vice which most predominates in thee,
and let that be the one with which thou preparest to fight.
More than this : consider well also in what way thou oughtest
both to run and to fight : "I, therefore, so run, . . . I so
fight," not merely "I run," not merely "I fight," but "so."
This is the right rule for making progress : not to take up the
business, so to speak, in the abstract, but in its several parts.
"So it becometh us to fulfil all justice,"[2] not merely "it
becometh us to fulfil" it, but "so to fulfil it."

II. Consider secondly, that the end which the Apostle had
set before him in his race was, without a doubt, to draw souls
to Christ ; that it was for this that he traversed so great a
multitude of countries without ever resting.   And yet the
principal means which he employed for the attainment of his
end was to make war upon his body by afflicting, mortifying,
scourging it—for all this is implied in the word *castigo*, which
bears the same meaning as *contundo*, and this is not accomplished without wounds and blows ; as though it was not
enough for him that his body should undergo intense fatigue,
but he must also inflict suffering on it.   But who can fail to
marvel exceedingly at this?   Would not one have thought
that every one who was moved by pity for all those nations
hastening to perdition, would have exhorted the Apostle to
spare himself for their good, not to undermine his health, not
to enfeeble his strength, not to shorten his life in this way?
And yet he was of a contrary opinion.   The means which he
considered best adapted to the attainment of his end was the

[1] St. Matt. x. 36.   [2] St. Matt. iii. 15.

mortification of the flesh : " I chastise my body." " I chastise,"
not " I slay ; " for this mortification must be exercised in such
measure as to conduce to the end, but still there must be
chastisement ; it must not be depreciated as though it were
a virtue belonging only to beginners.   " I chastise my body"
even after so many years of the spiritual life, not " I chastised
it " merely at the beginning of that life.   I chastise it in the
midst of labours and journeys and preachings, and all those
other noble works of charity which might well be thought,
of themselves alone, sufficient for my salvation.   This is
what he said.   And what dost thou say—thou, who art continually cherishing and pampering thyself under pretence of
preserving thyself for the greater glory of God.   Dost thou
suppose thyself to be so much more necessary to the human
race than the Apostle was?

III. Consider thirdly, that what should make us most afraid
is what the Apostle says next, " Lest, perhaps, when I have
preached to others, I myself should become a castaway," as
though damnation must be the consequence of neglecting the
mortification of the flesh : " What will the lamb do, when the
ram fears and trembles ? "[1]   Art thou sure that the neglect
of this mortification will not result in the same misery to thee?
If thou hast even the slightest suspicion that it may be so,
it ought to make thee very anxious.   This is why the Apostle
here says " lest perhaps," because the matter is so immensely
important, namely, salvation.   What will it avail thee to save
the world if thou art lost after all?   " What doth it profit
a man if he gain the whole world, and suffer the loss of his
own soul ? "[2]   Dost thou imagine that it is impossible for thee
to be sent to Hell after having sent many others to Heaven?
If this were true, the Apostle would not have said, " Lest
perhaps when I have preached to others, I myself should
become a castaway."   Whoever saved more people than he
did?   And yet he did not trust himself as one who having
been confirmed in grace could make sure of salvation.   Oh,
how greatly is even a slight risk to be feared, when the risk
is that of reprobation !   " At this my heart trembleth, and is
moved out of its place."[3]

IV. Consider fourthly, that this reprobation is always
possible because it is accomplished within ourselves.   It is
of God that we are chosen for the enjoyment of His glory:

---

[1] St. Augustine, Sermon xxi. *De Ver. Apost.*   [2] St. Matt. xvi. 26.
Job xxxvii. i.

it is of ourselves that we are rejected; and therefore the Apostle does not say "lest perhaps I should be sent away" (*reprobus evadam*), but "lest I should become a castaway," because every one is the maker of his own misery: "Destruction is thy own, O Israel."[1] But if our reprobation is accomplished within ourselves, who is there who ought not to fear greatly? It is this which is so marvellous, that the Apostle is actually afraid of being lost after labouring so much for God, and for this reason afflicts and mortifies himself, and that thou imaginest thy salvation to be within thy grasp whilst thou art leading a life of indulgence in thy own comforts! Wouldst thou have me believe that thy flesh is in greater subjection to the spirit in the midst of these comforts than the Apostle's was in the midst of his sufferings? I cannot think it. Listen to the words which he utters to the confusion of those who so readily imagine themselves to have become impeccable: "I chastise my body, and bring it into subjection." He does not say, "I keep it in subjection:" and this is a proof that the rebellion of the flesh is experienced even by those who are perfect, and that to the very end.

ELEVENTH DAY.

You are of your father, the devil; and the desires of your father you will do (St. John viii. 44).

I. Consider first, that there are four ways in which a man is commonly said to be the son of another, even although he is not directly begotten by him. The first is by nature, and it was in this way that the Jews boasted of having Abraham for their father: "Abraham is our father."[2] And Christ did not deny the truth of this, but only replied, "If you be the children of Abraham, do the works of Abraham." The second way is that of adoption, and it was in this sense, in the natural order, that Moses refused to be the son of Pharao's daughter, who had adopted him: he "denied himself to be the son of Pharao's daughter."[3] And in the supernatural order, all the just are thus actually called the children of God: "He hath predestinated us to the adoption of children through Jesus Christ."[4] The third way is that of teaching, and it was

[1] Osee xiii. 9.   [2] St. John viii. 39.   [3] Hebrews xi. 24.   [4] Ephes. i. 5.

of this that the Apostle spoke when he said to his Corinthian disciples, "I write not these things to confound you, but I admonish you as my dearest children,"[1] because he had brought them to the faith of Christ. The fourth way is by imitation, and in this sense the same Apostle said to the Ephesians, "Be ye therefore followers of God, as most dear children."[2] He added "most dear," because it is resemblance which usually most endears children to a father. When, therefore, in the last passage which I am proposing for thy meditation, Christ tells the perverse Jews, and all poor sinners in them, that the devil is their father, He does not mean to assert that they are his children by nature or by adoption, which are the first two kinds of sonship, but that they are so by the two others, namely, teaching and imitation. For it is he who gives them the most exact instruction and laws of evil, both which they, his guilty children, are quick to learn. If, then, there were no other reproach that could be with truth uttered against sinners but this, "You are of your father, the devil," how great a reproach would it be! A man who had the public hangman for his father could not appear without shame in the company of his fellow-citizens. And canst thou, who hast the devil for thy father, dare to present thyself among so many of God's servants? It is too evident that thou dost not understand the infamy of thy father.

II. Consider secondly, that in order to show themselves the devil's true children, sinners strive to be as much like him as possible in all things. And therefore Christ says, "You are of your father, the devil; and the desires of your father you will do." Not merely his "works," though that would be enough, but his "desires," so greatly do sinners aim at resembling their father interiorly as well as exteriorly. And so it is that often, when wicked men cannot sin in act, they at least do all they can by sinning in thought: giving way to carnal desires, to hatred, rage, rancour, and malignity without end. There is, indeed, another meaning which our Lord may have intended; for if thou observest, He did not say, "The desires of your father you do," but "you will do." And why did He say so? To show that these wicked children exert themselves even to go beyond their father. Therefore, when the devil is unable to injure the world in any way but by desire, they come to his aid by carrying it into execution. How much cockle would the devil sow among men, if he

---

[1] 1 Cor. iv. 14.   [2] Ephes. v. 1.

could! How much slaughter would he accomplish, how many murders would he commit, how much impurity would he spread even in the most strictly enclosed convents! But the wretched fiend cannot do all this, because God has tied his hands. Therefore, where the father's power is insufficient, his children step in and do "the desires of their father," by sowing the cockle he desires to see sown, by committing those murders, those impurities, which are, even in his eyes, so surpassingly hideous, that he would not dare to debase his nature by them. And do not imagine that the devil has to use compulsion in order to induce these unhappy children of his to commit these great sins. Not at all. They do it of their own free choice; and this is why Christ did not say, "The desires of your father you do," but "you will do," because it is their own will which leads them to it. And do they not show by this that they do these things because they are what they are? They do them, according to their character of children, and their deeds are the more infamous, because they are voluntary. Canst thou, therefore, conceive of any children in the world more wicked than these that I have described to thee? How terrible would it be if thou wert one of them!

III. Consider thirdly, if this is so, how much better it is to forsake so terrible a father, and exchange him for one Who is deserving of honour, nay, of the highest possible honour, since, just as by learning evil from the devil and imitating him in it, thou art the child of the devil, so too, if thou wouldst learn good from God and imitate Him in it, thou wouldst at once become the child of God."[1] Nay, there is more than this : for if thou becomest the child of God in the two ways above mentioned, thou wilt also become His child in another way, namely, that of adoption (since that of nature is exclusively reserved for Christ), and by this blessed adoption thou wilt be so exalted as to be partaker of the same grace and glory as belong to His Son by nature: "If sons, heirs also : heirs indeed of God, and joint-heirs with Christ." Would it not, then, be the very height of folly to refuse to be numbered among the sons of God in order to continue one of the sons of the devil? Yet this is what thou doest every time that thou refusest to turn from sin : "Behold, I should condemn the generation of Thy children."[2] This is what thou sayest to God : thou sayest that thou refusest to be His child, in order

[1] St. John i. 12.   [2] Psalm lxxii. 15.

to continue the child, not of a hangman, which would be no disgrace whatever in the sight of God, but of one who is a traitor, a rebel, and a renegade to God; one whom He has banished for ever from His presence, as guilty of high treason. Dost thou not think that this is the most enormous audacity that could by possibility be shown to one Who is God? But if the reverence thou owest Him is not sufficient to move thee, at least be moved by the additional motive of thy own ruin. And to this end, do thou think a while on the difference which there will be on the Day of Judgment between those who will then appear as children of the devil! "We fools," so will these miserable beings say on beholding those others: "We fools esteemed their life madness"—because they loved poverty, desired suffering, and longed for contempt—"and their end without honour," because they very often met with the contempt they preferred. And now—oh, what a difference! "Behold how they are numbered among the children of God" —with Whom, therefore, they will dwell, and enjoy Him for all eternity—"and their lot is among the saints."[1] Think of this, and see whether it is to thy advantage to choose to be one of the devil's children, when thou mightest be one of the children of God. And dost thou know how these wretched children of the devil, of whom we are speaking, are described in the Sacred Scriptures? They are called "children of Hell." "You make him the child of Hell two-fold more than yourselves," the meaning of which plainly is, that they, too, are doomed at last to the same inheritance which is that of their father in the infernal abyss.

TWELFTH DAY.

See, therefore, brethren, how you walk circumspectly, not as unwise, but as wise; redeeming the time, because the days are evil (Ephes. v. 15, 16).

I. Consider first, that the days of this life are given to us by God for a very important service, which is to transact the weighty business of our everlasting salvation. Still it is impossible to deny that, for the most part, they suit our purpose ill. For they are few, unstable, uncertain: and even of these few we have, in spite of ourselves, to give up a large portion to the necessities to which we are made subject by

[1] Wisdom v. 5.

original sin. Therefore, just as an instrument, when it does not serve its purpose well, is called bad, so, too, are our days said to be "evil." They are so called because there is but a small portion of them that can be employed as it should be : "The days of my pilgrimage are few and evil,"[1] which is the passage here alluded to by the Apostle. And yet, who is there who values that little good which our days contain, at the infinite price which it deserves? Many lose it in wicked, more still in useless ways ; and there are very few indeed who employ it altogether for the purpose for which it is given to us. This, then, is what the Apostle here means : that thou shouldst value time highly, and spend it all in the best possible way. Examine thyself, and see whether, on the contrary, thou hast not the bad habit of losing it.

II. Consider secondly, that the Apostle takes for granted that thou hast, at all events, lost much time in the past, as is the case with most persons, and therefore he tells thee in this passage to redeem it : "Redeeming the time." . But how is this possible when it is lost? By making good, in the short time that remains to thee, all the losses which thou hast suffered in the long time that thou hast lost, by growing in the love of retirement, by amending thy actions, by increasing thy prayer, by redoubling the fervour with which thou doest penance : "My eyes prevented the watches."[2] This is what is done by travellers who have lost several hours of the day in amusing themselves idly on the road. They make up for it by greater speed on the morrow. It is the same with labourers and artisans, and all persons who have incurred any loss through waste of time—they take the more pains to redeem it. And yet theirs is only a temporal loss. How, then, should it be with thee, who mayest easily have incurred one that is eternal? Look down into Hell, and ask there, what one of the damned would do, if it were in his power to return to earth and live the past over again. Dost thou think he would say that, like thee, he would indulge in sleep, in amusements, in idle talk and jesting, and hearing news? Far from it. Who can say how earnestly he would promise to labour, so as to do a great deal in a short space? How is it, then, that thou dost not apply thyself to the thought of thy own case? Art thou less bound by gratitude to God, because, instead of having taken thee out of Hell, where thou hast deserved to be by thy sins, He has graciously preserved thee from falling into it?

[1] Genesis xlvii. 9.   [2] Psalm lxxvi. 5.

Redeem the time, therefore, all the more; because it was, in the majority of cases, thy own fault when it was lost, through thy despising it, wasting it, or at least not defending it from robbers, that is, from those who stole it from thee for nothing.

III. Consider thirdly, that it is not only things which have been lost that are said to be redeemed, but those also which are in danger of being lost. Thus we say that a man redeems his life when he buys it for money of assassins who are standing ready, sword in hand, to take it from him. This, too, the Apostle enjoins on thee when he would have thee redeem thy time; he bids thee be careful to save it and buy it from assassins. And may not all those who rob thee of thy time be truly called thy murderers? Whenever they rob thee of time, they rob thee also of so much life, and that not temporal only, but eternal. Now, lift thy eyes from Hell to Heaven, and ask there how each one of the blessed would employ the time of which thou makest so light, now that he has at length come to know how much glory, greatness, and joy might be accumulated, even in one brief moment. If there could be sorrow in Heaven, surely this would be the way in which that sad intruder made his entrance—that there was no longer any time in which to merit. Wilt thou, then, who hast this precious time, allow thyself to be robbed of it? For this one thing, Heaven itself might envy thy state, that thou art able to merit: "Whilst we have time let us work good."[1] Let no one, therefore, rob thee of thy time—"Observe the time"[2]— all the more because, if it is lost, the evil is two-fold: the profit which ceases, and the loss which ensues. The profit which ceases is the gain which thou mightest be accumulating in Heaven by a good use of the present time, and which thou dost not so accumulate; and the loss which ensues is the punishment which thou wilt incur by this neglect of thy capital: "He hath called me against the time."[3]

IV. Consider fourthly, who these assassins are, from whom thou hast to redeem thy time for the future, so as not to lose it. They are the same who have so often robbed thee of it in the past--the men amongst whom thou art living. These may be divided generally into two classes, friends and enemies. The former very often try to rob thee of thy time by inviting thee to idle amusements; do thou redeem it, even at some cost, even by letting them think thee uncourteous. The latter try to rob thee of it by persecuting thee; they excite disputes

[1] Galat. vi. 10.   [2] Ecclus. iv. 23.   [3] Lament. i. 15.

about thee, they distrust and importune thee, almost endeavouring to compel thee to lose much time in defending thyself. Do thou redeem it from them also, even though it should be with some considerable loss of reputation or property. " Lose something," so was St. Augustine often wont to say, " so as to redeem time which thou mayest spend for God."[1] This is what is meant by " redeeming the time." And how wise is the man who acts in this way ! Yet there are few who understand this ; most men consider every other temporal good more valuable than time, whereas time is really more valuable than every other temporal good, because the eternal good may be purchased at last by a man of sound mind without anything else ; but without time it is impossible.

V. Consider fifthly, that just as thou art very careful not to fall into the hands of assassins, in order not to be obliged to redeem thy life at a great cost, so, too, oughtest thou to act in order not to be obliged to redeem time. Therefore does the Apostle say, in the first place, " See how you walk circumspectly." Because this is the first thing that thou hast to do, to walk circumspectly, so as to be on thy guard against those who would fain rob thee of time ; when thou art not able to do this, then to redeem it. But observe that he does not merely say, " See that you walk circumspectly," but, " how you walk : " because thou art also to study the best ways of avoiding encounters. This is the conduct of wise men, whereas fools run headlong into the power of assassins. Therefore, after saying, " See how you walk circumspectly," the Apostle adds immediately, " not as unwise, but as wise." The unwise are those who do not even recognize an evil that is present : " The senseless man shall not know."[2] The wise are those who foresee even a future evil, and so escape it : " A wise man feareth, and declineth from evil." This should be thy course—to foresee the opportunities of losing time, which may be given thee by so many, and to avoid them dexterously. And in this matter (which is more important than perhaps thou thinkest) do not regard the practice of the generality of people, because the number of the foolish is infinite, and such are those who do not value time, but live in idleness ; they are all foolish, foolish to the last degree : " He that pursueth idleness is very foolish."[3] Rather regard that which thou wilt be glad to have done when the hour of death comes. How wilt thou then

---

[1] *Hom.* x.   [2] Psalm xci. 7.   [3] Prov. xii. 11.

rejoice in the time which thou hast spent well, and how bitterly wilt thou weep over that which thou hast not so spent! But what will this avail? Thou wilt not be able to redeem it then —"Time shall be no more"[1]—for if even the days of our life serve us so ill to do the good we ought to do that they are called "evil," that of our death will not be able to serve us at all, for which reason it is called night instead of day: "The night cometh, when no man can work."[2] This, then, is the Apostle's meaning, when he says, "See how you walk circumspectly, not as unwise, but as wise, redeeming the time, because the days are evil."

THIRTEENTH DAY.

Now is the judgment of the world: now shall the prince of this world be cast out: and I, if I be lifted up from the earth, will draw all things to Myself (St. John xii. 31, 32).

I. Consider first, that there were two most happy results gained by the Death of Christ: one was the despoiling the devil of the dominion he possessed over the whole human race; and the other was the investing our Lord with that dominion. But do not suppose that this was done by chance or caprice. It was done by the just sentence passed by God, as Supreme Judge, in a most equitable judgment made by Him between Christ and the devil. And therefore, when He drew near to the time of His Death, Christ said, "Now is the judgment of the world;" and then added these words: "Now shall the prince of this world be cast out: and I, if I be lifted up from the earth, will draw all things to Myself." Oh, what devout meaning wilt thou be able to draw from these sublime words for thy soul's good, if only thou payest heed to them! Do thou, therefore, penetrate deeply into their meaning.

II. Consider secondly, that man allowed himself voluntarily to be conquered by the devil, by consenting to evil. And therefore he was, by a just sentence, condemned by God to a miserable slavery under the cruel tyrant whom he had chosen. Never could wretched man have freed himself from it through all time by his own power; on the contrary, yielding one after another to all the temptations presented to him by the devil, he would only have gone on continually adding sin to sin, till

---

[1] Apoc. x. 6.   [2] St. John ix. 4.

at length, after death, he would have gone to suffer the punishment awaiting him in everlasting fire. Hence it is as St. Augustine says,[1] that this dominion of the devil over man was just in itself, though exercised by that perfidious spirit with a most unjust intention. But if it was just, it was so not because it was his due by any title of right reason. It was just, only because it had pleased God to bestow it upon him ; as the power of the executioner over the criminal is just, because he has received it from the sovereign. Therefore, God, had He so pleased, might have pardoned man, and by His will have released him from the power of the devil, just as He had, by His will, given him into his power, without thereby doing the slightest wrong to the devil ; in the same way that every king may, at his pleasure, release a criminal from the hands of the executioner, without doing any wrong to the latter. But it did not please God to proceed in this manner— "The Lord is the God of judgment"[2]—and therefore He chose the matter to be transacted, if I may so speak, not in the tribunal of mercy, but of justice. For this reason He decreed that Jesus Christ, His own most holy, innocent, and spotless Son, the only one of the children of men who was not subject to the yoke of the devil, should come into the world to take upon Himself the payment of their debts. Unexpectedly, the devil beheld in the world a Man thus perfectly holy, and audaciously set himself to exercise over Him the same sovereignty, and with the same arrogance, which he exercised over the rest of mankind who were his subjects. He insolently ventured to approach Him in the desert, to tempt Him even to idolatry ; he persecuted Him, attacked Him, plotted against Him, and succeeded in getting Him condemned to death with savage fury, just as though He had been a sinner who deserved it. The perfidious fiend accomplished all that he desired through his great audacity in stirring up the Jewish people to put Christ to death. Christ most justly appealed against him to His well-beloved Father—"Arise, O God, judge Thy own cause"[3]—and His Father heard Him, as was right. Sentence was given against the devil, who heard it, with unwilling ears, fulminated from Heaven : "Thou hast caused judgment to be heard from Heaven."[4] And as he had unjustly sought to exercise his dominion over Christ, he was at once deprived even of that which had been granted to him over the rest of

[1] *De lib. Arb.* l. iii. c. 10. [2] Isaias xxx. 18.
[3] Psalm lxxiii. 22. [4] Psalm lxxv. 9.

men ; and it was declared that that dominion was justly due to Christ, Who had, moreover, made abundant satisfaction for the sins of the whole human race, not to the devil, who merely strove to multiply them by abusing his power, just as it was, to do things which were unjust. This, then, is what Christ intended to signify when He said, on the threshold of His Passion, "Now is the judgment of the world." He intended to signify that now, at length, the hour was approaching in which it would be decided to whom the sovereignty over all mankind (expressed by this word, "the world") belonged : to Him, Who had done so much to save it, or to the devil, who persecuted Him so ferociously. What dost thou say to this, thou who thinkest that thou hast escaped scot-free, as one may say, from the slavery of the devil? See, on the contrary, how this was effected at the cost of unspeakable torments inflicted by the devil on the Son of God, just as though He had been a vile man like thyself: "Tempted in all things like as we are, without sin."[1] Wilt not thou, then, at least, apply thyself to show that gratitude to the Son of God which is due from thee, by making war on the devil, who would fain, if it were possible, plot against Him even on His throne in Heaven?

III. Consider thirdly, that from this most just sentence there followed, in the first place, that despoiling of the devil which, as has just been said, took place when he was deprived of the dominion which had been granted to him over the whole of mankind subject to sin. This was signified by Christ in the words which follow : "Now shall the prince of this world be cast out." In words of sublime antonomasia, the devil is called "the prince of this world" in several passages of the Sacred Scriptures : "The prince of this world cometh, and in Me he hath not anything."[2] "The prince of this world is already judged."[3] And why is he so called but because of the authority which had been given him over the guilty world? "He is king over all the children of pride."[4] Now, of this authority he was deprived by direct sentence, especially on account of the outrages he had offered to Christ. And therefore Christ said that it was high time that so wicked a prince should be at last driven, not indeed from the world (for this He did not choose to do for good reasons of His own), but from his sovereignty : "Now shall the prince of this world be cast out." That is, out of his government, his dominion, his kingdom. Hence it follows that those

[1] Hebrews iv. 15.   [2] St. John xiv. 30.   [3] St. John xvi. 11.   [4] Job xli. 26.

who, notwithstanding, remain under the power of the devil, as is the case with so many idolaters, Jews, heathens, and Mahometans, as well as bad Catholics, do not remain in his power because the devil has the same authority over them that he would have had if Christ had not died for them, but because they are so foolish as to choose to remain thus, acting like the vilest of slaves, that is to say, those who are slaves by their own choice. For although it is quite true that men could never have escaped out of the devil's hands but for the grace merited for them by the Death of Christ, yet since they have that grace, the case is altered ; and if they choose, they might all escape from him : "Now we are loosed from the law of death, wherein we were detained."[1] And therefore, if the devil is their friend by keeping them subject to him, he is so merely because they make him their friend by choosing to obey him rather than Christ. What thinkest thou of this audacious rebellion, which is committed by so many men? Is it not right that thou shouldst both deplore and detest it, and endeavour utterly to destroy it, at least so far as thou art able? How will it be then if, on the contrary, thou art one of those who conspire to strengthen it?

There still remains to be considered the other result, which followed from the sentence given by the Eternal Father in favour of Christ, by which He was invested with the sovereignty of which the devil was deprived, a result expressed by Christ in these words : "And I, if I be lifted up from the earth, will draw all things to Myself." But, in order to consider them at greater leisure, as is right, we will keep them for the following day, a very suitable one, being the feast of the Exaltation of the Holy Cross.

FOURTEENTH DAY.

EXALTATION OF THE HOLY CROSS.

And I, if I be lifted up from the earth, will draw all things to Myself
(St. John xii. 32).

I. Consider first, that in the Sacred Scriptures we very commonly find the expression "all things" used to signify "all men." Thus it is said in one passage : "All (*omne*) that

[1] Romans vii. 6.

the Father giveth Me shall come to Me," meaning "every man." And again : "The Scripture hath concluded all (*omnia*) under sin,"[1] because man is in himself a compendium of all things. When, therefore, thou hearest Christ here say, "I, if I be lifted up from the earth, will draw all things to Myself," know that by the expression "all things" He does not mean to signify the types of the Old Testament, or the prophecies, or the prodigies and the commotion of the elements at His Death, as some of the saints have, however, learnedly explained these words ; neither does He merely intend to signify all the different races of men, as Jews, Greeks, Romans, and the like ; but He really means all men individually, and this is plain from the original text, where these terms are expressly used. But how is it true that when our Lord died upon the Cross He drew all men to Himself in the manner above mentioned, that is, individually ? This is what thou shouldst now try to understand, in order to draw thence those consequences which will, doubtless, conduce greatly to thy profit. Pray to God, therefore, to vouchsafe to make thee rightly enter into it.

II. Consider secondly, how Christ declared that by His Death (which He called an exaltation, for many reasons mentioned in the meditation for the 3rd of May, but especially because it was to take place from a high station, such as was the tree of the Cross), He would draw all men individually to Himself, because, the devil being despoiled of the dominion he had hitherto enjoyed over them, and Christ being invested with it, as was shown in the preceding meditation, it followed as a consequence that all must also individually belong to Christ, if not in fact (by reason of the contumacy of many of them), at least by right. This is the solution of the difficulty which has been mentioned. Nevertheless, it seems difficult clearly to understand how Christ declared so boldly that He was about to draw all men to Himself, when there were many who would, though by their own choice, refuse to go to Him, and who would consequently have been indeed merited, but not drawn by Him. Still, if thou considerest attentively, thou wilt see that Christ spoke with the most exact truth. All men are divided, as is well known, into two parties, those who are devout and those who are indevout to Christ. He spoke truly with regard to the former class when He said that by virtue of His Death He would draw them to Himself, because, by virtue

---

[1] Galat. iii. 22.

of His Death, He was to make them all His followers. Nor did He speak less truly as to the indevout, because, by virtue of that same Death, He was to make them all subject to Himself, at least on the Day of Judgment, by bringing them trembling to His feet, not, indeed, as His followers, for of that they would be quite unworthy, but as criminals dragged on by executioners : " For we shall all stand before the judgment-seat of Christ," not only "all" in kind, but "all" individually ; for it is written : " I live, saith the Lord, every knee shall bow to Me."[1] It cannot be denied that in acting thus He would draw the former by love and the latter by force to Himself. But, in spite of this difference, it remains perfectly true that all would be drawn : " All flesh shall come to Thee."[2] But, oh ! what a difference there is between these two ways of being drawn ! Do thou here enter into thyself and see whether it can ever be to thy profit to keep at a distance from Christ. Sooner or later thou wilt have to be brought to His feet, as thou hast heard, either by love or by force, either as His follower or as a criminal : there is no escape. Is it possible that thou wouldst rather be dragged as a criminal than run to Him as His follower ? What folly would this be ! Rather declare to God that thou wouldst rather die than subject thyself to be drawn in so shameful a way : " Draw me not away with the wicked ; and with the workers of iniquity destroy me not."[3] " Draw me not," by summoning me to Judgment ; " destroy me not," by condemning me after Judgment.

III. Consider thirdly, that after rightly understanding this explanation thou mayest think that those who, after the Death of Christ, continue indevout to Him are more truly said to be drawn to Him than those who are exceedingly devout to Him. For the latter, rightly considered, are not drawn : they go of themselves ; they who are drawn are those who have to be dragged by violence, as the wicked will be at the Last Day. But do not rest in that idea. For although both classes are truly said to be drawn, yet this can be said with greater truth of those who go to Him from love (though they are, indeed, drawn in the noblest way) than of those who go to Him by force, and the reason is that those who go from love follow the most powerful impetus that there is, that of their own will : " Each one is drawn by his own pleasure."[4] Still it should be observed that men are not drawn like brutes, but in a way

[1] Romans xiv. 11.　　[2] Psalm lxiv. 3.　　[3] Psalm xxvii. 3.　　[4] Virgil.

befitting their state, that is, the state of free agents, for which
reason God says: "I will draw them with the cords of Adam"
—according to some, "the cords of men"—that is, with
the cords with which I drew Abraham, Isaac, and Jacob
to Myself, and that was always "with the bands of love."[1]
There are, doubtless, many ways in which He does this; but
they may, after all, be reduced to three heads: persuasion,
favours, and sympathy. All these three are exceedingly
powerful, and they were all used with marvellous force by
Christ on His Cross, in order to draw multitudes of men to
Himself, though it is, indeed, true that He greatly strengthened
them by the interior virtue of that grace which He alone can
give. The first way of drawing man is by force of persuasion,
and this is two-fold, by word and by act. He who has the
gift of persuasion by words draws people by thousands, with
gentle violence, to himself. Still more does he do so who has
also the gift of persuasion by act, which is a language understood by all: "Speak . . . with all authority."[2] The second
way is by force of benefits, and this also is of two sorts, those
benefits which are already bestowed, and those which are yet
to be bestowed. The former class draws men by means of
gratitude, and the latter still more by interest: "He that
maketh presents . . . carrieth away the souls of the receivers."[3]
The third and last way is that of sympathy, and this, too, is of
two kinds. The first and broadest is the sympathy which
is born of likeness, for like always desires like: "Every
man shall associate himself to his like."[4] The second and
closer kind springs from a strong natural inclination such as
straws have for amber, iron for the loadstone, flames for the
heavens, and all things for their centre, to which they tend of
themselves with far greater velocity than that with which they
could be dragged by ropes in any other direction: "They are
sunk to the bottom like a stone."[5]

IV. Now fourthly, if thou wouldst know more particularly
how Christ drew so many followers to Himself from the Cross,
and drew them efficaciously, consider well all these three ways.
(1) He drew them by force of persuasion, for not only did
He, so to say, charm such multitudes by His preaching that
they were not able to leave Him, so great was their desire of
hearing Him—"Lord, to whom shall we go? Thou hast the
words of eternal life"[6]—but He added to this the persuasion

[1] Osee xi. 4.   [2] Titus ii. 15.   [3] Prov. xxii. 9.
[4] Ecclus. xiii. 20.   [5] Exodus xv. 5.   [6] St. John vi. 69.

of example, dying naked on a Cross between two thieves, with such humility, patience, peace, and resignation that even His executioners were smitten with love of Him, and came down from Calvary quite changed from what they were when they went up: "They returned striking their breasts."[1] (2) He drew them by force of benefits, both past and future. Past, because He had set them free from the slavery of the devil: "I will gather them together, because I have redeemed them."[2] Future, because He had thrown open to them the gates of Paradise: "I will give you a good gift, forsake not My law."[3] (3) And, lastly, He drew them also by force of sympathy; for on the Cross Christ truly showed Himself to be both God and Man, seeing that He endured death, as Man, and triumphed over it as God. Therefore He drew men to Himself as Man by the weaker kind of sympathy which comes from likeness, and as God He drew men to Himself by that more powerful sympathy which draws all things to their centre. For if the only centre of men's hearts is God, how is it possible for them to know Him and not be attracted to Him? If, then, all these excellent ways of being drawn are so powerful when used singly, think what must be their force when they are all combined. And it is thus that Christ used, and uses, and will use them till the end of the world for the good of those who will apply themselves to look upon Him with an eye of faith as He hangs on His Cross. This being so, did He not speak most truly when He said, "And I, if I be lifted up from the earth, will draw all things to Myself"? How sad would it be, therefore, if, notwithstanding, He should not yet have succeeded in drawing thee by any of these ways! If thou art not conquered by His words, be conquered at least by His example. If His example does not conquer thee, let His benefits do so, those which He has done and those which He intends doing to thee. If these do not move thee, then, at least, yield to that ruling instinct, which ought to be enough of itself alone to attract thee to Him, not merely because He is like to thee, but still more because it is in Him alone, as thy true centre, that thou wilt have peace: "These things I have spoken to you, that in Me you may have peace; in the world," which is out of the centre, "you shall have distress."[4] And if thou hast not yielded to any of these things, taken separately, do so, at least, to all of them combined.

[1] St. Luke xxiii. 48.
[2] Zach. x. 8.   [3] Prov. iv. 2.   [4] St. John xvi. 33.

FIFTEENTH DAY.

So let your light shine before men, that they may see your good works, and glorify your Father Who is in Heaven (St. Matt. v. 16).

I. Consider first, that this injunction was in the first place addressed by Christ to all the Apostles, and in them to all who were successively to succeed them both in the office of government and in that of preaching. And therefore He bade them let their light, that is their teaching, so shine before men, that it may be seen to be combined with works conformable to it, and thus give every one an occasion of perpetually praising God : "So let your light shine before men, that they may see your good works," that is, "that they may see that they are good, and glorify your Father Who is in Heaven." There cannot be a doubt that by "light," the teaching of the Gospel is plainly meant : "Now I send thee," our Lord said to Paul, "to open their eyes," (that is, the eyes of unbelievers), "that they may be converted from darkness to light."[1] When, therefore, it is manifest by this bright light, that he who sheds it upon the people himself practises what he preaches, how are they all enkindled to give praise to God ! But when the contrary is evident, how, on the other hand, are they scandalized by seeing the very doctrine which they hear extolled in words, condemned, so to speak, by the works of those who preach it ! And no wonder, for either that doctrine can be put into practice by weak mortals, or it cannot. If it cannot, of what use is it to teach it ? If it can, why do not those who teach it practise it ? This is the way the mass of men reason ; and it is a reasoning which, whatever its value may be, has so much power over their minds, that they are more disposed to act as their teacher acts, than to act as he teaches. Therefore, the Psalmist says : "To the sinner God hath said, Why dost thou declare My justices, and take My covenant in thy mouth ?"[2] He does not say "to the penitent," for one who has repented may preach with great fervour : indeed, he ought to do so, in order to atone for the offence he has given to God by the honour which he procures Him from others, and thus the penitent David said to God : "I will teach the unjust Thy ways."[3] Neither does he say "to the man who sins," because such a one, who sins sometimes through frailty, ought not immediately to leave off preaching, as unbecoming in him, but

[1] Acts xxvi. 18.   [2] Psalm xlix. 16.   [3] Psalm l. 15.

rather to gather fresh strength from that very preaching to rise
again manfully, and to show that he knows how to make use,
as a remedy for his own maladies, of the medicine which he
offers to others : " Of the fruit of a man's mouth shall his
belly be satisfied."[1] But he says " to the sinner," because the
man who is deliberately leading a bad life is most strictly
bound to keep silence : otherwise, there can be no doubt that
the better his words are, the worse will his works be ; for he
will show, all the more plainly, that he treats as a mere fable
that law which he explains so well and practises so ill. Art
thou in a state to lay down the law to others ? This is the
great obligation laid upon thee—to live as thou speakest : " So
let your light shine before men, that they may see your good
works, and glorify your Father Who is in Heaven." Christ
does not here require from every preacher that all his works
should be singularly perfect, for that would be to close the
mouths of countless numbers. But if He does not require
them to be perfect in kind, He requires them to be, at all
events, good ; since it is not becoming that one who reproves
others should himself be worthy of reproof.

II. Consider secondly, that, in the second place, our Lord
addressed this injunction to all those who bear the name of
Christian, still more that of religious, or of one consecrated to
the Divine service in any special way ; and He made it a duty
to all to act in such a manner as that their actions may correspond to a name so glorious not only in the sight of God, but
of men, and may thus be a still stronger motive for them to
praise Him : " So let your light shine before men, that they
may see your good works, and glorify your Father Who is in
Heaven." The suitability of this beautiful word " light " to
the name of Christian is sufficiently plain from Scripture :
" You were heretofore darkness, but now light in the Lord."[2]
But how does this name profit the many persons whose works
agree so ill with it ? The effect on those who see this can only
be to make them calumniate the law which these men profess.
For this reason it was always inculcated on Christians, even in
the earliest ages of the Church, not only to be good, but to
show that they were so : " Let your modesty be known to all
men."[3] And the reason was, that otherwise the accusations
made against Christians would become a reproach to Christ ;
whereas, when the justice of all their actions is clearly manifest,
it is of necessity that all who see the children living so

<hr>

[1] Prov. xviii. 20.   [2] Ephes. v. 8.   [3] Philipp. iv. 5.

virtuously will praise their Father, and therefore Christ here said so expressly, "That they may see your good works and glorify your Father Who is in Heaven:" not "your God," but "your Father," that hence might be deduced the strict obligation all Christians are under of honouring so good a Father by their lives. This, then, is what our Lord had especially in view in this injunction: to prevent scandal, and to encourage all men to give edification, taking care that this edification is not from a motive of self-glorification, but in order to glorify God; for He did not say "that they may glorify you," but "that they may glorify their Father Who is in Heaven." Hast thou this right intention in thy actions? If not, thou art, in truth, a most foolish and ungrateful son, who wilt therefore deserve to be condemned by those three heathen sons of a certain Diagoras Rhodius, who, being crowned on the theatre of the Olympic games for their prowess, all three at once took the garlands from their heads, and put them on that of their father, who was there present. And if thou wouldst have an example of the light itself, take it from the stars, of which it is written, that "they were called" to appear in the darkness; and with great promptitude "said, Here we are; and with cheerfulness shined forth to Him that made them."[1]

III. Consider thirdly, that there are some persons who are so far removed from these pious sentiments, that they actually abuse these words of Christ to give a colour to their own arrogance, regarding only those which come first: "So let your light shine before men, that they may see your good works;" and leaving out those which follow: "and glorify your Father Who is in Heaven." Thou wilt see that such persons cannot do the least good without making a display of it; so that if they present a chalice of ordinary value, a vestment, or an altar-frontal to a church, they must needs have it marked with their family arms; and so with everything else. Whatever good they do they are careful to join with it the utmost possible glory which may accrue from it, not to the Christian name, but to their own, which they put forward prominently wherever they go, engraving it on shining metal or solid marble. If, then, these persons attach themselves merely to the first part of this maxim of Christ without going on to the second, let them, at all events, pay attention to the way in which He speaks. It is true that it is said, "So let your light shine before men, that they may see your good works," but it is not

[1] Baruch iii. 35.

said "that they may see the good works to be yours." What
need, then, to seek further, to cover them with confusion?
This being so, thou shouldst, indeed, endeavour with all thy
might that thy works may be seen to be good, but not that
these good works may be seen to be thine. These are two
very different things : the former being safe from the danger
of ambition, and the latter subject to it ; because the first gives
all men an occasion of praising God, and the second of praising
thee. I say this, because the language of men nowadays is so
perverted. Formerly, for example, if a holy man was seen to
restore sight to a blind person, every one began at once praising
God : " All the people, when they saw it, gave praise to God."[1]
Now, on the contrary, they are more apt to extol the holy man,
because people have lost that keen sense which teaches that
God is the Author of all our good : " Every best gift and every
perfect gift is from above."[2] In our days, therefore, we have
to go to work very cautiously when we want to show not only
that our works are good, but that they are ours. It is true
indeed that, speaking generally, this ought not to be purposely
concealed, for that would be putting the light under a bushel,
which was blamed by Christ when He said, " Neither do men
light a candle and put it under a bushel, but upon a candlestick, that it may shine to all that are in the house ; "[3] but
neither ought it to be anxiously displayed ; for that would be
not merely putting the candle on a candlestick, but under the
eyes of those who are not looking for it, and have no desire to
see it. And this is what seems to be the object of those
persons who are bent on leaving a memorial of whatever little
good they do in the world, by some sort of sign or inscription.
Their object is to force others to notice them, a thing never
recommended by Christ, Who did indeed say, " So let your
light shine before men, that they may see your good works,"
but not " that they may be compelled to see them." Hence,
some preachers of the Gospel have sometimes gone so far as
to condemn with great vehemence the custom just mentioned,
even though it is so general in the Church. Their reason for
condemning it is not that it is absolutely improper to leave to
posterity some honourable memorial of the good done by their
pious forefathers, but because very often such a memorial is
not left on account of the good which has been done for a
pious object, but the good has been done for the sake of
leaving the memorial. It is true that it is not very easy to

[1] St. Luke xviii. 43.　　[2] St. James i. 17.　　[3] St. Matt. v. 15.

decide when it is better to conceal the good that is done, and
when not to do so; and therefore the following meditation
shall be devoted to this purpose, because a knowledge of this
is a great help in acting under all circumstances with that
liberty of spirit without which it is very difficult to feel joy in
well-doing.

SIXTEENTH DAY.

*Take heed that you do not your justice before men, to be seen by them,
otherwise you shall not have a reward of your Father Who is in Heaven*
(St. Matt. vi. 1).

I. Consider first, that the word "justice" is here used as
a general term by which to denote all the good works which
were directly afterwards divided by Christ into three kinds—
alms, prayer, and fasting—because they may all be reduced
to these heads. Neither is this to be wondered at, because
fasting is opposed to the concupiscence of the flesh; alms to
the concupiscence of the eyes; and prayer, which teaches us
our nothingness, to the pride of life. Fasting sets us in order
especially with regard to ourselves; alms with regard to our
neighbour; and prayer with regard to God. And thus, fasting
helps us to continence, which is especially a virtue of the
concupiscent part; alms to compassion, which belongs to
the irascible; and prayer to devotion, which belongs to that
which is called the rational part. And although all these
three good works, like all others, always contain in themselves merit, satisfaction, and impetration, yet nevertheless
fasting has a very special power of meriting, alms of satisfaction, and prayer of impetration. This being premised,
take notice how our Lord here speaks. He does not say
simply, "Take heed that you do not your justice before
men," but immediately adds, "to be seen by them," because
there is no harm whatever in thy fastings, alms, or prayers
being seen by other people: the harm is in doing them for
the purpose of their being seen. Neither is it wrong when
thou lettest them be seen for the sake of the glory which God
may get from them. The mischief is when thou lettest them
be seen for thy own glory. And therefore Christ purposely
said not, "Take heed that you do not your alms before
men, that they may see them," but "to be seen of them;"

for here lies the danger which requires to be carefully guarded against, that of having for our object not the showing of the works, but of ourselves. And how anxiously does the devil strive to get the little good which is now done in the world done in secret, just as it was in the days of the early persecutors, through fear of whom the Christians betook themselves to caves and catacombs. He knows the power of good example in inflaming men to do good, and therefore he does his utmost to remove it out of sight. And why was it, thinkest thou, that he once stirred up so fierce a war against holy images? It was because by the sight of them the faithful were mightily incited, some to martyrdom, some to devotion, some to penance, and others to different generous acts of virtue. And now that the devil is no longer able to make war, in our days, upon the dead images of men who were dear to God, he makes it upon their living images, that is to say, their bright examples. He endeavours, under various specious pretexts, to hide them entirely from the open light, so that they may not be a reproach to sinners or an encouragement to the timid. Thou thinkest that whenever thou concealest thy good works, it is done from a spirit of humility; but very often it is a temptation of the enemy, who is jealous of the good which thou mightest produce in others by not concealing them: "They are all dumb dogs, not able to bark."[1]

II. Consider secondly, that, speaking generally, there are two sorts of good works, those which are ordinary, and common to all Christians who desire to live in the due observance of their state, whether it be lay, ecclesiastical, or monastic; and of this sort are the usual penances of such a state, frequent Confession and Communion, assisting devoutly every day at Divine Office, and other things of the same sort, a failure in which is always considered an imperfection; and those which are not ordinary, but exceptional. As to these latter, the saints advise our doing them, for the most part, very secretly, to avoid admiration; but not so with regard to the former. On the contrary, they say that it is better to do them with all the openness which the strictest persons are in the habit of showing in such a state. And with good reason: for in such a state thou art either a private or a public person; if the latter, that is, a prelate, a prince, or a Superior, not only art thou right in liking this publicity, but thou art bound to it,

[1] Isaias lvi. 10.

because thy life has to be the rule for others: "In all things show thyself an example of good works."[1]   And if thou art a private person, it is still better to affect publicity than secrecy, not only on account of the good which, as has been said, will result from it to others, but still more on account of that which will result to thyself, because thereby thou declarest what thou art.   Why is it that thou sometimes thinkest well to do good secretly?   Is it through fear of vainglory?   Not so; it is to avoid pledging thyself, because thou thinkest that by being enrolled in a certain confraternity, by confessing and communicating every week, thou wilt no longer be free to accept the invitations of thy friends to accompany them to plays, races, or banquets, or because thou dost not like to be pointed out by those who see thee in such places.   But is it not better to make a generous resolution in this matter? "How long do you halt between two sides?"[2]   Thou wouldst rather not declare to whom thou belongest, to God or the world; but I tell thee that it is better to declare thyself.   For so long as thou dost not declare thyself to be on God's side, it will often happen that very wicked things are asked of thee, to which thou wilt consent out of human respect; whereas when once thou hast thus declared thyself, no one will venture to tempt thee any more.   It is enough, therefore, in all thy actions, always to maintain the same pure intention of pleasing God alone.   And this Christ meant to express when He said, with regard to almsgiving, "Let not thy left hand know what thy right hand doth;" with regard to prayer, "Enter into thy chamber;" and with regard to fasting, "Anoint thy head and wash thy face, that thou appear not to men to fast."   Certainly it was not His meaning by these words to forbid these actions being publicly performed, seeing that He Himself often performed them publicly.   But He meant to show, by a figurative way of speaking, that even when publicly performed they should be done with that purity of intention with which they are done by one who uses all the means just mentioned to conceal them.   Besides, if thou wouldst know how pleasing this liberty of doing well openly has always been to God, remember how He said to Abraham that He would have granted a general pardon to all the inhabitants of the wicked city of Sodom, if only He had found fifty just men amongst all those sinners: "If I find in Sodom fifty just within the city, I will spare the whole place for their sake."[3]   And

---

[1] Titus ii. 7.   [2] 3 Kings xviii. 21.   [3] Genesis xviii. 26.

observe, He did not say simply *in civitate*, but *in medio civitatis;* because, as some think, it was possible that amongst all those thousands of wicked men there might have been at least fifty who led a good life in secret, but that there certainly were not so many who were bold enough to do this openly."[1] And these last are the just men who have power to appease God, not those who merely keep themselves in His grace, but those who declare that they do so : " In the midst of the Church will I praise Thee ; "[2] " In the midst of many I will praise Him."[3]

III. Consider thirdly, that the value of this exterior declaration is in proportion to the purity of the interior intention, already mentioned, of not seeking our own glory, but the glory of God, in all that we do. When, on the contrary, this is wanting, who can doubt that the most admirable declaration is, after all, valueless in His eyes? Therefore Christ said : " Take heed that you do not your justice before men to be seen by them," that is, " in order that you may be seen by them ; otherwise you shall not have a reward of your Father Who is in Heaven." For how is it possible that thy Heavenly Father should reward thee in Heaven for the good which thou hast, indeed, done openly, but not for love of Him? Rather will He leave thee to be rewarded by men, whose approval thou hast chosen to prefer to His. And therefore observe that our Lord does not here say *a Patre vestro*, but *apud Patrem vestrum*, because He does very often give thee an earthly reward for the good which thou hast sometimes done from vanity, on account of the use which may very likely have resulted from it to the world ; but He will not give thee a heavenly reward. To gain this, the intention must have been altogether spiritual and holy ; because in Heaven it is not the mere material part of the work, or the husk, which is rewarded, but the formal part, which is the substance. Who, then, can express the importance of this intention ? But, then, does every act of vanity which may unfortunately be mixed up with such actions, in other respects so pleasing to God, take away their merit? Surely not. This is the case only when the act of vanity is sufficient, like a worm within the fruit, to vitiate the good work. I will explain myself, for fear of troubling thy mind in this matter. Either the desire of pleasing people

(which is the act of vanity) is antecedent to the good work of
which we are speaking (as, for example, to public almsgiving),
or it is concomitant or subsequent. If it is subsequent, it
cannot take away its merit, because such an act of vanity is
no more than a worm which would fain spoil the work by
gnawing it on the outside, but which finds it already perfect,
and therefore safe. If it is antecedent, it certainly does destroy
the merit when the object proposed in giving alms is only that
of pleasing men, for then the worm is in the heart of the
action. It is true that sometimes this very desire of pleasing
men may be subservient to the greater service of God, as in
the case of princes and prelates, who seek to gain the love of
their subjects by abundant alms, in order to be able to make
them more devout to God. In such a case, as the act is
legitimate, it is not like the worm we are speaking of, and
therefore cannot of itself in the least damage the value of the
work, which we suppose has the glory of God for its ultimate
object. Lastly, if the act of vanity is concomitant, it may or
it may not take away the merit of the work. It takes it away
when the work was begun in order to please God, but when,
before its completion, there has been a change of object, and
it is continued rather for the sake of pleasing men—in such a
case the worm is in time to spoil it. The merit is not taken
away when, although the work is not continued for the sake
of pleasing men, yet at the time of performing the work—
of bestowing, for instance, some charitable alms—the giver
indulges in a vain satisfaction which arises in his heart at
being surrounded by many persons who are noticing him ;
because, although this vain satisfaction does certainly amount
to a venial sin, yet it is pre-supposed to be an act entirely
distinct from the ultimate object for which the work is being
done, that is to say, the glory of God ; and so the worm is
entirely outside, because although the alms in question is
combined with the act of vanity, it does not arise from it.
In such a case, therefore, thou shouldst not cease from giving
alms, even publicly, from fear of vanity, but merely oppose
that vanity by repelling and repressing it, or at least by turning
thy thoughts in another direction. Do this, and the reward is
untouched. See, then, in the last place, how truly fatherly is
the conduct of thy Heavenly Father. He does not require of
thee what is impossible or unsuited to thee. He would but
have thee act as a submissive child, that is, He would have
thee prefer His approval very far before that of His servants.

SEVENTEENTH DAY.

THE HOLY NAME OF MARY.

Hail, Mary, full of grace (St. Luke i. 28).

I. Consider first, that although the Archangel Gabriel, when he saluted the Blessed Virgin by saying "Hail" (which was a word signifying greatness and announcing joy), did not immediately pronounce her name, as the Church's custom has since been, yet he implies it in declaring her "full of grace." For if Mary was full of grace, why was she so? Because of the closeness of her union with that Ocean which is the source of grace, a union which actually placed that Ocean under her authority as subject to her. If, then, Mary was full of grace, it was precisely because she was Mary, which, according to the most received etymology of that glorious name, means "Mistress of the sea." Observe, therefore, that in this address the Angel did not allude to any particular time; thus, he did not say, "Hail, thou that wert, that art, or that shalt be full of grace," but he said absolutely, "Hail, full of grace," in order thereby the better to include all times. Thus, beyond a doubt, there are three plenitudes of grace of which he meant to speak: that which she has received in the past, that which she was receiving in the present, and that which was reserved for her in the future. And if thou wouldst know what they are, they are those which all the holy Doctors have agreed in recognizing in her—the plenitude of sufficiency, of super-abundance, and of pre-eminence. The first made her full in herself; the second made her full in herself and full for others; and the third made her full in herself, and full for others, and full above all others put together. Art thou struck with wonder at the greatness of these plenitudes? That wonder will cease so soon as thou callest to mind that she is Mary, that is, the "Mistress of the sea." She has the ocean under her authority: "The sea is His, and He made it."[1] What wonder then that she is so rich? She is like a city which is mistress of the sea, and easily has the sovereignty over those who are not so: "Art thou better than the populous Alexandria? . . . the sea is its riches."[2] The thing at which thou mayest justly wonder is that, being so poor thyself, thou dost not make thy fixed abode in this city.

[1] Psalm xciv. 5.   [2] Nahum iii. 8.

II. Consider secondly, that the first plenitude, that of sufficiency, began in our Lady at the first moment of her conception, and so the Angel did not say to her "filled with grace," but "full of grace," in order to remove any idea that there ever was a moment in which she was without it. This made her full in herself, causing her to be, in the first place, full of grace as to every part of herself, which is what is termed the plenitude of the subject. She was full in her understanding, full in her affections, full in her appetites, full in her senses, and full in all the different portions of her soul, which were always most perfectly subject to God. Secondly, it caused her to be full as to the absence of every obstacle opposed to grace, none of which found a place in her. For she alone, among all the saints, was always free from the slightest stain, from disturbance of mind, from ignorance, from imprudence, from repugnance to anything good, from vain imaginations, from the slightest leaven of concupiscence, from anything whatever which could retard her flight to the highest sanctity. And thus, being free from all such obstacles, she was more capable of grace. Thirdly, it caused her to be full as to her actions, which were always performed by her with the plenitude of virtue, of vigour, and by perfect correspondence with the great lights bestowed on her by God. Fourthly, it caused her to be full as to all the kinds of grace which make man perfect in himself, and in which she was so rich from the beginning. These are the grace which makes pleasing (*gratia gratificans*), that grace by which God, from the beginning, had more complacency in the soul of the Blessed Virgin than in that of any other pure creature: habitual grace, that is, the grace which sanctifies; and actual grace, which is the grace that sustains; the infused virtues, both theological and moral, which in her were not divided, as in the other saints, of whom one was especially remarkable for faith, another for humility, another for obedience, and so on; but they were all combined. And lastly, the gifts of the Holy Ghost, which are the habits by which actions become heroic; their fruits, which are the beautiful actions proceeding from them, and their beatitudes, which are these actions in their highest perfection. Fifthly, it caused her to be full as to her office, that is, full of that particular kind of grace which became her who was predestined to be Mother of God, and consequently "Mistress of the sea," as her name teaches thee; and this is a grace which not only embraces all those which have been enumerated, but raises them into an order higher than any which our minds can conceive; for there is in the dignity of Mother of God something infinite, giving her, as it is said, a kind of affinity with God Himself. Such is the plenitude of sufficiency, which the Blessed Virgin possessed in herself from the beginning. But this did not prevent her continually increasing and augmenting it day by day; seeing that she was always *Viatrix*, without ever being weary. Nevertheless, she is said to be "full of grace," because, although the word "full," when applied to an ordinary vessel, such as a basin or pail, denotes a limited quantity, this is not the case when applied, for instance, to a vast lake, which is almost a sea. Thou oughtest to rejoice exceedingly in this plenitude, for it is impossible that one so full in herself should not delight in pouring her abundance on others; just as a nurse whose breasts are full seeks for an infant desiring milk: "Come over to Me, all ye that desire Me, and be filled with My fruits,"[1] that is, "with the milk of My breasts."

III. Consider thirdly, that the second plenitude, that of superabundance, began in our Lady from the moment when she conceived the Eternal Word in her most pure womb, rendering her superabundant in herself; for all that plenitude of sufficiency, which had till then been shut up in her soul as in its river-bed, burst its barriers and overflowed even her body, which was made worthy to become the habitation of the Most High, nay, even to supply from its substance the first material which He required in order to clothe Himself with human flesh, and afterwards what was needed for nourishment and growth in His Infancy. She was also made superabundant for the good of others, not merely because at that instant she entered on the possession of all the graces (*gratis data*), which made her perfect for the good of others, such as gifts of tongues, prophecy, power of working miracles, unbroken health, and the rest, which were, beyond a doubt, combined in her in an eminent degree, although she availed herself of them so little; but still more because, at that same instant, she received a far more sublime possession, that of Mediatrix between men and God, in virtue of which she has gained those glorious titles which she now bears of Healer of our diseases, Restorer of our world, and direct Dispenser of the treasures which descend upon us from the hands of God, because in that instant she became in reality what is declared to us by her glorious name—she became Mary, she became "Mistress of

[1] Ecclus. xxiv. 26.

the sea," so that she was able to dispose of Christ with that readiness and confidence with which a queen-mother who is greatly beloved disposes of him who is indeed her king and lord, but yet her son. If, then, there was reason for thee to rejoice in the plenitude of sufficiency, from the hope of receiving benefits from the Blessed Virgin, there is surely reason for still greater joy in this plenitude of superabundance, since in it she received her office of benefactress.

IV. Consider fourthly, that the third plenitude, that of pre-eminence, rendered our Lady not merely full in herself, and full for others, but full in such a manner that she surpassed all the blessed spirits who ever have been or ever shall be. This plenitude began in her at least towards the end of her life, but it is probable that it began before then. For all are agreed that at the first instant of her sanctification she was gifted by God with greater grace than that which was in the highest of the Seraphim, and this is a grace beyond words. Now to this grace, being possessed as she was of all that plenitude of sufficiency which has been described, she instantly corresponded in act, and thus operating with all her virtue and all her strength she merited at the least (according to the most received doctrine of theologians) the increase of all that grace which had been so graciously bestowed upon her, and thus she at once doubled her capital. In the next place, since in her this capital never lay idle (as some have asserted), not even in her sleep, she multiplied it in the course of her seventy-two years by the fresh interest which she made it produce, through the Divine help with which she was strengthened, not only every hour, but every minute and every second, in such a way that it is impossible for our minds to conceive the greatness of the treasures which she thus accumulated. For if, by any action which she performed, she became each time twice as rich as she was before, think how rich she must have been towards the end of her life! Now if to this grace, thus increased as a reward by her exact correspondence, we add that which Jesus Christ must have graciously conferred upon her by way of free gift on various occasions of extraordinary solemnity, such as His Incarnation, Nativity, Resurrection, Ascension, and the like, what an unfathomable and incalculable abyss these multiplied graces are! The famous benediction which Jacob gave to his son Joseph on account of his wonderful growth, may far more truly be applied to our

Lady: "Joseph is a growing son."[1]   "The Almighty shall
bless thee with the blessings of Heaven above, with the
blessings of the deep that lieth beneath, with the blessings of
the breasts and of the womb."[2]   Far truer indeed are those
words of her.   She is blessed "with the blessings of Heaven
above," that is, the plenitude of sufficiency which God poured
into her heart from the first moment of her conception in her
mother's womb.   She is blessed "with the blessings of the
deep," that is, the plenitude of pre-eminence, which, in comparison with all the spirits of the blessed, renders her an abyss,
and an abyss of profound depth, "the deep that lieth beneath,"
so far do her riches exceed all their vast treasures put together.
But what was the source of these two plenitudes?   That which
the Patriarch purposely put in the last place, keeping the order
of dignity, not of time, from the "blessings of the breasts and
of the womb;" from her being the Mother of God, from her
having carried Him in her womb, given Him birth, nourished
Him, brought Him up, and lastly exercised over Him, as
Mother, the authority revealed to us by that august name
of Mary, which teaches us that as such she was "Mistress of
the sea;" and of what sea? of the deep sea? nay, but of that
which is deepest of all, whence flow all the streams which
make us rich?

EIGHTEENTH DAY.

*Blessed is he that understandeth concerning the needy and the poor; the
Lord will deliver him in the evil day* (Psalm xl. 1).

I. Consider first, that, in the opinion of the most exact
commentators, the word "needy" here signifies the man who
has nothing, and is therefore in the extremity of want; and
the word "poor" the man who has little, and who therefore
is indeed in want, but only ordinary want.   And both these
epithets may be truly applied to Christ our Lord, for we see
that He had little, and that He had nothing—little in life,
nothing in death: little in life, because He passed His days
laboriously in a carpenter's workshop; nothing in death,
because He died naked upon a Cross.   Therefore it was no
redundancy of expression when He said of Himself, "I am

[1] Genesis xlix. 22.   [2] Genesis xlix. 25.

needy and poor,"[1] for He was both the one and the other
at different times. Now, to come to our point, according to
the Psalmist, he who in the first place "understandeth concerning the needy and the poor" is he who, seeing Christ our
Lord poor in life, and naked in death, does not stop there,
but goes on to understand that He is God. The man who
acts thus does not allow himself to be guided by sense, but
by faith: "Blessed are they that have not seen and have
believed."[2] But how few are there who act in this way!
Therefore, in these words of his, "Blessed is he that understandeth concerning the needy and the poor," David intended
to express the same thing which Christ did, when He said,
"Blessed is he, whosoever shall not be scandalized in Me."[3]
And what, thinkest thou, is the reason that so many are
ashamed to follow Christ in His profound abjection? because
such unhappy persons do not attain to the understanding of
more than they see: they do not understand "concerning the
needy and the poor." They do not penetrate so far as to see
that under that abjection is truly hidden their whole good.
Do thou endeavour to understand this as far as possible; for
at the hour of death thou wilt see how greatly it will profit
thee. Hear what the Psalmist says: "Blessed is he that
understandeth concerning the needy and the poor; the Lord
will deliver him in the evil day." This "evil day" is, without
doubt, the day of death: "Why shall I fear in the evil day?"[4]
And in that day, which is so absolutely called "evil," because
it is so to the majority of men, who is he who will be specially
protected by his God? He who has been faithful to Him at
the foot of the Cross; for none will have better proved his
love than such a one. Happy wilt thou be, if at that hour
thou art able to take thy crucifix in thy hand, and to say to
Him with truth that thou hast not been ashamed to follow
Him even in such a state of abjection.

II. Consider secondly, that Christ so highly esteemed
poverty, that now that He is no longer able to practise it in
His own Person, seeing that He has ascended in glory into
Heaven, He chooses, at least, to do this in the person of
others: and therefore He has plainly declared that He is
hidden under the form of every poor man whom we see:
"As long as you did it to one of these My least brethren, you
did it to Me."[5] And so, if, when He was upon earth, He

[1] Psalm lxix. 6.   [2] St. John xx. 29.   [3] St. Luke vii. 23.
[4] Psalm xlviii. 6.   [5] St. Matt. xxv. 40.

was a beggar in His own person only, now that He is in Heaven, He is a beggar in the persons of all. Who, then, is it, who in the second place, "understandeth concerning the needy and the poor"? It is he who, when he sees any poor man whatsoever reduced to a state either of extreme or of ordinary want, has come to understand clearly that Christ is hidden beneath the poor rags of that wretched man, and is hence moved to succour him, if he can ; and if he cannot, to respect him, to pity him, to console him, or at least to answer him gently, as though he were Christ Himself in Person. The man who acts in this manner is called blessed, because he has the real merit of that eminent virtue which regards the poor. But what great merit is it possible for thee to gain, when thou doest good to them from a mere instinct of natural compassion? This is an act of which even idolaters are capable. But thou dost gain a very great merit when thou doest good to them from that motive of faith which has just been mentioned, namely, to honour Jesus in them ; for then the act which would be natural passes into a higher order, higher than the heavens are above the earth, and becomes supernatural. And this is why so glorious a reward as deliverance from all the evils which otherwise await us at death, is promised to compassion towards the poor when it is practised in this admirable manner : "Blessed is he that understandeth concerning the needy and the poor : the Lord will deliver him in the evil day."[1] Not, indeed, that this compassion is sufficient of itself alone to save a man, but because these evils are either of sin or of penalty. If of penalty, this compassion is able to do away with them by the way of satisfaction—"Redeem thou thy sins with alms"[2]—and if they are of sin, it is able to keep a man free from them by the way of merit, as in the case of the innocent, in whom it is frequently the means of preserving grace—"Alms shall preserve the grace of a man as the apple of the eye"[3]—or to drive them from him by the way of right disposition, as in the case of penitents, for whom it very often procures at the hour of death that true repentance and resolution of amendment which otherwise they could not merit : "To the penitent He hath given the way of justice."[4] Do not say that these fruits were common to the virtue of compassion to the poor even when no one went so far as to recognize Christ in them : for, I reply, that though this was

---

[1] Psalm xl. 1.
[2] Daniel iv. 24. [3] Ecclus. xvii. 18. [4] Ecclus. xvii. 20.

the case it was by no means so in the same degree as at present.

III. Consider thirdly, that, in the third place, he "understandeth nothing concerning the needy and the poor," who does not require them to come to him to declare their wants, but who has these so much at heart, that he thinks of them himself, and makes the first advances to relieve them. The man who does this is also called "blessed." For in this passage either thou understandest by the "poor," Christ in His own Person, as was explained in the first point, and in this case there certainly is no great merit in waiting for Him to ask of thee in express terms some work for His glory or good pleasure, but in thyself discovering what He requires of thee: "The mind of the just studieth obedience."[1] For the love that thou bearest to Christ should extend so far as, if possible, to foresee and forestall His requirements. This was the way He acted towards thee when, without any prayer of thine, He went so far as even to die upon a Cross for thy salvation: "The Lord hath heard the desire of the poor."[2] Or by the "poor" thou understandest Christ in the person of the poor, as was also explained in the second point, and then too it is most certain that thy merit does not consist in waiting for the poor man to weary thee with his importunity. Thou shouldst be quick to perceive and relieve his misery of thyself, especially when his shame is so great that he would fain be understood even without speaking: "If I have denied to the poor what they desired"—not "what they requested," but "what they desired"—and "have made the eyes of the widow wait."[3] Dost thou not think that the man who acts thus will at his death reap a reward in proportion to his merit? "The Lord will deliver him in the evil day." What is the evil from which He will deliver him? There is no need to describe it; God knows what it is. And therefore if thou hast known how to discover what Christ desired of thee both for Himself and for His poor before He asked it, do not fear that at thy death He too will not know how to discover what thou desirest of Him, even though thou dost not express it in words.

IV. Consider fourthly, in the last place, that he is said to understand "concerning the needy and the poor," who watches over their needs like one who is their protector, provider, or advocate, and who thus takes up their cause just as though it were his own: "I was the father of the poor. . . . I broke the

[1] Prov. xv. 28.   [2] Psalm x. 17.   [3] Job xxxi. 16.

jaws of the wicked man, and out of his teeth I took away the prey."[1] Certainly he who acts thus may well be said to be blessed, and blessed beyond all others, because in this way not only does he himself do good to the poor, but he also resists the evil which, but for him, would be done to them by others : "The ear that heard me blessed me, . . . because I had delivered the poor man that cried out, and the fatherless that had no helper."[2] Here then, as has been already often said, thou beholdest thy Lord poor, both in Himself and in His poor. If thou wouldst be blessed, this is what thou must do : take His interests to heart in both ways, deliver "the poor man that cries out." See how many wrongs He receives every day in His own Person from proud men, who disdain the humility which He practised, and how many in the persons of those beggars who represent Him to thee. It is for thee then to take up arms to defend Him to the utmost from those who oppress Him, sure that by so doing thou wilt gain His favour, or to use terms more exactly applicable to the case, His protection : "Blessed is he that understandeth concerning the needy and the poor : the Lord will deliver him in the evil day." And when it is said that the Lord will deliver thee at thy death, is not this the same as saying that He will do battle for thee with thy proud infernal enemies, and not suffer them to triumph over thee? "Thou hast delivered me according to the multitude of the mercy of Thy Name from them that did roar, prepared to devour."[3] Is it not right that, in order to gain so mighty a deliverer, thou shouldst now use all thy knowledge, all thy mind, in His cause?

V. Consider fifthly, that the Psalmist does not say, "Blessed is he that understandeth concerning the poor and the needy," but "who understandeth concerning the needy and the poor." Do not suppose that this is without a deep meaning. It might seem reasonable that he should have reversed the order for the sake of the gradation. For if, as was said at the beginning, by "needy" is understood one who is in extreme want, and by "poor" one who is in what may be called ordinary want, it is certain that a man begins by having little, and so by being poor afterwards goes on to have nothing, and so to be needy. But here thou shouldst observe that the man who is in extreme want more easily finds one to help him than he who is in ordinary want. And therefore he is called "blessed" who not only "understandeth concerning the

[1] Job xxix. 16, 17.   [2] Job xxix. 11, 12.   [3] Ecclus. li. 4.

needy," but "concerning the poor," by rightly understanding
the obligation he is under to spend his superfluity upon the
poor, not only in their extreme, but also in their ordinary
wants.   This may be the reason that in so many other passages
of Scripture it has pleased God to join these expressions in the
same way : " I command thee to open thy hand to thy needy
and poor brother ;"[1] " Behold this was the iniquity of Sodom,
thy sister, she did not put forth her hand to the needy and to
the poor ;"[2] " That grieveth the needy and the poor ;"[3] " They
afflicted the needy and poor ;"[4] " You that oppress the needy
and crush the poor ;"[5] and there are many similar texts, to
show that those whom God commends to thy care are not
only those miserable ones who have none of the means of
subsistence, "the needy ;" but those too who have little, "the
poor."   And if this is so, how will it be possible for those
to be saved, who spend their income on dogs and horses,
rather than on the poor, unless they see that they are actually
perishing for want ?   Alas ! this is not "understanding concerning the needy and the poor ;" but only "concerning the
needy."   And yet Christ dwells alike under the persons of
both ; so that, in this sense also, He cries aloud, that all may
know it : " I am needy and poor."

NINETEENTH DAY.

Be not overcome by evil, but overcome evil by good (Romans xii. 21).

I. Consider first, that he is said to be overcome by another
who is drawn by that other to himself; for which reason it is
said that the loadstone overcomes iron, and not that iron
overcomes the loadstone, because it is the iron which is
attracted by the loadstone, and not the loadstone by the iron.
Thus much being premised, this is the first meaning of these
admirable words of the Apostle: "Be not overcome by evil,
but overcome evil by good."   It is, not to allow thyself to be
drawn by the enemy to do what thou oughtest not to do, but
to draw the enemy to do that which ought to be done by him.
In this way thou overcomest him ; and is it not certain that
whenever thou art offended, thou oughtest not to be angered

[1] Deut. xv. 11.   [3] Ezech. xvi. 49.
[2] Ezech. xviii. 12.   [4] Ezech. xxii. 29.   [5] Amos iv. 1.

or enraged, or to desire, in defiance of God, to revenge thyself, but to leave vengeance to Him alone, as thy Sovereign? "Revenge to Me, I will repay, saith the Lord."[1] If, therefore, thou sufferest thyself to be drawn by thy enemy to do what thou oughtest not to do, then thy enemy has already overcome thee : whereas, if thou art not disturbed, or irritated, or angered, as he would have thee be, but on the contrary, by doing him some great kindness, thou bringest him to lay aside his wrath, and to confess his fault in offending thee, and to humble himself for it, then thou overcomest him by drawing him to do that which he ought to do. And how canst thou possibly prefer being overcome to overcoming, since it is natural to every adversary always to do his utmost to overcome the other: "Be not overcome by evil, but overcome evil by good." Not "evil by evil," for this is a victory which even brutes succeed in gaining, but "evil by good," because this is worthy of a man. It is so noble a victory, that if it is possible to draw any distinction as to the perfection of those which Christ gained on earth, this would certainly be the greatest of all. For when He was dying on the Cross, it was of this that He thought, of drawing to Himself the very men who had nailed Him to that Cross. So that instead of crushing or destroying them, as He might have done, He deluged them with such an abundance of grace, that He brought great numbers of them to go down from the mountain, either struck with compunction or covered with confusion to such a degree, that they even went through the streets striking their breasts like public penitents : "They returned, striking their breasts."[2] Oh, how much nobler an act is this than that of one who avenges himself ! And so thou seest in all history, both sacred and profane, how much more glorious those are who overcame their enemies in this way, than those who suffered themselves to be overcome by them—that is, to be drawn to commit brutal or barbarous actions, by which they might succeed in rendering evil for evil. And if thou art unable, by all the kindnesses thou doest to thine enemy, to overcome him in such a manner as to draw him to do what he ought to do, yet thy victory will not be the less glorious, because thou wilt have done what ought to have succeeded in overcoming him. At any rate, if thou hast not overcome him, as the loadstone overcomes iron, by drawing it to itself, thou wilt have overcome him as gold overcomes lead, as the pearl overcomes sea-weed, as purple overcomes frieze,

[1] Romans xii. 19.   [2] St. Luke xxiii. 48.

as the cedar overcomes the cornel-tree, that is to say, by infinitely surpassing him in value, which is the other most ordinary mode of overcoming. By offending thee he did a base act of wickedness, and by pardoning his offence, and showing him kindness, thou doest an heroic act of Christian virtue ; and is not this really to overcome him ?

II. Consider secondly, the second sense of these words, which is not to suffer thyself to be overcome by the devil, or by those men who, as his colleagues or allies, would fain seduce thee to evil ; but rather to gain the victory over them all. The devil is frequently called " the wicked one " in the Scriptures by antonomasia : " Then cometh the wicked one, and catcheth away that which was sown in his heart." [1] For he was the first to bring evil into the world ; and not satisfied with that, he still goes on incessantly exciting and promoting it by means of the men who are his followers, and who, from their likeness to him, are often called "the wicked." "The wicked man is reserved to the day of destruction." [2] Now with regard to the devil, it is certainly true that thou wilt never be able to overcome him by drawing him to what is good; because he is so obstinate in his wickedness as to be inflexible, but thou canst at least not suffer thyself to be overcome by him whenever he seeks to draw thee to evil ; and, more than this, thou mayest overcome him by doing a good greater than the evil to which he instigates thee. In the first place, thou mayest not suffer thyself to be overcome by him, because, although there is no power on earth equal to his—" There is no power upon earth that can be compared with him"[3]— nevertheless, he cannot abuse that power so far as to violate thy free-will, but only to deceive and seduce it if thou art not on thy guard—"Cast thyself down "[4]—so that if thou wouldst not be overcome it is in thine own power to prevent it. All that is necessary is not to consent. And therefore the Apostle did not say, *Ne vincaris*, but *Noli vinci a malo*. And in the second place, thou mayest also overcome him by doing a good greater than the evil to which he instigates thee, because, for the very reason that the devil tempts thee to vainglory, for example, thou canst make a contrary act of humiliation : when he tempts thee to hatred, thou mayest make an act of charity; when he tempts thee to harshness, thou mayest make an act of gentleness ; when he tempts thee to gluttony, thou mayest

[1] St. Matt. xiii. 19.
[2] Job xxi. 30.　　　[3] Job xli. 24.　　　[4] St. Matt. iv. 6.

make an act of severe abstinence, and so of the rest. This is
more than not to suffer thyself to be overcome by him, that is,
not to be drawn by him to evil ; it is to conquer him, because
it is to do a good greater than the evil which he demanded.
This was what Job did, who, when he was so closely assailed
by the devil to insult God by uttering audacious words, not
only did he not suffer himself to be overcome by him, but
overcame him by breaking forth, on the contrary, into words
which gave the greatest possible honour to God : "The Lord
gave and the Lord hath taken away ; blessed be the name of
the Lord."[1] Then, as to the men whom the devil uses as his
ministers, thou must not be satisfied with so little ; but when
they would fain pervert thee, by drawing thee to evil—for
example, to profane amusements—thou oughtest to make every
effort to draw them to good : for example, by inducing them
to visit churches, convents, and retired oratories for penance.
This is the most glorious victory of all, and it is to this that
thou shouldst aspire. Consider, as an instance of this, what
St. Bernard did with his brothers. They wanted to draw him
out of religion, and to take him back to the world, and he
drew them from the world, and persuaded them all to follow
him into religion. Do thou do the same in thy measure with
thy companions, whenever they incite thee to evil : "They
shall be turned to thee, and thou shalt not be turned to
them."[2] When a fire is choked by great bundles of wood, it
does not overcome the oppression by merely not suffering
them to put it out, but by making them blaze, and so changing
them into fire.

III. Consider thirdly, that "evil" sometimes also signifies
in the Scriptures the irregular appetite which is within us :
"Evil is present with me."[3] Not that it is evil in its nature
(which cannot be truly said), but because it inclines us to evil,
which is why it is sometimes even called sin : "Now if I do
that which I will not, it is no more I that do it, but sin that
dwelleth in me."[4] And having premised this, thou hast here
the third sense of these words : "Be not overcome by evil, but
overcome evil by good," that is, do not let thyself be overcome by this sensual appetite of thine, but overcome it ; for
although it is true that it is very powerful, nevertheless, if thou
choosest, thou art lord over it by means of the sufficient aids
of grace which God grants thee for this purpose. Is it not a

[1] Job i. 21.
[2] Jerem. xv. 19.   [3] Romans vii. 21.   [4] Romans vii. 20.

great shame, then, if when thou art able to overcome it, thou
art contented to be overcome by it almost continually? "Lust
shall be under thee, and thou shalt have dominion over it:"[1]
this is the sublime command which thou hast received from
God, and according to which thou hast to act. This appetite
is "under thee" when thou overcomest it, instead of letting
thyself be overcome by it, not suffering thyself to be "overcome by evil." Thou "hast dominion over it" when thou not
only dost not let thyself be overcome by it, but overcomest it,
"overcoming evil by good," by accustoming even it to take
delight in those pleasures which are not of the flesh but of the
spirit. Think how some of the saints have even attained to
rejoicing in shame, delighting in infirmities, and being full of
rapture in the severest exercises of penance: "I exceedingly
abound with joy in all our tribulation."[2] And how did they
do this? In no other way but by accustoming their appetite
to love that which contains the true good. And this is the
way to conquer: "This is the victory which overcometh the
world, even our faith."[3]

TWENTIETH DAY.

*If you live according to the flesh, you shall die ; but if by the spirit you
mortify the deeds of the flesh, you shall live* (Romans viii. 13).

I. Consider first, how terrible the punishment is, with
which God threatens every one who chooses to live, not in the
flesh (for in this world it is impossible to do otherwise), but
according to the flesh, and from this we not only can, but
ought to keep ourselves. He threatens him with death: "If
you live according to the flesh, you shall die." And consider,
on the other hand, how great is the reward which He promises
to every one who shall not indeed destroy this flesh (for this is
neither required nor permitted), but mortify it. He promises
life: "If by the spirit of life you mortify the deeds of the flesh,
you shall live." It rests with thee, therefore, to choose which
thou wilt: "Behold I set before you the way of life and the
way of death."[4] It is altogether in thy power to take whichever
road thou choosest, that which leads to life, or that which
leads to death ; but consider well, before beginning the journey,

---

[1] Genesis iv. 7.
[2] 2 Cor. vii. 4.   [3] 1 St. John v. 4.   [4] Jerem. xxi. 8.

for it is far less easy to retrace one's steps half-way than not to enter upon the road.

II. Consider secondly, what that death is with which he is threatened who lives according to the flesh, that is, who daily pampers and gratifies it in all things, and yields to it in everything it desires. It is the most utter death that can be imagined : a death of sin, a death of nature, and a death of damnation. This is the death which God announces to each one of these miserable men in the words : " And if you live according to the flesh, you shall die." The first death is that of sin, because it is the first in order which they incur by their mode of life. The second death is that of nature, which, as it had its origin from the death of sin, so too is it nourished and accelerated by it, especially in those who are given up to pleasure, amusements, and gaiety, and thereby hasten their corruption. The third death is that of damnation, which is the consequence of the death of sin, and which follows immediately upon the death of nature, and this death is everlasting. " He that joineth himself to harlots," that is, who begins to live according to the flesh, after the manner of sensual persons, "will be wicked : " this is the first death to be expected, that of sin. " Rottenness and worms shall inherit him :" this is the second death, that of nature. "And his soul shall be taken away out of the number :"[1] this is the third death, that of damnation. All these deaths are incurred, one after another, by the man who indulges his own flesh excessively. Nay, how often do they all strike him at once ! At one and the same moment a man sins, and dies, and is cast into Hell. How, then, canst thou think it for thy profit to choose a life which leads to so terrible a death ?

III. Consider thirdly, on the other hand, what that life is which is promised to those who "by the spirit mortify the deeds of the flesh." This too, like the death we have just spoken of, is three-fold. There is the life of nature, which is the first in order of lives, just as the death of sin is the first in order of deaths ; secondly, the life of grace ; and thirdly, the life of glory. The man, therefore, who knows how to mortify his flesh gains, in the first place, the life of nature, because he lengthens his days : " He that is temperate shall prolong life."[2] He gains the life of grace, because it is by mortification that it is both acquired and preserved ; and lastly, he gains the life of glory, because it is by mortification that it is increased in

[1] Ecclus. xix. 3.   [2] Ecclus. xxxvii. 34.

the next world, and anticipated in this by those foretastes of heavenly consolation which are granted on earth to those only who mortify themselves. See, then, how beautiful a thing mortification is. It is the true love of ourselves. The foolish world thinks that the man who deliberately sets himself to mortify his flesh, hates it. Quite the contrary: rather there is no one who loves it so much, because there is no one who so much seeks its true good. Who will say that the sick man who exposes his flesh to be cut or burnt, however severely, by the surgeon, is indifferent to it? On the contrary, he loves it much more than the timid patient, who cannot be induced to bear the suffering. And why is this? Because the man who does not subject it to suffering gives it death, and the man who does so subject it gives it life. Our case is precisely similar; and if so, how canst thou be afraid to accustom thyself to mortify thy own flesh? If thou dost not mortify it, thou givest it death, and not only temporal, but eternal death; and if thou dost mortify it, thou givest it life: "If you live according to the flesh, you shall die; but if by the spirit you mortify the deeds of the flesh, you shall live." Is it possible that thou wouldst be one of those who prefer to give it death? Oh, how foolish a love is this which thou showest it!

IV. Consider fourthly, that, as the Apostle says, "If you live according to the flesh, you shall die," it might seem that by strict contrast he ought to have said, "If you live according to the spirit, you shall live." But he did not say this; he only said, "If by the spirit you mortify the deeds of the flesh." And why? Because in this world there are countless numbers who do really live altogether according to the flesh, but none who live altogether according to the spirit. A purely spiritual life, such as this would be, is not to be found in this world: it is reserved for us in Heaven, when the flesh will never rebel in the slightest degree against what the spirit requires of it. But if we cannot here live altogether according to the spirit, as has been said, we can at least, by the spirit, repress and restrain the assaults of that flesh, which is continually and eagerly striving to rebel against that which it ought to obey, not only in Heaven, but on earth, that is, the spirit; and therefore the Apostle merely said, "If by the spirit you mortify the deeds of the flesh, you shall live." He did not say, "If you mortify the flesh," because all are not equally able to mortify, afflict, punish, and discipline their flesh, although this is certainly most useful in keeping it obedient;

but all are equally able to mortify its deeds, that is, its rebellions, appetites, affections, and unreasonable motions, nay, rather all are bound to do so. There are three modes of life in the world which thou mayest imagine : one is the life of persons living altogether according to the spirit, and this is not to be hoped for here, for it would be the life of an angel ; the second is the life of those who live altogether according to the flesh, and this thou must shun with all thy might, for it is the life of a brute ; the third is the life of those who "by the spirit mortify the deeds of the flesh," and this is the life which is here enjoined on thee, for it is the life of a man, who occupies a middle place between brutes and angels. When this mortification is in an ordinary degree, it is merely that of a reasonable man, that to which, at the least, every Christian is bound ; when it is in an exalted degree, it is that of a spiritual man, and it is to this that thou shouldst aspire if thou hast not yet attained to it : "Always bearing about in our body the mortification of Jesus, that the life also of Jesus," that is, the life of spiritual persons, "may be made manifest in our bodies."[1] It is not the life of Epictetus, Seneca, Xenocrates, or any other pagan philosopher which ought to be made manifest in thy treatment of thy body, but that of Jesus Christ : "The life of Jesus."

TWENTY-FIRST DAY.

ST. MATTHEW THE APOSTLE.

The Spirit breatheth where He will, and thou hearest His voice, but thou knowest not whence He cometh and whither He goeth : so is every one that is born of the Spirit (St. John iii. 8).

I. Consider first, that as he that is born of the flesh by natural generation is like him who begot him according to the flesh, although he is not at once equal to him in perfection, but only becomes so after he is grown up—"That which is born of the flesh is flesh "[2]—so, too, he who by supernatural regeneration is born again of the Spirit is also like Him Who regenerates him according to the Spirit, that is, the Spirit of God, although he, too, is not equal to Him, but falls far short, above all until the time comes when he shall have attained

[1] 2 Cor. iv. 10.   [2] St. John iii. 6.

maturity in Heaven: "That which is born of the Spirit, is spirit."[1] Hence it is that the actions of a truly spiritual man have, as such, something Divine in them, which Christ meant to express in these words which thou art now preparing to meditate. For, as the inspirations of the Spirit of God have three special characteristics, represented to us by the blowing of the wind, first that He "breatheth where He will;" secondly, that He makes His voice clearly audible—"and thou hearest His voice;" and thirdly, that at the same time His ways are secret—"thou knowest not whence He cometh and whither He goeth;" so, too, does the spiritual man, in his operations, acquire a mode of action conformable to this by means of the virtue which he receives in corresponding with the inspirations above mentioned: "So is every one that is born of the Spirit." Such is the true explanation of this passage. But in order that thou mayest better understand it when reduced to practice, I propose to thee in a special manner the Apostle St. Matthew, who not only corresponded most admirably with Divine inspiration, but also showed in a rare degree what great things the Spirit of the Lord can do in a heart of which He has taken full possession.

II. Consider secondly, that it is said, in the first place, "The Spirit breatheth where He will," because He has an absolute freedom of operation in the inspirations which He vouchsafes to grant us; He is subject to no law, bound by no obligations, compelled by no necessity: "Dividing to every one according as He will."[2] And so He went in search of this Matthew, contrary to all that might have been thought, for he neither expected this call, nor asked for it, nor desired it, nor merited it, but, on the contrary, opposed powerful obstacles to it, being satisfied to sit at his place of business occupied with the basest usury: "He saw a man sitting in the custom-house, and He said to him, Follow Me."[3] And observe how Matthew himself received a similar freedom of operation immediately that he allowed the Spirit to take possession of him; for at once laying aside his own interest he prepared to follow the Lord Who called him to Himself; he did not go by compulsion, like a slave, nor tempted by reward, nor terrified by punishment; he went because he chose to go. This strange interior movement made him utterly disregard the idle remarks and babbling of men, and with a noble courage in the face of all his unbelieving companions and

---

[1] St. John iii. 6.   [2] 1 Cor. xii. 11.   [3] St. Matt. ix. 9.

countrymen, he prepared to follow Jesus : " And he arose up and followed Him." Now, such is the mode of action of every truly spiritual man—"So is every one that is born of the Spirit"—he acts not as a slave, but as a free man : "Where the Spirit of the Lord is, there is liberty."[1] It is enough for him to know the will of God ; he fulfils it at once, triumphing completely over all considerations of human respect which come in the way. How is it with thee? is this thy state? or art thou held back by a thousand obstacles which prevent thee from acting with freedom as thou oughtest to do in all that concerns the service of God? Remember that there is nothing more hateful to the Spirit of the Lord than compulsory action : " The Spirit breatheth where He will."

III. Consider thirdly, that it is said, in the second place, that the Spirit speaks to the heart in such a way that it is impossible for thee not to hear Him : " And thou hearest His voice." Thou mayest, indeed, not recognize that voice as His by persuading thyself that it is not God Who speaks, but a very different spirit ; thou mayest resist His inspirations, rebel against them, in a word, refuse to receive them, as was the case with so many obstinate Jews who heard the powerful preaching of our Lord, but thou canst not close thy ears, so as not to hear Him. Therefore it is not said "thou recognizest His voice," but "thou hearest" it. It is true that when He pleases the Spirit of God enters the heart by His voice in a manner at once so sweet, so powerful, and so penetrating, that when thou hearest it thou canst not but willingly yield to it. For as every voice has its particular accent, by which those who are familiar with it distinguish it from all others, so, too, has this Divine voice, although it is only to be discerned by the hearing of the soul ; nevertheless, it is certain that by it it may be at once known Who it is Who speaks, that there can be no doubt about it. This was the case with the Apostle St. Matthew, to whom the Spirit of God spoke in such a manner that not only did He make him hear His voice but also recognize it, so that any one would have been very foolish to tax the Apostle with imprudence in following Christ as one of Whom he had no knowledge. And here, once more, observe how the Apostle, being already made like Him Who called him, caused his voice also to be most plainly heard. For, in the sight of all present, he rose up from the place of custom and attached himself to Christ, clearly showing that

[1] 2 Cor. iii. 17.

he was suddenly changed into another man, no longer grasping nor covetous, nor desirous of the things of this world, but regarding them with a noble contempt. And thus it may be said that he made his voice clearly heard by those around when he gave an example sufficient to move them all, thus reproving the unbelief of those who, after so many miracles, were still unwilling to follow that Lord Whom he had followed at the first call, though surrounded by so many difficulties and hindrances: "At the hearing of the ear he has obeyed Me."[1] And here reflect that it is the same with every truly spiritual man: "So is every one that is born of the Spirit." He is easy to recognize; thou hast only to see him, and by the gravity of his demeanour, by his tranquillity, modesty, humility, obedience, and even tenour of life, thou at once feelest thyself urged to virtue, although by a language that is mute—"thou hearest His voice"—so that thou mayest indeed refuse to imitate, but not to hear him. Dost thou speak to all men with a voice like this? If, then, any one objects that thou seemest to be spiritual, but art not so in reality, that thou art a hypocrite, or a self-interested person, do not let this trouble thee; it is enough for thee to speak, and if any one chooses to say that thy speech proceeds from a human not from a Divine spirit, no matter, the voice at least is heard: "Thou hearest His voice."

IV. Consider fourthly, that it is said in the third place, that although the voice in which the Spirit of God speaks is very plainly heard, yet that no one can tell "whence He cometh and whither He goeth." It cannot be known "whence He cometh," because sometimes the Divine inspiration comes from accidentally seeing a corpse in some church, sometimes from hearing a sermon, from meeting a particular person, from reading a pious book through curiosity, so that it is very difficult to trace the way that He takes: "Who among men is he that can know the counsel of God?"[2] Neither can it be known "whither He goeth," for who is there who can positively foresee what God would have of us when He calls us to a better life? He intends to make one a martyr, another an anchorite, another an apostle, another an admirable example of patience in the midst of countless troubles; and so no one can possibly foresee what will be his last end: "Who can think what the will of God is?"[3] Who would ever have thought that when our Lord might have called a publican to

[1] Psalm xvii. 45.   [2] Wisdom ix. 13.   [3] Wisdom ix. 13.

Himself in so many other places, and in so many other ways, He would choose to do this as He was passing through the public streets, and when the man was sitting at the place of custom, occupied in counting money and making bargains and exchanges ; that is just when it seemed most difficult for him to hear and follow this call ? And in the same way who would ever have thought that of a publican He would have chosen to make so great an Evangelist ? And yet so it was, that in this also it might be verified that " His judgments are incomprehensible" as to knowing "whither He goeth," and " His ways unsearchable "[1] as to knowing "whence He cometh." But now see how Matthew came to act in a like manner so soon as he yielded to the inspiration sent him by God. He followed Christ, and although certainly he was neither able nor desirous to conceal that he followed Him, yet did he not for that reason reveal what were those pure intentions of his in following Him ; on the contrary, he left every one to think just what he pleased about him. One would say that he had left his business in consequence of failures, another from a changeable disposition, another through want of skill ; he was satisfied to have God alone for a witness of that high end for which he thus spurned the world. And such, in truth, is the conduct of every truly spiritual man : "So is every one who is born of the Spirit." He is never anxious to be considered so, although he does not deny it in his actions, and therefore he conceals his manner of life from every one except from him whom he has chosen to be to him on earth in the place of God : "This is our glory, the testimony of our conscience."[2] How, then, can it be that thou art guided by a true spirit in the ways thou art following, if thou revealest these to any one from a motive of vanity ?

TWENTY-SECOND DAY.

Thou hast made Me to serve with thy sins, thou hast wearied Me with thy iniquities (Isaias xliii. 24).

I. Consider first, who those men are of whom God here complains that they make Him "serve with their sins." Generally speaking, they are all those who by sinning abuse the gifts which they have so abundantly received from God as the Author of nature. They abuse their liberty, intellect,

[1] Romans xi. 33.   [2] 2 Cor. i. 12.

knowledge, riches, health, power, beauty; in a word, they abuse the powers which the honourable state in which God has placed them puts at their command, to do evil if they choose. But in a more special manner they are those who, by sinning, abuse the gifts which they have received from God as the Author of grace. Such are those ecclesiastics who often seem to desire that the immunity bestowed on them by their sacred character should become impunity. Such are those who take the bread of the poor to give it to relations who have more than enough already, or to spend it on horses and dogs. Such are those who are induced to make a traffic of the benefices which it is theirs to bestow. Such are those who, so to speak, sell the sacraments by refusing to administer them unless they are induced to do so by self-interest. Such are those who desire to hold churches from ambitious motives, or cures from avarice, or to preach for the sake of gaining not souls, but money. If, as is probable, thou art not of the second class, how likely is it that thou mayest be of the first! Think, therefore, whether it is reasonable that the God Whom thou oughtest to serve so zealously should serve thee so disgracefully, if the expression may be allowed. And yet it is certain that this is so, since He says with His own mouth that He is continually serving thee in thy sins, not of His own will; and therefore He does not say, "I have served thee;" but against His will, wherefore He says, "Thou hast made Me to serve." Nevertheless, He is forced to serve thee, because, although He has bestowed on thee abundant gifts, in order that thou mayest use them to glorify Him, thou, on the contrary, employest them all, or nearly all, in offending Him, because thou usually employest them for the purpose of bringing thy sinful designs to a more successful issue: "I have strengthened their arms, and they have imagined evil against Me."[1] Is it not then with great reason that thy God complains thus tenderly of the outrage thou doest to Him? "Thou hast made Me to serve" (what could He say more sorrowful?) "with thy sins."

II. Consider secondly, that if all sinners grieve God by obliging Him, so to speak, to serve in their sins, hardened sinners go further, and actually weary Him. Not that it is possible for God to endure weariness in anything, for which reason He does not say *laborare me fecisti*, as He did *servire*, but that, if this were possible, He would endure it. These

[1] Osee vii. 15.

hardened sinners do not fail, so far as they are concerned, to give Him abundant matter for it, and therefore He says *præbuisti mihi laborem.* Now, if thou wouldst understand in what this weariness consists, it consists, according to the opinion of the saints, in three things. (1) In the patience with which God continually bears all those insults which are the more intolerable the more constant and obstinate they are : "I am weary of bearing them."[1] (2) In the longanimity with which He waits for the repentance of those who thus insult Him, and not only waits but even invites, encourages, admonishes, and urges them : "I am weary of entreating thee."[2] (3) In the goodness with which He even defends them from the devils, who would fain at once bear their souls to Hell, which they have well deserved. And, therefore, these very words which thou art here meditating, "Thou hast wearied Me with thy iniquities," are rendered in the Septuagint : "I have defended thee in thy iniquities." Enter then seriously into thyself, and consider whether thou art one of those who cause this great weariness to their God, and if so, how is it possible thou art not even aware of it ? "You have wearied the Lord," so said Malachias to his hardened countrymen, and they did not hesitate to answer him boldly : "Wherein have we wearied Him ?"[3] to such a degree of blindness do sinners finally come, if they will not repent.

III. Consider thirdly, that if these words, on which thou hast been meditating, were applicable to every age, much more are they applicable to ours, now that our Lord has become Incarnate, to suffer so much for man's salvation. Set, therefore, before thine eyes Jesus crucified for thee, and contemplate Him in that state of nakedness, suffering, dishonour, and desolation. Then indeed wilt thou fully understand the meaning of these words : "Thou hast made Me to serve with thy sins ; thou hast wearied me with thy iniquities." For did He not serve thee to the utmost in thy sins, when, to save thee from them, He did not refuse to take the form of an abject slave ? "He emptied Himself, taking the form of a slave."[4] And did He not endure grievous weariness, when for thy sake he consented to live by the labour of His hands as a poor boy in a workshop : "I am poor, and in labours from my youth."[5] And yet all this is little compared to that which He has since done for thee, by placing Himself as

---

[1] Isaias i. 14.   [2] Jerem. xv. 6.
[3] Malach. ii. 17.   [4] Philipp. ii. 7.   [5] Psalm lxxxvii. 16.

a shield to save thee from the arrows of the wrath of God, which thou hadst so justly incurred : "And bringing forth the shield of His ministry" (*servitutis*), as it is written of Him in a figure in the Book of Wisdom, " He withstood the wrath,"[1] not merely by prayer, as Aaron did, but by submitting to be scourged, buffeted, wounded and tortured, from head to foot. Wherefore, whereas the Latin translator of the Septuagint, quoted above, did not even say, "I have defended thee in thy iniquities," some of the saints have translated it still more emphatically : "I have put forth a shield for thee in thy iniquities," understanding this passage as it has been just explained, in the most literal sense of Jesus, Who, for thy sake, made Himself a living mark for the wrath of God. And if this is so, how is it possible that thou art not covered with confusion at the sight ? It is at least certain, that in order to correspond in some degree with the kindness of thy Lord, thou art bound not only to cease from offending Him, as thou hast hitherto done, but, in addition, to serve Him with the greatest possible fidelity, and that not only in things which cost thee no labour, but in those also which appear to thee most grievous. Oh, how apt art thou to be restrained by sloth from labouring for love of Him ! If, therefore, thou couldst shake it off, think frequently of these words which God speaks to thee with His own mouth, "Thou hast made Me to serve with thy sins ; thou hast wearied Me with thy iniquities," and if necessary have them written at the foot of thy crucifix, either as a constant reproof or a constant remembrance. If thy Lord, as has been said, has served thee to such an excess in thy sins, which, after all, are the same as thy irregular desires, is it not right that thou shouldst now serve him in the fulfilment of His Divine will, which is so perfectly holy ? And if He has laboured for thee to such an excess in thy iniquities, that is, not only in thy actual, but thy habitual sins, is it not right that thou shouldst labour continually for Him in the propagation of His glory ?

[1] Wisdom xviii. 21.

TWENTY-THIRD DAY.

If any man think himself to be religious, not bridling his tongue, but deceiving his own heart, this man's religion is vain (St. James i. 26).

I. Consider first, that, speaking broadly, all these persons are religious who have given themselves to the service of God in an especial manner; for such persons have added to those general obligations by which they are bound to Him other rules and customs of their own. But, in a more restricted sense, those persons are religious who have consecrated themselves to the service of God by the solemn vows of poverty, chastity, and obedience, because such persons have still further bound themselves to God by stronger ties than any that exist in the world, by adding counsels to precepts, and that not for a time only, but constantly, that is, for their whole lives. Now, there is no doubt that it is necessary for all those who are serving God in an especial manner, to know how to bridle the tongue. But if among these persons there are any for whom it is more necessary than for others, they are certainly those to whom this name of religious, so dear to God, is more especially appropriate. For either these religious are wholly given to the contemplative, or to the active life, or to both together, at once learning from God, and teaching men, which is the most perfect kind of religious order. If they are wholly given to the contemplative life, it is easy to see how important it is for them to know how to bridle the tongue, for it is silence which prepares the soul to acquire the gift of contemplation: "I will lead her into the wilderness, and I will speak to her heart,"[1] and it is silence which preserves this gift when it is acquired: "He shall sit solitary and hold his peace, because he hath taken it up upon himself."[2] If they are wholly given to the active life, thou seest that they also should be very careful in bridling the tongue, because, though it is true that they are not bound to silence, like contemplatives, but are obliged to have intercourse with many of their neighbours, yet they are bound to speak without scandal and with great accuracy, which is perhaps more difficult than to be silent: "In the multitude of words there shall not want sin."[3] And if, in the last place, they combine both lives, as is the happy lot of those to whom David alluded in the words, "They shall publish the memory of the

[1] Osee ii. 14. [2] Lament. iii. 28. [3] Prov. x. 19.

abundance of Thy sweetness,"[1] they must know both how to be silent at the right time in order to possess themselves of that sweetness, and to speak at the right time in order to communicate it to others—"There is a time to keep silence, and a time to speak"[2]—and this belongs only to men of great judgment: "He that refraineth his lips is most wise."[3] To what degree hast thou hitherto attained in the government of thy tongue, in thy state of life? If thou dost not yet know how to govern it, listen to what the Apostle of the Lord here says to thee. He says that thou hast no right to glory in the name of religious, because all thy religion is vain, that is, it is devoid of the profit which it ought by its nature to produce, both in thyself and in others: "This man's religion is vain."

II. Consider secondly, that the tongue is like a vicious colt, that no one can succeed in taming it perfectly, unless he is more than man: "The tongue no man can tame."[4] It requires a very excellent gift of grace to prevent it from ever breaking bounds in some way: "Who is there that hath not offended with his tongue?"[5] Therefore the Apostle does not here say, "If any man think himself to be religious, not taming his tongue, this man's religion is vain," but he says only, "not bridling his tongue." For if he cannot attain to taming it in such a manner that it will never, so to speak, make a false step either through inconsiderateness or imprudence, he can at least attain to making it fear the bridle. This bridle is the dominion of reason, which, as it rules all the other members of the body, to keep them obedient to itself, so likewise should it rule the tongue; indeed, this is more necessary for the tongue than for the other members, since it is more troublesome than the rest to govern well. And the reason is that the other members, for the most part, run into only one kind of sins: the mouth into intemperance, the eyes into vanity, the ears into curiosity, the touch into immodesty, and so of the rest; but the tongue runs into every kind of sin, for which reason it is called "a world of iniquity."[6] I say it is not even contented with those sins which belong to it, such as haughty boasts, lies, calumnies, oaths, perjuries, backbitings, and the like, but it also appropriates those which do not belong to it, such as murders, thefts, frauds, uncleanness; since it is beyond a doubt that it is the tongue which is often so bold as

---

[1] Psalm cxliv. 7.　　[2] Eccles. iii. 7.　　[3] Prov. x. 19.
[4] St. James iii. 8.　　[5] Ecclus. xix. 17.　　[6] St. James iii. 6.

to teach these sins, to advise and command them before they are committed, and then to defend them when they are committed. So that if any one desires to keep himself from sins of the tongue, it is not enough to bridle it alone, but it is necessary to have really conquered all the passions—pride, anger, covetousness, envy, impurity—which are the passions that urge thee to say what should not be said: "I said, I will take heed to my ways, that I sin not with my tongue."[1] And this is another comprehensive reason why the man who does not bridle his tongue, cannot boast of being religious— "If any man think himself to be religious, not bridling his tongue, but deceiving his own heart, this man's religion is vain " —for this is a clear sign that he has not yet overcome his passions. Dost thou desire that thy tongue should obey the bridle? Take care at the same time to conquer those passions which most of all embolden it to rebel: "When the wood faileth, the fire shall go out."[2]

III. Consider thirdly, that men of irregular life sin with the tongue in a very different manner from men of spiritual life. The former see that they do wrong in speaking as they do, but take no pains to check themselves; on the contrary, they sharpen their wits in order to make their tongues more ready to say whatever is prompted by anger, hatred, ambition, and insolence, instead of reason. The latter, in order to speak freely, endeavour, in the first place, to deceive themselves, by persuading themselves that in such and such circumstances it is right to speak as they are doing. Observe, therefore, that the Apostle here says, "If any man think himself to be religious, not bridling his tongue, but deceiving his own heart, this man's religion is vain," because this is peculiar to religious who do not desire to keep their tongues in check : they deceive themselves with arguments which are frivolous rather than solid. If they want to break the silence so necessary to interior recollection, they begin by saying within themselves that the bow which is too much stretched breaks, and that by slackening it, it can be drawn with greater force afterwards. If they want to say words of self-praise, they try to persuade themselves in their heart that their only object is to gain credit, so as to reap more fruit from their work. If they want to blame the orders of Superiors, they encourage themselves to it by saying to themselves that it is not right to flatter them as others do, and so they call their complaints either the

---

[1] Psalm xxxviii. 1.   [2] Prov. xxvi. 20.

magnanimous love which they bear to truth, or their zeal
for correction, for charity, or for the honour of God. Do
thou, therefore, hold it as certain in thy own case, that if,
being bound in spirit, thou allowest thyself liberty of tongue,
thou deceivest thy own heart ; and to this certainly it is necessary to apply thy attention first of all. Begin by setting right
the erroneous opinions which lurk within it, and be very sure
that they are so many pretences invented to disguise thy
passions. At least set thyself to examine them with some
particular care, and do not be ready to trust to first appear-
ances, because it is precisely this which is to deceive thyself :
it is to throw dust into thine own eyes, to flatter and delude
thyself, and to be ready to approve the reasons suggested
by inclination, but not to examine them : "Be not seduced."
And why ? Because, no matter under what pretext they lurk,
unprofitable words always do harm : "Evil communications
corrupt good manners."[1]

TWENTY-FOURTH DAY.

*I say to you, that there shall be joy before the angels of God upon one
sinner doing penance more than upon ninety-nine just who need not penance*
(St. Luke xv. 7, 10).

I. Consider first, that according to the Hebrew idiom,
this positive phrase, "There shall be joy," has the force of
a perfect comparative, as also in the following passage : "It
is good to confide in the Lord rather than to have confidence in man ; it is good to trust in the Lord rather than
to trust in princes."[2] And so in our text, "There shall be
joy," means "there shall be greater joy." It is not indeed
here said, if we observe carefully, that a converted sinner is
more highly esteemed in Heaven than ninety-nine just who
have no need of penance, but only that there is greater joy
over him. For as to the esteem, it would be greater if that
converted sinner gave himself to God with such fervour of
spirit as actually to love Him more than all those innocent
persons who are here spoken of, but this case, such as was
perhaps that of St. Mary Magdalene in her conversion, is
exceedingly rare. And in this passage our Lord does not
intend to speak of what occurs in some exceptional conversion,

[1] 1 Cor. xv. 33.   [2] Psalm cxvii. 8, 9.

but of what occurs in all conversions, by their nature, and therefore the comparison which He makes is merely between an ordinary penitent, "one sinner doing penance," and ninety-nine ordinary just persons, "who need not penance;" not between an exceedingly fervent penitent and ninety-nine luke-warm just persons. Premising, then, that all these just persons together are, speaking generally, more highly esteemed by God than one penitent, still the penitent is the cause of greater joy, because joy has reference not so much to the esteem which a man has of anything in itself, as to the acquisition of the thing, particularly when this was either hopeless or difficult. So it was that when that loving father prepared a feast of such unusual abundance on the return of the prodigal son, the only reason which he assigned for this was that he had recovered him after so many years, that it was like seeing him come back from death to life: "It was fit that we should make merry and be glad, for this thy brother was dead, and is come to life again; he was lost, and is found."[1] How is it that this does not inflame thy heart with the greatest love of God? For what motive could He have in rejoicing so greatly over thy recovery, but the very high value He sets on thee, I do not say in comparison with the many just persons who are better than thee, but at least speaking absolutely? Would He not be equally happy without thee, equally great, equally glorious? What reason then can He have for rejoicing so much at thy return from sin to grace, except that thou art really dear to Him? This is a thing so wonderful, that it certainly could not be believed if God Himself had not actually sworn it with His own mouth, and therefore, as thou seest, He does thus swear it here: "I say unto you." "Oh, happy we on whose behalf God Himself swears."[2]

II. Consider secondly, the reason why God not only rejoices so much at the conversion of a sinner, that all the angels also rejoice: "There shall be joy before the angels of God." As though such a festival in Heaven were never private, but always public. There are three reasons for this: the angels consider it first as it concerns God, secondly as it concerns men, and thirdly as it concerns themselves. As regards God, the angels see how much glory (accidental glory of course) He gets from the conversion of men to penance, and therefore it is impossible but that they too should rejoice at it infinitely, because of their burning love for God. As

[1] St. Luke xv. 32.   [2] *Tert. De Pecc.*

regards men, it is certain that there is no envy in the angels, but that, on the contrary, their only desire is to have many with them to share their happiness; and for this reason also they rejoice exceedingly when they see those who have unhappily lost their right to that happiness regain it. Lastly, as regard themselves, there is also a plain reason for their joy, for since it is the office of the angels to labour for the salvation of men, as it is written: "Are they not all ministering spirits sent to minister for them who shall receive the inheritance of salvation?"[1] How can they but feel a sensible delight when they see the fruit of their labours? "What is our crown of glory? are not you in the presence of our Lord Jesus Christ?" said the Apostle to the Thessalonian converts whom he had gained to God; and so, too, it is evident, do the angels say. And thus, it may be, that the most zealous preacher in the world never rejoiced so greatly at saving many souls from sin, as each one of the angels always rejoices, the more so as the devils are always endeavouring to thwart them in making these glorious conquests, and therefore it is an immense joy to the angels to see themselves victorious and triumphant over their old enemies in this great battle: "That great dragon was cast out who seduceth the whole world. . . . Therefore rejoice, O Heavens, and you that dwell therein."[2] Think, at all events, how easily thou mayest this very day cause all these blessed spirits the greatest joy that they can ever feel, not indeed essential, for that is always the same, but accidental, and this by quitting a state which, moreover, must issue in thy everlasting destruction.

III. Consider thirdly, that these angels who in another place are called the angels of men—"Their angels in Heaven always see the face of My Father"[3]—are here, on the contrary, called the angels of God: "There shall be joy before the angels of God." But if thou observest closely, there is no contradiction whatever here, but rather an exact conformity, for this is said to point out fully the two parts of their office, namely, serving God, and acting as ministers to men. The angels serve God in three ways: by contemplating Him constantly, loving Him ardently, and vying with each other in praising Him incessantly: "All the angels stood round about the throne, . . . and they adored God, saying, Amen."[4] They serve men also in three ways: by purifying, illuminating, and

[1] Hebrews i. 14.
[2] Apoc. xii. 5—12.
[3] St. Matt. xviii. 10.
[4] Apoc. vii. 11.

perfecting them. They purify them from their faults, and this is the service which they render especially to beginners in the ways of God: "And one of the Seraphim flew to me, and touched my mouth, and said, Behold thy iniquities shall be taken away and thy sin shall be cleansed."[1] They illuminate them by instructions, and this is the special service which they render to those who are more advanced: "I am come to teach thee what things shall befall thy people in the latter days."[2] And they perfect them by the powerful supports of grace: "And behold an angel of the Lord touched him, and said to him, Arise, eat, for thou hast yet a great way to go."[3] These two parts of the angelic office, which consists in serving God and in labouring on behalf of us men, were wonderfully typified, as is well known, in the famous ladder on which Jacob saw the angels doing nothing but "ascending and descending;"[4] for this is the whole of their office: "You shall see the heaven opened, and the angels of God ascending and descending upon the Son of Man."[5] If then thou desirest not only to rejoice the angels by thy conversion, which is but a small thing, but also to imitate them, as is right, in their office here, it is plainly set before thee what thou hast to do. To ascend by the exercises of contemplation, to admire, love, and praise God, and to descend, by the works of an active life, to help all thy neighbours by purifying, illuminating, and perfecting them, according to their various states: "Whether we be transported in mind, it is to God, or whether we be sober, it is for you."[6] Thus thou wilt be, if not an angel, at least a man of angelic life, that is, wholly given up to the service of both God and man.

TWENTY-FIFTH DAY.

An obedient man shall speak of victory (Prov. xxi. 28).

I. Consider first, that the noblest action which it is possible for a man to do upon earth is that which many persons esteem least of all, namely, that of conquering self, for it is this which more than all others makes him act as what he is, that is, as a man. If thou observest tigers, leopards, panthers, lions, and other wild beasts, thou wilt see them indeed perform

[1] Isaias vi. 6.   [2] Daniel x. 14.   [3] 3 Kings xix. 7.
[4] Genesis xxviii. 12.   [5] St. John i. 51.   [6] 2 Cor. v. 13.

actions of the greatest courage in conquering animals stronger than themselves, but thou wilt never see them perform so exalted an action as this of conquering themselves. They always do that to which they are violently impelled by the appetite which predominates in them, whether of rage, greed, lust, or cruelty. This great act of conquering self is reserved on earth for man only, and of all the virtues it is obedience which produces this act: it makes thee conquer thyself in those things in which, according to the lower appetite, thou wouldst be least inclined to do so, and so makes thee truly act like a man, that is, like a reasonable creature, not a brute. There is therefore no reason to wonder why it is written in the Word of God that the obedient man only is allowed to boast of his victory: "An obedient man shall speak of victory." For any victory that a man gains merely by his strength in conquering others, is a victory which he has in common with brutes, and therefore he ought not to glory in any victory of this kind. He should glory in that alone which he gains through conquering himself by obedience, because such a victory not only shows that he is strong, as brutes are, but also that he is free, which cannot be said of him who, in order to satisfy his unrestrained desires, refuses to act in the manner which God has declared to him by the mouth of His servants. Say, then, to thyself, that if there were no other inducement for thee to obey perfectly, promptly, and gladly, this alone ought to be enough: to know that by obeying thou art performing so noble an action as this which we are speaking of. Therefore thou seest that he who obeys his Superior, not in a single instance, but habitually, nor from the desire of reward, which is ambition, nor from the fear of punishment, which is cowardice, but because it is right to obey, is called *vir*, because he is not only a man, but a man of no common order, a man who deserves beyond all others that noble name of *vir*.

II. Consider secondly, that all the many victories which are gained in the spiritual life may be summed up in that chief victory which a man gains over himself, in order to do what is commanded him. And therefore the Wise Man, according to the true reading of the Vulgate, did not choose to say, "An obedient man shall speak of victories," as several doctors would have it, but he used the singular number, "victory," because he who subjects his own will as he ought to that of his Superior, which is the especial victory of an obedient man, has no other enemies to fear. He has conquered them all in conquering himself: "Thy seed shall possess the gates of their enemies."[1] Such was the glorious reward given by God in the person of Abraham to all those who should truly imitate him in his obedience. The three powerful enemies of men are, as is well known, the flesh, the world, and the devil. Now, with regard to the first, the man who has not conquered the flesh, which is the vilest part of himself, cannot attain to a daily conquest over the will, which is the noblest part. And, therefore, whenever we see a truly obedient man, we may boldly declare that he is a chaste man, because we may be sure that he who has done that which is greater, has also done that which is less. Besides, this is a special reward which God, as the saints tell us, is accustomed to grant to such a man, namely, the subjugation of the flesh : " Let him who desires to have the lower part subject to himself, be himself subject to his Superior."[2] And, as a proof of this, we see that so long as our first parents did not transgress the commandment given to them in the earthly Paradise of not eating the forbidden fruit of the tree of knowledge, they never experienced any rebellious movement of the flesh, but only after they had transgressed that command. And therefore the saints also tell us that God, on the contrary, gives to the disobedient the sting of the flesh which makes them fall in a most shameful manner, in order that he who will not give an honourable obedience to his lord (as he is who occupies the place of God towards him on earth) may see his own servant shamefully refuse to obey him : "Thou who dost not obey thy lord art tormented by the servant."[3] Then, as to the second enemy, the world, a truly obedient man has no reason to fear it, because it is under his feet. For what is it which is most highly valued in the world? It is the glory of ruling others, and this is the very thing which the obedient man does not care for. Therefore, not only does he submit himself to those who are his Superiors, either in talents, rank, or office (as is done even by the followers of the world), but he also submits himself to those who are his inferiors, a thing which is never done in the world, except occasionally from motives of interest, for which reason St. Peter wrote : " Be ye subject to every human creature for God's sake," which is the only motive from which a truly obedient man thus acts ;

[1] Genesis xxii. 17.
[2] St. Augustine, in Psalm cxliii. [3] St. Augustine, *Ibid.*

for he who does not do it from this motive cannot be called obedient, but self-interested, and so he has not conquered the world. And lastly, as regards the devil, it is only the obedient man who can say that he is sure of having conquered him. Others may hope this, but they cannot be certain of it, because whoever is guided by his own judgment in doing well is liable to countless illusions and snares of the devil, from which he is free who never follows his own judgment, but that of his Superior : "The son that keepeth the word shall be free from destruction."[1] See how, in the great victory which thou gainest over thyself by obedience, thou conquerest all thine enemies ; and therefore, in the conflict which thou art entering, when thou resolvest to lead a spiritual life, do thou set thyself anxiously to attack any one in particular of the three enemies above mentioned. Set thyself to attack thine own will, which is the ruling part of thee : "You shall not fight against any, small or great, but against the king only."[2] Here fix thy eyes, here aim thy blows ; for in conquering this one enemy thou wilt gain a complete triumph.

III. Consider thirdly, what is inferred when it is said that "an obedient man shall speak of victory." Does it mean that he is to proclaim everywhere by sound of trumpet the glorious victory which he has gained by conquering himself, and with himself his fiercest enemies? No ; because we are plainly told that all victory is to be ascribed to God : "Thanking God, Who hath given us the victory."[3] What we are meant to understand is that the obedient man may speak of his victory to God Himself, by thanking, praising, and honouring Him ; and he may also speak of it to the saints, begging them all to supply what is wanting in him of the honour due to God. And if he desires also to speak of it to men, in order to instruct them how to gain a like victory, to encourage or console them, or for any such reason, he may do this because he will know how to do it. Some persons would fain give admirable instructions on self-conquest, merely because they have read them in books, but have never, or scarcely ever, put them in practice. Let all such persons hold their peace, because it is not said that a wise, an eloquent, or a learned man "shall speak of victory," but "an obedient man." Speculative science learnt from books is of very little use in speaking solidly on spiritual subjects : it is practical science which is of use. He who speaks without this is like a blind

[1] Prov. xxix. 27.   [2] 3 Kings xxii. 31.   [3] 1 Cor. xv. 57.

man talking of colours. "Let them that sail on the sea, tell the dangers thereof: and when we hear with our ears we shall admire;"[1] but if we hear a man talk of tempests who has never ventured many paces from the shore, we laugh at instead of admiring him. This, then, is another legitimate sense of these words, "An obedient man shall speak of victory," that it is lawful for any one to speak of the right manner of self-conquest, but not till he has himself practised it by the exercise of that perfect obedience which, above all things, enables a man to learn that practice. How easily mayest thou presume in spiritual matters in which thou hast as yet not begun to exercise thyself, or at least very superficially. "He that hath no experience knoweth little,"[2] because it is impossible for the man who has not first learnt to know things in himself to recognize them well in the persons of others.

TWENTY-SIXTH DAY.

Be instructed, O Jerusalem, lest My soul depart from thee (Jerem. vi. 8).

I. Consider first, that just as the first process of removing the roughness or dross from marble, metal, or wood, is called purifying it, so too is a soul said to be purified when it, too, is undergoing the first process of being free from whatever hinders it from receiving a good rule of life, that is, of being freed from those ill-regulated desires, or from those sentiments which are like vile dross overlaying it. This, without doubt, is the main labour, and therefore, to all those who courageously undertake it, there is promised so great a reward that it is even said, "They that instruct many to justice shall shine as stars for all eternity."[3] Now, this is the trouble which God is willing to take with thy soul. He desires to purify thee, that is to say, to cleanse thee from all that vile dross which He sees in thee—the love of pleasure, money, and glory, but, above all, that which is the source of all evil in thee, the high esteem of thyself: "Teaching, He instructed them in what they are to learn, that He may withdraw a man from the things he is doing, and may deliver him from pride."[4] It is true that He would have thee willingly submit to be thus

[1] Ecclus. xliii. 26.
[2] Ecclus. xxxiv. 10. [3] Daniel xii. 3. [4] Job xxxiii. 16, 17.

purified by Him, and to yield thyself willingly to the hand of
the great Sculptor, Who when He strikes thee does it for thy
good.   And therefore He says to thee, in the passage which
I am here proposing for thy meditation, "Be instructed,
O Jerusalem, lest My soul depart from thee." This is a work
which cannot be done entirely by thyself or entirely by God,
but by God and thyself together.   It is necessary for thee
therefore to let Him work and not to hinder Him, because He
does not deal with thee as with marble, or metal, or senseless
wood, but as a free spirit, which can either consent or refuse
to take the form to which God desires to mould it.   Therefore
"be instructed;" do not be of the number of those of whom
it is written: "They have turned their backs to Me, and not
their faces; when I taught them early in the morning, and
instructed them, and they would not hearken to receive
instruction."[1]

II. Consider secondly, that the chisel which God uses in
the process of which we are speaking is that of tribulation.
When He uses this, then He is said to do His work with a
strong hand.   "He hath taught me with a strong arm that I
should not walk in the way of this people,"[2] because nothing is
of more avail in removing the excessive love of ourselves,
which is the vilest of our dross, or to fill us with compunction
and bring about our conversion, than a great humiliation sent
us by God: "Thou hast chastised me, and I was instructed."[3]
Therefore, in this case especially, it is necessary to leave God
entire freedom of action—not to complain of Him, not to
show anger or impatience, but to accept with resignation all
those master-strokes which He sees it expedient to give thee,
otherwise thou incurrest the danger of His taking His hand
from thee, and leaving thee in thy rude state, and letting thee
walk, as thou desirest to do, "in the way of the people," that
is, in the way which leads to destruction.   Oh, if thou knewest
how great a benefit God confers on thee whenever He humbles
thee by some severe tribulation!   Thou canst not understand
it now, but I would fain hope that the day will come when
thou wilt see clearly that but for that sickness, that disgrace,
that misfortune, that adversity, which seemed so intolerable
at the time, thou wouldst surely have been lost.   Would not
those sheep be very foolish which should complain, when
scattered here and there over the Apennines, because their
shepherd brought them together again by using his stick?

[1] Jerem. xxxii. 33.   [2] Isaias viii. 11.   [3] Jerem. xxxi. 18.

That is the very time when they should be specially grateful to him, because he has shown himself so anxious to save them from falling over the precipice, and this is what God does when He sends us tribulation: "He hath mercy, and teacheth, and correcteth, as a shepherd doth his flock."[1]

III. Consider thirdly, that it is quite possible that God may not forsake thee, in spite of thy want of submission to His sweet will, in thy troubles, but that thou forsakest Him; and is not this of itself sufficient cause for great watchfulness on thy part? Therefore He says, "Be instructed, O Jerusalem, lest My soul depart from thee." It is true that in the Vulgate the word *forte* is added ("lest perchance"); but what then? Thou oughtest to be full of fear at even the slightest danger, when so great a matter is at stake, for the danger is that of thy damnation. What does God mean when He says, "Lest My soul depart from thee"? He means: Lest I should remove from thee that special protection, that affection, that love which I have extended to thee; for this is what is signified by the expression, "His soul:" "My soul delighteth in Him."[2] For it is true, that although He will never so far forsake thee, because of thy indocility, as to refuse thee the sufficient grace necessary for salvation, yet He will refuse thee that efficacious grace which He is bound by no law to give thee; He will refuse thee that special assistance, those extraordinary aids which are a purely gratuitous gift of His Heart. And therefore He says to thee, "Lest My soul depart from thee." And oh, how terrible a threat is this, enough to terrify not merely a beginner in the ways of God, such as I am here supposing thee to be, but even a saint! And yet it is to this threat that thou riskest becoming subject, by showing thyself so little conformed to the will of God in the time of adversity, which is the instruction here spoken of, even in a literal sense. For by such conduct thou art in danger of God's ceasing to send thee trouble, and consequently of His taking from thee that special kindness which He shows thee when He desires to purify and cleanse thy soul, with no other object than that of preparing it by this means for the reception of His grace. Therefore St. Jerome says, on this passage, that whenever thou art in any trouble which greatly distresses thee, thou shouldst always be ready with this verse, and think in thy heart that God is saying to thee, "Be instructed, O Jerusalem, lest My soul depart from thee." If thou scornest His instruction, it

[1] Ecclus. xviii. 13.   [2] Isaias xlii. 1.

may be that He will not depart from thee all at once; but
what will that avail thee, if He gradually goes farther and
farther away from thee, till He leaves thee altogether? And
this is the meaning of the word "depart."

TWENTY-SEVENTH DAY.

*Be not conformed to this world, but be reformed in the newness of your
mind, that you may prove what is the good, and the acceptable, and the
perfect will of God* (Romans xii. 2).

I. Consider first, that in this passage thou oughtest to
understand by the will of God those things which He would
have us do; just as when He says, "Teach Me to do Thy
will."[1] Now, there are three degrees of the things which God
desires of us. Those that are good: such, for example, as not
to hate our neighbour; those which are better, such as not
only not to hate, but to love him; and those which are best
of all, such as not only to love, but to do good to him. The
first belong to beginners, the second to those who are more
advanced, and the third to the perfect. Those, therefore,
which belong to the first class are here called "the good-will
of God;" those of the second, "the acceptable will of God;"
and those of the third, "the perfect will of God." The first are
called "good," because they are all works which are right in
the sight of God; the second are called "acceptable," because
they are works which are more than usually pleasing to Him;
the third are called perfect, because they are works which are
entirely conformed to His own. The thing, then, which the
Apostle here desires is that thou, for thy part, shouldst prepare
thyself to "prove" all these works in such a manner as to be
able one day to possess them all, which will be when thou
really givest thyself to perfection. See how far thou art, it
may be, from this, if thy works barely belong to the first class;
and bewail thy coldness.

II. Consider secondly, how accurately the Apostle speaks
when he bids thee prove these works: "That you may prove
what is the will of God." He does not say that thou art to
know them, but to prove them; for it is not enough to have a
speculative knowledge of things, even of the most sublime

[1] Psalm cxlii. 10.

perfection : there must be a practical knowledge also. And how are they known? By proving them : "That you may prove." Of what use is any amount of speculative knowledge in the works of virtue, if thou dost not reduce it to practice? The devil has so much knowledge, that it is for this very reason that he has his name ; because "demon," in Greek, means the same thing as *sciens* in Latin. And yet, as St. Augustine observes,[1] this word "demon" is always used in a bad sense in the Sacred Scriptures ; for of what avail is it to the devil to have a knowledge of so many good things when he does not practise them? Nay, this very thing makes him worse : "To him who knoweth to do good, and doth it not, to him it is sin."[2] Neither is the Apostle here satisfied with the doing these things which are spoken of, in any kind of way ; he would also have thee do them with relish, for this is the force of the word "prove." If the palate of thy soul is good, thou wilt soon see the difference between the food of beginners and that of the more advanced, and again between this last and the food of the perfect. We have in the Gospels three instances of our Lord feeding men. He fed them the first time with barley bread, the second time with wheaten bread, and the third time with the Heavenly Bread of the Blessed Sacrament. Now, consider that there is the same difference between the savour of the works belonging to those different states as between the savour of those three kinds of bread. If thou dost not distinguish it, it is because thou art of the number of those who have never yet proved it—"If so be that you have tasted that the Lord is sweet"[3]—thou art still eating the barley bread.

III. Consider thirdly, what is the way to acquire a palate capable of well distinguishing the increasing pleasure which there is in doing, not only that which is better, but that which is best. It is to reform the mind. And the reason is, that spiritual pleasures are perceived, not by the corporeal, but by the intellectual taste ; wherefore the Apostle here says, "Be reformed in the newness of your mind." The word translated mind is *sensus*, which thou wilt see has the same meaning here as *ratio*. For our reason is that interior sense which judges of spiritual things, just as touch, sight, hearing, and the other exterior senses judge of the things within their cognizance, that is, of material things : "I wished, and under-

---

[1] *De Civ. Dei*, lib. ix. cap. 19. [2] St. James iv. 17.
[3] 1 St. Peter ii. 3.

standing was given me."[1]   Now, it cannot be denied that this reason was originally given by God in perfection, which is why the Wise Man, speaking of our first parents, says, " He filled their hearts with wisdom" (*sensu*) ;[2] but afterwards, in consequence of sin, it was gradually so perverted that at length it became inveterate in judging of things wrongly : "Thou art grown old in a strange country."[3]   And this is precisely why the Apostle said, " Be reformed in the newness of your mind." For we must go back to the first rule of judgment given to us by God in the state of innocence, which is attained by virtue of the grace imparted to us by Christ our Lord, now that we have sinned.   For what other reason did He come into the world ?   He came to reform the judgment of the old man, and bring it back to its original newness.   Therefore if thou hast not set in order the judgment of thy mind, thou hast done nothing at all, for there is the root of the evil : "Woe to you that call evil good, and good evil, that put darkness for light, and light for darkness, that put bitter for sweet, and sweet for bitter."[4]

IV.   Consider fourthly, that what is principally required in order to take a new form is to lay aside the old one: "Stripping yourselves of the old man with his deeds, and putting on the new."[5]   Now, this old form is the same thing as the form of the world, and therefore, in the first place, the Apostle says, "Be not conformed to this world."   According to the judgment of the world, virtuous men are deprived of all pleasure, spiritual men still more so, and the saints most of all : "How very unpleasant is wisdom to the unlearned."[6] And why does the world judge so perversely ?   Because the only goods of which it has any knowledge are those which are perceived by the senses, such as pleasures, riches, and glory; and these it appreciates.   But what thou hast to do is utterly to lay aside the esteem of all these three goods which the world worships, to know that they are false, and useless, and unstable ; and in this manner thou wilt prepare thyself to receive that form which Jesus bore on earth, to destroy that which He found there.   Hear how He speaks against those who are given up to sensual pleasures : "Woe to you that now laugh ;" against those who are absorbed in riches : "Woe to you that are rich, for you have your consolation ;" and against those who are devoted to their own glory : "Woe to you when men shall bless you."[7]   Are not these three woes sufficient

[1] Wisdom vii. 7.   [2] Ecclus. xvii. 6.   [3] Baruch iii. 11.
[4] Isaias v. o.   [5] Coloss. iii. 9.   [6] Ecclus. vi. 21.   [7] St. Luke vi. 26.

instantly to destroy all thy love for the world? And all these three are uttered against those who love it, not only with their bodies, but also with their hearts: "Woe, woe, woe to the inhabitants of the earth."[1]

V. Consider fifthly, that if thou couldst really leave the world, not merely with thy heart, but also with thy body, then indeed thou wouldst prepare thyself to taste those higher pleasures which belong to such as fulfil the will of God with perfection. But because this is not possible for all, see how wisely the Apostle here spoke when he said, "Be not conformed to this world." He did not say, "Do not remain in this world," because many persons are compelled to remain in it; neither did he say, "Do not use this world," because many even of those who do not live in it are obliged to make use of it at times in order to provide what they require, at least for their necessary sustenance, such as food, clothing, and the like. He said, "Be not conformed to this world," because this is in the power of all. If, then, thou choosest to remain in the world, be it so; but hear in what manner thou art to live there. As Lot lived in Sodom, Job in Hus, Joseph in Egypt, Tobias in Ninive, Daniel in the splendid palace of Babylon, and others like them, who never conformed to the customs of the wicked nations among whom they lived, but were like fish which remain in salt water without ever contracting any saltness: "Having your conversation good among the Gentiles."[2] Thou wilt say that this is difficult; I grant it, and therefore, for those who are able to do so, it is better to leave the world. But if it is difficult, it is not therefore impossible to many persons, by the grace of God. If it were so, the Apostle would never have said, "Be not conformed to this world." But since he has said this, it is a proof that to do, or not to do it, is in thy own power. If thou findest it difficult, endeavour to lessen the difficulty as much as possible by continually asking God for His holy grace, by frequent confession and Communion, by reading some spiritual book every day, by frequenting religious houses, by loving to visit churches, and by completely abandoning all bad habits. Be constant in employing these means, and so wilt thou please God, and succeed in not being conformed to the world, which judges so foolishly: "I know where thou dwellest, where the seat of Satan is, and thou holdest fast My Name, and hast not denied My faith."[3]

[1] Apoc. viii. 13.   [2] 1 St. Peter ii. 12.   [3] Apoc. ii. 13.

TWENTY-EIGHTH DAY.

No man putting his hand to the plough, and looking back, is fit for the Kingdom of God (St. Luke ix. 62).

I. Consider first, that in order to understand rightly the intention of Christ in these terrible words, it is necessary, in the first place, to know with what object He uttered them. It was to reject a certain young man who had voluntarily offered himself as his perpetual disciple—"I will follow Thee, Lord" —but who first wished to obtain permission to acquaint his family with his purpose, in order to settle his domestic affairs : "But let me first take my leave of them that are at my house ;" without which permission it appears he was unwilling to carry out the resolution he had taken, as is shown by the conjunction "but," which is here strongly adversative. The only answer which our Lord gave to this young man was this severe saying : "No man putting his hand to the plough, and looking back, is fit for the Kingdom of God." By the words, "Kingdom of God," is understood either the Kingdom of Christ in Heaven, which is the kingdom of enjoyment, or His Kingdom on earth, which is the kingdom of labour ; and as this man is simply said not to be fit for the Kingdom of God, it follows that he is not fit for either of these kingdoms. Is not this a declaration to inspire the utmost horror, unless, indeed, it is explained in the most mitigated sense that is possible ?

II. Consider secondly, that he who "puts his hand" to the perfect following of Christ, such as this young man proposed to embrace in imitation of the Apostles, undertakes a great work, and one which consequently requires great love for our Lord, great courage and great attention ; and, therefore, Christ explained it by comparing it to what is certainly the most laborious occupation in agriculture, namely, ploughing, which therefore requires courage and attention : courage, because in a vast field it is a vast work ; and attention, because it cannot be performed at the same time as anything else, such as digging, sowing, or reaping, because all the furrows require to be drawn perfectly straight, which cannot possibly be done by a man who keeps constantly looking back. And this admirably explains our Lord's first meaning in this passage. Because the perfect or apostolic following of Him is a very great work, and one which requires all a man's energies ; for which, therefore, no one is fit who has not very great courage in undertaking it, and very great attention in accomplishing it. Now this young man had not great courage, because he could not make up his mind to abandon his family interests for Christ's sake, with that resolution which had been shown, not only by James and John when they left their nets, but by Matthew when he quitted the place of custom ; neither did he manifest that close attention which should distinguish such a follower of Christ, since at one and the same time he contemplated following Him, and leaving Him, though for a time only, on account of his domestic affairs. And therefore Christ declared that he who acts thus is not fit for the apostolate : I say for the apostolate, because the continuation of the metaphor here requires that after our Lord had said, "No man putting his hand to the plough, and looking back, is fit for the Kingdom of God," *excolendo* (cultivating) should be prefixed to the two last words to complete the proposition which is left imperfect. This is the mildest interpretation that can be given of the declaration here made by Christ. But it is enough to show how great an evil it is to be too much attached to worldly interests, being of itself sufficient to prevent so exalted a good as that of becoming an apostle.

III. Consider thirdly, that besides this more perfect following of Christ, there is another less perfect, to which every Christian is bound ; and it seems as though Christ purposely chose not to complete the declaration we are considering that it might be applied, as the saints apply it, to all persons in due proportion, with regard to their various shortcomings in following Him. And it is just this which is so terrible, for what Christ would have us understand generally is, that whoever is not firm in carrying out the good resolutions he has made, but breaks them, either through instability, or pusillanimity, or sloth, or attachment to the worldly interests which draw him to themselves (which was the stumbling-block of this poor young man), is no more fit to merit the enjoyment of God in His heavenly Kingdom, than he is to labour manfully for Him in His earthly Kingdom. Ask thyself now what thou canst justly say of thyself in this matter. Art thou as strong as thy Lord requires?

IV. Consider fourthly, that Christ says, in the first place, "No man putting his hand to the plough." He does not say "who hath put," nor "who shall put," but "putting," in order to show that not only the man who is not strong in carrying out

the good which he has undertaken is not fit for the Kingdom
of God; but also the man who is not strong to undertake that
which he has resolved. He who puts his hand to the plough
is the man who is firmly resolved to work; the man who works
is already ploughing. When, therefore, by a special vocation
sent to thee by God, thou hast resolved on something for His
service, set about it immediately. Do not delay, do not pro-
crastinate, do not turn back to hear what is said about it by
worldly persons, by thy companions and friends; for by so
doing thou wilt incur a grave risk of not fulfilling thy vocation
by reason of the obstacles which come in the way of all great
works. And, on the other side, who can tell that thy very
salvation may not be associated by God with the fulfilment of
this vocation, in the providential order of events which He
decreed for thee when He predestinated thee in His love? It
may well be, that in the case of this unhappy young man, the
refusal to serve our Lord in the apostolate involved his damnation, and this, not because he did not serve Him in the
apostolate, but because, refusing to serve Him in that manner,
he did not serve Him in any other, but remained in the bonds
of this world. And thus our Lord intended in this passage,
in the first place, to blame those who do not correspond with
Divine inspirations with that promptitude which belongs to the
strong, but interpose other business not necessarily evil in
itself, as irresolute persons do. How does it stand with thee
as to this promptitude of correspondence?

V. Consider fifthly, that in the second place Christ says
"looking back." He does not say "turning back," nor "going
back," but "looking back;" because no more is needed to
make thee unfit for the Kingdom of God than a mere glance
at earthly things, especially when, as in the case of this young
man, that glance is caused by an affection to them. When
thy Lord calls thee to the east, that is to say, to eternal things,
wilt thou at that very moment fix thine eyes on the west, that
is on temporal things? Oh, in what danger art thou of letting
thyself be so fascinated with them, as to think it impossible to
live without them! Therefore it is better to cut these bonds
than to untie them, for the latter is far more difficult than the
former: "Flee ye from the midst of Babylon, and let every
one save his own life."[1] He does not say "go out," but
"flee." Thus, in the second place, Christ here intended to
reprove those who choose to look with affection on that

---

[1] Jerem. li. 6.

which they have already abandoned in intention. Why all these excuses of desiring to make a good disposition of thy property? Thy Lord loves thee more than thy goods. No matter to whom they pass, do thou hasten to Christ, for very great is the danger of delay: "He that is in the field, let him not go back to take his coat."[1]

VI. Consider sixthly, that in the last place our Lord says of him who acts thus that he "is not fit for the Kingdom of God." He does not say that he shall not obtain it, but that he is not fit to obtain it, because it is possible that even one who looks back, after putting his hand to the plough, may come to be saved by sincerely repenting of his fault; but He says that he is not fit, because he has not in himself those dispositions which the Kingdom of God requires. The Kingdom of God demands men who are determined, firm, constant, and despisers of all that is most valued in the world. And how can such men, as we are speaking of, be of this sort? They are not fit for that Kingdom of Christ which is one of labour, because they are slothful men, and so they are equally unfit for that Kingdom of Christ which is one of enjoyment, because labour must necessarily precede enjoyment: "Because of the cold the sluggard would not plough, he shall beg therefore in the summer, and it shall not be given him."[2]

VII. Consider seventhly, that if these words of our Lord are so severe a blow to all those who are slothful in putting their good resolutions into practice, they are not only a blow, but a thunder-bolt to those who are not afraid to abandon them. For, if only to look back is at all events a sign of perdition in one who has put his hand to the plough, what must it be to withdraw his hand from it, in order to turn back? And do not suppose that only that man turns back who does so in person, in act, by doing the works of the world, such as apostates, who are "vessels of wrath, fitted for destruction."[3] He, too, turns back who merely does so in desire, because such a man has already repented of having put his hand to the plough, and so, in the sight of God, it is the same thing as though he had taken his hand from it. Be therefore always strong in heart, as well as in act, to serve God as thou hast resolved. "Our heart hath not turned back:"[4] this is the plough, do not take thy hand from it, whatever is the consequence; there is too much at stake, nothing less

[1] St. Matt. xxiv. 18.
[2] Prov. xx. 4. [3] Romans ix. 22. [4] Psalm xliii. 19.

than eternity. "She hath been in the field from morning till now, and hath not gone home for one moment:"[1] this is the conduct of one who desires to gain the favour of his Lord.

TWENTY-NINTH DAY.

ST. MICHAEL THE ARCHANGEL.

He hath showed might in His arm, He hath scattered the proud in the conceit of their heart. He hath put down the mighty from their seat, and hath exalted the humble (St. Luke i. 51, 52).

I. Consider first, that our Lord God has in all ages severely chastised pride. But never was this chastisement more severe than when first it came into existence, that is to say, in the empyrean Heaven. It was there that this wretched passion had its first origin in the mind of the rebel angels, and in the same instant it was hurled from the highest Heaven into the depths of Hell. Now, the words which I here propose for thy meditation refer principally, not only in a moral or mystical, but also in a literal sense, to that terrible act of justice wrought by God on all those glorious spirits whom He not only cast down from their lofty thrones, but condemned as the vilest of slaves to chains and torments, even creating the deep dungeon of Hell for their abode. Well will it be for thee if the contemplation of so terrible a doom produces in thee a true hatred of that vice which was its cause. It is certain that when Christ saw His disciples were somewhat uplifted by the wonderful works which they had done, although they did them by the power of His Name—"They returned with joy, saying, Lord, the devils also are subject to us in Thy Name"—the only means which He took to rebuke and humble them was to remind them that it was through pride that Lucifer fell even from Heaven, a fall swift as an arrow, terrible, destructive, and irrevocable: "And He said to them, I saw Satan like lightning falling from Heaven."[2] Do thou, therefore, learn to profit by this warning: "If God spared not the angels that sinned, but delivered them drawn down by infernal ropes to the lower Hell unto torments,"[3] how will it be with thee, vile worm of the earth, if ever thou displayest a pride like theirs?

[1] Ruth ii. 7.　　　[2] St. Luke x. 17, 18.　　　[3] St. Peter ii. 4.

II. Consider secondly, that these rebel angels are here called as if by name "the proud," because never were there in any age spirits prouder than these. It is enough to say that they all suffered themselves to be seduced by their wicked leader, Lucifer, to aim at a power so great as to be made, of themselves, like God: "I will be like the Most High."[1] I say of themselves, because no sooner were all the good angels admitted as a reward of their fidelity to the Beatific Vision, than they all obtained that likeness which belongs to it, but they did not desire to obtain it of themselves. If they desired to obtain it (as is probable), seeing that it was proposed to them as a reward by God, they desired to obtain it as a pure gift of grace, not of nature. The bad angels alone were so proud as to trust to their own strength to enable them to attain such greatness: "Thy heart is lifted up with thy strength."[2] And so it is said that they aimed at being equal with God—"Thy heart is lifted up, and thou hast said I am God"[3]—because they aspired at being able to be like God: their own beatitude. Now, these proud ones the Lord scattered in the thoughts they had conceived in their heart, for this is the meaning of these words, "He hath scattered the proud in the conceit of their heart;" for the word translated "in the conceit" is *mente*, so that the sense of the passage is, "He hath scattered the proud in the mind of their heart," that is, "by reason of the counsels of the heart, the thoughts of the heart, that which they meditated in their heart;" for "the mind of the heart," rightly considered, is the same thing as the designs framed by the will within itself. See, therefore, how truly God has scattered them in these machinations. These audacious spirits hoped to mount the very throne of God Himself, amidst splendour equal to His, and instead they found themselves at a distance from Him, utterly unlike Him, altogether hideous, tormented in the thickest darkness of Hell: "Thou saidst in thy heart, I will ascend into Heaven, . . . but yet thou shalt be brought down to Hell into the depth of the pit."[4] Do thou, for thy part, learn from this passage in what the great evil of pride consists. It does not consist in aiming even at the highest position; for what higher position can there be than that which we desire in Paradise? We desire that which Lucifer and his followers promised themselves; we desire that we too may be made like, if He did not

---

[1] Isaias xiv. 14.
[2] Ezech. xxviii. 5.　　[3] Ezech. xxviii. 2.　　[4] Isaias xiv. 13, 15.

deceive us when He said that in Heaven "we shall be like Him," for as He sees Himself in Himself, which is that in which His beatitude consists, so too shall we see Him in Heaven, and that not as we see Him here on earth in some image distinct from Himself, but "we shall be like to Him, because we shall see Him as He is."[1] But there is this difference between us and Lucifer, that Lucifer aspired to attain to this likeness by his own power, as St. Thomas teaches in several places, whereas we hope to attain it solely by grace. And in conformity with this principle there is nothing to forbid thy aiming at the most sublime sanctity, the highest degree of purity, poverty, or obedience—nay, even to the gift of the loftiest contemplation; there is no pride in all this: "Be zealous for the better gifts."[2] But at the same time, do thou always remember the great maxim, that of thyself thou canst do nothing: "Not that we are sufficient to think anything of ourselves, as of ourselves, but our sufficiency is from God."[3] Beg of Him to assist thee continually by His most holy grace; have recourse to Him, commend thyself to Him, confess thy weakness to Him at every step, and then aim as high as thou wilt, even, as Lucifer did, to be like God, and still thou wilt not be proud like him, but, on the contrary, truly humble, that is to say, at the same time both unpresuming and generous.

III. Consider thirdly, that these angels of whom we are speaking, who had thought themselves able to attain by their natural strength a sublimity of greatness which is natural to no mere creature, because it consists in becoming, by means of the Beatific Vision, if not equal, at least like God, in the glory which belongs to Him, were as a just punishment not only excluded from that greatness, which is only attainable by grace, but were also deprived of that which they did possess by nature; for after the words, "He hath scattered the proud in the conceit of their heart," by not suffering them to attain the supernatural beatitude which they had foolishly promised themselves by their own strength, there follows: "He hath put down the mighty from their seat," by further depriving them of the natural beatitude, which was already theirs. The devils are here called "mighty" in irony; not that they are not powerful, and exceedingly powerful by their nature, but because they foolishly promised themselves a still greater power, by imagining themselves able to soar on their own wings even to the throne of God. Now, all these "mighty"

---

[1] 1 St. John iii. 2.   [2] 1 Cor. xii. 31.   [3] 2 Cor. iii. 5.

ones, not only failed to reach that throne, but were also shamefully cast down from their thrones into Hell, amidst foulness, gall, stench, and darkness: "How art thou fallen from Heaven, O Lucifer, who didst rise in the morning?"[1] And how did God make this punishment more severe to them? By giving their thrones to men who were so inferior to them, in order that those proud spirits might rage with envy at the sight. And so it is not said that God "hath put down the seats of the mighty," but that "He hath put down the mighty from their seat," because the thrones of those angels are destined to the men who render that submission to God which was refused to Him by the original occupants of those thrones. Do thou for thy part learn from this that the particular virtue which will place thee on those angelic thrones is humility: "He hath put down the mighty from their seat, and hath exalted the humble," that is, those especially who do not boast that they possess anything of themselves; for just as by the "mighty" those are here meant who imagined that they were able to do by their own strength much more than they were really able to do, so, on the other hand, we must understand by the "humble" those, above all, who acknowledge before God that of themselves they can do nothing: "I am the man that see my poverty."[2]

IV. Consider fourthly, that this scattering and dethroning of the rebel angels was all done by means of His great Angel St. Michael. It was he above all others whom God employed as His captain-general, to conquer that vast army of the angels who were thus overthrown, just as He now employs him to defend His Church against the same angels, who now strive to overthrow it, and as He will employ him at the end of the world to oppose the furious assault which Antichrist will make when he vainly attempts to do what Lucifer failed to do in Heaven, namely, to be acknowledged by all as God: "So that he sitteth in the temple of God, showing himself as if he were God."[3] There it is said that in defeating the angels marshalled against Him, God "showed might in His arm," because he employed His arm, that is, St. Michael, to defeat them. It is he certainly of whom God has always made use as His chief minister, and therefore it cannot be denied that He made use of him as of His arm: "With the arm of Thy strength Thou hast scattered Thy enemies."[4] It is true that

---

[1] Isaias xiv. 12.
[2] Lament. iii. 1.   [3] 2 Thess. ii. 4.   [4] Psalm lxxxviii. 11

by the expression "the arm of the Lord" in the Sacred
Scriptures our Lord Jesus Christ is often signified : "To whom
is the arm of the Lord revealed?"[1]   But Jesus Christ is by
nature the arm of God, because He does the same things as
the Father, not only morally, as a prime minister does with
regard to his sovereign, but also naturally—"I and the Father
are one"[2]—whereas St. Michael is only metaphorically the
arm of God, because he is His prime minister.   However this
may be, it is certainly to him that thou shouldst have recourse
in all circumstances, but above all in time of temptation.
For St. Michael can truly be called the arm of God, especially
for this reason, that it is he whom God has employed, employs,
and will always employ to put the devils to flight : "Michael
and his angels fought with the dragon."[3]   It is true that all
the angels always fought together with him in that great battle,
but their leader was St. Michael, for which reason, as thou
seest, the rest are spoken of as "his angels," because they are
all subject to him.

THIRTIETH DAY.

ST. JEROME.

I sat alone, because Thou hast filled me with threats (Jerem. xv. 17).

I. Consider first, that the meaning of these words will at
once come home to thee by representing to thyself St. Jerome,
sitting on a stone on the lonely shore of a river overhung by
rugged rocks, with a volume of the Scripture before him,
clothed in a ragged garment, with a worn countenance, and
a breast bruised and bleeding from blows, and with scarcely
any feeling of a living man left but the dread with which from
time to time he turns to listen to the awful trumpet summoning
him to Judgment.   The words of the text are those spoken
by Jeremias in terror of the decisive threat which he had heard
uttered by the mouth of God when He had finally determined
on the destruction of Jerusalem.   But what is the destruction
of a city to that of a universe?   And, therefore, how much
more applicable are these words to the case with which we
are concerned.   Do thou endeavour to impress them firmly
on thy mind, for how greatly wilt thou be indebted to God,

[1] Isaias liii. 1.   [2] St. John x. 30.   [3] Apoc. xii. 7.

if thou too art able to say to Him with truth: "I sat alone, because Thou hast filled me with threats." Consider that the Prophet does not say that he is filled with any kind of terror, but with that terror which arises from threats; for there is also a terror arising from possible evil, whereas that which is caused by threats, arises from impending evil. And such is the terror which St. Jerome experienced when he said: "I am stained with the foulness of my sins; day and night I am overshadowed by the terror of having to pay the last farthing." Fear will be so general on the Last Day that it will be felt not only by just men, but even by angels and archangels, those spirits who are called strong by nature: "The powers of Heaven shall be moved."[1] But their fear will be very different from that of sinners, and this is the fear which ought to be thine. There are three kinds of fear which are inspired by a very great evil; namely, astonishment, bewilderment, and agony. The fear of the angelic spirits will be that of astonishment: they will consider that evil of the coming Judgment, as one which even their sublime intellects are incapable of thoroughly grasping, and at the thought of which they will be as it were thunderstruck. The fear of just men will be the fear of bewilderment: they will consider that evil as one which might very easily have befallen themselves but for the abundant grace bestowed on them by God, and so, being hardly able to believe in their own exemption, they will learn how infinitely greater an evil it is than they understood while in this world, and this thought will strike them with confusion and bewilderment. The fear of sinners will be the fear of agony: not only will they understand how immense and unheard of that evil is, but also that it is ready to fall upon them, and their state at this thought will be like that of one in his agony. This is the fear which with good reason ought to be thine when thou thinkest on the Last Day; and therefore consider to what a state it ought to bring thee, namely, a state of agony, because it is the fear of an evil which must be followed by everlasting death, unless thou strivest earnestly to escape it. The Prophet, at all events, said that he was full, even to overflowing, of this fear—"I sat alone, because thou hast filled me with threats"—so that like an overflowing vessel, he communicated this fear to those with whom he held intercourse. Such, too, was the fear of St. Jerome; his ears, and head, and heart, and tongue were full of the fear of

[1] St. Matt. xxiv. 29.

Judgment, so that, as though unable to contain this fulness, it overflowed in all his writings. Oh, how easily wouldst thou too be filled with this fear, didst thou but set thyself steadily to consider the terrible evil which may threaten thee on that day: "The Lord hath not called thy name Phassur, but fear on every side."[1]

II. Consider secondly, the effect which this fear produced in the Prophet; it was to make him retire from the company of men: "I sat alone, because Thou hast filled me with threats." And the same fear produced the same effects in St. Jerome, for it was the terror of Judgment which made him flee into the wilderness. The Prophet fled thither from pure fear; the Saint both from fear and for safety, because he thought that he could refrain more easily, at a distance from men, from those sins of which he would have to give an account on that terrible day. Dost thou think that if thou hadst a right fear of the Last Judgment thou wouldst indulge so freely in intercourse with men? What is the fruit of such intercourse but falls, and spiritual infirmities, and diseases, even such as are mortal? This was the experience of one who said, though late: "Lo, I have gone far off, flying away; and I abode in the wilderness."[2] How is it that thou dost not apply to thyself so profitable an example? At times, indeed, thou dost retire into solitude—"Going far off, flying away"—but thou dost not "abide" there, for scarcely canst thou bring thyself to continue half a day without being tired. Not so with the Prophet: for he said, "I sat alone;" and not so St. Jerome. He too "sat" in his wilderness; not that he remained in idleness there, for, on the contrary, he attained a very advanced age while occupied in speculating, writing, singing the Divine praises, and giving admirable replies to all those who consulted him as a living, universal oracle from all parts of Christendom; but that he made the wilderness his continual dwelling-place, refusing all the invitations he received from the most distinguished personages, even of Rome, because he was wholly absorbed in one single occupation, that of preparing for Judgment.

[1] Jerem. xx. 3.   [2] Psalm liv. 8.

OCTOBER.

FIRST DAY.

We have the more firm prophetical word, whereunto you do well to attend, as to a light that shineth in a dark place, until the day dawn, and the day-star arise in your hearts (2 St. Peter i. 19).

I. Consider first, how great a revelation that was with which St. Peter was favoured on Mount Tabor, when, together with the two holy brethren, James and John, he beheld the glory of our Lord Jesus Christ in His Transfiguration. And nevertheless he desired to show the faithful that he esteemed the Sacred Scriptures more than that revelation, since he here says expressly that he prefers the former to the latter: "We have the more firm prophetical word." He said "more firm," not that that revelation also was not firm, as every truth of faith is, but because that which is manifested to us in any revelation, however sublime, ought to be valued by us only just so far as it is in accordance with what we have learnt from the Sacred Scriptures, from which such revelations derive certainty, not in themselves, but relatively to us. Therefore it was with deep meaning that Christ ordained the apparition of both Moses and Elias in His Transfiguration, in order to show us that it is from the Books of the Law represented by Moses, and from the Books of the Prophets represented by Elias, that every revelation ought to receive uniform testimony in order to be genuine. Learn hence to form a due esteem of the Scriptures, and to prefer them to those ecstasies, sweetnesses, consolations, delights, and gifts of prayer which make thee think that thou hast already reached the heights of Tabor. What does thy spirit tell thee? That when thou enterest into prayer, thou seest Christ in His unveiled glory, and that thou too mayest exclaim, as St. Peter did, "Lord, it is good for us to be here." Do not think it, unless thou seest Moses and

Elias with Christ, that is, unless all that thou seest is in conformity with what is either imposed by the injunctions or taught by the dogmas of the Sacred Scriptures; for that which is thus heard is more to be relied on than that which is seen in a revelation: "We have the more firm prophetical word."

II. Consider secondly, that the Scriptures are compared by St. Peter to "a light that shineth in a dark place." The word translated "dark" is not *tenebroso*, but *caliginoso*, because where a light shines there is not darkness, but neither may there be clear light; and this is the case with us. Unbelievers who are without this light are in a darkness of ignorance that may be called palpable: "The Gentiles walk in the vanity of their mind, having their understanding darkened, being alienated from the life of God through the ignorance that is in them."[1] It is only we who are not in darkness: "But you, brethren, are not in darkness."[2] But if we are not in actual darkness, we are in "a dark place," because, although we have indeed a light, it is only like that of a lamp, which cannot entirely dissipate, but only diminish mental darkness; for the Sacred Scriptures do indeed take from us that ignorance of an evil diposition which belongs to one who takes falsehood for truth, as is the case of unbelievers; but they do not in the same way take from us that simply negative ignorance which belongs to one who does indeed know the truth, but only partially, as we do. "Now we know in part,"[3] because that which we know of God here is as nothing to that which we shall know of Him in Heaven, where, instead of the light of the lamp, we shall have the light of the sun: "When that which is perfect is come, that which is in part shall be done away."[4] This is the first reason why the Sacred Scriptures are called "a lamp," because they cannot entirely dissipate the darkness of the mind: "We cannot find Him worthily."[5]

But there is another reason why the Scriptures are called "a lamp," namely, to show us that they ought always to be ready in our hand wherever we go, to give us light at every step: "Thy word is a lamp to my feet."[6] And again, they are called a lamp, to show us that if we would have them give us a good light, we must follow them with very great reverence by keeping to their true sense, and never making them follow us by forced or distorted interpretations: "No prophecy of

[1] Ephes. iv. 18.   [2] 1 Thess. v. 4.   [3] 1 Cor. xiii. 9.
[4] 1 Cor. xiii. 10.   [5] Job vii. 32.   [6] Psalm cxviii. 105.

Scripture is made by private interpretation."[1] Do thou, on
thy part, dispose thyself so as to value the inestimable benefit
bestowed on thee by God, in enlightening thy darkness by
this beautiful lamp to direct thee in all thy ways ; and take
shame to thyself if thou art foolish enough to seek for light
from profane writers, from politicians and poets, like a child
who has lost its way, and follows a glow-worm, instead of
seeking it from this sacred lamp, this infallible and unfailing
lamp, the only one which God has placed upon the candlestick : "The lamp shining upon the holy candlestick."[2]

III. Consider thirdly, that the early Christians hardly ever
took their eyes from this blessed lamp, so attentively did they
meditate on the Sacred Scriptures, reading them again and
again, comparing different passages, and applying them. And
so thou seest that the Apostle did not here deem it necessary
to exhort them to this noble study, but only took occasion to
praise them for so diligently practising it : "We have the
more firm prophetical word, whereunto you do well to attend,
as to a light that shineth in a dark place." If, then, the
Apostle here says to his disciples "you do well," is it possible
that there can be directors so opposed to him as rather to say
to them "you do ill"? Observe, therefore, what attention
ought to be paid to the Sacred Scriptures—just that which
would be given to a lamp which was our only guide in a dark
cavern : as to "a light that shineth in a dark place." Oh,
how intently does every one keep his eyes fixed on such
a lamp, when going through dark and dangerous places, at the
risk of losing his way at every step. So also should we do :
"His lamp shined over my head, and I walked by His light
in darkness."[3] Or the comparison may be taken from sailors,
whose eyes are fixed in the darkness of the night on the lantern
placed on high, which is their only guide to the distant
harbour. And this also the Sacred Scriptures are to us.

IV. Consider fourthly, that it is not only for a short time,
as some are ready to grant, that this beneficent lamp should
be attentively regarded, but till the end of life—"till the day
dawn "—that is, till the night of this world is. at length
followed by that blessed day which alone merits the name,
because its light will be undimmed : "Till the day dawn, and
the day-star arise in your hearts." The word translated
day-star (*lucifer*) has a two-fold meaning : it may signify the
star, which is the bearer of light *in spe*, or the sun, which is

the bearer of light *in re*. It seems probable that the second *lucifer* is here referred to, not only because in the Syriac version it is expressly said "till the sun arise in your hearts," but also because if the first *lucifer* were intended, it would seem more correct to have said "until the day-star arise in your hearts, and the day dawn," rather than "until the day dawn, and the day-star arise in your hearts," because the star which announces the day precedes the dawn of the day itself. We may add that the use of the lamp does not cease with the coming of the first *lucifer*, which bears the light *in spe*, because at that hour it is still dark, its use ceases only with the coming of the second *lucifer*, which bears the light *in re:* "Her lamp shall not be put out in the night."[1] See, then, for how long a time thine eyes are to be fixed on the lamp, that is, on the light which is given thee by the Sacred Scriptures. Even till thy departure from this world—"till the day dawn"—because so long as thou art here, it can never be day, or at least full day, for thee. It is true indeed that by applying thyself to prayer, and attaining thereby to very lofty heights of contemplation, elevation, ecstasies, and visions, some *lucifer* may arise in thy heart, but only that one which bears the day *in spe*, never that which bears it *in re*. It will be the star, not the sun; and who would think that he no longer needs a lamp, because the star is risen which never brings with it full day, but only announces it? He must wait for the sun, which for thee will be the Beatific Vision; when that appears, the lamp will be no more, or at least will shine no more: "The light of the lamp shall shine no more at all."[2] And is it possible that thou dost not long with all thy soul for the dawning of this glorious sun, which will not only shine exteriorly, as the material sun, but in the inmost recesses of thy heart; for which reason it is not said "till the day dawn and the day-star arise on your hearts," but "in your hearts." It will be a sun which will make thee become, as it were, a second sun, like unto that which thou beholdest: "We know that when He shall appear, we shall be like to Him, because we shall see Him as He is."[3]

---

[1] Prov. xxxi. 18.   [2] Apoc. xviii. 23.   [3] 1 St. John iii. 2.

SECOND DAY.

THE HOLY ANGEL GUARDIANS.

For He hath given His angels charge over thee, to keep thee in all thy ways: in their hands they shall bear thee up, lest thou dash thy foot against a stone (Psalm xc. 1).

I. Consider first, that the first word in this passage should excite in thee the greatest confidence. That word, which is translated "for," is *quoniam*, and it does not give the reason for the preceding, but for the subsequent part of the text. The following, therefore, is the construction of the passage : "Because He hath given His angels charge over thee, to keep thee in all thy ways, therefore in their hands they shall bear thee up, lest thou dash thy foot against a stone." If, then, thou wouldst know why it is that the angels who are appointed thy guardians render thee so unwearied, so constant, and so loving a service, it is because God has commanded them to do so : "Because He hath given His angels charge." It matters not that thou hast no merit of thy own. It is enough merit in their eyes that they have received a command from God to assist thee. It is true, indeed, that they take pleasure in assisting thee for other reasons, such as their love for thee, and their hatred of the devil, and their desire to repair the ruin of Paradise ; but still their principal motive is to obey the Divine command. How, then, canst thou excuse thyself, when the motive which is sufficient for the angels is not sufficient for thee in the affairs of thy life, that is to say, the knowledge that a thing is God's will?—*Deus mandavit.* And dost thou inquire further? Most certainly this is not the teaching of an angel, but of the devil : "Why hath God commanded you that you should not eat of every tree of Paradise ?"[1]

II. Consider secondly, that thou art to contrast with the dignity of Him Who gives the command the vileness of a miserable creature like thee, on whose behalf it is given : "God . . . over thee." There is no proportion between the two. It is a God of infinite majesty Who is so concerned for thee, wretched worm that thou art. It is true that commentators understand the words "over thee" to refer to a just man, not to a sinner. Not that every sinner has not also a good angel who attends him as his guardian, as even Antichrist will have,

[1] Genesis iii. 1.

but because this Psalm treats of a just man who has placed his whole confidence in God : " He that dwelleth in the aid of the Most High." And it is such a just man whom God specially commends to the angels—he who most trusts in Him, because it is of him that He takes the greatest care : " He that dwelleth in the aid of the Most High shall abide under the protection of the God of Heaven." If, then, thou desirest that God should issue to His angels in thy behalf a more efficacious and express command than thou canst even desire, this is what thou must do—put thy whole trust in God.

III. Consider thirdly, who they are who receive this command. They are the angels, most glorious spirits, all of whom are princes of exalted rank, though some are greater than others ; and they are all, moreover, very well fitted for their office of guardians by the wonderful power which is theirs by nature, and by their great wisdom and holiness. From this thou oughtest to infer how highly God esteems thee, since He gives thee these sublime spirits for thy guardians : " He hath given His angels charge over thee." Who is there that is not confounded at such wonderful words? But do not suppose, because " angels " are spoken of, that every man has several, instead of one, for his especial guardians. That is the privilege of princes, prelates, and other important personages, who, as they require a two-fold prudence, one inferior for the government of themselves, and one superior for the government of others ; so, too, according to the teaching of the schoolmen, they have two guardians, an angel of a lower choir, who assists them as private persons, and one of a higher choir, who also assists them, but only in their public character. Nevertheless, he says "angels," not "angel," to all in general, because although one angel only is assigned to every one as his particular guardian, yet there is no one who has not at the same time the services of many others, such as are the angels to whom is committed the general charge of nations, kingdoms, cities, and of all important communities subject to God, in which it is just that God should have His own servants, as great sovereigns have theirs : " Upon thy walls, O Jerusalem, I have appointed watchmen."[1] What, then, dost thou think of this excess of condescension in these exalted personages, who act as thy servants at the very time when thou acknowledgest them so little, and showest them so little obedience and honour? Art thou not covered with confusion at the sight ?

[1] Isaias lxxi. 12.

IV. Consider fourthly, what the commandment is which the angels have received : " To keep thee." And from whom ? From all who lie in wait for thee, but especially from those whom thou art least able to be aware of thyself, namely, the devils, by whom thou art continually surrounded, although thou canst not see them. What, then, would become of thee, but for thy good Angel Guardian, who is continually either repelling or restraining them, or enabling thee to escape from their attacks by means unknown to thyself. This command, "to keep thee," is not therefore restricted to one kind of danger only, but includes numberless kinds, which concern both the body and the soul. For which reason is added, "in all thy ways." The word "way" in the Scriptures sometimes means the law of God : "I have run the way of Thy commandments, when Thou didst enlarge my heart."[1] Sometimes it means the actions of men : "Direct my way in Thy sight ; "[2] and it also means this mortal life itself, which is like a way leading us to the end of our journey, that is, to our future home : "Envy not the man who prospereth in his way."[3] Now, in all these ways the angels are charged to keep us, according to the peculiar needs of every one of them, although indeed each of these ways branches into several others. The law has many commandments ; operation has many acts ; life has many periods, many occupations, many cares, many states, all differing in kind. Who, then, can say how great is the assistance, always opportune and ready in every season, which thy Angel Guardian renders thee in all these ways ? And yet thou art not even mindful to give him thanks at the end of the day for all the countless favours he has done thee ! Thou wilt say that thou dost not know them. But why so ? Because he does them without revealing them to thee ? Is this a reason for esteeming his favours less highly, that they are done without ostentation ? Nay, but this is the best way of doing favours : "After thou hast given, upbraid not."[4]

V. Consider fifthly, that after contemplating the commandment, thou hast to consider the perfection with which the angels fulfil it ; for they do so not fully only, but superabundantly. The commandment which they have received is, "to keep thee," that is, to defend thee from the innumerable dangers to which, but for them, thou wouldst be every moment

---

[1] Psalm cxviii. 32.
[2] Psalm v. 9.   [3] Psalm xxxvi. 7.   [4] Ecclus. xli. 28.

exposed. Therefore this commandment would be satisfied, by their keeping at thy side to guide and direct thee. And yet, not contented with this, see how they take thee in their arms, and thus make thee safe: "In their hands they shall bear thee up, lest thou dash thy foot against a stone." Think, therefore, that thy good Angel is as one who, being appointed by thy Father to be thy guide in a dangerous journey, now over precipices, now through defiles or rivers, and now amidst rugged rocks, is not content with holding thy hand to keep thee from falling, but even carries thee to prevent thy stumbling in the most perilous places. This is why it is said, "In their hands they shall bear thee up, lest thou dash thy foot against a stone." It is not said "lest thou fall," but even "lest thou dash thy foot against a stone." These hands of the angel are the two faculties by which he governs thee, the understanding and the will; because by putting his executive power into action by means of these two alone, he is able to do everything. The stones are the obstacles and hindrances which meet us in any of the three ways which have been mentioned. And thy feet are thy affections, especially the two, love and fear, to which they may all be reduced. For all that man does, whether by thought, word, or work, is done from the love of attaining some good, and the fear of losing it, or else from the fear of incurring some evil, and the love of avoiding it. It is these two feet which guide thee everywhere. And therefore the reason why thou dost not make a false step with one or other of them, is because the angels carry thee even, as it were, on the palms of their hands, that is to say, they lift thee from the earth, so that, despising what is temporal, whether good or evil, according to the world's judgment, thou mayest love no good and fear no evil, but that which is eternal.

VI. Consider sixthly, that when the devil suggested to our Lord to cast Himself from the lofty battlements of the Temple, he quoted this text which thou art meditating in order to induce Him to do so, by this beautiful promise of angelical aid being always at hand. But he quoted it as his followers the heretics do, by distorting the Scripture from its true sense and falsifying it. In the first place, it was not true that Christ was spoken of in the text which says, "He hath given His angels charge over thee," for no angel was ever commanded by the Father to guard Him. For in what way would such a guardianship avail Him, as to His Soul or as to His Body?

Not as to His Soul, for that was always in a state of blessedness, and therefore He needed an Angel Guardian even less than the inhabitants of Paradise. Not as to His Body, which had a far better guardian than any angel, namely, the Eternal Word. And therefore the angels were indeed bound to serve Him, to obey Him, to worship Him, to manifest Him to the people, but not to assist Him : "You shall see the angels of God ascending and descending upon the Son of Man"[1]— ascending, when they went to Him to receive His messages, "descending" when they left Him to deliver them, as His humble servants, to men. In the next place, the evil one did not quote the whole text, for after having spoken these words, "For He hath given His angels charge over thee" (which according to the letter were not really spoken of Christ), he went on to the concluding ones : " In their hands they shall bear thee up, lest thou dash thy foot against a stone," omitting the intermediate part : "To keep thee in all thy ways." And it is evident that he omitted them out of malice, as not suiting his purpose. For even admitting that Christ had to be assisted by the angels, what kind of guardianship was promised in these words? To be kept from those falls to which He should expose Himself by wilfully casting Himself down? Certainly not ; but only "in all His ways," and those ways, moreover, which became Him. " He hath given His angels charge over thee, to keep thee in all thy ways," but not in wilfully incurred falls. What madness, then, to cast Himself down on the faith of assistance which was assumed, not promised ! But it was of little avail for the deceiver to pass over the words in question, when those which he immediately quoted were utterly inapplicable : " In their hands they shall bear thee up, lest thou dash thy foot against a stone." A man may stumble accidentally, even when he is walking prudently and cautiously ; but how can that which is said of a man who stumbles be applied to one who casts himself down from the top of a high building? It is one thing to trip accidentally against a stone, and another to expose oneself wilfully to a violent blow ; but the devil, who thought to deceive Christ by distorted quotations of Scripture, was himself deceived. For, on the one hand, Christ did not confute such foolish interpretations, in order to treat the devil as heretics, who sin through malice, ought to be treated, that is, by refusing to argue with them. On the other hand, Christ despised these interpretations in

[1] St. John i. 51.

two ways: first by deed, that is, by not choosing to do anything on the strength of them ; and secondly by word, that is, by quoting correctly another plain text, which overthrew the devil's false interpretations. For such was the text in which it is forbidden to tempt God by desiring to force Him to work miracles without necessity : "Thou shalt not tempt the Lord thy God."[1] From which text it may also be tacitly concluded, for the instruction of all, that no one ought ever to place himself in a position of danger, involving no good result, on the strength of the command which the angels have received from God, to give very great help to just men, because it is not meant by that commandment that they are to help these just men in all the dangers to which they expose themselves with or without reason, but only in those to which they expose themselves as just men.

THIRD DAY.

ST. FRANCIS BORGIA.

The Kingdom of Heaven suffereth violence, and the violent bear it away
(St. Matt. xi. 12).

I. Consider first, that the word translated "violence" is only said of a thing which is taken from a man against his will, as is the case also with what is stolen ; with this difference, that when a thing is stolen from a man, he suffers the wrong against his will because he does not know that it is taken from him ; but when anything is violently taken from him, it is against his will because, although he knows it, he is unable to prevent it. Now, in this sense it cannot be said that any one either steals Paradise, or takes it away violently, because God gives it freely to all : "God will have all men to be saved, and to come to the knowledge of the truth."[2] Nevertheless, Christ here used this manner of speaking, because, in consequence of the general corruption of the human race, things had come to such a pitch that it seemed as though Heaven were no longer intended by God for any but a very small number, namely, for His chosen people, Israel. These were His own people, "a peculiar people, and a privileged people, the people whom He

[1] St. Matt. iv. 7.   [2] 1 Timothy ii. 4.

protected : so much so, that the primary intention of Christ in coming into the world was to preach to them only : " I was not sent but to the sheep that are lost of the house of Israel."[1] Who, then, could at that time hope for Paradise, except that people? It appeared as though any others who might aspire to enter were desiring a thing to which they had no claim. But what then? all this was to be changed in time, chiefly because of the perversity of that people in rejecting the preaching of Christ. And, therefore, Christ here said that Paradise was no longer to be restricted, as had hitherto seemed to be the case, to one people only ; but that it was to be exposed, if I may say so, to a general assault. So that whoever should push forward to make it his own, no matter who he might be, Jew, Greek, Roman, Arab, Armenian, it should be possible for him to gain it, provided only that all these know the right way of proceeding, as was the case with the centurion, the woman of Canaan, and many other Gentiles, who being also united to Christ by a lively faith, were not only saved equally with those Jews to whom the preaching of Christ brought salvation, but in many instances so far surpassed them in zeal as to take their place : " Many shall come from the east and the west, and shall sit down with Abraham and Isaac and Jacob in the Kingdom of Heaven, but the children of the kingdom shall be cast out into exterior darkness."[2] This, then, is the first meaning of these words : " The Kingdom of Heaven suffereth violence, and the violent bear it away." The meaning is that Paradise was no longer to be restricted, as it had formerly seemed to be, to one people only, but that it was to be carried by assault. Do not, therefore, be afraid, no matter whether thou art noble or ignoble, freeman or slave, priest or layman, learned or ignorant, it matters nothing. Strive vigorously, and thou shalt be saved : " The Kingdom of Heaven suffereth violence." How often hast thou been told that Heaven is made for the poor ! " Amen, I say to you, that a rich man shall hardly enter into the Kingdom of Heaven."[3] And yet consider the great Saint whom we celebrate to-day, St. Francis Borgia. He was not only richly and nobly born, but a chief dignitary, a prince of exalted rank : and yet to what a high degree of sanctity did he attain ! Well, then, may it be said of him according to this first sense, not that he had Paradise, but that he took it by violence. Do thou learn to do likewise and

[1] St. Matt. xv. 24.
[2] St. Matt. viii. 11.   [3] St. Matt. xix. 23.

it shall be thine : "For that which every one had taken in the booty was his own."[1]

II. Consider that this bearing away implies violence : "The people of the land have committed robbery."[2] This, therefore, is the second sense of the words, "The Kingdom of Heaven suffereth violence, and the violent bear it away," namely, that it is violence which will win Heaven for thee. To whom, then, is this violence to be done? To God and to thyself. To God, by means of prayer ; for although He gives thee Paradise most willingly, yet He chooses for thy good to act as though thy hand had to take it by violence : "Because of his importunity He will give it to him."[3] And there is no other way in which this violence is ever said to be done to God, but the way of prayer : "Do not withstand Me, for I will not hear thee."[4] Then, thou art to do violence to thyself by complete self-abnegation. Of this sort are those movements which are in violent opposition to the natural appetite, such as closing thy eyes when thou art inclined to look on some fair woman ; shutting thy ears when thou art inclined to listen to licentious conversation ; bridling thy tongue when thou art inclined to give way to angry, impatient, proud, presumptuous, treacherous, or calumnious words. In such cases thou wilt do thyself that violence which is required in the case we are considering. Observe the violence which a soldier exercises in making an assault. He does violence to himself in pushing forward, and he does violence to the enemy who stands ready to repulse him on the walls. So, too, shouldst thou do, if thou aspirest as a valiant assailant to take Heaven by storm. We have an admirable example of this in the Saint of to-day, who did such violence to God that his prayer was almost incessant, even in the midst of the continual occupations he was engaged in ; and such violence to himself that it was enough for him to perceive that his senses desired some human consolation to refuse it to them.

III. Consider that this bearing away implies speed : "In haste they caught the word out of his mouth."[5] Hence, when a river runs very quickly, we say that its course is rapid (from *rapere*): "As the torrent that passeth swiftly (*raptim*) in the valleys."[6] This, then, is the third sense of these words : "The Kingdom of Heaven suffereth violence, and the violent bear it away," namely, that if thou knowest

[1] Numbers xxxi. 53.   [2] Ezech. xxii. 29.   [3] St. Luke xi. 8.
          [4] Jerem. vii. 16.   [5] 3 Kings xx. 33.   [6] Job vi. 15.

how to practise the requisite violence, thou wilt win Paradise instantaneously. How was it that the good thief won it by violence on the Cross? Because he made it his own in a few moments. It is true that the violence in his case was so extraordinary, that it is always considered a miracle; still, do not be afraid, even if thou art far advanced in years. In that case, thou hast to do all the more energetic violence both to God and to thyself: to the latter by self-abnegation, to the former by continual prayer. So shalt thou too attain in a short time to so high a place in Paradise that it will hardly be gained by another after many years. This was done by St. Francis Borgia, who did not bear the yoke of religion from his youth, but was advanced in years when he took it upon himself. And yet he is a greater Saint than many who have borne it from childhood.

IV. Consider that this bearing away also implies publicity. For it is this which beyond everything distinguishes robbery from theft: that the latter is done secretly, and the former openly. This shows thee who those are, in the fourth place, who "bear away" Paradise by force. They are those who not only choose to do so with violence and with speed, but also in the face of all men, caring nothing for what foolish people say of them, knowing that all rapine makes a certain noise: "Every violent taking of spoils is with tumult."[1] Such are the men who openly profess in the sight of the world to give themselves to prayer, and to practise strict self-denial. Those, on the contrary, who do the same things, but secretly, as though to avoid making a stir, should rather be said to steal Paradise than to take it by violence. And many will one day be seen to have been thieves, most happy thieves indeed, and whom no one ever suspected to be so. Such persons are represented to us in the Gospel by the woman with the issue of blood, who, approaching Christ under cover of the crowd, and seeming to touch Him, not from piety, faith, or confidence, but by mere accident, obtained her cure by the most artful theft of which we read. Not so the two lepers, who no sooner saw Christ than they began to lift up their voices from afar off; not so the centurion and the woman of Canaan, and, above all, the blind man of Jericho, who the more the crowd strove to silence him, the more loudly he cried, begging for light. All these are so many figures of those who do not steal Heaven, but take it by violence, and of this number was the

[1] Isaias ix. 5.

Saint of to-day. He did indeed set himself for a time to play
the thief, when he concealed his intention of becoming a Saint
beneath a rich and courtly dress. But after a while he took
courage, threw off the mask, and played the part of a violent
assailant, spurning all worldly pomp; and in order to trample
it under his feet, dressing not only soberly but shabbily, and
not shrinking even from sometimes appearing in public with
one of the vilest of animals on his shoulders. What then can
be said of thee, who not only lackest the courage to take it
by violence, but even the prudence to steal it as a thief?

V. Consider lastly, that if thou hast not courage either to
steal Paradise in the manner that has been described, or to
take it by violence, thou oughtest not even then to despair,
for Paradise will still be thine, if at least thou wilt consent to
be driven into it. For dost thou not know that the majority
of those that are saved are the poor, the suffering, the afflicted,
the persecuted, and many others whom God drives into
Heaven by means of various tribulations? These are they of
whom it is written that they are "compelled to come in."
For although it is true that they suffer all these ills against
their will, yet if they bear them submissively, not only is it
just as possible for them to reach Heaven, as for those who
steal it or take it by violence, but even to gain a higher place
there. If then thou art at least among these, thou shalt be
saved. Observe what happens in a dense throng, trying to
enter a church: those who are borne along by the force of
the crowd get in as well as those who obtain entrance by
violence; nay, very often they work their way higher up. And
the same thing happens in the case we are considering. If,
therefore, thou art of feeble courage, take care to let poverty,
troubles, afflictions, sickness, and, above all, the severe persecutions which, so to speak, are heaped upon thy shoulders,
make up for that vigour of voluntary action in which thou art
deficient: "Through many tribulations we must enter the
Kingdom of God."[1] Thou desirest, it may be, to receive that
Kingdom as a gift, but this is impossible, for "the Kingdom
of Heaven suffereth violence, and the violent bear it away."

[1] Acts xiv. 21.

FOURTH DAY.

ST. FRANCIS OF ASSISI.

The things that were gain to me, the same I have counted loss for Christ. Furthermore, I count all things to be but loss, for the excellent knowledge of Jesus Christ my Lord, for Whom I have suffered the loss of all things, and count them but as dung, that I may gain Christ (Philipp. iii. 7).

I. Consider first, what great things light is able to do in a soul. Those things in which the Apostle, like a merchant buying pearls in the dark, placed all his gains, that is to say, his greatest delight, riches, and reputation, appear to him, when seen by that light, not as gains but as losses, just as would be the case with a man who thought that he had been buying pearls, and afterwards perceived that he had been buying glass instead : "The things that were gain to me, the same I have counted loss for Christ." These things were the Judaic observances, which he had once applied himself ardently to learn, which he had professed and defended, so that he had acquired from his nation great applause as a zealous Israelite. Now, these things, contemplated by the light of faith which he had gained by the teaching of the Gospel, he clearly saw to be distinct losses, both because they brought in no gains (seeing that they help no one to the love of Christ), and because they entailed actual loss by taking it away from those who had already obtained it (seeing that it was no longer lawful to practise those observances). The same thing happens to every one who possesses a light similar to that of the Apostle. How does he wonder at himself if, like others, he once loved to follow the low maxims of worldly men, and to value like them vain emulation, precedence, points of honour, titles, followers, applause, dignities, and everything which he has now abandoned to follow Christ. If in such a case thou dost not thus wonder at thyself, what must we say? Surely this, that thou art not yet living in the light we are speaking of : "The sun of understanding hath not risen upon us."[1]

II. Consider secondly, that not only did the Apostle count as loss the things which he had formerly counted as gain, but advancing still further, he went so far as, for the same reason, to count as loss all other things which were not Christ, such as

---

[1] Wisdom v. 6.

noble birth, eloquence, learning, talents, and other like gifts,
however brilliant ; so that whoever values these things, must
either not aspire to follow Christ, or must abandon Him.
And it is this which the Apostle intends to express, when he
goes on to say: "Furthermore, I count all things to be but
loss." For this "furthermore" (*verumtamen*) has the force of
"nay, rather" (*quinimo*), which is an adverb implying a sort
of correction of oneself, as though one perceived that what
had been said was insufficient. So that it is as though he
added : "Nay, rather, I count as loss, not only those things
that were gain to me, but all things." But how did the
Apostle go on to form so bold a judgment, when he had
against him the torrent, if I may say so, of the whole human
race, who set the highest value on these goods ? He did so
by means of the excellent knowledge which he had acquired
in the school, not of Gamaliel, or of the Platonists, the
Peripatetics, or the Gymnosophists, but of Christ, the Son of
God: "For the excellent knowledge of Jesus Christ, my
Lord." All the knowledge which comes to us from Christ is
"excellent knowledge," as cannot be doubted, because it far
surpasses all other kinds which are not His. But if among
those which are His there is any one which may be said to
excel the rest, it is certainly the knowledge which teaches that
he who does not forsake all his possessions, all his friends,
cannot be a follower of Christ : "Every one that doth not
renounce all that he possesseth, cannot be My disciple."[1]
This is pre-eminently the "excellent knowledge," because it
is that which is least understood and least practised : to make
oneself naked, so as to desire nothing on earth but Christ
only. But it was well understood and practised by the
Apostle, as may be seen by the life he led in all the journeyings and persecutions that he endured in order to bear the
Name of Christ to unbelieving nations. This is the knowledge which thou shouldst strive to acquire, this "excellent
knowledge." When once thou hast acquired it, be very sure
that not only wilt thou count as distinct losses all the goods
which were formerly possessed by thee, but all which could by
possibility be possessed—"all things ;" yes, all without exception : "Furthermore, I count all things to be but loss, for the
excellent knowledge of Jesus Christ, my Lord." But the
difficulty lies in really attaining this knowledge, that is, in
really attaining to the conviction that if thou shall give up all

[1] St. Luke xiv. 33.

possible goods which can be offered thee in the world, for the sake of possessing Christ, naked on the Cross, He by Himself will be able to make up for all, nay, to satisfy thee more than all. Oh, how great a treasure is Christ, Who, when gained, is an equivalent for so much! And yet thou art ready to part with Him for any earthly good whatever, like children who are willing to give a diamond for a nut!

III. Consider thirdly, how the Apostle therefore concludes that for Christ's sake he has given a comprehensive rejection to all such temporal goods; but observe further his manner of speaking: "For Whom I have suffered the loss of all things," that is, "I have refused, rejected all things, and count them but as dung that I may gain Christ." Could he possibly have spoken of them more contemptuously? First, he says that he had cast them from him, and yet he did not choose to say, *Omnium detrimentum feci,* which would imply that the loss was a misfortune; but, to avoid seeming to mean that he had incurred any sort of injury by rejecting them, he said, *Omnia detrimentum feci,* that is, "I treated all things as though they were a loss," by dealing with them as men do with dangerous things, that is, casting them away. But, because there are other goods which the Apostle had not rejected because he never possessed them, such as posts of authority, tributes, thrones, suites of attendants, he added that he regarded all the goods in the world as dung, both those which he had possessed, and those which by possibility he might have possessed: "For Whom I have suffered the loss of all things, and count them but as dung," that is, "For Whom I treated all things that I possessed as loss, and count all things which I might possess but as dung, that I may gain Christ." And why so? Because he perceived the immense difference that there was between all the goods of the world together, and Christ only. Therefore all these worldly goods are compared to dung in several other passages of the Sacred Scriptures, and that with great reason. Because such goods belong either to the concupiscence of the flesh, that is to lust, and are like pestilential exhalation in the harm that they do to those who are at a distance by their evil fame, and to those who are near by bad example—"The beasts have rotted in their dung"[1]— or else they belong to the concupiscence of the eyes, that is, to avarice, and are justly compared to all that is most loathsome, because very few men can engage in the pursuit of

[1] Joel i. 17.

worldly wealth and escape contamination—"The sluggard is
pelted with the dung of oxen. Every one that toucheth
him will shake his hands"[1]—or else they belong to the pride
of life, that is, to ambition, and the comparison designates
well the quickness with which they putrefy: "Fear not the
words of a sinful man, for his glory is dung and worms."[2]
In the first text the reference is to beasts of burden, because
these are vile animals, as lustful persons are; in the second,
mention is made of oxen, because these are slothful animals;
and avaricious persons, although they are so greedy of gain,
yet to avoid enduring some great fatigue which they fear to
go through in striving after heavenly riches, declare themselves satisfied with earthly ones; in the third, the idea of
putrefaction is superadded, because all the glory of ambitious
men decays instantaneously: "The name of the wicked shall
rot."[3] Some of the Fathers, indeed, tracing the word *stercora*
to its origin, are of opinion that the allusion is rather to the
entrails of an animal, which are thrown to the dogs as only
fit for them. For those who are living in the light of truth,
when they see Christians who might aim at heavenly things
caring so much for earthly ones, it is as though they saw so
many dogs running round the shambles, fighting and quarrelling
with each other, and for what? to get a larger share of the
foul garbage from slaughtered animals lying on the ground.
Or, by the words of this passage thou mayest understand, as
some writers do, a dunghill, which every one is careful to place
at a distance from his house. What, then, can be thought of
persons, Christian, Catholic, nay, even religious, who are so
forgetful of their holy calling as actually to vie with each other
in bringing it into their house? "They that were brought up
in scarlet have embraced the dung."[4] Is it possible that thou
art willing to be one of these foolish persons? See what a
difference! The Apostle abandoned all earthly goods, treating
them as dung to gain Christ; and yet there are multitudes
who either abandon Christ, or are indifferent to Him, for the
sake of gaining the goods of this world which are "but
dung." How different are these merchants! to which dost
thou belong?

IV. Consider fourthly, that the great Saint of Assisi most
surely showed himself, like the Apostle, a wise and prudent
merchant, since utterly rejecting the goods which he possessed,

[1] Ecclus. xxii. 2.
[2] 1 Mach. ii. 62.   [3] Prov. x. 7.   [4] Lament. iv. 5.

as well as those which he might by possibility possess, when his father took him before the Bishop and disinherited him, he surrendered even his clothes, thus significantly declaring, that he was resolved to follow Christ in nakedness, in order to be able to follow Him the more quickly and the more freely, so as one day to make Him all his own. And did he not attain his end? Observe, therefore, how the Apostle here speaks. He said that he despised all things, even as dung. And why? "That I may gain Christ." He did not say, "That I may gain the love of Christ"—as it might seem that he would have said—nor the service of Christ, nor the fellowship of Christ, but "that I may gain Christ," because he would have nothing less than Christ wholly. And oh! how fully did he attain this. For he became, as it were, one and the same person with Christ, so that he was not afraid in the end to break forth into those wonderful words: "I live, now not I, but Christ liveth in me."[1] And to this the great Seraphic Patriarch, St. Francis, also attained. Observe him well, and thou wilt acknowledge that it is scarcely possible to see a difference between him and Jesus Christ: he was despised like Christ, poor like Christ, wounded like Christ, and an observer of the whole Gospel teaching, according to the letter of the words of Christ. But it is not possible to attain to this by the mere ordinary science to be learnt from the Gospel. The knowledge requisite for it is that "excellent knowledge" here spoken of.

FIFTH DAY.

Why seest thou the mote in thy brother's eye, but the beam that is in thy own eye thou considerest not? (St. Luke vi. 41).

I. Consider first, how unreasonable it is to observe closely even the small defects of thy neighbour, to criticize and condemn them, when thou thyself hast defects infinitely greater, and not only so, but greater in the same kind. It is with this that Christ here reproaches thee in the words: "Why seest thou the mote in thy brother's eye, but the beam that is in thy own eye thou considerest not?" A beam is, doubtless, greater than a mote, but it is not therefore of a different kind, since at the beginning it also was a mote, that is, a small germ,

[1] Galat. ii. 20.

which increased little by little, till it became a beam. Thou, then, seest this little germ in thy brother's eye, that is, the slight anger that has just arisen in him; and canst not see the beam in thy own eye, that is, the anger which has grown to such a height that it has become hatred. This, certainly, is a marvel of iniquity. Thou mayest say that it is much easier to see others than thyself; but it was precisely to do away with this frivolous excuse that Christ did not here say, "Why seest thou the mote in thy brother's eye, but the beam that is in thy own eye thou seest not?" but He said, "The beam that is in thy own eye thou considerest not," or, as the Greek text plainly has it: "Thou regardest not, watchest not." For if thou art not able to discern thy own defects with the same bodily eyes with which thou so easily seest those of others, thou oughtest to discern them with the eyes of the understanding. Before setting thyself to judge or to condemn thy neighbour, reflect carefully within thyself for a while whether a similar fault, such as anger, ambition, pride, intemperance, may not also be in thee, and even to a greater extent —"Before judgment, examine thyself"[1]—and by this means thou wilt abstain from showing so much zeal in regard to thy neighbour, because thou wilt acknowledge that thou art in a far worse state than he is. For if thou dost not do so in such a case, what injustice can be imagined more unbecoming and shameless?

II. Consider secondly, that Christ here gives the disgraceful name of hypocrite to one who acts in such an unseemly way: "Hypocrite, cast first the beam out of thy own eye, and then shalt thou see clearly to take the mote from thy brother's eye." Not only is such a man a hypocrite, but the worst of hypocrites, because he not only endeavours, as all hypocrites do, to appear better than others, when he is not so, but he endeavours to do this when he is really worse than others, and the means that he takes are not only alms, fastings, disciplines, or prolonged prayers, such as were practised by the Pharisee in the Temple, but the depreciation of his neighbour, whom he is bound to esteem better than himself, haughtiness, arrogance, pride, and the desire of behaving as a Superior, not merely in the lower act of commanding, but in the greater one of reproving. Dost not thou think that such a hypocrite is the most detestable possible? What, then, if thou art compelled to recognize thy own lineaments in his? Is there no other way in which thou

[1] Ecclus. xviii. 20.

canst gain credit, but by exercising towards others that zeal which they might more justly exercise towards thee! This is a way of gaining credit contrary to all reason.

III. Consider thirdly, that even if thou wert actuated by a right zeal, and not by pride, in desiring to condemn the smaller faults of thy brother without first looking to thy own, still thou art not only guilty of what is unreasonable, as has just been explained, by usurping an authority which does not at all belong to thee, but also of what is useless. Wherefore, when Christ here said, "Why seest thou the mote?" He signified by the word "why," "To what end, for what purpose dost thou see?" as in another passage of the same chapter: "Why call you Me Lord, Lord, and do not the things which I say?"[1] so that the meaning is, "What profit must thou derive from this zeal which thou showest with regard to thy brethren, without first thinking of thyself?" Thou canst derive none for thyself, because even if thou shouldst succeed in removing every possible mote from the eyes of others, this would avail thee nothing so long as the beam remained, all the time, in thine own. Thou wouldst be lost in spite of all the good thou hadst done to others by correcting and converting them, according to the declaration of Christ Himself: "He that shall break one of these least commandments, and shall so teach men, shall be called the least in the Kingdom of Heaven."[2] He does not say "shall be the least in the Kingdom of Heaven," because for such a one there will be no place, not so much as a corner, in Paradise; but He says "shall be called," because, no matter how highly he may be esteemed in the world as a great man, he will be despised in Heaven: "The just shall laugh at him, and say, Behold the man that made not God his helper, but trusted in the abundance of his riches," that is, of the riches of teaching which filled his sermons, of the number of his auditors and followers, of his reputation, "and prevailed in his vanity."[3] Is it not far better for thee, then, to employ for thy own profit the time and labour which thou devotest to the profit of others? There is a beam in thine eye, and it does not even move thee to tears! That is to say, thou bearest about with thee a grievous sin, and yet thou art not concerned or grieved, and instead of being in any degree anxious about thyself, thou occupiest thyself with persons less guilty than thou art. What evident madness is this! "Thou that teachest another teaches not

[1] St. Luke vi. 46.   [2] St. Matt. v. 19.   [3] Psalm li. 9.

thyself."[1] In the next place, as thou canst not derive any profit for thyself from such conduct, so neither canst thou do so for thy brethren. For dost thou not perceive that instead of profiting by the zeal that thou displayest concerning their faults, they will laugh at it? It is certain that if, with a beam in thy eye, thou art able to see the motes in theirs, much more will they, having motes in their eyes, be able to see the beam in thine. And if this is so, how clear it is that they will mock at thy zeal, and say within themselves, "Physician, heal thyself."[2] And not only will they mock at it, but they will be greatly scandalized by it, seeing that thou wouldst fain act as a judge at the very time that thou art a culprit. What, then, art thou to do? "Before judgment prepare thee justice."[3] If thou sincerely desirest to be of use to thy brethren by judging them, "first cast the beam out of thy own eye:" get rid of the sin that thou hast committed, grieve over it, detest it, change thy way of life—"prepare thee justice." Then, indeed, that will be esteemed a true zeal, which otherwise is esteemed either as pride or temerity, or presumption, such as it would be to desire to wipe the dust from another person's face with soiled hands: "What can be made clean by the unclean?"[4]

IV. Consider fourthly, that if thou dost not first seek to amend thyself, thou art doing what is not only both unjust and useless by occupying thyself with others, but also what is exceedingly injurious to thyself: "Wherein thou judgest another, thou condemnest thyself."[5] Dost thou not see, that if thou wishest to play the part of judge when thou art a criminal, thou provokest God's anger against thyself? It is true, indeed, that those persons whose office it is to judge, such as princes, prelates, and ministers, ought not to cease from the exercise of that office even when they are conscious of greater guilt than that which they judge. But it is not allowable for one who has not this office to take it upon himself, not even so far as to give a simple reproof, such as it is the province of preachers to administer. Every one who proposes to reprove others, either privately or publicly, for leading a bad life must first reform his own: "The priests were purified, and they purified the people."[6] If he does not do this, it is certain that he is guilty of a sin of presumption if his guilt is known to himself alone; and not only is he guilty of presumption, but of giving scandal, if it is known both to

---

[1] Romans ii. 21. [2] St. Luke iv. 23. [3] Ecclus. xviii. 19.
[4] Ecclus. xxxiv. 4. [5] Romans ii. 1. [6] 2 Esdras xii. 30.

himself and others. Is not this to provoke the anger of God
very greatly? If thy sin is one of presumption, God will justly
confound thee as a proud man who would fain conceal his
wickedness by reproving it: "God shall strike thee, thou
whited wall."[1] And if thy sin is, in addition, that of giving
scandal, then God will justly condemn thee as a seducer, who,
while professing a desire of converting souls to Him, art really
perverting them like those persons who are spoken of as "false
apostles," that is, deceitful workmen, transforming themselves
into the apostles of Christ."[2] Therefore, although it is, no
doubt, a pious act to desire to warn thy neighbour of those
straws which are in his eyes, that is, those beginnings of sin,
which perhaps they may not be able to discover of themselves,
yet do thou first cleanse thine own eyes of those massive
growths, which have, so to speak, taken deep root there: that
is to say, cleanse them of those sins of thine, which are not
only grievous, but inveterate; otherwise, thou wilt make
thyself only the more displeasing to God by acting the part
of a judge whilst thou art a criminal: "These that say, Depart
from me, come not near me, because thou art unclean, these
shall be smoke in My anger, a fire burning all day."[3]

SIXTH DAY.

ST. BRUNO.

*I will stand upon my watch, and fix my foot upon the tower, and I will
watch to see what will be said to me, and what I may answer to him that
reproveth me* (Habacuc ii. 1).

I. Consider first, that whoever sets himself to observe what
was the intention of the great Patriarch, St. Bruno, in founding
his glorious Institute upon the most inaccessible and solitary
heights about Grenoble, will come to the conclusion that he
learnt it from the prophetical passage here set before thee for
meditation. The first thing which was his object was this, to
be very watchful over himself: "I will stand upon my watch."
But, as this can only be accomplished when a man is encompassed, like a soldier, by numerous defences, he adds: "I will
fix my foot upon the tower." In the next place, being thus
well defended, both exteriorly and interiorly, what did the

[1] Acts xxiii. 3.   [2] 2 Cor. xi. 13.   [3] Isaias lxv. 5.

Saint propose to do? To stand upon his watch-tower like a sentinel, considering very carefully what Christ would come to ask him at the hour of death concerning all the works, words, and thoughts of his whole life; and what he would have to answer: "I will watch to see what will be said to me, and what I may answer to Him that reproveth me." For this holy man, terrified by the sight of that Parisian doctor who came out of his coffin, and cried out thrice that he had appeared before his Judge, had been examined and condemned, took occasion from this to retire with his devout companions to those caves which were at that time so far from all human society, and there set himself to consider the affairs of his soul. If, in the same way, thou meditatest for thy good on these words of the Prophet, how greatly will it be for thy salvation! And do not say that these words were spoken by him, according to their literal meaning, with reference to Christ's first coming into the world, as is shown by what follows: "It shall appear at the end, and shall not lie; if it make any delay, wait for it; for it shall surely come, and it shall not be slack," because, as thou well knowest, Christ's first and second coming are frequently spoken of indifferently.

II. Consider secondly, that the first thing thou hast to do is to guard thyself both exteriorly and interiorly: "I will stand upon my watch," this is interior watchfulness, "and I will fix my foot upon the tower," this is that which is exterior. As to the former, thou art to say, I will keep guard over myself— "I will stand upon my watch"—and will never allow any one to approach to force a way into my heart: "With all watchfulness keep thy heart, because life issueth out from it,"[1] that is, "life and death." Thy heart is like a citadel, on which depend the spiritual life and death of thy soul. There are three enemies bound together in a terrible alliance, who ardently desire to gain possession of it: the world which is all around, the flesh beneath, and the devil above thee. The world assails it by vanity, the flesh by pleasure, and the devil overcomes it by wickedness. See, therefore, how necessary it is to be on thy guard, above, beneath, and on every side. Thou shouldst defend thyself against the world by the love of poverty; save thyself from the flesh by the love of purity; and protect thyself from the devil by having recourse to God, in the first place, by prayer, and in the next place, by obedience to him who occupies his place on earth: "With all watchful-

[1] Prov. iv. 23.

ness keep thy heart." It is true, indeed, that this watchfulness cannot be of the same sort in every one, but varies according to the state of each individual. For which reason the Prophet does not say, "I will stand upon the watch over myself" (*custodiam mei*), but "upon my watch" (*custodiam meam*). There are different kinds of watchfulness requisite for a virgin and a married woman, a priest and a layman, a regular and a secular, a workman and a contemplative. And therefore it is that thou oughtest to say, according to what thy state of life requires, "I will stand upon my watch," that is, the most strict and exact watch which is demanded of me. Thou hast only to consider the matter to know what this is.

III. Consider thirdly, that no fortress, however strong and well guarded, is ever safe, unless it has, in addition, external defences. And therefore the Prophet adds : "And I will fix my foot upon the tower" (*munitionem*). What is the defence (*munitio*) here spoken of? It is, so to speak, the barrier, stockade, or rampart, which prevents every one who pleases from having access to thee, and without which the fortress may easily yield to some sudden surprise, which thou art not able to perceive in time. In order, therefore, to watch thyself well, thou shouldst be careful to allow no communications in thy house which are either suspicious or superfluous. It will, however, be of little avail not to allow the approach of such communications, if thou goest outside thy defences in quest of them. Therefore the Prophet purposely says : "I will fix my foot upon the tower," or rather "defence" (*munitionem*). Why "upon"? Was it not enough to say "within"? No ; because thou oughtest to remain within thy outworks, as a man does who stands sentinel on the summit of a bastion, to see whether any one is approaching from a distance : "I am upon the watch-tower of the Lord, standing continually by day ; and I am upon my ward, standing whole nights."[1] How important are all these considerations to one who is resolved to save his soul ! See what exertions are everywhere made to defend a fortress from hostile attacks. And yet it is but a temporal death that the fire and sword of the enemy can inflict. How, then, can any labour seem too great to guard thy soul from these enemies who are able to inflict a death that is eternal?

IV. Consider fourthly, that, while thus on guard, thou art not to be idle. For, not only hast thou thus to keep off all

[1] Isaias xxi. 8.

hostile attacks, which, in itself, is no small thing, but thou wilt
have the opportunity of carefully considering the only thing in
the world which is of importance, that is, thy last hour.   Dost
thou not know that thy Lord will speedily come to require
from thee a strict account of thyself?   What, then, art thou
about, that thou dost not at once set thyself to consider what
He will say to thee, and to resolve what thou wilt answer
Him?   This is the business with which thou oughtest to be
occupied, incomparably more than with any other ; and, therefore, thou art indeed foolish and senseless if thou only givest
it a passing thought from time to time.   Do not act thus.
Hear the words of a holy man : "I will watch to see what will
be said to me, and what I may answer to Him that reproveth
me."   He said not only "I will think," but "I will watch,"
because the kind of thought that is requisite is a careful,
undistracted thought, one as concentrated in kind as that of a
very sublime contemplation.   Oh, how different wouldst thou
soon become from what thou art, if thou didst but apply
thyself, not to the mere thought, but to this "watching" of the
Judgment !

V. Consider fifthly, that since it is of Judgment that he is
here speaking, the Prophet might have said with perfect
justice : "I will watch to see what will be said to me, and
what I may answer to Him that judgeth me."   Yet he preferred
to say, "Him that reproveth me," and he had a deep purpose
in this.   For thus, by a single word, he succeeded in expressing
all that is terrible in this Judgment.   The word *arguere* has
three meanings in Scripture.   Sometimes it means "to reveal:"
"Wine shall reveal (*arguet*) the hearts of the proud."[1]   In
this sense God will "reveal" the sinner in Judgment, because
this revelation which He will make of him will be two-fold :
the first to himself alone in the Particular Judgment—"I will
reveal (*arguam*) thee, and set before thy face,"[2] that is, "I will
set thee against thyself"—and the second before the whole
world in the General Judgment.   Sometimes *arguere* means to
convince by argument : "Why have you detracted the words
of truth, whereas there is none of you that can convince
(*arguere*) me?"[3]   And in this sense God will "convince" the
sinner in Judgment, by forcing him to acknowledge that if he
is lost he has only himself to blame : "Shall He reprove," or
convince, "thee for fear," like one who is able to bring forward
only weak arguments, "and come with thee into Judgment?"[4]

[1] Ecclus. xxxi. 31.   [2] Psalm xlix. 21.   [3] Job vi. 25.   [4] Job xxii. 4.

No; God will convict the sinner by general arguments drawn from the common graces bestowed on thee for thy salvation, and by particular arguments drawn from special graces. At other times *arguere* means to confound by reproof: "Them that sin reprove before all, that the rest also may have fear."[1] And in this sense God will reprove the sinner, rebuking him for all the wickedness he has committed against every law: "Behold the Lord cometh . . . to execute judgment upon all, and to reprove the ungodly for all the works of their ungodliness, whereby they have done ungodly, and for all the hard things which ungodly sinners have spoken against God."[2] Again, *arguere* means to condemn after Judgment: "And some indeed reprove, being judged" (that is, "being condemned"), "but others save, pulling them out of the fire."[3] And in this last sense will God "rebuke" every sinner, when He condemns him to everlasting fire: "Rebuke me not, O Lord, in Thy indignation" (that is, "punish me not in Hell"), "nor chastise me in Thy wrath"[4] (that is, "punish me not in Purgatory"), which is a very general interpretation of the passage. Now, is there not here sufficient matter of meditation for a lifetime? In the first place, thou hast to think of all that God will say to thee when He "reproves" thee in all these four ways which have been mentioned, namely, by setting thy sins before thy face, by convincing, confounding, and condemning thee, and, in the next place, thou hast to think of what thou wilt answer in each of these cases. And this being so, hast thou not good reason to come to the same conclusion as the Prophet, and the great St. Bruno also, to his great profit, did: "I will stand upon my watch, and fix my foot upon the tower, and I will watch to see what will be said to me, and what I may answer to Him that reproveth me."

———

SEVENTH DAY.

I am the Vine; you the branches: he that abideth in Me, and I in him, the same beareth much fruit; for without Me you can do nothing (St. John xv. 5).

I. Consider first, that just as the branches are in need of the vine, whilst the vine has no need of the branches, so, too,

---

[1] 1 Timothy v. 20.
[3] St. Jude 15.   [2] St. Jude 22, 23.   [4] Psalm xxxvii. 1.

does the case stand with regard to Christ and ourselves. We may cut as many branches from the vine as we please, and it will remain as vigorous as before, with the power of producing new ones. But the branch which is cut off loses all the vigour which it once had. This, therefore, it is which Christ especially meant to teach us in the passage we are about to consider, when He said: "I am the Vine: you the branches." He meant us to understand that, on the one hand, He has no more need of any of us—"What doth it profit God if thou be just?"[1]—and that we, on the other hand, are in as great need of Him as each branch of the vine on which it grows. Oh, if thou didst but enter thoroughly into this truth, that thy need of Christ for all the good that can come to thee is extreme, and that He has no need whatsoever of thee, how truly wouldst thou annihilate thyself before Him, and desire with all thy heart to abide in Him as a strong branch, amidst storms and tempests of every kind! "Who shall separate us from the love of Christ?"[2]

II. Consider secondly, what is the meaning of what is here called "abiding" in Christ, as the branch abides in the vine. It is to abide in Him in such a manner that He may transfuse His strength into thee. It is to be constant in loving Him, as He Himself declares a little further on in the words: "Abide in My love." There are branches which are actually cut off from the vine, and there are branches which are united to it; but amongst the latter there is this difference: some are united to it in a living, others in a dead manner. Those that are united to it by a living union are those that draw from it sufficient sap to bear fruit; those that are united to it in a dead manner are such as draw no sap from it, and therefore are poor, dry, and withered, and if not already dead, like the branches which are cut off, are at least very near death. The same thing occurs in the case we are considering: some are cut off from Christ, their Vine; and these are heretics— "Because of unbelief they were broken off"[3]—others are united to Him, and these are the faithful. But of these, some are united by faith only, others by both faith and charity; and these latter are said to be united to the Vine by a living union, because the union is reciprocal, so that they act by the union of the Vine with them: "He that abideth in charity abideth in God, and God in him."[4] Those

[1] Job xxii. 3.
[2] Romans viii. 35. [3] Romans xi. 20. [4] 1 St. John iv. 16.

whose union is one of faith only are said, indeed, to be united to their Vine; but it is a dead union, because the Vine is not united to them, as He says, " I love them that love Me,"[1] and therefore, as the Vine does not transfuse into them that life-giving sap, without which no branch can ever bring forth fruit to everlasting life, such branches, if they are, notwith-standing, united to the Vine, are not so united in a living manner. This is the state of such of the faithful as are living in mortal sin. And how wretched is this state! They are in Christ; but in what manner? Alas for them! in such a manner that Christ is not in them; at least, not as the author of grace: they are, and they are not, in Him, which is as much as to say that these unhappy persons are in Him as sickly branches which are on the point of withering away. How then is it possible for thee to live contented, if thou art unhappy enough to recognize thyself as one of these branches?

III. Consider thirdly, how Christ, as the true Vine, acts. " I am the true Vine:" and therefore, good and gracious as He is, He never ceases, so far as rests with Him, to transfuse into His branches the vital sap, if only they are not cut off from Him by sin. Do not wonder, then, that He said on this point, " Abide in Me and I in you," that is, " Abide in Me, and so abide in Me that I may abide in you," for this is the meaning of the saying. All that He desires is this reciprocal union of us with Him, and Him with us, and therefore it is that He makes it a command; but why does He command it unless because this union, on His side, is continual? If it were possible for us to be united to Him without His being also united to us by charity, it would be an unwise, unmean-ing, and unprofitable command. But since no such command ever proceeded from His lips, we must understand that if this Divine Vine does not infuse sap into us, the fault is ours; it is we who keep it divided, severed from ourselves: "Your iniquities have divided between you and your God."[2] What, then, should we do, but confess the misery of our condition and grieve over it? This very grief, indeed, on the part of the branches comes from the Vine: so that, if thou feelest any sorrow for thy sins, if thou art full of confusion, if thou art troubled, if thou beginnest to experience some degree of compunction for thy fault in keeping at a distance from thee Him Who alone can give thee every good, because He is thy

[1] Prov. viii. 17.   [2] Isaias lix. 2.

Vine, know that even this is of His goodness. He, even
though divided from thee, excites thee by His preventing
grace to consider how thou mayest be again united to Him:
so greatly does He desire to abide in thee, although, after all,
thou art but a branch, and He the Vine, Who has no need
whatever of thee, since, without thee, He has so many others:
" It stretched forth its branches unto the sea, and its boughs
unto the river."[1]

IV. Consider fourthly, that the branch receives from the
vine not only the power of producing fruit, but also the act of
production itself; because in every such production of fruit
which thou seest gradually developed from the branch, there
is the continual concurrence of the vine, acting, by its
strength, together with the branch, and fructifying it. Thus
does Christ act by virtue of grace when He is in us, not only
giving us the power to do works meritorious of eternal life—
" As the vine, I have brought forth a pleasant odour "[2]—not
only the power of doing them more easily, as even Pelagius
at last admitted, or of doing better or greater works; but
most positively supplying the act itself, precisely as the vine
enables the branch to produce grapes. For which reason
Christ here so emphatically said, " Without Me you can do
nothing," hereby showing that He was speaking not only of
the mode, but the substance of fructification. It is impossible
to do anything without Him. Does not this show thee still
more vividly the necessity of union with the Vine? Oh, if
thou wouldst often repeat within thyself these Divine words,
" Without Me you can do nothing," how deeply wouldst thou
enter into the sense of thy own nothingness!

V. Consider fifthly, that as there is no flower of truth from
which any one desiring, like the spider, to suck poison cannot
do so, so too there have been persons who have deduced a
palpable error from these glorious words of Christ, by attri-
buting the production of our good works so entirely to grace,
that nothing is open to free-will; as though Christ, while
giving us the power of producing fruit, took from us the act
of producing it. But how can this be? It would be little
glory to the vine if, by itself alone, it produced grapes. Its
chief glory is, that it imparts to the branches also the power
of concurring and co-operating in producing them. For this
reason Christ here said: " He that abideth in Me and I in
him, the same beareth much fruit." He did not absolutely

[1] Psalm lxxix. 12.   [2] Ecclus. xxiv. 23.

deny the power of the branch to produce grapes. What He did deny was its power to produce them of itself, that is, to produce them otherwise than by virtue of the vine: "As the branch cannot bear fruit of itself, unless it abide in the vine, so neither can you, unless you abide in Me." But is this a legitimate conclusion? The branch cannot bear fruit *unless* it is in the vine: therefore it cannot bear fruit *when* it is in the vine. The most uneducated peasant would see that such a conclusion is absurd. So that, although we consider the grapes to belong, as its fruit, to the vine, which is the principal agent in producing them, yet we do not hesitate to speak of them as also belonging, as its fruit, to the branch: "And going forward as far as the torrent, they cut off a branch with its cluster of grapes, which two men carried upon a lever."[1] If, then, the grapes may properly be said to be produced by the branch, although it is only a secondary agent, why cannot our good works be properly called ours, according to the words, "Give her of the fruit of her hands."[2] Such is the love borne to us by God, that He has been pleased to make His gifts our merits. And therefore, although He is the Vine, yet He is not a Vine which forces us to act; although we are His branches, He only enables us to act—"To bring forth fruit to God"[3]—because He deals with us as reasonable branches, which is what we are.

VI. Consider sixthly, that, this being so, we are under all the more obligation to Him, because, on the one hand, He gives us the power of acting, and therefore infuses grace into us, and, on the other hand, He does not deprive us of the merit of our works: on the contrary, He chooses them to be imputed to us, and that we should receive praise, credit, recompense, and a crown for them; and therefore He leaves us our free-will: "There is a wise man that is wise to his own soul, and the fruit of his understanding is commendable."[4] It is, indeed, true that even the good use of our free-will is wholly His gift, and that therefore we have no right ever to glory in anything but in Him: "He that glorieth, let him glory in the Lord;"[5] but it is not the less true that if we do not make a good use of it, the fault is ours, because we do not allow this Vine to act in us as He desires to do; for either we entirely reject the sap which comes from Him, or, if we receive it, we turn it into fruit that is either worthless or

---

[1] Numbers xiii. 24.   [2] Prov. xxxi. 31.
[3] Romans vii. 4.   [4] Ecclus. xxxvii. 25.   [5] 2 Cor. x. 17.

poisonous : "You have turned judgment into bitterness, and
the fruit of justice into wormwood."[1]   Do thou, then, always
keep in mind these two axioms : That if thou doest well, the
grace both to will and to do comes from God ; and that if
thou dost not do well, the fault is in thyself, who resistest grace
by thy free-will, and so, like many others, givest God reason
to complain : "You have chosen the things that displease
Me."[2]   Thus thou wilt keep the mean between two opposite
rocks, on one of which thou wilt otherwise make shipwreck.
The man who denies grace, proudly persists in ascribing his
good works to himself ; the man who denies free-will, blas-
phemously ascribes his misdeeds indirectly to God.   Do thou
avoid both these rocks, of which it is hard to say which is the
more terrible, and whilst acknowledging that all good comes
from God—"Without Me you can do nothing"—continue
always earnestly to beg of Him the power to do it—"From
Me is thy fruit found"[3]—and understanding that He will not
thus work in thee without thy doing thy part, be careful to
correspond and co-operate with His grace by means of self-
conquest : " Act like a man, and take courage, and do."[4]

EIGHTH DAY.

*If any one abide not in Me he shall be cast forth as a branch, and shall
wither, and they shall gather him up, and cast him into the fire, and he
burneth* (St. John xv. 6).

I. Consider how it might have been supposed that all the
thoughts which Christ desired to bring before thee in the
preceding meditation by means of the figure of the vine with
regard to its branches, or of the branches with regard to the
vine, could have been just as well expressed by taking any
other fruit-tree as a figure, such as an apple, pear, peach, or
cedar, with its branches.   But this is not so : the tree which He
chose was the fittest for the purpose ; first, because no other
renews its branches so readily as the vine.   No matter how
freely it is pruned, even so far as to strip it of all its boughs,
in a few months it puts forth new ones, and even in greater
number than before.   And for this reason there is no other

[1] Amos vi. 13.
[2] Isaias lxv. 13.   [3] Osee xiv. 9.   [4] Paral. xxviii. 20.

tree which equally shows how little Christ needs us when we
leave Him : " He shall break in pieces many and innumerable,
and shall make others to stand in their stead."[1] In the next
place, no other tree shows in an equal degree the great advantage which its branches derive from not being separated from
it : and this in two ways, both because there are no branches
more valuable when united to the tree from which they spring
than those of the vine, so sweet and wholesome is the fruit
they produce, and because there are none more worthless
when cut off from it. The branches of other trees, even
when cut off, are capable, when wrought by a skilful hand, of
serving some good use ; but not so those of the vine. There is
nothing to be done with them but to burn them in the fire :
" Son of man, what shall be made of the wood of the vine,
out of all the trees of the woods that are among the trees of
the forests ? Shall wood be taken of it to do any work, or
shall a pin be made of it for any vessel to hang thereon ?
Behold it is cast into the fire for fuel."[2] This was the first
object which Christ had in view when He employed this figure
of the vine ; and thou canst see how well it applies. Does not
thy heart tremble within thee when thou thinkest of the
alternatives between which thou art placed? Perfect happiness, if thou art willing to remain united to Christ by charity,
and utter misery if thou persistest in remaining separated from
Him : " One of two things must be the fate of the branch :
either to abide in the vine, or to be cast into the fire ; if it is
not in the vine, it must be in the fire."[3]

II. Consider the first punishment by which Christ began
to show the unhappiness of the man who is separated from
Him. It is that of being banished from His gracious providence : " If any one abide not in Me, he shall be cast forth as
a branch." When branches are cut off from the vine, they
are, in the first place, cast outside the vineyard, where they are
no longer worthy to remain together with those that are tended
there. And such will be the lot of all bad Christians at their
death : they will be cast out of the Church, that is, out of the
congregation of the faithful, with whom they will no longer
be partakers of any such good, either of grace or of glory,
through all eternity : " The angels shall go out, and shall
separate the wicked from among the just."[4] And when once
the branches have been cast out of the vineyard, never again

[1] Job xxxiv. 24.
[2] Ezech. xv. 2—4. [3] St. August. in Ezech. xi. [4] St. Matt. xlii. 49.

will they be allowed to enter it. What will become of thee if thou art one of them—a useless branch, an evil branch? Then, indeed, wilt thou shed those inconsolable tears which thou dost not shed now: "There shall be weeping and gnashing of teeth, when you shall see Abraham and Isaac and Jacob, and all the prophets in the Kingdom of God, and you yourselves thrust out."[1]

III. Consider the second punishment of the branches, signified by the expression of being withered: "If any one abide not in Me, he shall be cast forth as a branch, and shall wither." When the branch is cast out of the vineyard, it becomes at length so dry and withered that it loses every drop of moisture which belonged to it so long as it was united to the vine. Alas for the wretched sinner! if, while yet living, he had any good thing from Christ, he is entirely deprived of it by the sentence of damnation. While he lived he had, at least, the habit of faith, by which he might be said to be in some way united to Christ, though only in a dead way; and even if this was lost, as is the case with heretics, there might be left in him at least some habit of moral virtue which was, as it were, like that superficial verdure which remains on the branches for a short time, even after they have been cut off: some inspiration, some instinct, some feeling of remorse urging him to return to his former state. But after death there is an end of all that is good: "My strength is dried up like a potsherd."[2] For he that is lost will not retain the least drop of the sap which was once infused into him by the Vine, not one of those good inclinations, not one of those gifts and qualities, not one even of those talents, freely bestowed, by which he was often the cause of good in others, while he was at no pains to help himself: "Take away the pound from him."[3] Is it possible that, seeing the danger thy soul is in of falling into this fatally withered state, thou art not filled with terror? "They shall be consumed as stubble that is fully dry."[4]

IV. Consider the third punishment of the branches, which is signified by the expression of binding in bundles: "They shall gather him up." Three terrible evils are hereby included. The first is the deprivation of liberty to do good; for branches, endowed with reason, as men are, even when separated from Christ, their Vine, may still one day be again united to Him;

---

[1] St. Luke xiii. 28.
[2] Psalm xxi. 16.   [3] St. Luke xix. 24.   [4] Nahum i. 10.

but after receiving the sentence of damnation this will no longer be the case ; they will lose all power of the kind : "Bind his hands and feet, and cast him into the exterior darkness."[1] This expression of binding the hands and the feet shows that these miserable persons will no longer have power either to do good by their works, or to aspire to do good by affection. The second evil is the society of the wicked ; because, as the branches, when cut off, are bound together in close bundles, so will it be with the damned, who will be bound "into bundles to burn," in which state they will be capable of nothing but of crushing and reviling each other ; and thus the proud, the sensual, the covetous, will all be bound in separate bundles. These will be the "many mansions" which there are in Hell as there are in Paradise, namely, the different bundles, varying according to the different degrees of punishment of the senses. The third evil is that of subjection to their tormentors : for, as the branches, when gathered into bundles, are unable to escape from the hands of the servants whose business it is to cast them into the furnace, and when cast into it to turn them this way and that with their pitch-forks, so neither will the damned be able to escape from the hands of the devils : "They shall be gathered together," that is, all these bundles, "as in the gathering of one bundle into the pit, and they shall be shut up there in prison."[2] Now, then, see what will be the end of that liberty which now urges thee to depart from Christ.

V. Consider the fourth punishment of the branches, signified by the expression of being thrown into the fire : "They shall cast him into the fire." It has been already told thee that the branches must either bear fruit or be burned : there is no alternative. The damned, therefore, being no longer capable of bearing fruit, as they are entirely cut off from Christ, their wills are, by their own act, hardened and utterly bent to do evil, nothing remains for them but ever-lasting fire, and that, too, such a fire as dry branches are cast into, namely, one that consumes utterly. Therefore it is not said, "They shall condemn him to the fire," which might be said of one who was partially burned, as when torches are applied to the sides, breast, or feet of malefactors ; but it is said, "They shall cast him into the fire," because they will be entirely thrown into it, like dry vine-branches, without sparing any part : "Behold, it is cast into the fire for fuel ; the fire

[1] St. Matt. xxii. 13.   [2] Isaias xxiv. 22.

hath consumed both ends thereof,"[1] that is, "both parts (*utramque partem*), both body and soul. If thou dost but scorch thy little finger, thou criest out : what will it be, then, to be burned not merely by, but in the fire, like the dry branches which thou so often throwest in it, one on the top of another ; and what sort of fire will that be? A fire which continually consumes thee with agony, and yet never so consumes thee as to reduce thee to ashes.

VI. Consider the fifth punishment of the branches, signified by the expression "he burneth"—not "he shall burn," but "he burneth"—to show the strength, fierceness, and activity of the fire which so rapidly burns the branches. A slow fire does, indeed, burn them also, but gradually : a fierce fire burns them instantaneously. So will it be with the damned in the fire of Hell : "As the vine-tree among the trees of the forests which I have given to the fire to be consumed" (not merely *ad comedendum,* but *ad devorandum*), "so will I deliver up the inhabitants of Jerusalem."[2] It is true that the activity of the fire is greater in proportion to the combustibility of the material to be burnt, and therefore the branches being, as has been seen, so exceedingly dry, it is plain that there will be no delay in their bursting into flames in the fire of Hell : "They shall cast him into the fire, and he burneth." There may also be a further meaning intended by God in these words, "he burneth," namely, to show that the fire of Hell is a fire which burns continually, just as when it first began—"A fire that is not kindled shall devour him "[3]—because it is a fire which will go on burning in the same way for all eternity. Is there, then, any profit to the branches in forsaking the Vine? "Abide in My love." Oh, with what good reason does Christ remind thee not to depart from Him, no matter what storms assail thee ! All the good in thee depends on thy loving, obeying, honouring Him, and faithfully advancing His glory. And if the thought of the good which will be the result of thy abiding constantly as a firm branch in Him, thy loving Vine, is not enough to move thee, let the consideration, at least, of the evil which will be the consequence of forsaking Him suffice to do so.

[1] Ezech. xv. 4.   [2] Ezech. xv. 6.   [3] Job xx. 26.

NINTH DAY.

They have said to thy soul, Bow down, that we may go over : and thou hast laid thy body as the ground, and as a way to them that went over (Isaias li. 23).

I. Consider first, that although there have been personages, even of very high rank, who have paid service to barbarous sovereigns so as even to allow themselves to be trodden upon as footstools for their pride, yet that when this has happened, it has been a matter of compulsion, as was the case when the Emperor Valerian was overcome in battle by the haughty Sapor, King of Persia.   But when was there ever a man found willing, when absolute master of his actions, to render so degrading a service to another?   Yet it is just such a service which thou hast so often rendered to the devils who have been thy tempters : "They have said to thy soul, Bow down, that we may go over."   And what answer hast thou made?   None, indeed, in words ; but by thy deeds thou hast shown thyself but too willing to yield to their solicitations : "Thou hast laid thy body as the ground, and as a way to them that went over." See, then, by this passage, that the devils who tempt thee have no power at all to trample on thee, unless by thy consent. "They have said to thy soul, Bow down, that we may go over."   Why "they have said"?   Because they can never force thee to bow down against thy will.   They may, indeed, show thee their desire, they may urge, incite, and persuade thee, but they cannot force thy inclination.   And accordingly it is not said, as to thyself, that "thou hast been compelled to lay thy body as the ground," but "thou hast laid" it, because it has been with the fullest consent of thy will that thou hast chosen to gratify them.   And see how easily it has been gained : thou hast not even waited for them to press thee with either very urgent or frequent persuasions.   They had only to ask thee.   "They have said," and "thou hast laid," so ready was thy consent to comply with their suggestion.   Dost thou not blush for thyself on reflecting that thou, a Christian, as much higher in dignity than the devils as the son of a king is superior to his slaves, hast stooped to a degree of abasement and degradation surpassing all belief?   What shame oughtest thou to feel for voluntarily placing thyself beneath the feet of those devils whom it was for thee to trample under thine ! "Tread thou, my soul, upon the strong ones."[1]

[1] Judges v. 21.

II. Consider secondly, that in the very act of tempting thee, the devils acknowledge thy dignity by telling thee to bow down : "Bow down, that we may go over." When a man bows down, it is because he chooses to descend from a high to a low position. And this is precisely what the devils want to do with thee when they tempt thee to evil; they want thee to stoop so as to set a high value on earthly goods, and to seek to obtain them without a thought of the heavenly goods for which thou art born. But now, observe their malice; it is quite certain that the devils always wish thee to do what is worst, they would fain have thee prostrate on the earth beneath their feet, laying "thy body on the ground," and yet all that they ask of thee is to "bow down." For this is their universal rule, to ask for what is only a beginning of evil, which seems no great matter—a look, a laugh, a mere inclination for the forbidden fruit, as in the case of Eve; so sure are they that if they can obtain that small matter they can obtain everything, because it is so exceedingly easy for every one to go on from a small to a great thing in sin. "They have said to thy soul, Bow down that we may go over," and because thou hast not courageously resisted this demand, see to what a pitch of baseness thou art fallen : "Thou hast laid thy body as the ground, and as a way to them that went over." Thou wert not even able to content thyself with doing merely the amount of evil which was asked of thee, namely, to bow down towards the earth, or rather, thou wert able, but not willing to do so ; and therefore didst thou go on from bowing down to falling prostrate—in other words, to the very depths of sin.

III. Consider thirdly, that even this was not enough for thee, but thou wentest on so far as to continue lying on the ground so as actually to make thyself, for the service of the devils, like the earth, and that vile earth, moreover, which is trodden underfoot. And therefore observe that it is not said, "Thou hast laid thy body on the ground," but "as the ground;" and in order that it may be clear what kind is intended, it is immediately added, "And as a way to them that went over." No doubt a grassy piece of ground, such as that of a meadow, a courtyard, a field, is also "the ground," but it is ground of a particular sort, set apart so as not to be trodden on by every one indiscriminately ; it is only in the public roads that this is allowed with impunity. And thou hast actually chosen to reach such a depth of degradation as to make thyself a public road, on which thy enemies are free to pass up and down at their pleasure, and to thy great confusion ! Such is the state to which sinners come in the end ; they lay "their body as the ground," by each actual sin they commit, and they lay it "as a way" by habitual sin.

IV. Consider fourthly, that it is to the state of an habitual sinner that the devils really wish to bring thee, desiring, as they do, never to cease trampling on thee for all eternity ; and yet that at the beginning they ask nothing but a mere step : "Bow down, that we may go over." Is it not, then, madness to allow thyself to be deceived by their vile arts? Commit this one sin, they say, and go to confession afterwards. This seems, doubtless, to be merely asking thee to take one step. But grant their request, and thou wilt see the consequences. Soon this one step which thou hast granted will become the traffic of a public road, permanent and continual, which will make thy slavery last as long as the road. And wouldst thou grant them this? Do men give a right of way to their sworn enemies, traitors, tyrants, highwaymen? And such, if thou didst but rightly know them, are the devils who tempt thee.

TENTH DAY.

Man knoweth not whether he be worthy of love or hatred ; but all things are kept uncertain for the time to come (Eccles. ix. 1, 2).

I. Consider first, that the Preacher is not here speaking of any class of men, whether just or sinners, but of the just only ; because sinners know quite certainly that they are worthy of hatred, whereas the just do not know certainly that they are just ; and therefore neither do they know certainly that they are worthy of love. And the reason of this difference is, that sin is altogether our own work, so that we may easily know whether we are guilty of it. But infused and inherent grace, which renders us just, is altogether the work of God, and an interior and imperceptible work, so that we cannot know that we are possessed of it. We do indeed know that it follows infallibly when preceded by the right dispositions ; but who can assure us of possessing these? The only two channels which convey sanctifying grace to our souls are Baptism and Penance : the first cancels original, the second actual sin. The first requires the intention of the person who administers it, which is hidden from us; the second requires, in addition, on our part, a detestation of sin, which includes true repentance and a sincere resolution of amendment. And who can give us the assurance that we possess this in a sufficient degree? It is indeed written, "When thou shalt seek the Lord thy God thou shalt find Him;" but it is also added, "Yet so, if thou seek Him with all thy heart, and all the affliction of thy soul."[1] Here is the uncertainty, and therefore, "No man knoweth whether he be worthy of love or hatred, but all things are kept uncertain for the time to come," that is to say, "for the time when Christ shall sit in judgment." Till then we must always live in serious uncertainty about ourselves, not even knowing whether we shall be saved. When therefore, it is here said, "Man knoweth not whether he be worthy of love or hatred," two kinds of hatred are meant. First, that simple hatred with which God regards the sinner when He sees him in a state of sin, and although He is angry with him on account of that sin, yet bears with him; and this is the hatred of indignation. And secondly, that consummate hatred with which God regards him when He sees him in a state of sin, and is not only angry with him on account of that sin, but also allows him to die in it, and so to be damned; and this is the hatred of reprobation. Is it possible that thou art not afraid at finding thyself in so terrible an uncertainty as this? "Fear and trembling are come upon me, and darkness hath covered me,"[2] that is, "because darkness hath covered me"—"fear" as to the present, "trembling" as to the future.

II. Consider secondly, how it might seem at first sight that God deals very hardly with us in keeping us all our life in such dark uncertainty, when, if He pleased, He might so easily deliver us from it. Yet, perhaps, there is no way in which He could have so well provided for our needs. For if we were certain of being now in a good state, how easy it would be for us to grow proud, and so to fall from it; and if we were certain of being in a good state at last, how easily we should grow careless in this life! Thou mayest say that it would be possible for God to give us, at the same time, such an abundance of grace, that we should incur no danger of the kind. This is true; but since He does not do so, it is a proof that He has very good reasons for it. When was

[1] Deut. iv. 29.   [2] Psalm liv. 6.

it ever known that a physician received instructions from his
patients as to the way of treating them? It is he, not they,
who is the judge as to the best course. And what great benefit
have even the greatest saints derived from this painful but
blessed uncertainty! Nay, it is this very uncertainty which
has given to the Church her greatest saints. For it is owing
to this that they not only kept themselves in a continual state
of the greatest humility before God, in Whose hands they
knew that their destiny lay, but that they also kept themselves
very humble before men, whom they might rightly esteem
better than themselves. And yet how many persons there are
whom thou despisest, because thou foolishly considerest thyself
their superior! "Man knoweth not whether he be worthy of
love or hatred; but all things are kept uncertain for the time
to come." And yet thou boldly esteemest thyself better than
many who are perhaps, in the sight of God, in a far higher
state of grace than thou art, and who may also be one day in
a far higher state of glory. What mad presumption is this!
"I saw the wicked buried," says the Preacher, "who also,
when they were yet living, were in the holy place, and were
praised in the city as men of just works."[1] And yet, all the
while, they were wicked. Alas for thee, if these words should
one day be able to be recorded of thee, at thy burial! Be
humble, therefore, whilst as yet things are uncertain.

III. Consider thirdly, that this uncertainty, if thou considerest it aright, is so profitable, that even if it were in thy
power to put an end to it by receiving from God at this very
moment the positive assurance of thy future salvation, I would
almost say that thou oughtest not to desire to do so. And
why so? In order that thou mayest rest wholly, and with
great confidence, in the mercy of God: "Behold, God is my
Saviour; I will deal confidently, and will not fear."[2] Oh,
if thou didst but know, and rightly consider, the great merit
and advantage of so doing! There is, perhaps, no greater
honour that thou canst do thy God. This is why a great
servant of His, after having been tormented many years by
the anxiety arising from this terrible thought of his predestination, at length was inspired with such confidence as to
say, that if God were one day to place in his hand a written
assurance of his gaining Paradise, he would willingly tear it
before His face, in order to continue as before in a state of
complete dependence on His mere goodness. It is enough,

[1] Eccles. viii. 10. [2] Isaias xii. 2.

then, to have in thyself the witness of a good conscience,
telling thee that thou art not conscious of any sin which makes
thee God's enemy at the present moment, for otherwise thou
wouldst most certainly at once go and make confession of it;
it is enough that thou desirest to serve God to the utmost
of thy power in thy calling, to please Him, to advance His
glory; it is enough that thou art resolved to allow thyself
knowingly no sin, however small. If this is so, then believe
that thou wilt gain Paradise: "If our heart do not reprehend
us, we have confidence towards God."[1]   Thou mayest say
that God's judgments are very mysterious.   Certainly they
are: "Thy judgments are a great deep."[2] But what is thy
anchor in that deep? Always to have recourse to God, always
to commend thyself to Him, always to beg of Him never to
suffer thee to be faithless to Him.   Do this: and then no
possible certainty can be of so much value as the confidence
which thou placest in Him, for however great might be the
tranquillity produced by the former, thy merit would be pro-
portionately diminished: "Thy life shall be saved for thee,
because thou hast put thy trust in Me, saith the Lord."[3]

ELEVENTH DAY.

*Son, when thou comest to the service of God, stand in justice and in fear,
and prepare thy soul for temptation* (Ecclus. ii. 1).

I. Consider first, that to be tempted is common not only
to beginners in the ways of God, but to those who are advanced,
nay, even to the perfect.   For this reason even Christ Himself
condescended to suffer Himself to be tempted, in order that
no one might think it a dishonour. It might seem, therefore,
that instead of saying, "Son, when thou comest to the service
of God, . . . prepare thyself for temptation," it would have
been better to say "when thou hast come," so as to make
the rule applicable to all.   Nevertheless, the Preacher chose
rather to say "when thou comest;" for, if it is possible for
the advanced and the perfect to endure temptations, and very
severe ones, it is impossible for beginners not to endure them,
because the fury of the devil is fiercest against those whom
he sees in the act of escaping from his power: "It was told

[1] 1 St. John iii. 21.   [2] Psalm xxxv. 7.   [3] Jerem. xxxix. 18.

the King of the Egyptians that the people was fled, . . . and he took all the chariots that were in Egypt . . . and he pursued the children of Israel."[1] Besides, as thou seest, the intention of the Preacher in this passage is to exhort the just man to "prepare his soul for temptation," and therefore it was right to say, "Son, when thou comest," because the man who "has come" is either advanced or perfect in the spiritual life, and therefore it is taken for granted that he is already prepared for temptations, so as to know how to overcome them. Preparation is the act of beginners; and it is such persons who are here reminded not only to stand firm in the right resolution they have made of serving God, which is common also to the other two states, but always to fear, which more especially belongs to their own : "Stand in justice and in fear," because they are in greater danger owing to their inexperience. And it is not to be wondered that the word here used is "temptation," not "temptations," because the reason of this is to urge thee to prepare thyself not so much against every possible individual temptation as against that general practice of the devil to tempt those who have just left him for God, in the hope of winning them back to himself. If, then, thou art in the state of a beginner, say to thyself that the injunctions here given are addressed in a particular manner to thyself.

II. Consider secondly, that the first preparation which thou shouldst make against thy tempter, the devil, must be this, to learn what is his way of dealing with persons like thyself. And in what school canst thou learn this better than in that of the desert where the evil one was not afraid to attack Christ Himself, as though He had been a new soldier, because he had seen Him receive baptism from John, as though He were a penitent, and go into the wilderness? Observe therefore how, although the devil desires to lead thee into the greatest possible evil, he never asks it of thee suddenly, or all at once, but gradually, as he did of Christ, to Whom he suggested first a lesser, then a greater, and lastly, a still greater sin. He saw the weakness and hunger our Lord was suffering after His prolonged fast, and took occasion from this to urge Him to provide Himself with bread, not by means of robbery or violence, as many do, but only by working a miracle without necessity, which, if not a commendable thing, appears, at all events, but a small evil: "If Thou be

[1] Exodus xiv. 5.

the Son of God, command that these stones be made bread."
Foiled in his first attack, the devil thought the reason of this
'to be that Christ had attained to a great mortification of the
concupiscent appetites of the body ; and so, he went on to
make his next attack on the spiritual appetites, namely, those
of the irascible part, by tempting Him to the presumption
of showing confidence in the Divine assistance even when
wilfully exposing Himself to danger and injury by throwing
Himself, as though reckless of His life, from a great height:
"If Thou be the Son of God, cast Thyself down." Then
when he was unable to succeed in either attempt, he threw
off the mask, and by offering to make Him absolute monarch
of the world, sought to render Him so far beside Himself, as,
blinded by pride, ambition, and all the passions belonging to
the love of empire, to make Him for the sake of it fall at his
feet in worship : "All these will I give Thee, if, falling down,
Thou wilt adore me." In the first temptation, the devil
appeared in the form of a man naturally moved to compassion
by the suffering of another. In the second, he transformed
himself from a man into an angel of light, persuading to what
was evil indeed, but under the appearance of good, and supporting it even by the testimony of the Sacred Scriptures, which
he maliciously interpreted to suit his purposes. In the third
temptation, he threw off this angelic appearance, and showed
himself in his true character, as Satan ; so that while in the first
and second temptations, he said to our Lord, " If Thou be the
Son of God," because in both he concealed his evil intention,
he used no such form in the last, because the evil was no
longer veiled. In the first he appealed to the weakness which
he supposed must be even in Christ if He were merely man,
in the second to ignorance, in the third to malice : and thus
the first was a temptation to pusillanimity, as though there
were no possible way of support in that extremity of hunger
but by changing stones into bread ; the second temptation
was to presumption, as though God's help must always be at
hand even in a peril which was voluntarily incurred ; and the
third temptation was to flagrant rebellion, as though it were
lawful, for the sake of reigning, not only to trample on every
law of reason and religion, but also to invoke the aid of Satan :
" If justice may ever be violated, it may be violated for the
sake of reigning."

III. Consider thirdly, that, just as the devil dealt with
Christ, so, in due measure, does He deal with all who have

newly devoted themselves to the service of God, but more especially with a noble young religious who has left the world for the desert, in other words, to practise perfection in the cloister. First, the devil sets before him his lack of strength, and representing to him the austerity of the life on which he has entered, tries to persuade him that it is impossible to lead it for any length of time without a positive miracle ; and thus, like a man taking pity on his sufferings, he urges him to relax the severity of the discipline, and tempts him, in the first instance, to pusillanimity. Then, if the devil perceives that this young man, in the fervour of his soul, despises, nay, even rejoices in bodily suffering, he transforms himself from a man into a shining angel, and inflaming this spiritual fervour as much as possible, he urges him not to consider himself in any way, to ill-use, nay, even to destroy his body, as he may be certain of being able, by God's help, to endure any amount of punishment and torture ; he recommends him, relying on the ignorance for which he gives him credit, to attempt things that are beyond his strength, in order that at last he may sink under the burden : thus he tempts him to presumption. Then, in case he fails in both these assaults, the devil at length gives the rein to his fury, and throwing away the mask, no longer deals his blows in secret, but with despairing rage. He continually sets before this young man the life of enjoyment that may be spent in the world, the pleasures of liberty, the indulgences, the grandeur and pomp, the high connections and offices which may be attained in it, and making light of the wickedness of such a course, he insinuates to the unhappy youth that these things are well worth even such a price as apostasy : and so he tempts him to rebellion. Now, what oughtest thou to do in order to learn what the attacks are against which thou hast to guard thyself? The way is to be firmly convinced that the devil's plan is this : to intend the worst evil possible, but to attain it gradually. So that his earlier suggestions may be compared to the way in which a general calls upon the enemy's fortress to surrender, which he does, first very gently, then more peremptorily ; but his final attempt is like the shout of fury with which an army rushes with drawn swords to the charge. Do thou, therefore, prepare betimes for battle, and watch his every step : "He smelleth the battle afar off, the encouraging of the captains, and the shouting of the army."[1]

[1] Job xxxix. 25.

IV Consider fourthly, that if the first preparation, which is here necessary, is to know the arts which the devil is accustomed to employ in tempting all beginners, the second must be to learn the way to proceed in resisting him. This, too, has to be learnt in the desert, by studying the Divine method of our Lord, Who with great humility submitted to be assailed by the tempter in order to teach us how to parry his attacks. In the first place, considering this in the general, thou seest clearly that Christ did not set Himself to dispute with the devil, but repelled him, with authority, promptitude, and in few words. In the same way do thou, in the like case, refuse to argue with the idle imaginations in which the evil one lurks to attack thee; but taking thy stand on those principles of faith which thou hast already learnt, seek no further. Call to mind both the promises and the commandments of God, as Christ did, in order to oppose them both to the offers and the pretensions of the devil; and esteeming one plain and simple Divine saying better than any argument against him which may occur to thy mind, do not in anything, however small, concur with the intention of the devil, for nothing troubles him so much as this contemptuous treatment. Then pass on to each of these three kinds of temptation in particular; and in regard to the first, that to pusillanimity, which would fain make thee relax the severity of discipline, or provide thyself with clothing, food, or other human gratifications, in ways which are not allowed, say, "Not in bread alone doth man live, but in every word that proceedeth from the mouth of God." For if thou lackest one means of support, He will provide thee with another; it need not of necessity be that which the devil suggests to thee, it is enough for thee that it is provided for thee with exceeding love by God. For forty years the children of Israel were without ordinary bread, and God gave them manna instead. With regard to the second temptation, to presumption, which, on the contrary, urges thee (because of the faith in God which thou hast already shown) to perform penances beyond thy strength, or other acts of singular or indiscreet zeal, unsuitable to the state in which thou art placed, say, "Thou shalt not tempt the Lord thy God." For it is not lawful to take a dangerous leap, relying on extraordinary means of assistance, in order to reach a place which might be gradually attained by a safe path. And if thou art ignorant what this path may be, there are plenty of spiritual directors ready to act as faithful guides in showing it to thee: consult them. As to the third temptation, to rebellion; if the devil is so audacious as to assail thee with it, thou shouldst repel him instantly with a kind of authority and, as it were, anathematize him: "Begone, Satan; for it is written, The Lord thy God shalt thou adore, and Him only shalt thou serve." For although this temptation is the strongest of all, because of the great fascination with which thy mind may be deluded by those worldly goods all of which the devil shows thee at once—pleasures, riches, reputation, and even, were it possible, royal rank; yet observe that the devil shows thee, just as he showed Christ, the good but not the evil things of the world; he shows thee its joys but not its anguish; its lofty positions but not the falls from them; the roses but not the thorns. And thus he does, indeed, reveal to thee what is calculated to attract thee to the world, but he hides what would repel thee from it; he shows "all the kingdoms of the world, and the glory of them," but not "the misery of them." Besides, dost thou not see how evidently he lies to thee? He declares that he can give thee what is not his. His condition is so lamentable that he is always howling in fury, bound in chains of iron and bonds of fire, and yet he promises to make thee happy in this world if thou wilt adore him. Is not this a falsehood worthy of Satan? And can there be a better answer to drive him from thee than, "Begone, Satan!" since he is here so clearly seen in many ways to be what he is, malicious, false, insolent, and a sacrilegious usurper of the worship due to God alone? This, then, is what the Preacher desires of thee when he says: "Son, when thou comest to the service of God, stand in justice and in fear, and prepare thy soul for temptation." He would have thee know that the enemy is on the point of engaging in conflict with thee, and so learn the master-strokes which he will deal, and which thou must return: "The horse is prepared for the day of battle, but the Lord giveth safety."[1]

---

[1] Prov. xxi. 31.

TWELFTH DAY.

Man shall come to a deep heart, and God shall be exalted (Psalm lxiii. 7, 8).

I. Consider first, that these words are by some taken in a bad sense, which perhaps agrees best with the text; in this case the meaning is two-fold. For the word "heart" may signify the understanding, as in the passage, "Their foolish heart was darkened;"[1] or it may signify the will, as when it is said, "They made their heart as the adamant stone."[2] If we follow the first interpretation, then the meaning of these words is that when a man goes too far in investigating by the light of nature the mysteries of the Holy Trinity, Providence, predestination, grace, and the like, the explanation of which God reserves for us in Heaven, God, in such a case, recedes from him, as it were, by raising Himself so high above him, that at length He makes Himself imperceptible, as would be the case if a small boat were to give chase to a great vessel at sea. The boat, after proceeding some distance from the shore in pursuit, has either to turn back, and humbly confess its impotence—and this is what the most modest among philosophers have done, who, after long arguments, have come to the conclusion that the things of God surpass human capacity: "Behold, God is great, exceeding our knowledge;"[3] or else, if it is rash enough to continue the chase, not only will it altogether lose sight of the ship when far out at sea, but it will itself be lost; and that has been the fate of those audacious philosophers whose grand speculations have come to nothing—"They became vain in their thoughts"[4]—and who, not being able to comprehend Divine things, have ventured to deny them, even going so far as to say in their hearts that there is no God: "The fool hath said in his heart, there is no God."[5] And not only to say so in their hearts, but sometimes even in conversation and in public lectures, which is as though the boat we have spoken of should insolently deny that the ship is anywhere on the seas, because, the longer it follows her, the further she recedes from view. This, then, is the first interpretation of the words, "Man shall come to a deep heart, and God shall be exalted," namely, "Man shall come to a deep heart by seeking to comprehend the inscrutable things of God, and God shall be

---

[1] Romans i. 21.   [2] Zach. vii. 12.   [3] Job xxxvi. 26.
[4] Romans i. 21.   [5] Psalm xiii. 1.

exalted by withdrawing Himself from man," which is what the Preacher intended by the words, "I have said I will be wise, and wisdom departed farther from me, much more than it was."[1] If by "heart" thou understandest the will, then the meaning is this, that when man, with deep malice, such as is the characteristic of wicked politicians, endeavours, as it were, to place himself above God in his actions, God rises so far above him as to overrule matters by bringing to pass the very opposite of that which man had ventured to place against God —"He bringeth counsellors to a foolish end"[2]—just as He did in the selling of Joseph, the conspiracy of Aman, the plots of Achitophel, and, above all, in the counsel by which the Jews, in their madness, crucified Christ, and by so doing made the name of Jesus more glorious throughout the world by the very means whereby they intended to make it for ever infamous: "They have searched after iniquities, they have failed in their search."[3] This is another meaning of these words, "Man shall come to a deep heart, and God shall be exalted." They mean, "Man shall come to a deep heart by seeking to thwart God, and God shall be exalted by overruling man." How is it with thee? What sort of a heart is thine? Art thou humble both in understanding and in will? If not, rest assured that God will make a mock of thee, as He always does of those who, like those mighty men of Babel, would fain erect a tower, by which to climb higher than the clouds: "Man shall come to a deep heart, and God shall be exalted."

II. Consider secondly, that there are some who take these words in a good sense. Their meaning in this case is, that when a man applies his understanding, piously and not out of curiosity, to the contemplation of the greatness of God, then the more he understands of it, the more he knows that there remains to be understood: "He will yet far exceed."[4] For when men seek Him in a spirit of pride, God exalts Himself, and at the same time hides Himself in His own light. But when they seek Him in a spirit of devotion, God reveals Himself to them in the act of exalting Himself. So that He is always making Himself more known, but always from a greater distance—"All men see Him; every one beholdeth afar off"[5]—so that the soul is rapt in admiration at the greatness of that glory, and becoming like an eagle gazing on the sun, the more she knows of God, the more she confesses how

---

[1] Eccles. vii. 24.   [2] Job xii. 17.
[3] Psalm lxiii. 7.   [4] Ecclus. xliii. 32.   [5] Job xxxvi. 25.

little she knows; and the more she confesses this, the more she desires to know, so great is at once the splendour and the sublimity which she discovers in that glorious presence: "Man snall come to a deep heart, and God shall be exalted," that is, "Man shall come to a deep heart in contemplating the glory of God, and God shall be exalted by appearing more glorious." This is when the word "heart" is taken to mean the understanding." Now, if by "heart" is understood the will, the meaning is that the more man exalts himself to love God, the worthier does God appear of being loved still more; and so, in this case, God is not exalted in Himself, because He cannot become higher in Himself than He is, but He becomes so in the estimation of man, who, entranced by such marvellous goodness, is like the needle, which the more it perceives the attracting loadstone rise, the more also itself rises to meet it, although to effect this it has to overcome the weight of its own body by excessive efforts. This is the meaning here of the words, "Man shall come to a deep heart, and God shall be exalted," that is, "Man shall come to a deep heart by a great love of God, and God shall be exalted by showing Himself worthy of a still greater love." Oh, happy art thou, if thy heart is "deep" in this sense! But how is such a heart to be acquired? By understanding clearly that thou art made for God, and that therefore it is the lowest depth of baseness if thou art content to leave God, and so to remain always on the earth, like a worm condemned to crawl in the mire: "Arise, arise, put on thy strength, O Sion,"[1] so as to rise as high as possible.

III. Consider thirdly, that if thou dost not understand the meaning of ecstasy, it is here clearly shown thee what it is, because it rests entirely on these words taken in a good sense. There are two kinds of ecstasy: one of the understanding, the other of the will. The former is founded on admiration, the latter on love; but not any kind of admiration, nor any kind of love, but only the admiration and the love of one who has "come to a deep heart." The soul applies herself to the contemplation of the beauty and the goodness of her God, and perceiving both to be far greater than she has any power of conceiving, she is so overwhelmed with wonder, that at length she is carried out of herself by ecstasy, as the Queen of Saba was on beholding the great city of Solomon: "And when the Queen of Saba saw all the wisdom of Solomon, - - -

[1] Isaias lii. 1.

she had no longer any spirit in her."[1] This is the ecstasy of the understanding. Or else, the soul applies herself sincerely to the contemplation of the beauty and the goodness of her God, and then God touches the inmost depths of her heart, and draws her to Himself with so sweet a love, that, unable to endure that sweetness, she is, in a manner, compelled to go out of herself in order to unite herself to Him, Who calls her from afar to Himself, as He did the bride: "Arise, make haste, My love, and come."[2] This is the ecstasy of the will. In both of these "man cometh to a deep heart," whether by admiration or by love; for how deep must both this admiration and this love be! And in both "God is exalted" above man. For if God did not show Himself continually to surpass all the admiration and love of Him which is in the soul, she would remain within her own bounds, so to speak, as though capable of containing Him, instead of breaking them by her aspirations to rise above herself in order to fly after her only Good, Who is so high above her: "So that my soul rather chooseth hanging."[3] Her flying after Him shows that she sees Him continually rising higher than she can attain by her efforts: "Man shall come to a deep heart, and God shall be exalted."

IV. Consider fourthly, that, speaking generally, these two kinds of ecstasy go together, for it is impossible that when once the Divine Sun has entered a soul it should enlighten without inflaming, or inflame without enlightening her. And thus love is increased by admiration, and admiration by love. Still, the two are not so completely one that they cannot be separated, and the reason is that in order to love God upon earth, it is not necessary that knowledge should be in equal proportion to love. The heat which emanates from the Divine Sun may not unfrequently be greater than the light. And so it is possible for a poor simple old woman to love God more than He is loved by many eminent theologians, whose knowledge of Him certainly surpasses hers, as St. Bonaventure declared to Brother Giles. Therefore, when admiration is greater than love, the ecstasy is said to be in the understanding, and when love is greater than admiration, it is said to be in the will. It is not impossible that the former kind may co-exist with but little love, because it may be a pure gift of God; but this is rarely the case. And therefore it is more to be suspected, both because much that is natural may be

---

[1] 3 Kings x. 4, 5.   [2] Cant. ii. 10.   [3] Job vii. 15.

mingled with it, and because it is liable to delusions from the
devil, who may easily bring before the mind wonderful intellectual lights, which carry it away, and at the same time, by a
refinement of malice, may excite in the heart some slight
degree of deceptive love, that is, a love that is rather sweet
than solid. The ecstasy of the will is more sure, always
provided that the love is equally fervent in devotion and in
works. For if in prayer love is so powerful as to take the
soul, as it were, out of the body, as in death—nay, sometimes
even making the body itself, though apparently lifeless, rise
from the earth, in a manner contrary to nature, and follow
nothing but the sweetness of that overpowering joy which fills
the soul, how is it possible that such a love can be feeble in
works? It must, on the contrary, be so sublime in this respect
also, as to make the soul accomplish the law of God, not in
an ordinary, but a heroic degree of perfection, so as not only
to despise for His sake impure pleasures, deceitful gains, false
glory, and everything that is in the least degree contrary to
the Divine will, but even generously to embrace all kinds of
suffering, to rejoice in poverty, to triumph in persecution,
showing plainly that he who does this no longer lives to
himself, but to God alone—nay, that by a kind of ecstatic life
he lives in Him as a drop of water in the ocean, which is so
lost in it as to be no longer itself: "Seek the things that are
above, . . . mind the things that are above, . . . for you are
dead, and your life is hid with Christ in God."[1]   Therefore,
whenever this ecstasy, which may be called vital, is not to be
found, every other kind, whether of the intellect or the will, is
entirely illusory.

THIRTEENTH DAY.

Let none of you suffer as a murderer, or a thief, or a railer, or a coveter
of other men's things. But if, as a Christian, let him not be ashamed, but
let him glorify God in this name (1 St. Peter iv. 15, 16).

I. Consider first, how St. Peter, desiring in this passage to
enumerate some of the crimes which are most severely punished
by law, has selected those which are accounted the most
disgraceful, because they are injurious to our neighbour, for
which reason he says: "Let none of you suffer as a murderer,

[1] Coloss. iii. 3.

or a thief, or a railer, or a coveter of other men's things." A "murderer," as every one knows, is one who has injured his neighbour as to his life; a "railer" is one who has injured him as to his reputation; a "coveter of other men's things" is one who, if he has failed in actually inflicting these injuries on his neighbour in act, has, at least, endeavoured and planned to do so, by the attempt. So that "a coveter of other men's things" means, strictly speaking, one who tries to seize upon them, or, at any rate, who looks on them with the intention of so doing; because the law cannot go so far as to take notice of the desire, but only of the attempt. Now, all these crimes are especially disgraceful, because, since the stability of a state depends on the maintenance of mutual justice among men, it is right that any one who infringes it should be punished, not only by such penalties as may be escaped by flight, but by infamy which attaches to the culprit everywhere: "The nations have heard of thy disgrace."[1] Therefore, says St. Peter: "Let none of you suffer as a murderer, or a thief, or a railer, or a coveter of other men's things;" for this being to suffer as an unjust, so also is it to suffer as a disgraced man. Far otherwise is it with regard to suffering as a Christian. To be a Christian is an exceedingly glorious thing; and consequently it is also exceedingly glorious to suffer as a Christian, for the evil of disgrace does not consist in receiving, but in meriting disgrace: "If you suffer anything for justice' sake, blessed are ye."[2] It is very evident that it is an exceedingly glorious thing to be a Christian, because to be so is not only to observe justice, but charity to all men: "By this shall all men know that you are My disciples, if you have love one for another."[3] So that not only is it not to injure one's neighbour as to his life, like the murderer, but even to save it, if needs be, at the cost of one's own; not only is it not to injure him in his goods, like the thief, but to bestow them upon him, even to the stripping of oneself to clothe others; not only is it not to injure him in his reputation, like the detractor, but to increase it, even by yielding to him, sometimes, the credit due to oneself. Surely, there can be no greater glory than this: and therefore there can be no greater glory than to suffer for being a Christian, in other words, for observing so glorious a law. "But if as a Christian," that is, "if he suffer as a Christian, let him not be ashamed," because, although punishment is in itself a disgraceful thing, it ceases to be so when it is not merely un-

---

[1] Jerem. xlvi. 12.　　[2] 1 St. Peter iii. 14.　　[3] St. John xiii. 35.

reasonable, but contrary to reason : " Be ashamed at your own ways, O house of Israel," saith the Lord.[1] He does not say, " Be ashamed at what others think of you," but "at your own ways." How, then, would it be if thy action were the very opposite of what has been described ? Wouldst thou not have reason to fear both meriting and suffering the disgrace ? Observe the prudence of the Apostle. He does not say that a man is not to be grieved when he has to suffer some shame as a Christian : all that he says is that he is not to "be ashamed" of it. He does not forbid his being grieved, knowing that even in the saints, nature claims as much as this as its right ; but he forbids his being ashamed, because he also knows that if nature claims its right in him, it should be in moderation ; and therefore, if some pain is excited in him on account of the evil he is enduring, even if it is as a Christian that he endures it, there ought not to be any shame excited, because in wise men shame should only be felt for what is blameable ; although, so long as their virtue is weak, there may be some little shame for the blame also, for which reason they need encouraging, from time to time, not to fear it : " Fear ye not the reproach of men."[2] If, therefore, not one of all those who suffer innocently ought ever to be ashamed of that suffering (as even philosophers maintain), but to despise it with a generous courage ; how much less should he be ashamed who suffers as a Christian, that is, who suffers to maintain the honour of Christ, who suffers for the sake of piety, modesty, charity, or to prevent error from taking root among the people. It is the part of a Christian to spurn the glory of this world as false and frail, and to glory in the hope of that which will be his in the next world, as a child of God : " We glory in the hope of the glory of the sons of God."[3] What reason, then, has he to be ashamed because he is unjustly deprived of glory here, when that glory of which he is thus unjustly deprived here will serve so much to increase that which will be justly bestowed on him in Heaven ? The more he is blamed the less reason is there for him to be ashamed : " If as a Christian, let him not be ashamed."

II. Consider secondly, that therefore it is not enough for St. Peter that the man who suffers as a Christian should " not be ashamed." He would have him also glorify God, and that too in this very name of Christian : " Let him glorify God in this name." But what is the meaning of glorifying Him in

[1] Ezech. xxxvi. 32.   [2] Isaias li. 7.   [3] Romans v. 2.

this name? Does it mean glorifying by standing firm, in the midst of disgrace, in the name of Christian? Certainly: but this is not enough. It means also glorifying Him by doing nothing, in that disgrace, which is unbecoming one who bears that name. If, when thou sufferest any insult which is offered thee as a sincere, upright, religious, or zealous Christian, thou complainest bitterly of those who have offered it to thee, if thou art violent and angry, if thou secretly wishest some evil to them by which it may be shown that Heaven, in punishing them, defends thee, can it be said that thou sufferest that insult "as a Christian"? Dost thou "glorify God in this name"? Surely not: for the law of Christ teaches thee to pray for those who calumniate thee, to bless them, to do good to them, to love them, as much after as before the most terrible injuries; that is, to love them as thyself. This, then, is what thou hast to do: then will it be well with thee; then, indeed, wilt thou glorify thy God as thou oughtest to do; for what greater glory can He get from His servants than this: to see that for love of Him they are ready to love those by whom they have been not merely offended but ill-used, and why? Only because, with a true Christian heart, they were resolved to save the honour of Christ. And so, although, doubtless, He was infinitely glorified by all the martyrs, yet most of all was He glorified by those who even in the midst of their torments rendered good for evil to their tormentors, sometimes expelling devils from the bodies of those savage men, sometimes restoring their sight, saving their lives, or making them their heirs, as the great Bishop St. Cyprian did the executioner when his sword was raised to strike his head from his body. Try to imitate as far as is possible in thy state these great examples: "Be not overcome by evil, but overcome evil by good."[1] In this way, while suffering as a Christian, thou wilt not only glorify God simply as he does who ceases not while suffering to declare himself a Christian, but thou wilt glorify Him in a more noble way, that is, by acting as a Christian even in that painful position.

[1] Romans xii. 21.

FOURTEENTH DAY.

For I have always feared God as waves swelling over me, and His weight I was not able to bear (Job xxxi. 23).

I. Consider first, how greatly mistaken all those are who think that it belongs rather to sinners than saints to fear the anger of God. Could any man be holier than Job was, whether in prosperity or adversity? And yet, listen to what he says of himself: "I have always feared God as waves swelling over me." No terror can be equal to that of sailors in mid-ocean, when they are assailed by storms and whirlwinds, and see the waves threatening to overwhelm and sink their vessel. What a scene of agitation and deafening tumult and shrieks and groans must that be! And yet Job says that he "always feared God as waves swelling over" him, that is, not merely as a tempest which might possibly arise, but as one which had already broken forth. And so far from this being incompatible with sanctity, it is perfectly in conformity with it; for it is precisely this which gives its force to sanctity. What is sanctity? Is it not a universal contempt of all earthly things? And the special source of this contempt is the seeing God, as it were, like a mighty tempest ready to burst overhead. And just as sailors in such circumstances give no thought to banquets, or glory, or gain, or amusements, but only to that which alone is important, the saving of their lives, so neither do the saints in the case we are considering think of anything but saving their souls. It may be that hitherto thou hast lived in a state of excessive attachment to the goods of this miserable world. What is this a sign of? It is a sign that thou always regardest God as a calm sea, from which there is no danger of shipwreck. But if thou regardest Him as a sea in storm, thou wilt see that henceforth thou wilt be able to think nothing but of saving, if it must be, only thy bare life by clinging to a plank: "Being mightily tossed with the tempest, the next day they lightened the ship "[1]—so quickly is everything thrown overboard when once a violent storm has arisen.

II. Consider secondly, that sailors in a storm are not satisfied with making light of all that they possess if only they may not perish, but that they cry so piteously to Heaven, that never at any other time have they wept or prayed so vehemently. So, too, is it with the saints in the case we are con-

[1] Acts xxvii. 18.

sidering. For which reason Job said, "I have always feared God as waves swelling over me," to show that he had all his life commended himself to God with the earnestness and fervour of one who sees angry waves breaking over him : "As overflowing waters, so is my roaring."[1] It is true, indeed, that just as sailors, for all their prayers to Heaven that their lives, now all but lost, may be saved, do not for that reason neglect to do all in their power to help themselves, by rowing, weighing anchor, lightening load, and cutting away all that is possible ; so, too, do the saints act in the present case, and so does Job, under a metaphor, imply that he also had acted—"For my heart doth not reprehend me in all my life"[2]—so strictly had he always observed his duties that he was able to speak in this way. How, then, is it with thee, who either dost not commend thyself to God, or else, while so doing, dost not act in conformity with what thou desirest of God in so commending thyself? This proves that hitherto thou art far indeed from having learnt what it is to fear Him, as in a tempest.

III. Consider thirdly, that if at times thou dost think of the anger of God, and picture it as ready to burst forth like swelling waves, over whom is it that thou imaginest them descending ? It is always over other men's ships, never over thine own. What wonder, then, that thou art not afraid ? Far different is it with the saints. They all say, as Job did, "I have always feared God as waves swelling over me," not over others, but "over me ;" for as they have a mean opinion of themselves, they are persuaded that if they were to presume to defy God, He would at once destroy them, as a small boat would be destroyed which should dare to defy the storms and whirlwinds which are breaking over it. Thou thinkest that thy sins will be easily borne with by God, not because thou hast a high esteem of the Divine mercy, but because thou hast a high esteem of thyself. Thou conceivest thyself so well endowed with graces, merits, or talents, that greater regard is due to thee than to the mass of mankind, even as to the sins that thou committest. But what gross pride is this ! "Their iniquity hath come forth as it were from fatness."[3] If thou hadst ever so little humility thou wouldst exclaim, with far greater reason than Job, "I have always feared God as waves swelling over me," so justly wouldst thou see thyself deserving of a prompt and speedy punishment, ready to fall upon thee like waves swelling on high, as the same Job in another

[1] Job iii. 24.　　[2] Job xxvii. 6.　　[3] Psalm lxxii. 7.

passage declared still more clearly in the following words: "I feared all my works, knowing that Thou didst not spare the offender,"[1] that is, not "any offender," for it is written in another passage of Holy Scripture, "I will spare them, as a man spareth his son that serveth him,"[2] but "me, the offender," so worthy did Job esteem himself of inexorable punishment for even the smallest sin: "If I be wicked, woe unto me."[3]

IV. Consider fourthly, that this may appear to thee a servile fear, and therefore unsuitable for great saints, who ought to keep themselves from sin in order not to offend, displease, and dishonour their God, not for fear of being instantly punished by Him for the first slight offence they may commit against Him. But thou reasonest in this way because thou hast not yet well considered the words of holy Job. Listen to his words: "I have always feared God as waves swelling over me." He does not say "the chastisements of God," but "God." It is one thing to fear God's punishments, and another to fear God, Who is able to punish us, nay, Who is ready to do so, as swelling waves are ready to break. The former is the fear of slaves, the second of sons, who take occasion from the equity and severity of the king, their father, to enhance the obligation they are under of living in obedience to him, and who, therefore, on the one hand, are ready to kiss the rod when he thinks fit to punish them, and, on the other, are careful not to excite his anger in the least degree. And why is this? Because, the more power a king possesses, the more he ought to be reverenced by his subjects: "Who shall not fear Thee, O King of nations?"[4] Such a fear is not servile, but reverential, and is ascribed even to the angels: "The pillars of Heaven tremble, and dread at His beck."[5] And this, too, is a characteristic of the saints, who, therefore, are said in countless passages of Scripture to fear God, to fear His greatness, His justice, His anger; but I cannot recall a passage in which they are said to fear His punishments, except, indeed, in the sense of acknowledging themselves worthy to be punished, as when St. Augustine, in the height of his charity, said, "I fear everlasting fire." This is the fear which Job displays in this passage, when He says, "I have always feared God as waves swelling over me, and His weight I was not able to bear," because, when he contemplated the great

[1] Job ix. 28. [2] Malach. iii. 17.
[3] Job x. 15. [4] Jerem. x. 7. [5] Job xxvi. 11.

power by which God was able to overwhelm him in an instant, even as a boat which is the sport of the tempest, he humbled himself utterly in His mighty presence, he abased and annihilated himself, and declared himself unable to shake off or to bear so great a weight, even as a little boat is unable to do so when overwhelmed by the ocean breaking over it : " His weight I was not able to bear," that is, His power, His might, which is so great that, like an enormous weight, it not only overcomes, but crushes every one on whom it falls. Even our Lord Jesus Christ was not exempt from this fear, nay, it was greater in Him than in any other, for of Him alone is it written that He was full of it—" He shall be filled with the spirit of the fear of the Lord "[1]—because He alone had that fear which is due to Him. He who fears God does not fear Him as good and gracious, but as the rigid chastiser of the wicked. And thus it was that Christ feared God ; not that it was possible for Christ Himself to be punished, but because Christ, as Man, acknowledged that absolute sovereignty which is in God for the chastisement of every one who rebels against Him, and hence He humbled Himself before Him with a feeling of reverence proportionate to that supreme sovereignty. And it may be that this was why Christ said in His Gospel : " Fear ye Him, Who after He hath killed hath power to cast into Hell."[2] He might, had He pleased, have said " fear Hell," since to do so is right. Nevertheless, He chose rather to say, " Fear Him, Who after He hath killed hath power to cast into Hell," to teach us that the chief object of our fear is not the chastisement, but the Chastiser. Do but exercise thyself in the true love of God, and thou wilt find how great will be thy joy in acknowledging Him to be worthy of the greatest fear that is possible : " The fear of the Lord is honour and glory and gladness and a crown of joy."[3] There is nothing in which a holy soul experiences greater consolation than in abasing and annihilating herself, and never is she more sensible of this self-abasement and annihilation than when she pictures herself as standing in the presence of a God Who is able to overwhelm her in an instant, as a shell is swallowed up in a furious sea : " I have always feared God as waves swelling over me, and His weight I was not able to bear."

---

[1] Isaias xi. 3.   [2] St. Luke xii. 5.   [3] Ecclus. i. 11.

FIFTEENTH DAY.

ST. TERESA.

I will cry like a young swallow, I will meditate like a dove
(Isaias xxxviii. 14).

I. Consider first, how eagerly the hungry young swallow in its nest opens its beak and cries to its mother, to show her its need of food. If thou observest well, thou wilt see that of all young birds there is none that, in proportion to its size, opens it so widely. For which reason it certainly may most suitably represent to thee the earnestness with which thou shouldst daily invoke God, when, while reciting thy vocal prayers, thou askest of Him that which will be for thy greatest spiritual good, since this should be the only food which thou desirest: "I will cry like a young swallow." But what avails the labour of the tongue in much asking, if it alone asks? There must be the union of the mind with the tongue: "If I pray with the tongue, my spirit prayeth" (that is "my breath"), "but my understanding is without fruit." What is it then? "I will pray with the spirit, I will pray also with the understanding."[1] While, therefore, thou criest to God like a hungry young swallow, thou shouldst meditate like a thoughtful dove, sighing from the depths of her heart: "I will meditate like a dove." But what does the word "meditate" here mean? It means to consider the thing that thou askest of God, and to endeavour to enter thoroughly into the sense, the force, the object of the words which thou addressest to him, in short, into everything that may make thy requests more forcible. Is it not a cause of great shame that thou hast recited the *Pater noster* for so long a time, and yet hast not arrived at the comprehension of its meaning? If thou wouldst know in a few words the origin of so great an evil, it is when thou recitest it, thou dost indeed cry like the swallow, but dost not meditate like the dove.

II. Consider secondly, that speaking generally, to meditate is simply to think attentively. Hence it is sometimes employed in a bad sense: "He hath devised (*meditatus est*) iniquity on his bed."[2] Now, however, we use the word exclusively in a pious sense. For example, then, there are three ways in which thou mayest think of the petitions of the *Pater noster* which are all day in thy mouth. Thou mayest think of them without

[1] 1 Cor. xiv. 14, 15.   [2] Psalm xxxv. 5.

any kind of consideration of their meaning, which is the mere act of thinking. Thou mayest think of them with a consideration of their meaning, but with the object of extracting from them some ingenious idea, as is done with words which are not sacred, and this is merely to study them. Lastly, thou mayest think of them with a consideration on their meaning which arises not from curiosity, but from the desire to excite within thyself sentiments of devotion, and this is what we now call to meditate. Observe what takes place among the flowers of thy garden : thou seest flies, butterflies, and bees hovering over the same roses, but for very different ends. The flies merely pass from one rose to another, which is like mere thinking ; the butterflies fly to them and rest on them in order to suck their ordinary nourishment from them, and this is like mere study ; the bees fly to the roses and alight on them in the same way, but it is in order to extract thence their sweetest and choicest juice, from which to make honey : and this, thou mayest understand, is to meditate. And so it is that meditation is indeed also study, but not that of the understanding only, but of the understanding and the will. This is what thou shouldst do when thou recitest the *Pater noster :* endeavour certainly to understand as well as possible the deep meaning of the prayers thou art addressing to God, but in order, at the same time, to benefit thy soul by those affections, now of confidence, now of confusion, compunction, or love, from which is formed the choice honey which we call devotion. When thou appliest both thy understanding and thy will to the matter of which thou art treating with God, then mayest thou be truly said to meditate ; just as the dove is said to meditate when she thinks and mourns at the same time : "We shall lament as mournful doves" (*columbæ meditantes*).[1]

III. Consider thirdly, that this kind of study, although it is adapted to the true nourishment of thy soul, may appear tedious to thee, and therefore thou mayest say that there is no good in meditation, that contemplation is far better ; since, on the one hand, the same and even more fruit is gathered from it than would be gathered from meditation, and, on the other hand, this is done without labour, and without giving the soul any of those occasions of distraction or dryness, which are so frequent in meditation, which is very laborious in its nature. But if thou shouldst speak in this way, thou wouldst in truth show thyself but ill versed in the school of prayer, seeing that

[1] Isaias lix. 11.

thou wouldst be in error as to its first rudiments. For what is the difference, in general, between meditation and contemplation? According to the universal opinion, the difference is that contemplation is undoubtedly a meditation also, but a mature, advanced meditation which is no longer made by protracted reasoning, as was formerly the case, but merely by a glance, which, far from giving trouble, infuses the greatest joy, although this joy is greater or less according to the degree of love which has been acquired. But how canst thou expect to attain all at once to the understanding by a mere glance at as a whole, that which thou hast not first taken the trouble to study in its separate parts? It is enough for the bride to hear her Spouse spoken of to be dissolved in sweetness: "He is all lovely, such is my Beloved."[1] But why is this? Because she has first spent much time in dwelling on all His lineaments individually with a special delight in each one: "His head is as the finest gold: His locks as branches of palm-trees: . . . His eyes as doves."[2] And thou wouldst fain have the highest gifts of love in time of prayer without having first laboured greatly in meditation to acquire them! In what a delusion art thou living! No one can deny that the sweet fire of Divine love is enjoyed in contemplation, but this fire is enkindled in meditation: "In my meditation a fire shall flame out."[3] Be not, therefore, ashamed of doing as he did, who said, "I will cry like a young swallow, I will meditate like a dove;" otherwise, when occasions of self-conquest come, thou wilt perceive that the prayer thou delightest in is a plant rich indeed in leaves and blossoms, but not in fruit, because its roots are not deeply struck within thee.

IV. Consider fourthly, that the doctrine which has been here given thee is drawn from the maxims of St. Teresa, the Saint whom all now regard as so great a teacher in the sublimest school of prayer. She appeared in the Church in the last century like a gentle swallow heralding the coming spring. For it was not only in her time, but by her counsels and co-operation that there was a revival of the great Order of Carmel, which after being, as is said, the first to arise, and to become a garden of chosen contemplatives in which God took delight, had, in the course of time, been, as it were, wholly overwhelmed by a cruel winter. Then, having completely finished her work, she disappeared, transformed into a dove, in the likeness of which she was seen by some in her passage,

---

[1] Cant. v. 16.   [2] Cant. v. 11.   [3] Psalm xxxviii. 4.

perhaps to signify the lofty place she was to occupy in Heaven. Now this great Saint, who has given rules suited to every stage of prayer which any one can be in, always herself practised and taught to others that of which I have been speaking to thee, namely, not to aspire to higher flights till thou hast well tried thy wings. As to her own practice, she always cried out like a humble swallow from her nest, acknowledging her own misery, and imploring the Divine mercy; and she also meditated like a dove, for it was her usual custom to begin her prayer by meditating on some point of the Passion, according to the wise counsels she had received on this subject from a holy man, and then she freely left her soul in the hands of God, like a vessel which is put out to sea by manual exertion, and when once in deep water, is borne along by the wind. And, desiring to teach her daughters the noblest possible form of prayer, she, in her *Way of Perfection*, declared this to be the *Pater noster*, the favourite meditation of so many holy Doctors before her, and of so many after her. Do thou, therefore, choose this Saint as thy advocate, to teach thee how to practise both these profitable parts of which we have been speaking: that of the swallow, so as to commend thyself fervently to God, and that of the dove, so as to meditate carefully at the same time. And since there is no better means of doing both these things than this prayer, the *Pater noster*, I am going to propose it to thee as the subject of meditation for several mornings, according to the simplest and most profitable interpretations which I have gathered, after examining, if I am not mistaken, most of the books which treat professedly of it. And this I have done in order that thou mayest always remember, when reciting it, that these two things are necessary for saying it well—an ardent desire, and loving attention: "I will cry like a young swallow, I will meditate like a dove."

SIXTEENTH DAY.

Thus, therefore, shall you pray: Our Father Who art in Heaven, &c.
(St. Matt. vi. 9).

I. Consider first, that if the son of the king to whom is committed the direct administration of the kingdom, were, with his own lips, to dictate to thee the petition which thou wert to present to the king, his father, it is very certain that thou wouldst not go in search of any other more suitable for the attainment of thy desire. Now this is precisely what that famous prayer is which we commonly call the *Pater noster*, and which I now wish to propose for thy profitable meditation. It is a petition to be presented to God the Father, a petition taught by Christ with His own mouth : by Christ, Who is not only the Son of so great a King, and the Son charged with the direct administration of the Kingdom, but a Son Who acts as our Advocate with the King ; an Advocate so loving that we cannot doubt that He desired to teach us how to act aright, and so wise, that neither can we fear that He was unable to do what He desired. Think, then, whether there can be any petition more sure of success than this. And yet how often and eagerly dost thou affect others, to the neglect of this, which surpasses them as far as the ocean surpasses all rivers, even those which came forth from Eden ! If this is thy case, thou deservest to hear Christ say to thee : "You transgress the commandment of God for your tradition."[1] Stir thyself up, then, to the constant use of this prayer, so as to be able to use it aright ; prepare thy mind to understand its value by first looking at it in general, as one does on first entering a splendid palace, and then going on to examine it in detail.

II. Consider secondly, that the first thing required to make a prayer efficacious in attaining its object is, that that object be a right one : "Prayer is the asking of God those things that are right." For if unreasonable or foolish petitions are not presented even to an earthly sovereign, how much less should they be addressed to the King of Heaven ? Now, this prayer of our Lord is a prayer that is right in every way ; for there are two things which may be rightly asked of God. The first, that He will give us that which is truly good, and this is what is properly termed prayer ; the second, that He will save us from what is truly evil, which may more properly be termed deprecation. And these are the two things which are here asked. As to the good, indeed, we are not to be satisfied with asking of Him our own only, but His also ; nay, His even than our own. And because His good can be only His more extrinsic glorification, it is this which we ask of Him when we say, "Hallowed be Thy name." As to our good, it is of three kinds : heavenly, spiritual, and temporal. The first is to be asked for absolutely, and this we do when

---

[1] St. Matt. xv. 3.

we say, " Thy Kingdom come." The second is to be asked according as it may most conduce to our attainment of the heavenly good. And this we do when we say, " Thy will be done," &c. The third is to be asked in so far as it does not hinder, but furthers our spiritual good, and this we do when we say, " Give us this day our daily bread." Next, as to evil, we should ask God to save us from all which is opposed to the good of which we have been speaking. Now, as to the first, that which belongs to God, it is in danger of no contrary evil, because no one can lessen it in the slightest degree : " If thou sin, what shalt thou hurt Him ? "[1] Nay, God gains glory from the dishonour which is done Him by the reprobate, as well as from the glory rendered Him by the elect ; because His omnipotence is shown at the same time equally in rewarding the latter as in punishing the former. So that, as regards Himself, we do not ask that He may be delivered from any evil, since He is liable to no such fatal necessity. We do but ask Him to save us from the evil which is contrary to our good. And therefore, since sin alone is opposed to the heavenly good (which is the attainment of Paradise), we say, " Forgive us our trespasses." And since temptation is, in its nature, opposed to our spiritual good, we say, " And lead us not into temptation." And since all adversity is, in its nature, opposed to our temporal good, we say, " But deliver us from evil." If, then, thou observest attentively, thou wilt here see the most perfect rectitude in the things asked ; and if this is so, how is it possible to doubt that God will grant them ? " He that speaketh right things shall be loved."[2]

III. Consider thirdly, that for prayer to be sure of its object, it should not only be right, but duly ordered, because prayer is the interpreter of desires, and therefore, what should we think of one who should listen to him who desired more earnestly what is much less to be desired, or who desired less earnestly what is much more to be desired ? See, therefore, how well Christ has arranged the petitions which we should present to God in our prayer. He has done so according to the order which we ought to observe in our desires ; since it is most natural for every one to ask first for what he most desires. And so thou seest that, as regards good, He first makes us ask for that which concerns God, and afterwards for our own. And even with regard to our own good, He makes us ask

[1] Job xxxv. 6.   [2] Prov. xvi. 13.

---

first that which is heavenly, then for the spiritual, and lastly for the temporal. The heavenly is our end, that is, His Kingdom, and therefore it comes first; the spiritual is the means for attaining that end, namely, the fulfilment of His will, and therefore it comes next; the temporal is the help which makes those means easier, that is, our daily bread, and therefore it comes last. In the same way, as to evil, He makes us first pray to be free from sin, which is contrary to spiritual good. And seeing that this is so, oughtest thou not to conceive a great confidence of being heard, since God sees, in this manner, that thou art not only right, but in right order in thy desires? What then canst thou fear? "To the just their desire shall be given."[1] But who can be more just in his desires than the man who not only desires that which ought to be desired, but desires it, too, in the order in which it ought to be desired? "He set in order charity in me."[2] This is the choicest and sweetest melody which thou canst offer to God—the harmony of desires. And is not this just what is meant by a well-ordered prayer?

IV. Consider fourthly, that a prayer, in order to be sure, should be conceived with great confidence; for we all know by experience how much it inclines us to grant just requests when we see that the petitioner has great confidence in our affection, and therefore makes them with courage, tenderness, and brevity, whereas we are disinclined to do a favour to one who asks in a contrary way. And yet we are all an unkindly race. How much more will this, then, be the case with God, Whose chief glory it is to love to do good? "Thy life shall be saved for thee because thou hast put thy trust in Me."[3] Now, see that these requests are made to God with courage, tenderness, and brevity, the three requisites which characterize a petition made with confidence. They are made with courage, which is seen by the expressions we use, *Sanctificetur, Adveniat, Fiat, Da, Dimitte, Ne inducas, Libera,* all of which are such as might seem peremptory, if they had not been dictated to us by Christ, to teach us that one who asks right things of God should never ask them with hesitation, as we do when asking requests of men: "Let him ask in faith, nothing wavering."[4] They are made with tenderness; for since this tenderness has its source in that sweetness of charity which is exercised towards God and man, we here say "Father," to express our charity towards God, and "Our Father," to express our charity towards man; and not only do we add, "Forgive us our debts, as we also forgive our debtors," but we ask for all men that which we ask for ourselves, praying always in the plural, like a full choir singing together. And they are also made with brevity; for all that is asked could not possibly be asked in a more succinct or speedy manner. This, too, shows very great confidence, for the habit of circumlocution which is practised towards earthly sovereigns is a plain sign of want of confidence. For which reason Christ said, in reference to this subject, "When you are praying, speak not much." He did not say "pray not much, do not ask for much," but "speak not much," and "speak not much as the heathens," who believed that they were able to prevail with the gods by their eloquence: "For they think that in their much speaking they may be heard." It is desire, not words, which makes God hear us: "The Lord hath heard the desire of the poor."[1] And a desire may last as long as we please; indeed, if we ought to pray always as Christ commanded, it ought to last always.

V. Consider fifthly, that the confidence which is required in prayer must never be founded on our opinion of our merits, but wholly on the goodness of God. And therefore, in order that prayer may be sure of success, it is requisite, lastly, that it should proceed from a heart filled with the spirit of humility —"The prayer of him that humbleth himself shall pierce the clouds"[2]—because, according to our manner of speaking, it is able to penetrate into the inmost sanctuary of the Most High. And this humility is wonderfully displayed in the prayer which Christ here teaches us. For true humility consists in complete distrust of self, as being altogether miserable, and in expecting all good from God. And whoever uses this prayer manifests this; for not only does he show that he expects every possible good from God alone, but also that it is from Him alone that he expects deliverance from all evil, past, present, and to come, to which humility also makes him take for granted that he is liable. With great reason, then, did our Lord say, "Thus therefore shall you pray," for this is the right way of praying so as to be heard. He did not say, "In these words shall you pray," so as to exclude various other prayers, such as those which are piously recited daily by the Church, who desires to elevate the souls of the faithful by a variety of forms, but He said "thus," to teach us that other forms, in order to be

_____
[1] Psalm ix. 17. [2] Ecclus. xxxv. 21.

efficacious, must all be in conformity with this one, both as
to the nature and order of the petitions, and as to the confidence and humility with which they are made. And therefore
it is the opinion of St. Augustine, that all good prayers must
be able to be resolved into this one, which, however, must
also be esteemed better than others, seeing that it is the rule
for all : "Thus shall you pray." There is no rule fixed for
praising God, because there is no limit to the praises which
are due to Him : "Exalt Him as much as you can, for He is
above all praise."[1]   But there is a fixed rule for praying to
Him, because all petitions must be kept within the limits here
prescribed by Jesus Christ, Who therefore here says, "Thus
shall you pray," but neither here nor elsewhere ever said,
"Thus shall you praise."

SEVENTEENTH DAY.

Father.

I. Consider first, how great a marvel this is, that a vile
man when making his supplication to God, can with truth call
Him Father, and not only can, but ought to do so. It is a
thing that would seem impossible to be done, unless Christ
had commanded it. The priest, therefore, when about to
recite the *Pater noster* at the altar with the people, as a public
minister, always prefaces it by this preamble : "Instructed by
Thy saving precepts, and following Thy Divine institution, we
presume to say : Our Father," &c., by way of declaration that
such a mode of address ceases to be presumptuous when
preceded by a command. Do thou, too, before opening thy
lips to say "Father" to God, awaken in thyself a deep feeling
of confusion at seeing what thou art in regard to God, a vile
worm, hideous, defiled, and sinful : "And now, O Lord, Thou
art our Father and we are clay."[2]

II. Consider secondly, that all men may call God their
Father, inasmuch as they are His work, formed by His hands,
and in His likeness ; inasmuch, also, as they are protected,
cared for, and fed every day by His fatherly love : "Have we
not all one Father ?"[3] But we, who are the faithful, look
higher when we call God Father. We call Him so because of
that great supernatural election, which we possess in our state

[1] Ecclus. xliii. 33.   [2] Isaias lxiv. 8.   [3] Malach. ii. 10.

of grace. Hence, although God is, in the broadest sense, the universal Father of all men, He, nevertheless, gives but common gifts to the rest of men in the world, as Abraham, who was a father, and a rich father, when he sent away his son Ismael, only gave him a loaf of bread and a bottle of water, which he put upon his shoulder. He keeps the inheritance for us, who are His faithful, as Abraham did for Isaac. See, then, with what affection thou shouldst always say this word when thou sayest "Father" to God. And this affection should be two-fold, the affection of a son in the order of nature, and that of a son in the order of grace. In the former character thou owest Him thy whole being, and therefore thou art more bound to be His, with all that thou doest, than the tree, with all its leaves, flowers, and fruits, is bound to be for the advantage of the owner who planted it. And in the latter character, not only dost thou owe Him thy whole being, but His also, which He has begun to impart to thee with the intention of making thee one day altogether like Him in glory, as thou art now like Him in grace. Consider, then, at this point, what feelings should be thine when thou sayest to God, "Father."

III. Consider thirdly, that the blessed saints of the Old Testament were also the adopted children of God, as we are, through the grace which, from the beginning of the world, was given to all those who had faith in the future coming of Christ. And yet those saints very seldom called God their Father, except in the character of Creator. And the reason is, that although they, too, were truly His adopted children, yet they did not venture to call themselves so, because they were still in the condition of servants, just as children when they are little, are subject to a strict tutor, which the Law was to them : "As long as the heir is a child, he differeth nothing from a servant, though he be lord of all."[1] With the coming of Christ, "when the fulness of the time was come," we issued from the state of servitude : we are "not servants, but sons." So that now we not only are the adopted children of God, as those who were before Christ's coming, but we are called so : "They shall be called the sons of the living God."[2] And therefore it was the will of Christ, that as, by His favour, we are free to call ourselves the children of God, so, too, are we free to call God, Father ; and this is what the Apostle also meant when he said : "Because you are sons, God hath sent the Spirit of

[1] Galat. iv. 1.   [2] Romans ix. 26.

His Son into your hearts, crying, Abba, Father."[1] What
thinkest thou, then, of thy state? Is it not one which deserves
to be highly esteemed? Thy state is the same as that of
Jesus, with the difference that He is the Son of God by nature,
and thou art so by adoption. Still, thou also art a true and a
mature son: "I have said you are gods, and all of you the
sons of the Most High."[2]

IV. Consider fourthly, that it is for this reason that Christ
would have us, whenever we say this prayer of our Lord, first
call God our Father in that nobler sense of which we have
just spoken, in order that we may always call to mind the
dignity of our state, and therefore, if we are sons, refuse to
degenerate from it so basely as to behave like servants, as is
the case with so many Christians who are unworthy of the
name they bear. Is it, then, right for such a one as thou art
to pursue madly after the miserable goods of this world, like
the sons of Mahomet or Melancthon? "The prince will
devise such things as are worthy of a prince."[3] It is a far
greater disgrace to thee, who art of so high a rank as to be the
son of God, to stoop so low as to look upon money, fame, or
sinful pleasures, than it would be for the son of an emperor
to collect the filth of the dunghills, to desire the management
of the sewers, or to steep himself in the foulness of putrefaction. And yet what dost thou not sometimes do for the
sake of worldly goods? Thou wilt go so far as to renounce
thy sonship, nay, actually to make thyself the slave of the
devil who offers them to thee, seducing thee with his deceitful
promises, and saying to thee as he said to Christ, the Son of
God by nature, in the hope of deceiving Him: "All these
will I give thee, if, falling down, thou wilt adore me." Why
dost thou not answer him, as Christ did, by an anathema?—
"Begone, Satan!" A son of God to make himself the devil's
slave! Oh, what a monstrous, unnatural insanity! What is
this but going into the country to feed swine? Surely thou
canst not venture to lift thy eyes any more to Heaven to recite
the *Pater noster*, unless thou first cast thyself with the prodigal
son at thy Father's feet, saying to Him with bitter tears:
"Father, I have sinned against Heaven, and before Thee, I
am not now worthy to be called Thy son."[4]

V. Consider fifthly, that our Lord would have us address
God by this name of Father whenever we say the *Pater noster*,

[1] Galat. iv. 6.
[2] Psalm lxxxi. 6.   [3] Isaias xxxii. 8.   [4] St. Luke xv. 21.

in order that we may call to mind not only the dignity of our
state, as has been just said, but also those deep obligations by
which we are bound to behave to God as sons. These may
be reduced to five : namely, to love Him, to honour Him,
to obey Him, to imitate Him, and lastly to submit to His
paternal correction : " Children, hear the judgment of your
father, and so do that you may be saved."[1] The first obligation is to love Him : " With all thy strength love Him that
made thee."[2] The obligation is fulfilled chiefly by the heart.
It is true that it can only be fulfilled in one way, that is, by
loving God for God, which is the love of a son ; it is not
fulfilled by loving Him for the gifts which we hope for from
Him, for that is the love of a hireling. The second obligation
is that of honouring Him : " If I be a Father, where is My
honour ?"[3] This obligation is chiefly fulfilled by words, that
is by words of praise, respect, and reverence towards God :
" The sacrifice of praise shall glorify Me."[4] It is true that
the honour that is pleasing to God is not that which is merely
exterior, but that which is both exterior and interior. Otherwise, what honour is it ? Not that which is paid by a son to
a beloved father, but by a courtier to a prince : " This people
glorify Me with their lips, but their heart is far from Me."[5]
The third obligation is that of obeying Him : " Thou shalt be
as the obedient son of the Most High."[6] This obligation is
fulfilled chiefly by actions, because it consists in the diligent
practice of His commandments : " I will do all things, father,
which thou hast commanded me."[7] It is true that this, too,
can only be fulfilled in one way, that is, by obeying from love.
He who obeys for fear of punishment obeys as a slave, not
as a son. The fourth obligation is that of imitating Him :
" Thou shalt call Me Father, and shalt not cease to walk after
Me."[8] This obligation can only be fulfilled by the whole man,
by heart, words, and works together, because it consists in
aiming at doing whatever is done for the love of God, with
the greatest perfection possible : " Be you therefore perfect,
as also your Heavenly Father is perfect."[9] Lastly, the fifth
obligation is that of submitting to His paternal correction :
" Persevere under discipline. God dealeth with you as with
His sons, for what son is there whom the father doth not
correct ?"[10] This is fulfilled by patiently accepting the chastise-

[1] Ecclus. iii. 2.　[2] Ecclus. vii. 32.　[3] Malach. i. 6.
[4] Psalm xlix. 23.　[5] Isaias xxix. 13.　[6] Ecclus. iv. 11.　[7] Tobias v. 1.
[8] Jerem. iii. 19.　[9] St. Matt. v. 48.　[10] Hebrews xii. 7.

ments which God sends us, such as poverty, sickness, disgrace, temptations, and by being convinced that He really sends them for our good: "He that loveth his son frequently chastiseth him, that he may rejoice in his latter end."[1] This is the conduct of a son, but to murmur is the conduct of a rebel: "In vain have I struck your children, they have not received correction."[2] This, therefore, is what Christ would have thee remember whenever thou sayest "Father" to God. He would have thee remember all the obligations thou owest Him as a son; but especially that in which thou art most wanting. What manner of son, therefore, wouldst thou be to God, if, unhappily, thou shouldst discover thyself to be wanting in all?

VI. Consider sixthly, that Christ has commanded us to call God Father in this prayer; because, as very great things are asked in it, as thou wilt see in time, He desired by these means to encourage our hearts by an infallible certainty of obtaining them. And where is there even an earthly father, who does not take pleasure in gratifying his son in what is right? How much more, then, the Heavenly Father Who is so much greater and better than they, that in comparison none of them deserves even that we should call him father: "Call none your father on earth, for One is your Father, Who is in Heaven."[3] But this is the principal reason which ought to excite in thee a very great confidence in God, the knowledge that thou belongest to Him as the effect does to its cause: "And now, O Lord, Thou art our Maker, and we all are the work of Thy hands."[4] Therefore, just as a statue, if it had reason, would count upon every possible good from the excellent sculptor who made it, as also every picture would from its painter, every palace from its architect, every utensil from its artificer, much more may we certainly count upon every good from God: "Cannot I do with you as this potter, O house of Israel, saith the Lord?"[5] I say much more, because other agents may be prevented in numberless ways and by various defects from successfully accomplishing their aim, however dear it may be to them. But it is not so with God, for He is liable to no defects whatsoever. Not to impotence, because His hand conquers all things: "The hand of the Lord is not shortened," like the hand of one who is paralyzed or stiffened, "that it cannot save."[6] Not

[1] Ecclus. xxx. 1.　　[2] Jerem. ii. 30.　　[3] St. Matt. xxiii. 9.
[4] Isaias lxiv. 8.　　[5] Jerem. xviii. 6.　　[6] Isaias lix. 1.

to ignorance, for His mind sees all things: "All things are naked and open to His eyes"[1]—"naked," because He sees them from without, like a naked body; "open," because He also sees them within, like a body laid open to anatomy. Neither can there be in Him any defect of perfect good-will, because His Heart loves all: "For Thou lovest all things that are, and Thou didst not appoint or make anything hating it "[2]—"appoint" being said of the decree which is called that of intention, and "make" of the decree of execution. If, therefore, God, even when prayer is not made to Him, must yet do good to us merely by reason of His being our cause, how much more will He do so when earnestly entreated? This is the foundation whereon thou hast to rest that hope which is not confounded: the knowledge that God is thy Father by so many titles; and therefore this word "Father" seems to be placed in this prayer as the foundation of the whole of it, and of each one of its parts, just as though it were repeated in every petition: "Father, hallowed be Thy name; Father, Thy Kingdom come; Father, Thy will be done," and so on. It is this word "Father," I repeat, which is the key-note of the whole.

EIGHTEENTH DAY.

Our Father.

I. Consider first, that the only son of his father has hopes of obtaining from him much more in proportion than one who has numerous brothers. But do not suppose because this is true of an earthly father, that it can be so of the Heavenly Father. The number of God's children may be so great as to surpass that of the sands of the sea, yet, for all that, not one of them has any reason to hope less for himself on that account, because He is a Father Who is abundantly bountiful to all: "If the number of the children be as the sand of the sea, a remnant shall be saved."[3] Do not, therefore, be discouraged when thou hearest that in this beautiful prayer of our Lord thou art not to say to God, "My Father," like an only son, but "Our Father," as one of many brethren, because, notwithstanding, He is entirely ready to hear thee, as though He were the Father of none of these, but of thee only.

[1] Hebrews iv. 13.   [2] Wisdom xi. 25.   [3] Romans ix. 27.

Indeed, He will hear thee more gladly when thou sayest to Him "Our Father," instead of "My Father," because it is a mark of thy not doubting His power, as so great a Father, to do good to all, while doing it to thee—nay, it is a declaration on thy part that He thinks of all, provides for all, feeds all, and takes equal care of all: "He hath equally care of all."[1] This is the first reason why Christ would have us, His faithful, here say "Our Father," and not "My Father," that we may show that high esteem of our Father, which all the rivers would have of theirs if they were able to say to the ocean, "Our Father." Is it this kind of esteem which thou showest when thou sometimes fanciest within thy heart that God does not think particularly of thee, because He has, at the same time, so many others to think of. This is to fear that His Heart is less wide than the ocean, which is as much bound to consider one of the countless rivers great or small, which are derived from it, as to consider them all together.

II. Consider secondly, that, in the next place, Christ would have us here say "Our Father," not "My Father," to give us an opportunity of remembering that we are brethren, vying with each other, as it were, in doing good to one another. Thou, when thou settest thyself to pray, art more ready to pray for thyself alone than both for thyself and others. Nay, when thou prayest for thyself it is with great feeling and fervour, and when thou prayest for others it is, as a rule, but languidly. But this is a serious error. Is it that thou art afraid of being the loser if thou prayest also for others, and not for thyself only? On the contrary, then is the very time that thou gainest most for thyself. For when thou prayest only for thyself it is possible that thou mayest be influenced merely by self-love. But whenever thou prayest also for others, and especially for such as are not bound to thee by any tie but this of Christian brotherhood, it is certain that thou art then influenced by pure charity, and therefore, by thus making thyself dearer to God, thou preparest thyself also to receive from Him more abundantly that which thou askest of Him in one brief moment for thyself, for which reason the Apostle said to his Roman converts, "God is my witness . . . that without ceasing I make a commemoration of you always in my prayers." See how important the Apostle thought this point—he went so far as to swear it. Unless, indeed, he did so, because to pray earnestly for others is so

[1] Wisdom vi. 8.

rare amongst men as hardly to be believed by any one without an oath. Further still, by praying for others together with thyself thou also showest greater love for God than by praying for thyself alone, because it is a sign that thou desirest the number of those who serve Him to be great ; and by praying for others thou also honourest Him more, because thou showest thy esteem not only for Him, but for all who bear His likeness. Thou also obeyest Him more, because thou showest care not only for thyself, but for all those who are commended to thee by Him. Thou also imitatest Him more by displaying a love like His own which pours itself like a golden shower abundantly over all. How much more then dost thou merit by praying for others together with thyself? Thou hereby provest thyself God's true son : " Be ye therefore followers of God, as most dear children " (and such are those who most resemble their Father), "and walk in love."[1] Do not wonder, then, that Christ would have thee here say "Our Father," not "My Father." He would have every one prepare himself to obtain his requests with greater ease, by practising so many beautiful acts of virtue as these are which he offers to God as perfumes collected in a mass of heavenly sweetness.

III. Consider thirdly, also, that by making this prayer thus universal for all, Christ has taken away all pride from men, for what noble or prince could ever despise any one in the world if he remembered that we are all children of one Father ? " Hath not one God created us ? Why, then, doth every one of us despise his brother ?"[2] He has taken away envy, because every one thus endeavours to promote the good of others as his own. He has taken away inequality, because every one endeavours also to obtain as much good for others as for himself. He has taken away enmity, because how can one who has not first acknowledged his neighbour as his brother offer for him such sublime petitions as these are without being rejected by God as an impudent liar, having honey on his lips and poison in his heart ? "They blessed with their mouth, but cursed with their heart."[3] And lastly, He has established a means of wonderful power for taking Paradise by storm, since He has united together the forces of all His faithful into one body. A great number of feeble soldiers, if they were to fight singly, would be despised, but when collected together, they become formidable : "The children of Israel, pursuing in one body, defeated all that they

---

[1] Ephes. v. 1, 2.　　[2] Malach. ii. 10.　　[3] Psalm lxi. 5.

could find."[1] And therefore Christ would have the faithful, when praying daily together, not pray each for himself, but all for each and each for all—"Pray one for another, that you may be saved "[2]—so that the attack made upon Heaven may be of immense power : "If one fall he shall be supported by the other."[3] Dost thou then not value or practise so beautiful a method of prayer ? "Curse ye the land of Meroz, said the angel of the Lord, curse the inhabitants thereof, because they came not to the help of the Lord, to the help of His most valiant men."[4]

IV. Consider fourthly, that although the mutual help which we give one another by praying in this way is of immense virtue in obtaining every possible good that we can ask of our Heavenly Father, yet doubtless the mighty help which our Elder Brother, Jesus Christ, is pleased to give us every day by making Himself our most loving Advocate is of far greater virtue, since He of Himself alone is able to do far more than all of us together. Now we merit this help of Jesus Christ much more by saying to God in this prayer of our Lord, "Our Father." And why? Because by so saying we show to Christ a beautiful act of reverence, respect, and worship by leaving it to Him to say "My Father." Strictly speaking, Christ only has the right to say "My Father" to God : "My Father, if this chalice may not pass away, but I must drink it, Thy will be done."[5] And the reason is, that this is the privilege of the Only Begotten, Who alone can always say in the house, "My Father." Where there are many brothers it is right for them to say to Him "Our Father," particularly when speaking to Him all together. But Christ is the Only Begotten of God the Father, and therefore He alone can speak as such : "He shall cry out to Me, Thou art My Father."[6] We are not even second or third begotten, because, as St. Augustine observes, we are not begotten at all—we are created, and therefore we ought to say to God, "Our Father." For, though we are adopted into the very same sonship which belongs to Jesus Christ, it is through Him we are thus adopted. He gave us that wonderful example of the great and only Son in His Father's Kingdom, desiring to have many brethren as co-heirs in that Kingdom, and therefore asking His Father to adopt us so that we might sit with Him on His own throne. But this very fact is a reason

---

[1] Judith xv. 4.   [2] St. James v. 16.   [3] Eccles. iv. 10.
[4] Judges v. 23.   [5] St. Matt. xxvi. 42.   [6] Psalm lxxxviii. 27.

why every one of us should, out of gratitude to so good a Brother, leave to Him alone that great honour which He merits by nature, of saying to God, "My Father," and not seek to claim it, even by grace; especially since we cannot, being many, speak of ourselves as God's only sons, even in the order of grace. It is not, however, forbidden to thee, when praying privately in thy chamber, ever to say to thy Father, "My Father," by an impulse of affection, just as a son may do so in a house where there are many brothers. But, in the case we are considering, remember that thou never canst do so by right, on account of the deep reverence due to Jesus Christ, Who always said, "My Father," when speaking to God, and when speaking of Him to men always said, "My Father," or else "Your Father," but never "Our Father." Whenever, therefore, thou, a vile worm, desirest to say to God, "My Father," I would point out to thee that thou shouldst, as it were, ask leave of Christ each time to be free to do so, and thus treat Him with that respect and reverence which are due to Him as the Only Begotten. For to whom among men (with the exception of Christ alone) "hath God said at any time : Thou art My Son, to-day have I begotten Thee?"[1]

NINETEENTH DAY.

Who art in Heaven.

I. Consider first, how evident it is that so noble a Father as has been described in the last two meditations cannot possibly be an earthly one. Nevertheless, in order to make a more marked distinction between Him and all earthly fathers, we must add, after saying "Our Father," these words, "Who art in Heaven." And for what reason? Is it in order to gain His good-will (as men sometimes do) by so fair and magnificent a preamble? Surely not : such arts are altogether superfluous in addressing the Father. We are to do this in order to stir ourselves up to remember that we are speaking to a Heavenly Father, and that therefore we should not ask of such a Father anything earthly, at least not as the final object of our prayers, but only that which is worthy to be asked of so great a Father. "Seek the things that are above, . . . mind the things that

[1] Hebrews i. 5.

are above, not the things that are upon the earth."[1] Dost
thou not think that thou wouldst be doing a great dishonour
to the sun, if it were endowed with reason, to ask of it wild
berries, rushes, and other wild and common plants? It would
certainly be right to acknowledge them as coming from it,
seeing that they are, in their way, of use to men. Nevertheless,
supposing prayers to be addressed to the sun, thou wouldst
ask for flowers, corn, gold, pearls, rubies, and choice diamonds,
which are more properly its gifts. And so, although all gifts,
including those which are temporal, come from God, nevertheless, if thou wouldst behave to Him in a manner becoming
His dignity, thou wilt not ask of Him those goods which dogs
and horses, if they could speak, would ask of Him. Thou
shouldst ask of Him those only which it is His glory to give,
namely, spiritual goods, because the others, if it is well for
thee to have them, will be given to thee by Him without the
asking : "All these things shall be added unto you."[2] In
what manner, then, dost thou pray to this great Father?
Dost thou behave to Him as to a Heavenly Father, which He
is ? or dost thou venture to ask frivolous things of such a
Father, as if thou wert a pagan—foolish things of so wise,
wicked things of so holy, a Father? This would be doing
Him a greater wrong than thou wouldst do to a king if thou
wert earnestly to beg him to cover thee with filth.

II. Consider secondly, that it would have been sufficient
to remind us that we should ask of a Heavenly Father those
goods only which are suited to His dignity, if when praying
to Him, we were to add "Heavenly" to the word "Father,"
that being His usual title : "Behold the birds of the air, . . .
your Heavenly Father feedeth them."[3] Nevertheless, it was
the will of Christ, that here, instead of saying "Heavenly,"
we should say "Who art in Heaven." And why? That we
may lift up our souls more vigorously above this low part of
the world in which we are living, and may soar, as it were,
instantaneously to that higher part which is the empyrean
Heaven, above which we are wont to represent God to ourselves as dwelling in His royal palace : "To thee have I lifted
up my eyes, Who dwellest in Heaven."[4] Every one knows
that God is present everywhere alike : "Whither shall I go
from Thy Spirit?" He is in the plains, the mountains, the
seas, and the lowest depths of the earth : "If I descend into

[1] Coloss. iii. 1, 2.
[2] St. Matt. vi. 33. [3] St. Matt. vi. 26. [4] Psalm cxxii. 1.

Hell Thou art present."[1] Still, more strictly speaking, it is
said that He is in Heaven—" He that dwelleth in Heaven
shall laugh at them "[2]—for just as our soul, although it is
in every living part of our body, no matter how mean, yet
is said to be, in an especial sense, in the heart and in the
head, because it is there that it performs its noblest operations :
in the heart, as the principle of animal ; in the head, as the
principle of intellectual life. So, although God is always
altogether present in every part of the universe, however low,
He is nevertheless more properly said to be in Heaven—" He
that dwelleth in Heaven "—in the ethereal and the empyrean
Heaven, because it is there that His chief operations are
performed : in the ethereal Heaven as Lord of the natural,
in the empyrean as Lord of the supernatural order : " His
dwelling is above."[3] Our soul, indeed, is contained in the
body, as in a receptacle, but God is not contained : on the
contrary, He contains in Himself even those vast spaces in
which, according to our gross manner of thinking, He is
contained, and which He far exceeds by His more vast immensity : " Thy magnificence is elevated above the heavens."[4]
This, then, is what thou hast to aim at, first of all, whenever
thou beginnest to pray, namely, to lift thy mind from the
earth by a lively faith, raising it as high as thou art able, not
only to those regions whence thy great Father, as Sovereign
of the natural order, sends down all those pure and favourable
influences which flow upon us from the skies ; but also to those
where, as Sovereign of the supernatural order, He makes blessed
all the angels, archangels, and elect souls, who encircle Him
like a glorious crown, for it is this Heaven which is the abode
prepared for thee, also, by thy good Father, if thou dost but
desire it. This is why, whenever Christ prayed, He, too, was
wont to lift up His eyes to heaven—" Lifting up His eyes
to heaven, He said, Father, the hour is come "[5]—to teach
us how much more we miserable creatures, when we pray,
should remember that our Father listens to us from above,
and so should detach our minds from earth when we are
addressing Him, since at present we cannot so detach our
bodies. This is the first reason why our Lord would have us
here say, not "Our Heavenly Father," but " Our Father Who
art in Heaven," in order to excite our faith more vividly at
the beginning of our prayer by the thought of His local

[1] Psalm cxxxviii. 8.　[2] Psalm ii. 4.
[3] Deut. xxxiii. 27.　[4] Psalm viii. 2.　[5] St. John xvii. 1.

presence, to believe that God is in Heaven hearing us, as our beloved Father, on His royal throne : " The Lord's throne is in Heaven."[1]

III. Consider thirdly, that Christ would have us here say "Our Father Who art in Heaven," rather than "Our Heavenly Father," not only in order to excite in us a more lively faith, as has just been said, but, also, together with faith, to excite in us the hope which is so very necessary to one who prays. It is certain that hope, by its nature, always tends to those things which are difficult, high, and great, since we do not hope for things that are easy : "What a man seeth, why doth he hope for ?"[2] We rather consider them as already ours. As, then, in this sacred prayer of our Lord, thou art about to ask great things of thy Father, it will help thee immediately to think of Him as in the highest heavens, because thou wilt then at once understand that by merely thence holding out His hand to thee He is able to raise thee to a very lofty place with Him : " He sent from on high, and took me, and received me."[3] And do not say that, being so high above thee, especially among so many other men, and so many, too, who are greater than thyself, amongst whom thou art living ; since this, on the contrary, is a ground for hope that He will not lose sight of thee, because His dwelling is on high : " He dwelleth on high, and looketh down on the things . . . on earth."[4] Just as, because the sun is so high in the heavens, there is no one in the world who can fear that he will not share in his favours, every one enjoys them as much as though he were alone in the world. He looks down as attentively on a little flower as on innumerable palm-trees, cedars, cypresses, and planes, amongst which it is more lost than a pigmy would be in a nation of giants : "The sun, giving light, hath looked upon all things."[5] Why, then, shouldst thou, poor and mean as thou art, fear lest God should not discern thee in the crowd of more important persons? He discerns thee more clearly even than the sun discerns the flower amongst numerous tall trees. And, just as the sun is not prevented from communicating himself wholly to the little flower by communicating himself to all the countless thousands of plants which grow on the earth, as though that flower were the only object he had on which to pour down the light of his beams, so, too, is it with regard to God and thyself, provided only that thou

---

[1] Psalm x. 5.   [2] Romans viii. 24.
[3] Psalm xvii. 17.   [4] Psalm cxii. 5.   [5] Ecclus. xlii. 16.

interposest no obstacle between thyself and Him. And therefore, when thou sayest to Him, "Who art in Heaven," have great confidence, for it is not without reason that He would have thee think of Him as on high, and not as enclosed within a temple or shrine, as was once the case with the foolish Jewish people, who consequently supposed that they were obliged to seek Him in the Tabernacle whenever they wanted to pray. Thou mayest always find Him in Heaven—"He hath set His tabernacle in the sun"[1]—in a place which is manifest, vast, and lofty; where He hears thee wherever thou mayest be, in plains, mountains, lakes, seas, gardens, or woods. All that thou hast to do is to call upon Him: "I will cry to God the Most High."[2] For being, as He is, not in the sun, but infinitely higher than the sun, there is no more danger of His not having His eye upon thee everywhere, than there is of the sun not shining upon thee: "The Lord hath looked from Heaven; He hath beheld all the sons of men."[3] Further still, just as it follows, from God's dwelling being so high, that He easily sees all that He chooses, as we do from a lofty tower, so also does it follow that He has power over all. And why? Because no one has authority over Him. Thy Father is "in Heaven," and certainly He is there as its ruler. Why, then, dost thou fear the fatality of adverse influences, like the heathen, who consequently thought all prayer useless? The very contrary is commanded thee: "Be not afraid of the signs of heaven, which the heathens fear."[4] Thy great Father is where He holds under His dominion all those causes which we call favourable, all the intelligences, spheres, stars, and lower powers; and therefore none of them can withstand the execution of His Divine decrees, if it is His will to save thee: "All things are in Thy power, and there is none that can resist Thy will if Thou determine to save Israel," said Mardochai in his affliction.[5] And just this, only more briefly, dost thou say in the words, "Our Father, Who art in Heaven."

IV. Consider fourthly, that, besides faith and hope, this form of words is calculated also to excite charity within thee, for it is impossible that when, with a little reflection, thou here expressest what that place is in which thy great Father dwells, thou shouldst not infinitely rejoice in the happiness which is so justly His. It is never said that a king is in any city in which he is living *incognito;* it is only said of one where he is

[1] Psalm xviii. 6. [2] Psalm lvi. 3.
[3] Psalm xxxii. 13. [4] Jerem. x. 2. [5] Esther xiii. 9.

known, loved, valued, and courted by a loyal people, as he is, above all, in his metropolis. Doubtless, thy Father is the Universal King of the whole world, and He is so everywhere, on earth as much as in Heaven. But on earth He may be said to live *incognito*, so little does He here receive of the homage due to His sovereign Majesty ; and therefore it may almost be said that He is not here. Where, then, is He? In Heaven above, where He is really honoured as He deserves ; where "all" know "Him, from the least of them even to the greatest."[1] When, therefore, thou sayest to Him, "Who art in Heaven," what oughtest thou to understand by this word "art"? Thou oughtest to understand "art known, art loved, praised, glorified, and exalted." How greatly shouldst thou rejoice in so saying! It is true that, considering the great distance there is from the earth in which thou art living to Heaven, there will come over thee a desire for the wings of the dove, so that thou mightest fly thither to thy beloved Father ; and because this is impossible, thou wilt grieve and say in thy heart, "Who will grant me that I might know and find Him, and come even to His throne?"[2] But do not be troubled at this, for it is only the effect of charity, which will therefore all the more move God to hear thy prayer. A little child, who is the son of a great king, and who sees his father sitting on a splendid throne, would fain climb the steps leading to it, so as to be clasped in his arms, but not being able to do so, he begins to weep. And these very tears bring consolation, for they compel the father to come down from his throne to embrace him. So will it be with thee. Those tears which thou sheddest at seeing thy Father so high and thyself so low will draw Him down lovingly to thee, to unite Himself to thee till the hour comes when He will call thee to Him, as a son of mature years, to sit with Him and reign with Him on His own throne.

————

TWENTIETH DAY.

Hallowed be Thy Name.

I. Consider first, that, as thou callest thy God "Father" in this prayer, thou art bound to show, in all the petitions that thou addressest to Him, that thou behavest like a true

[1] Jerem. xxxi. 34.　　　[2] Job xxiii. 3.

son. But what can a wise, respectful, and loving son first of all desire, save what is for the honour of his father? And so thou, too, art to ask, in the first place, of thy Heavenly Father that only which is for His honour: "Hallowed be Thy Name." And this is the very noblest petition that we make in this sublime prayer of our Lord. When making it, therefore, let us strip ourselves of all self-interest, let us love God for His own sake, not for any advantage to be gained by loving Him. This petition is placed at the very beginning, to teach us that it must be the final end of all the rest which follow. If we pray to God that His Kingdom may come, that His will may be done, that He will give us our daily bread, forgive us our sins or keep us from them, and lastly, deliver us from every evil—what is the final end for which we ought to ask all these things? Is it our own good? Certainly not; it is His. This is the conduct of a son, not to act like marshes having their source in the sea, which slothfully keep, to swell themselves, all the water they receive from it; but like rivers, which give it all back as a tribute. Thou seest, therefore, that to make this petition aright would require the heart of a seraph, who loves God for His own sake, and only takes delight in loving Him because his doing so redounds in the end to the honour of God. Thou art not a seraph, but thou canst compel thyself to be one: and how? By sending this petition up to God like a dart from thy soul, in all thy daily actions. "Hallowed be Thy Name:" this is the dart of love, which, whether it bear to God the offering of the most precious or the most insignificant thing that thou hast, is equally pleasing to Him: "Thou hast wounded My Heart" (in the same manner) "with one of thy eyes," that is, with something of great price, "and with one hair of thy neck," that is, with something of no value.[1]

II. Consider secondly, that it is certain that God is not capable in Himself of receiving any good, because He is rich in all things. There is only one way in which He can receive it outside of Himself, and that is, by increase of His glory, which, as it was the final end for which He made the world, according to the words, "Every one that calleth upon My Name, I have created him for My glory, I have formed him, and made him,"[2] that is, "created him" by the creation of the soul, "formed him" by the formation of the body, and "made him" by that noble combination resulting from the

---

[1] Cant. iv. 9.   [2] Isaias xliii. 7.

union of soul and body; so, too, is it His will that this
should be the final end of everything that we do, just as every
artificer justly chooses His glory to be the final end of all the
good which his work produces to others. Ought we, then,
never to do anything for our own glory? God forbid that
we ever should. All must be to the greater glory of God:
"To Thee, O Lord, justice," that is, glory; "but to us confusion of face."[1] And it is this glory, which is so justly due
to Him as to be called "justice," that we here ask of Him,
because He only can cause it to be rightly given to Him. We
do not, therefore, ask it by the name of glorification, though
we might very properly do this, but by that of sanctification,
because that is the glory most pleasing to God: "Holy, Holy,
Holy, Lord God of hosts." And when this is said by all on
earth as it is in Heaven, there is nothing more that can be
said: "All the earth is full of His glory."[2] Now, it must be
premised that this expression "to sanctify" has two meanings
in Scripture: the first is to make holy, the second to treat as
holy. In the first sense it is said that God sanctified the
Sabbath: "He blessed the seventh day, and sanctified it,"[3]
because He set it apart for Himself. In the second sense
it is said that God commanded us to sanctify that day—
"Remember that thou keep holy the Sabbath-day"[4]—because
He caused it to be reverenced as His day. Now, God's
Name cannot be sanctified in the first sense, because He
cannot be holier in Himself than He is, that is, infinitely
holy: "Holy is His Name."[5] It can be sanctified only in the
second sense. And in what way? In the same way as He
formerly would have the Sabbath, and as He now would have
the Sunday and all other days that are consecrated to God,
sanctified, that is, first, by refraining from profaning them by
servile, unbecoming, or wicked works, which is a merely negative sanctification—"Keeping the Sabbath from profaning it"[6]
—and, secondly, by various holy acts of religion, which is
positive sanctification. So, too, when we beg of God that
His holy Name may be hallowed, we first ask that He will not
suffer His Name to be profaned, that is, dishonoured or
mocked, as is done by so many unbelievers, who give that
Name to stocks and stones, and even to the foulest fiends of
Hell, nay, as is done by many of the faithful even, who blas-
pheme like devils; and then we beg of Him that His Name

may also be honoured by acts of religion, and above all, of immense adoration, love, and praise. Thou seest, then, how much more properly we address God when we say, "Hallowed be Thy Name," than if we said, "Praised, manifested, magnified, glorified be Thy Name." We say all in one word, and in those terms, also, which are most pleasing to God: "Sing to the Lord, O ye His saints, and give praise to the memory of His holiness."[1]

III. Consider thirdly, why it is that we do not here beg of God that He may be hallowed, but only His Holy Name may be hallowed. Ought it not to be desired that He should be glorified, in the ways that have been mentioned, still more in His Person than in His Name? Surely it ought: "The Holy God shall be sanctified."[2] But he who desires that God should be glorified in His Name, shows by that very desire that he also wishes Him to be glorified in His Person. And yet we do not here say, "Hallowed be Thou, Father," but "Hallowed be Thy Name," because a good son cannot endure to see his Father dishonoured, not only in His Person, but not even in the name He bears: "I will glorify Thy Name for ever."[3] Another reason may be that the praise which is rendered to every one usually corresponds to the name or reputation he bears. If he has the reputation of being munificent, he is praised for munificence; if of being gentle, he is praised for gentleness, and so on. Therefore, when we desire of God that His Name should be glorified in Him, or He in His Name, we do not merely desire that He may be glorified, but that He may be glorified according to every name that He possesses, whether it be Lord, or powerful, or provident, or just, or good, or gracious, or holy, and so on without end: "Let them know that the Lord is Thy Name,"[4] or that "powerful, provident, &c., is Thy Name." Hence the Psalmist was not satisfied with saying, "O ye children of God, bring to the Lord glory," but "glory to His Name," because he desired that God might be glorified according to every name belonging to Him: "According to Thy Name, O God, so also is Thy praise unto the ends of the earth."[5] It is true that, for all this, we ought not here to add any of these titles of God, powerful, provident, and the rest, but simply to say "Thy Name." And why? Because it ought to be enough to know that any name is God's, for us to desire that it may be

---

[1] Psalm xxix. 5.   [2] Isaias v. 16.
[3] Psalm lxxv. 12.   [4] Psalm lxxxii. 19.   [5] Psalm xlvii. 11.

glorified. When thou desirest that God may be glorified according to any of these names, powerful, provident, and the rest, it is possible that thy wish may have reference to the benefits which, as such, He has conferred on thee personally. But in this petition thou oughtest altogether to forget thyself, and so say to God, "Hallowed be Thy Name," that is, "because it is Thy Name," without thinking of anything else that might be added. This is to act like a son who knows that he loves his Father and his Father's Name as he ought to do: "All they that love Thy Name shall glory in Thee:"[1] not "in Thy gifts," but "in Thee."

IV. Consider fourthly, that loving sons not only desire most ardently that the name of their father may be glorified, but desire also that they themselves may glorify it above all others: "I will declare Thy Name to my brethren."[2] It might seem, therefore, that in order to show thyself a loving son to thy Father, thou oughtest to say, not, "Hallowed be Thy Name," but "may I," or at least, since thou art praying with others, "may we hallow Thy Name." But thou art grievously mistaken in so thinking. Thou oughtest always to say, "Hallowed be Thy Name." And why? Because this shows thee to be a loving son. It is quite true that such a son ought to desire to be the one who, more ·than all others, shall give glory to his Father, but he ought not to desire this above all things. What he ought to desire above all things is that his Father may be glorified, whether by himself or by others: by himself, if so great a favour may be granted him; if not, then, at least by others. Therefore it is not true that thou shouldst rather say to God, "may I," or "may we hallow Thy Name." Thou shouldst say precisely as is said here, "Hallowed be Thy Name," to show that the thing that thou askest is that which thou desirest above all things. And dost thou not know that it is impossible for thee to give glory to God in any special manner without its redounding to thy own great honour? "The glory of a man is from the honour of his father."[3] How easily mayest thou be grossly deceived by self-love, so as to seek thyself even while seeking thy Father's glory. How frequently, indeed, is this the case! Thou wouldst be the only one in the world to glorify God; the first to bring souls to true penance; the first in preaching, expounding, instructing, governing; the first to gain abundant palms of honour: thus acting like the disciples who, while yet rude

[1] Psalm v. 12.   [2] Psalm xxi. 23.   [3] Ecclus. iii. 13.

and untaught, desired to be alone in giving glory to the Name
of Christ by casting devils out of men's bodies by virtue of it :
" Master, we saw a certain man casting out devils in Thy
Name, and we forbade him because he followeth not with us."[1]
Remember how Christ immediately answered those disciples :
" Forbid him not, for he that is not against you is with you."
Exercise thyself, then, in saying continually to thy God, not
"may I," or "may we hallow Thy Name," but "Hallowed be
Thy Name." It is this which thou art to desire above all
things, not that thou mayest thyself glorify the Name of God,
but that the Name of God may be glorified by all ; and as
thou shouldst desire it, so too shouldst thou pray for it above
all things. It follows, that if, in spite of thy endeavouring to
the utmost of thy feeble strength to promote His glory as
much as all others, thy efforts are unsuccessful, this ought not
to distress or humiliate thee ; rather shouldst thou rejoice that
there are so many in the world younger and stronger than
thyself to supply thy deficiencies, and desire that they may
supply them : " Praise the Lord, ye children, praise ye the
Name of the Lord."[2] It might be thought, however, that it
would be preferable to say, " Hallow Thy Name," rather than
" Hallowed be Thy Name," because it is God only Who can
give to His Name the glory due to it. Nevertheless, it is not
so : for if it is the will of God to be glorified, it is His will to
be glorified by our means also, not by Himself alone ; and
therefore we ought to say *sanctificetur*, which is an abstract
expression including both Himself and us, for as we are not
able to do anything for the glory of God without Him, so
neither will God accept anything from us without our exertions.

TWENTY-FIRST DAY.

Thy Kingdom come.

I. Consider first, that, after the good of his father, every
good son may, nay ought, rightly to think of his own. But of
what kind of good ought he to think first? Surely of that
which he ought chiefly to love and value, which is, without
doubt, his inheritance. It is this which, above all, is due to
him as a son: "If sons, heirs also."[3] And it is this which,

[1] St. Luke ix. 49.   [2] Psalm cxii. 1.   [3] Romans viii. 17.

as a son, he should, above all, endeavour to secure. This, therefore, is the reason why, after having said to our Heavenly Father, "Hallowed be Thy Name," Christ would have us immediately add, "Thy Kingdom come;" because, if it is right that, after thinking of the glory of the Father, we should think of ourselves, there is nothing which we ought so earnestly to desire or seek after for ourselves, as to secure that inheritance in our Father's house which is the attainment of the final end of all His children. Neither is it to be wondered at that we may boldly ask this inheritance of God. For a heavenly inheritance is not like others. If a son in this world desires the inheritance prepared for him by his earthly father, the very fact of his desiring it makes him unworthy of obtaining it, because it is the same thing as desiring his father's death. Not so with the heavenly inheritance prepared for us by our Father above. For this inheritance is the same as the enjoyment of Himself: "The Lord is the portion," that is, the whole portion "of my inheritance."[1] It is to see Him, to be united to Him, to live in Him, and therefore it is the same thing to ask Him to admit us to our inheritance as to ask Him to grant that we may be with Him for all eternity. And is it possible that thou art not ravished by the thought of so magnificent an inheritance? Oh, how fair, how rich an inheritance it is! "My inheritance is goodly to me."[2] Is it not right to be asking for it continually? "Thy Kingdom come." Therefore it is not here said, "Let us come to Thy Kingdom," but "Thy Kingdom come," that is, "May Thy Kingdom come to us," because this is the way in which to speak of an inheritance; we ought never to pursue after our inheritance, but to wait for it to come to us.

II. Consider secondly, that when Christ would have us here pray for what is in substance our inheritance, He would not have us pray for it by that name, but by the name of a kingdom—"Thy Kingdom come"—in order that we may esteem that inheritance as we ought. Do not think that it is an ordinary good that we shall inherit when we inherit the Beatific Vision. We shall inherit a Kingdom that has no equal, for it is that Kingdom which belongs to God only, namely, beatitude: "Heirs of the Kingdom which God hath promised to them that love Him."[3] We cannot imagine greater blessedness on earth than that of a sovereign prince, because the state of a sovereign appears to us to be the only

<hr />

[1] Psalm xv. 5.   [2] Psalm xv. 6.   [3] St. James ii. 5.

one which contains in itself a perfect assemblage of all good. One who reigns has whatever he desires, whether it be money, or company, or followers, or pleasures, or music, or the chase ; for which reason God, with His own mouth, thus described the kingdom when He gave it to Jeroboam : "I will take thee, and thou shalt reign over all that thy soul desireth."[1] But that which we think constitutes the chief blessing of a sovereign is that he is absolute lord of all his people, and deals with them according to his pleasure. It is true, indeed, that this pleasure on earth is very imperfect. For where is there a king, however great, who is not without innumerable good things which he would fain possess, and who does not, moreover, meet with disobedience, stubbornness, rebellion, and countless acts of at least secret disloyalty from his people? To reign truly belongs to Heaven alone ; for we see that on earth even God Himself, although its true and universal King—"God is a great King over all the earth"[2]—nevertheless does not reign so absolutely as not to be subject to flagrant disobedience from great numbers. Nay, do not even His own children band together against Him with Satan, the king of darkness, and make war upon Him every day? He can be said to reign truly in Heaven only, where all the blessed pay Him that complete homage which is rendered to Him nowhere else, not even by any of the just. And He will reign there still more absolutely when, having utterly destroyed the kingdom of the devil, and finally trampled upon all the disobedient and rebellious, He will reign triumphantly with all His faithful children in everlasting peace : "Sion, Thy King shall reign."[3] And this, strictly speaking, is the reign for which we here pray when we say to God, "Thy Kingdom come," that sovereign beatitude in which we shall reign with Him for endless ages above the stars, when, possessing God, we shall possess with Him all that can be desired ; there we shall indeed "reign over all that our soul desires," and shall see not only all our own inferior appetites subject to us, but all the damned and all the devils whom Christ our Judge will put under us in these words : "Come, ye blessed of My Father, possess you the kingdom prepared for you from the foundation of the world."[4]

III. Consider thirdly, that it might seem as though, when asking this Kingdom of our Father we might say, "May our Kingdom come," because if it is, as has been said, the inherit-

---

[1] 3 Kings xi. 37.
[2] Psalm xlvi. 3.   [3] Isaias lii. 7.   [4] St. Matt. xxv. 34.

ance which belongs to us as children of God, we might think
that we had a right to demand as our own that which is
"prepared" for us. But such was not our Lord's will: He
would have us say to God, "Thy Kingdom," not "Our
Kingdom," because although it is true that Paradise will truly
be the Kingdom, not only of our Heavenly Father, but also of
us, His children, nevertheless, if we would ask for it in a holy
manner, it must never be as ours, but as His. This is the
conduct of a son of generous mind—to love his inheritance,
not for his own advantage, at least principally, but because by
means of it he can do greater honour to his father. Therefore,
when thou here sayest to thy God, "Thy Kingdom come,"
thou oughtest to think, above all, of that Kingdom which God
will then possess in thyself, when there will no longer be
anything of thyself left in thee which is in opposition to God,
or which can keep thee from Him, but when thou wilt be all
His in will, imagination, understanding, speech, and every part
of thee, even the least: "The Lord will reign over them in
Mount Sion, from this time, now, and for ever."[1] This is the
chief delight of the blessed in Heaven—not the being kings
themselves, but the seeing that God reigns over them, and
therefore when they thank Christ for the beatitude which He
has won for them by His Blood, they all say to Him in unison:
"Thou hast redeemed us to God in Thy Blood . . . and hast
made us to our God a kingdom and priests, and we shall reign
on the earth."[2] They thank Him first because they are made
a kingdom to God, that is, because God will reign over them
wholly, and afterwards they thank Him because they too have
been made kings, but kings who are also priests, as were all
the kings of the chosen people, that is, kings who should
continually offer to God in golden thuribles the incense of
eternal praise: "Thou hast made us to our God priests, and
we shall reign upon the earth," that is, "priests reigning upon
the earth," reigning over all things which they, together with
God, shall have to all eternity beneath their feet. Thus thou
seest that their first cause of rejoicing is that they are the
Kingdom of God, and the second that they also are to reign
with God. And thou, upon the earth, shouldst observe this
same beautiful order which the saints do in Heaven. They
rejoice incomparably more because they are God's Kingdom
than because they are kings; and this too shouldst thou
desire incomparably more whenever thou offerest to God this

[1] Micheas iv. 7.   [2] Apoc. v. 9, 10.

petition, "Thy Kingdom come," not so much that thou mayest reign with God as that God may, at the same time, perfectly reign over thee.

IV. Consider fourthly, that there are two classes of persons who can never say with a good countenance to God, as others can, "Thy Kingdom come." The first is that of obstinate sinners, and the second of those imperfect just persons whose hearts are unduly attached to their mortal life. The first class cannot say this, because when they ask of God that His Kingdom may come, they are, in fact, asking for their final damnation. God certainly will reign to all eternity over all, not the just only, but sinners also—"God shall reign over the nations "[1]—but in a very different manner. He will reign over the just in Heaven, and over sinners in Hell. And thus the just will be God's Kingdom, because He will reign over them as a loving Sovereign reigns over kings who, having been crowned by Him, will, in return, rejoice to vie with each other in placing their crowns before His glorious throne. And sinners will be God's Kingdom, because God will reign over them as a terrible Sovereign over slaves whom He has condemned to perpetual imprisonment, and who will hopelessly try to shake off the iron chains and fiery fetters under which they groan, desiring in their despair to find death, but in vain. This, then, is what hardened sinners are unconsciously asking for when they pray to God that His Kingdom may come; they are asking that that everlasting slavery may come, which is to be their portion in the abyss of Hell: "Woe to them that desire the day of the Lord."[2] Neither can these words be said by those imperfect persons among the just who are living in too great an attachment to their earthly life; for with what face can they pray to God that His Kingdom may come if their hearts are so ill-disposed as to be almost willing to resign Heaven for ever if only God would give them leave to remain with a good conscience in this world for ever? Whenever, therefore, thou recitest the *Pater noster*, think for a moment in what state thou art when thou askest of God that His Kingdom may come, and if thou art living in sin, tremble and fear at the thought of thy danger when that Kingdom draws near : "The Kingdom of God is come nigh unto you."[3] And if thou art still too much attached to this world, seek to be detached from it ; for how is it possible that thou canst be so much attached to a poor cottage, such as the earth is in

---

[1] Amos v. 18.   [2] St. Luke x. 9.   [3] Psalm xlvi. 9.

comparison with Heaven, as to shrink for fear of leaving it, from going to another country, distant though it be, to take possession of a vast kingdom, which is thine by right of inheritance? Even though thou wert, I do not say a noble or a prince in this world, but even one of its greatest kings, still it would be for thee to say, as Christ did, "My kingdom is not of this world."[1] He did not say "in this world," but "of this world," because He was truly King of this world, as well as of the next, and yet He made no account of the former, but only of the latter, and therefore He said that He was King of Heaven, not of earth, because it was not in His earthly, but His heavenly Kingdom that He found His consolation: "My Kingdom is not from hence."[2] If thou doest this, thou wilt accustom thyself to say these beautiful words, "Thy Kingdom come," in life and in death with the utmost affection to thy Heavenly Father: in life, with the sentiments of one who desires that the Kingdom of God may come to him also, as it does to so many here ; in death, with those of one who sees it approaching, and gives it the welcome which is due to it.

V. Consider fifthly, that even if thou hast not yet got rid of sin, or laid aside that excessive affection which thou bearest to the world, thou oughtest not to think on that account that the Lord's Prayer is either useless or unbecoming in thy present state, and so leave off saying it. First, because thou prayest in common with others, always in the plural, and therefore it is not unbecoming, because if thou art conscious of not being able to ask the gift for thyself, thou art performing an act of charity in asking it for others. Secondly, because thou prayest materially at least, if in no other way, and so performest an act, if not of charity, still, at any rate, of external religion, an act which is easy to pious persons, but tedious to others. And so this prayer is not useless even to thee, because, by virtue of that material good act, which is pleasing to God, thou mayest prevail with Him to give thee one day such grace as may bring thee altogether out of thy miserable state, and enable thee at length to say boldly, not for others only, but for thyself also, "Thy Kingdom come."

[1] St. John xviii. 36.   [2] St. John xviii. 36.

TWENTY-SECOND DAY.

Thy will be done on earth as it is in Heaven.

I. Consider first, that it is perfectly just for every son to
aspire to his inheritance, but on one condition, that he does
not make himself unworthy of it by the lack of respect which
he shows his father. He is, indeed, bound to merit it by
positive submission to his father's will in all things. Dost
thou not think it right, then, that after saying to our Heavenly
Father, "Thy Kingdom come," in which petition we ask for
the inheritance, we should immediately add, "Thy will be
done," thus showing our readiness to do whatever He pleases?
Nevertheless, we do not say to our Father "may we do Thy
will," in order not to ascribe to ourselves, by such a way of
speaking, more than is right. We say *fiat*, in order to testify,
by a more modest form of expression, on the one hand our
readiness, as free agents, to do the holy will of God, and on
the other, the need we have of His grace to enable us to do it.
It is true that when we say "Thy will be done," we do not
mean to imply merely that it is to be done by us, but both by
and in us. For a good son is bound not only to do whatever
his father commands in particular instances, such as when he
bids him, for example, come or go, or leave off amusing
himself; but he should also be willing to be disposed of in
general as his father pleases—to be placed, for instance, with
certain company, or at a certain court, or in a certain calling.
And this is what we signify our readiness for by our *fiat* with
respect to our Father Who reigns in Heaven. First, that His
will may be done by us—"Thy will be done by us"—that is,
that His precepts and counsels and most secret inspirations
may be performed and followed by us: "In the head of the
book it is written of Me that I should do Thy will."[1] Secondly,
that His will may be done concerning us—"Thy will be done
in us"—that is, that He may dispose of us in all our affairs,
both in prosperity and in adversity: "Yet not My will, but
Thine be done."[2] Dost thou think, then, that it is treating
thy God as a Father, and so meriting the inheritance which
He prepares for thee, if thou art so far from submitting to His
will in both these ways, as to fulfil it in neither? "He that
doeth the will of My Father Who is in Heaven, he shall enter
into the Kingdom of Heaven."[3]

[1] Psalm xxxix. 8, 9.　　[2] St. Luke xxii. 42.　　[3] St. Matt. vii. 21.

II. Consider secondly, that the first of these two wills here
mentioned is what is called the will of intimation, or the will
which is made known to us; for it is not a will by which God
decrees that a work required by Him shall be done, but only
a will of desire, manifested to us by precepts, counsels, and
other similar signs by which He discovers to us what He
desires: "He hath made known His wills to the children of
Israel."[1] The second is called the will of good pleasure, and
is that absolute will by which God has positively decreed to
dispose of us in one particular way rather than in any other,
without the possibility of any one ever resisting it: "All My
will shall be done."[2] Speaking strictly, it is said that we
obey the former of these wills, and that we are conformed to
the second. When, therefore, by the words "Thy will be
done," we mean "done by us," we pray to God to give us
a perfect obedience: "Teach me to do Thy will, for Thou
art my God."[3] And when we mean "done concerning us,"
we offer Him an entire conformity of our will with His: "Not
as I will, but as Thou wilt."[4] And do not say that this is not
a petition, but an act of resignation; for even those things
which God has positively determined to do for our salvation,
He has for the most part determined to do by our means,
and especially by the instrumentality of our prayers, which,
therefore, we have the intention of applying to that great end.
And when we mean both "by us" and "concerning us," then
we do both of these things, that is, we ask Him for a perfect
obedience, and we offer Him a perfect conformity. See, then,
what a sublime petition this is! It may be said to be a golden
epitome and compendium of all sanctity at once. For it is
certain that as the means for obtaining the inheritance prepared
for each one of us, the prompt employment in due season of
all the Christian virtues is necessary, as of so many different
and, so to speak, small coins which are in use. There are
patience, mortification, meekness, humility, chastity, charity,
fortitude, and so many more, that their number would certainly
exceed that of all the different coins which are in circulation.
But it is evident that it would be a tedious exercise to pray
for all these to God separately and by name as often as they
are required. What, therefore, has Jesus, the Infinite Wisdom,
done? He has comprised them all in one, but one which,
like a gold coin of immense value, is equivalent to them all—

[1] Psalm cii. 7.
[2] Isaias xlvi. 10.　[3] Psalm cxlii. 10.　[4] St. Matt. xxvi. 39.

that is, the doing of the holy will of God. So that when we say "Thy will be done," we seem, indeed, to be asking of Him one thing only, namely, that His will may be done ; but in reality we are asking Him an infinite number of things. For what is the will of God, but that we should practise all these virtues as the saints did ? "For this is the will of God, your sanctification."[1] This is what Christ has here made us ask. How is it, then, that these words are not continually on thy lips, when thou knowest them to be so powerful ?

III. Consider thirdly, how right it is that these words should always be in our mouths in the first sense of asking God for grace to do His will : "May He incline our hearts to Himself, to walk in all His ways."[2] It is right because of the honour we render to God by doing it, and it is right also because of the benefit that accrues to us from doing it. It is right because of the honour we render to God by doing it, for obedience is the first honour which every father requires of his children : "Why call you Me Lord, Lord, and do not the things which I say ?"[3] Thus, Christ declared of Himself that to do in all things the will of His Father was the chief end for which He came down from Heaven to earth : "I came down from Heaven not to do My own will, but the will of Him that sent Me."[4] How, then, would it be if thou wert, on the contrary, so ill-disposed, that after having readily performed some good work, such as going to a hospital, fasting, taking the discipline, because it suited thy humour, thou shouldst immediately lose all desire of performing it as soon as ever it is commanded thee? This certainly is not honouring thy Father. In the next place, it is right because of the benefit accruing to ourselves, for no son is more loved by a father than a very obedient one : it is such a son whom he embraces and on whom he lavishes most caresses and favours. So is it with God : "I have found David, the son of Jesse, a man according to My own Heart, who shall do all My wills."[5] Children who, on the contrary, are continually disobeying their father, are never in favour, but continually under his displeasure. Dost thou wonder why thou never livest in peace with God? Because of thy disobedience : "Who hath resisted Him and hath had peace ?"[6]

IV. Consider fourthly, also, how right it is that these words, "Thy will be done," should be continually on our lips

---

[1] 1 Thess. iv. 3.   [3] 3 Kings viii. 58.   [2] St. Luke vi. 46.
[4] St. John vi. 38.   [5] Acts xiii. 22.   [6] Job ix. 4.

in the second sense, namely, that of rejoicing that the will of God should be done in us: "It is the Lord, let Him do what is good in His sight."[1] And this for the same reasons. First, for the honour redounding hence to our great Father; for when we willingly give Him absolute dominion over us, we show how greatly we trust in Him, in His love, power, providence, goodness, and wisdom: "The Lord ruleth me, and I shall want nothing."[2] This is the greatest honour that He can receive from us. Sailors can show no greater honour to their pilot than by sleeping quietly in their berths when he is at the helm. But if they stand round him, full of anxiety to know why he steers more to the left than the right, they insult him and make him very angry in the end. Thou canst do no greater dishonour to God than by obliging Him, so to speak, to give thee an account of His government: "Why have we fasted, and Thou hast not regarded?"[3] Wouldst thou truly honour Him? Say to Him continually from thy heart, "Thy will be done," that is, "because it is Thine," for I desire no other reason. It is right also because of the great advantage we derive from it, for we are ignorant children who, unless we freely let ourselves be guided by our Father in all things, are in danger of losing our way at every step. A sheep wandering by herself through a wood, is trembling and full of fear. And why? Because, simple as she is, she understands her great need of guidance. When she follows in the steps of her shepherd she is at rest. So, too, with us. If we desire to walk safely on earth, we must let ourselves be guided by God like simple sheep. This alone will take away all fear from us: "And I am not troubled, following Thee for my pastor."[4]

V. Consider fifthly, that to desire what is the will of God in every way, whether by us or in us, is so important a work that we should endeavour to perform it in the most perfect manner possible. Therefore Christ commanded us, whenever we say to our Father, "Thy will be done," always to add, "on earth as it is in Heaven." It is doubtless impossible for the will of God to be valued and adored on earth by every one as it is in Heaven, where love is in proportion to knowledge. Nevertheless, we shall aim high in order to reach the highest possible mark: "I show you yet a more excellent way."[5] And this to see what is done in Heaven, where both these wills of God, His revealed will and His will of good pleasure,

---

[1] 1 Kings iii. 18.   [2] Psalm xxii. 1.   [3] Isaias lviii. 3.
[4] Jerem. xvii. 16.   [5] 1 Cor. xii. 31.

are accomplished. His revealed will is done specially by the
angels, who, as indefatigable messengers of their God, are
always ready with outspread wings to hasten whithersoever He
sends them : "Bless the Lord, all ye His angels, you that are
mighty in strength, and execute His word." And how is this
will done by them? Promptly? Exactly? This is not
enough : it is done out of pure obedience—"Hearkening to
the voice of His orders"[1]—that is, not merely "so soon as
they hear His voice," as some have explained it, but "for
this one end, that they may hear or obey His voice," which
is the interpretation especially insisted on by Bellarmine as
most conformable to the original ; for the angels obey from
no interest of their own : they obey simply for the sake of
obedience. The will of good pleasure is done continually,
not only by the angels, but by all the blessed. And how is
this also done? It is done with all their soul, that is, with the
perfect assent of the understanding, which is resolved to think
that which God wills in every matter is the very best ; and it is
done with the perfect assent of the will, which is resolved to
choose it as the best. We upon the earth sometimes obey
God with promptness and exactitude, but we obey Him at the
same time for the advantage accruing to us from obedience.
This is not the obedience of the angels. And we are sometimes also conformed to the Divine will, but at the same time
we wish, if it were possible, that God's will were different.
This is not the conformity of the blessed. The blessed not
only desire whatever God wills, but they desire it in such a
manner that they would not, if they could choose, have God's
will different from what it is. Thus it is that the will of the
blessed is so transformed into the will of God as not to be
distinguishable from it : "He who is joined to the Lord is one
spirit."[2] It is a consequence of this, that although the blessed
are not all equal in their beatitude, they are all equally contented. The reason is that all of them, as loving sons, not only
do not desire a portion of the inheritance in the least degree
greater or less than that which their Eternal Father has from all
eternity appointed to each one of them, but they cannot even
desire that it were His will to appoint such a portion. This
is a thing which it may be that thou art not able fully to
understand here, because nature here overcomes grace by its
motions ; but thou wilt understand it in Heaven, when grace
overcomes nature. And just as God cannot desire ever to

[1] Psalm cii. 20.   [2] 1 Cor. vi. 17.

have willed anything different from what He has willed with regard to any one of the blessed, so neither can the blessed, who have one spirit with God, desire it. It is, then, this short saying, "Thy will be done," which makes the perfect bliss of Paradise. As, therefore, if it were possible for self-will to depart from Hell, it would be Hell no longer—"If there were an end of self-will, there would be no Hell"—so if it were possible for self-will to enter Paradise, it would be Paradise no longer, because there would no longer reign there that perfect peace which comes from there being in Paradise one only and single will, that is, the will of God: "Thou shalt be called My pleasure in her."[1] Wouldst thou know why it is that thy heart, instead of being a little paradise of pleasure and peace, is so often a hell of confusion? It is because self-will is there: "Israel shall be confounded in his own will."[2]

TWENTY-THIRD DAY.

Give us this day our daily bread.

I. Consider first, that as every father justly requires due reverence from the sons whom he would make his heirs, so, in order that they may easily render it to him, he ought, on his part, to take thought for their daily food, more especially when he is a father who is exceedingly rich, and when they have nothing. But where is there a father so rich as our Heavenly Father? and where are there children who, without Him, are so poor, or rather so destitute as all of us? Therefore, in order to give thee the assurance that this thy great Father will not fail to provide thee with all the nourishment necessary for thee, Christ, after the first three petitions which can be perfectly granted in Heaven only, invited thee to beg that nourishment of Him, not that thy Father is not very ready of Himself to give it, but to accustom thee to acknowledge that everything comes to thee after all from Him alone. There are two kinds of food, corporal and spiritual, the latter to support the life of the soul, the former that of the body. And as an earthly father shall provide his children, as far as he is able, with both of these, giving them as to the body meat, clothes, and shelter, and whatever else is necessary for life,

---

[1] Isaias lxii. 4. [2] Osee x. 6.

and as to the soul whatever is necessary for leading a good life : much more should the Heavenly Father do so. Hence it is that these words, "Give us this day our daily bread," are explained by some of spiritual food, because the bread, which is here called "daily" by one Evangelist, is by another called "supersubstantial ;" and they are explained by others of corporal food, because the bread here called "supersubstantial" by one Evangelist is by another called "daily ;" and, lastly, they are taken by some in both senses, because the word admits of both meanings in the Greek root from which it is derived. It will be safest for thee to adhere to the opinion of these last, and to understand by "bread" both spiritual and corporal food, both because a good father is bound to provide both of them, and a good son to desire each, and because all the words of which this petition is made up are applicable to both. Do thou beg of God light to understand it all aright, so that when thou askest of Him this bread thou mayest not do so as beasts would ask for their food.

II. Consider secondly, these words first in their higher sense, that which refers to spiritual food. And here thou hast to observe that this food is expressed by the word "bread :" first, because the chief of all kinds of spiritual food is that of the Eucharist, which is especially designated by that name —"This is the Bread that came down from Heaven"[1]— and next, because this same name is commonly used to express all other similar kinds of food, as, for example, the Word of God, the consolations, lights, and tears that accompany prayer, and, above all, those helps of grace which are called actual, and which by their strengthening influence enable us to do the will of God with ease, and to rest in it. These helps are asked of God only by the name of bread, not that they are not in themselves full of sweetness, but because we ought not to ask them of God for that reason, but only because they have the power of strengthening and confirming the soul : "That bread may strengthen man's heart."[2] And by so doing God has here, in the first place, removed that excessive taste which some have for feeding their souls on spiritual sweetnesses ; let us be content with bread. The next thing to be noted is the word "our," which is added that we may not desire to appropriate the bread of others, but be satisfied with our own, that is, with that which belongs to our state. It may be that thou seest with displeasure the daily

[1] St. John vi. 59. [2] Psalm ciii. 15.

Communions which are permitted to another by the same Spiritual Father who refuses them to thee. Thou enviest the higher gift of prayer which thou discernest in another, the lights, nay, even the raptures, ecstasies, revelations, and wonderful communications of grace which God does not grant to thee, either because thou art not worthy of them, or because they are not suited to thy state. This is not to desire thy own bread only. Be satisfied with that which God thinks well to give thee as suited to thee, and never complain that He gives to others bread which is made of finer flour; say to Him "our bread." In the third place, thou sayest "daily bread," that is, that which is eaten daily; not that all these kinds of food are necessarily to be taken every day, but because they are usually taken, at least by desire. Such is, in an especial manner, the Most Holy Eucharist, which thou mayest receive daily, if not sacramentally, at least spiritually, as Christ did Himself, Who for three-and-thirty years desired it only: "With desire I have desired to eat this pasch with you before I suffer,"[1] not any pasch, but "this," that is, the one in which He instituted Holy Communion, and, as is most probable, was the first to receive It, so making in Himself a worthy abode for Himself: "The children are partakers of flesh and blood, and He also Himself in like manner hath been partaker of the same."[2] So that if, instead of calling this "daily," thou preferrest to call it "super-substantial bread," thou seest why it has that name. It is because it is intended to nourish the noblest substance of man, namely, his soul. The next words to be noticed are "give us," and from these thou shouldst take great confidence in asking food of so good a Father. The word used is *da*, not *dona*, a word which is not used in speaking of food. Food is not given as a present, but as a matter of course, especially by a father. Only it must be remembered that this is not a reason for living in idleness; for although it is true that a rich father gladly gives food to his children who have nothing of their own, yet this is not the case when he sees them standing idle, and doing nothing to help the family. And is it just, that God should feed thee, even daily, with the food of the Eucharist, and give thee spiritual joy and tears, and extraordinary aids, while thou on thy part renderest Him no service? There is surely great inconsistency in demanding food and not labouring in any way: "If any man will

[1] St. Luke xxii. 15.   [2] Hebrews ii. 14.

not work, neither let him eat."[1]   Fifthly and lastly, the restrictive phrase "this day," that is, "for this day," is added in order to do away with all excessive anxiety in thinking of the future.   Often thou losest heart, and dost not apply thyself in earnest to the spiritual life for fear of soon losing those helps which make its beginnings so sweet.   Do not do this ; think only of the present day, "this day," as Christ teaches thee by this *hodie*: to-morrow thou shalt think of to-morrow, but how dost thou know that thou wilt live to see it ?   "Be not solicitous for to-morrow."

III. Consider thirdly, in the same way, that these words which we are here meditating may be easily applied to the food which is intended for our bodily sustenance.   It is called bread, because if we are not to seek after even spiritual, how much less after carnal delights, when the body which we would pamper becomes food for worms in a few days after death ! It is true that according to the Hebrew expression the word means not only bread, but all kinds of food—"Call him that he may eat bread"[2]—and even everything that is in any way necessary for the support of life : "He that taketh away the bread gotten by sweat is like him that killeth his neighbour."[3] But it is asked for by the name of bread to remind us that as we are not in the habit of eating more than enough of it (for very few persons would eat it from gluttony), so, too, should we act with regard to all temporal goods which we ask of God, using them in moderation : "Use as a frugal man the things that are set before thee."[4]   It is called "ours," because we ought to be content with asking only that bread which is ours—"We will eat our own bread"[5]—for there are too many persons in the world who desire that of others ; and if this is unlawful even as to spiritual bread, which does not diminish, no matter among how many it is divided, how much more is this the case as to corporal food, which is so limited ?   It is called "daily," to teach us that no one ought to act like certain insatiable rich men, who do not, indeed, rob others, but who busy themselves in heaping up what would suffice for the support of many families who have hardly enough to eat : "That hoard up silver and gold, and there is no end of their getting."[6]   This is not to desire food, but a large income.   And if thou askest how the bread, which means common food for the body, can be called not only "daily,"

[1] 2 Thess. iii. 10.   [2] Exodus ii. 20.   [3] Ecclus. xxxiv. 26.
[4] Ecclus. xxxi. 19.   [5] Isaias iv. 1.   [6] Baruch iii. 18.

but "supersubstantial," it is to teach thee for what end thou
shouldst ask this food of thy great Father.   It is not merely
for the preservation of thy body, which is the inferior substance,
but that thy body, being preserved and strengthened by this
food, may be subject to the soul, which is the superior
substance, here called *supersubstantia*.   In asking for this
bread we say *da*, not *dona*, because even these corporal goods,
if they are asked for only as food, and food intended for so
good a purpose as that of subjecting the body to the soul,
should be asked for with confidence.   Dost thou fear that
God will deny needful nourishment to one who is His son,
as thou art, when He gives it even to brutes?   "He giveth
to beasts their food, and to the young ravens that call upon
Him."[1]   Oh, how greatly dost thou wrong Him by not
trusting Him, and by providing thyself with this nourishment
in unlawful ways!   Be satisfied with meriting that He should
give it thee, by behaving as a son who does not live in idleness, and remembering that He has innumerable ways of
supporting thee.   We are told to say "this day" in praying
for our corporal nourishment, first, because it is taken for
granted that thou art to have recourse to God every day to
beg for it, as well-conditioned children, who do not go about
the house helping themselves from the larder, but ask their
father for it; and next, because thou art to ask for it this
day without anxiety for the morrow, as the children do of
whom we have been speaking.   If they were to ask for the
provisions of next day they would show a lack of confidence
in him, as though he would not be the same father on however
many days they appealed to him.   The manna was given day
by day to the children of Israel, yet it never failed them for
forty years.

IV. Consider fourthly, that this petition, salutary as it is,
may become a rock on which two very different classes of men,
the rich and the poor, are equally in danger of striking.   If
thou art rich thou art in danger of thinking that the daily use
of this prayer is unnecessary for thee, seeing that thou art
provided not only for days, but for almost innumerable years:
"Soul, thou hast much goods laid up for many years."[2]   Thy
cellars and barns are full, how then canst thou require to say
to God, as the poor do, "Give us this day our daily bread,"
whether thou understandest by "bread" the spiritual or corporal
food which thou hast in abundance?   But thou art greatly

[1] Psalm cxlvi. 9.   [2] St. Luke xii. 19.

mistaken, for if thou hast much thou mayest lose much, and that in an instant. Therefore, as thou mayest very easily every day lose all that thou possessest, so, too, shouldst thou every day pray to God to preserve for thee at least as much as is sufficient for thy honest maintenance. Neither oughtest thou to change the forms of petition because thou art rich, by saying to God, "preserve," or "continue," instead of "give." For every moment that He preserves thy possessions for thee, so that no harm may come to them, He really gives them to thee ; and therefore thou art bound to come before thy God daily, as a poor beggar, to beg of Him bread for thy support. If thou art poor thou art in danger from the contrary quarter, that is, the danger of not troubling thyself to labour for thy daily bread, but only asking it of Him Who is certain to give it thee if thou dost ask it. But is not this folly? No father, as we have said, intends when he gives food to his children to encourage them in idleness, but to remove them from the danger of it by giving them strength to work. Do not say, either, "But if I work to gain my daily bread, what is the use of asking for it?" Because if thou didst not ask for it thy work would be useless. God might send down upon thee hail, rain, and storms to destroy thy work, and so thou wouldst indeed labour, but without profit. When, therefore, thou sayest to God, "Give us this day our daily bread," in whichever of the two senses that have been explained thou usest the petition, whether for spiritual or corporal needs, thou art not by it to ask exemption from the universal law which declares, "In the sweat of thy face shalt thou eat bread,"[1] but to ask that thy labours may be so far successful as to provide thee with the means of living; for it is of little use for thee to plant and water the tree unless God give it the interior sap from Heaven: "Neither he that planteth is anything, nor he that watereth, but God that giveth the increase."[2] Thou seest, therefore, that whether thou art rich or poor thou hast always alike to say to God these words, "Give us this day our daily bread," for it is in virtue of them that thy food is given thee.

[1] Genesis iii. 19.   [2] 1 Cor. iii. 7.

TWENTY-FOURTH DAY.

And forgive us our debts.

I. Consider first, that a father, who merits the greatest honour, both in himself and on account of his extreme care of his children, not only in providing them with a rich inherit- ance, but with food which is both suitable and unfailing to support them till they receive it, would certainly deserve from all his sons such reverence that not one of their number would displease him on any account. But it is impossible, at least morally, that this should happen, so great is the corruption of human nature. And therefore Christ, Who well knew that notwithstanding our obligations to our Heavenly Father we should be insane enough to displease Him repeatedly and grievously, chose to join together the preceding petition for daily bread and this one for the pardon of debts by the word "and," to point out to us the close connection there is between the innumerable favours we receive from God and the equally innumerable acts of ingratitude with which we repay them. Still we must take courage, for after this sadly significant "and," Christ goes on immediately to teach us how to beg of God so needful a pardon with an infallible certainty of obtaining it, if we ask it with all our hearts. For what would be the use of teaching us to ask it, if by asking we could not receive it? "Ask and it shall be given you." Imagine, therefore, that in this beautiful prayer we have hitherto been speaking to our Father as innocent children, seeing that, after praying for the glory of His great name in that fervent petition, "Hallowed be Thy Name," we have asked of Him (as was right) first our promised inheritance in the words, "Thy Kingdom come," then the interior merit for obtaining it, in the words, "Thy will be done," and, thirdly, the means, both exterior and interior, for doing so, in the words, "Give us this day our daily bread." We now begin speaking to Him as guilty but penitent children, because no father should think only of his children who are well, but also of those who are sick. Indeed, it is the chief joy of a true father to win back his erring children. This was shown by the father in the famous parable, who rejoiced more at the return of the prodigal than in all the service he received from the good son : "Let us eat and make merry, because this my son was dead and is come to life again."[1] Do thou, therefore, conceive a great confidence in remembering that when thou sayest to God, "Forgive us our debts, as we also forgive our debtors," it is to a Father that thou sayest it.

II. Consider secondly, that we are truly debtors to another when we have either taken from him or withheld from him anything that is his by right. But what is the right of God over us as our Father? It is that, as good sons, we should, in all circumstances, prefer His pleasure to our own. Whenever, therefore, we fail to do so, we are debtors to God on a large scale, that is, both in the guilt and in the penalty, according to the degree of the sin committed. These, then, are the two great debts of which thou beggest forgiveness of God whenever thou sayest "Forgive us our debts." It is neither the guilt alone nor the penalty alone which thou askest Him to remove from thee, but thou prayest Him, as a tender Father, to remit both: first, indeed, the guilt, as one who is truly penitent desires, and then the penalty. It is true that thou canst only beg Him to forgive thee these debts in the appointed way, and therefore, as to the debt of guilt, if thou desirest to obtain its remission by these words (which have not the power of conferring it in themselves, as the sacraments do, but only by way of impetration), it is necessary for thee to have in thy heart at the time both the true sorrow and the true purpose of amendment which are requisite. And, as to the debt of penalty, it is necessary for thee to give to God due satisfaction, both by confessing the sin thou hast committed to him who occupies His place on earth, and by performing the penance imposed on thee for it. Does this seem a great thing to thee? If so, thou dost not understand what these debts are. The debt thou hast contracted by the commission of the smallest venial sin is so great that, if all the saints and all other pure creatures most pleasing to God were to desire to make condign compensation for it by their homage, and were even to come down from Heaven to offer solemn sacrifices for thee in this valley of tears, to fast, to take the discipline, to pray incessantly for thee, they could not succeed in making that compensation to all eternity? And why? Because God hates the smallest venial sin that is committed in the world more than He loves all the united worship of His pure creatures. Is it so great a thing when all the children of a house join together in reverencing and

[1] St. Luke xv. 23, 24.

honouring their father? They are only doing their duty;
indeed, they always fall short of it : but if one of them offends
him, he sins greatly against him to whom he is bound, and
so there is no proportion between the cases. The debt of
penalty is also so great that it can never be understood but
by one who is now in Hell or in Purgatory in order to pay it
to the last farthing. Canst thou think it a great thing that
God, for the forgiveness of thy debt, requires thee to retract
from thy heart the sin thou hast committed, to confess it,
secretly, but sincerely, to a priest, and to perform some
penance imposed on thee for the good of thy soul? Return
thanks to Jesus Christ, that, by making satisfaction for thee by
His works, which are of infinite value, He has also obtained
for thee full remission. And remember that however much
thou mightest do, it would all be nothing. When, therefore,
thou sayest to God, "Forgive us our debts," think of what
thou art saying. Do not think thou art asking of God what
costs nothing. It is true, indeed, that the pardon which thou
obtainest by this prayer costs thee nothing; but how much
did it cost Jesus, the Son of God, when He made Himself a
sacrifice for the good of all? "He gave Himself a redemption
for all."[1]

III. Consider thirdly, that this great petition was directed
by our Lord to two principal ends : to save men both from
presumption and despair, those two tremendous precipices,
one of which is the danger of the just, the other that of
sinners. There are some who may reach such a pitch of
audacity in this world, as to imagine that they have nothing to
ask their Father's pardon for—"I have never transgressed thy
commandment"[2]—and there are others who may be so filled
with terror as to have no hope of forgiveness : "My iniquity
is greater than that I may deserve pardon."[3] Both these
dangers, then, are provided against in this beautiful prayer.
This prayer was, in the first instance, enjoined on the Apostles,
and then, in them, to all the faithful without exception :
"Thus shall you pray." And it is commanded to be recited
every day, for which reason it is called the Daily Prayer; it is
to be said in public, in private, in every part of the world.
Let no one, then, presume in himself, for however holy he may
be he is bound to say to God, not only for others, but for
himself (as has been taught by numerous Councils), "Forgive
us our debts." The Blessed Virgin alone could not say it for

---

[1] 1 Timothy ii. 6.   [2] St. Luke xv. 29.   [3] Genesis iv. 13.

herself, but only for others, or if she said it for herself it was only because, like our Lord, she in her charity esteemed the debts of all mankind as her own. But who besides has there ever been in the world who could exclude himself from the long list of debtors? "If we say that we have no sin, we deceive ourselves, and the truth is not in us;"[1] not only, as St. Augustine observes, "humility is not in us," but not even "truth." It is indeed, possible, that at some particular moment of time in which thou recitest this prayer, thou mayest be free from every kind of debt in consequence of having just gained a Plenary Indulgence by which all thy sins, even the very smallest, have been done away. But how canst thou know this certainly, unless an angel were to come down from Heaven on purpose to reveal it to thee? So that even in this case thou shouldst still pray in the same manner, because thou art certain that the debt has been incurred, and not certain that it has ever been cancelled : "Be not without fear about sin forgiven."[2] In the next place, as no one who says the Lord's Prayer ought to presume, so neither ought any one to despair, provided only that he does not merely recite it with his lips (as even parrots have been taught to do), but from the bottom of his heart. For how could all men be alike commanded to say to God, "Forgive us our debts," if there could possibly be debts so enormous that it was impossible for them to be forgiven, even though this petition were offered with heartfelt sincerity? This is not so. Ask, and then thou mayest be firmly convinced of obtaining thy request : "I forgave thee all the debt, because thou besoughtest me."[3] And hereby two foolish heresies are also confuted, that of Jovinian, who said that the grace of Baptism made men impeccable, and that of Novatus, which was the direct reverse, and taught that whoever lost the grace of Baptism by sin could never regain it. Both are utterly false. Our Lord has charged baptized persons to say daily, "Forgive us our debts," and therefore it is possible both to fall into sin and to obtain forgiveness of it after Baptism.

IV. Consider fourthly, that here thou mayest ask whether a sinner who is not resolved to do penance can say this prayer, since every one who says the words, "Forgive us our debts," ought to say them, as is taught by the decisions of Councils, not only for others, but for himself. But, I would ask, what is it that such a sinner has the intention of asking in these

---

[1] I St. John i. 8.   [2] Ecclus. v. 5.   [3] St. Matt. xviii. 32.

words? Is it that his debts, whether of guilt or penalty, may be forgiven him in spite of his maintaining an obstinate determination of persevering in his sinful life? If so, he would be making a prayer as audacious as it is sacrilegious, and therefore there can be no doubt that he ought to leave off making it, for he would be praying contrary to our Lord's intention, which was that we should pray for the remission of our debts, not for impunity. But if, notwithstanding his hardened determination to sin, he does not intend to ask of God that his debts may be forgiven him in his present state of a sinner who is resolved not to make satisfaction, but that He would give him the dispositions for abandoning that state, then he may pray, not only without sin, but with profit, because he is not asking for a remission at the present time, which is incompatible with the state he is in, but for one in the future, when it will not be so. Therefore, at the very least, however great a sinner thou mayest be when saying the Lord's Prayer, thou shouldst desire to cease one day to be a sinner. And surely, unless thou hast become a fiend in human shape, thou canst not think this is much to do. If thou wilt not do it, then take to thyself that terrible text from the Proverbs, " He that turneth away his ears from hearing the Law, his prayer shall be an abomination."[1] Observe that it is not said, " He that heareth not the Law," which is the case with every sinner who does not practise it, but, " He that turneth away his ears from hearing " it, as is the case with those hardened men who stop their ears, like the adders, because they do not wish to practise it.

TWENTY-FIFTH DAY.

As we also forgive our debtors.

I. Consider first, that if there is anything of which the father of a numerous family should be desirous, it is that all his children should live together in peace. " Behold how good and how pleasant it is for brethren to dwell together in unity."[2] "Good," because it is advantageous, and "pleasant," because it is delightful. And whereas peace reigning in a family makes it a paradise, when the scene is changed by the disturbance of that peace, it is turned into a hell. Hence,

---

[1] Prov. xxviii. 9.   [2] Psalm cxxxii. 1.

while our Heavenly Father is perfectly satisfied that all the
other petitions of the *Pater noster*, no matter how comprehensive they are, should be offered to Him unconditionally,
He has not chosen that this should be the case in this one
petition, in which we apply to Him for the forgiveness of sins.
He would have us ask for it, indeed, but on this condition,
that we, too, forgive our brethren : "Forgive us our debts as
we also forgive our debtors." This word "as" is not, however,
employed by God to signify the rule of proportion for the
remission which we beg of Him. Alas for us, if this were so !
The debts which He forgives us far exceed any which we
forgive, or can forgive, our neighbour. We do but forgive the
hundred pence, as in that beautiful parable in the Gospel, and
He forgives us ten thousand talents. There is no comparison
between the two. Then, as to the manner of this forgiveness.
God forgives us with unbounded love, and we forgive our
neighbour with one that is restricted ; God forgives promptly,
and we reluctantly ; God, so heartily that He buries our debts
in the depths of the sea, never to rise again—"He will cast
all our sins into the bottom of the sea "[1]—and we so feebly
that the debts are always, as it were, just below the surface, so
hard do we find it to blot them from our remembrance. This
"as," then, is not set by God as a rule, but a condition ; it is,
therefore, not a thing to be fulfilled, but that is already fulfilled,
or in the act of fulfilment. So that we are not to say, "Forgive
us our debts as we will forgive our debtors," but "as we
forgive," that thou mayest not act like a cheat, who, if he
receives the favour before fulfilling the annexed condition,
either fails to fulfil it or delays doing so. But if, after all,
thou wouldst take this "as" not merely for a condition, but
for a rule (as some of the Fathers seem to do), it must not
be thought to be in any way a rule of perfection, but only of
proportion. It is not a rule of perfection ; for what are we,
vile worms of earth, that we should lay down the law to God
as to the perfect way of doing His works? We ought to take
the law from Him, not give it to Him : "Be you therefore
perfect, as also your Heavenly Father is perfect."[2] But it is
a rule of proportion, because, in proportion to the love with
which we forgive our brethren, God will forgive us. If we do
nothing but what we are strictly bound to do, namely, to
forgive injuries, God will deal with us in like manner. If we
not only forgive, but repay them by extraordinary, special,

[1] Micheas vii. 19.   [2] St. Matt. v. 48.

and superabundant acts of kindness, so, too, will God treat us: "With what measure you mete it shall be measured to you again."[1] See, therefore, how full of meaning is this little word "as;" a whole day would not be enough to exhaust it.

II. Consider secondly, that though this word "as" is used here with perfect propriety, it might have been thought sufficient to leave it to be understood as a tacit condition, not needing to be expressed. For if it is taken as a condition absolutely necessary for obtaining pardon of God, that condition was already sufficiently imposed by Christ in the words, "When you shall stand to pray, forgive, if you have aught against any man, that your Father also Who is in Heaven may forgive you your sins."[2] Or, if it is taken as a rule of proportion, this also was fully implied in the saying: "With what judgment you judge you shall be judged."[3] What, then, was the use of emphasizing this "as" so forcibly that it should be impossible to recite the Lord's Prayer, even once only in a lifetime, without clearly, deliberately, and expressly declaring to God that we forgive? Of what use was it? Of infinite use; for when in this prayer thou beggest God to forgive thy debts, either thou art in dispositions to forgive thy debtors, or not. If thou art so disposed, then, by immediately adding, "as we also forgive our debtors," thou receivest a very strong inducement to do so very completely, because this word "as" then comes before thee as a rule, and reminds thee that in the same proportion with which thou forgivest thou shalt be forgiven. If thou art not so disposed, then this word compels thee to enter into thyself, because in that case this word is put before thee as an absolutely necessary condition, and reminds thee that unless this condition is fulfilled, it is not only useless but foolish to expect pardon. Further, with what confusion oughtest thou to be covered, if, although thou daily recitest the Lord's Prayer, both in public and private, thou findest that thy conduct is the very opposite of what thou declarest it to be to God! If thou wert to present a memorial to thy sovereign, in which he were to discover a falsehood, especially of such a kind as rendered it fraudulent, thou wouldst certainly be so overwhelmed with shame that if thou hadst any sense of honour thou wouldst be ready to sink into the earth. And yet thou dost not hesitate to beg of God so often to forgive thee because thou forgivest thy neighbour, while all the while this is false. If thou actest in this way thou deservest that

[1] St. Matt. vii. 2.   [2] St. Mark xi. 25.   [3] St. Matt. vii. 2.

whenever thou comest to these words of the *Pater noster,* "as we also forgive our debtors," all the devils should surround thee and cry out, "Thou liest! it is false! We know that for many months thou hast not even returned the greeting of such a one, not to speak of those kind and courteous offices which are due to all in token of real peace; and canst thou talk of forgiving?"

III. Consider thirdly, that in order to avoid these just reproaches thou mayest say that thou wilt resolve, when thou sayest the Lord's Prayer, to pass over these inconvenient words, which so evidently reveal thy falsehood. Dost thou imagine that this is an original idea? Consult Cassian, and thou wilt see that formerly some persons who were superstitious rather than devout in their prayers did the very same thing. Take care never to imitate them. Canst thou suppose that thy Heavenly Father, to Whom thou art speaking, is either so forgetful or so distracted as not immediately to perceive thy omission? He knows what thou art leaving out, and why thou art doing so. And do not say that thou actest in this way out of reverence, to avoid lying to a God of so great Majesty. For if this reverence impel thee to refrain from falsely saying to Him that thou forgivest when thou dost not forgive, why then does it not far more impel thee to obey Him by forgiving? This is not reverence, but shame at seeing thy miserable state, which thou lackest the courage to forsake. This, therefore, is what thou shouldst do. Say the words, and say them all, as is right; and if in the act of saying them thou art so weak as not to be able as yet thoroughly to change thy heart, at least desire to change it. In this way, if thou dost not forgive at the time, thou wilt at least have some intention of forgiving, and so thou wilt not lie when saying these great words, "as we also forgive our debtors;" and this is not merely because thou sayest them in common with all (for if that were sufficient excuse, all the saints would not speak so strongly with one voice against the man who says them, however seldom, without practising what he professes), but also because, if thou hast not yet reached the goal of forgiving like others, thou art at least on the way to do so. The worst of all would be neither to have this desire nor to wish to have it. In such a case, what can I say? That thou shouldst altogether abandon the practice of saying the *Pater noster,* since it is not right to mutilate it? God forbid! But I do say that when thou sayest it, thou shouldst declare to God that thou art no longer worthy to do so except in the common name of Christendom, because thou canst not yet ask pardon of thy sins in thy own person, not having granted it, for love of Him, to thy neighbour.

IV. Consider fourthly, that although to forgive is a necessary condition for obtaining God's forgiveness, it is not, therefore, as some have thought, a sufficient condition. For if at the same time thou dost not abandon evil habits, restore his good name or his goods to one whom thou hast robbed of either, and do whatever else is commanded thee by the law of thy God, it is certain that He will not forgive thee thy debts, even although thou mayest forgive those of thy debtors. For there is this difference in the Scriptures between what are called affirmative promises—such as, "He that believeth and is baptized shall be saved," and those which are negative, like the converse, "He that believeth not shall be condemned"[1]—that the latter are to be understood without limitation, and therefore it is certain that not to believe is enough to condemn any one; but the former are always understood with this limitation, that the promise holds good if certain necessary things are not neglected. And thus thou perceivest that it is not enough for salvation merely to believe and to be baptized, as certain heretics of our time maintain; but that it is also necessary to act as a believer and a baptized person if life is prolonged. So, too, in our case: if thou dost not forgive thy debtors their debts, it is plain that God will not forgive thee thine; for this is an instance of a negative declaration: "If you will not forgive men, neither will your Father forgive you your offences," and therefore it is without limitation. But the fact of thy having forgiven thy debtors is not enough to make God forgive thy debts, because this is an affirmative declaration: "If you will forgive men their offences, your Heavenly Father will forgive you also your offences,"[2] and therefore it is to be understood with the aforesaid limitation, that thou on thy part fulfillest what remains. Unless thou art chaste, sincere, temperate, and virtuous, it is evident that to forgive is not enough of itself for thy salvation. Do not think, however, on this account, that Christ makes promises which are more magnificent than real when He repeats so often and in so many ways that the means of obtaining forgiveness of God is to grant it to our neighbour: "Forgive, and you shall be forgiven."[1] For though it is certain that to forgive one's neighbour is not a work sufficient of itself to gain God's forgiveness, yet it is one so dear in itself to God that He is often moved by it to change men's hearts, sometimes even miraculously, as in the case of St. John Gualbert, so as to fill them with compunction, to convert them, and to make them accomplish with ease everything else which is necessary to obtain God's pardon. Whereas the contrary act has sometimes caused God to drive away from Him one who was on the very point of gaining the palm of martyrdom, as was the case with the unhappy Sapritius. How anxious, then, oughtest thou to be to satisfy thy Father in this matter! He is above all things desirous, as a good Father, to see peace in His house. Woe to those quarrelsome brothers who dispute and rage together. They can hope for no good from Him; for just as it is right to encourage peaceable children, so is it necessary to repress turbulent ones: "For God is not the God of dissension, but of peace."[2]

TWENTY-SIXTH DAY.

And lead us not into temptation.

I. Consider first, that a good resolution is the touchstone by which repentance is tried before God accepts it as fine gold. If, therefore, we desire to appear before our good Father full of sorrow for our offences against Him, we must show Him truly that we have made a firm resolution not to repeat them, for this is the test: "To depart from injustice is to offer a propitiatory sacrifice for injustices."[3] And this cannot be better done in our case than by begging Him to keep us Himself far from everything that might lead us into fresh prevarications; because we can, indeed, abstain from putting ourselves into fresh occasions of sin, as he did who said, "I shall keep myself from my iniquity,"[4] but we cannot prevent them from finding us out of themselves. Do not suppose, however, that, when we here say to God, "And lead us not into temptation," we are asking Him never to allow us to be tempted in any way. In the first place, this would be

[1] St. Luke vi. 37.
[2] I Cor. xiv. 33.
[3] Ecclus. xxxv. 3.
[4] Psalm xvii. 24.

impossible, seeing that our very life is a battlefield: "The life of man upon earth is a warfare."[1] Secondly, it would not be to our advantage, for the enduring of temptation is of immense value to those who know how to profit by it: "My brethren, count it all joy, when you shall fall into divers temptations."[2] Thirdly, it would not be fitting, for it appears contrary to all reason to desire to be dispensed from all fighting, and yet to be crowned: "But this every one is sure of that worshippeth Thee, that his life, if it be under trial, shall be crowned."[3] What we ask, then, is that we may never be so tempted as to fall in the temptation, as birds, deer, and other animals fall into the snare, and are caught in it: "And lead us not into temptation." And so, in fact, we ask of God to be preserved not from every kind of temptation in general, but from those in particular in which He foresees that we shall fall, either from being enticed by pleasure, like birds, which allow themselves to be tempted into the net by a grain of millet, or from being overcome by suffering, like deer and other wild game, which run into the snare from not being able any longer to resist the fierce pursuit of the huntsmen. This may be gathered from our manner of here addressing God, when we say to Him, "Lead us not." In other temptations, which issue in good to us we do not yield, but stand firm and unseduced by the snare that is laid for us: it cannot, therefore, be said that He leads us into these. That expression may be used of those temptations which are injurious to us; not, indeed, that He ever positively makes us fall into them, but that He suffers us to do so. And, as thou knowest, this way of speaking is used in Scripture with regard to God: it is a human way of speaking. It is said that God hardens our hearts when He foresees that if He does not give us timely and efficacious assistance we shall be hardened, and allows us to be hardened: "Thou hast hardened our heart that we should not fear Thee."[4] It is said that He blinds our eyes, when He allows them to be blinded; it is said that He makes our ears dull, when He allows us to make them dull; it is even said that He makes us go astray from His precepts, when He allows us to do so: "Thou hast made us to err, O Lord, from Thy ways."[5] And so, in the present case, it is said that God makes us abide in a state of temptation when He allows this: "Thou hast brought us into a net."[6] This, therefore, should

1 Job vii. 1.   2 St. James i. 2.   3 Tobias iii. 21.
4 Isaias lxiii. 17.   5 Isaias lxiii. 17.   6 Psalm lxv. 11.

properly be thy meaning when thou addressest these words to God, "Lead us not into temptation," that He will not suffer thee to come into that temptation which He sees will entrap thee. And so, speaking strictly, thou here askest for two things, which may, indeed, be reduced to one, but which are nevertheless two: the first is that thou mayest not fall when temptation comes, that is, mayest not consent, and by this thou askest to be kept from sin; the second is that He will not allow any temptation to come into which He foresees that thou wilt fall, and by this thou not only humbly confessest thy weakness, but also thy determination not to fall.

II. Consider secondly, that there are two kinds of hurtful temptations, some interior, others exterior. The first have their source in the innate concupiscence which is within us; the last come to us from external objects. The first are said to arise from the flesh, which, by its interior assaults, aims at two things, to deter us from the good to which the soul would otherwise tend, and to incite us to evil: "Every man is tempted by his own concupiscence, being drawn away and allured:"[1] "drawn away" from good, "allured" to evil. The second class of temptations are said to come from the world, which aims, indeed, at the same end as the flesh, namely, to deter us from good and to allure us to evil, but does not, like the flesh, do this in one way only. The flesh attacks us only by way of flattery, as Dalila did Samson; the world employs both the way of flattery and that of persecution, as Saul did with David. It uses the former by setting before us all its sensible goods, and the latter by representing to us contempt, imprisonment, crosses, and horrible torments. It is true that both these most dangerous tempters would be less formidable, if they had not a powerful infernal ally; for not only are we tempted by the world and the flesh, but also by the devil, who has a share in both exterior and interior temptations: in those which are interior, by incessantly instigating the flesh to flatter the spirit, saying to it, as was said by the Philistines to Dalila, "Deceive him;" and in those which are exterior, by aggravating the world's rage or cunning, as the case may be, and by stirring it up to injure the good, as he stirred up Saul to injure David: "An evil spirit troubled him."[2] And so the devil, by himself alone, can do nothing. His power is in the degree to which he is able to excite the flesh and the world against thee. This being so, thou oughtest to convince thyself

[1] St. James i. 14. [2] 1 Kings xvi. 14.

firmly of this rule : that thy principal care should be directed
to defending thyself against the flesh; because this is an
interior temptress that never, leaves thee for a single instant,
not only holding thee in its arms, as Samson was held by
Dalila, but dwelling in the innermost recesses of thy being.
Thy second care must be to defend thyself from the world,
which encloses thee all around, so that thou hast to dread its
assaults on whichever side thou turnest, as when David was
persecuted by Saul in town and country, in caves and in
deserts. Thy third consideration should be to defend thyself
against the devil, who, if thou art on thy guard against the
flesh, as Samson should have been, and against the world, as
David was, will have very little power to prevail against thee.
And do not think that if thou takest great care to protect
thyself from these three cruel tempters, it is needless to say
continually to God, "Lead us not into temptation;" because,
however carefully thou mayest guard thyself, very great need
hast thou still of God's help, so many and so fierce are the
temptations that may surprise thee at any moment without thy
perceiving them : "Watch ye and pray, that you enter not into
temptation."[1] It is not enough to watch, prayer too is necessary; just as one man who keeps timely watch, and another
who calls to his neighbours for help, both escape the danger of
thieves.

   III. Consider thirdly, how great is thy folly, if, without
even waiting for the three malicious traitors to set themselves
to catch thee in their snares, thou walkest into these of thy
own accord : "Will the bird fall into the snare upon the earth
if there be no fowler?" asks the Prophet Amos, as though the
case were impossible. And yet it is one which happens every
time that thou dost not wait to be tempted, but goest of
thyself to meet the temptation : thou fallest into the snare
when there is no fowler. And when is this? Whenever thou
voluntarily placest thyself in any grave occasion of sin. Thou
oughtest to know that, in such a case, thou offerest a vain
prayer to God when thou sayest, "Lead us not into temptation," for art thou not mocking God when thou askest Him
not to let thee fall into temptation, whilst thou art inviting it
by thy own will? If, therefore, thou observest well, this is not
a prayer intended to save us from those snares in which a man
entangles himself through curiosity, caprice, or idleness ; but
from those which surprise him against his will, as was the case

[1] St. Mark xiv, 38.

with those which were laid for David : "The snares of death prevented me."[1] For it is an infallible rule, that he who knowingly lets himself into a snare, like Samson, will be caught fast in it : "He hath thrust his feet into a net. . . . The sole of his foot shall be held in a snare."[2] Who would pity birds, if they had sense to discern the net, and did not avoid it ? We pity them, because they are simple little creatures that do not know whither they are going when they fly so merrily to the snare : "The bird maketh haste to the snare, and knoweth not that his life is in danger."[3] Who pities the man who stirs up a wasp's nest ? who pities the man who attacks vipers, or provokes panthers in their dens ? "Who will pity any that come near wild beasts ?"[4] No one, surely : and this is what thou doest when thou goest in search of temptation ; thou art "coming near wild beasts," thou art provoking thy tempters ; and then, if they spring upon thee, thou expectest God to save and preserve thee ! I will tell thee when thou mayest make this prayer with great confidence of being heard even when it is not the occasion of sin that has sought thee, but thou the occasion : it is when thou art induced to seek it by a good object, that is either by an obligation of duty, or a disposition of obedience, or a law of charity, as when Judith set herself to go voluntarily into the tent of the wicked Holofernes, and could yet say boldly to God, "Give me constancy in my mind that I may despise him, and fortitude that I may overthrow him,"[5] because she was going to deliver her people. But with the exception of such cases, how canst thou beg of God to preserve thee from a temptation which thou art seeking ? "He that loveth danger shall perish in it."[6] He is not said to love danger who places himself in it from an honourable motive, but he only who does so wantonly. If, therefore, thou wantonly seekest temptation, which is thy net, trifling and playing with it, do not ask of God to preserve thee from entering into it ; for this is asking Him for miracles, merely in order that thou mayest take thy pastime with impunity. And if so, is it not, still further, to ask Him not to let thee fall into temptation, and at the same time to tempt Him ? "Thou shalt not tempt the Lord thy God."[7]

---

[1] 2 Kings xxii. 6.   [2] Job xviii. 8, 9.   [3] Prov. vii. 23.
[4] Ecclus. xii. 13.   [5] Judith ix. 14.   [6] Ecclus. iii. 27.   [7] St. Matt. iv. 7.

TWENTY-SEVENTH DAY.

But deliver us from evil.

I. Consider first, that in the two preceding petitions all that we have been doing is to entreat our Heavenly Father to deliver us from evil, and therefore they, like the petition we are about to consider, are called by commentators "deprecations," which differ from prayers in this, that the latter have for their object the attainment of good, and the former the averting of evil: "Hear my prayer (*orationem*), O Lord, and my supplication (*deprecationem*)," *oratio* referring to the good to be obtained, *deprecatio* to the evil to be averted. In the words, "Forgive us our debts," we have asked to be absolved from past sins and from the penalty we have incurred by them; in the words, "And lead us not into temptation," we have asked to be preserved from future sins, and from the penalty we may incur by them. Why, then, do we add, "But deliver us from evil," as though we had asked nothing of all this? We add in these words as much more as though we said "deliver us from all evil," for besides the deliverance from sin and the penalty answering to it, there remains to be asked also the deliverance from many other evils which we call temporal, thorns to which even those who by their innocence are, as it were, virgin soil, are subject; these are evils which are certainly like thorns, not merely in their power of tormenting, but in their manner; for some are natural, such as ignorance and infirmities; others are planned, like the private persecutions we endure, and like wars, seditions, and schisms; and others again, such as we call accidental, like fires, floods, failures, tempests, earthquakes, famines, and many other similar evils, from which our good Father loves to deliver us, lest these terrible scourges should so overwhelm our hearts as to prevent their bringing forth due fruit to the glory of God, but from which for the most part He loves to deliver us by virtue of our prayers: "If My people . . . being converted shall make supplication to Me . . . then will I hear from Heaven, . . . and will heal their land,"[1] which is the reason why so many prayers are ordered by the Church every day for this end. And thus, in effect, these three last petitions by which we ask all that is good for ourselves answer to the other three immediately preceding

[1] 2 Paral. vii. 14.

them. When we beg of God that He will forgive us our debts we ask to be delivered from that which is directly opposed to the attainment of our inheritance, that is, the beatitude of Heaven, namely, from sin and its penalty incurred by us. Therefore the petition, "Forgive us our debts," answers to that other, "Thy Kingdom come." When we pray to God not to lead us into temptation, we ask to be delivered from that which directly prevents our doing the will of God, and desiring that it may be done in us, namely, from those temptations to which God foresees that we should yield if He allowed them to assault us. And therefore this petition, "And lead us not into temptation," answers to "Thy will be done." Lastly, when we beg of God to deliver us from every evil, we pray to be delivered from that which interferes with the supply of our daily food, both spiritual and temporal, namely, from the innumerable adversities to which human life is subject. And, therefore, this petition, "But deliver us from evil," corresponds to that other, "Give us this day our daily bread." Or thou mayest take this last petition as an epitome of all the others, and then when we say to God, "But deliver us from evil," it is as though we tacitly asked Him to grant us all the good that we have prayed for in the preceding petitions, and to prevent our falling, as He might justly allow us to do, into the contrary evil. Hence, whenever we say to God, "But deliver us from evil," we should make this petition with exceeding humility, knowing that we deserve not one particular evil or another, but every evil, simply because it is evil.

II. Consider secondly, that if we restrict this petition to those evils, either natural, planned, or accidental, to which, as we have said, all, even the most innocent, are subject in this life (which seems the best interpretation), thou art not to think that God only delivers us from them by preventing our being assaulted by them, as it is said, for instance, that He delivered just Lot from the destruction appointed for the wicked cities: "He delivered Lot out of the destruction of the cities in which he had dwelt."[1] This kind of complete deliverance from every kind of evil cannot be found in our valley of tears. If, therefore, this were thy desire whenever thou shouldst say, "But deliver us from evil," thou wouldst be simply asking to go to Paradise, where there is neither hunger, thirst, weariness, nor any other evil, whether planned

[1] Genesis xix. 29.

or accidental, or merely natural: "The creature also itself shall be delivered from the servitude of corruption."[1] If, then, thou wouldst ask for such a deliverance from evil as is suited to our miserable life, where glory is to be gained by suffering, do not ask only that which is complete, but that which God, in His most wise providence, prefers. Dost thou suppose that He has no other ways of delivering us but that one which is least suited for us? On the contrary, He has three ways, all of which are more glorious than that one. The first is by alleviating the evil by those consolations which make it easy to bear; thus He dealt with Jacob, to whom, when fleeing from his brother's anger, He appeared so many times to console him with glorious promises, and showed Heaven opened in a vision. The second is by compensating the evil with a good which counterbalances it, as when He made Daniel, in his sorrowful captivity, find favour with the princes who kept him prisoner. The third is by changing the evil itself into a greater good. This was Joseph's case, when his being sold was the cause of his prosperity. When, therefore, thou here sayest to God, "But deliver us from evil," thou art not, so to speak, to desire to tie His hands by asking Him absolutely not to send thee some particular kind of evil, because thou art ignorant which will be for thy greatest advantage—"Remember that thou knowest not His work"[2] —but only to beg Him to deliver thee from it in that way which He sees to be most beseeming His glory. If, therefore, it is His pleasure to deliver thee from such an evil by not sending it to thee at all, blessed be His Name: "I will give glory to Thy Name . . . because Thou hast delivered me from them that did roar, prepared to devour."[3] If this is not His will, then leave it to Him to deliver thee as seems best to Him: "Deliver me in Thy justice."[4] He may so console thee in that evil that thou wilt scarcely feel it, as in Jacob's case. This is to take from the evil all its power of tormenting: "I exceedingly abound with joy in all our tribulation."[5] He may counterbalance it by giving thee an equivalent good, which will make thee forget it, or count it as nothing, which was Daniel's case. This is to take from the evil both its power of tormenting and of injuring: "Afflicted in few things, in many they shall be well rewarded." Or He may turn the evil into good, as He did with Joseph, which is the peculiar art of

[1] Romans viii. 21. [2] Job xxxvi. 24. [3] Ecclus. li. 4.
[4] Psalm xxx. 2. [5] 2 Cor. vii. 4.

His Divine wisdom, by which He makes even affliction turn into joy, and injury into profit : "You thought evil against me, but God turned it into good."[1] See, therefore, in what manner God is to be addressed. We may not say, "Deliver us from tribulation," but "deliver us from evil," because tribulation is often made the cause of far greater good than the merely not suffering tribulation would be ; and if so, it is not for thy advantage to ask Him to deliver thee from one particular tribulation which thou dislikest, but only from evil —"The Lord keep thee from all evil"[2]—otherwise thou art in danger of acting like those who foolishly confuse evil with good and good with evil : "Woe to you who call evil good and good evil." Remember, too, that the greatest good to be gathered from the evils of this world is to accustom oneself to know how to bear them with tranquillity : "Tribulation worketh patience." If, therefore, God grants thee this good together with the evil thou art suffering, seek for no other, this alone is enough to justify the saying that thou art free from all evil.

III. Consider thirdly, that since petitions so sublime are made in this holy prayer of our Lord to the Eternal Father, it might have seemed proper to end it in the manner so customary with the Church, "through our Lord Jesus Christ Thy Son," instead of by the simple "Amen," which cannot impart the same force as the other conclusion, which would bring in the thought and the merit of Jesus, to make the prayer more acceptable to God. But Jesus Himself, the author of the prayer, ordered otherwise, and would have it concluded only by an "Amen." Neither should this surprise thee : first, because if (as many doctors hold) He was in the habit of frequently reciting this prayer aloud, together with the Apostles, it appears not quite in harmony with this custom to name Himself as interceding for those things which He was asking of His dear Father for Himself also, although not for Himself in His own character, but as Head of the mystical body which He deigned to form with His disciples. A second reason is that the Father at once recognizes the words, the style, the sense, and the language of His Son, so that it was superfluous for us to mention the Son in those petitions which are made to the Father, not only by His command, but also in His words. Thou mightest inquire also with what object Christ concluded His prayer

---

[1] Genesis l. 20.   [2] Psalm cxx. 7.

with the word "Amen;" but this, too, has a devout meaning.
"Amen" is a Hebrew word so rich in meanings that it has
never been translated into Latin, from the impossibility of
finding an equivalent in that language. Nevertheless, we
may briefly say that when it stands at the beginning of a
phrase it has the force of an affirmation; thus, when our
Lord was about to speak of a truth of great importance,
He was often in the habit of saying, "Amen, I say unto
you," which was not an oath, as has been vulgarly believed,
but an affirmation. When it is not at the beginning, but the
end, then it has two meanings: one that of confirming,
approving, or accepting what has been said, and the other
that of desiring it in addition. And so, when the curses
fulminated against transgressors and the blessings invoked
upon observers of the Divine precepts were formerly read
out, the assembled people were to answer "Amen" to each
of them. When the "Amen" was given in answer to the
curses, the intention was to confirm, approve, and accept
them unanimously; when it was in answer to the blessings
there was the further intention of desiring them: and so we
read in the Psalm, "Blessed be the Lord, the God of Israel,
from everlasting to everlasting," and then it is added, "And
let all the people say Amen, amen."[1]   The meaning was to
express a surpassingly eager and vehement desire of this, such
as was expressed by the great Bishop St. Cyprian, who, on
hearing sentence of death loudly pronounced against him,
replied as loudly, "Amen;" an "Amen," indeed, of immense
value! When, therefore, we say "Amen" at the end of the
Pater noster, what is it that we intend to say? It is "So be
it." "The Lord fulfil all our petitions."[2]   Lastly, then, this
word avails for the recollection of our minds, so that if we
happen to have wandered or been distracted while offering
any of these seven petitions to God, we may supply for our
faults by this word, which should be understood as added to
each one in particular, although, to avoid frequent repetition,
we are content to put it at the end only, as a signature or seal
to the whole; and yet how little thou thinkest of this "Amen"!

IV. Consider fourthly, that this "Amen" is of use even
to utterly ignorant persons. For although there ought not to
be any one in the Church so rude and untaught as not to
know quite well what it is that is prayed for in all the petitions
of the *Pater noster*, nevertheless, such persons are constantly

[1] Psalm cv. 48.   [2] Psalm xix. 7.

to be met with. And therefore every such person who knows, even in a confused way, that everything which the Church asks of God is perfectly right, does, by this "Amen," unite his intention to that of the others whose minds are wiser and loftier than his. And if he does this with lively faith he obtains what is asked equally with others, just as the peasant does who presents to the King a petition, the force of which he does not understand, but who declares that he earnestly desires that its contents, so far as they have been explained to him by practised and well-informed persons, may be granted. And this is the point concerning which St. Paul gave orders when he said that the public prayers in the church were not to be said, at least not all of them, in a low voice or in an unknown tongue, in order that the ministers (who are in the place of the ignorant) might be certain when to direct them to answer "Amen." "Else, if thou shalt bless with the spirit, how shall he that holdeth the place of the unlearned say Amen to thy blessing? because he knoweth not what thou sayest."[1] Besides, thou oughtest not to think that the prayers which are approved by the Church are of no use to thee, even when thou dost not understand them. It is enough for thee to know how to say "Amen" with her ministers, so long as thou sayest it from thy heart. They are not useless in inclining God to hear them, for although thou dost not understand the value of the jewels which thou offerest Him, as an expert lapidary would do, yet He understands it, and therefore He will accept them, as He accepts the prayers of little children: "Out of the mouth of infants and of sucklings hast Thou perfected praise."[2] Neither are they useless in terrifying the devils, just as the words of an incantation spoken with the right intention by an enchanter avail to terrify serpents, even when he does not perfectly understand their meaning.

TWENTY-EIGHTH DAY.

Thus, therefore, shall you pray: Our Father, &c.

I. Consider first, that although whoever is satisfied with saying "Amen" with the ignorant in the *Pater noster* does not lose the good of this Divine prayer, nevertheless it is

[1] 1 Cor. xiv. 16.   [2] Psalm viii. 3.

quite a different kind of fruit which is gathered by one who understands it well, and not only recites it in the usual way, which is merely to run through the petitions with the tongue, but who lets his mind pause over each one of them, as bees do on flowers, dwelling on them, meditating on them, and endeavouring, as it were, to extract from them their sweetest honey. Therefore, as the word pray has a two-fold meaning, the more restricted one of asking by supplication—"Pray for them that persecute you"[1]—and the wider one, which we call making our prayer—"He went up into a mountain alone to pray"[2]—it is reasonable to think that when Christ here said to His disciples, "Thus, therefore, shall you pray," He did not intend to say to them merely, "You shall ask thus," but also, "When you ask you shall meditate thus." As then in the somewhat extended explanation of the *Pater noster* which has been given, thou hast already seen the intention, the teaching, and the method of this beautiful prayer, not only in general but in detail, it will be easy for thee to feed thy soul upon it daily, and to use its petitions at one time as the remedy of thy ills, at another as means of strength and consolation, as though they were so many beautiful ejaculatory prayers, collected into a quiver, which every one may send to God according to the strength of his arm. There are three classes of persons in the ways of God, beginners, proficients, and those who are perfect. Beginners, who have just abandoned sin, when they say, "Our Father Who art in Heaven," should say the word "Father" with sentiments of great confusion, combined with great confidence; proficients, with those of confidence and love; the perfect, with those of love and admiration. And in the same way every one should gather for his profit from all the petitions, the few best adapted for him, just as is done with regard to their pasture by different animals, those that have just begun to eat grass, those that are older, and those which are most experienced. Of which class art thou? To whichever thou mayest belong, it will be useful for thee to know the method which should be practised by each according to his state.

II. Consider secondly, that if thou belongest to the state of beginners, thou oughtest to see what is the vice which has most power over thee, and to make choice accordingly of that petition which most conduces to its speedy overthrow. If pride is thy predominant vice, declare frequently to God that

[1] St. Matt. v. 44.   [2] St. Matt. xiv. 23.

it is to His Name, not to thine, that glory is due; pray therefore that His Name may be glorified: "Hallowed be Thy Name." If it is avarice, protest to Him that thou wilt no more value those goods on which the kingdom of worldlings is founded, but only long for those which belong to His: "Thy Kingdom come." If it is envy that torments thee, say to God that the cause of it is that thou dost not understand that the will of God should be to every one the sublime law in which he rests; pray that His will may be accomplished: "Thy will be done on earth as it is in Heaven." Be content that that Divine will should honour, enrich, advance whomsoever it chooses; thou, as a poor beggar, deserving nothing, wilt only ask of God what He gives as alms. If it is gluttony that troubles thee, declare to Him that thou art not worthy even of mere daily bread, because thou hast so often made thy belly thy God, by taking food for the sake of satisfying thy appetite, but that it is as a favour that thou askest Him for this bread—"Give us this day our daily bread"—and that therefore thou no longer askest for it to please a false god, but only that thou mayest have strength to serve the true God. If thou art passionate, and it seems to thee a hard thing to restrain thy anger, say often to God, "Forgive us our debts, as we also forgive our debtors," for thou wilt triumph over it by the repetition of this prayer and protestation. If thou hast contracted irregular habits through the vice of luxury, which make thee afraid of relapsing easily, go on in the same way saying continually to God, "And lead us not into temptation," because this is a temptation which seldom besets any without fault of their own. And if, lastly, thy negligence in spiritual exercises exposes thee to be overcome by sloth, beg of God frequently to preserve thee from evil, that is, from this sloth, which is called the source of all evil: "But deliver us from evil." Thou hast, indeed, good reason to strive for a complete deliverance from this evil, which is the parent of many others: "Idleness hath taught much evil."[1]

III. Consider thirdly, that if thou belongest rather to the state of proficients, thou shouldst see what is the virtue to which thou thinkest thyself most attracted or most adapted, and without neglecting the others, cultivate that, making use of it, as it were, as a foundation on which they are to rest, as richly embroidered hanging forms a ground for gold, rubies,

[1] Ecclus. xxxiii. 29.

or pearls. If thou art conscious of possessing a lively faith, thou shouldst desire that this light of faith which God has given thee may increase, and shed itself over others, so that all may vie in seeking God's honour only: "Hallowed be Thy Name." If the hope of future glory makes thee very courageous to do and to suffer great things for God, say to Him that it is not an earthly but only a heavenly reward that thou desirest: "Thy Kingdom come." If charity has set up its standard in thy heart, and desires to reign absolutely, so that self-love may altogether die, and the love of God altogether live in thee, say to Him as often as thou canst, "Thy will be done on earth as it is in Heaven." If thou delightest to act with the prudence which aims, both in temporal and spiritual needs, at being neither too careless as to the present nor too anxious as to the future, accustom thyself to say frequently, "Give us this day our daily bread." If thou lovest to see justice equally practised, and would not have it exercised strictly in the houses of others and indulgently in thine own, as many do, occupy thyself in saying, "Forgive us our debts as we also forgive our debtors." If thou rejoicest that thy irregular appetites, more particularly those which arise from the rebellion of the flesh, should be under the restraint of temperance, take pleasure in repeating, "And lead us not into temptation." And if it is thy delight to bear adversities with fortitude, or even to court them for the love of God, ask Him to keep thee from evil—"But deliver us from evil"—not, indeed, from the evil of much suffering, which is an evil thou knowest how to bear, but from that only which is really evil in it, namely, that of suffering with impatience.

IV. Consider fourthly, that if thou art so happy as even to have advanced some distance in the higher state of the perfect, it is impossible but that when thou thinkest of God, the Supreme Good, thou shouldst be enkindled to wish Him every possible good. But what good can we wish Him Who is the Supreme Good? Not knowing, therefore, how to give vent to thy love, thou wilt at least desire that all may join with thee in loving Him, and since there are so many ungrateful men who never even think of praising Him, though they receive so many great graces from Him continually, thou wilt call upon even the inanimate creation, the woods, seas, and mountains, to supply for them, praising Him and crying, "Hallowed be Thy Name." But the more thou desirest to praise God, the more thou wilt see that He is beyond all praise. And therefore there will arise in thy heart a very earnest desire of going to that Heaven where alone He is worthily praised : " Thy Kingdom come." But what will it avail thee to be so enamoured of this prospect, like him who said, " I desire to be dissolved," when the time has not yet come ? Thou hast still to remain an exile upon this earth, where every one offends thy God, instead of going where every one is ceaselessly employed in praising Him. There can be but one consolation for thee, to say to God, " Thy will be done." But still this will, indeed, enable thee to live, but it cannot prevent thee from languishing with desire. And in this liquefaction of the will, in order to incorporate and plunge it for ever in that of God, like the wills of the blessed in Heaven—" On earth as it is in Heaven "—thy spirit will so faint within thee, that thou wilt be compelled from time to time to beg some support from Him : " Give us this day our daily bread." It is true that thy greatest support will not be from the tokens of love which God will give thee by visiting thee when thou retirest for prayer, nor from lights and illuminations, nor from that " bread of tears "[1] with which He may feed thee : it will come from that only which thou art allowed to receive at the holy altar. Therefore, as the Heaven of the blessed is where they have the King of glory present with them, so, too, will thine be where the King of glory is present—invisibly indeed, but personally. And although thou wilt find Him there every day, yet every day thou wilt desire to return to Him, so abundantly will He there pour into thy soul His gifts and His sweetness. But the more these gifts and this sweetness grow in thee, so much the more does thy obligation of loving Him grow. And this is thy greatest suffering, the knowledge of thy shortcomings in discharging this obligation. The only way of relieving thyself then will be to say to God, " Forgive us our debts as we also forgive our debtors," so that if there should be no one who injures or hates thee, thou mayest almost desire that it were so, could such a desire be lawful, in order to be able by returning him good for evil, to do him what God, to thy shame and confusion, is continually doing to thee. Yet this sorrow would be more endurable if, loving God so little, thou wert at least certain that the time would never come when thou wouldst displease Him, even slightly, again. But who can

[1] Psalm lxxix. 6.

give thee such an assurance? There will come into thy mind all the cunning arts of which Satan makes use, and thou wilt think how easily he may deceive thee also—nay, is it certain that he has not already deceived thee by making thee believe that thou lovest God when thou dost not love Him? Then thou wilt have doubts about all that is good in thee, interior recollection, lights, illuminations, and even the union of thy soul with God, and thou wilt think that He speaks to thy heart to make thee aware of the delusion in which thou art living, saying with a voice of deep reproach, "And canst thou profess to love Me?" This will strike thee with sorrow so deep that thou wilt be ready to lose all confidence, and all that thou canst do is to entreat of Him not to suffer thee to be lost in so terrible a storm: "And lead us not into temptation." Yet here there is a light which will shine upon thee like the star so dear to mariners. It is the thought that it must be thy joy upon earth only to suffer for God. Say to Him, then, that thou art willing that He should send thee whatever temptations He sees right for thee, tribulations and labours and interior trials if He please, though thou findest these the heaviest of all, if only amidst all these trials He will deliver thee from the only evil in the world which it is unlawful to desire even for love of Him, namely, to be separated from Him for an instant: "But deliver us from evil." And in this confidence thy heart will be at rest, so that seeing thyself, as it were, safe in port, thou wilt be constrained to say, "Amen."

TWENTY-NINTH DAY.

You are they who have continued with Me in My temptations, and I dispose to you as My Father hath disposed to Me, a Kingdom, that you may eat and drink at My table in My Kingdom (St. Luke xxii. 28, 29).

I. Consider first, that it seems strange that when Christ was promising the Apostles so great a Kingdom as Paradise, He should have nothing to say of that Kingdom but that in it they should eat and drink as much as they chose at His table. Is there nothing to be done in Heaven but to eat and drink? On the contrary, there is no such thing there. "The Kingdom of God is not meat and drink,"[1] as the

[1] Romans xiv. 17.

Apostle said to reprove the intemperate Cerinthus, who sought to teach the contrary in the Church. There will be no desire of meat and drink in Heaven—"They shall no more hunger nor thirst"[1]—and if so, what enjoyment would there be in making use of food? It would be like continuing a remedy when the disease is cured. The intention of our Lord in using this form of expression was to explain to the Apostles, who were as yet gross in their minds, the beatitude of Heaven under the lively figure of a banquet, which is familiar to all. A banquet is a feast of the most delicate viands, it is joyful and splendid, and it gives all the guests an opportunity of thoroughly satisfying their hunger. So, too, but in a far higher sense, will it be with the beatitude of Heaven : " I shall be satisfied when Thy glory shall appear."[2] Raise, therefore, thy cleansed and purified imagination from what is material, and picture to thyself in Heaven a banquet indeed, but a spiritual banquet, which is that promised to men by God, far different from that of a Mahometan paradise.

II. Consider secondly, that it is possible for a king to entertain a great number of nobles very splendidly in his royal banqueting-hall without necessarily seating them at his own table. That is a still higher honour, which certainly was not done by Assuerus to the immense multitude whom he invited in " Susan from the greatest to the least."[3] This favour was reserved for a few of the personages of most distinction, " Who saw the face of the King, and were used to sit first after him."[4] When, therefore, Christ, just before His Death, told His Apostles that as though by will He disposed His Kingdom to them, that is, assigned and appointed it, with the express declaration that they should always sit with Him there at His own table—" I dispose to you a Kingdom, that you may eat and drink at My table"—He doubtless intended by this to assign to each of them a more distinguished honour than that which should be enjoyed in their degree by all the other guests, but at different tables. The true meaning, therefore, of this form of expression is that the Apostles were to be the nearest of all the blessed to their Lord, and were to sit at His table in His Kingdom, just as at the Last Judgment they were to sit on thrones of dignity like His own, judging mankind with Him. And therefore after Christ had said, " I dispose to you a Kingdom, that you may eat and drink at My table,"

---

[1] Apoc. vii. 16.
[2] Psalm xvi. 15.   [3] Esther i. 5.   [4] Esther i. 14.

He immediately added, as though in further explanation of an honour which was not to be shared by all : "And may sit upon thrones judging the twelve tribes of Israel." How is it, then, that thou art so little in the habit of venerating these blessed Apostles, even upon the days specially assigned by the Church for their worship? It is they who are to judge thee together with Christ on the Day of Judgment, and who are now in Heaven His familiar friends and favourites, near to Him in every sense, and yet thou thinkest so little of them ! It is impossible to say what good they may continually obtain for thee, if only thou hast recourse to them in time. And why? Because they are seated so high in Heaven. Those who have the power of obtaining favours from the sovereign are those, as a rule, who sit at his table. And this, too, Christ meant His Apostles to understand when He said they should sit at His table in Paradise ; He meant that they would be also the most prompt there in doing His will : "He was the King's guest, and was honoured above all his friends."[1]

III. Consider thirdly, what was the reason why Christ told His Apostles He would exalt them so highly. It was because they had been faithful to Him in His labours and trials, and had not turned away from Him like others who, for fear of the anger of the Jews, either ceased to follow Him at all, or only followed Him secretly : "You are they who have continued with Me in My temptations." Oh, how beautiful a thing it is not to forsake the Master in time of adversity ! Many love to be near Him at His table : "He is a friend, a companion at the table ;" but few keep at His side in His poverty : "He will not abide in the day of distress."[2] It was, then, because the Apostles, on the contrary, were faithful to Christ in His poverty—"Abiding in the day of distress"— that He decreed, when He should reign in His Kingdom, to receive them as "companions at His table." For this is the universal rule, that he who would rejoice with Christ must first have suffered with Him : "As you are partakers of the sufferings, so shall you be also of the consolation."[3] Observe now how marvellous is the antithesis : "You are they who have continued with Me in My temptations, and I dispose unto you a Kingdom." What disproportion can be greater than that which there is between these opposite terms, "You with Me," "I to you"? Is it because servants so mean have shown a little fidelity in suffering for so good a Lord, that the

[1] Daniel xiv. 1.   [2] Ecclus. vi. 10.   [3] 2 Cor. i. 7.

Lord is to make them almost His equals in the Kingdom? Yet this is what Christ here says, "You are they who have continued with Me in My temptations: and I dispose to you a Kingdom, that you may eat and drink at My table in My Kingdom." And how do I "dispose" it? "As My Father hath disposed to Me," that is, "I dispose My Kingdom to you as My Father hath done to Me: with the same love, the same dignity, the same essential beatitude, which consists in seeing the face of God, with this difference, that the Father has disposed it to Me by nature, and I dispose it you by grace." What thinkest thou then of such a kind of gift as is here bestowed by Christ? Does it not fill thee with an ardent desire of bearing Him company, of clinging to Him, of keeping at His side whenever He bears His Cross? His sufferings, persecutions, and want are what He here calls "temptations," because the Father, so to speak, tried Him by their means, not that He might know Him, but that the world might know Him by them, which is the reason why the temptations of Christ were not common, but grievous, universal, and varied: He was "tempted in all things."[1] Certainly the Apostles never suffered them in any equal degree; they did but share in them, for which reason Christ did not say to them, "You are they who have borne My temptations with Me," but only, "You are they who have continued with Me in My temptations." And yet He gives so great a reward for so little! How foolish art thou if thou dost not serve so good a Master.

IV. Consider fourthly, that it seems not a little strange that Christ should say to the Apostles that they had been so faithful to Him in His trials, when they so easily forsook Him in His Passion: "All leaving Him, fled."[2] Nevertheless, thou oughtest to observe first, that when Christ said this they had not yet forsaken Him in this way, because He said it when He was about to rise up from the Last Supper to go to His Death, and therefore neither was Judas the perfidious traitor present; for when the Supper was half over, he left the Cenacle, in order to carry out His infamous bargain: "Having received the morsel, he went out immediately."[3] This should teach thee that Christ judges us only by our present state of justice. The Apostles to whom He spoke had all been faithful to Him up to that time, and He therefore spoke of them as still faithful. It is true that they were very shortly to

[1] Hebrews iv. 15.   [2] St. Matt. xxvi. 56.   [3] St. John xiii. 30.

forsake Him, as He showed that He knew when, a little later,
on His way to the Garden, He declared to them that they
would all be scattered away from Him like sheep upon the
mountain when they see their shepherd struck to the earth by
a sudden whirlwind: "All you shall be scandalized in Me
this night: for it is written, I will strike the Shepherd, and
the sheep of the flock shall be dispersed."[1]   And besides,
if they were all to be dispersed then, they were all to return to
Him after that dispersion, with their whole hearts, like penitent
sheep to their shepherd when, the storm being over, he has
risen from the earth.   And because Christ thinks no more of
the sins which have been bewailed with sincere sorrow, He
here spoke to the Apostles in a manner which showed that
those sins had not prevented His accomplishing His sublime
designs on their behalf. Besides, it is well known that a person
who leaves another, and returns in a short time, is not considered to have left him by the law: "A woman, who returns
to her husband in a short time, is not said to have left him."
And therefore Christ, knowing that the Apostles would return
to Him immediately after their flight, chose to speak of them
here as though they were not to leave Him at all.   If thou
too art ever so unhappy as to forsake thy Lord, make no delay
in returning to Him: "Delay not to be converted to the
Lord."[2]   And be of good courage; for notwithstanding that
thou hast left Him, He will treat thee as though thou hadst
always faithfully persevered in His service: "You are they
who have continued with Me in My temptations, and I dispose
to you, as My Father hath disposed to Me, a Kingdom, that
you may eat and drink at My table in My Kingdom."   Thou
mayest say that thou canst not hope to sit in Paradise at so
rich and splendid a table as the Apostles; but what then?
should that prevent thy being satisfied with thy portion?
"Blessed is he that shall eat bread in the Kingdom of God."[3]

[1] St. Matt. xxvi. 31.   [2] Ecclus. v. 8.   [3] St. Luke xiv. 15.

THIRTIETH DAY.

Let every man be slow to anger. For the anger of man worketh not the justice of God (St. James i. 19).

I. Consider first, how trivial an excuse it is to say that thou canst not help being easily angered, because thy temper is so hot. If it were a valid one, St. James ought not to have here declared it as a universal law to all, to be "slow to anger;" but should have taken care to distinguish between differences of sex, state, and temperament. Since, then, he excepts no one from this law, it is a proof that every one may triumph over nature by the help of grace, as David did, who although he was naturally of a very warm and sanguine disposition, more so, perhaps, than thou art, yet attained to perform acts of such heroic meekness to Saul, Semei, and others of his enemies, that towards the end of his life he prayed to God to show mercy to him on account of this: "O Lord, remember David and all his meekness."[1] Why, then, is it that thou, too, canst not conquer thy nature in the same way? Because thou art not willing to fight. Imitate that holy King who said, "I will pursue after my enemies and overtake them, and I will not turn again till they are consumed."[2] Aim at repressing those irregular motions of anger which prevail in thee. Do not let a day pass without examining thyself particularly concerning them, so as to accustom thyself not to make light of them. Whenever thou findest thyself in the act of yielding to them, immediately make a contrary act of submission or of excuse for the person who angers thee, as thou thinkest best. Accuse thyself of each one of them to God every evening, with the intention of confessing them with contrition in due time, make acts of sorrow and purpose of amendment, and beg of God, above all, not to let thee fall so continually. Do all this perseveringly, and thou wilt see that in time thy rebellious nature will be overcome, so that at last thou wilt be able to say with David, "The Lord hath done great things for us, we are become joyful."[3] What dost thou suppose the Apostle meant when he said, "Let the peace of Christ rejoice in your hearts"?[4] He desired that one day there should be in thy heart, as the fruit of this peace, the triumph of one who has gained the victory: "Let the peace of

[1] Psalm cxxxi. 1.
[2] Psalm xvii. 38. [3] Psalm cxxv. 3. [4] Coloss. iii. 15.

God rule in your hearts," as the reading of the text is according to some. This, then, is a proof that the peace of God, which is the same thing as gentleness of manners, affability, kindness, meekness, is able to conquer its enemy, that is, anger.

II. Consider secondly, that anger is not one of those passions that are called vices, like gluttony, sloth, pride, envy, and many others : it is a natural passion, which is common to all, even holy men. And therefore, speaking absolutely, to be angry is not a sin. We know that our Lord Himself was more than once angry with the Pharisees : He "looked round about on them with anger."[1] And with the desecrators of the Temple He was so angry that He made a scourge of cords, and drove them out with His own hand. The sin is in being unreasonably angry; that is, either with some one with whom one ought not to be angry, or before one ought to be angry, or more than one ought to be angry. Thou art to think, therefore, of anger as of a soldier given us by God to fight according to reason. If it does not attempt to act without reason, if it obeys and honours reason, it is a good soldier ; it is a bad one when it despises reason. And this is why St. James does not here say that thou art never to be angry, but only that thou art to be "slow to anger." For, however good a soldier anger may be, yet thou art not to employ it constantly, but only in cases of extreme need : because this soldier is no less fiery than good, and therefore it is not so easy to control him after he has been summoned to the aid of reason, as it was not to summon him. Most frequently anger is like Joab, who received very strict orders from David his lord concerning the audacious Absalom, which was to take him indeed, to guard and bring him to the King, but not to slay him—"Save me the boy Absalom"[2]— but when he had captured him, he thought he knew better than David, and insisted on thrusting three lances into the heart of the rebellious son, in order the better to secure the kingdom to the father. So it is with anger : when once it has arms in hand, it very readily exceeds the limits that were prescribed to it by reason, esteeming them too narrow. Therefore St. James tells thee to be slow in using it—"slow to anger"—which means "slow to make use of anger," for most people do not know how to hold it in check. How often, in thy own case, hast thou been moved by zeal to condemn some scandal thou hast seen or heard of, and in the end thou hast gone so far as to speak disrespectfully of some Superior who was bound to prevent it, and failed in doing so. Not only is anger evil when it exceeds the orders given it by reason, but also when it does not wait patiently for them, but acts as St. Peter did when he asked our Lord in the Garden whether he should take up arms—"Lord, shall we strike with the sword?"[1]—and then, without waiting for the answer, drew it, and "struck the servant of the High Priest." What is to be done in such a case? The anger must be repressed immediately—"Suffer ye thus far"—otherwise, if thou dost not do so, thou committest sin, for thou art giving it vent before hearing reason.

III. Consider thirdly, that as St. James said that every one should be "slow to anger," he might have said that he should not be quick, more particularly as this was the expression of the Wise Man in Ecclesiastes: "Be not quickly angry."[2] But yet St. James was not contented with saying this: he will have thee be not merely not quick, but slow to anger. For in the Old Law there was a somewhat greater concession to certain natural human tendencies, and the reason is that there was not yet that strength of grace which Christ merited for us by His Death, in the New Law. The Old Law, too, gave much greater opening for anger, because it was altogether a law of threatening and storm and punishment, and therefore there was frequent need for the exercise of anger. Whereas the New Law is a law of love. And so thou seest that when the two sons of thunder, James and John, desired that fire might come down from heaven on the Samaritans, who had refused to receive Christ—"Lord, wilt Thou that we command fire to come down from heaven and consume them?"—our Lord reproved them, telling them that they did not know what spirit influenced them in this desire—"Turning, He rebuked them, saying: You know not of what spirit you are"[3]—thus implying that the days of Elias were already passed: "The Son of Man came not to destroy souls, but to save."[4] So that, if formerly it was enough not to be quick to anger—"be not quickly angry"—it is now necessary to be even "slow to anger." Do not, therefore, think that thou art living like a perfect Christian, if thou art easily angered, even with a just anger; for zeal (which is the same thing as just anger, which cannot endure to see iniquity triumph in the world) ought indeed to be strong,

---

[1] St. Luke xxii. 49.
[2] Eccles. vii. 9.   [3] St. Luke ix. 54, 55.   [4] St. Luke ix. 56.

but at the same time gentle, as Christ's was. Therefore it was written that neither sadness nor perturbation should ever be seen in Him—"He shall not be sad nor troublesome"[1]—for sadness is characteristic of one who has not power to attain his end, and perturbation, or excitement, of one who attains it, but with disturbance of mind. And thus thou seest that Christ, in His greatest fervour, when He drove out the desecrators of the Temple, displayed a zeal which was as gentle as it was strong. It was very strong, because it attained its end, and very gentle both as to the action itself, the means by which it was affected, and the manner of doing so. As regards the action, He did not think of killing, striking, or crushing those sacrilegious men, but only of putting them to flight. As to the means, He made use of nothing more than a scourge of small cords; and, as to the manner, He did it with so much modesty and majesty, that none of those whom He drove out could help reverencing Him; and with so much propriety and gentleness, that the bystanders, instead of feeling horror at the action, immediately surrounded Him, and entreated Him to cure their infirmities: "And there came to Him the blind and the lame in the Temple, and He healed them."[2] How often dost thou think that thou art displaying zeal when thou losest all gentleness on seeing or hearing of the faults of others! It is no such thing: it is thy natural anger, which has succeeded, by specious pretence, in deceiving reason, and which will never rest till it has extorted from it by fraud and force full freedom of action, as though it were true zeal.

IV. Consider fourthly, that there are two parts in zeal: the one is to punish the wrongs done to God, and the other to prevent them. It punishes them by blaming their author, by reproving, rebuking, and even severely mortifying him; it prevents them by admonishing him privately, by praying with him, by suffering for him, offering penances for him to God. Thou art very ready to perform the first office of zeal, which belongs to a Superior, and exceedingly negligent as to the last, which is common to all. What does this show? It shows that what thou thinkest true zeal is not so, but thy own anger, passing by the name of zeal, if, indeed, it is not ambition and pride assuming its character. Do first the humble offices of zeal, and then thou mayest with more reason trust thyself when thou art impelled to those which are more showy.

[1] Isaias xlii. 4.   [2] St. Matt. xxi. 14.

THIRTY-FIRST DAY.

The anger of man worketh not the justice of God (St. James i. 20).

I. Consider first, what is the reason given by St. James to persuade thee to be slow to anger, as has been explained in the preceding meditation, which was left incomplete, in order to divide between two days the matter which might easily be found too much for one. The reason is that anger never does any good : "The anger of man worketh not the justice of God." At first sight these expressions may seem to thee exaggerated, but weigh them well, so that, seeing their justice, thou mayest learn deep reverence for the Divine words. It is certain that all the good at which anger aims by its operations, may be reduced to one kind of justice, that is, retributive justice. If thou observest closely, thou wilt see that its object is vengeance ; and this vengeance is not always desired on a just title, or for a just end, or in a just manner, or in just circumstances. This being so, it follows that, in these operations, either reason must prevail over anger, or anger over reason. If anger prevails over reason, it is true that the actions are attributed to anger as the principal agent, and that, therefore, there may be some excuse for them as being due rather to excitement and impulse than to advertence ; but they can never be works of justice, for that is never justice in which all the rules of justice are not observed. In this case, therefore, St. James spoke well when he said that "the anger of man worketh not the justice of God," because not only does it not work justice, but it works against justice. If, on the other hand, reason prevails over anger, it is true that the works are works of justice, but they are not such as can be attributed to anger, which is only the secondary agent ; they must be attributed to reason, since, as we all know, operations of every kind are attributed to the principal agent : to the general, not to the soldiers ; to the prince, not to the magistrates ; to the master, not to the servants ; to the architect, not to his workmen. So that, in this case also, St. James spoke divinely when he said, "The anger of man worketh not the justice of God," because it is not "the anger of man" which then "worketh the justice of God :" it is the reason of man making use of anger. And if so, how evident it is that it is right for thee to be "slow to anger," even when it appears to be moving thee for a good object, and in a good manner, since it is not in it but in reason that thou art to have thy resources; the meaning of which is that in every matter, even in one most deeply affecting the glory of God, thou oughtest to regard chiefly, not the zeal, the impulse of which thou art conscious within thyself, but rather that which concerns the duty of reason, otherwise thou wilt often think thyself to be acting zealously when, in reality, thou art acting furiously.

II. Consider secondly, why it is that St. James was not satisfied with saying, "The anger of man worketh not justice," but chose to add "the justice of God." The reason is that human justice, in order to be right, must resemble Divine justice as much as possible. This being so, even if it were the anger of man that "worketh justice," it is, at all events, impossible that it can be such a justice as the anger of God works, either as to its manner of working, or as to the act. It cannot be so as to the manner: for the anger of God, if it may be called so, is not a passion, like the anger of man, but simply the will to punish those who merit punishment. And therefore it always effects its justice with calmness, serenity, and the utmost tranquillity, because this will does not occasion the smallest change in God: "Thou, being master of power, judgest with tranquillity."[1] Whereas, as we know, the anger of man is a passion, and an exceedingly violent one, which is always accompanied by a great perturbation of the blood, and much commotion about the heart, sending up vapours to the mind, which are very apt to cloud it: and therefore it must always cause great disturbance of soul and body: "My eye is troubled with wrath, my soul and my belly."[2] Thou seest, then, that, as to manner, the anger of man cannot work a justice like that of God, because it cannot do it with tranquillity. Neither is this possible as regards the act, because the anger of God being simply that will to punish which has been spoken of, it leaves Him room for the exercise of mercy, when He so pleases, at the same time that He is exercising justice; for which reason the justice of God is always combined with great mercy: "Will He in His anger shut up His mercies?"[3] The anger of man, far from admitting of compassion, rejects it, as its direct opposite, until it has been indulged to the point which it considers just: "Anger hath no mercy, nor fury when it breaketh forth," not "anger that is checked," but that "breaketh forth."[4] And therefore the

[1] Wisdom xii. 18.
[2] Psalm xxx. 10.   [3] Psalm lxxvi. 10.   [4] Prov. xxvii. 4.

anger of man, even as regards the act, cannot work a justice like that of God, that is a merciful justice, but one which is full and complete. So that it is always most true that the "anger of man" (and not merely *hominis*, but *viri*, that is, of a man who is very excellent) never "worketh the justice of God" in any way whatever. The only exception is that of Jesus, Who was, indeed, true Man, but true God also. If, then, it were possible, a man ought to desire that he might do all his works of justice as God does. But since, on account of his imperfection, he could rarely do them with great vigour, but most frequently in a languid, lax, or constrained manner, he is at liberty to call in the aid of anger, only it must be as seldom as possible: he must be "slow to anger," that is, "slow to make use of anger," in order that his justice may be as much as possible like that of God, calm and merciful—merciful as to the act, calm as to the manner.

III. Consider thirdly, that every Superior, particularly every Religious Superior, who ought to be a pattern of perfection to others, should inscribe these words in his cell, "The anger of man worketh not the justice of God," so that, having them continually before his eyes, he may be sure not to forget them. He is bound to endeavour that his justice may as much as possible be like that of God (since he is in His place), and therefore he should take care never to allow his mind to be clouded, or his authority rendered harsh by anger. It is very seldom that a religious subject is benefited by a punishment which he sees to be given by his Superior from human anger, that is, with a degree of discomposure and severity which denotes passion. But when he sees that he is punished, indeed, but not with anger, he is really benefited: "Mildness is come upon us, and we shall be corrected."[1] And this is the case, when it is evident that when the Superior punishes, it is only that he may not incur the sin of Heli by neglecting the duty of chastising his erring sons, and therefore it is in a gentle manner and with prudent mortifications that he does so. Now, to do this is only difficult at the time when he is moved by anger. Therefore, if thou art a Superior, never impose a punishment at such a time, but wait till thy anger, no matter how just, is calmed down. Neither do thou urge in opposition, that Phineas, Moses, and Matathias went so far as to slay sinners in the heat of anger. For first it is to be observed that they did so in cases of grievous scandal,

[1] Psalm lxxix. 10.

necessarily requiring a prompt remedy, such as might only be applied by striking great terror. In the next place, it must not be supposed that they acted in these cases merely by the light of natural reason. They acted by a clear light from Heaven which showed them that such was the Divine will in this instance: and therefore theirs was not "the anger of man." It was the anger of a higher spirit, inciting them to actions which are to be admired, indeed, but not imitated, certainly not by such as we are. If thou art, however, not a Superior, but only a subject, thou art, none the less, to submit even to what seems to thee an unreasonable anger in thy Superior, because this is the obligation of thy state; and not to be angry thyself even when thou knowest that thou art being punished by "the anger of man." "Not revenging yourselves, my dearly beloved; but give place unto wrath."[1] And thou givest place to thy Superior's anger, when thou leavest it free course, and dost not oppose it.

[1] Romans xii. 19.

NOVEMBER.

FIRST DAY.

FEAST OF ALL SAINTS.

Blessed are they that dwell in Thy house, O Lord; they shall praise Thee for ever and ever (Psalm lxxxiii. 5).

I. Consider first, what must be the beauty of Paradise, since it is the house of God! The greater the prince, the grander and more sumptuous should be the house in which he dwells. And therefore what house can there be superior to that which is the habitation of the greatest of princes, "the Lord of lords"? A perfect house has five qualities—great size, excellent arrangement, beauty, splendour, and pleasantness. And where shall we find all of these combined except in the house of God? Wouldst thou know its size? It is too great for the weakness of thy mind to imagine, far less comprehend it: "O Israel, how great is the house of God!"[1] Its arrangement is indicated by Christ Himself when He said, "In My Father's house there are many mansions,"[2] for it is well known that what most prevents confusion in the houses of great persons is the variety of apartments. As to beauty, a glance, even from so far off as our earth, is enough to enchant the heart: "I have loved, O Lord, the beauty of Thy house."[3] The next thing is splendour, and whosoever would have true riches must go there for them: "Wealth shall be in His house."[4] All that is found elsewhere is poverty, not riches. Lastly, as to pleasantness, dost thou not know that this Divine house may rather be called a garden of sweet delights than a house? It is for this reason that it has the name of Paradise: "Thou wast in the pleasures of the Paradise of God."[5] See,

---

[1] Baruch iii. 24.   [2] St. John xiv. 2.
[3] Psalm xxv. 8.   [4] Psalm cxi. 3.   [5] Ezech. xxviii. 13.

therefore, how justly the Psalmist exclaims, "Blessed are they that dwell in Thy house, O Lord!" What other house can make thee blessed merely by dwelling in it? Thou art often more unhappy in the houses of great men or of princes than in thy own, because thou art free in thy own and a slave in theirs. It is the peculiar privilege of the house of Beatitude to make thee blessed in whatever part of it thou dwellest. Such is the house of the Lord. Is it possible that thou art not already ravished with this house, seeing that wherever thou art in this world, thou hast no better dwelling than a mere cabin? "They dwell in houses of clay."[1]

II. Consider secondly, that although every possible good is in the house of God, and therefore every one who dwells in it is blessed, this, nevertheless, is not the reason why the Psalmist feels this pious envy of its inhabitants, when he says, "Blessed are they that dwell" there. This would have been too low a motive for so noble a mind as his. If he envies them when he calls them blessed, it is because they are never occupied in anything but in praising God: "Blessed are they that dwell in Thy house, O Lord; they shall praise Thee for ever and ever." He does not say "they shall see Thee," but "they shall praise Thee." And it is in this way that thou, too, art to purify thy desire of Heaven, and render it more perfect. If, in desiring it, thy last end is the enjoyment of God, it is thy own good that thou desirest; if thy end is to praise Him, it is His: and this is perfection. Therefore, just as, when thou fearest Hell, thou shouldst do so, at least principally, that thou mayest not be obliged to curse God eternally, for this is the quality which chiefly ennobles thy fear; so, in desiring Heaven, thou shouldst do so that thou mayest eternally bless Him, and where, but in Heaven, can this be? In this world we cannot praise God continually, as we ought, because we are often obliged to intermit His praises in order to lay our wants before Him. But in Paradise there is no want of anything, and so there is nothing to do but to praise God: "They shall praise Thee for ever and ever." And besides, even if it were possible for us always to praise Him here, we should not know how to do it, for which reason we occupy ourselves here far more in praising His works than Himself: "Generation and generation" (that is, those which succeed each other on the earth) "shall praise Thy works."[2] But we shall know very well how to do it in Heaven, and

[1] Job iv. 19.   [2] Psalm cxliv. 4.

therefore the Psalmist says of those who dwell there, "They shall praise Thee for ever and ever"—not "Thy works," but "Thee." A man who sees a beautiful palace, a beautiful painting, without knowing the artist, praises the work, but one who is well acquainted with the artist praises him. In this world we do not know God directly in Himself, but only in His works, and therefore our praise is not so much of Him as of the beautiful works of His hands. In Heaven we shall know Him as He is in Himself—"We shall see Him as He is"[1] and therefore in Heaven our praise will be rather of Him than of His works. And so, although the blessed will praise God greatly for all those exterior goods which He enjoys, such as the glory He receives from the works of creation, justification, sanctification, and even from the punishment of the reprobate, still they will praise Him yet more for those that are interior, that is, for being what He is, blessed in Himself alone, Eternal, Immense, Infinite, and Incomprehensible: "According to Thy Name, O God, so also is Thy praise."[2] And herein lies the excellence of their praise, because the exterior goods which God enjoys concern the good of the blessed also, but not so the interior; these, in their nature, concern Himself alone. And therefore, just as the excellence of the love of the blessed consists in loving God for His interior, rather than exterior goods—"For the Almighty Himself is above all His works"[3]—so, too, will the excellence of their praise consist in praising Him more for the former than the latter: "They shall praise Thee for ever and ever."

III. Consider thirdly, that it may be difficult for thee to understand how it is that the blessed will never weary of this incessant occupation of praising God, but this comes of thy measuring their love by thy own. When the blessed grow weary of loving God, then they will grow weary of praising Him. But who can ever grow weary of loving Him Who is all good? And on what account, thinkest thou, can this weariness come? It cannot be on account of Him Who is praised, for if the blessed had any one but God to praise, then, I grant, they might be weary in the end, because any one else may indeed be worthy of praise, but of a limited praise. But since it is God Whom they have to praise, there is no danger of their growing weary, even though they praise Him "for ever and ever," because they will be continually finding fresh reasons for praising Him: "Blessing the Lord,

[1] 1 St. John iii. 2.   [2] Psalm xlvii. 11.   [3] Ecclus. xliii. 30.

exalt Him as much as you can, for He is above all praise."[1]
Neither can this weariness ever come from those who
praise; because, as the blessed love God far more than
themselves, so do they love to praise God even more than to
see Him. The Seraphim, of whom Isaias had a vision, covered
their eyes with their wings in God's presence, while their lips
were incessantly singing, "Holy, Holy, Holy." And why was
this? It seems to me that it was to show that they would
rather cease seeing than praising Him. And so the blessed,
who have attained to a most excellent love of God, rather than
cease praising Him, would be willing even to cease seeing
Him, preferring to renounce their beatitude rather than His
praise. Therefore, as it is impossible for the blessed ever to
grow weary of being blessed, so, far more impossible is it for
them to weary of rendering to God those praises which they
prize even more than their beatitude. With great reason,
then, did the Psalmist say to God, "Blessed are they that
dwell in Thy house, O Lord; they shall praise Thee for ever
and ever," for it is in this that all their blessedness consists—
in praising God: "This people have I formed for Myself;
they shall show forth My praise."[2]

SECOND DAY.

THE COMMEMORATION OF THE FAITHFUL DEPARTED.

*It is a holy and wholesome thought to pray for the dead, that they may
be loosed from sins* (2 Mach. xii. 46).

I. Consider first, that this thought, which invites thee to-day
to pray with especial affection for the dead, is a holy thought:
"It is a holy thought to pray for the dead." It is holy because
it is founded on an act of charity, which is the chief of virtues.
What does charity require? Is it only that those members
who are in health should help those who are sick? No; it
requires them to go further, and to help those also who are,
indeed, in health, but in bonds: "Remember them that are
in bonds, as if you were bound with them."[3] Now, it is quite
certain that the faithful departed who are in Purgatory are
members of the Church equally with those who are living.

[1] Ecclus. xliii. 33. [2] Isaias xliii. 21. [3] Hebrews xiii. 3.

They are, doubtless, in health, because they are in grace; but they are bound, because they are not capable of helping themselves in their necessities, seeing that death has put an end to the time appointed by God for each one of them to merit in: "The night cometh, when no man can work."[1] It is, therefore, a holy thing that the faithful who are living, especially such as are healthy members, should bestow some help on those who are dead: "That the members may be mutually careful one for another."[2] How, then, canst thou see those poor souls in the flames, and bound in those flames, without being moved to some degree of pity for them? Thou art not worthy of being a member of so glorious a body as the Church, which is bound together by charity: "Bear ye one another's burdens, and so you shall fulfil the law of Christ."[3]

II. Consider secondly, that this succour which is given to the dead makes the reciprocal communication of the Church perfect with regard to all its members: "By charity of the spirit serve one another."[4] This communication may be divided into four parts: that from the living to the living, from the dead to the dead, from the dead to the living, and from the living to the dead. There can be no doubt that the living assist each other in the Church, for we pray for each other every day in this world: "Pray one for another, that you may be saved."[5] It is equally certain that the dead help one another; for this was shown when the dead Eliseus raised the other dead man who was cast into the same tomb with him; and we know that the saints in Heaven pray for the saints in Purgatory, and especially for those who are buried in their churches, as is taught by St. Augustine.[6] It is certain also that the dead help the living, for the favours are innumerable which we receive from them by their many loving apparitions to us; neither is there any city which has not a patron in Heaven to do for it what Jeremias was seen doing for Jerusalem in the days of the Machabees: "This is he that prayeth much for the people, and for all the holy city, Jeremias, the Prophet of God."[7] It is very right, therefore, to complete the communication between all the members of the Church, that the living should also help the dead, so that nothing may be lacking to perfect the charity which she professes: "Stretch out thy hand to the poor" (so that thou who art living mayest

---

[1] St. John ix. 4.   [2] 1 Cor. xii. 25.
[3] Galat. vi. 2.   [4] Galat. v. 13.   [5] St. James v. 16.
[6] Lib. ii. *De curâ pro mortuis.*   [7] 2 Mach. xv. 14.

help the living), "and restrain not grace from the dead" (so that thou who art living mayest help the dead).[1]

III. Consider thirdly, that this thought of praying for the dead is not only "holy," but "wholesome." No one can doubt that it is salutary to pray for the dead, for it is especially ordained for their benefit; not, indeed, for that of the dead who are condemned to Hell, for these are members already cut off from the whole of the mystical body of Christ, but for that of the dead who are suffering in Purgatory, and who, although they are no longer wayfarers, in the sense of journeying along the way, are so as being as yet detained from the goal of the journey, which is the glory of Paradise. And therefore, if we cannot any longer help them to merit, as we could when they were wayfarers like ourselves, we may help them very greatly to attain the reward of their merits, now that they have ended the journey without being as yet comprehenders. And therefore, salutary as the thought which moves thee to pray for the dead is to them, it is still more so to thyself, because, if it hastens their glory, it increases thine; so that if thou art in a state of grace, thou meritest and acquirest riches by praying for them—"Thus thou storest up to thyself a good reward for the day of necessity"[2]—whereas they do not merit, but only enter into possession of the recompense which they formerly merited also. And dost thou not know, too, how grateful those holy souls will be to thee when they have come to glory? It may be that they will obtain for thee, by their powerful intercession, a degree of glory to which, but for it, thou wouldst never have been worthy to attain. We know that merely to bury the bodies of the dead is esteemed a work very profitable to those who perform it: "Blessed be you to the Lord, who have showed this mercy to your master Saul, and have buried him; and now the Lord surely will render you mercy and truth."[3] How much more, then, to send their souls to Paradise, and deliver them from those bonds which detain them in a dungeon which resembles that of Hell, even though it is less terrible!

IV. Consider fourthly, what the bonds are which detain these souls from glory. They are their sins, which are pardoned indeed, but for which satisfaction has not been made, and therefore it is said, "It is a holy and wholesome thought to pray for the dead, that they may be loosed from sins." Thou seest what is done to the body by cords, chains, fetters, and

[1] Ecclus. vii. 36, 37.   [2] Tobias iv. 10.   [3] 2 Kings ii. 5, 6.

other cruel bonds. The same evils are brought upon the soul by sins : "He is fast bound with the ropes of his own sins."[1] So that, when thou sinnest, thou weavest with thy own hand these ropes which bind thee so fast, and in two ways, by making thee guilty of the sin and liable to its penalty. The souls that are detained in Purgatory are indeed free from the former bonds, but not as yet from the latter ; and so it is said that "it is a holy and wholesome thought to pray for the dead, that they may be loosed from sins." It is not said "that they may loose themselves from sins," for it is in this life only that a man may, by God's grace, shake off all the bonds which fetter him—"Loose the bonds from off thy neck, O captive daughter of Sion "[2]—but it is said "that they may be loosed from sins," because they are in want of some one to loose them. Can it be that, seeing them in so sore a plight, thou art not moved to give them succour? And observe that their bonds are bonds of fire, so that there is no time even for unfastening them : they must be burst asunder.

V. Consider fifthly, how this deliverance is to be effected. In two ways: by the way of grace or by the way of justice. The former includes Masses and prayers ; the latter fasting and alms. By the way of grace may be interposed on behalf of the dead the public intercession of the whole mystical body of the Church, and this is done in the Most Holy Sacrifice of the Mass ; and also the private intercession of its members, which is done by the prayers offered up by each one in particular. By the way of justice the debt due by the dead to the Divine justice may be both paid and redeemed. Fasting, which includes all other afflictions of the flesh, has power to do the former, and alms the latter. It is true, indeed, that all these works which are applied either to the payment or the redeeming of the penalties left owing by the dead, are, after all, accepted by God by way of suffrage, as it is called, because there is, so to speak, no proportion between the punishment inflicted on the dead by the Divine justice and that which is accepted by the Divine justice from the living in its stead. In our world the tribunal of this Divine justice is exceedingly gentle, like the civil or canonical courts in which the punishments awarded are very mild : "He doth not now revenge wickedness exceedingly."[3] In the next world the tribunal of God's justice is very terrible, like the criminal court, the proceedings of which are of the utmost severity:

---

[1] Prov. v. 22.   [2] Isaias lii. 2.   [3] Job xxxv. 15.

"Amen, I say to thee, thou shalt not go out from thence till thou repay the last farthing."[1]   Therefore, whether it is in payment or redemption that God accepts the punishments belonging to a very mild tribunal instead of those belonging to a very severe one, it must always be a grace. He can accept them if He pleases, and He does so please; but He is free not to accept them if such is His pleasure. What, therefore, is to be done? We must pray continually that He may be pleased to accept them. This, then, is the reason why it is here simply said, "It is a holy and wholesome thought to pray for the dead, that they may be loosed from sins." It might have been said, to visit churches, to fast, to take the discipline, and do all possible good works; but this is not said because, after all, all these are comprised in one word—to pray for the dead.   Do, therefore, all that thou canst on their behalf: visit churches, fast, take the discipline, give alms; but at the same time always beg of God that He may be pleased in His mercy to accept the little that thou art doing, because it is always very much less than they owe, and that little thou shouldst always unite to the Blood of Christ, which prays so much better than thou canst do.   If thou doest this, then have great confidence, for it is one of the especial glories ascribed to that Precious Blood, that it opens the door of deliverance to so many imprisoned souls that are consumed with a burning thirst of seeing God which they are unable to quench: "Thou also, by the Blood of Thy Testament, hast sent forth Thy prisoners out of the pit wherein is no water."[2]

VI. Consider sixthly, that there are some in that fire who, although they died in grace, yet were not sufficiently careful in their lifetime to make satisfaction for the sins they committed, saying that they would do penance for them afterwards in Purgatory; who did not value that reciprocal communication of merits of which it is so easy for the faithful to partake, and never thought about the dead, did not love and succour them, or even fulfil their pious bequests with promptitude.   And in this way they failed to merit the grace which God bestows when He is pleased to accept our prayers on behalf of the dead.   What is to be done, then, if thou wouldst help them?   Thou must pray with great fervour, for this is a case in which it is not enough merely to pray (*orare*) for the dead, but so to pray as to gain thy request (*exorare*).   Will it be so easy, thinkest thou, for the dead to benefit by a kindness

[1] St. Matt. v. 26.   [2] Zach. ix. 11.

which they never themselves showed to others? This would not be fitting, for even mercy itself will be in some proportion with merits, according as there has been more or less disposition to practise it: "All mercy shall make a place for every man according to the merit of his works."[1] Can there be a doubt, therefore, that thou oughtest to pray with redoubled fervour for those who have so small a claim on the treasures which are so largely lavished on behalf of the merciful? Think, then, how it will be with thee if thou showest no mercy to the dead: this is enough of itself to make thee incapable of meriting it.

THIRD DAY.

The fear of the Lord is the beginning of wisdom (Psalm cx. 10).

I. Consider first, that "the beginning of wisdom" may have two meanings: it may mean that which is its beginning, as to its nature, or it may mean that which is its beginning as to its effects. For example, in the art of building, there are its beginnings as to its nature, and these are the rules on which that art is essentially founded, that is, what we call the rules of architecture; and there are its beginnings as to its effects, and these are the foundations which are laid by that art, after digging out the earth, because it is out of them that the building begins to rise. So, too, as to wisdom, the chief of all arts, which proposes as its end to aim always and in all things at pleasing and glorifying God. When, therefore, thou art here told that "the beginning of wisdom is the fear of the Lord," do not take the word "beginning" in the first sense, because, according to that, the beginning of wisdom is the rule of the faith, by which we must be guided so as not to go wrong. It must be taken in the second sense, because the fear of God is the first effect proceeding from wisdom, when it begins to operate in the hearts of the just. For thou oughtest not to think that by wisdom is here intended that of which the only end is to know God, namely, speculative wisdom; but that, of which the end is rather to serve Him perfectly, namely, practical wisdom. Now, when this wisdom begins to operate, as such, in the heart of the just man, the first thing it does is to make him fear that God Whom it means gradually to make

[1] Ecclus. xvi. 15.

him love, and that very deeply, for "the fear of the Lord is the beginning of His love."[1] And because it is upon this foundation that it proceeds to erect its edifice, therefore it is said, "The fear of the Lord is the beginning of wisdom." Dost thou see, then, what the fear of the Lord means? It means the foundation of the whole spiritual building: and if so, what will become of thee if it gives way from weakness? The building will fall to pieces: "Unless thou hold thyself diligently in the fear of the Lord, thy house shall quickly be overthrown."[2]

II. Consider secondly, that by "the fear of the Lord" is not meant a servile fear, a fear which makes Christians act like slaves, and keep themselves from offending God only because they know that if they offend Him they will not go unpunished. This fear is good in itself, because it is that of which it is written, "The fear of the Lord driveth out sin;"[3] still it is not that of which the Psalmist is here speaking when he says, "The fear of the Lord is the beginning of wisdom," because he is speaking of an interior principle; whereas a servile fear, which may co-exist with sin, before it has driven it out, is, with regard to the actions that proceed from Divine wisdom, an exterior principle, as it were, which prepares for them the heart in which they have yet to begin—"For he that is without fear cannot be justified"[4]—it is not an interior principle of those actions which have already begun. The fear which is here spoken of is that filial fear, which is the interior principle, or beginning of these actions—"the beginning of love"—by which the just man, acknowledging how worthy God is in Himself of being valued and loved above all things, subjects himself entirely and reverently to Him, as a son to a father, for fear of offending Him. If, then, thou desirest to know whether Divine wisdom has really begun its beautiful operations within, and not merely outside of thee, see what sort of fear rules thy soul in thy relations with God: is it a filial or a servile fear?

III. Consider thirdly, that this filial fear is not generally at the beginning very perfect in all: because a man after his conversion does not readily leave off thinking of the penalty annexed to sin, indeed, he thinks of it too much with the fear belonging to his state, which is called initial. But as wisdom gradually makes the esteem and love which are due to God

---

[1] Ecclus. xxv. 16.
[2] Ecclus. xxvii. 4.   [3] Ecclus. i. 27.   [4] Ecclus. i. 28.

more perfect in the heart, it also gradually purifies the fear which it aroused in it, so that, when charity is perfect, initial fear becomes chaste, that is, altogether unmixed with the thought of punishment. This is precisely the fear which is intended in the words, " Perfect charity casteth out fear,"[1] that is, the fear of punishment, not servile fear, which, though not evil in itself, is considered as having already done its part, which is that of disposing for the work, *timor extra sumptus,* but even initial fear, which is a part of the work, *timor intra sumptus.* This is the fear which I say is " cast out " by perfect charity; for the more a man is lost in the love of God, the less does he think of his own dangers and losses : he thinks of God only. Thou oughtest, then, to represent wisdom to thyself as making use of the fear of punishment which has been described, much as a princess uses linen thread for her embroidery, merely to prepare the work, that is, just so far as is necessary to fix the cloth or silk on which she intends to work her embroidery in gold, which is the fear of guilt, but no further. And so Divine wisdom does indeed use it, but not permanently; for in proportion as it makes its work perfect in the heart of the just man, which is already given to God, it takes away this fear. The fear which it leaves in the heart is the chaste fear which always remains—" The fear of the Lord is holy, enduring for ever "[2]—and this is the fear of guilt, which, so far from ceasing, increases continually. For the more any one advances in the love of God, the more anxious does he become to do nothing which may displease or dishonour Him. Art thou one of those who art not in the least fearful of offending Him ? It is a very plain sign that thou hast not yet attained to esteem and love Him perfectly. It is well, indeed, to trust that thou wilt not offend Him, but at the same time to fear. Nay, thou oughtest to fear even that thou mayest suddenly be lost after suddenly offending Him as thou hast done, for this would be strictly just ; but do not fear it with the fear of a slave, but of a son, who cannot dread anything as fatal or terrible in the thought of banishment from his home, but separation from his father. A feeling of terror such as this is no way incompatible with chaste fear : " I said in the excess of my mind, I am cast away from before Thy eyes."[3]

IV. Consider fourthly, that since the fear of God always remains in the heart of the just man, it may be difficult to

---

[1] 1 St. John iv. 18.   [2] Psalm xviii. 10.   [3] Psalm xxx. 23.

understand how it can be called "the beginning of wisdom."
It would seem rather to be its beginning, continuation, perfection, and chiefest honour : "The fear of the Lord is a crown
of joy."[1]   It might be thought, therefore, that Job spoke more
rightly when he said that the fear of God is the very essence
of wisdom—"Behold the fear of the Lord that is wisdom"—
than the Psalmist when he said, "The fear of the Lord is the
beginning of wisdom."   But thou wilt not reason in this way
if thou understandest aright what the beginning is which is
here spoken of.   It is, doubtless, the beginning of the whole
of man's life when well regulated, for being, as this is, the
whole of the work of wisdom : "The love of God is honourable
wisdom."[2]   But it is not any kind of a beginning : it is a
beginning of the nature of a root.   The root is, indeed, as it
were, the foundation of the tree, but it is a vital foundation
which not only supports the tree, but nourishes, increases,
beautifies, enriches it, and gives it all the good it possesses :
"The root of wisdom is to fear the Lord."[3]   And, therefore,
as we say truly that the root is in reality the whole tree,
although speaking strictly it is its beginning, so, too, is the
fear of God said to be the whole of wisdom—"To fear God
is the fulness of wisdom"[4]—that is, the whole of man's life
regulated by wisdom.   Thou seest into how many branches a
well-regulated life spreads : how many leaves, flowers, and
fruits it produces.   All of these have their source in the holy
fear of God as their true origin.   If this were to fail, how
quickly should we see them all wither away !   It is not, then,
that the just man does no other good works besides fearing
God, for they are innumerable.   He does works of justice,
humility, obedience, mercy, purity, prudence, piety, fortitude,
and countless others—"He that feareth God will do good"[5]—
but the beginning of them all is the holy fear of God.   And
what kind of beginning ?   A beginning which is always united
with them, supplying vigour to each of them, however numerous
they may be.   As I said, it is a beginning of the nature of a
root—"The root of wisdom is to fear the Lord"—and therefore
the other virtues are called its branches, which are never
wanting where it is present : "The branches thereof are long-
lived."   See, then, how glorious a thing it is to continue in
the fear of God.   "Blessed is the man to whom it is given to
have the fear of God:"[6] there is no one to be compared to

[1] Ecclus. i. 11.   [2] Ecclus. i. 14.   [3] Ecclus. i. 25.
[4] Ecclus. i. 20.   [5] Ecclus. xv. 1.   [6] Ecclus. xxv. 15.

him on earth. It is not, indeed, enough merely to have it; it must be kept in a vigorous condition : " He that holdeth it, to whom shall he be likened ? "[1] For the deeper a root strikes the better it is.

V. Consider fifthly, that, no doubt, thou art exceedingly desirous of knowing whether this holy fear of God, from which proceeds all good, is in thee : " The root of wisdom is to fear the Lord." But do not wonder that thou canst not know it, at least not with a certain conviction. It is a root, what wonder then that it is out of sight ? God keeps it hidden for our own good—" To whom hath the root of wisdom been revealed ? "[2]—for so is this fear more perfectly preserved by the continual fear of not having it. Therefore, just as the more the root is covered with earth the more vigorous it is, so is it in the case we are considering. It is true that if the fruits belonging to this root are never wanting, they show very plainly, in him, that the root is living ; otherwise how could they be nourished or increased ? If thou refrainest from evil through human respect, for the sake of advancing thy interests, or gaining credit, or, at least, of not suffering in the opinions of men, most certainly thou canst have no security of possessing aright the holy fear of God ; for the branches have their root elsewhere : " Thy root and the nativity is of the land of Chanaan,"[3] which is corrupt human nature. But if thou so refrainest purely in order not to offend thy God, do not fear ; for although thou dost not see in thyself the root of which thou wouldst fain have a proof, it must be in thee, and the more deeply hidden the better.

FOURTH DAY.

ST. CHARLES.

I can do all things in Him Who strengtheneth me (Philipp. iv. 13).

I. Consider first, what boldness the Apostle showed when he said these words; he showed that, in a certain manner, he thought himself omnipotent : " I can do all things." And yet, because he so thought himself, not in his own strength, but that of the God Who alone could make him so, he was

[1] Ecclus. xxv. 15.   [2] Ecclus. i. 6.   [3] Ezech. xvi. 3.

not proud, but courageous. Humility consists not in thinking
we can do nothing for God ; for if so, the idle, the timid, the
pusillanimous, the slothful, would be the humblest men in the
world. Humility consists in thinking that we can do nothing
of ourselves. Sometimes it appears impossible to thee to
overcome thy predominant vice, to escape certain dangers, to
perform certain works of penance, to fulfil some duty of thy
state perfectly ; and thou restest in this belief, as though it
provided thy humility with pleasant nourishment. But if thou
considerest rightly, this is not humility, but sloth : "The
slothful man sayeth, 'There is a lion in the way, and a lioness
in the road ; I shall be slain in the midst of the streets."[1]
Nay, if we look closely, it is rather pride wearing the mask of
humility. Thou regardest thyself as though thy whole well-
being depended on thy natural strength, and therefore thou
losest courage as if thou wert told to slay lions and lionesses
by thy unassisted arm. Take off thine eyes from thyself, and
fix them on God alone ; endeavour to believe and understand
thoroughly that thou art to do everything in His strength Who
chooses to make use of thee, foolish, vile, and feeble as thou
art, for this very reason, to show that it is He Who is the
Author of the works He bids thee do. And thou, what canst
thou fear ? No matter then how many come forth to terrify
thee, not lions and lionesses only, but hosts of infernal fiends,
thou art certain of victory : "Though I should walk in the
midst of the shadow of death, I will fear no evils, for Thou art
with me."[2] Dost thou think that there was any fear in the
heart of the Apostle when he said, " I can do all things in
Him Who strengtheneth me " ?

II. Consider secondly, that the Apostle did not say, " He
Who strengtheneth me can do all things in me," but " I can
do all things in Him ; " not that he did not perfectly understand that the glory of the work is due to the Chief Worker,
as he showed when he said, " Not I, but the grace of God
with me,"[3] but because he was ready to confess his own
ability : not, indeed, ability which depended on his own
natural powers, for if this had been the case, he would simply
have said, " I can do all things," but ability which depended
on Him Who infused a supernatural vigour into those powers :
" By the grace of God I am what I am."[4] And here, too, if
thou observest rightly, he ascribes the whole glory to the Chief

---

[1] Prov. xxii. 13 ; xxvi. 19.
[2] Psalm xxii. 4.   [3] 1 Cor. xv. 10.   [4] 1 Cor. xv. 10.

Worker, not saying, "I can do all things with Him Who strengtheneth me," but "in Him," to show that not only were his actions done in union with God, but in the strength of God. The thought that ought to give thee courage to do great things is not that thou hast to do them with God, for in that case, as far as thy part is concerned, thou mightest lose heart, as though a pigmy were charged with half the labour of moving some great mass of stones together with a giant; but that thou hast to do these great things in the strength of God, as if the giant infused into the pigmy his own vast force, so as to share with him the labour of moving that weight: "They that hope in the Lord shall renew" (or "change," *mutabunt*) "their strength."[1] Not only will their natural strength be increased, but it shall be changed into a strength which is supernatural; for whereas originally they could do nothing but what was in the natural order, they will pass, when strengthened by the confidence they have put in God, into the supernatural order, and do things which are beyond nature.

III. Consider thirdly, how anxious the Apostle was to show that God not only worked in him, but that He made him work by infusing into him, as it were, His own omnipotence. It seems, therefore, as though he purposely avoided saying, "I can do all things in Him Who governs, supports, or succours me," preferring rather to say, "Who strengtheneth me," in order to show that although what he did was certainly by the power of grace, yet that he did it as being strengthened, not compelled to act: "Take courage, and do."[2] Strengthening implies that the person who is strengthened concurs, according to his ability, in the work for which he is strengthened. And so we do not say that a chisel is strengthened to carve, a brush to paint, or a pen to write. One who is strengthened must have such a share in the work, that it can be ascribed to him in a certain measure. Thus, when the tongue of Balaam's ass was loosed so as to speak with a human voice, it might, indeed, be said that it was made to speak by the Angel, but not that it was strengthened to speak. This, then, is the effect of grace in man: it strengthens them, that is, it invigorates, refreshes, and helps them: "I am thy God; I have strengthened thee, and have helped thee."[3] And this shows that they, too, on their part, perform those actions spontaneously for which they are divinely strengthened; for

[1] Isaias xl. 31.
[2] 1 Esdras x. 4.   [3] Isaias xli. 10.

it cannot be said that one who does nothing himself is strengthened to do it, but rather that he is made to do it. Do not expect that God will ever compel thee to act by virtue of grace, in the same way that Balaam's ass was made to speak; thou hast to concur by thy free-will in such a manner that the work may be ascribed to thee also, but by virtue of the strength granted thee: "The Lord stood by me, and strengthened me, that by me the preaching may be accomplished."[1] Could it be more clearly expressed than the Apostle here expresses it?

IV. Consider fourthly, what those things are of which the Apostle meant more particularly to speak when he said, "I can do all things in Him Who strengtheneth me." He meant contempt, poverty, journeyings, and all the various trials which met him in preaching the Gospel, which, though they seemed insuperable by human strength, he declared that he did not fear in the strength of grace. So that thou perceivest that the strength did not take away the suffering, but enabled him to suffer courageously. And if thou desirest to see a Saint of our own day, who might truly say with the Apostle that he was confident of being able to "do all things," that Saint is certainly the glorious St. Charles. He seems to have been given by God to our times for this very purpose, to show men of refinement and delicacy how much is possible to the weakness of nature when strengthened by the power of grace; for being, as he was, of very noble birth, brought up in luxury and accustomed to command, he brought himself, although of a very frail constitution, to endure great labours in preaching, travelling, giving audiences, making visitations, conducting processions, studying, holding synods, carrying out reforms, and even ministering to the plague-stricken. And to all these labours he added numerous penances of all sorts, hunger, thirst, watching, and subjected his virginal flesh to afflictions and punishment, not uncertain and intermittent, as thine are wont to be, but incessant. Dost thou not think that, under such an accumulation, he suffered so as to know that in himself he was a man weak as others? And yet he never lost courage. And why? Because he knew that grace can do all things in one who does not place obstacles in its way: "I can do all things in Him Who strengtheneth me." Do thou, too, have confidence in thy God; and at the end of thy life thou, too, wilt be able, as St. Charles was, to say, "His

[1] 2 Timothy iv. 17.

grace in me hath not been void; but I have laboured more abundantly than all they: yet not I, but the grace of God with me."[1]

FIFTH DAY.

Dissemblers and crafty men provoke the wrath of God, neither shall they cry when they are bound (Job xxxvi. 13).

I. Consider first, that some persons make it their whole study to feign the virtues which they do not possess, or, if they do not go so far as this, to conceal their vices by cunning. The former are here called "dissemblers," and the latter "crafty men," and both are said to "provoke the wrath of God." Not only is it said that they "merit" it, for this is common to all who sin, even through lack of knowledge, as was the case with the holy King Josaphat, when he made a friendship and alliance with the wicked Achab, merely for the destruction of unbelievers: "Thou helpest the ungodly, and thou art joined in friendship with them that hate the Lord, and therefore thou didst indeed deserve the wrath of the Lord, but good works are found in thee because thou hast taken away the groves out of the land of Juda."[2] But it is said that they "provoke the wrath of God," because these accursed hypocrites never sin through want of knowledge, most of them being remarkably acute, but through malice, and therefore, when they sin, they not only merit the wrath of God, as all sinners do, but they provoke it; for, trusting to their crafty manner of proceeding, they are so bold as to show that they do not fear it, saying, as they do sometimes, to mask their guilt, that they wish God may strike them dead, crush, destroy, and deprive them of all happiness, if the thing of which they are accused is true. "They ask of Me the judgments of justice," that is, the judgments which they ought to fear greatly, and not to provoke; "they are willing to approach God," by coming to the sacraments frequently, by intruding into congregations and monasteries, as though they, too, were really just men instead of deceivers: "As a nation that hath done justice, and hath not forsaken the judgment of their God."[3] What art thou thinking of (supposing that thou hast ever been tempted by the devil to act in a similar way)?

---

[1] 1 Cor. xv. 10. [2] 2 Paral. xix. 3. [3] Isaias lviii. .

Dost thou imagine that thou canst possibly deceive the eyes of God as thou deceivest those of men? It is easy so to act before them that the grave of an adulterer who has died in sin may seem to be an altar, so richly may it be adorned with choice marbles and porphyry outside; but God, Who sees the inside, knows what is there: "Man seeth those things which appear, but the Lord beholdeth the heart."[1]

II. Consider secondly, that it is commonly believed that there are very few hypocrites now in the world; but this is not true: there are only too many. And how many persons, even if they do not feign virtues which they do not possess, at any rate make a show of being rich by setting out their scanty store of wares at the entrance of their shops. These also are "dissemblers," for they pretend to do more good than is the case: "Feigning long prayer."[2] And how many others are there, who, if they cannot entirely conceal their vices, because they are so flagrant, manage to gild them over with a thousand excuses, and never take the blame of them to themselves, but act like a clever thief, who, even when caught with the stolen goods in his possession, knows so well how to twist and turn things in his own favour, that he triumphs in the end, and the court acquits him, and arrests an innocent man in his place. Well, indeed, are they called "crafty." The "prudent" (or "crafty") man "saw the evil," either of blame, or dishonour, or anything else which threatened him, "and hid himself," in order to elude the punishment he deserved. "The simple," when he least thought of it, "passed on and suffered loss,"[3] by being taken in the other's stead. Thou seest, then, that both these classes of men are hypocrites in the strictest sense of the word. For there are four kinds of hypocrisy described by Doctors of the Church. To feign a good falsely, and to conceal an evil which is real, to magnify an existing good, and to excuse an existing evil. Dost thou not think that such persons abound everywhere, to the great injury of that holy simplicity which is compelled to suffer banishment on earth? God grant that thou mayest not thyself be one of this unhappy number, or at least, that thou mayest not be in danger of becoming so from thy anxiety to appear better than thou art in every way, from exaggerating what is good, or concealing what is bad in thee: "Why dost thou endeavour to show thy way good, to seek love?"[4] It is true that by so doing thou

[1] 1 Kings xvi. 7.
[2] St. Luke xx. 47.　　[3] Prov. xxii. 3.　　[4] Jerem. ii. 33.

mayest sometimes gain commendation and applause, like swans which have white plumage and a black skin ; but what will that avail thee if at the same time thou provokest the wrath of God against thyself? "Dissemblers and crafty men provoke the wrath of God ;" so that those swans, that enjoy the false praise of being clean birds among men, will be numbered by God among the unclean.

III. Consider thirdly, that these wicked men, whether "dissemblers " or "crafty," are said to "provoke the wrath of God," because, by exciting it, they not only bring upon themselves grievous punishments, but hasten the time of their infliction. It is the nature of God to be exceedingly slow to punish: "The Lord waiteth that He may have mercy on you."[1] And so thou mayest observe that, in the case even of some very wicked men, He puts off punishing them till after death. But He very seldom does this with hypocrites. He punishes them, as a rule, in this life also, because, if pride of every sort is extremely displeasing to Him, it is most of all so when it aims at affecting a holiness which does not exist. What prince would ever allow the circulation of bad coin to go on for a long time in his dominions? And if this is never permitted with regard to any kind of money, least of all is it as to gold ; because the more valuable the good coin is, the greater is the injury done to the public by debasing it. The same thing may be said in the case we are considering. And if God rarely leaves those long unpunished who falsely lay claim to noble birth, knowledge, judgment, or power which they do not possess, still less does He leave unpunished those impious hypocrites, who falsely lay claim to holiness ; but just as they have attained that height of praise and applause at which they have been aiming by the deception of years, He suddenly unmasks their secret iniquity in the way they least suspect, and brings them to confusion by unexpected disgrace, and sometimes also by other shameful punishments, such as condemnation, imprisonment, or public deposition from their honourable positions. "Be not a hypocrite in the sight of men," says Ecclesiasticus, "and let not thy lips be a stumbling-block to thee," by pretending to perfections thou dost not possess, or by colouring over those imperfections which on certain occasions thou art bound to make known, "lest thou fall " into some deep pit, "and bring dishonour upon thy soul " at the very time that thou wast most highly thought of ;

[1] Isaias xxx. 18.

"and God discover thy secrets," not only in the next world, but in this, "and cast thee down in the midst of the congregation,"[1] by bringing on thee the torment of a tremendous fall, as of an image hurled from the niche where it did not deserve to be placed. Does not this detestation which God has for simulated goodness, move thee also to hate it exceedingly? "Dissemblers and crafty men provoke the wrath of God." Let this suffice to make thee resolve to be, on the contrary, perfectly simple and sincere in all thy doings.

IV. Consider fourthly, that if these scourges with which God chastises impious hypocrites and dissemblers, served to correct them, it could not be truly said that those unhappy men, by bringing them on their heads, provoked God's wrath. For, in that case, their punishment would undoubtedly be the greatest mercy to each one of them. The misery is, that most frequently these chastisements do but act as a simple punishment, which does not avail to bring those deceivers to repentance. And therefore it remains most true that they "provoke the wrath of God," because it is not the wrath which causes the penalties belonging to the next world to be cancelled in this, but rather that which causes them to begin here. And this is what is meant by the words, "Dissemblers and crafty men provoke the wrath of God, neither shall they cry when they are bound." For thou oughtest to think that when He sends upon these criminals the punishments which have been mentioned, the intention of the judge in thus putting them to the torture, is to make them confess the hypocrisy of their lives, and also of the ecstasies, revelations, raptures, and visions that they have feigned, if, unhappily, they have gone so far in wickedness. But they are so jealous of the reputation they have enjoyed for so many years, that they continue in their obstinacy—"They shall not cry when they are bound"—that is, they will not confess their sin, they will not ask for pity or pardon, or if they do so in secret in their hearts, they will not do so aloud so as to be heard by all those whom the miserable men have deceived so long: "They shall not cry." And so they will go to Hell rather than acknowledge that they have borne a false character for holiness. "Even when severely chastised, sinners refuse to make confession, because they have formerly been considered holy in the opinion of all; and although they know that they will be sent into everlasting punishment, they desire to continue in the

---

[1] Ecclus. i. 37.

judgment of men such as they have always laboured to appear."[1] See, then, what may be the end of this fatal love of seeming what thou art not, especially as regards virtue. If thou art destitute of merit, never seek to appear well furnished with it, and if thou knowest thyself to be full of faults, strive not to conceal, but to amend them. "The wicked man imprudently hardeneth his face," as Judas did when he sought to conceal his treason by a kiss; "but he that is righteous correcteth his way,"[2] as St. Peter did, whose abhorrence of his sin caused him to weep bitterly for it during his whole life.

SIXTH DAY.

He that hath looked into the perfect law of liberty and hath continued therein, not becoming a forgetful hearer, but a doer of the work, this man shall be blessed in his deed (St. James i. 25).

I. Consider first, that the final end which so many human legislators have aimed at in their laws has been the happiness of the cities, families, and individuals who should observe them. But none of them have ever been able to attain their object : "They that call this people blessed shall cause them to err ; and they that are called blessed shall be thrown down headlong."[3] And the reason is, that because those laws are not capable of bestowing eternal life, neither are they capable of blessing, but only of condemning. It is the observance of the law of Christ alone that can bestow beatitude. And so thou seest that when He went up into the Mount with His disciples, and opened His sacred lips to promulgate it, He began by announcing this beatitude : "Blessed are the poor —the meek," and so on. No doubt this language occasioned the greatest astonishment, being so opposed to the opinions of the whole of mankind, who up to that time had considered beatitude to consist in directly contrary things, in riches, glory, greatness, and prosperity : "They have called the people happy that hath these things."[4] So that it is not without reason that St. James here says, "He that hath looked into the perfect law of liberty . . . this man shall be blessed in his deed," in order that no one may imagine that it is

---

[1] St. Gregory on this passage.   [2] Prov. xxi. 29.
[3] Isaias ix. 16.   [4] Psalm cxliii. 15.

possible for him ever to attain beatitude by conformity to any law but that of Jesus Christ. Do thou endeavour to understand this most important truth, for upon it rests the foundation of that noble building, our Christian life.

II. Consider secondly, that this law of Christ is called "the perfect law of liberty." It is called so as differing from the Jewish law, which was a law of servitude—"Engendering unto bondage"[1]—and also, because the Jewish law never was able to make any one perfect: "The Law brought nothing to perfection."[2] And this for two reasons: first, because it lacked the perfection of the end, which was eternal life, to which the Law, in itself, was not able to bring any one, but only to dispose him for it; and secondly, because it lacked also the perfection of the means, namely, the three entirely new Evangelical Counsels, by which every one now may arrive so rapidly at perfection, that any uneducated man may aspire to it: "And he shall go before Him . . . to prepare unto the Lord a perfect people."[3] Nevertheless, if there is any one part of the law of Christ which may be said to be more perfect than another, it certainly is those eight marvellous sentences called by Him beatitudes, which, in reality, are only so many maxims of virtue, but of virtue practised in a heroic degree, that is, in a degree which is more Divine than human, so that they alone are able to render man blessed. And this, if thou considerest rightly, is what St. James means when he says, "He that hath looked into the perfect law of liberty . . . this man shall be blessed in his deed." He may certainly be said to allude particularly to the practice of these sublime maxims, since it is these which more particularly make thee blessed. How is it, then, that up to this time, it may be, thou hast never tried to understand them properly?

III. Consider thirdly, that there are two ways in which man may be called blessed: in possession, and in hope. He is blessed in possession when he attains to the glory of Paradise —"Blessed are they that dwell in Thy house, O Lord"[4]— then he is perfectly blessed. He is blessed in hope, when he has a firm and well-founded expectation of attaining to that glory—"Blessed is he whom Thou hast chosen and taken to Thee: he shall dwell in Thy courts"[5]—then, also, he is blessed, but imperfectly. Now, it is certain that the Eight Beatitudes of the Gospel cannot give thee, on this earth, that

[1] Galat. iv. 24.   [2] Hebrews vii. 19.   [3] St. Luke i. 17.
[4] Psalm lxxxiii. 5.   [5] Psalm lxiv. 5.

beatitude which is perfect, because they cannot make thee blessed in possession; but, at least, they give thee that which is imperfect, because, in a very special manner they make thee blessed in hope. They are the clearest of all marks of pre-destination, and therefore they cause thee to hope for the glory of Paradise with the greatest confidence and the firmest foundation that is possible, consistently with remaining within the limits of hope: "We are saved by hope."[1] Is it possible that thou art not enamoured of them?

IV. Consider fourthly, that there is between these two beatitudes, of hope and of possession, an intermediate kind, so to speak: it is that which not only disposes thee to attain the glory of Paradise by way of merit, but which begins to give thee a foretaste of it. And this precisely is the peculiar characteristic of these eight great maxims of virtue. For when they are rightly practised, that is, in a heroic manner, they make thee begin to taste on earth that unalterable spiritual sweetness which belongs to the saints in Heaven. And, therefore, St. James here says, "He that hath looked into the perfect law of liberty . . . this man shall be blessed in his deed." He does not say "on account of his deed," because every just man, who does any meritorious work, will be blessed because of it if he does but persevere; but he says "in his deed," which can only be said of holy men, because, acting as they do in a heroic manner, they are not only blessed "on account of their deed," that is, by their works, but also "in their deed," that is, in their works: so great is the happiness they feel in acting in this Divine manner. And therefore, in a certain sense, it may be said that these eminently just persons are blessed in possession, even on earth, because, if they are not yet plunged in the ocean of the joys of Paradise, they are, at least, beginning to taste its streams. It is certain that they are more than blessed in hope, because their expectation of being one day plunged in those joys far surpasses that of other just men, just as he who sees flowers budding on the tree has a much stronger hope of the fruit that he desires than one who sees leaves only. Why, then, wilt thou be satisfied with leaves, when it is in thy power also to enjoy the flowers which are so sure an earnest of fruit?

V. Consider fifthly, that if thou desirest to partake of the great advantages which these beatitudes produce, it behoves

[1] Romans viii. 24.

thee to fulfil two preliminary conditions mentioned by St. James. The first is, to understand thoroughly what are the virtues which are formed by this perfect law : and to do this, we must look into that perfect law, which does not mean giving it a superficial glance, as men do to the notices at the corners of the streets ; for that would be looking "at," not "into" the law. It means looking thoroughly into it, examining it, considering it, contemplating it attentively. For this purpose thou hast here these beatitudes divided into an equal number of separate meditations for the next eight days, in order that thou mayest discover their true meaning, in that precise degree which is for thy profit. Hast thou remarked the difference that there is between the sailor and the astronomer? They both contemplate the stars attentively : but the astronomer does so out of curiosity to learn as much as possible concerning their attitude, their appearance, their aspects, and their motions ; whereas the sailor's object is only to direct his course aright. And it is this second rule which thou oughtest to follow in meditating. The other condition is, when thou hast thoroughly understood the glorious truth taught by Christ, to set thyself very courageously to put it in practice, being firmly persuaded that it will not benefit thee in the least to court holiness, so to speak, all day long, if thou dost not espouse her. And this is what St. James means to infer when he says, "He that hath looked into the perfect law of liberty and hath continued therein, not becoming a forgetful hearer, but a doer of the work, this man shall be blessed in his deed." To "continue" in the law is a form of expression used in the Sacred Scriptures, signifying a continual, persistent, solid observance of it : "Cursed be he that abideth not in the words of this law, and fulfilleth them not in work."[1] And it is this kind of observance which is required in the case before us for becoming blessed. So that, when thou art told that "Blessed are the poor, the meek," and so on, there is always this condition understood, if not expressed : "If they have continued in that perfect law of poverty, meekness, and the rest." Otherwise, it is very certain that thou wilt not be blessed, even if thou hast espoused this perfect law, if thou resolvest on repudiating it after a few days. For dost thou suppose that Christ is like many human teachers, who consider themselves sufficiently appreciated by their hearers, when they perceive

[1] Deut. xxvii. 26.

that they have fully understood the good instructions they have received either in medicine, morals, or canon law, even though they may be indifferent to acting according to them? The very contrary is the case : if thou dost not put in practice the teachings of Christ, it will be the same as if thou hadst purposely forgotten them. For which reason, the man who does not practise them is not merely called "a forgetful hearer" by St. James, but one who has "become a forgetful hearer," because his case is not that of one who simply forgets, but of one who forgets intentionally.

SEVENTH DAY.

Blessed are the poor in spirit, for theirs is the Kingdom of Heaven (St. Matt. v. 3).

I. Consider first, that there are two sorts of poor on earth —the poor by necessity, and the poor by choice—both of whom are, indeed, capable of attaining the Kingdom of Heaven, but neither of whom are the happy persons to whom it is so distinctly here promised by our Lord, and whom He calls "the poor in spirit." For if thou considerest those who are merely poor from necessity, how can they lay claim to that Kingdom by right of the poverty they endure, who endure it certainly, but unwillingly? And if thou considerest those who are merely poor by choice, how can they lay claim to it either, who have, indeed, made themselves poor, but who have done so out of pride, like certain of the ancient philosophers? Those, then, to whom the Kingdom of Heaven is here so expressly promised are those who are poor, not only by choice, but in spirit. And these are, in the most literal sense, those who, following the movement of the Holy Ghost Who has incited them to do so, have embraced evangelical poverty according to the counsel of Christ, by entirely, exactly, and perpetually renouncing everything belonging to them. I know that those also share in this beatitude who, although they are rich, have their hearts so detached from riches as to be ready, if it were in their power, to make themselves as poor as St. Francis for the love of Jesus. Still, they share it in a remote sense, in the same way that those belong to the army of martyrs who have gone among the fiercest barbarians to find a Decius or a Diocletian, but without doing so. For such persons are not truly poor; and though in desire they may be poor, still they are not "poor in spirit:" in other words, they are poor in affection, but not in act, having, indeed, a spirit of poverty, but not also poverty of spirit. Those only have true poverty of spirit who have truly left everything for Jesus, without any hope of ever possessing it again, or any desire to do so, so that they are able to say with St. Peter, "Behold, we have left all things, and have followed Thee."[1] This is the most probable, because the most exact, interpretation of the passage, and it is that which is given by St. Jerome, St. Basil, St. Bernard, St. Ambrose, and others of the ancient Fathers, as well as by most modern commentators. Think, then, how excellent is the state of those good religious who are so poor and destitute that thou, perhaps, art ready to despise them with pride. It is the state of men to whom is reserved the glorious Kingdom of Paradise, which is promised to them by that title because there is no other so great. Oh, how much higher than thyself, it may be, wilt thou, on the Day of Judgment, see those whom now thou dost not deign to admit into thy presence!

II. Consider secondly, that although the Kingdom which is here said to be destined to these evangelical poor is in the future, yet Christ was not content to say, "Blessed are the poor in spirit, for theirs shall be the Kingdom of Heaven." He said, "Theirs *is* the Kingdom of Heaven." And why was this, except to show how infallibly certain they may be said to be of attaining it, so many aids does this holy poverty afford of refraining from evil and of doing good? But besides this, have not these blessed poor already paid the full price required by Christ for this Kingdom when He said, "Amen, I say to you, that every one that hath left house, or brethren, or sisters, or father, or mother, or wife, or children, or lands, for My Name's sake, shall receive an hundred-fold, and shall possess life everlasting."[2] Therefore, just as when a man has paid the full price demanded by the sovereign for receiving a governorship, a county, or a marquisate, he may be said to be the owner of that governorship, county, or marquisate, although he may not yet be installed in it, so, too, may he be said to own Paradise who has in the same way paid the price so expressly fixed for it by Christ. All that remains to be done by one who has brought himself to such a state of true poverty

[1] St. Mark x. 28.   [2] St. Matt. xix. 29.

for Christ's sake is to persevere in it, and not to become again attached in that state to the things of this world, its comforts, grandeur, glory, or distinctions, all of which are foreign to his state. What would this be but gradually to take back the price which has been paid, and so to forfeit the right which was once possessed to the Kingdom? How blessed, then, is the man who, in the state of poverty, continues truly poor, and behaves as such, and professes himself such through all his life! How sure is he of winning Paradise! And thus thou seest how it is that evangelical poverty constantly persevered in is a sign of predestination. It is, indeed, the most evident sign there is. For though it cannot be denied that all the beatitudes are also signs of it, as will be seen when we consider them separately, yet they are not so plain to us. For who can make sure of possessing the requisite meekness, sorrow, justice, mercy, purity, or grace? All these virtues are chiefly interior, and so, although, doubtless, they too are the price at which Paradise is bought, yet they do not so well enable us to recognize their perfection or value. But to have left all for Christ, and to act like the poor, and to profess to be poor, is a thing which is palpably evident : how great, then, is the certainty it is able to give thee! How is it that thou art still insensible to the charms of this beautiful beatitude? Is it not foolish to be able to share in it by exercising one resolute act of the will, and not to care about doing so?

III. Consider thirdly, why our Lord made poverty the first of the beatitudes. It was in order to do away with the chief obstacle to the salvation of men—riches : "Amen, I say to you, that a rich man shall hardly enter into the Kingdom of Heaven."[1] For although poverty is also the cause of many evils, as it is written, "Through poverty many have sinned,"[2] this is only when it is hated, not when it is loved. Indeed, when it is loved it brings with it very great advantages ; for, if there were no other, it is a powerful aid in becoming humble, mortified, modest, to be which, at least perfectly, is almost impossible in riches. And therefore the Spirit of God never urges us to get riches, but to be indifferent to them : "If riches abound, set not your heart upon them."[3] Besides this, poverty makes a man more free and unencumbered to follow Christ all over the world ; and therefore He made it the foundation of the Apostolate : "Every one of you that doth

[1] St. Matt. xix. 23. [2] Ecclus. xxvii. 1. [3] Psalm lxi. 11.

not renounce all that he possesseth cannot be My disciple."[1]
And not only so, but this beatitude is also the foundation of
all those that follow. For if thou observest closely, thou wilt
see that poverty is the greatest help in attaining the virtues
belonging to them. It is easier for one who is poor to be
meek, to mourn, to sacrifice himself as a victim to justice, to
have a merciful or a pure heart, and to preserve a deep peace
amidst all the disturbances of the world, being in a state which
is troubled about nothing. And so Christ made this the basis
of the other beatitudes : poverty espoused in this world for
the pure love of God. Oh, if thou didst but know the beauty
of this bride, what love wouldst thou feel for her! Consider
how well the Son of God knew it, so that, not being able to
espouse her in Heaven, He came down to earth to do so:
"Being rich, He became poor."[2] And if thou art no longer
able to espouse poverty, at least do not depreciate or despise
her, or prefer riches to her—riches which, as though to bring
shame on our Lord, are now considered the chief beatitude
by worldlings.

IV. Consider fourthly, that that gift of the Holy Spirit
which is called fear, corresponds to this beatitude, which is
the first pronounced by Christ. For he who greatly fears God
and is afraid both of His judgments and His punishments,
and above all, of those evils which may threaten him from
God in the next world, will be most eager to strip himself of
everything which is the principal cause of them to the majority
of persons. And such are the riches which they love : "Riches
kept to the hurt of the owner."[3] See how sailors act in a
storm. They hasten to lay hands upon the most valuable
things—wool, linen, silver of great price—and cast them
without hesitation into the sea, so great is their fear of being
lost if they should be bold enough to spare their merchandise
in the face of a raging sea : "The mariners were afraid, and
they cast forth the wares that were in the ship into the sea,
to lighten it."[4] This is what they also do who are really afraid
of being lost in the far more terrible sea of the Divine justice
stirred to wrath : they hasten to save themselves by casting
away their riches, that fatal freight which threatens to sink the
ship. What, then, can be said of all those rich men who,
instead of lightening their ship in the storm, show the greatest
anxiety and eagerness in loading it heavily? "They heap

[1] St. Luke xiv. 33.
[2] 2 Cor. viii. 9.   [3] Eccles. v. 12.   [4] Jonas i. 5.

together silver as earth."[1] They have no fear: if they had, dost thou think they would be so mad as to cast themselves away to save their wealth, instead of casting away their wealth to save themselves?

———

EIGHTH DAY.

Blessed are the meek, for they shall possess the land (St. Matt. v. 4).

I. Consider first, that, speaking strictly, the meek are those who are quick to repress the motions of anger, that is, of that heat which inclines us to feel resentment against those who have offended, are offending, or have the intention of offending us. It is true that this readiness to repress such motions may arise, if we observe closely, from three causes. First, from a purely natural light, which shows thee the deformity and disorder of anger whenever it acts not in conformity with reason, but in defiance of it. This certainly is a virtue, but a moral virtue such as was common to many even of the heathens, such as Socrates, Antigonus, Anaxagoras, and others, who were meek merely because they were ashamed of displaying anger. Secondly, from strict obedience to the law of Christ, which so expressly and emphatically forbids all revenge. This is, indeed, a Christian, but a common Christian virtue, because it does not prevent thee from suffering intensely when thou art obliged to repress a movement of anger. Thirdly, from a great love of God, which makes thee gladly endure every personal offence for His sake, and also from a great hatred to thyself which prevents thy feeling it. This last is not merely a Christian, but heroic virtue; and therefore it is this which is here spoken of, because it is this which makes thee truly meek. Do not therefore imagine that when Christ said, "Blessed are the meek," He meant to call all the meek who are in the world blessed. He meant those who are possessed of this sublime and solid meekness which has been just described, for it is in this that true peace is found. Dost thou desire to know whether this great gift is thine? Thy own heart can best answer the question, for it is quite possible that it may be a little Etna filled with a fire of which only itself is conscious. Oh, how often dost thou pretend to a meekness which thou dost not possess! "Blessed are the meek:" not those who

[1] Zach. ix. 3.

have the art of appearing meek, but those who are so in reality; and of these there are but few in the world.

II. Consider secondly, that the meekness here spoken of is a special sign of predestination. First, because it makes thee like Christ, Who set so high a value on this virtue, that He took His title from it: "Tell ye the daughter of Sion, Behold thy King cometh to thee, meek."[1] Secondly, because it preserves thee from innumerable dangers of sin by preserving thee from anger, which is a sin leading to many others—"He that is easily stirred up to wrath shall be more prone to sin"[2]—both because of the object of the angry man, which is that vengeance which is sweeter to a man than honey, and also because of the mad violence with which he is transported in his desire for it: "Thou destroyest thy soul in thy fury."[3] Thirdly, because it bestows on thee a wonderful disposition for receiving the grace which will make the practice of virtue easy to thee by preserving thee in great peace: "To the meek He will give grace."[4] And so, when Christ here said, "Blessed are the meek, for they shall possess the land," He did not mean by "the land," this earth, which is also possessed by the furious and violent; He meant that land where there is no place for those proud men, His empyrean Heaven. But He called it by the name of "land," because, just as with the Jews the brazen serpent signified the Saviour on the Cross, the sea signified Baptism, the manna the Eucharist, and every other type signified, however obscurely, its antitype; so did the land, which has been so often promised, signify Heaven to them. "I said, Thou art my hope, my portion in the land of the living."[5] "They that wait upon the Lord, they shall inherit the land."[6] "Such as bless Him shall inherit the land."[7] And, most appropriately of all to our subject: "The meek shall inherit the land."[8] It is to this passage that our Lord alluded in His discourse, only that whereas the Psalmist said "inherit," Christ said "possess," because those who were still minors could, indeed, inherit the heavenly beatitude, but not enter on its possession. However that may be, dost thou see the way to win Paradise? It is by yielding. Thou art accustomed to see that this land wherein thou dwellest is always won by means of disputes, quarrels, contentions, and sharp struggles. Do not suppose that the

[1] St. Matt. xxi. 5.   [2] Prov. xxix. 22.
[3] Job xviii. 4.   [4] Prov. iii. 34.   [5] Psalm cxli. 6.
[6] Psalm xxxvi. 9.   [7] Psalm xxxvi. 22.   [8] Psalm xxxvi. 11.

land which is on high is won in the same way. It is won by meekness, that is, by yielding to all. "Blessed are the meek, for they shall possess the land." This is the other reason why Christ made use of this expression: to make His language more admirable.

III. Consider thirdly, why it was that, having said in the first place, "Blessed are the poor," Christ added in the second, "Blessed are the meek." It was because the first thing which is necessary for the poor is to prepare to be despised. For it is the way of the foolish world to value men, as metals and marbles are valued, for their splendour: "The rich man spoke, and all held their peace; the poor man spoke, and they said, Who is this?"[1] It is necessary therefore for him who has resolved to leave all for God, to arm himself in the first place with very great meekness, in order to be able to endure the mockery and insults which await him. It is true that this is still easier for him, if he is willing to see it, from his freedom from all obligation to observe the vain points of honour of the world. And so, immediately after poverty, our Lord adds meekness, because it is more unseemly for one who is poor, more especially poor in spirit, to be haughty, unbending, disputatious, and arrogant. If then thou art poor by necessity, thou hast to despise the being despised; if thou art poor by choice, thou hast to love it, because thou art bound to love everything which is necessarily connected with the state which thou hast chosen.

IV. Consider fourthly, why it is that the love of contempt will, more than anything else, help thee to conquer that many-headed hydra, anger. The reason is that it kills it at a blow. Love contempt, and thou wilt be meek. To prove the truth of this, consider who it is against whom thy anger is most violently excited. Is it all those who are gravely offended with thee? Certainly not: for if thou knowest that any one has reason to be offended with thee, as, for instance, when a sovereign, a father, or an official punishes thee for some fault which thou hast committed, in such a case thou wilt indeed beg for mercy, thou wilt be grieved and troubled, but not angry. It is when thou knowest thyself to be despised that thou art angry. And so, if any one offends thee through ignorance or thoughtlessness, thou art not angry, at least not seriously angry, only so far as thou thinkest that the other has failed in his duty of considering what he was doing. Thy

[1] Ecclus. xiii. 28, 29.

anger is greater if any one offends thee when carried away by a fit of passion; but even this does not excite thy greatest anger. This happens when the person who offends thee does so intentionally, shows it openly, publishes it, and glories in it: for such a one evidently despises thee. Do, then, as I say: love to be despised, and then thou wilt not be in the least angry at seeing thyself despised. But thou dost not love this: thou mayest, indeed, sometimes despise thyself by speaking depreciatingly of thyself, but thou canst not endure to be despised, not even by the same words which thou hast used in speaking of thyself. If such conduct is to despise oneself, it certainly is not to love being despised, as is necessary for meekness. Therefore, think often of thy many offences against God; and thus, conceiving a holy hatred of thyself, thou wilt not only love to be despised, but will wonder that all men do not despise thee.

V. Consider fifthly, that the gift of piety corresponds to this second beatitude. And this is not wonderful, for piety is of very great help in acquiring meekness. What is piety but the virtue which disposes us to acknowledge God as our Father, and to consider and treat Him as such with true reverence? Now, if thou so acknowledgest Him, dost thou not also know that He governs thee with His special providence, that He succours thee, and loves thee; and that, therefore, whenever He allows any misfortune to befall thee, it is all for thy greater good? How is it then that thou art so easily disturbed by every such casualty? This is a want of piety, because it is a want of reverence to thy Father. If any one offends, or mortifies, or calumniates thee, why is it in his power to do so? Because thy Father permits it. And yet thou art angry, as though thy Father knew nothing about it! When any Semei mocks thee, do thou say as King David did, "Let him alone, and let him curse, for the Lord hath bid him curse David, and who is he that shall dare say, Why hath He done so?"[1] This is an act of true piety, which will help thee very greatly to acquire the virtue of meekness.

[1] 2 Kings xvi. 10.

NINTH DAY.

Blessed are they that mourn, for they shall be comforted (St. Matt. v. 5).

I. Consider first, that although this word "mourning" has come to have a very wide meaning, yet it referred originally to that kind of a sorrow arising from the loss of a good. Go through the Sacred Scriptures, and thou wilt see that this is always inferred in the expression, "mourning apparel, days of mourning, house of mourning," and the like. And in our own days, speaking of a childless couple who had long asked a child of God in vain, we should say that they were in great grief, but we should not use the word "mourning." But if we spoke of parents who had lost, or were on the point of losing, a child, then we should do so, because these parents in their state of mourning refrain altogether from the pleasures and amusements which the childless parents, who are not in mourning, are free to enjoy. Having said this, thou wilt understand who they are whom Christ here meant to call blessed, when He said, "Blessed are they that mourn:" they are those who mourn for a lost good. But, then, are all those blessed who weep for the dead, or for a lost inheritance, or for a dignity from which they have been deposed? No; because the reason given by Christ does not apply in their case. He said, "Blessed are they that mourn, for they shall be comforted." But these persons whom thou hast suggested cannot be the mourners here intended. For, though they should weep a very ocean of tears, they can never avail to recover what they have lost: tears will neither raise the dead nor restore money and dignities, and therefore they do not make those who shed them blessed, because they can give them no comfort; rather do they increase the sorrow of these mourners by consuming them unavailingly day by day. Those who are blessed because of their tears are they who weep for the losses brought upon them by sin, for it is they only who can thus regain what they have lost, and who will therefore one day be comforted. Now, these losses are two: there is the loss of the goods of grace, and the loss of the goods of glory. It is these persons, then, whom Christ here especially called blessed when He said, "Blessed are they that mourn, for they shall be comforted:" those who mourn for the losses which have been just named, and whose one desire, therefore, is to repair them by hearty penitence. How is it, then, that thou so keenly regrettest every little worldly good which is taken from thee? Keep thy grief for nobler objects, keep it to bewail that which was lost by thee in a moment by sin, the goods of grace and of glory; otherwise it will be not useless only, but injurious.

II. Consider secondly, that this just mourning is a sign of predestination: "Blessed are they that mourn;" because it infallibly brings with it the reparation of these two heavy losses which are being lamented, those of the goods of grace and of glory: "Blessed are they that mourn, for they shall be comforted." There are three ways of comforting one who is grieving for a good that he has lost: first, by urging him to bear his loss with calmness; secondly, by giving him a good in some degree equivalent to that of which he has been deprived; and thirdly, by restoring to him the very good which he has lost. He whose consolation is of this last kind is a comforter indeed. It was thus that Christ comforted the widow of Naim. And so when He here said, "Blessed are they that mourn, for they shall be comforted," He certainly could not intend the first kind of consolation, for that would be very wrong. A man can never be told to bear with calmness the losses either of grace or of glory which he has incurred by sin; on the contrary, he should be told never to cease lamenting them. Neither can the second kind be intended, for the world contains no good which is in any way, however remote, equivalent to those which are lost by sin, namely, grace and glory. Our Lord, therefore, could only have intended to speak of the third kind of consolation, that is, the true one. And the certainty of being one day thus comforted involves the fact of predestination. Therefore Christ said, "Blessed are they that mourn, for they shall be comforted," or "shall receive consolation," as some versions have it, for the sake of greater clearness, because that true consolation of which He spoke is altogether future. It cannot, indeed, be denied that this blessed mourning brings with it, even in this world, a very great consolation. But it arises entirely from the pleasure which is taken in the flower as an earnest of the fruit: "We became like men comforted."[1] And it can never be perfect here, because there must always remain a doubt as to whether the fruit will set. There must always be a good deal of alarm troubling the confidence of having recovered the grace of God which has been lost by sin: "Who can say,

---

[1] Psalm cxxv. 1,

My heart is clean?"[1] And there is still greater fear as to the certainty of persevering to the end in that grace even when it has been recovered. The only perfect consolation, then, is that which will come from the ripe fruit, and this fruit will at last be found in Heaven, and it is this which is the consolation here promised by Christ, not only because in Heaven every penitent will certainly regain those goods of grace and glory for which he is mourning, but also those temporal goods of which he deprived himself that he might live in a state of mourning, such as pleasure, glory, friendship, greatness, comforts, and the like, which are incompatible with sorrow of heart. Oh, with what lavish interest will all these goods be then restored! How, then, canst thou, who art now living a life of mourning as a penitent, fear that thou shalt not be comforted? "Comfort is hidden from my eyes."[2] Take courage, for thy mourning shall be followed by the only true comfort, that comfort which, as I have said, will restore the very same good which was lost: "I, I Myself will comfort you."[3]

III. Consider thirdly, why Christ, after laying the foundation by the words, "Blessed are the poor in spirit," added, first, "Blessed are the meek," and then, "Blessed are they that mourn," as it stands in the Vulgate, which it is always best to follow. The reason is, that just as poverty is the best disposition for meekness, as was shown in the last meditation, so is meekness the best disposition for mourning, and therefore it rightly precedes it. We may add that Christ intended by these three beatitudes which have been explained, to set in order the whole of the old man in himself. And therefore He would have him, in the first place, trample on all those goods that are beneath him, which are the external goods signified by riches; and then, passing on to what is interior, He would have him first moderate the irascible part by meekness, and next the concupiscent by mourning, because when once the ardent and ambitious movements of anger are pacified, then is the time for thinking quietly of one's own soul, and for lamenting its unhappiness, by depriving oneself, for that purpose, of all pleasures, whether impure or imperfect, which are ill-suited to a mourner. See, therefore, how thou mayest really know whether thou art living in a state of mourning, namely, by the marks which are characteristic of that state.

---

[1] Prov. xx. 9.
[2] Osee xiii. 14.   [3] Isaias li. 12.

IV. Consider fourthly, that the first of these marks are those belonging to the concupiscent part, to which mourning is directly antagonistic. One who is a true mourner is with difficulty persuaded to take even a little food, so great is his disinclination to it; how impossible, then, is it for him to think of feasting or banqueting, or even of choice dishes! Fasting goes hand in hand with mourning: "Anna wept and did not eat."[1] A mourner has done with vain society, with theatres and balls, and such-like idle amusements which gay people frequent: "A tale out of time is like music in mourning."[2] How, then, canst thou pretend to live in mourning if thou art devoted to such things? Next come those marks which concern the irascible part, which always aims at superiority, and therefore is incompatible with mourning. A mourner does not court glory, he tramples on it. It is in this state that he is humble towards all men ; he has recourse to all, asks help of all, thinking himself more wretched than any: "As one mourning and sorrowful, so was I humbled."[3] Again, I ask, what sort of mourning can thine be if thou art not able to think of so many ways of exalting thy name? If thou wert a true mourner thou wouldst abase thyself lower than the sorrowful Mephiboseth, who, when David lavished honours upon him, answered, "Who am I, thy servant, that thou shouldst look upon such a dead dog as I am?"[4] Thirdly, and lastly, come the marks which concern exterior goods, or what are commonly called the goods of fortune, pomp, grandeur, presents, prodigality. How unbecoming to mourners is rich apparel! Such persons lay aside all gay dresses and ornaments, and choose black garments : "Tearing his garments, Jacob put on sackcloth, mourning for his son a long time."[5] How is it with thee? Hast thou in thy mourning entirely banished every kind of vanity? Look at the houses of mourners, notice the bare walls, the unornamented furniture, the hard beds : all this shows true mourning, and if thy conduct is the reverse thou art no mourner. This, then, may teach thee what Christ meant when He said, "Blessed are they that mourn." He meant those whose hearts are detached from whatever is incompatible with mourning.

V. Consider fifthly, that this beatitude answers to the gift of knowledge, because it is this more than anything else which will impart to thee that superhuman compunction that will

---

[1] 1 Kings i. 7.   [2] Ecclus. xxii. 6.
[3] Psalm xxxiv. 14.   [4] 2 Kings ix. 8.   [5] Genesis xxxvii. 34.

make thee blessed : " He that addeth knowledge addeth also labour " (*dolorem*).[1] Why is it that there are so many Christians who do not sorrow for their losses, enormous as they are? Because they are ignorant. They do not know what those goods of grace and glory are which they have lost. And so the loss of all these troubles them far less than the loss of a horse or a hound. Far otherwise is it with one who has a deep knowledge of those goods. How does he grieve when he sees that he has lost them ! " My tears have been my bread day and night, whilst it is said to me daily : Where is thy God ? "[2] This, then, is the right way of spending the time of mourning : to go into the heart of the matter, that is, our sins, and when this is done, mourning alone seems but a small thing, we shall go on to weep and lament, and even to punish ourselves mercilessly. And this is, lastly, the meaning of living, as some describe this state, not merely in "mourning," but in "mourning, weeping, and lamentation." If thou thinkest that this is an unmeaning accumulation of words, thou art greatly mistaken. On the contrary, these expressions describe the three stages through which a true penitent has to pass : mourning (*luctus*), weeping (*fletus*), and lamentation (*planctus*). The first describes the deep grief which is contained in the heart, the second the tears in which that grief vents itself, and the third the acts of self-chastisement and violence to the body which accompany those tears. Such is the opinion of many great doctors. And so thou seest that in the Sacred Scriptures joy is contrasted with mourning, as Solomon says, " Mourning taketh hold of the end of joy."[3] Laughter is opposed to weeping : " A time to weep and a time to laugh ; "[4] and dancing to lamentation : " A time to mourn (*plangendi*), and a time to dance."[5] This, then, is what thou hast to do if thou desirest to lead the life of a perfect penitent : first, let thy heart be filled with a deep and abiding compunction for the grievous sins thou hast committed ; then go on to weep for them bitterly before God, if thou art worthy of the gift of tears ; and if not, at least desire so to weep. And next, be careful to afflict thy flesh, according to thy ability by penances suited to thy strength, such as hair-shirts, chains, disciplines, and the like : " Make thee mourning as for an only son, a bitter lamentation,"[6] such as has been here described. Do not imagine that it is an ordinary kind of

[1] Eccles. i. 18.   [2] Psalm xli. 4.   [3] Prov. xiv. 13.
[4] Eccles. iii. 4.   [5] Eccles. iii. 4.   [6] Jerem. vi. 26.

mourning which makes the mourner "blessed." It must be such as cannot refrain from bitter weeping, since, as is the universal opinion, the beatitudes enumerated by Christ are, indeed, in other words, the virtues befitting a Christian, but they must be those virtues practised in a heroic degree.

TENTH DAY.

Blessed are they that hunger and thirst after justice, for they shall have their fill (St. Matt. v. 6).

I. Consider first, that by the word "justice" we must here understand all the operations of the just man, in other words, every kind of virtue: "Blessed are they that do justice at all times."[1] And it is worthy of note that our Lord was not satisfied with saying, as the Psalmist did, "They who do justice," but went further, and said, "They who hunger and thirst after justice;" because it is not enough always to do well, there must be a very great and ardent desire of always doing better. And therefore this beatitude applies alike to the beginner, the proficient, and the perfect, all of whom, like persons continually hungry and thirsty, should never be satisfied. Beginners can never expect to be of that happy company of the blessed if, directly after their conversion, they set themselves with a feeble will to the practice of virtue, like men who sit down to table without an appetite. They must, on the contrary, apply themselves to it with the determination, if possible, of becoming saints, and not say, as some do, If only I can enter Paradise, it is enough for me, no matter what may be my place there; for this is a very foolish way of speaking. Neither can proficients cherish such an expectation if, when they have reached a certain point, they think they may stop there; there is no standing still in God's service: "He that is holy, let him be sanctified still."[2] Therefore, they ought to be continually aiming at higher and higher perfection, as though they were but just beginning: "When a man hath done, then shall he begin."[3] Neither can those even who are perfect enter into this blessed company, if they are satisfied with well-doing themselves, and do not strive to the utmost of their power to induce others to do well; for the

[1] Psalm cv. 3.   [2] Apoc. xxii. 11.   [3] Ecclus. xviii. 6.

hunger and thirst of justice is not confined to our own personal
good, but extends to that of others. And the reason is, that
spiritual is not like corporal punishment, of which the more
thou bestowest on others the less there is for thyself; but,
on the contrary, the more spiritual nourishment thou takest
thyself, the more thou hast to give to others. And so the
hunger and thirst with which thou art consumed do not excuse
thee from throwing open thy barns and cellars to thy neighbours. Rather shouldst thou invite even those at a distance
to satisfy their wants abundantly: "Come, eat My bread, and
drink the wine which I have mingled for you."[1] Thus will it
be evident that thou hast a real hunger and thirst of justice—
thirst, as to those virtues which are easier, like drink; hunger,
as to those which are more difficult, like meat—and thus
shalt thou, too, be blessed: "Blessed are they that hunger
and thirst after Justice." How, then, canst thou be said to
have this hunger and thirst, if every little act of virtue which
thou performest seems to thee a great thing?

II. Consider secondly, that this hunger and thirst is a sign
of predestination; for it will procure for thee a very high place
in Heaven: "Blessed are they that hunger and thirst after
justice, for they shall have their fill." That it will take thee
to Heaven is beyond a doubt; for if Christ declares that thou
shalt "have thy fill," He can refer to no other place, since on
earth thou art to be always hungering, always thirsting. And
the reason is, that here it is impossible for thee ever to attain
to a sufficient degree of justice; indeed, it is only when thou
learnest by thy progress how far thou still art from any high
degree of justice, that thou canst really be said to have
attained to it. This is what St. Augustine says: "That man
has made great progress in this life who has learnt, as he
advances, how far he is from perfect justice."[2] It can therefore be in Heaven only that thou wilt have thy fill, because
there only is perfect justice: "I shall be satisfied when Thy
glory shall appear."[3] But it is also certain that this hunger
and thirst will procure for thee a very high place in Heaven.
For satiety, of whatever kind, must always be in proportion to
desire. The amount of nourishment which would be amply
sufficient for a man who felt some slight need of support and
refreshment would not be so for one who was suffering very
great hunger and thirst. And therefore, when our Lord
assures thee that thou shalt have thy fill of justice, supposing

thy desire of it to be very keen, it is certain that the fare which will be thy portion must be far richer than that which is prepared for those who are more indifferent. And this is the same thing as to have a high place in Heaven: "He hath filled the hungry with good things," not merely "He hath refreshed," but "filled" them; because in Heaven the man whose justice is greater will be more rewarded than he whose justice is less. How is it, then, that thou dost not use all possible means to excite in thyself a hunger and thirst which are so advantageous to thee? If thou desirest to acquire it, drive out all evil humours, endeavour to bear a little regular fasting from those sensual or sensible pleasures in which thou indulgest to excess. Begin, instead, to make trial of those that are spiritual; give thyself to frequent prayer; enter deeply into the contemplation of the beauty of justice; see how profitable it is, how sweet, how glorious. This will excite in thee so great a hunger and thirst after justice, that thou wilt languish with desire at the thought that on this miserable earth thou wilt never be able entirely to satisfy it.

III. Consider thirdly, why Christ put this beatitude in the fourth place. The saints tell us that it was because, having in the preceding beatitudes withdrawn man from evil, from affection to the goods which are beneath him by leading him to a complete renunciation of all that is his, from affection to a state of superiority by repressing the irascible part by means of meekness, and from affection to corporal pleasure by repressing it also by sincere mourning; having, I say, done all this, there remained the work of moving him to good, according to that great law: "Turn away from evil, and do good." Therefore He began this by giving him a keen hunger and thirst for what is good, because the first necessary condition for well-doing is the desire to do so. It is true that for any virtue to reach the height of a beatitude, it must be, as has already been often said, not merely in an ordinary, but a heroic degree. And so Christ was not contented with any kind of desire, even though it were a desire of justice; for He did not say, "Blessed are they that desire justice," but He would have this desire resemble that of a man pining with hunger and thirst, which is the strongest desire that can be felt, and therefore He used this very expressive phrase: "They that hunger and thirst." The Prophet said of the Jews who were besieged in Jerusalem, "They have given all their precious things for food to relieve the soul," not "to sustain"

it, for that was more than they dared to hope 'for, but only
"to relieve" it. And in like manner thou oughtest to take no
thought for any earthly thing, when it is a question of feeding
thy soul with this precious food of justice, which is of so
much greater value. This will show that thou really hast the
desire which Christ intended, the desire of a man pining with
hunger and thirst. And if none of the means which have
been mentioned suffice to give thee such a desire, do thou at
least long to experience it. Desire the desire : "My soul
hath coveted to long for Thy justifications at all times."[1] Act
like a sick man who has, indeed, no appetite, but who would
give anything to gain one, and yet it is not in his power to
gain it by merely desiring to do so ; whereas, if only thou
desirest this keen hunger and thirst of justice of which we are
speaking, the very desire will be the beginning of it.

IV. Consider fourthly, that the gift of fortitude corresponds
to this beatitude. And the reason is that it is no ordinary
energy that is needed to overcome the difficulties in the way
of satisfying this strong and vehement desire of justice, but
courage. See to what dangers a man will expose himself to
satisfy a fierce hunger or to quench a burning thirst : he will
even face an armed host, as was done by the besieged in
Bethulia. Fortitude, therefore, is needed ; without it nothing
will avail. "Desires kill the slothful :"[2] because a slothful
man has the same desire of perfection as another, but not the
same courage in setting himself to attain it. It may be that
this is what keeps thee back from the good which thou
mightest be always doing—a want of courage. Thou art
afraid of opposition, of the talk and mockery of men, of the
dangers which may often meet thee in this life. Therefore,
granting that thou hast a very great desire of doing well, still
it is necessary to add to it fortitude : "The hand of the industrious (or the strong, *fortium*) getteth riches."[3]

---

[1] Psalm cxviii. 20.   [2] Prov. xxi. 25.   [3] Prov. x. 4.

ELEVENTH DAY.

ST. MARTIN, BISHOP.

Blessed are the merciful, for they shall obtain mercy (St. Matt. v. 7).

I. Consider first, that those whom our Lord here calls blessed are not only those who actually do works of mercy, either corporal or spiritual, but also those, who being prevented from doing them by lack of talent, or strength, or means, or opportunity, would, at least, rejoice to do them, if it were in their power. Therefore, He did not say, "Blessed are they that do mercy," but "Blessed are the merciful," in order that none might be excluded from this beautiful beatitude but those who do not desire to share in it; for although it is true that mercy includes a readiness of will to succour the needy, it can only be done when there is the power. St. Augustine gives this definition of mercy: "Mercy is the compassion for the misery of another in our own heart, by which, if it is in our power, we are impelled to succour him."[1] And therefore, if there are any who are unable to practise any kind of works of mercy, let them be consoled, for if only they show mercy in desire it is enough: "According to thy ability be merciful; if thou have much, give abundantly; if thou have little, take care, even so, to bestow willingly a little; for thus thou storest up to thyself a good reward for the day of necessity."[2] And what is this "good reward," but one equal to that which is given to those whom Christ here calls "merciful." It is true that from this it also follows that the man who does not practise mercy, when it is in his power, is not merciful; because mercy, whenever it can be practised in act, ought never to end, so to speak, in mere leaves of compassion and condolence, like wild vines, but ought to bear fruit: otherwise, of what avail is it? "If a brother or a sister be naked and want daily food, and one of you say to them, Go in peace, be you warmed and filled, yet give them not those things that are necessary for the body, what shall it profit."[3] And therefore God is not merely called *Misericors*, but *Miserator*, a title often given Him by the Psalmist, because it would be of little use to us that He is, by His nature, disposed to help us abundantly, if He did not actually do so. Now, in order for

---

[1] *De Civitate Dei*, lib. ix. cap. v.
[2] Tobias iv. 8—10.   [3] St. James ii. 15, 16.

this mercy to be practised in the sublime degree which is requisite for every beatitude, it must have three qualities, like those of the sun: it must be extended to all men, so as to do good even to enemies; it must be extended to all circumstances, so as to do good in every sort of necessity; and it must be practised with entire disinterestedness, according to the words: "When thou makest a feast, call the poor, the maimed, the lame, and the blind; and thou shalt be blessed, because they have not wherewith to make thee recompense;"[1] for otherwise it would not be mercy, but a bargain in the guise of charity. If this is so, how is it with thyself? Thinkest thou that thy name is in this noble catalogue of the merciful? But how is it possible if thou art so cruel, that instead of succouring thy neighbour, when occasion offers, in those wants of his which thou perceivest, or, at least, of compassionating him, thou very often either despisest him, or art angry with him, or reproachest him, or even revilest him freely?

II. Consider secondly, that mercy is a special mark of predestination, not merely because this can be proved in many other ways, but because of the promise made by Christ in these words, to which I would have thee confine thy attention: "Blessed are the merciful, for they shall obtain mercy." It is true that they do not expressly say that the merciful shall obtain mercy from God, but only that they shall obtain mercy, which may be equally understood of men, who generally show kindness to those who are kind themselves. But what sort of mercy is it, after all, that men can show thee? It is one that is very imperfect, which may, indeed, relieve thee in some misery, poverty, or danger, but which can never make thee blessed. It is only God Who can show thee a mercy which will make thee blessed. And not every kind of mercy which is shown thee by God even can do this, but only that by which He allows thee to die in grace. It is of this therefore, most certainly, that Christ meant to speak when He said, "Blessed are the merciful, for they shall obtain mercy," because it is remarkable that when men practise works of mercy, God usually gives them grace either to forsake sin in time, or to keep themselves from it, and so to be saved at last: "Alms," whether spiritual or corporal, "maketh to find mercy."[2] Now thou canst understand why it is that at the Last Day Christ will declare to the elect that He rewards them on account of the works of mercy they have practised,

[1] St. Luke xiv. 13, 14.   [2] Tobias xii. 9.

rather than of any other of the many virtues for which they have been distinguished, such as chastity, obedience, humility, mortification, or even on account of their having bravely suffered death for God. The reason is, not that the elect will be more rewarded in Paradise for those works of mercy than for their other sublime qualities, but because it was those works by which especially they disposed themselves to receive from God the grace of being chaste, obedient, humble, or mortified, and, in some cases, even of dying a martyr's death. And therefore Christ will on that day make particular mention of these works, as being the root from which sprang so many fruits. So, too, on the other hand, will He reproach the wicked with their neglect of these works, because it was owing to that neglect that that efficacious grace was denied to them, by virtue of which they would have been preserved from the sins into which they afterwards fell, or have been recovered from them; for just as "alms maketh to find mercy"—that is, is the means by which he obtains that efficacious grace which God would not otherwise be bound to give us—so, on the contrary, the neglect of alms prevents mercy being found: "For the iniquity of his covetousness I was angry, and I struck him, and he went away wandering in his own heart."[1] What, then, art thou doing, thou who so greatly desirest mercy of God? Do not imagine that, because everlasting salvation is called mercy on account of the grace on which its first origin depends, therefore it is not thy part to gain it. Listen to what our Lord says: not that the merciful shall receive, but that they "shall obtain mercy." This, then, is a proof that, as a rule, even mercy is not bestowed by God as a gift, but as a reward, although, indeed, a reward so superabundant as never to cease being mercy. If, then, it is bestowed as a reward, how canst thou rely on receiving it as a gift?

III. Consider thirdly, why Christ put this beatitude in the fifth place. It was because, having in the preceding one incited man to do good, not only in himself, but in others, by works of justice which are those to which each person is particularly bound, He then went on, in the present one, to urge him still further to do that sort of good also to which he is not, otherwise, so strictly bound. And these must necessarily be those which are called works of mercy, works of superabundance, of supererogation: "The lips of many shall bless him that is liberal of his bread."[2] So that when thou

---

[1] Isaias lvii. 17.   [2] Ecclus. xxxi. 28.

givest to a poor man in very great want merely of thy superfluity, either in clothing, lodging, feeding him, or doing him
good in any other way, this is not, strictly speaking, showing
him any kind of mercy: it is merely giving him what belongs
to him. It is truly showing mercy when, in such a case, thou
not only givest him thy superfluity, but that which is strictly
necessary for thyself, and, like St. Martin, dividest thy cloak
for the poor man. In the same way, as to the spiritual works
of mercy, do not think that thou art showing mercy to thy
neighbour when thou correctest him merely because of thy
position towards him, whether that of father, master, parish
priest, or bishop, because this concerns justice: it is mercy
that thou showest when such correction is in no way obligatory. So, too, it is not a work of mercy to teach a person
who pays thee for doing so, to comfort one who maintains
thee, or to counsel one who recompenses thee; but it is a
work of mercy when no obligation binds thee to any of these
things, but thou art moved purely by charity. See, then, to
what degree thou art to rise, if thou desirest really to make
one of that happy number whom our Lord here calls blessed:
thou must do more than thou art bound to do by the obligations of thy state, as the Apostle understood to be his own
case when he said, "I most gladly will spend and be spent
myself for your souls."[1] Otherwise, speaking strictly, thou art
just indeed (since thou art exact in spending thyself in every
way to which thou art bound), but thou art not merciful.
When thou spendest thyself, not only in ways to which thou
art bound, but beyond them also, then thou art merciful.

IV. Consider fourthly, that the gift of counsel corresponds
to this beatitude, for no one uses it more nobly than one who
shows mercy to his neighbour. He who does so gains much
at little cost; and no counsel can be more clear-sighted or
prudent than this: so that it was with great reason that
Daniel asked King Nabuchodonosor to accept it: "Wherefore,
O King, let my counsel be acceptable to thee; and redeem
thou thy sins with alms."[2] It is true that to forgive an injury,
particularly a very sore and grievous one, is a work of mercy
which costs corrupt nature something. But what is it in
comparison with the immense gain which comes of such
forgiveness? By it not only dost thou move God to forgive
thyself, but thou actually obligest Him to do so, because He
has expressly given thee His word for this: "Forgive, and you

[1] 2 Cor. xii. 15.   [2] Daniel iv. 24.

shall be forgiven."[1] And what proportion is there between
those offences which God forgives thee, and those which thou
forgivest thy neighbour? The evil which the latter brought
upon thee was only transitory; and that which the former
brought upon thee was eternal. And if he gains much at little
cost who performs so difficult a work of mercy as this of
forgiveness, how much more may the same be said of one who
speaks a few words to teach, or console, or counsel, or reprove
his neighbour, or who gives a little money to relieve him when
he is afflicted by some grievous temporal distress? They are
indeed of the number of those of whom the Preacher spoke
when he said, "There is that buyeth much for a small price;"[2]
they give earth, and gain Heaven. Is it not a wise counsel to
consider such an exchange seriously? See, then, what is the
title justly merited by those who do not give themselves up,
during their lifetime, to these corporal and spiritual works of
mercy which are so dear to God. They merit the disgraceful
name of fools: "Thou fool, this night do they require thy soul
of thee: and whose shall those things be which thou hast
provided?"[3]

TWELFTH DAY.

Blessed are the clean of heart, for they shall see God (St. Matt. v. 8).

I. Consider first, that by the heart of man, taken, as it is
here, not in a material but a metaphorical sense, is sometimes
signified in the Sacred Scriptures the understanding: "Their
foolish heart was darkened;"[4] sometimes the memory: "Mary
kept all these things, pondering them in her heart;"[5] and
sometimes the will: "How good is God to Israel, to them
that are of a right heart."[6] At other times we must understand
the union of all these three powers as they should be exercised
especially in meditation: "He will give his heart early to
resort to Him that made him, and he will pray in the sight of
the Most High." When, therefore, thou shalt have attained
to purity in all these three noble powers, at one and the same
time, then thou wilt be of that happy number whom Christ
with His own lips here calls blessed: "Blessed are the clean
of heart." But what is the meaning of purity in these powers

of the soul? It is easy to know this. Grain is pure from which the chaff has been cleared, a fruit is pure when stripped of its rind, cloth is pure when cleansed from stains, gold is pure when separated from its dross. And so, when thou hast taken from these three powers all that makes them less simple and true in their nature, they will be pure: "Cleanse thy heart from all offence."[1] The understanding should be cleansed by purifying it from false doctrine, mischievous curiosity, precipitate counsels, and malicious judgments. The memory should be cleansed by making it forget the persons, conversations, and comforts which were forsaken when the soul came out of Egypt, and everything the remembrance of which is apt to divert it from its God. And the will should be purified not only from sins, even light ones, but also from all inclination to them, from irregular intentions of pleasing others rather than God alone in our actions, from carnal affections, bodily appetites, and even those secret motions which the rebellious senses are always ready to excite: "Let us cleanse ourselves from all defilements of the flesh and of the spirit, perfecting sanctification in the fear of God."[2] Truly may a man who has attained so far as this say that he is "clean of heart." Thou wilt answer that there is no one in the world who can do this, at least perfectly: "Who can say, My heart is clean?"[3] I grant it; but neither can any one on earth attain to loving God perfectly with his whole heart, and yet this is expressly commanded—"Thou shalt love the Lord thy God with thy whole heart"[4]—in order that every one may see the goal of the great race he has to run, and so may endeavour to approach it as closely as possible. And it is the same in the present case. How is it, then? Dost thou think that thou approachest that purity, the perfect idea of which is here set before thee? The nearer thou comest to it the more blessed art thou. God forbid that thou shouldst be one of those who think themselves "clean," when they have not yet so much as thought of washing themselves: "A generation that are pure in their own eyes, and yet are not washed from their filthiness."[5]

II. Consider secondly, that this cleanness is also a sign of predestination, and an immediate sign, seeing that it is the disposition which most directly prepares the soul for seeing God. What is the most direct disposition in a mirror for

[1] Ecclus. xxxviii. 10.   [2] 2 Cor. vii. 1.
[3] Prov. xx. 9.   [4] Deut. vi. 5.   [5] Prov. xxx. 12.

perfectly reflecting the sun? It is freedom from every stain.
So is it with man. When the powers of his soul are thus
cleansed, there is no obstacle to prevent the full splendour of
God from being shed over them all. But we know that,
speaking generally at least, such a vision of God cannot be
attained on earth: "Man shall not see Me, and live."[1] It
must then be reserved for the enjoyment of Heaven. And
this is what Christ here meant when He said, "Blessed are
the clean of heart, for they shall see God." If He had said
"they shall contemplate, consider, understand," He would
have said what is indeed true, but what is applicable even to
dim and clouded mirrors, such as men in this world always
are. But since He intended to speak of what can be attained
only in Heaven, where these mirrors are all bright and shining,
He said "they shall see." Now, see how well it is worth the
pains to aim earnestly at obtaining this purity, which more
than anything else prepares thee for seeing God. But how wilt
thou obtain it? By cleansing thy heart in just the same way
as mirrors are cleansed, by wiping, scouring, and washing it.
The heart may be said to be wiped by frequent examination
into the sins that have been committed, and by the repentance and purpose of amendment which accompany a thorough
examination. Scouring represents the more painful works of
satisfaction which are practised in addition, and the heart is
washed by frequent recourse to the fountains of the Saviour,
that is, to the holy Sacraments of Confession and Communion.
It is true that even these must derive all their power from the
faith which leads thee to make use of them, and therefore it
is to faith more especially that the purification of the human
heart is ascribed in the Sacred Scriptures: "Purifying their
hearts by faith."[2] But this very fact shows that this cleanness
of heart is a sure sign of predestination; for as the merit of
faith consists in firmly believing what is not seen, so the
reward which answers to it will be to see clearly what thou
hast believed.

III. Consider thirdly, why Christ put this beatitude in the
sixth place. It is because man, being set in order by the
former beatitudes, both as to himself and his neighbour—as
to himself by the first three, and as to his neighbour by the
two next—it was most right that he should go on to set
himself in order as to God; and therefore the first thing is
this cleanness of heart, which is so necessary for every one

[1] Exodus xxxiii. 20.   [2] Acts xv. 9.

who desires to converse intimately with Him : " Be clean, you
that carry the vessels of the Lord."[1]   Besides, since in the
beatitude immediately preceding this, the works of mercy were
greatly exalted, it might easily have been thought that these
alone were amply sufficient for salvation, as has been asserted
by some.   And therefore Christ took occasion to warn men
that if the heart is impure it will not avail it to be tender : it
must be clean also.   Dost not thou know that there are many
who live like brutes, and never trouble themselves about it,
because they are in the habit of giving some daily alms to the
poor ?   They make a boast of Christ's words to the Pharisees
who were so stained with vice : " That which remaineth give
alms, and behold all things are clean unto you."[2]   But such
persons grievously distort this text ; for although I am ready
to grant that it was not spoken ironically, as some have
thought, yet it must be remembered that while these Pharisees
were exceedingly particular about washing their bodies carefully every day, they had no scruple whatever in having
their consciences always stained with rapine, fraud, deceit,
and grievous wrong done to the poor.   And therefore Christ,
without forbidding these external ablutions, commanded them
to add those which were interior, by giving more frequent
alms, which would purify them from their past extortions, and
then, indeed, they would be altogether clean.   This is what
our Lord meant when He said, " All things are clean unto
you."   He meant that they were to be cleansed thoroughly,
and not to be like persons who wash a dish very carefully
outside and not at all within.   Doubtless it is true that alms
avail to cancel sin, as the Angel said to old Tobias, " Alms is
that which purgeth away sins," but it does so only by way of
disposition ; and so, if thou art so unhappy as to be steeped
to the lips in carnal sins, give alms by all means, which will be
a great help in getting grace from God to come out of that
slough in which thou art lying.   But it is one thing to give
alms in order to get grace from God to come out of the mire,
and another to do so in order to get leave to lie in the mire to
the end, and yet be saved at last.   This would not be a desire
to cancel sin, but to encourage it by alms.   And who could
possibly expect such an extravagant thing ?

IV.   Consider fourthly, that the gift of understanding
corresponds to this beatitude ; and this gift consists in a
sublime light from God, which elevates the mind to com-

[1] Psalm lii. 11.   [2] St. Luke xi. 41.

prehend the Sacred Scriptures, and to explain them in the truest sense: "Then He opened their understanding, that they might understand the Scriptures."[1] This gift, therefore, belongs to the clean of heart on two grounds, which mutually help each other: first, because purity of heart is an assistance in understanding the Sacred Scriptures; and secondly, because the understanding of the Sacred Scriptures helps to increase purity of heart. That purity of heart is a help to the understanding of the Sacred Scriptures is very certain, indeed it is not only a help but a necessity. What man in his senses would pour a precious balsam into a dirty vessel? The very first thing he does is to cleanse it. And so, too, does the Holy Spirit. He will not pour the sense of the Sacred Scriptures into a vessel that is impure. For if there is occasionally a man of bad life to be found who, notwithstanding, is very learned in interpreting the Sacred Scriptures, do not imagine that he does so generally by an infused gift; he does it by collecting the interpretations of different persons, which he searches out in the works of holy writers. And we may observe that the Psalmist first says, "Blessed are the undefiled in the way, who walk in the law of the Lord," and then, "Blessed are they that search His testimonies."[2] It is also true that the understanding of the Holy Scriptures helps to increase purity of heart, for they may be compared to the River Pactolus, whose waters not only cleanse but enrich the soil; and while the streams of all human knowledge bring with them much mud and impurity, that is, the vices of arrogance, ambition, envy, and temerity, which they leave behind them, this stream, far from bringing them with it, carries them away, leaving on the soil it flows over an abundance of gold sufficient to make the soul rich in virtue. Thus thou seest that those saints who have been most learned in the Scriptures, have also been among the most eminent. Neither is this to be wondered at, for it is written that "the consumption abridged" (such as are the many precepts of perfection condensed into so small a volume as that of the Sacred Scriptures) "shall overflow with justice."[3] Do not, then, think the time wasted which thou spendest in learning and meditating on these texts which I am proposing for thy study, for they have the power not only of watering thy soul, but of making it overflow with holiness.

[1] St. Luke xxiv. 45.   [2] Psalm cxviii. 1, 2.   [3] Isaias x. 22.

THIRTEENTH DAY.

Blessed are the peaceful (pacifici), *for they shall be called the children of God* (St. Matt. v. 9).

I. Consider first, that admirable definition of peace given by St. Augustine in two words when he called it "the tranquillity of order." An order such as is found in a state which is, indeed, well constituted, but disturbed by the frequent rebellions which occur in it, is not sufficient for peace, because it is an order which lacks tranquillity. Neither is such tranquillity as exists in a state which is quiet, but ill-constituted by reason of the lack of subordination in its government, sufficient to form a peace that is lasting, because it is a tranquillity which lacks order. True peace requires both order and tranquillity. Having remarked this, thou canst see who those are of whom our Lord is particularly speaking when He says, "Blessed are the peaceful." It is certain that the wicked can never deserve this title, for if they sometimes possess tranquillity, as happens even to the greatest sinners, they do not possess order, seeing that there is the most utter disorder in them, the inferior part commanding, and the superior obeying : "There is no peace to the wicked, saith the Lord."[1] Neither is it ordinary just persons who are here meant, for if there is order, there is not tranquillity in them, because the order is continually being disturbed by the frequent rebellion of the passions which are still bold enough to rise up in revolt: "We have looked for peace, and behold trouble."[2] Those eminently just persons are the truly peaceful who, having already mortified their passions, make them altogether subject to the will, as their mistress, while the will itself is subject to God, not only obeying Him promptly and exactly, but allowing itself to be guided by Him in everything, as a child by a most loving Father ; and so, no matter what befalls them, they are always the same, always happy, always cheerful, always contented. Truly are these souls peaceful : "Much peace have they that love Thy law ;"[3] for in them there is in truth "the tranquillity of order." There is order, because the powers of the soul are in perfect subordination ; and there is tranquillity, because this subordination is very difficult to disturb, not that even the greatest of saints are not at times liable to some disturbance of their affections—"There is no just man upon

[1] Isaias xlviii. 22.   [2] Jerem. xiv. 19.   [3] Psalm cxviii. 165.

earth that sinneth not"[1]—but that this disturbance is very slight. We all know that a trifling tumult made occasionally in a kingdom by some ill-disposed person, especially when it is soon quieted, in no way destroys the tranquillity of the community, and so does not destroy its peace : much less is peace destroyed in the souls we are speaking of by the exterior confusion arising from diabolical suggestions, for who would ever say that the peace of a commonwealth was destroyed because the dogs barked continually in the city? How, then, is it with thee in this respect? If thou hast not true peace, learn, at least, what is requisite for obtaining it : namely, a solid and well-regulated order of all the powers of thy soul by means of the perfect subordination with which it depends on the Divine will : "Submit thyself then to Him, and be at peace."[2]

II. Consider secondly, that this peace of which we have been speaking is an eminent sign of predestination, because if all who possess it are the children of God, it is evident that all will likewise receive the inheritance, which is nothing less than everlasting life : "If sons, heirs also." And so, too, Christ says, " Blessed are the peaceful, for they shall be called the children of God." They are, therefore, called by the lofty title of children of God, because their conduct is that of children. Servants also submit to their masters, but it is because they are obliged to do so; they submit out of fear, unwillingly, and reluctantly, whereas sons submit to their father out of reverence, with alacrity and love. And this is the conduct of those eminently just persons of whom we are speaking ; they willingly let themselves be governed by God as He pleases, and therefore they are His children : "Whosoever are led by the Spirit of God, they are the sons of God," not those who resist His Spirit. Neither shouldst thou wonder that Christ did not say, "Blessed are the peaceful, for they are the children of God," but "for they shall be called the children of God," because, in Hebrew, "to be called" has frequently the force of "to be." "My house shall be called the house of prayer."[3] And besides, in this passage, "they shall be called" is more forcible than "they shall be." For, what dost thou think that Christ implied by this expression? He implied that not only shall the just persons in question be the children of God, as ordinary just men are, by right of their supernatural adoption, but they shall be recognized as such by

---

[1] Eccles. vii. 21.   [2] Job xxii. 21.   [3] Isaias lvi. 7.

all men, as gold is known by its lustre. In the same way it was said of Christ, "He shall be called the Son of the Most High,"[1] not that He was not to be the true Son of God, and His Son by nature, but because He was to be so in such a way that it could not be doubted by any one unless he purposely closed his eyes as bats do to the light of day, so great were to be His holiness, wisdom, science, and gentle sweetness to all men. We will suppose that thou art a child of God, because thou art just; but is thy life such that every one who knows or sees thee is at once convinced that thou art so? This is the most certain sign of it that thou canst show, the entire abandonment of thyself into the hands of thy Father, which is the most perfect subjection that thou canst practise towards Him. But how can this be seen in thee, who art so easily disturbed by everything that happens? Peace has been compared to a river flowing in an even stream, where the bed is full, never to a torrent: "Oh, that thou hadst hearkened to My commandments: thy peace had been as a river."[2]

III. Consider thirdly, why Christ put this beatitude in the seventh place, that is, after cleanness of heart. The reason is, that it was necessary to observe this rule in setting the just man in perfect order with regard to God. The first thing is to cleanse him by purity of heart which, in its nature, signifies only a negative, although a very high perfection; and then to incite him to an entire union of that heart with God, signified by the word peace, which means positive perfection. Purity is the right disposition for seeing God; union for loving Him. And therefore it is so great a good first to see, and then to love Him; because union must be preceded by purity, not purity by union, which is what is pointed out by St. James, when he says, "The wisdom that is from above first indeed is chaste, then peaceable."[3] Here, then, thou seest man at the highest point of heroic perfection which he can hope to attain on earth. For if perfection consists in loving God, it is certain that he loves Him most who is conformed to His holy will in all things with most imperturbability and courage; and therefore, for such a one peace is reserved. "Being justified, therefore, by faith," what must we do, if we would be, not just only, but saints? The Apostle answers: "Let us have peace with God."[4] I know that there is also an esteemed interpretation which understands the word *pacifici* as referring to those

---

[1] St. Luke i. 32.
[2] Isaias xlviii. 18.   [3] St. James iii. 17.   [4] Romans v. 1.

who labour to reconcile to God those sinners who have revolted from Him. But these are not merely *pacifici*, or "peaceful," but also *pacificatores*, or "peacemakers," which it is not in the power of all men to be. What Christ here said therefore, if we keep close to the Vulgate, was merely, "Blessed are the peaceful;" not that the peacemakers are not blessed also, and very highly blessed, seeing that they do the very same work which God's own Son came into the world to do, but because, as in all the preceding beatitudes He chose to put only those virtues to which it is in the power of all to attain, if they please (as it is easy to see by going through them), it seemed more suitable to do the same in this one also. It may be added, that in no other passage are those who make it their business to bring about peace called *pacifici*, but *pacificantes:* "Rich men in virtue, living at peace (*pacificantes*) in their houses."[1] And so, if thou art living in the seclusion of a cell, if thou art sick, or in any other way hindered from being a "peacemaker," this will in no way exclude thee from this beatitude, if only thou art "peaceful" in all thy troubles.

IV. Consider fourthly, that the gift corresponding to this beatitude is that of wisdom, because, since peace consists, as has been said, in the tranquillity of good order, it is evident that this cannot be attained without that gift, because it is wisdom to which it belongs, in every line, to establish order, to maintain it when established, and to restore and renew it when it has been disturbed. And so thou mayest observe that it is the wise men of a State who are entrusted with the care of watching over the order which should prevail in it; and the same holds good in the army, in medicine, and even in mechanical arts; because no one can form a judgment concerning them except a man who is wise in those matters, that is, who is acquainted with the things relating to them as their final cause: "As a wise architect, I have laid the foundation."[2] We must also remember that the wisdom, which is a gift of the Holy Ghost, is that sublime wisdom which knows the First Cause, that is, God, and according to which it is guided in every matter in order to be right. So that this wisdom is not such as is acquired by many through either study or sagacity. It is a wisdom infused in us by the same Spirit Who teaches us, from time to time, the practical knowledge of what is most pleasing to God in the existing circum-

[1] Ecclus. xliv. 6.   [2] 1 Cor. iii. 10.

stances, so as to move us to do it. And this is therefore the wisdom the love of which ought to inspire thee, which thou shouldst ask of God with all thy heart, for it is not the most learned, the most studious, or the most eloquent who possess it, but the most highly favoured by God in prayer: "I called upon God, and the spirit of wisdom came upon me." [1] And so it is, that a poor ignorant old woman may possess it in a higher degree than the most erudite and sublime preacher. Be diligent, therefore, in always begging God to enlighten, assist, and instruct thee in all thy affairs, and thou wilt see with what wisdom thou wilt succeed in maintaining good order within thy soul, so as always to be subject in all things to God, as is necessary for the enjoyment of a deep peace in Him.

FOURTEENTH DAY.

Blessed are they that suffer persecution for justice' sake, for theirs is the Kingdom of Heaven (St. Matt. v. 10).

I. Consider first, that if all the gold which thou seest in the houses of great men, on their dress, and in their furniture, were thrown into a heated crucible, it would be seen that a great deal which every one has supposed to be fine gold is in truth artificial. So is it with regard to virtues. How many there are that are false, even among those who have the reputation of being very spiritual persons! But because they have never as yet been subjected to the test of a severe persecution, they have the credit of being real. No wonder, then, that Christ, after having, as it might seem, brought the whole man to perfection, as to himself, his neighbour, and God, in the seven preceding beatitudes, added this one more: "Blessed are they that suffer persecution for justice' sake." He would have thee not trust so easily to thyself in thinking that thou art truly poor in spirit, meek, contrite, a true lover of justice, truly merciful, clean of heart, and peaceful; but He would have thee wait till the time comes when thou wilt have to encounter some sharp persecution through thy resolution to practise some one or other of these virtues. Then it will be seen by thy constancy whether those virtues of thine are sterling metal or base. This beatitude, therefore, is not so

[1] Wisdom vii. 7.

much a new one, as a test, or also the crown of the rest. For
the height of perfection is not to do all the good contained in
those beatitudes, but to do it and to receive evil in return.
This is the real test of every virtue : "If, doing well, you
suffer patiently, this is thanksworthy before God."[1]   Impress
on thy mind, then, that this is the crown of all the beatitudes :
"To suffer persecution for justice' sake," to be mocked,
insulted, calumniated, plotted against, hunted to death for this
reason, because thou art resolved to act like a Christian who
is faithful to Christ.   So marvellous a truth as this is incomprehensible to thee.   Thou esteemest thyself blessed when, on
the contrary, all the good that thou doest brings thee good in
return.   But Christ's judgment is the reverse of this : He
would have thee esteem thyself blessed when all the good
that thou doest brings thee evil in return, and very serious
evil, such as is properly signified by the word "persecution,"
which means a violent assault such as aims at depriving thee
of rest, property, reputation, and life ; which, moreover, is not
transient, but indefatigably persistent.   That is not considered
to be tried gold which is taken out of the fire almost as soon
as it is put in, but that which becomes more brilliant the
longer it remains there.   So is it with true virtue : "Thou
hast tried me by fire, and iniquity hath not been found in
me."[2]

II. Consider secondly, that it is needless to inquire
whether this beatitude is a clear sign of predestination,
because, as it pre-supposes in itself all the merits of the
former beatitudes, so also does it pre-suppose all their rewards.
I know, indeed, that there have been those who have passed
at once from being idolaters to become martyrs : in other
words, to triumph over the greatest persecutions in the world.
But this is as great a miracle in the order of grace, as it would
be in the order of nature for a dwarf to be changed into a
giant.   Ordinarily, to endure grievous persecution patiently
requires a long practice of all those virtues which Christ has
here comprised in this glorious compendium of sanctity, as
these seven beatitudes may be termed.   I say, to endure them
patiently, because this is implied in the word "suffer," which
has not a merely passive meaning, as in the passage, "I have
suffered many things this day in a dream because of Him,"[3]
but one which is both passive and active, as in that other

[1] 1 St. Peter ii. 20.
[2] Psalm xvi. 3.   [3] St. Matt. xxvii. 19.

passage, "Have you suffered so great things in vain?"[1] because it is not a compulsory, but a voluntary suffering which is intended, like that of the Christian martyrs. And to such suffering the Kingdom of Heaven is expressly promised, as it was to poverty in the first beatitude, in order to maintain the due relation between the merit and the recompense. For there are in the idea of a kingdom two very high qualities— riches and dominion, and it is promised to the poor especially with reference to the former, and to the persecuted with reference to the latter, unless, indeed, thou holdest with St. Bernard, St. Bernardine, and others, that the poor of Christ are placed by Him in the same rank as the martyrs, and that therefore Heaven is said in the same terms to belong to both. Neither shouldst thou wonder that it is said that it "is," not that it "shall be" theirs. For it is not the fruits appertaining to the glory of Paradise which are here spoken of, as in the other beatitudes, but only the right to that glory. And this is not future, like those fruits, but present. He who is poor for Christ's sake, and He who is persecuted for Christ's sake, are considered in Paradise as already lords of a kingdom of which they are not in possession. And yet thou shrinkest from the danger of bringing thyself into this condition.

III. Consider thirdly, that we cannot say of this beatitude that any one particular gift corresponds to it, because they all do so. The fear of God corresponds to it, because the first defence against every temptation that can overtake thee is the fear of offending God, if thou art conquered by it. Piety corresponds to it, because that gift adds reverence, respect, and filial love to fear. Knowledge corresponds to it, because it teaches thee that to stand firm in persecution is the chief good, as the chief evil is the yielding to it. Fortitude corresponds to it, because it is this which gives thee the courage to despise persecution. Counsel corresponds to it, because it enables thee to adopt the means most calculated to ensure victory. Understanding corresponds to it, because it enlightens thee to have recourse to God in order to beg help of Him. And lastly, wisdom corresponds to it, because it enables thee to behave in this conflict with a calmness which belongs not to a beginner, but to an experienced general. When Dalila desired that Samson might be overcome by the might of the Philistines, who were persecuting him so fiercely, she plucked out seven of his hairs, which, according to the Fathers, were

[1] Galat. iii. 4.

a figure of these Seven Gifts of the Holy Ghost. If, therefore, thou art shamefully conquered by every persecution which is excited against thee in the service of God, look well to see whether the reason is not that the devil has dealt with thee in like manner, and pray continually that God may make thee worthy of possessing these gifts in the high degree requisite for obtaining that sublime beatitude which is the crown of all the rest : "This every one is sure of that worshippeth Thee, that his life, if it be under trial, shall be crowned."[1]

FIFTEENTH DAY.

Blessed is the man whose help is from Thee : in his heart he hath disposed to ascend by steps, in the vale of tears, in the place which he hath set (Psalm lxxxiii. 6, 7).

I. Consider first, that if it were by thy own strength that thou hadst to attain the virtues constituting all the beatitudes which have been the subjects of the late meditations, thou mightest have good reason to be afraid, since of thyself thou canst do nothing. But thy confidence must be placed in God ; and then, what canst thou fear? "Blessed is the man whose help is from Thee," so says the Psalmist ; for he who has God to help him may trust firmly to attain to even the loftiest height of perfection, such as is comprised in these beatitudes. It is true that God does not forbid thy having recourse, in addition to His help, to a good Spiritual Father to guide thee in this important journey. And therefore the Psalmist does not say, "Blessed is the man whose help Thou art," in order that thou mayest not expect always to receive immediate help from God, but "Blessed is the man whose help is from Thee," to teach thee that God often chooses to help thee by means of others. Yet, even so, thou art still blessed, for after all it is God from Whom thy help always comes, although it may not always come to thee directly from Him. Indeed, it is generally by means of others that He chooses to help thee, this being required by that sweetness of disposition which is the way of His Providence. And so, when the wise old man Tobias was told by his son that he did not know the right way to Rages, he did not tell him to set out trusting in

[1] Tobias iii. 21.

God's love for him to guide him in the right road, but he told him to inquire for some one to show it to him: "Seek thee out some faithful man to go with thee for his hire."[1] This is a most important direction. Do not venture upon so serious a journey as that of the spiritual life in the expectation that God will help thee in person: "Blessed is the man whose help is in Thee," not "whose help Thou art," to teach thee that thou art not to look for this. Pray, therefore, to God that, as He sent an angel to guide the young Tobias, so too He may send thee, if not an angel, at least a man of the most evangelical spirit that can be found.

II. Consider secondly, that since thou art to receive help so sublime from God, as has been said, thou mayest think that thou wilt instantly attain to the high perfection which is the object of thy desires; but this is a very great mistake. Thou shalt, indeed, attain to it, but gradually. This is why, as thou seest, the Psalmist does not say, even of a just man who is thus helped by God, "In his heart he hath disposed to fly," but "to ascend by steps," for such flights are granted to very few. And this is the principal reason why so few come to be saints, because most people would fain be caught up, like St. Paul, to the third Heaven, and this is not the will of God. He would have us ascend, not fly, that we may have greater merit in the violence that we do to ourselves when we conquer ourselves little by little, as in climbing up a high mountain: "Come, and let us go up to the mountain of the Lord."[2] What merit would Elias have had if the holy Angel who encouraged him to walk to the top of Mount Horeb had, so to say, given him wings to fly thither? His merit was in the perseverance he had to exercise while walking, day and night, incessantly, along so dangerous, solitary, and long a road as that which led to the top of the mountain. Do not imagine, therefore, that thy Spiritual Father, even though he were an angel, can give thee wings to carry thee without any trouble to the heights of sanctity. It is much that he can strengthen thee, as the Angel strengthened Elias, so that thou mayest attain them, if only thou choosest to do so, but in a human way, that is to say, step by step. This is what thou art here told, "In his heart he hath disposed to ascend by steps;" steps are not flights, nor even leaps.

III. Consider thirdly, that by these steps which the just man has disposed in his heart, thou mayest well under-

---

[1] Tobias v. 4.   [2] Isaias ii. 3.

stand, as some do, the beatitudes on which we have just been
meditating, for they are truly steps, which are so admirably
disposed, that one leads up to another. Poverty of spirit,
which consists in a great contempt of those exterior goods
which hinder thee from running the way of perfection swiftly,
disposes thee also to despise thyself, and to mortify thy
passions, especially those which are the strongest and most
violent, and so from poverty thou ascendest to meekness.
The mortification of thy passions disposes thee to enter more
calmly into thyself, so as to reflect upon all the evil thou hast
done, and to weep bitterly over it; and so from meekness
thou ascendest to the compunction which Christ calls mourning. This sorrow for all the evil thou hast committed disposes
thee to make amends for it by as many good works; and so
from mourning thou ascendest to an ardent desire of justice.
The desire of doing well is an excellent disposition for doing
still more than thou knowest thyself to be strictly bound to
do, and so thou ascendest from the ardent desire of justice to
the practice of works of pure mercy—that is, of superabundance
and supererogation. And this practice of virtue beyond what
is of strict obligation disposes thee to receive from God greater
grace than He would otherwise be bound to give thee for the
cleansing thy soul from every stain; and so from the works
of mercy thou ascendest to the greatest cleanness of heart to
which it is possible to attain in the body. The cleansing of
thy soul from all stain disposes thee to union with God, and
so from cleanness of heart thou ascendest to that sublime
peace in which that man rests who has reached the summit of
perfection. If then these steps are so well disposed, as thou
seest, would it not be a strange temerity to desire to fly straight
from the first to the last? We must "ascend by steps."

IV. Consider fourthly, that it is, doubtless, a work of
labour to ascend in this way to the top of a very high
mountain, such as that of perfection. But do not let this
dismay thee, for the joy will be in proportion to the labour.
And therefore, as the steps are according to merits in the
beatitudes, so too are they according to their rewards, which
are proposed by Christ in most exact order, so that each one
not only always contains in itself the good of those which have
preceded it, but even surpasses it. Thus thou mayest observe
that the first reward promised thee by Christ, when He says
that the Kingdom of Heaven is thine, is certainly very great;
but this is not enough, for thou mightest object that many

persons, even in this world, have a kingdom, and yet do not enjoy it for want of a sure and solid possession. And so Christ adds, in the second place, that thou shalt possess His Heavenly Kingdom, not as though it were one founded on the treacherous waters, as might be the empire of the seas possessed by some famous pirate, but as a kingdom founded on the solid earth. And because many possess such a kingdom who yet have no consolation in it, on account of the grievous troubles they have there, Christ goes on to add, in the third place, that in thy kingdom thou shalt be comforted. And because there are many who do indeed receive comfort, but not full comfort, in their kingdom, for lack of many pleasures which they still would wish to have, Christ adds, fourthly, that in thy kingdom thou shalt be, not comforted only, but satisfied. And because many persons may, perhaps, be satisfied with happiness in their kingdom, but only in proportion to their limited capacity, He goes on to add, in the fifth place, that in thy kingdom thou wilt receive a good which very far exceeds anything which thou canst desire as coming within the bounds of thy merit, because not justice only, but mercy, will be exercised towards thee to that end. And because many have, in their kingdom, a good exceeding their merit, but not the highest good, which is to see God, Christ tells thee, in the sixth place, that in thy kingdom thou shalt see Him clearly. And because, lastly, thou mightest here object, that even to see God is a less thing than a perfect resemblance to Him would be, Christ adds, in the seventh place, that thou shalt be like God in thy kingdom, just as a child is like his father, which is the most perfect resemblance possible. Dost thou not think, then, that Christ too has admirably disposed His steps in these rewards? Do not, therefore, think it hard that thou hast to dispose them in thy merits also.

V. Consider fifthly, that thou frequently resolvest to make these steps of merits in thy heart, but without going on to dispose them, because thou dost not consider rightly within thyself what are the means to enable thee to ascend by them most rapidly. And, therefore, hear what the Psalmist says : " Blessed is the man whose help is from Thee, in his heart he hath disposed to ascend by steps ; " he doth not say " he hath proposed," but " he hath disposed." Thinkest thou that God will act in thee independently of thyself? This is a great mistake ; if He did so, it would not be helping thee, but doing the whole work. When, therefore, the Psalmist says of the just man, "Blessed is the man whose help is from Thee," he shows the great power of the grace which strengthens him; when he says, "In his heart he hath disposed to ascend by steps," he shows that, notwithstanding, thy co-operation is necessary. Do thou, therefore, do thy part. Begin to exercise thyself in a more special manner in these beatitudes, according to the order which thou seest to be prescribed by Jesus Christ: dwell upon their meaning, admire and esteem them, examine thyself concerning them; and when thou judgest thyself to have made some progress in one, go on to the next; and in this way thou wilt perform the obligation thou art under to dispose the steps by which thou art able to ascend.

VI. Consider sixthly, that in doing this there are two very necessary cautions, which thou shouldst constantly keep in mind. The first is, that this ascent by steps is to be made "in the vale of tears," where no beatitude can ever be perfectly acquired on account of the innumerable miseries, distractions, troubles, and temptations which assail us here. If, therefore, it seems to thee that thou dost not attain perfection, do not lose courage. Go on more and more bravely in thy journey from the valley to the mountain, and in time thou wilt arrive there. The great misery is to turn back half-way up the mountain through base slothfulness, and to rush back to the depths of the valley. The second caution is, that this ascent is to be made by every one "in the place which He hath set," that is, as St. Augustine explains the passage, "in the place which God has appointed for him," namely, in his own state of life. Do not, therefore, imitate those persons who, when they cannot attain perfection, always lay the blame on the state in which God has placed them, and so are always unsettled and disturbed, and would like to be continually changing their calling, house, or order. What a mistake this is! There have been great saints in every state of life, and if thou art not a saint in thine, blame thyself, not thy state. I do not say that, supposing thee to be now in a position to make a good election of a state of life, thou art not to make the best that is possible according to thy condition; but I do say, that when once the election has been made, thou shouldst persevere in it. For although it is true that two things are requisite to bring thee to perfection, the grace of God and thy co-operation with that grace, as has been already said, nevertheless it is not in thy co-operation that thou shouldst place any reliance whatsoever, but only in the grace that God is pleased to grant thee. And if so, what is the object of wandering about? "Trust in God, and stay in thy place,"[1] since He can give thee His grace as easily in one place as in another.

———

SIXTEENTH DAY.

And it shall come to pass at that time that I will search Jerusalem with lamps, and will visit upon the men that are settled on their lees, that say in their hearts, The Lord will not do good, nor will He do evil (Sophonias i. 12).

I. Consider first, that by Jerusalem is here meant every Christian soul which has been chosen by Christ for His abode, but which is ungrateful to Him. And therefore He tells such a soul not to be at ease, for that "at that time," that is, the time appointed by Him for calling it to account for the evil it has done, He will search it very closely: "I will search Jerusalem with lamps." Thou knowest that the woman in the Gospel who was very anxious and diligent in seeking for the lost groat, "lighted a candle." And God would have thee infer that He will be equally careful, from this most proverbial expression in which He declares that He will use a lamp in the search which He will make into all thy works. He may also have intended to say that a light is used for two especial reasons in seeking for things: to see them when they are in the dark, or to distinguish them when they are too small to be easily perceived. Both these objects are here implied in these words of God. Thou art easy about a grievous sin that thou hast committed because, if it is interior, it is in the depths of thy heart, and, if exterior, it is concealed in the darkness either of secrecy or of oblivion. And thou art easy about a light sin, because thou thinkest it will escape notice. But how canst thou trust to such thoughts, when God has told thee that He has lamps to discover what He pleases? "I will search Jerusalem with candles." If thou desirest that God should not make use of this searching light in thy case, do thou first make use of it thyself, for it is written that "if we would judge ourselves we should not be judged."[2]

II. Consider secondly, that one lamp is sufficient to find things by, even in the darkness of night. And yet God does

———

[1] Ecclus. xl. 22.   [2] 1 Cor. xl. 31.

not say, "I will search Jerusalem with a lamp," but "with lamps," to show thee that He has not one only, but many ready to search thee, so clearly will He have all things displayed when He judges thee. The first and greatest light which He will use is that which is uncreated, that is to say, His Divine wisdom, which sees all things, knows all things, and distinguishes all things: "There is no creature invisible in His sight;"[1] and of all the lamps that He uses this is the most formidable. The rest are all created; and the chief among them will be the angels, both good and bad, who, by reason of their spiritual nature, can go everywhere, and discover everything more readily than the brightest torches: "Who makest Thy angels spirits and Thy ministers a burning fire."[2] The Lord will summon them as witnesses on that day to all that thou hast done. The second lamp will be the vivid light of reason which illuminated thee, according to the words: "The light of Thy countenance, O Lord, is signed upon us."[3] And by this light, which thou now doest thy utmost to dim, thou wilt on that day see clearly all the foulness of thy sins: "The spirit of a man is the lamp of the Lord, which searcheth all the hidden things of the bowels,"[4] that is, of the memory, which will preserve the image of everything which has taken place in thee, whether thoughts, words, or works. The third lamp is the law inspired by the mouth of God Himself, which has been so often recalled to thy mind by wise preachers or directors, or by spiritual books, and which thou hast despised: "The commandment is a light, and the law a lamp."[5] And this, too, will very clearly show thee thy shortcomings. The fourth lamp will be the sun, which witnessed thy sins by day, and the stars which witnessed them by night, nay, even the earth, air, water, plants, in short, every creature of which thou madest use to sin, will on that day be made use of by God to discover thy sin: "The heavens shall reveal his iniquity, and the earth shall rise up against him."[6] Lastly, the fifth lamp will be the example of Christ, and of countless multitudes of His faithful saints, in comparison with whom thou wilt be seen on that day to be still more faulty: "Elias the Prophet stood up as a fire, and his word burnt as a torch."[7] What wilt thou do, then, amidst the blaze of all these brilliant lamps? Wilt thou be able to conceal one single sin? Whither wilt thou turn?

---

[1] Hebrews iv. 13.   [2] Psalm ciii. 4.   [3] Psalm iv. 7.
[4] Prov. xx. 27.   [5] Prov. vi. 23.   [6] Job xx. 27.   [7] Ecclus. xlviii. 1.

Whither wilt thou flee? Where wilt thou hide thyself? Ah! it is easy now to pretend to be what thou art not really; but then it will be no longer possible; then it will be all over with those whose beauty, like thine, is but on the outside: "All are cut off that are wrapped up in silver."[1]

III. Consider thirdly, that if God will bring forth all these lamps to discover the most secret sins even of Jerusalem, that is to say, of every soul which is holy either in conduct or by profession, much more will He do so, we must think, to discover those of a wicked soul. And, therefore, with regard to these, He changes His manner of speaking, and says only that He will visit them: "I will visit upon the men that are settled on their lees." Neither should this surprise thee, for as to these unhappy souls their wretched condition is so evident that a mere glance is enough. The first thing, therefore, is to take notice who they are whom God here describes as "settled on their lees," or, as it is in the Hebrew, "set fast, congealed" (*coagulatos, congelatos*). They are hardened sinners, that is, such sinners as find peace in the turbid joys of this world, in pleasure, gain, or glory. It is these who are "settled" in them, for those sinners who are disturbed by frequent troubles, such as sickness, calumnies, or contrarieties, are not so much so. They fall into them and rise out of them, like wine which is stirred up with its dregs, whereas those who "settle" on them are those who prosper in them, like wine which is left undisturbed on its dregs. These, then, are the sinners whom God will especially visit on the Last Day, that is, whom He will trouble, torment, put under His feet, and punish as they deserve: "I will visit upon the men who are settled on their lees." God's visitations, when the word is used in a bad sense in Scripture, are the calamities which He sends: "Behold, the Lord will come out of His place, to visit the iniquity of the inhabitants of the earth against Him;"[2] but these visitations, when made in the life of sinners, are like those of a physician, for the purpose of healing: "Thy visitation hath preserved my spirit;"[3] when made in the next life they are those of a judge, for punishment: "In the Day of Judgment He will visit them; He will give fire and worms into their flesh," that is, "fire" externally, "worms" internally: "that they may burn and feel," that is, "burn" with the power of sense, and "feel" with the pain of loss "for ever."[4] So that

[1] Sophonias i. 11.
[2] Isaias xxvi. 21.   [3] Job x. 12.   [4] Judith xvi. 20.

these sinners, who prospered in their iniquity, not having been visited by God as a Physician here, will be visited by Him as a Judge in the Last Day. How earnestly, therefore, shouldst thou beg of God to visit thee so soon as thou hast sinned, because, if He delays doing so, alas for thee! "What will you do in the day of visitation, which cometh from afar?"[1]

IV. Consider fourthly, that there would be very few sinners in the world who would "settle on their lees," for any length of time, if they did not succeed in shaking off the fear of this "visitation which cometh from afar." For after saying, "I will visit upon the men that are settled on their lees," God immediately adds, "that say in their hearts, The Lord will not do good, nor will He do evil." Surely such persons are not to be met with among Christians? Alas, but too often they are atheists, who, since they cannot dwell amongst us except in disguise, "say," but only "in their hearts," either that "there is no God,"[2] or that, if there is, He has something else to do than to look so closely into our affairs: "He considereth not."[3] Nay, there are many amongst us who whisper these sentiments, thus revealing them to those at least with whom they are most intimate. If thou frequentest the company of astute courtiers, of worldly-wise and politic persons, thou wilt see how little they seem to believe that God will render them good for good or evil for evil. If they did believe it they would not give others the evil counsels they do, as being profitable for their advancement, neither would they so often follow them themselves, and endeavour to attain eminence by deceit or treachery. It is because they do not believe it that they act as if their own reason were the only God. Pray Him, therefore, to show thee speedily that there is a God, by chastising thee whenever thou sinnest: "Correct me, O Lord, but yet with judgment and not in Thy fury, lest Thou bring me to nothing."[4] For nothing so forcibly impresses on us the truth of the great visitation of our sins which He will make at the Last Day, as those lesser visitations which He makes in this life, just as nothing more tends to atheism than to be at the same time both sinful and prosperous.

[1] Isaias x. 3.
[2] Psalm xiii. 1. [3] Job xxxv. 14. [4] Jerem. x. 24.

SEVENTEENTH DAY.

Gladly will I glory in my infirmities, that the power of Christ may dwell in me (2 Cor. xii. 9).

I. Consider first, by how many troubles the Apostle was overwhelmed during those thirty-six years that he had spent in Christ's service: imprisonment, scourgings, stonings, accusations, plots, revilings, banishment. And yet we are never told that he prayed earnestly to God to be delivered from any of these evils. It was only from the "sting of the flesh" that he so earnestly begged deliverance: "Thrice I besought the Lord that it might depart from me;"[1] for in Scripture language "thrice" signifies very frequently. Not that he yielded to the temptation, for by the grace of God he so chastised his body as to have it subject to him—"I chastise my body, and bring it into subjection"[2]—and therefore the spirit which was allowed to tempt him had not power to do more than buffet him, that is, it was rather a humiliation than a stumbling-block: "There was given me a sting of the flesh, an angel of Satan to buffet me."[3] And yet, no sooner did our Lord tell the Apostle that it was better for him to be subject, like other men, to those infirmities consequent on the rebellion of concupiscence through sin, which we have contracted from Adam—"My grace is sufficient for thee, for power is made perfect in infirmity"—than he changed his mind, so as even to say that he loved to glory in his weakness: "Gladly will I glory in my infirmities." And why? Certainly not for love of them, but because it was these infirmities which had established the power of Christ in him: "Gladly will I glory in my infirmities, that the power of Christ may dwell in me." Such is the most legitimate and literal sense of this passage. Do thou learn from it that thy glory is not to consist in being favoured by God beyond the mass of men, in being exempt from temptations, even from such as are impure and humiliating. It is to consist in gathering from them the good which God designs them to do to thy soul: "Because thou wast acceptable to God, it was necessary that temptation should prove thee."[4]

II. Consider secondly, what this "power (or virtue) of Christ" is, which the Apostle saw was so greatly strengthened in him by these infirmities. Surely it was that virtue which

---

[1] 2 Cor. xii. 8. [2] 1 Cor. ix. 27. [3] 2 Cor. xii. 7. [4] Tobias xii. 13.

was the distinguishing characteristic of Christ: humility in His own Person, and meekness towards others. And this is what He was most desirous of teaching mankind, who were so utterly ignorant of this new doctrine: "Learn of Me, because I am meek and humble of heart."[1] This may therefore be truly said to be "the virtue of Christ," that is, the virtue most preached and most practised by Christ. Now, the "sting of the flesh," as it is here called, was very useful in keeping the Apostle humble in himself, because having, as he had, so many grounds to boast of the favours showered upon him by our Lord, it did for him the office of the slave who had to walk before the triumphal chariot of the Roman conquerors, to remind them continually, in the midst of all the shouts and acclamations around them, that they too were men, like others, formed of the dust of the earth: "Remember that thou art a man." And what was the result of this humility which the Apostle always had in himself? It was to make him always meek to others, so that, compassionating their faults in a merciful spirit, he excused and bore with them, and while treating them as a physician, did so as a physician subject to the same infirmities. Would that thou also knewest how to get the same good from thine, namely, to be humble and meek! Then, indeed, like the Apostle, thou too mightest begin even to glory in them, that is, to hold them in that estimation in which men hold the gifts and endowments in which they take pride: "If I must needs glory, I will glory of the things which concern my infirmity."[2] These infirmities are like so many windows which admit the sun into thy room, that is, the light which both warms and illuminates. It illuminates thee by that low esteem of thyself, which is the light that thou needest most of all; and it warms thee by charity to thy neighbour, which is the heat of which thou art most in want. If, then, the good they produce in thee is so great, how canst thou despise them? Dost thou not see that if these useful windows art kept closed, thou wilt be in the dark, and mightest easily esteem thyself to be what thou art very far from being? Bear, therefore, with the admonitions they give thee: "A grievous sickness maketh the soul sober."[3]

III. Consider thirdly, that thou mayest think that if thou standest in need of being admonished and reminded of thy vileness, still the admonitions need not be so very intimate as those given by the sufferings of sense, which at almost

---

[1] St. Matt. xi. 29. [2] 2 Cor. xi. 30. [3] Ecclus. xxxi. a.

every moment remind thee harshly of thy sad condition.
The Apostle had such a reminder on account of the wonderful
revelations which had been granted to him: "Lest the
greatness of the revelations should exalt me, there was given
me a sting of my flesh, an angel of Satan to buffet me." But
as thou hast no such grounds for pride, it seems all the harder
to be subject to such a sting. Thou shouldst remember,
however, that it is not always those who have something to
be proud of who are the proudest. I grant that thou hast
no grounds for being proud, but look well to it, lest thou
shouldst be so notwithstanding. And if so, if thou knowest
that thou art often proud without any occasion, how would it
be with thee if such an occasion should arise? "He that is
glorified in poverty, how much more in wealth?"[1] If God
bestows on thee a few tears in an ordinary kind of prayer,
a little sweetness in devotion, a gift of desire, thou thinkest
thyself at once in the third Heaven with the Apostle. This
may show thee that thou art in greater need than he was of
having thy own vileness brought home to thee in a painful
manner, since thou art not a conqueror, like the Apostle, and
yet art as full of thyself as though thy life was a perpetual
triumph. And whence arises thy lack of charity to thy neighbour but from an over-estimation of thyself? It is this which
makes thee so harsh in reproving, so bitter in censuring. Is
there not, then, amply sufficient reason why God allows in
thee some of those weaknesses which are found in souls so
far nobler than thine, to strengthen them? In such souls
they are allowed, as ballast is necessary in ships which fly
swiftly before the wind, in that they are allowed as a chastisement. Art thou both poor and proud? "The pride of thy
heart hath lifted thee up who dwelleth in the clefts of the
rocks."[2] Is there not, then, great reason for thee to be
ashamed?

IV. Consider fourthly, the great advantage there is in
being humble in oneself and meek towards others, since in
order to possess this virtue it becomes a gain even to have
been subject to the most humiliating temptations. And this
is not wonderful, since it is only on the humble and the
meek that Christ bestows great grace: "God giveth grace
to the humble."[3] "To the meek He will give grace."[4]
To the humble, because it is always necessary to practise
humility; to the meek, because it is necessary to practise meek-

[1] Ecclus. x. 34.   [2] Abdias i. 3.   [3] St. James iv. 6.   [4] Prov. iii. 34.

ness when occasion arises.   And this is the grace which
strengthens thee interiorly. The perfect strength of a Christian
is to do and to suffer : to do much, and to suffer much ; but
all to the glory of God, as was the case with the Apostle.
Our Lord gives the humble grace to do much, because that
man does much who, knowing that of himself he can do
nothing, has recourse to Christ, and puts all his trust in Him.
And He gives grace to suffer much to the meek, because that
man suffers much who, having determined to resent nothing,
lets himself be treated by all as they choose.   Was not the
Apostle right, therefore, when he exclaimed : " Gladly will
I glory in my infirmities, that the power (or virtue) of Christ
may dwell in me?"   He might have said *virtutes Christi*,
that is, the humility of Christ and the meekness of Christ ;
but he chose to say *virtus*, not only because these two virtues
are so closely connected as to be like one, but because what
he especially valued in both was that keen force, vigour, and
strength (*virtus*), which was to result in him both in doing
and suffering much for God.   The Christian virtues which
we possess should not be valued by us because they adorn
us, rendering us, for example, humble and meek, but because
by them we receive power to employ ourselves better for the
glory of God ; and so we ought not to love them as an end,
but only as the means of serving God, Who is our end :
" Thou art the glory of their strength."[1]

EIGHTEENTH DAY.

Turn away from evil, and do good, seek after peace and pursue it
(Psalm xxxiii. 15).

I. Consider first, that it is, doubtless, our sins of commission which will make the Last Judgment so terrible to all
men, but still more those of omission.   These, indeed, will
make it terrible beyond words.   And the reason is this : that
if any one has been guilty in his life of theft, adultery, murder,
envy of another, or the like, he at once sees his guilt, and
so can take steps for his amendment ; but who is there that
can be fully aware of all the good he has omitted to do in
his state of life, whether in regard to God, his neighbour, or

[1] Psalm lxxxviii. 18.

himself? "Who can understand sins?"[1]   And therefore
the Psalmist is not contented with saying, "Turn away from
evil;" he adds, "and do good:" for salvation consists in the
union of the two.   Thou art quite satisfied with thyself as
soon as thou thinkest that thou art wronging no one.   But
go on to ask thyself how thou art performing thy duties as
a religious, a preacher, a bishop, the father of a family, or
whatever they may be that are laid upon thee.   It is not
enough to refrain from evil in them, thou shouldst also do
good.   It is not enough for a rich man, for instance, not to
rob the poor in order to secure his salvation: he must also
clothe them.   Therefore, observe that our Lord declares that
on the Day of Judgment He will inquire particularly concerning what are called sins of omission, saying, "I was a stranger,
and ye took Me not in; naked, and you covered Me not;
sick and in prison, and you did not visit Me,"[2] because such
sins are least noticed.   They have two principal sources, sloth
and fraud.   The former is the sin of those who know the
obligations of their state, but do not fulfil them from dislike
to incur what is disagreeable: "The Levites were negligent."[3]
Fraud is the sin of those who affect ignorance of their duties
in order to avoid the remorse of conscience which overpowers
those who neglect them: "They practise deceits against their
own souls."[4]   Consider not only the evil which thou hast
committed, but the good which thou hast not done; for it is
not noxious plants only, but barren ones, that God will cast
into the fire: "Every tree that doth not yield good fruit shall
be cut down, and cast into the fire."[5]

II. Consider secondly, that as the Psalmist said, "Do
good," so, too, he might have said: "Do not do evil."   But
what he said was, "Turn away from evil, and do good;"
because all the hope that we have of not doing evil, even the
most grievous, consists in avoiding it and defending ourselves
from it.   Show me a man who does not keep at as great
a distance as possible from the occasions of committing sin,
and it is certain that, in the end, he will commit it.   Therefore, just as in battles, where strength is lacking, artifice must
be employed, so, too, is it in our case: "Turn away from
evil."   We must find out by-ways, arts, and subterfuges, in
order to avoid it: "A wise man feareth, and declineth from
evil; the fool leapeth over, and is confident."[6]   Do not say

[1] Psalm xviii. 13.   [2] St. Matt. xxv. 43.   [3] 2 Paral. xxiv. 5.
[4] Prov. i. 18.   [5] St. Matt. iii. 10.   [6] Prov. xiv. 16.

that to turn away from evil is not to conquer it, as a strong
man would; for he is esteemed strong who knows how to flee
from it: "A wise man is strong," for, if not strong, he does
as much as one who is; "and a knowing man stout and
valiant."[1]   Do not, therefore, wait till dangers come, but
forestall them by prudence, as men do when there is a
threatening of famine, or war, or any other evil far less serious
than sin: and so wilt thou do what is here called turning
away from evil.   "Let not the way of evil men please thee;"
for even that pleasure is a sin: "Flee from it, pass not by it;
go aside, and forsake it:" that is, "flee from it" in person;
"pass not by it" in thought; "go aside from it" if it meets
thee; "forsake it" if thou art in it.[2]

III. Consider thirdly, that if it seems so hard to thee to
"turn away from evil and do good," thou shouldst take
courage, because, even in this world, thou wilt reap much
fruit from so doing.   And what is this fruit?   "The peace of
God which surpasseth all understanding."[3]   It is this good
which all men long after: usurers seek for it in their money,
the proud in their dignities, the sensual in their pleasures.
But, oh, how far are these unhappy men from finding it!
"There is no peace to the wicked, saith the Lord God."[4]
No matter which way thou turnest, there is but one which
leads to it, that which is pointed out to thee by the Psalmist
in these words: "Turn away from evil, and do good."   Turning from evil takes away the pain of a bad conscience; doing
good, and that superabundantly, bestows, in addition, the joy
which is given by a good one, and so peace is gained: "The
work of justice shall be peace."[5]   It is true that, in this world,
there can be no perfect peace, because no one can ever attain
to doing good, nor even to turning away from evil, without
a struggle: "I see another law in my members, fighting
against the law of my mind."[6]   But this does not matter, for
the very struggle may be considerably diminished by subjecting
the flesh to the spirit by means of exterior and interior mortification.   And it is this which the Psalmist teaches thee when
he says: "Seek after peace, and pursue it."   If thou thinkest
that thou hast not yet obtained the peace that thou desirest,
be not weary either of seeking it though it be far off, or of
pursuing it though it flee from thee: for those who have
missed the way of peace, like worldlings who "have not

[1] Prov. xxiv. 5.   [2] Prov. iv. 15.   [3] Philipp. iv. 7.
[4] Isaias lvii. 21.   [5] Isaias xxxii. 17.   [6] Romans vii. 23.

known "[1] it, do indeed seek in vain for peace, no matter how eagerly they seek after it; but he who follows the way that leads to it approaches, even if he does not attain it: "I am become in His presence as one finding peace."[2] And it is far better to limp along the right way than to run in the wrong one.

NINETEENTH DAY.

And He said to all: If any man will come after Me, let him deny himself, and take up his cross daily, and follow Me (St. Luke ix. 23).

I. Consider first, how greatly those are mistaken who imagine that manly self-denial, mortification, self-discipline, and patient suffering are only binding on persons in religion, whose profession is perfection. These things are common to all. For this reason the Evangelist here declared that these great words, "If any man will come after Me, let him deny himself, and take up his cross daily, and follow Me," were not addressed by Christ to the Apostles only, but to others also. "He said to all:" that is, to those then living, to those who were to come after, to all Christians without exception, who are those here described by our Lord in the words, "If any man will come after Me." There were many at that time who went after Christ, but with what object? Some to listen to Him, others to admire Him, others to ask help from Him in their troubles; but none of these were His followers. His followers were those who flocked to Him in order to keep close to Him. Therefore He did not say, "If any man will come to Me," but "after Me;" for being a Christian consists in following Him as the true Lawgiver, Leader, and Head, and consequently in letting ourselves be guided by Him as He pleases. What is thy object in following Christ? Is it gain, or glory? If so, thou art no loyal follower. He must be followed because He is worthy of being followed. And so He said, "If any man will come after Me," not "after My gifts." If thou lovest Christ for the sake of thy own interest, especially temporal interest, He will disdain thy service. The Sichemites were all circumcised, with the determination of abandoning their idols, and nevertheless God did not accept this religious act, because their reason for performing it was

[1] Psalm xiii. 3.   [2] Cant. viii. 10.

aggrandizement: "We must circumcise every male among us, following the manner of the nation, and their substance and cattle and all that they possess shall be ours."[1]

II. Consider secondly, that this "will come," is not merely the future tense of the verb to come, but is *vult venire*, that is, "is willing to come," to show that He requires a good-will in all who follow Him. Those are acceptable servants who give their master a willing, not a compulsory service : "All the children of Israel dedicated voluntary offerings to the Lord."[2] Besides, since it is so noble a thing in itself to follow Christ, why should any one wish to wait till he is obliged to do so? A tacit invitation ought to be enough, such as is given by a sovereign when he causes it to be signified to his subjects that he is about to take the field. And thou knowest well how much poverty, persecution, and shame Christ first bore for thy sake : so that thou hast seen Him even die naked upon a Cross between two thieves. How, then, canst thou ask for more than a simple call to follow Him? Is it not a shame that when the devil sounds his trumpet every one hastens at the sound—"A man of Belial, whose name was Seba, sounded the trumpet, and all Israel followed him"[3] —but when Christ sounds His, hardly any one stirs? No wonder that when speaking to a vast multitude, "to all," He said, "If any man." He knew that many would be called and few chosen.

III. Consider thirdly, that the object of this invitation given by Christ is that which is put in the last place, the following Him : "Let him follow Me." And what is this following? It is that which must lead thee, if need be, to Calvary. For if thou examinest to see on what occasion Christ invited all to tread in His footsteps, it was not when He was on His way to the marriage of Cana, nor when He went up the Mount of Transfiguration, nor when He was entering the city in triumph. It was directly after He had declared that His terrible Passion was drawing near : "The Son of Man must suffer many things," &c. This, therefore, is what every one ought to set before him : that he must follow Christ so constantly, both in His doctrine, maxims, and in the imitation of His virtues, as to be ready to be crucified with Him, rather than consent to forsake Him. But do not suppose that this is an easy matter ; for Christ required, as a preparatory disposition for dying with Him on

[1] Genesis xxxiv. 22, 23.   [2] Exodus xxxv. 29.   [3] 2 Kings xx. 1.

the Cross, that every one should accustom himself to carry his daily cross, that is to say, the trial, tribulation, or affliction which is daily sent to him: "Let him take up his cross daily, and follow Me." Thou art sometimes inclined, when engaged in prayer, to think thyself very ready to give up thy life for Christ, and to go so far as to defy the very lions, like St. Ignatius, to say nothing of swords and scourges, and all the time thou findest the greatest difficulty in bearing some little act of discourtesy. This is to desire to be crucified with Christ without having first, like Him, gone forth to meet death bearing the Cross upon thy shoulders.

IV. Consider fourthly, how forcible these few words are: "Let him take up his cross daily." It is not said "let him bear," but "let him take up his cross," to show that thou art to embrace thy cross with readiness, alacrity, and gladness, not to wait till thou art compelled to carry it, like Simon of Cyrene. The word "cross" is used, because by it is signified every adversity which may meet thee, but the word is also used in preference to trouble, affliction, or any other, because it is made sweeter by reminding us that all we have to bear is less than Christ bore when for love of us He died upon the Cross. It is said "his" cross, because there are many persons who think themselves prepared to carry very heavy crosses, only not those which are given them to bear. But it is in this that all thy merit consists: not in desiring to bear the cross of another, but in being willing to bear thy own, which is just that suffering which the duties of thy state bring with them. The cross of sovereigns is the giving of audiences; that of bishops, making visitations; that of priests, saying the Divine Office with devotion; that of religious is solitude; that of married persons, bearing with one another, and so on. Every one thinks the cross of another easy to bear, and accuses him of being careless or lukewarm in his way of bearing it; and very few know how to bear their own properly. Lastly, there is the word "daily," to show that the carrying of this cross is not the business of only one day in the week, as is the case with some persons who wear a hair-shirt, a chain, or some such instrument of penance, but, on the contrary, that it must be the work of every day, so many are the contrarieties which are met with every day of human life in consequence of sin. See, therefore, whether thou art ready to receive thy daily cross with open arms, and hence draw the conclusion whether thou art ready, in case of need, not only to follow Christ in

ways which are moderately rough, but to go with Him even
to Calvary with the fidelity of a perfect disciple.

V. Consider fifthly, that there is nothing which is so great
an obstacle in every one to this cheerful bearing of the cross
as self-love. For just as Christ here made it a necessary
preliminary disposition for the perfect following of Him even
to Calvary, that a man should accustom himself to the daily
carrying of his own cross, so, too, did he make complete
self-abnegation a necessary condition for that daily carrying
of his own cross. This is the meaning of the words, "If any
man will come after Me, let him deny himself:" not only
must this abnegation concern his possessions, *sua*, but *se*,
himself. Would that thou couldst enter into all the deep
meaning of the words, "to deny oneself!" Our Lord does
not say that thou art not to be too self-indulgent to thyself;
He says that thou art to deny thyself, which is the same thing
as saying that thou art constantly to go against thy inclinations,
more especially when they are in the least opposed to the will
of God. If thou wouldst understand what it is to deny thyself,
see what it is to deny or renounce another. When thou hast
thus renounced some false friend whose treachery thou hast
discovered, though thou shouldst see him fall into the hands
of justice, be thrown into prison, bound in chains, condemned
to the gallows, thou wouldst not be moved to pity or aid him,
thou wouldst, on the contrary, be glad to see him enduring
the punishment due to his base treachery. Just so oughtest
thou to act in denying thyself, that is, if thou deniest that
part of thee which is the traitor, thy irregular concupiscence
from which spring so many appetites, some sinful, others
irrational, thou shouldst not even pity thyself for what thou
sufferest, but confess that it is justly thy due. And here it is
to be observed that it is not in thy power to root out thy evil
inclinations, for which reason Christ only commands thee
to deny them, that is, not to allow them to rule over thee:
"Let not sin reign in your mortal body, so as to obey the
lusts thereof,"[1] and this is always in thy power. If, then, thou
art not to allow them to prevail over thee even when they rise
up of themselves against thy will, still more hast thou to be
on thy guard against arousing or exciting them when they are,
so to speak, asleep. Yet what else art thou doing by pampering thyself with so many luxuries? Thou art stirring up those
very desires which thou oughtest to keep in continual sub-

[1] Romans vi. 12.

jection. Think, therefore, that the life of a Christian should always be such as thou hast been here told : to deny himself, so as to accustom himself to bear every daily cross which God sends him, so as to follow Christ faithfully, if needs be, even to Calvary : "If any one will come after Me," by being a Christian, "let him deny himself" in time of prosperity, "and take up his cross daily," but more especially in time of adversity, "and follow Me," even in time of furious persecution.

TWENTIETH DAY.

He that is faithful in that which is least, is faithful also in that which is greater ; and he that is unjust in that which is little, is unjust also in that which is greater (St. Luke xvi. 10).

I. Consider first, that no more serious mistake can be made in the spiritual life than that of desiring to do very great things for God which will never fall to our lot ; such as walking barefoot, like St. Pachomius, along rough roads strewn with stones and sharp thorns ; burying ourselves, like St. James, in tombs where it is impossible to stand upright ; or, like Guarinus, creeping on hands and feet through caves ; and, at the same time, neglecting to perform perfectly the little simple works of the Divine service which present themselves in the course of the day. What reliance can be placed, in that case, on such desires, however fervent ? None whatever. Sometimes they may even be exceedingly mischievous ; for they may lead thee to think thyself already very rich in virtues when thou art very destitute of them : "Thou sayest, I am rich and made wealthy, and have need of nothing, and knowest not that thou art wretched and miserable."[1] First, therefore, thou oughtest to exercise thyself with very great care in doing little things, and then aspire to great ones. And why so ? Because Christ says, "He that is faithful in that which is least," that is, in doing the least good, "is faithful also in that which is greater." To do a good work which is not only little, but even the least, is a sign of being able in time to do, not only what is great, but even greatest. It is true that it is not said, "He who does the least good," but "He who is faithful in " doing "that which is least," because it cannot be at once

[1] Apoc. iii. 17.

concluded from every little good work that thou doest that if
thou hadst a favourable opportunity, thou wouldst also do a
great one.   This can only be concluded when thou art faithful
in doing these little things, that is, when thou art in the habit
of doing all the good that thou canst.

II. Consider secondly, why this faithfulness in little things
is so important.   It is because a habit of suffering for God,
which is of a long standing, is the thing which most of all
helps us to bear willingly those trials which are especially
repugnant to human nature, such as libels on our character,
imprisonment, violence, a death of shame and suffering.   But
this habit cannot be formed by these great trials, which at
most occur once in a lifetime; and therefore it must be
formed by means of little things which are constantly happening.   This, then, should be thy daily study.   Not to pray that
thou mayest, like St. Ignatius, have to face wild beasts in
the amphitheatre, for this is not required of thee; but to arm
thyself with patience to bear such annoyances as that which
the gnats cause thee every day in thy cell.   Prepare thyself to
endure the sharp words which are addressed to thee when
thou least expectest them.   Strengthen thyself so as to pass
over the discourteous treatment, or to forget the unbecoming
expressions of thy neighbour.   Thus, indeed, wilt thou make
great progress: "He that is faithful in that which is least, is
faithful also in that which is greater."   Besides, how canst
thou hope to drink up, like milk, the floods with which the
ocean threatens to overwhelm thee, if thou art unable to digest
the bitter drops which, frequently as God gives them thee, are
yet so small?   It is precisely of these that thou art to make
thy store of merit, if thou wouldst set about it rightly.   The
bees which make the sweetest honey are not those which will
only gather it from the royal lily, but those which do not
disdain the little rosemary blossom, nor confine themselves to
the choicest aromatic herbs, but visit the broom, the humble
wild thyme and marjoram, because these common plants yield
much greater store than choicer flowers which are more rare.

III. Consider thirdly, that as there is no hope that the
man who despises a little good will ever perform a great one
when the occasion presents itself, so too, on the other hand,
is there much reason to fear that he who slights a little evil
will fall into a great one.   And so, as thou seest, Christ goes
on to say: "And he that is unjust in that which is little, is
unjust also in that which is greater."   He did not say, "He

that doth a little injustice," because that would be drawing a conclusion from a single act, but "He that is unjust in that which is little," because this is drawing it not from the act but from the habit; for that man is not called wicked who occasionally commits a wicked act, but the man who is in the habit of committing such acts. But if it is evident that thou art not in the least careful to keep thyself from slight offences, this is a fair ground for thinking that thou wilt not refrain from grievous ones should the opportunity occur. For if, as we have seen, habit is so strong in doing good, still more is this the case in doing evil, because to the force of habit is added that of nature, which is of itself more prone to evil than to good. Imagine a stream of running water which is directed into a level bed; it will gradually widen the space so as to be able to flow easily in it. But if thou guidest it to run down a slope, it will gradually become a torrent. The same is true in our case; and therefore how certain the consequence always is! A man allows himself to be seduced by avarice into petty sins, such as receiving little presents to which he has no right; he deceives, cheats, commits small acts of dishonesty when it is in his power to do so. Then, like another Judas, he lets himself be so blinded by his avarice, that quickly going on from little to great crimes, he will at length slay Jesus Christ, dishonour the priesthood, violate the sanctuary, and sell, if need be, the very sacraments. And this consequence, which is proved to follow on a habit of avarice, is the same with regard to sensuality, pride, ambition, and intemperance, when the habit is the result of multiplied acts, even though they may not be grave in themselves. A young calf, when carried on the shoulders even of a strong man for the first time, appears an unbearable weight, but let him do it the next day and the next, and so on uninterruptedly, and in time he will be able to carry it with ease when it has become an ox. Such is the force of habit even in laborious things. What then must it be in easy ones? and how plain it is that "he that is unjust in that which is little, is unjust also in that which is greater." He does not say "will be," but "is." For although, as all Scriptural commentators agree, the evil which is little is present, and that which is greater is future, still the latter is so near that it may be spoken of as present. And yet thou wilt not believe it, but art bold enough to give the lie to Christ! Take care that this little evil of thine does not drive thee to a great one, and that too with a fall which is irreparable.

David sinned through a sinful love of woman, but he was not disposed to that sin by a previous habit of conversing with them or looking at them too much. In one and the same instant he saw Bethsabee, fell in love with her beauty, and sinned with her. Solomon, his son, sinned through precisely the same love, but his sin followed on a habit of indulging in innumerable idle enjoyments, pleasures, and amusements with them, which, though not sinful, were excessive: "I made me singing men and singing women, and the delights of the sons of men."[1] And what was the consequence? At the first rebuke that David received for his sin, he repented so deeply that he continued to deplore it all his life, and his first sin of the senses was followed by no other; whereas Solomon went on from one sin to another with such fatal speed, that in his latter days he did not shrink from joining the women he had so long loved in worshipping their idols, rather than displease them.

TWENTY-FIRST DAY.

THE PRESENTATION OF OUR LADY.

Who is she that cometh forth as the morning rising, fair as the moon, bright as the sun, terrible as an army set in array? (Cant. vi. 9).

I. Consider first, with what good reason thou mayest exclaim on seeing this heavenly child on this day ascending the steps of the Temple so firmly, by herself: "Who is she that cometh forth as the morning rising?" and the rest. The Blessed Virgin is certainly that happy Dawn longed after for so many ages of the world by the holy Fathers. For, just as the dawn comes between the night and the day which it is so soon to bring forth, so did the Blessed Virgin come between the night of sin which ruled over the race of mankind and the day of grace which was to come, between the night of sorrow and the day of consolation, between the night of terror and the day of joy—the night of the Law and the day of the Gospel. Therefore, the word translated "cometh forth" is not *egreditur*, which better suits the day of her happy Nativity; but the word for to-day is *progreditur*, because already she is advancing, like the "morning rising," with steps both silent

[1] Eccles. ii. 8.

and firm : silent, because there are few in the world who know the progress she is already making in all virtues, so deeply are they sunk, some in the sleep of sin, others in that of ignorance ; and firm, because no one can ever impede that progress, so free is she from all that withholds others from virtue or retards them in its pursuit. Who can ever hinder the dawn from giving birth to the sun in due time? Now observe, that if Mary is on this day compared to the dawn, it is with reference to her dignity as Mother of God, to prepare herself for which she is now going to the Temple. And, therefore, she is not simply compared to the morning, but to the very early morning—"Who is she that cometh forth as the morning rising ?"—to show that the time is not yet come for her to give birth to her Child, but that she is to be gradually prepared for it by ever-increasing merits. Paradise rejoices in this Dawn, because soon it will see the renewal of intercourse between Heaven and earth, so long interrupted by that night whose darkness is already beginning to be dispelled. Earth rejoices, because it at length sees its hopes of salvation blossom anew, hopes which had not only drooped in that night, but were all but withered. And Hell is furious, because, just as thieves, assassins, and adulterers know that the dawn is their enemy—"If the morning suddenly appear, it is to them the shadow of death "[1]—so do the devils know that this child who has come into the world is their enemy. What, on the other hand, hast thou to do? Thou hast to remember that when the dawn appears, then is the right time to rise and praise God : "We ought to prevent the sun to bless Thee, and adore Thee at the dawning of the light."[2]

II. Consider secondly, that this child who is spoken of as the Dawn, because of the dignity of Divine Maternity for which she is being prepared—"Who is she that cometh forth as the morning rising ?"—is, at the same time, called "fair as the moon, bright as the sun." She is "fair as the moon," by grace; "bright as the sun" by glory. She is not called "fair as the sun," because the sun has its beauty from itself; she is called "fair as the moon," because the moon has her beauty from the sun. When, therefore, thou hearest that Mary is all fair : "Thou art all fair, O my love, and there is not a spot in thee;"[3] when thou hearest that in the first instant of her Conception she had a greater abundance of grace than any of the saints ever possessed

[1] Job xxiv. 17.　　[2] Wisdom xvi. 28.　　[3] Cant. iv. 7.

at the end of their lives: "Her foundations are in the holy mountains;"[1] when thou hearest that in her are united all the gifts of graces (even those known as *gratis data*), privileges, and merits, which are distributed among others: "My abode is in the fulness (*plenitudine*) of the saints;"[2] when thou readest that she, too, shares the lofty titles of Reparatrix, Redemptress, Mediatrix, Hope, Salvation, and Life, which belong, properly speaking, to the sun, that is Christ: "The light of the moon shall be as the light of the sun,"[3] do not be afraid, as though all this were to exalt her beauty too highly. No matter how highly it is exalted, there is no fear, since we know that there will, after all, always remain the same difference between her and Christ that there is between the sun and the moon. The beauty of Christ is in Himself, but Mary receives her beauty from Him. And is it not to the honour of the sun that he is able to impart his own glory to the moon? "A great sign appeared in heaven: a Woman clothed with the sun."[4] Next it is said that she is "bright as the sun," or rather, "elect as the sun," for the word in the Vulgate is *electa*, because her election to glory was not separate from that of Christ; but when He was decreed by God as "the First-born amongst many brethren,"[5] Mary also was decreed as the Mother of Christ, and was predestinated to so splendid a throne of glory in Heaven, that as Christ constitutes, of Himself alone, an order in the Heavenly Beatitude, higher than that of all the saints, as their King, so, too, does Mary, as their Queen: "The Queen stood on Thy right hand, in gilded clothing, surrounded with variety."[6] She "stood," she did not "sit," because it belongs to Christ to decree what graces are to be bestowed on men, and to Mary to beg for and distribute them. She is on the "right hand," not the left, because she has nothing to do with the terrible chastisements which are also fulminated by Christ, but only with His graces. She is "in gilded clothing," not "golden," because the two-fold robe of glory which adorns her in soul and in body is not hers by nature, as Christ's is, but is bestowed upon her. She is "surrounded with variety," because the different aureolæ belonging to all the choirs of prophets, apostles, anchorets, martyrs, and all the rest, meet together in her: "I live, saith the Lord, thou shalt be clothed with all these, as with an ornament."[7] Wilt thou not, then,

---

[1] Psalm lxxxvi. 1.   [2] Ecclus. xxiv. 16.   [3] Isaias xxx. 26.
[4] Apoc. xii. 1.   [5] Romans viii. 29.   [6] Psalm xliv. 9.   [7] Isaias xlix. 18.

admire and love the child who is one day to be so glorious?
The whole Church is in the habit of saluting her three times
a day, morning, noon, and evening : in the morning, to remind
thee of the great benefits she brought thee by her Divine Son,
when she came forth, "as the morning rising ;" in the evening,
to remind thee of the abundant graces that she has received
for herself and for others like the moon, which is fairest at the
full : " Fair as the moon ;" and at noon, to remind thee of the
glory she is now enjoying, so that, in union with her Son, she
may pour down from Heaven eternal light upon thy soul :
" Elect as the sun."

III. Consider thirdly, that this same child, so full of
sweetness, is nevertheless lastly here described as full of terror:
" Terrible as an army set in array." But thou hast no reason
to be afraid, for her terrors are not for thee but for thy
enemies. The devils know the power of the sighs and prayers
which from her very cradle she began to send to Heaven, and
how greatly do they therefore fear her ! The terror with
which she inspires them is as great as though she were a whole
army of Principalities and Powers prepared for battle. I say,
prepared, because Mary is not yet terrible "as an army engaged
in battle," but only "as an army set in array," that is, ready
for battle. She is not said to be engaged in battle, because
she has not yet taken the field to rout the hosts of Hell, as
she will one day do at the foot of the Cross on which her Son
is dying ; but she is said to be "in battle array," because she
is already preparing for the conflict. We know that a well-
ordered army may already be said to be partly victorious : it
has no need to make any effort to terrify the enemy, to
brandish arms, to fire shots ; this is not necessary, the mere
sight of it strikes terror. So was it with Mary while yet an
infant ; nay, so may it be said of her even now, for to put
to flight and vanquish all Hell, it is enough for her to show
herself : "She weakened him with the beauty of her face."[1]
Hence it is, that not Hell alone, but all its allies combined
with it, cannot endure the sound of her name. Hell has three
allies, heathens, Jews, and heretics ; and how do all these
three hosts also dread our Lady? For how often has she
conquered them merely by the force of her mighty name
invoked against them by Christendom ! And so the Church
sings : "Rejoice, O Virgin Mary, thou alone hast destroyed
all heresies in the world." And why so? Is it not because

---

[1] Judith xvi. 8.

she gave to the world the Sun which at once dispelled all the errors prevailing in it? Surely; but this is not the only reason. It is also because she gave special instructions, in the first instance, to the Apostles, who went forth to attack all those three hostile armies, and has ever since continued her protection to the princes, popes, and doctors who have made war on them either with arms, anathemas, or disputations. Is she not terrible, then, to all these three armies of the enemies of God? Yes, she is terrible to them as "an army set in array," because she has never need to marshal her forces against them; she is always "set in array." What, then, shouldst thou do? Thou shouldst hasten to her tents to place thyself in safety there, if thou art entirely given to the contemplative life; and if to the active, then go there to be enrolled in her ranks, and to fight for, or at least together with her.

TWENTY-SECOND DAY.

Wash thy heart from wickedness, O Jerusalem, that thou mayest be saved: how long shall hurtful thoughts abide in thee? (Jerem. iv. 14).

I. Consider first, how few there are who wash their heart from wickedness. Many cleanse it, because many purge it by confession from the sins with which they have soiled it; few wash it, because few purge it by confession in such a way as to leave nothing of their sins cleaving to it. And this is the meaning of washing the heart, not to leave in it even the affection to evil: "Wash thy heart from wickedness, O Jerusalem, that thou mayest be saved." When thou confessest, thou accusest thyself, for example, of having sought the vain esteem of men in thy actions so many times, but thou dost not go on to rid thy soul at the same time of the esteem thou cherishest for that esteem, by considering within thyself how foolish and unprofitable it is, and how little it deserves to be courted; so far art thou from doing this, that thou retainest so strong an inclination towards it, as almost to consider the man blessed who possesses it. So long as thou actest thus, thou cleansest, but dost not wash thyself from wickedness. Then, if thou examinest thy heart, thou wilt see how much affection thou retainest, not only for the vain esteem of men, but for worldly friendships, pleasures, honours, amusements, and everything that the world worships. If to wash the heart were as easy to all men as to cleanse it, these words would not be addressed to Jerusalem, that is, to a soul which is already consecrated to God: "Wash thy heart from wickedness, O Jerusalem, that thou mayest be saved: how long shall such hurtful thoughts abide in thee?"

II. Consider secondly, what it is that shows the heart not to have been washed from wickedness. It is the "hurtful thoughts" which abide in it. In the first place, I say "hurtful," not evil; for if evil thoughts abode in it, it is certain that the heart could not even have been cleansed. But if there are not evil, there are hurtful thoughts in it, thoughts which do not imply any grievous offence against God, but which may, nevertheless, gradually lead to it, such as are thoughts of worldly glory or pleasure. These, doubtless, proceed from the affection remaining in thy heart for such vanities, and therefore they are a sign that if it is cleansed, it is not washed. And, in the second place, I say "abide," not "pass." "How long shall hurtful thoughts abide in thee?" because hurtful thoughts often pass through the minds of all men, and this is no proof of any affection to evil: it is their abiding there that is a proof of this. And so the Prophet does not say to Jerusalem, "How long shall hurtful thoughts come to thee?" or "force their way into thee," or "enter thee;" he says "abide in thee," because here lies all the mischief: it is not the flies that come and go that corrupt the balsam, but those that settle on it: "Dying flies spoil the sweetness of the ointment."[1] How is it with thee now as to this matter of keeping thy mind free not only from evil, but from hurtful thoughts? Know that it is thy thoughts which principally show thy predominant affection: "Your sins have appeared in all your devices (*cogitationibus*)."[2] When, therefore, thou examinest thyself before confession, consider what thy thoughts are most apt to dwel' upon, and thou wilt know where thou most requirest washing.

III. Consider thirdly, how thou art to proceed in order to cleanse thy heart not merely from the stains of sin, but from all affection to it, which is the proper meaning of washing it. We all know that thou art to conceive a hatred for sin, but it must not be a weak, but a vigorous hatred. For if thou hast only a weak hatred for that which has a great power of making itself master of our corrupt nature, it will be extremely difficult not to relapse into love for it. If thou desirest never to love

---

[1] Eccles. x. 1.   [2] Ezech. xxi. 24.

it again as long as thou livest, then must thou hate it intensely. See what Queen Esther did to prevent herself from becoming attached to the royal diadem that she wore ; she abominated it : "I abominate the signs of my pride and glory which is upon my head in the days of my public appearance."[1] And why was this bitter hatred? Because she knew that if she had not hated it so intensely, she would have come to love it by degrees, and perhaps even to love it more than her duty, as was the case with the Israelites, who from retaining some affection to the flesh and leeks of Egypt after they had left it, fell into a sin which they had never been guilty of in Egypt, namely, an affection to its idols. Do not imagine, therefore, that it is a matter of superabundance or supererogation that is required of thee when thou art told to hate sin with an exceeding hatred. It is the only way never to have the least degree of love for it. Do not wonder, then, that the Prophet says, "Wash thy heart from wickedness, O Jerusalem, that thou mayest be saved ;" not only, "that thou mayest become holy," but "that thou mayest be saved," for this washing of the heart, that is, the conceiving a vehement hatred against sin, is necessary for our very salvation, not that it is not sufficient for salvation to have no love for sin, but that the man who does not hate it so greatly as even to abominate it will very soon come to love it : "I have hated and abhorred iniquity, but I have loved thy law."[2] See how much is needed even for the attainment of nothing more than the love of the law of God, which is so opposed to the senses ; we must not merely hate wickedness, but utterly abhor it.

TWENTY-THIRD DAY.

Then shall the King say to them that shall be on His right hand : Come, ye blessed of My Father, possess you the Kingdom prepared for you from the foundation of the world (St. Matt. xxv. 34).

I. Consider first, that when a man says "come," he signifies two points, the whence and the whither. And it is certain that our Lord will mean to express both these points when, after the Judgment, He will turn to the elect and pronounce over them all the sentence of everlasting beatitude,

[1] Esther xiv. 16.   [2] Psalm cxviii. 163.

and say to them "come," that is, come from labour to rest, from poverty to riches, from tears to smiles, from the battle to the crown, which you have merited by your victory. O what a joyful "come!" Truly, "going they went and wept, casting their seeds; but coming they shall come with joyfulness, carrying their sheaves."[1] And do not suppose this "come" to be a general form of invitation to all. For each order of the saints will be perfectly able to distinguish it as said to them in particular, with reference to their especial merits. Come, prophets, exiled for My sake; come, patriarchs, solitary for My sake; come, apostles, rejected and mocked at by the world for My sake; come, martyrs slain, monks despised, virgins who dedicated the flower of your youth to Me; and so of all the rest. And because there is nothing which the saints have so greatly desired as to be with their King, like faithful subjects, for this reason also will He say to them "come." Up to that time there will have been some united to Him by grace, some by grace and glory, and a very few by the fulness of glory which consists in the union of soul and body: "In my flesh I shall see my God."[2] Therefore will He say "come," because, until then, only a very few out of so many will have been called to Himself by Christ. Oh, how have the saints longed all their lives for this "come;" now they hear it. But how is he who desires to follow Christ on that day to merit it? By following Him now: "If any man will come after Me, let him deny himself, and take up his cross, and follow Me."[3] "If any man will come after Me" in glory, "let him follow Me" in humiliation. Dost thou think it just to follow Christ to the enjoyment and not to the conquest of His Kingdom? "To him that overcometh I will give to sit with Me in My throne."[4]

II. Consider secondly, how, out of all the titles relating to their glorification and joy, any of which He might give to the elect on that solemn occasion, Christ will choose that of blessed of the Father—"Come, ye blessed of My Father"—because this one title includes all the rest. With us men to say and to do are different things, and therefore when we bless any one, we intend either to praise him for some good that he possesses, or to pray for it for him. But with God to say and to do are the same: "He spoke, and they were made."[5] And so, when He wishes us a good He gives it to us, whether it be

1 Psalm cxxv. 6.　　2 Job xix. 26.
3 St. Matt. xvi. 24.　　4 Apoc. iii. 21.　　5 Psalm xxxii. 9.
MM　　　　　　　　　　　　　VOL. II.

grace, gifts, privileges, or every sort of virtue. Christ, therefore, would have all the blessed in that vast assembly know that all the good they have has come to them from their Father. And so He will say, "Come, ye blessed of My Father." And oh, how will they then break forth all together into the words, "Blessed be the God and Father of our Lord Jesus Christ, Who hath blessed us with spiritual blessings in heavenly places in Christ!"[1] The blessings given by the Father to the elect will have been most various: "He blessed every one with their proper blessings,"[2] but all of them were given with reference to this last blessing on the Last Day, which is called the everlasting blessing: "May the Eternal Father bless us with an everlasting blessing. Amen." And it is for this that they will all thank Him individually. Now observe, that Christ might very justly say to the elect on that day, "Come ye who are blessed of My Father through Me," because every blessing whatsoever was given them by the Father "in Christ," Who merited it for each one of them. But He will not say so; for by pronouncing on them that everlasting blessing to which all others led up, He will plainly show on that day that those others all depended on Him. Oh, well for thee if thou meritest that blessing for thyself! But in order to do so thou must practise the respect and reverence due to so great a Father: "Behold, thus shall the man be blessed that feareth the Lord;"[3] not, he "shall be blessed," but "thus," because this same Father has, indeed, other blessings with which He recompenses less dutiful children for any good thing that they do from time to time, but they are not like this blessing; they are carnal blessings, that is, such as are suited to their corrupt inclinations; they are not "spiritual blessings," they are "of the fatness of the earth," not "of the dew of heaven." Thou seest, therefore, that the elect on that day will thank Him for those blessings which alone were dear to them: "Spiritual blessings in heavenly places." And, oh, how right they will be! For by this is meant all the blessedness of Paradise.

III. Consider thirdly, that it is precisely the possession of Paradise which Christ will give to the elect on that day, when He says to them, "Come, ye blessed of My Father, possess you the Kingdom prepared for you from the foundation of the world." In doing this He might make use of other expressions, such as "enter into the Kingdom," "behold the Kingdom,"

---

[1] Ephes. i. 3.   [2] Genesis xlix. 28.   [3] Psalm cxxvii. 4.

"enjoy the Kingdom," but what He will say is, "possess you the Kingdom," and this for two reasons. First, to show the peaceful security in which the blessed will enjoy that Kingdom to all eternity; and, secondly, to denote ownership. This is implied in possession; we possess goods which belong to us as our own, not by being lent, rented, hired, or pledged, and also to which we have not merely the right (which we have to those that have been taken from us unjustly), but over which we have actual power, as a king has over the kingdom he governs; such will be the beatitude of the blessed, and therefore Christ will say to them "possess," as it is written, "He that hath overcome shall possess these things."[1] And if thou askest why, when Christ might call this beatitude by so many other names, such as reward, prize, exceeding joy, He will rather call it, on that day, by the name of Kingdom, I answer, it is because no other name so well denotes, not only the interior enjoyments which the blessed will have in the possession of God, but also the greatness, glory, and majesty which will also be theirs exteriorly in their dominion over the damned. And how will these last then rage with fury on hearing this word, Kingdom! To see for ever reigning above them those poor miserable beings whom once they deemed unworthy of a glance! "The saints of the Most High God shall take the Kingdom."[2] Joseph's brethren could not endure the innocent boy's fancying, even in a dream, that he was ever to reign over them: "Shalt thou be our King, or shall we be subject to thy dominion?"[3] Think, then, what will be the fury of the damned on that day, when they see reigning over them in reality not a brother, but strangers, rivals, and those whom they held in contempt: "Hath not God chosen the poor in this world, . . . heirs of the Kingdom which God hath promised to them that love Him?"[4] Now thou understandest why Christ will wait for that day to say to the elect, "Possess you the Kingdom prepared for you." It is because then only the blessed will, together with our Lord, have all their persecutors beneath their feet. "Judgment shall sit, that his power" (that is, Lucifer's and his followers') "may be taken away, and that the Kingdom, and power, and the greatness of the Kingdom under the whole heaven may be given to the people of the saints of the Most High."[5] Wilt thou say that thou dost not care for the Kingdom? Then

[1] Apoc. xxi. 7.   [2] Daniel vii. 18.
[3] Genesis xxxvii. 8.   [4] St. James ii. 5.   [5] Daniel vii. 26, 27.

thou wilt be damned ; there will be no intermediate state on that day—there will be nothing but the right hand or the left, the south or the north, Paradise or Hell.

IV. Consider fourthly, that the crowning point of the happiness of the blessed on that day will be the knowledge that this Kingdom was made on purpose for them. For, even after attaining so vast a good, there might possibly remain in them some degree of anxiety, some slight admixture of doubt, lest they should ever lose it, if, although possessed by them, it had, nevertheless, not been made for them. But when they hear not only that this great Kingdom is theirs, but that it was made for them, what doubt can remain? And, therefore, our Lord will say to them, "Possess you the Kingdom prepared for you from the foundation of the world;" not merely "the Kingdom," but "the Kingdom prepared for you." It is true, that this is not the only reason. Christ will also speak thus, in order that the elect may still more clearly see the great love borne to them by the Father, since, at the very time when He determined to create the world, He also determined to prepare for them this glorious Kingdom of the Empyrean—"He hath prepared for them a city"[1]—predestinating them to that height, both of grace, glory, and dignity, of which on that day they will take possession. Imagine, if thou wilt, the greatness of the praise they will give the Father, but be sure that thou canst never do so adequately. And now observe that the Kingdom is not said to have been "given," but "prepared" for them "from the foundation of the world," just as the prize is prepared for the runner, the recompense for the warrior, the reward for him who endures great labour ; for the fact of the Kingdom being prepared does not preclude the necessity of their really gaining it by their merits: "God hath prepared for them that love Him ;"[2] the Kingdom was "prepared from the foundation of the world ;" but it was after the foundation of the world that it was merited.

[1] Hebrews xi. 16.   [2] 1 Cor. ii. 9.

TWENTY-FOURTH DAY.

Then He shall say to them also that shall be on His left hand: Depart from Me, you cursed, into everlasting fire (St. Matt. xxv. 41).

I. Consider first, how different will be the words Christ will speak to the reprobate on the Day of Judgment from those which He will just have spoken to the elect. To the latter He said "come;" to the former He will say "depart;" and "depart from Me into everlasting fire." What a terrible expulsion! Compare in thy mind, as before, the two terms—whence and whither—"From Me . . . into everlasting fire"—and thou wilt be filled with terror. It will be surely no light evil to be banished from the beautiful face of God; but to be not only banished, but banished to burn in the most excruciating fire that can be imagined, and that, too, an everlasting fire; think what this will be! In every mortal sin committed by sinners there were two crimes: the turning away from God, and the turning to the creature; and therefore it is just that each should be punished according to its desert. The pain of loss corresponds to the turning away from God: "He shall not see the glory of the Lord."[1] And this will be intimated by Christ when He says to these unhappy wretches, "Depart from Me;" for it is just that those who have slighted the land of promise should not enter it: "I swore in My wrath that they should not enter into My rest."[2] The pain of sense answers to the turning to the creature: they "shall be tormented, day and night, for ever." And it is to imply this that He will add "into everlasting fire," for it is perfectly just that one who has disregarded God for the sake of gratifying his senses, his feelings, his body, should be tormented by his own passions, like so many furies, and should endure, in his senses, his feelings, and his body, not only the pain of fire, but of all the other torments answering to his past sins, which are inflicted by fire: "In measure against measure, when it shall be cast off, thou shalt judge it,"[3] that is, in the measure of punishment against the measure of sin. These torments, therefore, are all included under the name of fire, not only because the damned will have to suffer them in a prison of fire—"They shall cast them into the furnace of fire"[4]—but also because all the other torments, besides that of fire, will have

[1] Isaias xxvi. 10.
[2] Psalm xciv. 11.   [3] Isaias xxvii. 8.   [4] St. Matt. xiii. 42.

to be inflicted by an activity and ferocity which have the characteristics of fire: "They shall go out from fire, and fire shall consume them."[1]   Fetters of fire, swords of fire, arrows of fire, serpents of fire; in short, everything that thou canst imagine in Hell, not excepting the putrid breath which will be exhaled from the mouths of the damned—all will be of fire: "Your breath, as fire, shall devour you."[2]   Is it possible, then, that when it is in thy power to hear Christ say "come," on that day, thou preferrest to hear Him say "depart"?   Surely not. Surely thou wilt resolve to purchase that "come" at any cost! If there were nothing else to gain, is it not enough to escape the fire?   No prize is too high to pay for that.   And consider that this fire is everlasting!   "Night and day it shall not be quenched; the smoke thereof shall go up for ever."[3]

II. Consider secondly, that just as Christ will honour the elect with the title of "blessed," so, too, will He brand the reprobate with that of "cursed."   There is, indeed, this difference with regard to the elect and the reprobate, that the former have all their good from their Father, while the reprobate have their evil from themselves: "Destruction is thy own, O Israel; thy help is only in Me."[4]   No wonder, therefore, that the former are not only called blessed, but blessed of the Father, while the latter are simply called cursed.   Not one among us could ever gain, or be capable of gaining, Paradise without the Father, and so Christ will say to all who shall gain it, "Come, you blessed of My Father;" but every one of us could of himself disregard the gaining of Paradise, and so Christ will say to those who fail to gain it, "Depart you cursed," but not "cursed of My Father."   Not that, on that day, the curse as well as the blessing will not be pronounced by Christ, but only that it is not to be ascribed to the Father.   What father is there who is not far more ready to bless his children than to curse them?   If he curses them, it is because they compel him to do so by their rebellion: "They are cursed who decline from Thy commandments."[5] What dost thou say to this?   Think well of it; for it rests with thee, so long as thou livest, to merit the blessing which thy Father desires to give thee, or to lose it by thy own fault. But remember, that if thou dost not gain the blessing, it will be impossible for thee to escape the curse.   One or the other must be Thy portion.   A father either blesses his children, if

[1] Ezech. xv. 7.   [2] Isaias xxxiii. 11.
[3] Isaias xxxiv. 10.   [4] Osee xiii. 9.   [5] Psalm cxviii. 21.

they are good, by making them his heirs, or curses them, if they are bad, by disinheriting them : " Behold, I set forth in your sight this day a blessing and a curse : a blessing if you obey the commandments of the Lord your God, . . . a curse, if you obey them not."[1]   How unhappy, then, is the son who chooses the curse ! " He loved cursing and it shall come unto him ; he would not have blessing and it shall be far from him."[2]   The word is not *recedet*, but *elongabitur*, to show that when the unhappy man, having recognized his error, shall desire the blessing, he will no longer be able to pursue it : " For know ye, that afterwards, when Esau desired to inherit the benediction, he was rejected, for he found no place of repentance, although with tears he had sought it."[3]

III. Consider thirdly, that, in order to make it still clearer that the curse is not to be ascribed to the Father, Christ will say, when speaking to the elect on that day, "Possess you the Kingdom prepared for you ;" but, when speaking to the reprobate, He will not say, " Depart into the fire prepared for you," because the Father made Paradise before it was merited by any of His children, but not Hell. Hell was made by Him at the very moment when the rebel angels merited it. And therefore, this being so, it was made for the devil, not for men : and so, when speaking of it to men, Christ will say, " Which was prepared for the devil and his angels," not "for you." It is true, that after it was made, the Father used it in the same way for men as He had at first done for the devils, but this was not His primary intention. He used it for men, because as the great majority of them chose to obey Lucifer rather than God, it was just that they should go at last to dwell in the dominions of him whom they had chosen for their king. Besides, canst thou believe that if Hell had been made for us, the Father would have sent His Divine Son from Heaven to earth to deliver us from it at the cost of all His Blood? It was made only for the angels who rebelled against Him : " Prepared for the devil and his angels." And therefore thou seest that, after they had sinned, there was no help of any kind granted to them, as there was to us. What shame, then, will be thy portion, if thou losest the Kingdom which has been made for thee, and art cast into the fire which was not made for thee, but for thy foes, the devils : " I was the brother of dragons, and companion of ostriches."[4]

---

[1] Deut. xi. 26—28.
[2] Psalm cviii. 18.   [3] Hebrews xii. 17.   [4] Job xxx. 29.

IV. Consider fourthly, that Christ will first call the elect to the Kingdom by saying "come," and then drive the reprobate to the fire by saying "depart." There are three reasons for this: the first is to show how much more He rejoices in doing good than injury: "Is it My will that a sinner should die? saith the Lord God."[1] And this reason is on account of the kindness of the Judge. The second is to console the elect as soon as possible, and to do them honour in the presence of those enemies of theirs who either ill-treated or despised them in the world: "He that hath been humbled shall be in glory."[2] And this reason is on account of the dignity of those who are on the right hand. The third is to cause greater anguish to the reprobate, and to make them rage with envy at the sight of the glory of the elect, and the joy with which they will hear their sentence: "The wicked shall see, and shall be angry; he shall gnash with his teeth, and pine away."[3] And this reason is because of the confusion of those on the left hand. Now contemplate, for a moment, the different ways which will be taken: upwards by the elect, and downwards by the reprobate: "These shall go into everlasting punishments, but the just into life everlasting." Although, in truth, it is not correct to imagine any ways in this case—the everlasting separation between these great multitudes will be made in an instant—the elect, ravished by the love which lifts them up, will fly, like flames, to the heavenly sphere, at the same instant of time that the earth will yawn to swallow up the reprobate in its depths; such is the power which will be in the voice of Christ when He says to the former "come," and to the latter "depart."

TWENTY-FIFTH DAY.

ST. CATHARINE, VIRGIN AND MARTYR.

Wisdom will not enter into a malicious soul, nor dwell in a body subject to sins (Wisdom i. 4).

I. Consider first, that that man is wise, in every matter whatsoever, who knows how to judge of things according to their first and highest causes, and to order them accordingly. And such are the masters of every art whose merit is greater

---

[1] Ezech. xviii. 23.   [2] Job xxii. 29.   [3] Psalm cxi. 10.

in proportion as they are better able to form a high judgment of things in their art, and to order them : " As a wise architect, I have laid the foundation."[1] It is true that the first and highest cause, surpassing all others in every matter, is God. He, therefore, who is able only to judge of things and order them according to causes lower than God, no matter how high they may be, is indeed called wise, but only in that particular line, whether it be architecture, civil or canon law, medicine, anatomy, astronomy, or arithmetic : he cannot be called wise absolutely. That man alone can be so called who knows how to judge of things and order them according to their First Cause, that is, God : " Behold, the fear of the Lord, that is wisdom."[2] And the reason is that all those lesser arts are of small value unless there is, with them, the possession of that highest art which consists in the attainment of our Last End. All the rest must be subservient to this. And, therefore, be sure of this, that thou mayest be lost, even if thou art possessed of every one of the rest, if this one is not added to them : " For if one be perfect among the children of men, yet if Thy wisdom be not with him, he shall be nothing regarded."[3]

II. Consider secondly, that this sublime wisdom, of which we are speaking, is a special gift of the Holy Ghost, Who gives to souls a particular inspiration by which they judge of and order all things according to God, that is, according to the will, the pleasure, the glory of God, according to that which has most power to win for us the love of God ; and this is what the Apostle meant when he said, "The spiritual man judgeth all things,"[4] not that he is always able to judge of them by lower, or human rules, but that he is always able to judge of them by those higher rules which are Divine. Thou shouldst not wonder, therefore, at the Wise Man here saying that "wisdom will not enter into a malicious soul, nor dwell in a body subject to sins," because the Holy Ghost, Who gives this wisdom, abhors the malicious soul, a soul, that is, which is given up to pride, anger, envy, avarice, sloth, which are the capital sins most properly attributed to the soul ; and He also abhors the "body subject to sins," that is, to gluttony and luxury, which are the capital sins properly attributed to the body. And, therefore, how can the Holy Ghost infuse into such persons that wisdom, which is so great a gift of His?

---

[1] 1 Cor. iii. 10.
[2] Job xxviii. 28.   [3] Wisdom ix. 6.   [4] 1 Cor. ii. 15.

"The Holy Spirit will withdraw Himself from thoughts that
are without understanding."[1] Thou must first dispose thy
heart to receive so great a guest as the Holy Ghost by driving
out of it all sins whatsoever as the extreme of folly, and then
thou shalt be a partaker of His gifts, for He is not a Sovereign
Who sends His gifts by the hand of another, He always
brings them Himself: "A full Spirit shall come to me."[2]
"The Spirit of the Lord came upon him; the Spirit of the
Lord fell upon him; the Spirit of the Lord abode in him;"
such are the terms in which He is spoken of. And this, too,
is the reason why He is elsewhere likened to a violent wind,
because He not only sends down His favours like rain into
thy life, as an ordinary wind does, but actually brings them to
thee: "Suddenly there came a sound from heaven, as of a
mighty wind coming, and it filled the whole house where they
were sitting."[3] If, then, thou desirest this sublime wisdom,
which is His gift, beg Him to vouchsafe to come into thee:
"I called upon God, and the spirit of wisdom came upon
me,"[4] that is, "the Spirit, Who is the giver of wisdom."

III. Consider thirdly, that all the sins, both carnal and
spiritual, which have been noticed in the last point, contain in
themselves two disorders: the turning from God, and the
turning to the creature. Still, there is this difference between
spiritual and carnal sins, that in the latter there is more of
conversion to the creature, because they are all for the
gratification of the sensual appetite, which is not capable of
the joys whose source is in God, except, as it were, by
rebound, and, therefore, is less guilty in not appreciating them
as the spirit does; whereas spiritual sins have more of aversion
from God, because the spirit is perfectly capable of those joys
which have their source in God, and nevertheless slights them
to attach itself rather to those which come from the creature.
Hence, spiritual sins have more of the diabolic, and carnal
sins more of the animal, element. In the former, the soul
operates as the soul, by virtue of itself, and therefore, with
regard to these sins, it keeps the name of soul: "Wisdom will
not enter into a malicious soul." In the latter, the soul (as
though it were identical with the body) operates rather by
virtue of the body than by its own virtue; and therefore,
speaking of these sins, it is designated by the name of the
body instead of that of the soul: "Nor dwell in a body

[1] Wisdom i. 5.
[2] Jerem. iv. 12.   [3] Acts ii. 2.   [4] Wisdom vii. 7.

subject to sins." In the former, the soul operates, as mistress, by its own will, and therefore it is said to choose evil: "Wisdom will not enter into a malicious soul," that is, "a soul choosing evil" (*in malevolam animam*). In the latter, the soul is rather drawn to evil, like a slave, by the sensual appetites, as though it were rather a body than a soul; and therefore, in such sins, it is not said to choose evil so much, as to have been forced to choose it: "Nor will it dwell in a body subject to sins." Art thou not ashamed, then, if, when contemplating thyself from the lowest part to the highest, thou canst not say whether the noble or the base part of thee is the more polluted?

IV. Consider fourthly, as a thing of primary importance to observe, that it is said that wisdom will not "dwell" in a body subject to sins, and that it will not "enter" into a malicious soul. And the reason is, because as spiritual sins have in their nature more of what is called aversion from God than carnal sins have, they prevent even the approach to the mind of that wisdom which will have God to be the first rule in all things: "The beginning of the pride of man is to fall off from God, because his heart is departed from Him that made him."[1] And as carnal sins have more of what is called conversion to the creature, than spiritual sins have, although they have less of aversion from God, so it is true that at times they allow wisdom to approach the heart of man by some bright ray of faith which arouses him, or some terrible exhortation which alarms him, or some excellent example which smites him with compunction; but then, if they allow it to enter a little way, they do not, as a rule, allow it to remain long, because the heart has become so strongly attached to the creature it has turned to: "If the morning suddenly appear, it is to them the shadow of death; and they walk in darkness as if it were in light;"[2] because sensual men delight in the pleasures of sense, as though they were the true pleasures of wisdom, of which they are not capable. In a word, the man who is under the dominion of spiritual sins is extremely difficult to convert: "Wisdom will not enter into a malicious soul." The man who is under the dominion of carnal sins, if less difficult of conversion, is exceedingly weak in persevering: wisdom "will not dwell in a body subject to sins." Both are very grievous evils; it would be hard to decide which is the most grievous. What, then, if in thy case both spiritual and

[1] Ecclus. x. 14, 15.   [2] Job xxiv. 17.

carnal sins combine to do their worst, at least by the pernicious inclinations which thou hast never thoroughly eradicated from thy soul which is so cold in its love for God, or from thy body which is too much captivated by the pleasures of sense? If this is so, thou wilt be condemned to be one of those sons of Agar, who are incapable of seeking after any but earthly wisdom: "The children of Agar, that seek after the wisdom that is of the earth, but the way of wisdom they have not known."[1]   How unlike the Virgin-Saint of this day, who had so much wisdom, because she was the habitation of the Holy Spirit Who abode in her, as in His temple!

TWENTY-SIXTH DAY.

The Lord is my helper : I will not fear what man can do unto me
(Psalm cxvii. 6).

I. Consider first, that perhaps one of the greatest disquietudes which can distress thy soul in the spiritual life is to imagine what thou wouldst do if ever thou wert exposed to the terrible trial of having to lose all thy friends, property, reputation, relatives, and even life itself, rather than commit sin. Wouldst thou generously resist the attack and prefer to be burned, tortured, lacerated, torn limb from limb, or wouldst thou yield?   This is one of those thoughts which, as is well known, thou oughtest never to excite voluntarily in thy mind; because evils that are very vividly realized have a tremendous power of terrifying us, and so, by allowing thyself to dwell on this thought, thou wouldst foolishly place thyself in temptation. And, therefore, it is sufficient to represent to thyself such evils as are likely to befall thee, either to encourage thy soul, or to arm it against danger, since it was only with reference to these objects that the Apostle wrote: "Try your own selves, if you be in the faith."[2]   There is no question of imagining evils which are only possible.   But then thou wilt say, that if thou dost not represent these evils to thyself, the enemy of man, thy tempter, sometimes purposely represents them to thee, in order to endeavour to conquer thee by phantoms. If, therefore, thou wouldst know how to act on such occasions, I will tell thee

[1] Baruch iii. 23.   [2] 2 Cor. xiii. 5.

in a very few words. Drive him away from thee with this verse, which was hurled against him in the like case by St. Martin, and tell him that with the help of God thou art afraid of nothing: "The Lord is my helper: I will not fear what man can do unto me." Dost thou not see that these terrors are like the walls of fire which we hear of as surrounding enchanted palaces? If thou makest much of them they will cause thee to stop short in terror; if thou dashest at them, they yield at once as if they were walls of mist, that is, walls which are not to be climbed or thrown down like stone walls, but only to be passed through: "Through my God I will go through (*transgrediar*) a wall."[1]

II. Consider secondly, for thy encouragement, that it is not in any degree in thyself, but only in God, that thou art to put thy confidence—"The Lord is my helper"—and therefore the interior distrust of thy own strength which thou feelest does not betoken in thee any lack of resolution on thy part to act, in all circumstances whatsoever, as the glory of God demands; rather does it betoken a true and deep knowledge of thy own misery which justly makes thee fear, as far as thou art concerned, the worst that can happen. All that is wanting is, at the same time, to trust God as greatly as thou distrustest thyself, nay, far more, for His mercy will be always incomparably greater than thy demerits, and His power than thy weakness: "There are none that can resist His hand."[2] And besides, this feeling of self-distrust is a very valuable thing. It is far better than to think oneself firm and safe, because God loves to confound the presumptuous: "Thou humblest them that glory in their own strength."[3] And thus we see many persons who have had great confidence in themselves shamefully beaten when the struggle came—"They have turned back in the day of battle"[4]—while others, who trembled for fear, stood firm, because the very feeling they had of their own weakness urged them to seek help from God, to humble themselves, to watch and pray with the greatest fervour, lest they, too, should yield to temptation. And this was why the Apostle said, "When I am weak, then am I powerful."[5] Do not be distressed, then, at feeling that thou wouldst fall if thou wert to be confronted with a great temptation. It is sufficient if, despite this feeling, thou hast a strong confidence that thou wilt not fall, not indeed by virtue of thy present strength,

[1] Psalm xvii. 30. [2] Daniel iv. 32.
[3] Judith vi. 15. [4] Psalm lxxvii. 9. [5] 2 Cor. xii. 10.

which thou seest is very small, but of that which God will give thee, in proportion to thy need, when the time comes for helping thee.

III. Consider thirdly, also, for thy still greater encouragement, that it is not necessary for thee to have in thyself at the present time a spirit of fortitude great enough to overcome the terrible kind of temptations which have been mentioned, for God does nothing in vain; and therefore He is not in the habit of giving us the grace which is requisite for victory in great battles when there is no occasion for it. But this is no loss to us; if He does not give it now, He will give it when the time comes: "The Lord will give strength to His people."[1] Samson was the strongest man in the world; but dost thou think that he was always conscious of that surpassing strength? By no means; but when he met lions in the forest, or saw himself surrounded by enemies in the city, attacked and well-nigh overpowered by them, then suddenly he felt it come to him from Heaven. Therefore, whenever he did some wonderful deed, the Sacred Scriptures speak of him as surprised by the Spirit of the Lord: "The Spirit of the Lord came upon Samson, and he tore the lion."[2] "The Spirit of the Lord came upon him, and he slew thirty men."[3] "The Spirit of the Lord came strongly upon him; and as the flax is wont to be consumed at the approach of fire, so the bands with which he was bound were broken and loosed:"[4] showing that this supernatural strength was given to him as the occasions for using it presented themselves. Do not be afraid of anything which may now be suggested to thee by the enemy to make thee lose courage, but hope in God, Who, for this very reason, is called "a helper in due time;"[5] and when that time comes He will give thee the strength which thou lackest now: "The Spirit of the Lord shall come upon thee, and thou shalt be changed into another man."[6] Dost thou not remember what is written of the saints who had faith in God? That by that faith they "became valiant in battle,"[7] because, at the very moment when they were called upon to put forth strength for the honour of God they received it: "They recovered strength from weakness."[8] Hence it is that hope in God should not be chiefly founded on the grace which we have already received from Him, but on God Himself, Who, when there is

---

[1] Psalm xxviii. 11. [2] Judges xiv. 6.
[3] Judges xiv. 19. [4] Judges xv. 14. [5] Psalm ix. 9.
[6] 1 Kings x. 6. [7] Hebrews xi. 34. [8] Hebrews xi. 34.

need for it, will give us incomparably greater grace than He has yet given : "In Him will I put my trust."[1]

IV. Consider fourthly, that hope in God is a very different thing from presumption. Presumption goes beyond the limits of the laws He has laid down ; hope keeps within them. Now, the laws which He has laid down with regard to trusting in Him are these : that when we know ourselves to be poor in the grace required for great conflicts, we should desire it, pray for it, and meanwhile do our utmost to train ourselves in easier conflicts by the grace which God never fails to give in them : "Exercise thyself unto godliness."[2] I said advisedly "do our utmost," because, if even in these we often fall through weakness, we are not to lose heart, as though by losing lesser battles we became incapable of meriting the grace for those that are greater. To lose the battle is not always to lose merit ; that is done when we lose the battle through sloth, when we lose it because we do not choose to fight at all, but would have God conquer by Himself, for us, but independently of us. It is this which displeases God, for this is the pernicious confidence of presumption. Listen to the words of the Psalmist : "The Lord is my helper." But if He helps us, that implies our doing something on our side, otherwise it would not be true that He helps us, He would do the whole. If, then, it appears to thee that thou dost not feel that strength in thyself which would be necessary for the overcoming those difficulties which the enemy represents to thee as possibly coming upon thee one day from some man whom he has possessed, thou hast only to desire and pray for that strength, which is an easy thing to do, and meanwhile to exert that little measure of it which God bestows on thee according to what is required for the trials of the day ; for it is said even of Samson, whose prodigious strength was bestowed on him by God only for the object of defeating the Philistines, that from a child he gave good tokens of his future prowess in his own country : "The child grew, and the Lord blessed him ; and the Spirit of the Lord began to be with him in the camp of Dan."[3] First "in the camp of Dan," which was a contest in the lists ; and then "in the camp of the Philistines," which was the field of battle.

---

[1] Psalm xvii. 3.   [2] 1 Timothy iv. 7.   [3] Judges xiii. 24, 25.

TWENTY-SEVENTH DAY.

When I shall take a time, I shall judge justices (Psalm lxxiv. 2).

I. Consider first, that God is now giving thee convenient and ample time in which to do good, if only thou choosest to make use of it: "I gave her a time that she should do penance."[1] But observe that, as He gives it thee now, so will He one day take it from thee, and make it all His own. Think, therefore, that He does not here say, "When the time cometh, I shall judge justices," but "When I shall take a time," because, according to some, He means to show thee that at last He will take for Himself this time which is now thy own, so that thou wilt not have so much as a moment of it: "Time shall be no longer."[2] At all events, it is certain that He will take His own time, that is, the time He has fixed and appointed for judgment. And then how strict will be the account which He will demand from thee of the time which He now gives thee! "He hath called against me the time."[3] Think a little on the present time; how art thou spending it? on useful or on vain things? God gives it thee for the purpose of transacting the great business of gaining Paradise; and thou either despisest it, or undervaluest it, or only usest it for the purpose of earning damnation. Alas, for that ill-spent time! Thou wilt know its value when thy time is over, and God's has come. Dost thou imagine that this is a long way off? "Her time is near at hand."[4]

II. Consider secondly, that when He has taken this time, He will (as He here says) "judge justices." What does this mean? According to the Hebrew phrase, "to judge justices" signifies to judge equitably, rightly, according to the strict rules of justice: "Thou hast sat on the throne, Who judgest justice."[5] And therefore God intends to say that on that day there will be no more room for mercy, all will be given up to justice. But, according to the general consent of the Fathers, the expression "to judge justices," also means to judge those works which are just in themselves, to see whether they were done at the right time, with a right intention, in the right way, and with all the right circumstances. And therefore God here says, "When I shall take a time, I shall judge justices," to teach thee that in that day He will judge, not only what is

[1] Apoc. ii. 21. [2] Apoc. x. 6.
[3] Lament. i. 15. [4] Isaias xiv. 1. [5] Psalm ix. 5.

evil, but what is just : " He shall purify the sons of Levi," that is, the just, "and shall refine them as gold and as silver."[1] And, if this is so, which of us, miserable men, will be safe ? Examine thyself on this point, and thou wilt see that many of the works thou art continually doing are, very likely, just works. But, God grant that thou art also doing them all in the right way ! Every one knows that saying the rosary, reciting psalms, receiving the sacraments, hearing Mass, giving alms to the poor, are in themselves just works. But how dost thou do them? With what distractions, what want of application, what a variety of faults ! And yet, how is it written ? "They that have kept just things justly shall be justified,"[2] not "they that have kept just things," but "they that have kept just things justly." What constitutes the holiness of a man is not so much that which is merely material, as that which is formal in his works.

III. Consider thirdly, that there are many persons in the world who judge matters of justice of themselves, and assert that they are not bound in conscience to make due restitution in such and such circumstances, nor to make peace with certain persons, nor to break off certain habits, and who, if they would discuss the matter with others, more pious, or more learned than themselves, would soon see that they are so bound. Therefore, God would also here point out to us, that every one ought to be very slow to judge himself privately in these matters, because in due time he will see their importance : "When I shall take a time, I shall judge justices." There may be many things which thou considerest lawful which are not really so. Thou wilt not take the trouble to inquire of casuists or confessors, as is the right course. Thou art one of those who are, indeed, "a law to themselves,"[3] but a law made after their own fashion. Always remember, then, that everything will undergo a review ; and at the hands of what Judge ? One Who can distinguish true from simulated justice : "He shall not judge according to the sight of the eyes."[4] Accustom thyself therefore to be less ready to follow thine own judgment either as to the good which thou doest or that which thou omittest ; but take counsel. "Blessed is he that condemneth " (or judgeth) "not himself in that which he alloweth."[5]

[1] Malach. iii. 3.   [2] Wisdom vi. 11.
[3] Romans ii. 14.   [4] Isaias xi. 3.   [5] Romans xiv. 22.

IV. Consider fourthly, that just as God will judge in His own time the justices, or judgments, which thou falsely makest of thyself, so too will He judge those which are falsely made of thee, when thou art innocent, by hostile judges. If, then, thou now meetest with some wrong from men, do not be afraid or cast down, for the great God, of Whom it is written that He "executeth judgment for them that suffer wrong, and giveth food to the hungry,"[1] will judge unjust sentences also, so as to right the injured, the dejected, and the oppressed: "When I shall take a time, I shall judge justices." Only, be content to wait a while: "Wherefore, expect Me, saith the Lord, in the day of My Resurrection that is to come."[2] So soon as thou receivest an injury from some prince, or prelate, or court, thou wouldst fain see thunder-bolts fall from heaven to speak with tongues of fire on thy behalf, and very often thou art ready to say to God with Habacuc, "How long, O Lord, shall I cry, and Thou wilt not hear? Shall I cry out to thee, suffering violence, and Thou wilt not save?"[3] Wait for the time; dost thou not hear what God says, "When I shall take a time, I shall judge justices"? He will right thee, do not doubt it; but not now, not to-day, not to-morrow, not on the day which thou wouldst fix for Him, for it is not for the culprit to dictate the time to his judge, and that the Sovereign Judge. He will choose the time far better than thou canst; it would be very easy for thee to err by taking one day instead of another. He knows which is the right one; He has appointed the day when "He shall judge the world with equity."[4] Remember those afflicted souls who cried to God from under the altar: "How long dost Thou not judge and revenge our blood on them that dwell on the earth?" and who were told to "rest for a little time, till their fellow-servants and their brethren who are to be slain, even as they, should be filled up."[5] Dost thou think thyself alone in the world in suffering oppression from the violence or tyranny of those who can do what they will? Not so; thou hast many companions who have suffered even greater wrongs than thine from earthly judges. And the number of all these must "be filled up," in order that the triumph of Divine justice over all defective human justice may be both more complete and more glorious. Meanwhile, do thou endure in silence—"Rest for a little time "—for if God were to choose at this time to

<hr>

[1] Psalm cxlv. 7.   [2] Sophonias iii. 8.
[3] Habacuc i. 2.   [4] Psalm xcvii. 8.   [5] Apoc. vi. 11.

justify thy honour only, His glory would come short; whereas, by justifying thine and that of countless others at the same time, His glory will be complete.

TWENTY-EIGHTH DAY.

I will heap evils upon them, and will spend my arrows among them (Deut. xxxii. 23).

I. Consider first, that all the evils of this world, however bitter or painful they may be, do not altogether deserve the name, because they are never unmixed evils; there is always some admixture of good in them, which if it does not over-power, at least sweetens their bitterness: "Good is set against evil."[1] Not so in Hell. There all evils are entirely unmixed, and therefore, when God is here speaking of the damned, He says so emphatically, "I will heap evils upon them." He does not specify poison, or wounds, or burns, or any other of the tortures inflicted by executioners on criminals in this world, for then thou mightest imagine antidotes to poisons, balm for wounds, cooling applications for burns, and, at all events, death to bring relief to every kind of torment—that great relief which the thought that they will come to an end is in all evils. No; He says simply "evils," in order that, after imagining poisons, wounds, burns, and every possible kind of suffering, thou mayest not go on to add to these things any degree of good, for there is nothing of the kind for the damned: "Behold, I will watch over them for evil, and not for good,"[2] says God. He will "watch for evil," because He will make them know the bitterness of tears, but "not for good," because He will not allow them to experience relief from them. He will "watch for evil," because He will make them feel the distress of imprisonment, but "not for good," because He will not let them find rest in its seclusion. He will "watch for evil," because He will make them feel the darkness of night, but "not for good," because He will not let them know its silence, its sleep, its quiet, for so much as a moment. At least these poor wretches might hope that after millions and millions of ages there will be an end of their evils. But neither can this be, for if God ever intended

<hr />

[1] Ecclus. xxxiii. 15.   [2] Jerem. xliv. 27.

this end to come, then He would "watch over them for good"
as well as "for evil." If evils are to be unmixed evils, then
must he who endures them be perfectly certain that they are
to be everlasting: "I said, My end is perished."[1] Everlasting
tears, everlasting imprisonment, everlasting darkness, everlasting fire, and, above all, everlasting despair of ever seeing
that beautiful face of God for which they were made: they
"shall suffer eternal punishment in destruction from the face
of the Lord."[2] Does not this thought fill thee with terror?
When thou experiencest any evil in this world, thou consolest
thyself with the thought that, if long, it is not grievous, and if
grievous, not long. All such consolation is over in Hell, for
there there is no grievous evil that is not eternal, and there
neither is, nor can be, an eternal evil that is not grievous.

II. Consider secondly, that as the evils of this world are
not unmixed, so neither can they all meet together in one
man, however unfortunate or desperate his condition may be,
but they may be said to be scattered about the earth seeking
where to abide. The man who is subject to one of them is
exempt from another; indeed, many evils are the opposites of
each other, as cold and heat, wakefulness and stupor, which
cannot torment the same body at the same time. But in Hell
all evils, even those which are most contrary to one another,
will, by the power of God, be united for the torment of the
damned; and for this reason also God here says, "I will heap
evils upon them," because the evils which on earth are divided
among many persons, those even which are mutually opposed,
are in Hell all summoned by God to league together.
Although, indeed, God does not here say, "I will summon,"
but "I will heap," in order to teach us that it is not a question
of mere assembling, but of crowding and loading: "Assemble
together like the bruchus."[3] What possible evil is there that
cannot be found in a pit which God has for all ages made the
very centre of all evils? "The place of torments."[4] The
only good which could possibly be there left thee would be
existence, but even this will be changed for thee into the
existence of evil, if ever thou allowest thyself to be dragged
down there by the weight of thy sins, since it is a far less evil
not to exist at all than to exist in the midst of such evils.
What, then, wilt thou do? Wilt thou endeavour not to exist?
This is impossible: "There is no poison of destruction"

[1] Lament. iii. 18.
[2] 2 Thess. i. 9. [3] Nahum iii. 15. [4] St. Luke xvi. 28.

there.[1] The inhabitants of Hell must always exist, and exist always as criminals, always in tears, always in prison, always in the power of all the furies of Hell; and so God did not say, "I will heap evils against them," but "upon them," that thou mayest know that never to all eternity canst thou rise above these evils, they will always be above thee: "Let his confidence be rooted out of his tabernacle, and let destruction tread upon him like a king."[2]

III. Consider thirdly, that God says of the damned, "I will spend My arrows among them." By these "arrows" are meant the maledictions which He is continually hurling like thunder-bolts against the wicked by the mouth of His preachers and prophets, when they cry aloud: "If thou wilt not hear the voice of the Lord thy God . . . all these curses shall come upon thee and overtake thee."[3] Some, when they hear them, are terrified, and humbling themselves in compunction and contrition, ask pardon; and then God takes from these arrows their power of hurting sinners, and turns them against their enemies, the devils, who seduced them to evil: "When thou shalt be touched with repentance of thy heart, and shalt return to Him, . . . He will turn all these curses upon thy enemies and upon them that hate and persecute thee."[4] Others are so obstinate that, far from being afraid of these arrows, they sometimes despise them in their hearts, and sometimes even deride them openly, as though they were mere bravado, and say in their arrogance, "The evil shall not come upon us."[5] And it is against such persons that God here declares that He will show them in the end how completely and perfectly He will accomplish these threats: "I will spend My arrows among them." The saints observe that the arrows which God has sent against the body of man in consequence of sin are seven in number—hunger, thirst, cold, heat, weariness, sickness, and death. These are common to all who have sinned in Adam. But they are not "spent" perfectly on earth; they are arrows that have lost their feathers or their points, and so the wounds they inflict are not very deep. But in Hell, with what fulness of force will they fly to strike every sinner! "I will make My arrows drunk with blood;"[6] and therefore it is of the damned that God says most justly that He will "spend" His arrows on them because it is upon them that He will make them take full effect. Is it

[1] Wisdom i. 14.   [2] Job xviii. 14.   [3] Deut. xxviii. 15.
[4] Deut. xxx. 1—7.   [5] Jerem. v. 12.   [6] Deut. xxxii. 42.

possible, then, that thou canst know that, if thou sinnest, this terrible place awaits thee, and yet continue to sin, as though thou didst not believe it?

TWENTY-NINTH DAY.

He that feareth God neglecteth nothing (Eccles. vii. 19).

I. Consider first, that this word, *negligere*, has two meanings, to neglect and to despise. The Apostle used it in the former sense when he said to Timothy, "Neglect not the grace that is in thee."[1] And God used it in the latter sense by the mouth of the same Apostle when He said, "Because they continued not in My testament, and I regarded them not" (*neglexi*).[2] From this word *negligere*, used in the first sense, is derived *negligentia*, "negligence," and from the same word, used in its second sense, is derived *neglectus*, "neglect." Endeavour to have all this clearly before thy mind, not, as it may hitherto have been, indistinctly, and then thou wilt at once understand what the Wise Man here means to declare when he says, "He that feareth God neglecteth nothing." His meaning is, that he who fears God neglects nothing that is good as being superfluous, and also that he who fears God despises nothing that is evil, as being a light matter. It may be said that the whole machine of Christian perfection works, as it were, on these two hinges of salvation. What, on the other hand, is the cause of the ruin, and the irreparable ruin, of so many? It is from not keeping firmly on these two hinges. Since, then, consequences of so much importance may result from following out this excellent teaching, or from failing to do so, endeavour to enter into its sense as deeply as possible.

II. Consider secondly, that the Wise Man does not say that "he who feareth God omitteth nothing." For even the greatest saint sometimes omits to do some good thing that he might have done beyond what is usual. What he says is that he "neglecteth nothing." For if such a one omits to do anything good, he does so through frailty, through feebleness, not through that base vice called negligence. Negligence is the fault of those who not only do not do that greater good

[1] 1 Timothy iv. 14. [2] Hebrews viii. 9.

which they might do if they choose, but who do not wish to do it, being satisfied with doing what is barely sufficient to prevent their losing their master's favour. And oh, how great is this evil! It is nothing less than to lose, through demerits, those superabundant helps which God gives to those who are careful to please Him. And such helps are, as is well known, those strong wings which are called eagles' wings, on which thou wouldst see thyself borne by God in so short a time to the heights of perfection : " You have seen how I have carried you upon the wings of eagles, and have taken you to Myself."[1] While, for want of these helps, how many there are who become gradually more and more feeble, like persons whose income is so small that they can hardly subsist! And therefore the Wise Man says in another place, " He that neglecteth his own way shall die."[2] Oh, how sad is this saying! The word translated "shall die," is not *morietur*, because by this habitual negligence in the service of God thou dost not formally incur that terrible death of the soul, which is damnation ; it is *mortificabitur*, because, if thou dost not formally incur the death of the soul by this negligence, which does not always amount to a mortal sin, thou at any rate disposest thyself to incur it by the extreme poverty of spirit and lack of sustenance to which thou reducest thyself : " The slothful hand hath wrought poverty."[3]

III. Consider thirdly, what the virtue is which should be opposed to negligence. It is diligence, which consists, according to the saints, in three things : (1) In the study of all those means, however small, which are able to lead us more quickly to the perfection required of us in our state by God. This study is opposed to negligence, inasmuch as it is a fault in matters of election. (2) It consists in the prompt employment of these means. And this promptitude is opposed to negligence, inasmuch as it is a fault in execution. (3) It consists in the careful application of these means. And this is opposed to negligence, inasmuch as it is a fault as to attention. Examine thyself as to thy conduct, and thou wilt see that thou art often wanting in some one of these fruitful departments of diligence, even if, at times, thou dost not fail in all three. Do, therefore, as God bids thee, "cleanse thyself from negligence with the few," because there are few persons who are at the pains of particularly accusing themselves in confession of this negligence, whichever of the three kinds

[1] Exodus xix. 4.   [2] Prov. xix. 16.   [3] Prov. x. 4.

above mentioned he may be guilty of. The very most that
they do is to accuse themselves of it now and then in general
terms, which mean nothing. Very few really repent of it, and
resolve to correct it. But do not mind what the many do;
act like the few, for it is the few, not the many, who will be
saved : "Many are called, but few chosen."[1]

IV. Consider fourthly, that the Wise Man does not say
that "he that feareth God doeth nothing evil," for there is no
one, not even a great saint, who does not sometimes commit
at least a venial sin—"In many things we all offend "[2]—what
he says is "neglecteth nothing." For if a holy man commits
some venial sin, he does not make light of it, particularly if it
was a deliberate venial sin ; on the contrary, he grieves bitterly
over it. The man who makes light of it is he who thinks
it an evil of no consequence because it is venial. But is a
venial sin a light evil? Alas for thee if thou shouldst die in
so mad an error! Venial sin is the greatest evil there is, or
can be in the world, except mortal sin. So that to have one
single deliberate sin upon thy soul is a greater evil than to be
suffering from every kind of wound, fever, sickness, madness,
in the world, nay, greater than it would be to be tormented
by all the devils in Hell; so that it would not be lawful for
thee, in order to escape all these evils, to tell one single lie,
even in jest, to attempt a trifling theft, or to plan a petty
deception. Nor is this all ; but if it were possible for thee
by one such venial sin to bring over to the faith of Christ in
one day all the Jews, Turks, Tartars, heathen, in a word, all
the nations who have rebelled against Him, thou art not
free to commit it: and so, far from being pleased with thee
for what thou hadst done, God would punish thee with
the keen and sensible pain of Purgatory, which exceeds all
sufferings of this world. Knowing all this, canst thou still
venture to despise one single venial sin, and to say, "What
great harm is it?" "He that feareth God neglecteth nothing."
It is true that thou dost not go so far as to offend God
grievously by venial as thou dost by mortal sin, but thou
displeasest and dishonourest Him. How, then, canst thou
make light of such a sin on account of what it is in itself,
and say to thyself, like a graceless son, "So long as my Father
is not grievously offended by me, I am satisfied"? Besides,
it is true that venial sin is not a disease that of itself slays the
soul, like mortal sin, which is a disease that is so matured, full,

[1] St. Matt. xx. 16.   [2] St. James iii. 2.

perfect, and complete in its evil, as even to destroy the vital principle of the soul, which is charity; but still it is certainly the beginning of that disease. How, then, canst thou make light of such a sin on account of its effects, and say within thyself, like a madman: "So long as I do not commit mortal sin, I am not anxious about any other kind"? Dost thou make light of every disease that is not mortal? On the contrary, thou avoidest every kind as much as possible, knowing that all neglected diseases may bring thee into a condition in which thou mayest contract an incurable one. Why, then, dost thou act in a precisely contrary way in the case with which we are concerned? "He that feareth God neglected nothing;" both because of the intrinsic character of this evil, and because of its effects. When, therefore, thou hearest it said at any time that venial sin is a light sin, do not suppose that this is ever said absolutely speaking, but only relatively, that is, in comparison with mortal sin. Considered from any other point of view, it is the enormous evil which it has been just shown to be.

V. Consider fifthly, that it may seem as though the Wise Man might more properly have said, not "he that feareth," but "he that loveth God neglecteth nothing;" because, not to neglect any good as unimportant, and not to make light of any evil as light, may appear the conduct of one who greatly loves God than of one who greatly fears Him only. But this is a great error; for although, from one point of view, thy objection is valid, yet from another it was more fitting for the Wise Man to say "he that feareth," than "he that loveth God," in order that no one might suppose that to neglect nothing, whether good or evil, belonged only to certain great saints, who are on fire with the love of God. He would have thee know that this is the duty of all, even of those who have attained no higher than to fear Him, since it is beyond a doubt that God often visits small sins, both of omission and commission, with very terrible punishments, and those not only negative, such as consist merely in the withdrawal of favours, but positive, such as being consumed by fire, devoured by wild beasts, and the like, which are related in the Sacred Scriptures. If, then, thou neglectest to do a great deal of good which is in thy power, or makest light of doing evil, what does this show? That thou dost not love God? That is saying little: it shows that thou dost not even fear Him rightly. If it is of faith, that "he that feareth God" (which is the same as "whosoever feareth God") "neglecteth nothing," either good or evil, it follows that whosoever does neglect these things, does not fear God.

———

THIRTIETH DAY.

ST. ANDREW THE APOSTLE.

She is a tree of life to them that lay hold on her ; and he that shall retain her is blessed (Prov. iii. 18).

I. Consider first, that Heaven is our native country, and the earth, where we now live, a land of exile. Is it not certain, then, that we ought continually to long for the country of our inheritance? But, alas! what a gulf divides us from it: a terrible and stormy gulf, for such is this mortal life. Certainly we need a vessel to cross it. What shall it be? The fair vessel which God had provided us with for this purpose was innocence, in which we should have been borne to land safe and happy. But this bark was lost in the miserable shipwreck which all the descendants of Adam suffered together with him. Nothing is left now, but for every one to cling to penitence, which is for this reason called the plank of safety after shipwreck, and this is, in other words, the Cross of Christ. For in what does this consist but in suffering, in mortifying, punishing, and humiliating ourselves, in behaving ourselves always as wretched penitents unworthy of enjoying any good in the world? This blessed Cross is the bark in which is our only hope of salvation : "No one can traverse the sea of this world unless he is borne by the Cross of Christ," says St. Augustine. Do not wonder, then, at hearing it here called "a tree of life." If thou dost not cling close to it there is no hope ; it is certain that thou wilt sink, that is, that thou wilt go down to Hell to be the haters of the Cross of Christ, who are lost for ever : "The enemies of the Cross of Christ, whose end is destruction."[1]

II. Consider secondly, what a passionate struggle there is when a shipwrecked vessel goes to pieces in mid-ocean, and there is nothing left for the wretched crew but to cling to some plank. How the unhappy creatures strive with each

[1] Philipp. iii. 18.

other to get hold of it! How they push and struggle, and when at length they have seized it, how they clasp it! And why? Because it is the wood which is to save their life: "The tree" (or wood, *lignum*) "of life." Oh, how happy would it be for all Christians if they understood that this is what the Cross is for them: "A tree of life to them that lay hold" of it. Instead of leaving it to others, every one would try to be the first to seize it himself; but this truth is not understood! People look at what the Cross is in itself, a "contemptible wood."[1] And therefore every one, instead of striving to reach it, pushes it from him. Thou knowest that a piece of wood which before the shipwreck was of no value, is after the shipwreck sought after, caught at, taken by force, no matter from whom, and would not be parted with at any price. So, too, with the Cross. Regarded in itself it is "contemptible wood," but it is wood which is left to us after shipwreck, and this of itself is enough to give it great value: "The scandal of the Cross is made void."[2] It is no time for considering what is pleasant to nature. We must value the Cross by our state as a shipwrecked crew, and not merely "take" (*prendere*), but "lay hold (*apprehendere*) of it," that is, as it were, strive for it, so precious is it when it is the means of saving life, and that life everlasting. It is "a tree of life to them that lay hold on it." How is it with thee? Art thou leaving the Cross to others, or striving for it thyself? Take good heed, for as it is a very strong sign of salvation to prize the Cross, so, too, is it an evident sign of reprobation to despise and contemn it: "The word of the Cross to them that perish is foolishness, but to them that are saved it is the power of God."[3] It is true that we do not prize the Cross by merely honouring it as the mass of Christians do, or by preaching and lauding it, but by clasping it to our hearts. For it is "a tree of life," not to all, but only to those who embrace it: "to them that lay hold on it," not "to them that honour it," or that "worship it," or that "exalt it," but that "lay hold on it."

III. Consider thirdly, that, in order to be saved after shipwreck, it is not enough even to embrace the plank of wood. It must be held firmly. And so, too, it is here said of the Cross, that "it is a tree of life to them that lay hold on it, and he that shall retain it is blessed." He who only "lays hold" is not blessed, for if a man who is shipwrecked

---

[1] Wisdom x. 4.   [2] Galat. v. 11.   [3] 1 Cor. i. 18.

seizes a plank, and then loses hold, because he has not courage
to face the raging waves which dash over him, he will be lost
as surely as though he had never seized it. It is "he that
shall retain it" who is called "blessed," because it is he only
who is sure to come ashore. And so it is with the Cross.
Of what use is it to clasp it lovingly for a short time, if thou
art terrified by temptations into letting it go? Cling closely
to it; learn a lesson from shipwrecked men, who, made
courageous by the sight of the threatened danger, submit to
be lashed, flung this way and that, dashed about by the raging
sea without ever being forced to unclasp their arms from the
plank. So, too, oughtest thou to do, for everything depends
on this. There is no lack of crosses, for after the terrible
shipwreck of all mankind God would have no one lost for
want of a plank of safety. And therefore the difficulty is not
to find them, nor to take hold of them; it is to keep that hold
firm, courageously disregarding all the storms which are raging
round: "God forbid that I should glory save in the Cross
of our Lord Jesus Christ."[1] What wonder, then, that the
Wise Man here says, "He that shall retain it is blessed."
Oh, how many more there are who embrace the Cross than
who hold it firmly! And this is not surprising; the same
thing happens as to the planks in a shipwreck. And therefore
in the first part of the verse the plural number is used:
"It is a tree of life to them that lay hold on it," for which
the singular is substituted afterwards: "He that shall retain
it is blessed." Think well of this; do not merely consider
that thou hast embraced the Cross, as many do, but that it
must be held firmly to the end, which is the case with very
few: "With Christ I am nailed to the Cross."[2]

IV. Consider fourthly, that the words thou art here meditating were spoken, in the first place, in praise of the Divine
wisdom, but that, in a secondary sense, they are referred
by many of the saints, to the Cross. And there is a deep
meaning in this, because, at the present day, if thou observest
closely, the wisdom of Christians is reduced to this—the love
of the Cross of Christ: "I judged not myself to know anything
among you but Jesus Christ, and Him crucified."[3] And so
a man who has never learnt any rules of perfection, if he
stands firm in desiring nothing but the Cross, leaving to
others, if they will, comforts, pleasures, and dignities, and
choosing for his part that which the world abhors, is sure

[1] Galat. vi. 14.   [2] Galat. ii. 19.   [3] 1 Cor. ii. 2.

of attaining a very high place in Heaven. And this is where the Cross has a vast pre-eminence over other planks of safety in shipwreck, which, notwithstanding that they are all the word of life, do not always save thee. No matter how firm thy hold of them may be, they may still cast thee miserably on a desolate shore, where thou mayest find a more wretched death on land than would have been thy lot by sea. Far otherwise is it with the Cross. The Cross is sure to bring thee to Paradise. Do but cling closely to it, and there is no fear of its missing the way; it will land thee safe. "It may be that one of weak sight embraces this Cross," that is, an unlearned and ignorant man, who knows very little of what lies beyond seas, "and who cannot see far in the direction in which he is going, yet if he does not loose his hold it will carry him safely." So says St. Augustine.[1] And this is why the Cross is now our wisdom. Consider the glorious Apostle . St. Andrew. Not only did he rejoice and triumph at the sight of the Cross, and greet it with the utmost delight, but he said with perfect certainty, "Give me to my Master, that He Who by thee redeemed me may also receive me by thee;" because he knew that the Cross could not possibly bring him to any shore but that which he desired to reach.

[1] Tract. ii. *In S. Joannem.*

DECEMBER.

FIRST DAY.

Before prayer prepare thy soul, and be not as a man that tempteth God
(Ecclus. xviii. 23).

I. Consider first, that there are two ways of tempting God:
one express, the other implied. The former is when a man
neglects to do, on his part, what it is in his power to do,
merely for the purpose of trying what limit there is to the
goodness, the power, or the knowledge of his God in supplying
him with what he desires. The latter way is when a man does
not actually, and of set purpose, make this the object of his
neglect, but yet acts as though he did so. This being so, we
shall very rarely find a man who, when he neglects to prepare
himself for prayer, has the intention of trying whether God
will not, in spite of this, vouchsafe to communicate Himself
interiorly to him, as He does to the man who makes such
preparation. And therefore the Preacher does not here say,
"And do not tempt God," that is, in the way which is
expressed. But it is not uncommon to find one who neglects
to prepare himself as though it were his intention to make
such an experiment; and therefore the Preacher says, "And
be not as a man that tempteth God," that is, in an implied
manner. When, therefore, thou comest into the presence of
God, to pray without any preparation, is not this like staking
everything on a chance? But God will not have thee, on thy
part, neglect to do all that thy weak ability is capable of, even
in such a matter as this. If, then, thou findest thyself dry,
wandering, and distracted in prayer, do not be surprised. It
is thy own fault, for whereas it is in thy power to prepare
thyself for it, as so many faithful servants of God do, thou
nevertheless neglectest to do this, either through disinclination
or want of attention, and at the same time persuadest thyself

that God will not, on this account, fail to show Himself to thee in prayer with the same gracious countenance which is the reward earned by others at the cost of great diligence and preparation. Is not this an extraordinary kind of presumption, nay, even of irreverence and impiety? For if, when thou hast to speak to thy sovereign, thou thinkest carefully before-hand of what thou hast to say, how much more oughtest thou to do so when it is to God that thou art about to speak? "Speak not anything rashly before God."[1]

II. Consider secondly, that in this preparation there are two parts, the remote and the proximate. The remote prepa-ration is a pure and mortified life : pure, because it is by this that the understanding is prepared, like a polished mirror, to receive abundance of light ; mortified, because it is by this that the will is prepared, like an empty vessel, to be partaker of those spiritual delights which God refuses to him who does not sacrifice to Him those which are sensual. And the proxi-mate preparation is retirement, recollection, and above all, that which the saints so strongly inculcate, the arrangement beforehand of that which thou proposest to meditate upon for thy soul's good : "Before prayer prepare thy soul, and be not as a man that tempteth God." For is not this like tempting God, to begin to pray, after the manner of a vessel impru-dently launched on the waters, without rudder, pilot, or captain, to be borne wherever the wind may carry her? And what will become of thee if the wind does not blow? And canst thou expect that very wind to blow which is needed in thy particular case? That would be expecting God to be bound to work miracles. Therefore, always consider what duty is most binding on thee, or to what fault thou art most subject, and direct thy prayer accordingly. And if thou shouldst consider thyself so perfect as to be in no need of thinking about advancing to perfection, reforming thyself as to things in which thou art lax, reviving thy fervour, oh, how grievously thou art in error! "Be not afraid to be justified even to death," said the Preacher, and immediately afterwards he added, as a pressing reminder, "Before prayer prepare thy soul," to teach thee that the time in which thou art to make preparation for prayer is as long as that which is to be employed in thy justification.

III. Consider thirdly, that it is possible that thou mayest think thou art living in a continual preparation for prayer. I

[1] Eccles. v. 1.

answer that, in that case, the admonition here given by the
Wise Man is not for thee ; for doubtless, a man who is already
prepared stands in no need of preparation. But be very
careful that thou art right in what thou sayest. There are
some who are content to be like stocks and stones in time of
prayer, doing nothing at all ; and certainly it is very easy to be
in a state of continual preparation for such a kind of prayer
as this. But this must not satisfy thee. What thou art to do
is to resolve to exercise the powers of thy soul to the honour
of God, as the saints did. If, therefore, thou art not of the
small number of those whose hearts are always not only
habitually, but actually inflamed with the love of God, it is
certain that thou must first prepare the means of arousing it
by recollecting thyself before prayer ; for, as prayer is a mental
act, it is also certain that it consists not in habit, but in act.
Thou seest, then, how far the preparation which is here
enjoined on thee by the Wise Man should extend. It should
extend so far as this : that whenever thou settest thyself to
pray, thou shouldst not appear to be setting thyself to tempt
God. And that man does appear to tempt God who, when
aiming at a certain end, does not first have recourse to the
few means which are in his power for more easily attaining it.
But canst thou think that thou art doing this, when before
placing thyself in God's presence to transact with Him so
important a business as that of thy salvation, thy spiritual
progress or perfection, thou hast not thought particularly
about that which thou hast to ask for the attainment of so
great an end ? Thou wilt say, perhaps, that it is enough to
ask this of Him in a general way. But this is not the
teaching of Jesus Christ. The blind man of Jericho had
said to Him in a general way, "Jesus, Son of David, have
mercy on me," and yet Christ urged him to specify what he
desired—"What wilt thou that I should do to thee ? "—in
order to teach us, so the saints remark, that He delights in our
laying our wants before Him with great confidence and pre-
cision : "Lord, that I may see."

IV. Consider fourthly, that although, before entering on
thy prayer, thou shouldst set before thee the end to which thy
thoughts are particularly directed, whether it be the overcoming of a vice, or the acquiring of a virtue, or a more
generous following of Christ ; yet still, thou oughtest not to
tie thyself to it in such a way as not to follow His guidance
with full liberty of spirit, if, in the course of thy prayer, God

should be pleased to lead thee in some other direction? What is it that the Wise Man requires of thee? Merely that thou shouldst "not be as a man that tempteth God." Now, that captain is not tempting God who, in order to take in provisions, directs his course towards some particular country or port, and then, because the wind carries him to another equally good, lets himself be borne by it. But he would be tempting God if he were to steer for no particular country or port, but chose to go only to that one to which the wind might carry him. In this matter, therefore, there are two extremes to be avoided : the one, that of setting no end whatever before thee, when thou enterest on thy prayer ; and the other, that of keeping it before thee so rigidly as to be attached to it in spite of everything. It may happen, indeed, that thy prayer is sometimes fruitful, when there has been no preparation for it ; but dost thou not know that it would be far more fruitful if there had been that preparation? It is possible for medicines to be beneficial, even when taken at hap-hazard ; but the most salutary way is to take them on a system : "The skill of the physician shall lift up his head."[1]

SECOND DAY.

Brethren, labour the more, that by good works you may make sure your calling and election. For, doing these things, you shall not sin at any time (2 St. Peter i. 10).

I. Consider first, the greatness of thy folly if thou sufferest thyself to be tempted by the devil to say within thyself, as some do, Of what use is it to labour so much for my salvation ? If God has predestinated me to glory, I shall be saved without doing these things ; if not, then I shall not be saved, do what I may. This is madness. For, I ask thee, what wouldst thou do, if having called in a doctor when thou art suffering from some grievous sickness, he were to say to thee, Of what use is it, sir, to go through the trouble of taking so many medicines ? If God intends you to be cured, you will recover without taking them ; if not, then taking them will not make you recover. Would such a way of speaking please thee? I am well aware that thou wouldst reproach him as foolish or mad ;

[1] Ecclus. xxxviii. 3.

thou wouldst say that if it is God's will to restore thee to health, thou hast to believe that it is His will also to do this in the right way—that is, by the use of suitable remedies—and that therefore prudence requires thee to take them. Why, then, dost thou not argue in the same way in the present case? Nay, there is much greater need for thee to do so in the present case, because it is possible that it may sometimes be God's will to cure thee without the use of remedies of any kind; but it can never be His will to save thee without any kind of good works. On the contrary, it is most likely that He requires many such, and those too painful, difficult, and very hard, which is His ordinary way with most persons: "Strive to enter by the narrow gate."[1] Why, then, dost thou not prepare for them manfully? It is this which St. Peter desires to enjoin on thee, and on all who desire to win Paradise: "Labour, that by good works you may make sure your calling and election." He would have thee concur by thy good works in making thy predestination sure, not indeed as to its cause, which is the Divine decree, but as to its effect; for when God decreed without thyself to save thee, He did not decree to save thee without thyself; but He decreed thy salvation through those works which were to be done by thee for this end. When, therefore, thou unhappily neglectest to do them, thou hast great reason to doubt thy predestination; for it is a rule admitting of no exception, that without doing them no one can be saved: "If thou wilt enter into life, keep the commandments."

II. Consider secondly, that thou mayest here say that thou dost not understand how God's decrees can be infallible, if, notwithstanding, their taking effect or not rests with thee. But what does that prove? Such a difficulty is of no greater force in the matter of the salvation of the soul than in that of the recovery of health, of the preservation of life, of the gaining of victories, and all the other events which God has decreed in thy case, but decreed in such a way as to be dependent on thy free-will. As, therefore, although thou knowest certainly in the natural order that that will always happen to thee which is written above, thou dost not neglect for that reason to take medicine in order to be cured, to take food in order to live, to fight in order to conquer, and to take measures for procuring any other goods. So, too, although thou art sure of the same thing in the supernatural order, thou

[1] St. Luke xiii. 24.

-oughtest not to neglect to do all the good in thy power in order to secure thy salvation : "Labour that by good works you may make sure your calling and election." Are not the Divine decrees equally immutable in all cases? "Whatsoever the Lord pleased He hath done both in Heaven and in earth."[1] "In Heaven," that is, in the supernatural order; "in earth," in the natural order. Why, then, dost thou say in the one case, There is nothing to be done : if it is written in Heaven that I am to be saved, I shall be saved whether I do good works or not; and not also in the other, If it is written in Heaven that I am to be cured, I shall be cured whether I take medicines or not? This is being led astray by mere caprice : "Deceive not your souls."[2]

III. Consider thirdly, that although it is, in general, necessary to do good works in order to be saved, thou mayest nevertheless think that it is not necessary to do this or that particular good work, but that all that is requisite is to die in a state of grace. And so thou dost not perceive that St. Peter, when urging thee to make thy salvation sure, is not satisfied with saying, "Act (*agite*) so that by good works you may make sure your calling and election," but he says "labour" (*satagite*). And I answer that his saying *satagite*, and not *agite*, is a sign that salvation is a harder work than thou thinkest. Who has ever told thee that good works are only necessary in a general way for this end, and that it is not necessary to do any one in particular, when it is a question of enjoining good works? They are all necessary in themselves, although it may occur that after the omission of this or that one, God may, in His mercy, give thee time for repentance before death, and that so thou mayest be saved. But who can promise thee this? If, therefore, thou wouldst make thy salvation sure, and not trust it to the thread of a possibility, do not neglect so much as one of those good works, which are in themselves necessary to eternal life, but do them all. And the reason is, that if God has predestined thee to salvation, this predestination is not only by means of good works taken in general, but by means of those particular good works which He foresaw that it would be thy duty to do. So that, whenever thou, on the contrary, neglectest to do them, thy predestination becomes exceedingly uncertain; because it is certain that if thou wert to die in thy present state of a transgressor, thou wouldst be damned; and it is not certain that thou wilt

[1] Psalm cxxxiv. 6.   [2] Jerem. xxxvii. 8.

not so die. And this is what St. Peter means when, after saying, "Labour that by good works you may make sure your calling and election," he directly adds, "For doing these things" (that is, "for the end of making sure your calling and election"), "you shall not sin at any time." And why are we to keep ourselves from sinning (that is to say by a complete and perfect, namely, a mortal sin) even one single time (*aliquando*)? Because we know not what may happen after that one time. It is certain that the sin has been committed; it is not certain that conversion will follow: and therefore neither is salvation certain.

IV. Consider fourthly, that besides the intrinsic certainty which good works confer on thy predestination, there is also the extrinsic certainty which these works give to thee: a certainty which is not, indeed, physical, like the former, but only moral; yet still a very great one, because the greatest of all possible signs of predestination is an eagerness to do as many good works as are in our power. And the reason is, that although it is not beginning well, but enduring well, that wins the crown; yet if thou art always increasingly faithful and fervent, God will not fail to assist thee specially by His grace at the end of thy life, and to give thee the crown: for, generally speaking, He does not allow one who has for a long time done his utmost to lead a good life to fall away miserably at the last, and make a bad end: "Be thou in the fear of the Lord all the day long, because thou shalt have hope in the latter end, and thy expectation shall not be taken away."[1] This, too, is what St. Peter means to imply when he says, "Labour that by good works you may make sure your calling and election." He would have thee strive to attain the greatest of all moral certainties of predestination, that, namely, which depends on good works, especially those which are said to be superabundant and of supererogation, to which, according to some commentators, he here alludes by saying not merely "labour," but "labour the more," that is to say, more than is necessary; for it is not God's will to allow Himself to be surpassed in generosity by any one, but, on the contrary, as He shows Himself niggardly to the niggardly, so, too, does He show Himself liberal to the liberal and lavish to the lavish: "The Lord will reward me according to my justice."[2] Do not, therefore, be satisfied with merely doing some good work now and then, for even the reprobate do this;

---

[1] Prov. xxiii. 17, 18.   [2] Psalm xvii. 15.

but do many many such works, and do more and more every day. This is never done by them, but only by the predestinate, and those whose predestination is most clear. If, therefore, thou wouldst have the certainty of belonging not only to those who are called to glory, but to the elect, examine how far thou art anxious to do good every day. If thou doest many good works, be sure that God will keep thee from ever sinning, that is, mortally : " Labour the more that by good works you may make sure your calling and election. For, doing these things, you shall not sin at any time."

THIRD DAY.

ST. FRANCIS XAVIER, APOSTLE OF THE INDIES.

As arrows in the hand of the mighty, so the children of them that have been shaken (Psalm cxxvi. 4).

I. Consider first, that, according to the universal opinion of sacred commentators, by "children of them that have been shaken" are here intended all the just, but particularly the generous successors of the Apostles. Doubtless all the first are so called because they all alike acknowledge the Apostles to be their fathers, and to them this glorious title belongs by a two-fold right, both as shaking and being shaken. It belongs to them in the former or active sense, because in order to follow Jesus faithfully they shook off not only the yoke of the world, but every tie, every affection, everything, in short, belonging to the world, not even choosing to allow its dust to cling to their feet wherever it would have men live as enemies to Christ. And in the second, or passive sense, it is still more suitable to them, for the Apostles were shaken by innumerable persecutions, rejections, repulses, and ill-treatment from all men, like corn that is winnowed. Now, in both these senses all the just are called *filii excussorum*, that is, children of those who shook, and of those who were shaken, because they were begotten to God by the Apostles : " In Christ Jesus by the Gospel I have begotten you."[1] But still more are their generous successors their children, because they have made it their study to imitate them in this very particular of shaking

[1] 1 Cor. iv. 15.

off everything which came to them from the world, in order
the better to bring back souls to Christ, and also in that other
particular of allowing themselves to be shaken by the world.
Art thou one of this number? Examine thyself, and thou wilt
discover, it may be, how far thou art yet from being able to
make so high a boast.

II. Consider secondly, that the Psalmist predicts of these
nobler successors of the Apostles, that is, of their successors
in the great work of winning back souls to Christ, that they
should be as so many arrows in the hand of a mighty archer—
that is, Christ. For, at a single sign from Him or His Vicar,
they were to be ready to go with wonderful speed to the
farthest ends of the earth: "As arrows in the hand of the
mighty, so the children of them that have been shaken."
Neither were they only to be swift in flight like arrows, but
like them also to be straight in their aim, bold in their assault,
piercing to the depths of the hardest heart. And certainly if
there is one of this number who more than another resembles
such an arrow, it is the great Saint whose memory thou
particularly honourest to-day, St. Francis Xavier, who was so
worthy a son of the Apostles, and so illustrious an imitator of
their conduct as to merit the name, not only of an Apostolic
man, which many others have borne, but of an Apostle. Do
thou, if thou bearest any love to this Apostle, learn, after his
example, to desire to be, so far as the nature of thy state
allows, as an arrow in the hand of thy Lord Jesus—"In the
hand of the mighty"—so that, if thou art good for anything,
He may make what use He pleases of thee.

III. Consider thirdly, that St. Francis Xavier was an arrow
swift in flight. No sooner did he hear the will of God declared
to him by the mouth of his Father, St. Ignatius, than taking
with him only a cassock, a Bible, and a breviary, so as to be
free from all encumbrances, he hastened from Rome to Lisbon,
from Lisbon to Goa, from Goa to the Moluccas, and from the
Moluccas to Melinda, Malabar, Malacca, and many other
unknown nations of India and Japan; so that in ten years he
had travelled more than a hundred thousand miles, that is to
say, a distance equal to four times round the whole earth.
Thinkest thou, then, that if he could do so much in ten years,
he could have wasted many hours, as thou dost, in idle talk,
in sleep, in pleasure, or idleness? Oh, how unlike such an
arrow art thou, who clingest so closely to thy own comforts as
to lack the courage to leave thy country for the sake of serving

God, or even to quit some particular city, or community, or house in which thou findest thyself at thy ease! I pray thee declare to God that, like this glorious Saint, thou, too, wilt live in a state of detachment from all things: "Lo, here am I, send me."[1] Beg of Him to send thee whithersoever He pleases —"Shoot an arrow"[2]—since it is for the archer to decide concerning the arrow, and not for the arrow to decide concerning the archer.

IV. Consider fourthly, that an arrow goes, not only with marvellous rapidity, but perfectly straight to the point at which it is aimed. There is no danger of its diverging in the least degree from its course; indeed it almost seems to have an eye to see the goal, so directly does it strike it. So, too, did St. Francis Xavier. The conversion of the heathen, which he knew was the object for which he had been sent to the Indies, was always before his eyes, and so he went so straight to that end that he never lost sight of it. Nay, he would not even turn a few miles out of his way in order to go to his home and console his aged mother by the sight of him. If, then, he acted in this way in passing, so to say, from one work to another, judge whether he ever allowed himself to be diverted afterwards from his course by less holy feelings, by avarice, pride, ambition, or vanity. From the same motive this man, who made it his ordinary spiritual consolation to walk barefoot, even on thorns, and who humbled himself so far as to serve all his travelling companions like the meanest servant in galleys, inns, hospitals, or stables, when he saw that the conversion of souls required a different line of conduct, did not refuse the burden of ceremonious embassies, or to give audiences, and receive splendid lodging, retinue, and attendance, and this with such entire detachment of will, that he went back from all this pomp with increased delight to his mean occupations. Thus he showed himself learned with the learned, ignorant with the ignorant, weak with the weak, sorrowful with the sorrowful, and even a gamester with gamesters, so straight was his course. He never sought his own glory, but God's. Is this thy case? Alas! how great is the power of self-love over thee; it is this which turns thee in everything out of the straight course: Go "like an arrow shot at a mark,"[3] that is, go in a straight flight to thy aim.

V. Consider fifthly, that if thou watchest an arrow shot by a strong hand, not only does it seem as though it had eyes to

---

[1] Isaias vi. 8.　　[2] 4 Kings xiii. 17.　　[3] Wisdom vi. 2.

see the goal at which it is aimed, but almost as though it had a mind to attack and overcome every obstacle which intervenes, so determined is its course. So, too, with St. Francis Xavier. See what resolution was his! He did not allow himself to be intimidated by those seas which even at our day, after being so repeatedly navigated, are accounted formidable, whereas at that time they were hardly discovered. Nothing daunted him, neither precipices, nor torrents, nor chasms, nor cold, nor heat, nor pernicious climates. Numberless persons did all in their power to deter him from going to that terrible island of the Moors, but he persisted in going, without even taking with him any antidote against the poisons which were predicted to him as certain to be met with there. How often did he brave death face to face among the plague-stricken, and how often was he threatened with it by numerous bands who rushed upon him to stone him, without losing courage! It is enough to say that he attempted to enter China; and although that immense Empire was then defended by so many ramparts and walls as to prevent the entrance of any one, yet at least he died on its threshold. This, indeed, is resolution in God's service. How is it with thee? How easily art thou deterred by the least hindrance in thy way: "The arrow of Jonathan never turned back."[1]

VI. Consider sixthly, that all these qualities of speed, of straightforwardness, of resolution, would be of no avail to the arrow if, after all, it failed to pierce deeply wherever it is aimed; for it is with this object that it is used in battle, to penetrate the ranks of the enemy with violence, even at a great distance. And can it be doubted that St. Francis Xavier possessed this violence, a most holy violence, in the highest degree? "My arrow is violent without any sin."[2] To be convinced of this it is only necessary to glance at the number and kind of persons whom he converted by his preaching. As to the number, he baptized with his own hand more than a million and two hundred thousand idolaters, which gives an idea of how many more he caused to be baptized by others, in order to have leisure to make further conquests of souls: "Thy arrows are sharp; under thee shall people fall."[3] And as to kind, they were all sorts of the most barbarous persons. How wonderful, then, was this arrow of the Lord, which penetrated hearts of stone rather than flesh! He was indeed an arrow which, "like that of a mighty man, shall not return

[1] 2 Kings i. 22.   [2] Job xxxiv. 6.   [3] Psalm xliv. 6.

in vain."[1] Five kings fell, pierced by this arrow, and laid their crowns at the feet of Francis, to receive Holy Baptism at his hand. But what most shows the force of this arrow is to see the difference between St. Francis Xavier's converts and those of others, so much superior were the former in their fidelity and strength in keeping the promises made to God in their baptism. How sure a sign this is that the arrow went deep: "Thy arrows are fastened in me."[2] Not that this is to be wondered at, when he made use not of words only in converting men, but of numberless industries, suggested to him by his fervent spirit. How canst thou expect to have the power of penetrating into those souls whom thou too, it may be, art desirous of converting to God? "The arrows of children are their wounds,"[3] when thou dost not first suffer the arm of the Lord to take possession of thee, before going on to make deep wounds in others?

VII. Consider seventhly, that if thou observest, it is not in itself that the arrow has the power of flying with speed, straightness, determination, or violence; all this depends on the arm of the archer. Therefore it is written: "The sharp arrows of the mighty."[4] No matter of what kind the arrow may be that is shot by a weak arm, it is always blunt. It is when it is shot by a strong arm that it is sharp. For then it strikes so keenly on the quick as to seem to carry burning coals with it: "The sharp arrows of the mighty with coals that lay waste." So is it with us. Of ourselves we have no power to strike hearts, all our power comes from God. Only there is this difference between material arrows and us men, when God is pleased to make use of us as arrows—that the former cannot resist the arm which directs their course. According to the impetus given by it when the arrow leaves the bow, will their flight be swift, straight, determined, vehement in wounding. But with us it is different; it is but too possible for us to resist that mighty arm which makes use of us, because we are arrows possessed of free-will. And so it is not wonderful if we are like arrows shot by children. The reason is, that we do not suffer God to dispose of us with perfect freedom. St. Francis Xavier gave himself entirely up into God's hand, not merely not resisting Him, but co-operating perfectly with the strong impulse which he received from God, when it was His will to send him to the Indies, and this

[1] Jerem. L 9.
[2] Psalm xxxvii. 3.　　[3] Psalm lxiii. 8.　　[4] Psalm cxix. 4.

was why he did so much there for His honour. He was "an arrow in the hand of the mighty," that is, an arrow which never attempted to do anything of itself, but which let itself be entirely directed by God, without any reservation, though he should have to lay down his life in His service. Thou art indeed an arrow "of the hand of the mighty," but not yet, it may be, "in the hand of the mighty," because thou dost not leave thyself to be used with perfect freedom by God for His service, in the way most pleasing to Him.

———

FOURTH DAY.

As much as she hath glorified herself, and lived in delicacies, so much torment and sorrow give ye to her (Apoc. xviii. 7).

I. Consider first, that there are, so to speak, in corrupt man, two sources of all sin, the irascible and the concupiscible part. The latter causes him to spurn the dictates of reason and to aim at an inordinate desire of seeking his own gratification; the former causes him to aim at an inordinate seeking after his own pre-eminence. To the former are especially ascribed spiritual, to the latter carnal sins. The concupiscible part is the cause of man's letting himself be carried away by the irregular love of creatures, the irascible part is the cause of his turning his back (through his love of creatures) on God, Who has forbidden that love. Hence it is that, in Hell, the sources of all pains must correspond with these two sources of all sins, and so be divided into two : the pain of loss, and the pain of sense. The pain of loss corresponds especially with the excesses of the irascible part, in which was the origin of aversion from God. The pain of sense corresponds with the excesses of the concupiscible part, in which was the origin of conversion to creatures. And so by the pain of sense are expiated the irregular pleasures in which a man has indulged himself principally for the gratification of his body, and by the pain of loss is expiated his spiritual pride. Now, therefore, thou wilt easily understand the meaning of these words which are spoken to the devils by Divine Justice for the terror of every sinful soul : "As much as she hath glorified herself, and lived in delicacies, so much torment and sorrow give ye to her." By the words "hath glorified herself" are to be understood the sins most proper to the irascible part, that is, spiritual sins ; and by the words "lived in delicacies" are to be understood those most proper to the concupiscible part, that is, carnal sins. By the words "give her torment" is signified the pain of sense, answering more particularly to the sins of the concupiscible part ; and by "give her sorrow" the pain of loss, answering more particularly to the sins of the irascible part. Does not the blood freeze in thy veins at the thought of these pains which are infallibly prepared for thee also, if ever thou allowest thyself to be dominated by these irregular passions? O proud man, O self-indulgent man, see what will be the end of thy pride and thy pleasures !

II. Consider secondly, that as the pain has to be in proportion to the nature, so, too, must it be in proportion to the degree of the sin. And therefore it is here said : "As much as she hath glorified herself and lived in delicacies, so much torment and sorrow give ye to her." Now, in mortal sin, there are, as thou hast seen, two evils : the turning from the Creator, and the turning to the creature. The turning from the Creator is the turning from an Infinite Good ; and on this account mortal sin contains in itself, so to speak, an infinite malice. The turning to creatures is not only a turning to a finite good, but it is a turning to them by acts which are also finite. And therefore the turning from God corresponds more particularly with the pain of loss, which is, in a certain sense, an infinite pain, being the privation of an Infinite Good ; and the turning to creatures corresponds more particularly with the pain of sense, which is a finite pain, being a pain greater in one case, and less in another, according to the degree of this turning to creatures which has been finite in each individual ; so that the man who has loved those creatures with more disorder is more punished, and *vice versâ :* "According to the measure of the sin shall the measure also of the stripes be."[1] What, then, oughtest thou to think on hearing this "as much" and "so much"? That the torment which the pain will cause to the damned is to be no greater than the pleasure which they felt in the sin? Surely not ; for, on the contrary, it will be very much greater. For a very trifling pleasure they will suffer a torment exceeding that which was suffered by all the martyrs together. What thou oughtest to think is that by the words "as much" and "so much" is denoted not

[1] Deut. xxv. 2.

equality, but proportion: so that the man who has sinned more suffers more, not only in the pain of sense, but in the pain of loss; not that this latter does not deprive all equally of an equal good, namely, the Beatific Vision, but because the man in whose power it was to attain that good more easily will curse his folly with so much the more tribulation and anguish; and therefore it is said: "As much as she hath glorified herself, and lived in delicacies, so much," not only "torment," but "sorrow give ye to her." Do thou, therefore, who failest to learn from its guilt how great an evil sin is, at least acknowledge it from the penalty.

III. Consider thirdly, that as the pain is to be in proportion to the guilt in the degree of its severity, so, too, thou mayest think that it ought to be in the degree of its duration, and therefore thou canst not understand how this rule is observed—"As much as she hath glorified herself, and lived in delicacies, so much torment and sorrow give ye to her"—seeing that often the sin lasted but a moment, whereas the pain of the damned will have to last for ever. But, as to that, what court of justice is there on earth that does not punish a crime by a penalty which is longer in duration than that crime was? A murder is committed in an instant, and yet sovereigns are continually punishing it by driving the murderer not only out of their dominions, but out of the world. The reason is, that all punishments which have an end, seem, after all, contemptible to an audacious heart: it is those which are eternal that are really dreaded. And therefore, that the fear of Hell might be more capable of restraining the passion or the pride of man from sinning, it was right that the punishment of sin should be not only severe, but everlasting: "These shall go into everlasting punishment."[1] But if the duration of the pains of Hell, being eternal, surpass that of the sin, they do not surpass its gravity. There is no sin, not even the smallest, provided it be mortal, which does not contain in itself, as it were, an infinite gravity of malice, because it is committed against God. Therefore, as it cannot be punished by pain which is infinite in intensity, it is just that it should be punished by one which is, at all events, infinite in extent; all the more because, the sin not being retracted, it is just that it should be punished in the damned as long as it lasts, at least morally, and as long as it continues, by reason of the past act, to render them as truly bad, impure, wicked, hateful to

[1] St. Matt. xxv. 46.

God, and deserving of punishment as they were when actually sinning. Therefore is it said, "As much as she hath glorified herself, and lived in delicacies, so much torment and sorrow give ye to her," because, although, in the reprobate, the act of glorifying themselves and the act of living in pleasure are past, yet the desert to suffer for those acts, which have indeed been punished, but never sufficiently punished, is not past, but continually present in them. Do not reply that the damned repent of the evil they have done, saying within themselves, "We have erred from the way of truth,"[1] because their repentance is not from hatred of the sin that they committed, for which, on the contrary, as sin, they retain the utmost affection; it is only from hatred of the pain which torments them: "His soul shall mourn over him."[2] If, then, thou wouldst not be compelled one day to repent thus, with a repentance as unavailing as it is endless, do not delay longer repenting in the right way; for if in time past thou hast been occupied in providing thy body with pleasures, and thy soul with vainglory, thou knowest that now they must submit to torment and sorrow.

FIFTH DAY.

If any of you want wisdom, let him ask of God, Who giveth to all men abundantly, and upbraideth not, and it shall be given him. But let him ask in faith, nothing wavering (St. James i. 5, 6).

I. Consider first, that although it is true that every kind of wisdom is an ornament to man, yet he does not need every kind. So that if thou wert to ask of God the wisdom of St. Thomas, Albert the Great, or Alexander of Hales, thou couldst not very readily make sure of obtaining it. But if thou askest for the wisdom which thou needest in thy state of life, that is, the wisdom which consists in knowing how to guide thyself aright in doubtful cases which meet thee in thy affairs, thy ministry, the work of thy salvation, then do not doubt that thou wilt obtain it. Therefore St. James here says, "If any of you want wisdom, let him ask of God;" he does not say, "If any of you love or delight in wisdom," but "want wisdom;" for this is the wisdom which thou art sure of obtaining from God by asking for it, namely, the wisdom

[1] Wisdom v. 6.   [2] Job xiv. 22.

which thou needest; and especially that without which thou canst not make any good progress in the Divine service. Often thou dost not know how to act in it, and so thou art disquieted. Have recourse to God: "Desire of Him to direct thy ways."[1] This is the safe refuge. He will never fail to enlighten thee: "As we know not what to do, we can only turn our eyes to Thee."[2]

II. Consider secondly, that the knowledge of thy own unworthiness may deter thee from doing this; and therefore St. James says, to encourage thee, "If any of you want wisdom, let him ask of God, Who giveth to all men abundantly, and upbraideth not." If God limited the giving of this wisdom to some of His special favourites, thou mightest easily fear that thou wouldst not obtain it. But He gives it "to all men," that is, to all who ask for it, and gives it, too, "abundantly," "although His manner of giving it is so delicate, so veiled, so silent, that very often it does not appear to come from Him. And this is what the Apostle would have thee understand by the words "upbraideth not." When men of the world do thee any kindness, they do it in such a manner as to make their doing it very evident. Dost thou not see that in this way they are in truth reproaching thee with thy need of them: "He will give a few things, and upbraid much."[3] Not so God; God gives without even seeming to do so. Thou askest right counsel of Him, and He orders things so that it is given thee by a friend, as it were by chance, or that thou meetest it in a book, or receivest it in a light which darts into thy mind when thou least expectest it. This is giving without any sort of upbraiding, for it is giving, and at the same time making it possible for a man almost to ascribe to himself what really has been given by God. Do not take occasion, from the delicacy of thy Lord in showing thee kindness, to fall into so base an error, but know that all wisdom, no matter through what channel it is received, comes to thee from Him: "All wisdom is from the Lord God."[4]

III. Consider thirdly, the condition necessary for the certain obtaining of this wisdom. It is to ask it of God devoutly and perseveringly; devoutly, that is, on the strength of the promises which He has made thee in the Sacred Scriptures, and therefore the Apostle says, "Let him ask in faith;" and perseveringly, that is, never ceasing to ask,

[1] Tobias iv. 20.
[2] 2 Paral. xx. 12.　　　[3] Ecclus. xx. 15.　　　[4] Ecclus. i. 1.

although it may seem as though thou wert not heard. And
therefore the Apostle adds "nothing wavering." Nothing
so much leads thee to abandon prayer as seeing that thou
hast been asking for a long time without obtaining. Do
not act thus; but be firmly convinced that thou wilt obtain,
and adding confidence to faith, go on asking, "nothing
wavering," and thou wilt see in the end whether God's promises
fail. There would, indeed, be ground for discouragement if
it was on the merits of thy petitions that thou hadst to rely,
faulty and miserable as thou art. But thy reliance must be on
the word of God. What then canst thou fear? He who asks
in faith, that is, on the faith of that more than royal word, may
well "ask, nothing wavering."

IV. Consider fourthly, that it is to show that want of
perseverance in prayer proceeds from this uncertainty, that
St. James adds, "For he that wavereth is like a wave of the
sea, which is moved and carried about by the wind." When
waves are troubled by the wind, they are now carried straight
to the shore, then, as though their intention were changed,
they pause, and instead of going on in that direction, they
suffer themselves to be turned this way and that without any
rule. The same thing happens whenever confidence and
perseverance in prayer become wavering. The man who asks
for a time, and then leaves off, may be sure that he will obtain
nothing—"Let not that man think that he shall receive
anything of the Lord"—because God chooses that our confidence in Him shall be constant and enduring, and therefore
He chooses that we should continue to ask, even when He
does not grant our petitions. "Pray without ceasing,"[1] for
He very often delays giving graces for this very reason, that
is, in order to try whether we have a right confidence in Him.
What great merit would there be in thy prayer if, as soon as
thy lips were opened, thy requests were granted? This would
certainly be to ask "nothing wavering," but not "in faith."
The merit consists in reiterating thy petition when thou
seest it rejected, or the answer delayed, like the woman
of Canaan in the Gospel, who at last heard the words:
"O woman, great is thy faith; be it done to thee as thou
wilt."[2]

[1] Thess. v. 17.   [2] St. Matt. xv. 28.

SIXTH DAY.

The Spirit searcheth all things, yea, the deep things of God
(1 Cor. ii. 10).

I. Consider first, that just as when it is said that the Spirit
of God "asketh for us with unspeakable groanings,"[1] the
meaning is that He makes us ask ; so, too, when it is said
here that He "searcheth the deep things of God," what is
meant is that He makes us search them, because that which
is the gift of the Spirit should be ascribed to Him.   It is true
that everything is not ascribed to Him in the same measure,
because as the work of the Spirit is to spiritualize us—"that
which is born of the Spirit is spirit"[2]—so those qualities
which are especially His are those which are particularly said
to be imparted to us by Him.   The Spirit is agile, prompt,
swift, and diligent ; there is no sloth in Him : "The Spirit
goeth forward, surveying all places round about."[3]   He has a
wonderful force of impulsion : "His breath (*spiritus ejus*) as
a torrent overflowing even to the midst of the neck."[4]   He
cannot be checked or bound : "Who hath held the wind
(*spiritum*) in his hands?"[5]   He goes as it pleases Him, where
He chooses, as far as He chooses, when He chooses, with
perfect liberty : "The Spirit breatheth where He will."[6]   All
these, and other like qualities, are easily infused into us by the
Spirit of God : "So is every one that is born of the Spirit."[7]
Now amongst other qualities of the Spirit there is that of
penetrating with the utmost subtlety to search that which is
hidden, though it were in the depth of the sea ; and in this,
too, He enables us to imitate Him, making us "search all
things, yea, the deep things of God."   And yet thou art not
desirous to know anything of God, although thou art full of
curiosity to fathom the secrets of Nature, or those of princes,
dignitaries, or even of any of thy neighbours.   Thou art
entirely indifferent to what concerns God.   Look well to it,
for it is certainly a subtle spirit which makes thee so eager to
look into the doings of others, but it is not a holy one.   A
holy spirit is that of which it is written that it is "pure and
subtle :" "subtle,"[8] because it penetrates everywhere ; "pure,"
because it has no desire to penetrate where it may be defiled :
"In unnecessary matters be not over-curious."[9]

[1] Romans viii. 26.   [2] St. John iii. 6.   [3] Eccles. i. 6.
[4] Isaias xxx. 28.   [5] Prov. xxx. 4.   [6] St. John iii. 8.
[7] St. John iii. 8.   [8] Wisdom vii. 23.   [9] Ecclus. iii. 24.

II. Consider secondly, that the Spirit of God is said to be
"one and manifold:"[1] "one," by unity of essence; "manifold,"
by multiplicity of gifts : " There are diversities of grace, but
the same Spirit."[2] Now, as His principal gifts are seven in
number, so are we told that there are seven spirits which
He infuses into the heart of the just, according to these words:
" The Spirit of the Lord shall rest upon him ; the spirit of
wisdom and of understanding ; the spirit of counsel and of
fortitude ; the spirit of knowledge and of godliness, and he
shall be filled with the spirit of the fear of the Lord."[3] Now,
every one of these spirits searches, and makes us search, into
" deep things," but in a holy manner worthy of the Holy
Spirit : " The Spirit searcheth " (or causeth to be searched)
" all things."   If thou wouldst see this, examine each of those
spirits, which are those "seven spirits of God, sent forth into
all the earth,"[4] and thou wilt see that all of them make thee
desirous to search.   The last in order of enumeration, the fear
of God, is the first in order of action, because the first step in
holiness is taken by the fear of God : " By the fear of the
Lord men depart from evil ;"[5] and therefore it is from the
fear of God that we must rise gradually to His wisdom.   Now,
the spirit of fear searches the recesses of the conscience, so
that no unsuspected sin may lurk there, neither is it satisfied
with looking to grievous sins only, which have the power of
separating us from God, but finds out lesser ones also.   The
spirit of piety searches into the kinds of service due from a son
to a father, so as to pay them all most carefully to God, and
it also searches out the hidden miseries of our neighbour, both
corporal and spiritual, so as to be able to succour him in
them, no matter how deeply they may be hidden.   The spirit
of knowledge searcheth out sunk rocks, that is, the delusions
and deceits on which so many are dashed to pieces : " Making
shipwreck concerning the faith."[6]   The spirit of fortitude
searches into the dangers besetting the honour of God,
especially from all those who are wolves while seeming lambs,
and is not satisfied with defending the Church against her
persecutors, and not also attacking those who secretly undermine her.   The spirit of counsel searches into the remedies
suitable for all sick persons, more especially such as disdain
the physician, that is to say, those sinners who will not endure
admonition, and from the first adapts itself to their habits as

[1] Wisdom vii. 22.   [2] 1 Cor. xii. 4.   [3] Isaias xi. 2, 3.
[4] Apoc. v. 6.   [5] Prov. xvi. 6.   [6] 1 Timothy i. 20.

far as is possible without sin, in order to persuade those
miserable men later on to change their manner of life. The
spirit of understanding searches out the deep meaning of the
Scriptures to throw light upon them, not content with a
superficial examination, because the most valued treasures
cannot be found without digging deep. And, in the last
place, the spirit of wisdom searches into all which is most
deep concerning God, His essence, His attributes, His acts,
His titles, His persons, processions, decrees, in short, into the
most abstruse questions imaginable ; and in this way, as thou
seest, are these words most perfectly verified, that "the Spirit
searcheth all things, yea, the deep things of God ; " so that, by
means of this beautiful gift of wisdom, which is the chief of
all, a thousand truths have been discovered concerning God,
of which the ancient philosophers, whom the world so much
admires, were entirely ignorant : "Which none of the princes
of this world knew."[1] Do thou, according to the gift which
God has granted thee in the greatest measure, not content
thyself with what lies, as it were, on the surface of the water,
but go down into its depths, and there discern and find those
hidden treasures, which may be profitable either to thyself or
others ; for this examination is one of the principal effects of
the Spirit : "His eye hath seen every precious thing ; the
depths of rivers He hath searched, and hidden things He hath
brought forth to light."[2]

III. Consider thirdly, that as the good spirit goes everywhere to search for what may be most profitable to souls
which are faithful to God—it "reacheth everywhere"—so, too,
does the evil spirit go about everywhere to discover what may
do them the greatest harm. He, too, is therefore "one and
manifold : " "one" in the end he aims at, which is the ruin of
souls ; "manifold" in the means he employs. And therefore
he, too, has seven specially famous spirits, which are the
opposite of the Divine spirits. Pride is opposed to the spirit
of fear, envy to the spirit of piety, anger to the spirit of knowledge, sloth to the spirit of fortitude, avarice to the spirit of
counsel, gluttony to the spirit of understanding, luxury to the
spirit of wisdom, as thou mayest easily know for thyself, if
thou observest the hindrance placed by each of these vices in
the way of the practice of the gift opposed to it. These are
those seven spirits which always accompany Lucifer when he
goes about the earth—"I have gone round about the earth,

[1] 1 Cor. ii. 8.   [2] Job xxviii. 10, 11.

and walked through it "[1]—and which he will introduce into
thy house if thou dost not know how to keep it well closed, to
search for everything good that thou hast, and to rob thee of
it : "Thus saith Benadad, I will send my servants to thee, and
they shall search thy house, . . . and all that pleaseth them
they shall put in their hands and take away."[2] Then, as this
evil spirit, for all that he is thus manifold, has not always the
courage to attack the good spirit in open fight, he has in
readiness seven treacherous spirits, which enter by guile and
cunning when violence does not succeed. These are those
spirits to which Christ alluded when He said, "He taketh unto
him seven spirits more wicked than himself,"[3] because, as
simulated virtues are far worse than open vices, so, too, are
the seven simulations, which go about under the appearance
of good spirits, worse than the seven wicked spirits that have
been spoken of. They are simulated wisdom, understanding,
counsel, fortitude, knowledge, piety, and fear of the Lord.
All these simulations united form the deceitful spirit of
hypocrisy, and therefore it is to be observed that Christ said
seven times, "Woe to you, Scribes and Pharisees, hypocrites."
All of these are sent by the evil spirit to search out the just,
even in the cells of Carmel, Tabor, or the Thebaid, to enter
into their hearts, and if they succeed in gaining possession of
them, to induce them to make a show of the gifts of the good
spirit, which they do not really possess. Take care that thou
art not one of these miserable souls, for it is of these pretended
just ones that Christ said, "These shall receive greater damnation,"[4] that is, a two-fold Hell, one as due to their secret vices,
and the other to their feigned virtues.

SEVENTH DAY.

ST. AMBROSE.

*I will not accept the person of man, and I will not level God with man :
for I know not how long I shall continue, and whether, after a while, my
Maker may take me away* (Job xxxii. 21, 22).

I. Consider first, that all those exterior advantages on
account of which thou sometimes esteemest some persons

[1] Job i. vii.
[2] 3 Kings xx. 6.   [3] St. Matt. xii. 45.   [4] St. Luke xx. 47.

more highly than the just, such as great riches, high station, sublime knowledge, beauty of person, are a mask which prevents thy discerning what they are, although thou art in daily intercourse with them: that is, it prevents thy seeing that they are a miserable bag of corruption, like thyself. Is it possible that for their sake thou actually sometimes wouldst displease God? Never let it be so; say rather, like Job, "I will not accept the person of man." What does this mean? The word translated "person" (*persona*) signifies "a mask," and the meaning is: "I will not accept the mask of a man instead of the man himself;" I will not suffer myself to be deceived by the mask which he wears; I will regard neither his riches, nor his station, nor his knowledge, nor his attractive appearance; but whenever such a man urges me to offend God, I will cast him from me indignantly and without remorse. Oh, how profitable will it be for thee to keep vividly before thy mind that the world is like a stage, full of personages fair of aspect indeed, but dressed for show! Respect them as is right, but remember at the same time that they will very soon have to leave the boards, and to stand, naked, pale, and trembling, before God, to give an account of themselves all alike: "The Lord is Judge, and there is not with Him respect of person."[1]

II. Consider secondly, the importance of putting this rule in practice. It is so important, that if thou forgettest it thou wilt come to prefer a character of the stage to the Lord Whom he represents, and will cease to pay to God the honour and obedience which are His due. And why? For fear of displeasing a man who possesses the merest shadow of the Divine riches, sovereignty, wisdom, or beauty, which thou failest to reverence. Not only art thou forbidden to prefer any man to God, but even to think of comparing him to God: "I will not accept the person of man, and I will not level God with man." Think whether there can be a greater distance than between the Creator and the creature, the master and the servant, the king and the slave, man and God! Canst thou think of putting these two on a level? What fearful madness! No matter who is in question, however closely bound to thee by friendship or authority, or any other title to respect, God alone must prevail: "Who art thou, that thou shouldst be afraid of a mortal man, . . . and thou hast forgotten the Lord thy Maker?"[2]

[1] Ecclus. xxxv. 15.　　[2] Isaias li. 12, 13.

III. Consider thirdly, the motive with which thou shouldst encourage thyself to refuse to set any man before God, or even to put him on a level with God. It is that of thy impending death. Perhaps this motive seems to thee far-fetched or unconnected with the subject, but it is not so. It is this which was made use of by Job, when he said, "I will not accept the person of man, and I will not level God with man." And thou shouldst do the same when the occasion presents itself: "For I know not how long I shall continue, and whether, after a while, my Maker may take me away." How wilt thou fare if, when thou hast to appear before the judgment-seat of Christ, He should reproach thee with having made more account either of the friendship or authority of men than of His? How great will be thy confusion, thy bitterness, and anguish of heart! Will there be any hope that a single one of those men whom thou hast formerly so regarded and honoured will give thee any help at that tribunal by speaking, or pleading for thee, or by offering to bear the punishment which is thy due? Poor wretch, dost thou not know that thy fate for all eternity is in the hands of God alone? If thou dost know this, how, then, is it possible for thee to neglect Him for the sake of any other, or not to serve Him with due fidelity? Thou seest, therefore, how close the connection is between these two clauses: "I will not accept the person of man, and I will not level God with man," and "For I know not how long I shall continue, and whether, after a while, my Maker may take me away," remote as it might seem at first.

IV. Consider fourthly, that if ever there was a man in whose mouth these words were becoming, it was the great Archbishop whom thou honourest to-day, St. Ambrose. For remember how he withstood the Emperor Theodosius face to face—a prince so powerful, and, as a rule, so pious—when he refused to admit him into the Church on account of his cruelty in the massacre of Thessalonica. Think, then, that it was these words which encouraged him to that great act; indeed, it is very possible that he may have uttered them at the time, or if not these precisely, at least others with the same meaning. Do thou keep them in thy mind, in readiness for any occasion that may arise; and whenever thou desirest courageously to overcome human respect, say at once within thy heart, "I will not accept the person of man, and I will not level God with man." And if this is not enough to enable thee to gain the victory immediately, go on to say, "For I know not how long I shall continue, and whether, after a while, my Maker may take me away."

———

EIGHTH DAY.

THE IMMACULATE CONCEPTION.

Wisdom hath built herself a house, she hath hewn her out seven pillars
(Prov. ix. 1).

I. Consider first, that by the universal consent of the Fathers, the "house" here spoken of is the Blessed Virgin Mary, chosen from all eternity by the Word to be His glorious Mother. Therefore, observe how He speaks. Doubtless, as He was to come down to earth, He chose a house; but this house was not, so to speak, hired, as though He had appointed an ordinary woman to fulfil this great office of being His Mother, on the contrary, He made it, nay, rather, He built it (*ædificavit*), that is, He did not make it, like all other created things, almost without spending thought on what He was making—" He spoke, and they were made "[1] —but He made it with design, with study, plan, and architecture. And for whom did He build it? For Himself (*sibi*). He did not build it for any other to inhabit it, but for Himself alone, that is, that it might be His shelter, His dwelling-place, and, consequently, a house worthy of a God, and therefore this house never entertained any other; but as the Word chose to be the Child of Mary, so, too, did He choose to be her only Child. This being so, must He not have been careful to endow her with every perfection, prerogative, and advantage which could make her dearer to Him? No king spares expense when it is a question of building his royal palace, especially from the very foundations. And canst thou suppose that the Eternal Word could have acted differently? This is, indeed, the very reason why He here prefers to appear under His name of Wisdom rather than under any other—"Wisdom hath built herself a house"—in order to show that it was wisdom which He particularly exercised in building this glorious house, so as to remove far from it every defect,

[1] Psalm cxlviii. 5.

deformity, and disfigurement, and adorn it in so masterly a manner as to make it evident that it was intended to be a monument of His wisdom. If thou hadst no other standard for measuring Mary's inestimable privileges, suffice it to know that she is that house which was built by Wisdom, and built for Himself alone, and no other: "Wisdom hath built herself a house."

II. Consider secondly, what sort of a king he would be, who, having built himself a splendid palace, should suffer it to be inhabited before him by one who was a renegade, a traitor, or a rebel, so as to be polluted by the breath of such a wretch? So far from allowing this, he would not let a miscreant of this sort approach within a thousand miles of it. And can it be believed, that when the Eternal Word had made so fair a house as Mary is, and made it, too, expressly for Himself, He would nevertheless allow that traitor to Him, the devil, to go to dwell in it before Him, and not only to dwell in it, but to take possession of it by means of original sin? No reasonable man could believe so. For in what manner would it be possible for the Divine Word to allow the devil to take possession? By necessity, or by election? If by necessity, then He had not in Himself power enough to be able to prevent it; if by election, then He had not love enough for Mary to choose to do so. Is there any one who could possibly admit either of these enormous absurdities? "Wisdom hath built herself a house," and therefore it must be believed that He Who built it for Himself would keep it for Himself. And if He would not even allow the devil to approach it after Him, how would He possibly have suffered him to dwell in it before Him? It was the part of Wisdom to build this glorious house, and that of Providence to defend it from all hostile powers: "By wisdom the house shall be built, and by prudence it shall be strengthened."[1]

III. Consider thirdly, that in order that this house might be more stately, we are told that Wisdom, when building it, set up many pillars, both to support and embellish it: "She hath hewn her out seven pillars," that is to say, "many pillars," according to the frequent meaning of "seven" in the Scriptures. "The soul of a holy man discovereth sometimes true things, more than seven watchmen that sit in a high place to watch."[2] These pillars were the virtues with which our Lady's soul was adorned. And if any one should ask how

---

[1] Prov. xxiv. 3.   [2] Ecclus. xxxvii. 18.

many these virtues were, I answer all. For this, too, is the
meaning of "seven" in the Scriptures. The number seven
denotes totality. Although, indeed, if all the virtues are
reduced to their different kinds, they are, in the strictest sense,
seven. And therefore they are here said to be seven in this
sense also, not in number but in kind. These, too, are the
seven primary virtues, from which all the rest proceed. They
are the three theological virtues, faith, hope, and charity,
which are called superhuman or Divine virtues, because they
are virtues which belong to man in proportion as he has been
elevated to a participation of the Divine Nature ; and the four
cardinal virtues, prudence, justice, fortitude, and temperance,
which are called human, or moral virtues, because they are
proper to man even according to his natural state, when not
thus elevated. All of these, then, were not in the Blessed
Virgin as they are in us, that is wavering, but firm and solid,
for which reason they are called pillars—"She hath hewn her
out seven pillars "—because they never decayed, but on the
contrary, were at once established by the most permanent and
richly endowed confirmation in grace that is possible, that,
namely, from which all leaven of sin is excluded : "I have
established the pillars thereof."[1] What remains for thee, on
seeing these beautiful pillars, but to gaze on them with admiration ? Look on them attentively, and thou wilt see represented
in each one of them many incomparable actions of Mary,
some belonging to faith, some to hope, some to charity, and
others to the other virtues which have been above mentioned.
Admire, love, and embrace them with the kiss of a devout
heart ; and if thou desirest to show a true devotion, copy
them in thy soul. It is right to praise, love, and admire the
virtues of Mary, but above all to imitate them.

IV. Consider fourthly, that what is here particularly
ascribed to Wisdom is not the chiselling or erecting or
embellishing these columns, so much as the hewing them
out—"She hath hewn her out seven pillars "—and this is that
we may understand from what quarry He dug them, one so
rare and precious, that to search its inmost depths was a work
only to be accomplished by Uncreated Wisdom. This is why
even the virtues which are common to others were possessed
by Mary in so heroic and surpassing a degree, that they
constitute an order superior to that in which the rest of the
just possess them. But if this is so, who can suppose that

[1] Psalm lxxiv. 4.

she was included by God under the condition which He made with Adam, by which the happiness of all his posterity was made dependent on his obedience? when she was to possess faith, hope, charity, prudence, justice, fortitude, temperance, so far greater than those of Adam, and greater not only in act, but in habit, so that, by reason of the perfection which was proper to her virtues, the Blessed Virgin might be able the more easily to fulfil most exactly the law of her God. This is an argument of great probability for proving how justly Mary, through the grace of Christ, Whose worthy dwelling-place she was to be, was to be excluded from the common lot of having to depend on the fidelity of Adam : a lot desirable, on many grounds, for others, but on none whatever for her. Do thou congratulate her from thy heart on being chosen for such an office as that of being the Mother of her Lord. And if so many glorious prerogatives followed this as a consequence, thou mayest well believe that this one of her Immaculate Conception was among the number. Otherwise, what inconsistency of plan would it have been to bestow on her, in the second instant of her life, virtues so illustrious and sublime, so far transcending all laws even of grace, and at the same time to have allowed her to be a child of wrath in the first instant of her life? "As golden pillars upon bases of silver,"[1] so says the Lord, not "upon bases of clay."

NINTH DAY.

Cursed be the man that trusteth in man, and maketh flesh his arm, and whose heart departeth from the Lord (Jerem. xvii. 5).

I. Consider first, that although the confidence which thou hast placed in men has so often been deceived, thou dost not even yet amend. See, then, whether these words at least will not be sufficient to make thee withdraw thy confidence henceforward from them, and place it in God: "Cursed be the man that trusteth in man, and maketh flesh his arm." Thy hopes of finding help in men are founded on two motives, their fidelity and their strength; for if thou believedst that they were either unable or unwilling to give thee the good which thou desirest by their means, in spite of the large

[1] Ecclus. xxvi. 23.

promises they have made thee, thou wouldst not hope in them. Now, it is with reference to one who founds his hopes on their fidelity that the Prophet here says, "Cursed be the man that trusteth in man," and to one who founds them on their power, that he adds, "And maketh flesh his arm." Is it possible that thou canst give admittance in thy heart to a confidence which brings a curse to him who harbours it?

II. Consider secondly, that this word "to curse," may have three meanings, all bearing on our subject. (1) To foretell evil: "They are cursed who decline from Thy commandments."[1] (2) To imprecate evil: "Come and curse Jacob."[2] (3) To send down evil: "And looking back, he saw them, and cursed them in the Name of the Lord."[3] All these three meanings are here employed by the Prophet to show how justly cursed is the man who puts his trust in man.

1. He foretells evil, as a Prophet, to these persons whom he curses. For how can anything but sorrow be presaged of one who relies on the faith of man, which is so treacherous, or on the power of man, which is so weak? That the faith of men is treacherous is evident, for he is most inconstant by nature: "Every man is a liar."[4] He is as inconstant in will as in judgment, and as inconstant in judgment as in his appreciation. Do we not all know that his appreciations are as variable as the tints of the chameleon? A look changes them; a mere word spoken against thee turns a friend into an enemy: "A wicked word shall change the heart."[5] And who is there that cannot see how weak the strength of man is, when he is unable to save himself, much less others: "Put not your trust in princes, in the children of men, in whom there is no salvation."[6] If, therefore, there is "no salvation in them," how can there be salvation "from them"?

2. As the neighbour of such persons, he imprecates evil by cursing them; because, although it is never lawful to wish any one evil, as evil, it is lawful to wish him evil as being for his good: "Fill their faces with shame, and they shall seek Thy Name, O Lord."[7] And in this way the Prophet, as his neighbour, here desires that every one who relies either on the fidelity or the power of man may be deceived in his expectation, so that he may in this find a motive for henceforth

---

[1] Psalm cxviii. 21.   [2] Numbers xxiii. 7.   [3] 4 Kings ii. 24.
[4] Psalm cxv. 11.   [5] Ecclus. xxxvii. 21.   [6] Psalm cxlv. 2, 3.
[7] Psalm lxxxii. 17.

seeking love and help, not from men, but from God: "They were all confounded at a people that could not profit them."[1]

3. As the priest and directly commissioned servant of God, he brings evil upon them by cursing them, because he is the executor of the Divine will, and therefore he says, "Thus saith the Lord, cursed be the man that trusteth in man," to show that he is speaking in God's Name, not his own. And it must needs be that the fulfilment of this curse on these unhappy persons is immediate, because by it the evil is neither foretold nor imprecated, but directly sent. So it was that the fig-tree, which was thus cursed by Christ, "immediately withered away,"[2] because with God to say and to do are the same, there is no intervening time: "He spoke, and they were made." If, indeed, the only result of this curse were to render vain the favour of men, the evil it entailed would be but slight. The worst is, that it also entails the loss of God's favour, Who is justly incensed at seeing men preferred before Him. Dost not thou tremble at so terrible a malediction, which robs thee of everything, on both sides, of earth and of Heaven?

III. Consider thirdly, that it is not every kind of trust that is put either in the fidelity or the power of men that merits the curse of God, but that only which is contrary to the greater trust which ought to be put in the fidelity and power of God. For this reason, after saying, "Cursed be the man that trusteth in man, and maketh flesh his arm," the Prophet adds in conclusion, "And whose heart departeth from the Lord," because it is this which is the evil to be detested—the departing from God in his heart, which man is guilty of in such a case. When is it, therefore, that thy heart departs from God when thou trustest in men? In the first place, it is when, for the sake of gaining the favour of men, thou dost not hesitate to do something which offends God—to flatter, to detract, to lie, to transgress the laws of thy state in any way. And in the second place, it is when thou so trustest in the favour of men as not to trust at the same time far more in that of God, as it is thy duty to do, knowing and believing that men can only benefit thee just so far as it is God's will that they should do so. The first of these errors is trusting in men more than in God; the second is trusting in men as in God; and both are alike detestable: "Woe to them, for they have departed from Me."[3] Besides, dost thou not know

---

[1] Isaias xxx. 5.　　[2] St. Matt. xxi. 19.　　[3] Osee vii. 13.

that no man, however great he may be, can ever do thee any good unless he is moved by God to do it? "The heart of the king is in the hand of the Lord: whithersoever He will He shall turn it."[1] How canst thou ever dare, then, to offend God for the sake of gaining the favour of men, or how is it that, when seeking their favour, thou dost not far more seek the favour of God, on Whom they all depend, as the clay in the potter's hand? "Behold, as clay is in the hand of the potter, so are you in My hand, O house of Israel."[2] Though all men should fail thee, thou hast every possible good in God alone; but if God should fail thee, in whom canst thou hope? "On whom dost thou trust, that thou hast revolted from Me?"[3]

IV. Consider fourthly, therefore, how much better it is to trust in God: "It is good to confide in the Lord, rather than to have confidence in man;"[4] because, just as nothing but evil can be augured of him who trusts in man, so, on the other hand, every good may be augured of one who trusts in God: "Blessed be the man that trusteth in the Lord."[5] We may augur good of one who trusts in His fidelity, because the faith of God is not inconstant, like that of men: "God is not as a man that He should lie, nor as the son of man, that He should be changed,"[6] that is, "that He should lie" by an evil will, or "be changed" by an inconstancy of opinion. And we may augur all good of one who relies on the power of God; for what cannot he do who leans on the arm of Omnipotence? "Then shalt thou abound in delights in the Almighty;" not only "abound in good things," but "in delights," because thou shalt have not only what is necessary for satisfying thy desires, but what is superabundant. Why, then, dost thou not resolve to withdraw thy confidence henceforth from men, and to place it in God? It is in God alone that we can hope absolutely. In men we may indeed hope, but only as instruments of whom God makes use to do thee good. And therefore, if thou considerest well, all thy confidence must in the end resolve itself into confidence in God, from Whom everything comes: "Destruction is thy own, O Israel; thy help is only in Me."[7] Here, then, observe, for thy practical instruction, that it is not here said, "Cursed be the man that hath recourse to man," but "that trusteth in man;" neither is it

---

[1] Prov. xxl. 1. [2] Jerem. xviii. 6. [3] Isaias xxxvi. 5.
[4] Psalm cxvii. 8. [5] Jerem. xvii. 7. [6] Numbers xxiii. 19.
[7] Osee xiii. 9.

said "that useth flesh as his arm," but "that maketh flesh his arm," because it is not forbidden to have recourse to men in a right way in our necessities, nor to make a right use of the authority or assistance of men, that is, with a due subordination to the Divine will. What is forbidden is relying on men as the foundation of our confidence, for that should always be placed in God as the *primum mobile*, on Whom all the lower spheres depend.

TENTH DAY.

I have been delighted in the way of Thy testimonies, as in all riches
(Psalm cxviii. 14).

I. Consider first, that by God's "testimonies" are often meant in the Scripture those reasons which show us that our religion is the only one to be followed, such as the perfect fulfilment of all the prophecies concerning Christ, all the miracles and martyrdoms, and other like proofs which are clear not only after reflection, but at first sight: "Thy testimonies are become exceedingly credible."[1] Now, the Psalmist says that he has experienced the utmost delight in the way of these testimonies, as being that which leads to the discerning the true religion from those that are false. And to express this fully, he compares it to the delight which he experiences who knows himself to be the possessor of every sort of riches: "I have been delighted in the way of Thy testimonies as in riches." There are two kinds of riches, natural and artificial. The former are those which have the power of directly relieving the natural wants of man: such are food, clothing, beds, houses, carriages, and the like. The latter are the fortunes, which are the means of procuring these goods. Do not imagine, then, that the delight which David felt in thinking on all the many beautiful proofs which were to establish the truth of the Gospel faith was like that which the rich derive merely from the first kind of riches; for that is a delight which lasts only as long as the want to which it corresponds, so that when hunger and cold are at an end, it is more a trouble than anything else to take food or put on clothing. David's delight resembled that which the rich take in the second kind of riches, a delight which is lasting. Thou seest that it is not

[1] Psalm xcii. 5.

enough for them merely to know they are rich, they take pleasure in often thinking about it; and although they are not in want either of more food, or clothing, or anything of the same sort, yet they open their safes for the mere delight of seeing how full they are, they count their treasures, gaze upon their gold, and dwell in rapture on one jewel after another, for the pleasure of seeing how prosperous their condition is. This is what David did, with this difference, that in his case it was praiseworthy, and in theirs blameable: "I have been delighted in the way of Thy testimonies, as in all riches." There was no limit for him in this delight. How is it with thee? Is it possible that thou art satisfied with the mere knowledge that thou possessest the true faith? Dwell frequently upon the thought, and give hearty thanks for this blessing, above all when comparing thy lot with such vast multitudes who have no knowledge whatever of God, and who bear evident marks of being lost.

II. Consider secondly, that God's testimonies also mean in the Sacred Scriptures the commandments which He has given us in His most holy law; and they are so called because they testify to us most certainly what is the will of God: "My soul hath kept Thy testimonies, and hath loved them exceedingly."[1] And the Psalmist here says that as others delight in both kind of riches, so did he delight in the way of these testimonies, which is the way leading to grace and glory, without which it would avail us but little to have been born in Christendom: "I have been delighted in the way of Thy testimonies, as in all riches." And the reason was, that as to the first kind he was sure of never lacking anything he needed in that way; and that as to the second, he never should need them. He was sure, as to the first kind, that he should never lack anything necessary, because whoever faithfully observes God's law has God for his provider. There is no fear that God will ever cease thinking of the man who thinks of Him: "They that seek the Lord shall not be deprived of any good."[2] And as to the second kind, he did not want them, for by the keeping of God's law he had already attained to the state in which all irregular appetites are repressed. And of what avail are excessive riches, except to gratify those appetites. Do thou repress them, and thy happiness will be so great in seeing how many things thou thus ceasest to require, that thou wilt not have need to envy any rich man in the world.

[1] Psalm cxviii. 167.   [2] Psalm xxxiii. 11.

None of these men ever have as much as they want: "The rich have wanted, and have suffered hunger;"[1] they "have wanted" even as regards natural riches, because their avarice has become nature, and they "have suffered hunger" as to artificial riches, because the more avarice is fed, the greater is its hunger: "It never saith, It is enough."[2] What, then, oughtest thou to do, instead of putting thy delight in money? Thou oughtest to be indifferent to it. Beg of God to give thee grace to make more account of the least word of His than of all the treasures of the universe: "The law of Thy mouth is good to me, above thousands of gold and silver."[3] Esteem the law of God more than any other good—"Place thy treasure in the commandments of the Most High"[4]—and thou wilt see how thou wilt abound with delights: "I have been delighted in the way of Thy testimonies, as in all riches."

III. Consider thirdly, that by God's testimonies in the Scriptures is also understood the example of Christ, which was in all respects conformed to His Evangelical Counsels: "I am one that giveth testimony of Myself."[5] These are the words of Christ, and He spoke thus because the primal truth is like the light which alone has the power of giving authoritative testimony of itself in the world. Now, in this sense, to follow the way of the Divine testimonies is the same thing as to follow the way in which Christ walked. And canst thou not say with David in this way also, "I have been delighted in the way of Thy testimonies, as in all riches"? Nay, thou canst say so in this way more than in any other; for in this, not only wilt thou care nothing for excessive, superfluous, superabundant riches, which artificial ones are, but thou wilt rejoice to suffer the want of those which are natural; thou wilt rejoice in hunger, in cold, in every kind of discomfort to bodily delicacy, provided only that all these things are suffered by thee to please Christ: "I please myself in my infirmities, in reproaches, in necessities, in persecutions, in distresses for Christ."[6] Therefore, whosoever rejoices in being really like Christ, if he could have all the riches of the world without even having to go through the labour necessary for gaining them, would give them all up for the sake of the delight which he feels in merely thinking that he has nothing: "I have been delighted in the way of Thy testimonies, as in all riches;"

[1] Psalm xxxiii. 11. [2] Prov. xxx. 16. [3] Psalm cxviii. 72.
[4] Ecclus. xxix. 14. [5] St. John viii. 18. [6] 2 Cor. xii. 10.

not only "as in gaining all riches," but "as in all riches when already gained," so little does he value the harvest, not only when growing, but even when laid up in the barn! Dost thou feel this delight? If not, acknowledge that the fault is thy own. Thou dost not set thyself to penetrate with thy understanding the hidden treasures which are to be found in the imitation of Christ, and thou dost not practise thy will in preferring them to every other good. What wonder, then, if thou dost not taste this delight? For observe how the Psalmist speaks. He does not say, "The way of Thy testimonies has delighted me as all riches," but "I have been delighted in the way of Thy testimonies, as in all riches," for, as a wise man, he did not expect that the way of God's testimonies, in whatever sense it is taken, would give him any delight if he did not do what he could on his side in order to find his whole delight in them.

IV. Consider fourthly, how little all these doctrines are understood or are believed by the foolish world; and therefore, wherever thou goest, thou art always meeting people who, for the sake of making thee, so to speak, die of envy, will delight in parading before thee all the riches they have to show in their palaces, equipage, or dress, or in the intolerable pomp which they affect in public. What, then, shouldst thou do on witnessing a sight so unbecoming a Christian? Thou shouldst immediately turn to thy God, and say in thy heart, "I have been delighted in the way of Thy testimonies, as in all riches." This is the most wholesome correction which thou canst ever employ to prevent thyself from being infected by this same avarice, which, as it enters by the eyes, takes its name also from the eyes: "The concupiscence of the eyes." Dost thou see those palaces which rob the neighbouring houses both of air and light, those dresses showered over with jewels, that furniture laden with rich trimmings, those carriages, grooms, and pages, those splendid horses, so exactly matched? Do not let any of these steal away thy heart, but turn it at once away from them, and lift it up to God: "I have been delighted in the way of Thy testimonies, as in all riches." And in order to be able to do so with ease, take care to place thy delight in these three things: in those proofs which make evident the truth of thy faith, in the fulfilment of the Divine precepts which gives it life, and in the imitation of Jesus which makes it perfect.

ELEVENTH DAY.

Moab hath been fruitful from his youth, and hath rested upon his lees, and hath not been poured out from vessel to vessel, nor hath gone into captivity; therefore his taste hath remained in him, and his scent is not changed (Jerem. xlviii. 11).

I. Consider first, how injurious to virtue is attachment to the comforts which are more particularly enjoyed in a fixed place of abode. The Moabites had, from the beginning, been lavishly provided with these things, their country being a very fair and fertile one, not far from the desolated cities of the plain, from which they had learnt much licence : " Moab hath been fruitful from his youth." And thou seest to what this brought them, to live at ease in the midst of dregs—" He hath rested upon his lees "—because, never having been removed, their case was that of some strong, generous, excellent wine, which has not been changed from one vessel to another ; they went on decaying in their original vices, till at last they had to be treated like spoilt wine, cast aside as worthless. Thou mightest easily, if thou choosest, do a great deal for God's honour, since thou lackest neither strength, nor talent, nor ability, nor disposition for doing it, and yet thou dost nothing ; thou art "resting upon thy lees." And what is the reason ? Thou hast not "been poured out from vessel to vessel." Thou hast always remained in thy native place, or wherever else thou hast chosen to remain from habit, or friendship, or on account of powerful patrons, or the various comforts to be enjoyed there. And so it may be said of thee as it was of Moab, " He has gone into captivity." It is no wonder, therefore, if in the end thou losest all thy strength in the dregs to which thou hast become so much attached. Begin to leave God to dispose of thee as best pleases Him : " Lo, here am I, send me."[1] Shake off thy attachment to thy native place, and to thy relations, and to all the places where thou mayest live in comfort—" Furnish thyself to go into captivity, thou daughter inhabitant of Egypt "[2]—and thou wilt see how ready even thou wilt become for God's service.

II. Consider secondly, what the evils are which result from this attachment to ease ; there are two of them, and they are precisely those which wine contracts from its dregs when it is allowed to stand a long time on them without changing the

[1] Isaias vi. 8.   [2] Jerem. xlvi. 19.

vessel—it never loses its bad taste, and it never loses its bad smell : "His taste hath remained in him, and his scent is not changed." The bad taste is the evil inclination which a person acquires not to leave the place where he has chosen to take up his abode for some time. This inclination grows upon him, till at length it reaches such a pitch, that even when he does leave the place it is not possible for him to lose the inclination, just as wine when it has become infected by the dregs is not bettered by being put into a fresh cask, it never loses the taste : "His taste hath remained in him." And the bad smell is the bad reputation which in the long run is acquired by attachment to ease and comforts. For who can believe that such a man will begin in his old age to undertake labours to which he has not accustomed himself in his youth? "His scent is not changed." And therefore he who does not hasten to leave his dregs, renders himself quite useless for doing any good : "Wisdom is not found in the land of them that live in delights."[1] Thou thinkest, nevertheless, that thou hast already left those dregs ; as to this, others may indeed bear witness from the odour thou sheddest around, but thou oughtest to be certain about it from the taste that thou perceivest thyself.

III. Consider thirdly, that those ancient servants of God may well be compared to wine, not left upon the lees, but "poured out from vessel to vessel," of whom the Apostle says that they scarcely found any place of shelter or rest on earth, so continually were they driven away, mocked, or persecuted by all men : "They wandered about in sheep-skins, in goat-skins, being in want, distressed, afflicted, of whom the world was not worthy, wandering in deserts, in mountains, in dens, and in caves of the earth."[2] And who can say what great instruments of God's glory they were for this very reason? Does not this encourage thee to do for that same glory so much less a thing as to abandon the excessive comforts which thou art enjoying in peace? Take heed, for this peace is that of wine on the lees, a peace which leads, little by little, to corruption : "I will visit upon the men that are settled on their lees."[3]

---

[1] Job xxviii. 13.   [2] Hebrews xi. 37, 38.   [3] Sophonias i. 12.

TWELFTH DAY.

*Wash me yet more from my iniquity, and cleanse me from my sin ; for I
know my iniquity, and my sin is always before me* (Psalm l. 4).

I. Consider first, that no sooner did David, having repented
of his fault, say to the Prophet Nathan, "I have sinned against
the Lord," than the latter answered him: "The Lord also
hath taken away thy sin."[1] He must, therefore, have been
certain that he had obtained pardon for it. And yet for all
this certainty he never ceased to continue to pray for pardon,
not from any doubt as to its having been granted, but from
the desire of receiving it in fuller measure every day, as is the
case with those in whom, if sin has abounded, grace has not
only abounded, but superabounded: "Where sin abounded,
grace did more abound."[2] And therefore he asked pardon of
God, not only according to the greatness of God's mercy in
itself, by which every sort of sin is forgiven—"Have mercy on
me, O God, according to thy great mercy"—but he asked it
also according to the multitude of those various acts in which
God had exercised that mercy towards him—"And according
to the multitude of Thy tender mercies blot out my iniquity"—
for who can say how great and glorious and wonderful these
acts have been seen to be in regard to so many different kinds
of sinners? "His tender mercies are over all His works."[3]
And if thou preferrest to think that David, although certain of
forgiveness, continued to pray for it with so much earnestness,
in order to teach thee what thou oughtest to do, then I say to
thee that in thy case not only is there no such certainty, but
there may be a very great uncertainty. Dost thou imagine
that to dwell upon thy own sins is any hindrance to perfect
union with God? David says that he did so, not only often,
but continually: "My sin is always before me." He did not,
indeed, say "my adultery," but only "my sin," because it is
better, particularly in certain cases, not to call to mind in
detail the evil that has been committed, but only in general.
Nevertheless, it is true that he said "always," because, however
just and pious and perfect thou mayest be, nay, even if thou
art in a state of mystical sublimity, like David, still thou
oughtest to think seriously upon all thy miseries, and grieve
bitterly for them to the last day of thy life. Therefore this
psalm is called, "A Psalm unto the end," that is, "A Psalm

[1] 2 Kings xii. 13.   [2] Romans v. 20.   [3] Psalm cxliv. 9.

to be sung unto the end of the world," as Bellarmine explains it ; for if thou wert to live till the end of the world thou wouldst have to say " I have sinned " till the end of the world. " Remember and forget not how thou provokedst the Lord thy God to wrath in the wilderness."[1] " Remember " as to the present, and "forget not " as to the future.

II. Consider secondly, the difference that there is between bodily and spiritual diseases. It is enough for the physician to know the former ; not so as to the latter : these must be known also by the patient. Therefore the Psalmist, when begging of God an abundant supply of justifying grace, adds, as a reason for obtaining it, that he has on his side already complied with the necessary condition, that, namely, of knowing the gravity of the evil he had done : " Wash me yet more from my iniquity, and cleanse me from my sin : for I know my iniquity, and my sin is always before me." Do not object that it is not enough to know the evil that has been committed, but that it must also be grieved over, detested, that there must be a firm resolution of amendment ; for when any one says that he knows his sin as it ought to be known, he has said everything. Just as it is impossible to know clearly what is supremely good, and not to love it intensely, so, too, is it impossible clearly to know what is supremely evil, and, I will not say not to hate it, but not utterly to abhor it. Therefore, all that God requires as a condition of thy pardon is to know thy sins : " I am holy, saith the Lord, and I will not be angry for ever ; but yet acknowledge thy iniquity." He does not say "bewail" or "detest," He only says "acknowledge" or "know" (*scito*), because, if thou understandest what an evil thou hast committed in offending so good a God, it is impossible for thy eyes not to become two fountains of perpetual weeping.

III. Consider thirdly, that although the words sin, iniquity, and impiety are most frequently confounded together, yet, according to their strictest sense, they are intended to denote the three celebrated distinctions of transgression which are committed by man against himself, against his neighbour, and against God. The first of these is called simply sin, the second iniquity, and the third impiety : not that any sin which violates the order that the sinner owes either to himself or to his neighbour, does not also violate that which he owes to God as the Supreme Legislator, but because that sin is

[1] Deut. ix. 7.

properly called impiety which violates the order due to God, as Father or Lord, in the worship which is called religious. Now, in the case of David, he had certainly violated the order due to himself by means of the malice of his sin, and he had violated that which he owed to his neighbour by means of the wrong done to Urias, which was in every way so grievous; but he had not violated the order due to God as regards the worship of religion, because his sin had not been one of infidelity, or simony, or perjury, or blasphemy, or the like; and therefore he here speaks only of sin and of iniquity—there is no mention of impiety throughout the Psalm; although elsewhere, speaking of the general principle of impiety, which is, after all, contained in every sin, he did say, "I said, I will confess against myself my injustice to the Lord, and Thou hast forgiven the wickedness (*impietatem*) of my sin."[1] Because of the iniquity he begs of God to wash him: "Wash me yet more from my iniquity;" because of the sin he begs of God to cleanse him: "Cleanse me from my sin"—to wash him, that is, from the remains of the past, and to cleanse him from the danger of the future evil; and both to wash and cleanse him more and more continually: "Wash me yet more; cleanse me yet more." What canst thou say for thyself, who in thy time hast most likely committed not only sins against thyself and iniquities against thy neighbour, but, it may be, also enormous impieties against God; and yet if thou canst remember to have once asked pardon for them, thou thinkest that thou hast fully satisfied thy obligation? "Every night I will wash my bed."[2] See how many nights of tears one night's sin cost David!

IV. Consider fourthly, that, as to the iniquity, David had just before begged of God to do away with it: "According to the multitude of Thy tender mercies, blot out my iniquity." But, not contented with that, he still prays to be washed from it, that is, washed from all affection or attachment, however slight, to the past evil which might have been left in him by that iniquity which had lasted for nine months; and therefore he does not say "wash my iniquity" as he said "blot out," but "wash me from my iniquity," that is, "blot out the iniquity; wash the author of the iniquity." This is indeed the conduct of one who holds in abhorrence the stain upon his face: not only to efface the stain, but so to wash the whole of his face at the same time as not to leave on it the faintest

---

[1] Psalm xxxi. 5.   [2] Psalm vi. 7.

speck from that hateful stain. So, too, he prays to be cleansed from his sin: "Cleanse me from my sin," that is, to be cleansed from the malice of his will. And the reason is, that for him who is unclean, not only in act, but in the power of acting, it is not sufficient even to be washed; in a short time he will commit fresh impurities, just as the face, soon after being washed, becomes soiled again. And therefore David does not here ask to be merely washed, but cleansed from his evil will, which had become evil not only in act, by reason of the sin he had committed, but also in the power of committing still greater sin, and therefore he never ceased to fear. He feared because, ever since the original sin of Adam, the will is in itself inclined to evil in every one: "The senses of man's heart are inclined to evil from his youth." And he feared also because he had inclined it to evil still more by his own actual sin. Oh, if thou knewest how many bad dispositions are left in thy will by every sin (especially such as are neglected and of daily occurrence), thou wouldst certainly not put off repenting of it heartily, as thou dost sometimes for months together. So far from acting thus, thou wouldst not be satisfied even with knowing that thou hast repented, since it is possible for thee to be injured by forgiven sin even; not indeed in itself, but in its worst effects: "Be not without fear about sin forgiven."[1]

V. Consider fifthly, that the heart is washed and cleansed by contrition, confession, and the good works which are afterwards performed in satisfaction for the sins which have been committed. But all this is the part of the sinner, according to what is written: "Wash yourselves, be clean, take away the evil of your devices from My eyes."[2] And therefore it may seem as though David, instead of asking to be washed and cleansed, might more properly have declared to God his intention of washing and cleansing himself. But thou oughtest here to remember the custom of the Sacred Scriptures, which is, that when any actions of man are performed both by grace working within and free-will co-operating with grace, they are sometimes entirely attributed to God, sometimes entirely to man, thus showing how perfectly they work together: "Incline my heart to Thy testimonies," here is the work of grace; "Incline thy heart to know wisdom," here is the work attributed to free-will; "Lead me in the right way," here is the work of grace; "Direct thy

---

[1] Ecclus. v. 5.   [2] Isaias i. 16.

heart in the right path," here is the same work attributed to free-will; "Create a clean heart in me, O God," here is the work of grace; "Make you a new heart," here is the same work attributed to free-will. Therefore, all the prayers of man to God which are contained in the Sacred Scriptures are so many proofs of our need of grace, and all God's commands to man are so many proofs of the freedom of our will. But, indeed, if we consider aright, the Psalmist here does not intend, under the figures of blotting out, washing, and cleansing, to denote the dispositions by which the penitent prepares the way for sanctifying grace, namely, contrition, confession, and other good works, but grace itself; and therefore there is all the more reason for his asking it of God, since to God alone it belongs to give it: "I am, I am He, that blot out thy iniquities for My own sake, and I will not remember thy sins."[1] The "blotting out" of iniquities is the remitting to the sinner not only the guilt, but also the eternal punishment which he had incurred in the lofty decrees of Divine justice. The "washing" and the "cleansing" of the sinner is the infusing of sanctifying grace, which is able not only to purify him from past stains, but to preserve him from future ones. But who can do this save God? "Who can make him clean that is conceived of unclean seed? Is it not Thou Who only art?"[2] Sanctifying grace may go on continually increasing —and therefore "yet more" is added when washing and cleansing are spoken of—whereas the remission both of the guilt and of the eternal punishment is completely effected in an instant, and therefore these words are not added when speaking of blotting out. If thou, too, longest to be thus day by day washed and cleansed more and more by God's sanctifying grace, do thy part first, by washing and cleansing thyself by means of those dispositions which it is for thee to procure : "Wash thy heart from wickedness, O Jerusalem, that thou mayest be saved."[3]

VI. Consider sixthly, that if thou neglectest the performance of this duty, it is entirely because thy sin does not trouble thee, as David's troubled him during the whole of his life. Listen to what he says: "My sin is always before me," or rather "against me," for the word is not *coram*, but *contra me*, because his sin was always standing, as it were, in the act of throwing in his face the ingratitude with which he had treated his God for the sake of a vile, sensual pleasure: "Thy

[1] Isaias xliii. 25. [2] Job xiv. 4. [3] Jerem. iv. 14.

own wickedness shall reprove thee."[1]   Doubtless David might
have turned away his eyes from this unpleasant monitor, but
he would not do so because he esteemed the remembrance
of his sin most useful in exciting humility, compunction, and
caution : " After Thou didst show unto me I struck my thigh :
I am confounded and ashamed."[2] If thy sin does not occasion
thee equal, or at least similar trouble, examine well, and thou
wilt see that this proceeds from thy keeping it studiously at
a distance from thy thoughts, and lending an ear in preference
to the world, the flesh, and the devil : to the world, which
flatters thee in thy sin ; to the flesh, which makes excuses for
it ; and to the devil, who encourages thee to listen to the
world and the flesh rather than to the salutary reproofs of thy
conscience.   But how much better is it to be rebuked by a
wise man than to be flattered by all the fools in the world !
" It is better to be rebuked by a wise man than to be deceived
by the flattery of fools."[3]   Besides, do what thou wilt, sooner
or later thy sin will stand face to face with thee ; if not in
life, certainly in death : " I will reprove thee, and set before
thy face."[4]

THIRTEENTH DAY.

My mouth shall meditate truth, and my lips shall hate wickedness
(Prov. viii. 7).

I. Consider first, that being, as thou art, so inclined to
speak evil of thy neighbour that thou frequently makest
a boast of it, saying that thou art a man of an open, frank,
and candid disposition, to whom it is impossible not to speak
of things always just as they are, this noble passage of the
Wise Man may appear to be plainly on thy side.   But this
is a great mistake.   What is it that he says? " My mouth
shall meditate truth."   He does not say "shall utter, shall
speak," but " shall meditate," because every truth that rises
to thy lips must not be spoken out merely because thou art
a man of frank character, but only that which it appears right
to express after mature consideration.   Canst thou, then,
think it right to speak all sorts of evil of thy neighbour merely

[1] Jerem. ii. 19.
[2] Jerem. xxxi. 19.   [3] Eccles. vii. 6.   [4] Psalm xlix. 21.

because it is true? Thou art not willing to have evil spoken of thyself for that reason; why then dost thou permit thyself to speak evil of others on these grounds? "My mouth shall meditate truth." First weigh well the reasons which induce thee to utter a truth that is injurious to thy neighbour's character, and those which induced thee to keep silence; then, if the former do really, in the sight of God, preponderate over the latter, thou art free to utter it, lest thou shouldst be one of those who, even in a court of justice, betray the truth for the sake of a vile and worthless gain : "For a morsel of bread he forsaketh the truth."[1] But never speak without having thus first weighed the question. Dost thou really think it a matter for a man to boast of, that he is very free spoken? With many persons, this means that they are never able to bridle their tongues. And yet thou shouldst be ready to spend all thy richest treasures for the sake of learning how to do this : "Melt down thy gold and silver, and make a balance for thy words, and a just bridle for thy mouth."[2] "Make a balance," by considering whether the truth ought to be said or not ; and make "a just bridle," by learning how to govern thyself in saying it, supposing it is to be said, or to abstain from saying it, if it is not to be said. And besides, why is it that thou so readily persuadest thyself that this saying of the Wise Man is a justification of thy tendency to detraction? Because thou supposest that these words, *Labia mea detestabuntur impium*, mean "My lips shall hate the wicked man." Not so : they mean "My lips shall hate wickedness"—*Id quod impium est.* Such is the force of the expression. It is one thing to hate wickedness in the abstract, and quite another to hate it in this or that particular man. If then, even with regard to hating wickedness in the abstract, the Wise Man declares that he will first deliberate well as to what he has to say—"My mouth shall meditate truth"—what would he have said about hating it in the concrete, that is, in the person of another?

II. Consider secondly, that if *impium* in this passage does not mean the wicked man, but wickedness, it would seem that after having first said, "My mouth shall meditate truth," the correct antithesis would have been to add, "And my lips shall hate falsehood." And yet the Wise Man says "shall hate wickedness." And he says so, that thou mayest clearly understand what the truth is of which he is speaking. It is the truth concerning our law. Thou oughtest to know,

---

[1] Prov. xxviii. 21.   [2] Ecclus. xxviii. 29.

therefore, that in this passage the Wise Man is speaking in the person of Christ, the Eternal Wisdom. Now, it is certain that Christ was to come, in order to teach the world truth, that is, to teach what is the true end to which our affections ought to be directed, and what are the true means for attaining it. Therefore did He say of Himself, " I am the Way, the Truth, and the Life:" the Way, as regards the means; the Life, as regards the end; and the Truth, as regards both the means and the end. Here, then, thou seest the Truth placed between the Way and the Life, because it teaches the means to the man who needs only the means, and the end to him who needs also to know the end. As, therefore, Christ says, "My mouth shall meditate truth," so, too, does He justly add, "And My lips shall hate wickedness," that is, they shall hate all the falsehood, opposed to religion, that shall be in the world, for this is the most hateful kind of falsehood, that which is not only wicked and infamous, but impious. Of course, all falsehood is to be disapproved, as we all know; but that falsehood which is opposed to religion is also to be hated, that is, to be abhorred, detested, and kept at a distance like an abominable plague, the merest breath of which cannot be endured: "My lips shall hate wickedness." But if this is so, how is it that sometimes thou dost not scruple in conversation to go so far as to jest about this kind of falsehood, as though it were not so detestable in itself as it is represented? Is it possible that thou canst even commit so monstrous an evil? Whenever thou happenest to hear religious questions mooted, say immediately within thyself, "My mouth shall meditate truth, and my lips shall hate wickedness." Thou shouldst say, "My mouth shall meditate truth," in order not to be like those audacious persons who, having never in their lives studied anything beyond a few story-books, enter into conversations on the sublimest mysteries, such as immortality, predestination, providence, the agreement of grace and free-will, without having an idea what they are doing. And thou shouldst say, "My lips shall hate wickedness," because, whenever there is a discussion concerning any error which is contrary to the faith, thou shouldst instantly hate it, just because it is opposed to the faith, even if thou knowest no other reason for hating it. "My lips," not "my mouth," but "my lips shall hate wickedness," so ready oughtest thou to be to give expression to this hatred. Is it not a shame if at any time thou art heard speaking in praise of revenge, of hoarding money, of ambition, of luxury, and other things opposed to the doctrine of Christ? "My lips shall hate wickedness."

III. Consider thirdly, that if it is Christ Who is here speaking by the mouth of the Wise Man, it would seem that He might more properly have said, "My mouth shall speak," rather than "meditate truth," because Christ, Who is Eternal Wisdom, had no need of first thinking what He should say; it was enough for Him to open His mouth, He was always certain never to err. Yet He says, "My mouth shall meditate truth," in order to remind thee how many years He, the Eternal Wisdom, waited before opening His mouth. He did not wait all those years in order to think of what He had to say, but in order to show thee how much thou oughtest to think about it; for He condescended to perform countless actions, not on His own account, but for our instruction. Besides, dost thou not know that there are two ways of meditating? Things which are to be said may be meditated in the mind, and also in act; and this is the two-fold signification of the words, "Blessed is the man who shall meditate on the law of the Lord day and night."[1] It was in this second way that Christ also meditated the truth before teaching it, there being no need for Him, as there is for us, to meditate it in the other way. What precept did He ever give concerning poverty, humility, obedience, charity, meekness, modesty, or religion, that He had not first practised for a long while? "The government is upon His shoulder:"[2] because He first bore upon His own shoulders all the load, which He was afterwards, as a King, to lay on others. Dost thou act thus? Art thou not, rather, very ready to tell others what good works they ought to do, whilst thou art reluctant, or at least tardy, in first exercising thyself in them? What wonder, therefore, is it if thy words have no force? "My mouth shall meditate truth, and my lips shall hate wickedness." Wouldst thou hate wickedness in others in such a way as to be able to confound and crush it, and almost to banish it from the earth so soon as thou openest thy mouth? First meditate well, both in thy mind and by thy works, those Christian virtues by means of which thou art to overthrow it: "Before judgment, prepare thee justice."[3]

---

[1] Psalm i. 2.   [2] Isaias ix. 6.   [3] Ecclus. xviii. 19.

FOURTEENTH DAY.

None of us liveth to himself, and no man dieth to himself; for whether we live, we live unto the Lord, or whether we die, we die unto the Lord; therefore, whether we live, or whether we die, we are the Lord's (Romans xiv. 7, 8).

I. Consider first, that great kings have in their armies a regiment called the forlorn hope, the soldiers of which have devoted themselves to the service of their lord in such a manner as to regard their lives as in no way their own but his. Thus if the preservation of their lives will be to his advantage, they preserve them ; if not, they rush courageously for his sake into the thickest ranks of the enemy. Think now, that it was in such a band as this that the Apostle gladly enrolled himself when he said : "None of us liveth to himself, and no man dieth to himself ; for whether we live, we live unto the Lord, or whether we die, we die unto the Lord ; therefore, whether we live or whether we die, we are the Lord's." A soldier of the forlorn hope does not live to himself, because for him the end of living is not himself, that is, it is not the preservation of himself, but his object in preserving himself is the service of his lord, and therefore he does not "live to himself." Neither does he die to himself, because his object in dying is not any advantage or benefit which is to accrue to him after death, but only that which may accrue to his lord, and therefore he "dieth not to himself." This, too, is the conduct of those true servants of God who have devoted themselves entirely to Him. They are indifferent to both life and death ; but if they live, they are determined to live to Him, and if they die, to die also for Him : "Whether we live, we live unto the Lord, or whether we die, we die unto the Lord." How is it with thee? Think how many conditions and reservations thou makest ! Thou hast not the courage to live to God by detaching thyself from these comforts which make thee live to thyself ; and still less hast thou the courage to die for God, by exposing thyself to the danger of one day losing thy life for His honour. And yet, if it were so, what happiness would be thine ! Think over all the dangers to which, as a soldier of the forlorn hope, the Apostle exposed himself for the love of Jesus : "In perils of waters, in perils of robbers, in perils from my own nation, in perils from the Gentiles, in perils in the city, in perils in the wilderness, in perils in the sea, in perils

from false brethren."[1]   And thou art overwhelmed with terror
at one single danger to which thou hast not even exposed
thyself, but which meets thee on thy way: "O you of Israel,
that have willingly offered your lives to danger, bless the
Lord."[2]

II.   Consider secondly, that those "live to themselves"
who live according to their own judgment or fancy; and that
those "die to themselves" who die either in consequence of
the grievous disorders they fall into for the gratification of their
bodies, as it is written, "By surfeiting many have perished,"[3]
or through the excessive labours they undergo either for
the sake of ambition or of avarice.   Not so the servants
of God: "None of us liveth to himself, and no man dieth to
himself," this is what they say.   It is, indeed, a very base
thing to live to oneself, for brutes can do that; and it is a
miserable thing to die for oneself, for it would be difficult to
find even a brute that would do so.   If we are to live, we
should live to Christ, and if we are to die, we should also die
for Christ: "Christ shall be magnified in my body whether it
be by life or by death."[4]   What a noble sentiment!   Christ
cannot either increase or decrease in Himself.   He cannot
increase, because being true God, He is infinite in perfections;
and He cannot decrease, because He is indefectible.   He can
increase and decrease in others only, that is, in the greater or
less knowledge that others have of Him.   A man magnifies
Christ when he spreads abroad His Name as much as possible.
And he magnifies Him in his body, when he magnifies Him,
not only interiorly, but exteriorly.   If he magnifies Him by
employing his tongue, his feet, his eyes, his ears, his hands in
Christ's honour, then he magnifies Him "by life;" and if he
magnifies Him by losing tongue, feet, eyes, ears, hands, nay
life itself, for the love of Christ, then he magnifies Him
"by death."   This is what Christ's faithful servants have set
before them as their end: "Christ shall be magnified in my
body whether it be by life or by death."   But none have so
set it before them as those who do this unreservedly, those
true soldiers of the forlorn hope, who are able to say, as the
Apostle did, "To me, to live is Christ, and to die is gain."[5]
To live "is Christ" for them, because Christ is the principle
of their actions, and to die "is gain" for them, because they
account it a great gain to be able to give that life voluntarily

[1] 2 Cor. xi. 26.   [2] Judges v. 2.
[3] Ecclus. xxxvii. 34.   [4] Philipp. i. 20.   [5] Philipp. i. 21.

for Christ, which some day or other must be lost, whether they will or no. Canst thou possibly choose to be one of this ignoble, rather than of that noble band?

III. Consider thirdly, that, besides the natural, there is the civil life, which consists in the reputation which thou enjoyest, in thy employments, intercourse, friendships; and this, too, if thou art one of the forlorn hope of Jesus Christ, thou hast to give entirely to Him, so as to keep back nothing from spending and being spent altogether for His service. "Whether we live" this civil life also, "we live unto the Lord," for we must care nothing for our reputation except so far as it enables us to gain greater glory for God: "Not to us, O Lord, not to us, but to Thy Name give glory."[1] And in the midst of employments, intercourse, and friendships, we must study to please people for this reason only, that we may be able the more readily to win them to God: "I in all things please all men, not seeking that which is profitable to myself, but to many, that they may be saved."[2] And so, as to this civil death also, "we die unto the Lord," because, if we must lose all those things, and be left bereft of reputation, forsaken, hated, forgotten, be it so, if only we lose them for God: "We are delivered unto death for Jesus' sake;"[3] for did not Christ face both of these deaths, the natural and the civil, for the love of thee? What great matter is it, then, if thou, who art a wretched slave, diest for Christ, when thou knowest that He chose to die for thee? "The Mediator of God and men, Christ Jesus, gave Himself a redemption for all,"[4] that is, even for the most unworthy of men, of whom thou art: "He loved me, and delivered Himself for me."[5]

IV. Consider fourthly, that what most encourages soldiers of the forlorn hope not to care about themselves is to remember that they are not their own, but that they belong to the king for whom they are fighting. And this, too, should encourage thee in the case we are considering, only with still greater reason. Remember Whose thou art: "Whether we live, or whether we die, we are the Lord's." What king ever had so many titles of sovereignty over a man as God has over each one of us, who have been created, preserved, redeemed by Him? "Know you not that you are not your own? for you are bought with a great price."[6] Besides, the knowledge that we are "the Lord's," ought to inspire us with very great con-

---

[1] Psalm cxiii. 9.   [2] 1 Cor. x. 33.   [3] 2 Cor. iv. 11.
[4] 1 Timothy ii. 5, 6.   [5] Galat. ii. 20.   [6] 1 Cor. vi. 20.

fidence. And the reason is, that no earthly king can ever extend the same protection to his servants, living and dead, that God does to us: "Whether we live, . . . we are the Lord's," and therefore it belongs to Him to defend us from all those who attempt to do us an injury against His will; and "whether we die, we are the Lord's," and therefore it also belongs to Him to give us back the life which we have given for Him; for earthly monarchs cannot restore life to those who have given it for them, but God can and will do so: "Thou, indeed, O most wicked man, destroyest us out of this present life; but the King of the world will raise us up, who die for His laws, in the resurrection of eternal life."[1] What, then, prevents thee, I do not say from very gladly spending thy life to the honour of God, but even from losing it as His servant already pledged to a certain risk, knowing that to lose it is in reality to find it, nay, that it is never more surely found than when it is joyfully lost for Him? "He that will save his life, shall lose it," because the man who lives to himself will lose it, no matter how much he tries to preserve it, and will, perhaps, lose it the more quickly in consequence of the excessive pains he has taken to preserve it. "And he that shall lose his life for My sake shall find it," because he who has died for God has found his life in the very act of losing it: he has lost a temporal, he has found an eternal life.

FIFTEENTH DAY.

The riches of salvation, wisdom, and knowledge; the fear of the Lord is his treasure (Isaias xxxiii. 6).

I. Consider first, that, as there are bodily, so, too, are there spiritual treasures. The more the former are loved, the more do they endanger the eternal salvation of their owners; and therefore they are called the riches of perdition: "Thy money perish with thee."[2] The more the latter are loved, the more do they help their owners to ensure their salvation; and they are called "the riches of salvation." It is a characteristic of the former that the keeping of them brings with it no kind of good, rather every kind of evil, in consequence of the excessive affection for them which is contracted in keeping

[1] 2 Mach. vii. 9.   [2] Acts viii. 20.

them : the evil of guilt and the evil of punishment—"Riches kept to the hurt of the owner "[1]—and therefore they are riches of perdition. The keeping of the latter brings with it every kind of good, both that of grace and that of glory. Neither canst thou say that this good may also be bestowed by the former sort of riches, for when this is so, it is by being, not kept, but spent ; and therefore what kind of riches are those which only benefit thee when thou no longer possessest them ? Spiritual riches benefit thee when thou hast them in possession. And although these, also, can be distributed to others, like corporal riches, yet they are not lost by being thus distributed, like them ; on the contrary, they are multiplied, for thou becomest more especially rich the more thou impartest to others the riches God has given thee, whether by instructing the ignorant, reproving the sinful, counselling the doubtful, or comforting the afflicted. Who would have thought, then, that there are so many more seekers after the former than the latter ? See how anxiously and laboriously men busy them-selves continually in heaping up bodily riches : " There is but one, and he hath not a second ; no child, no brother, and yet he ceaseth not to labour, neither are his eyes satisfied with riches."[2]   And who is there that expends even the half of this study and labour on the acquisition of spiritual riches ? Do thou, at all events, remember, that it is possible for a man to receive corporal riches as a gift, for instance, by inheritance, but that spiritual riches cannot be had without labour : "The slothful hand hath wrought poverty, but the hand of the industrious getteth riches."[3]

II. Consider secondly, what these riches are, which are here called "riches of salvation." They are wisdom and know-ledge. Wisdom concerns our last end, that is, God; knowledge concerns the means which lead us to that great end. That man, therefore, possesses true wisdom on earth, who knows for what end he was created, and who does not set before him as his end either the favour of great men, or pleasure, or money, or rank, or glory, or any of those empty idols which the world worships. And he possesses true knowledge who not only knows his true end, but also is able to discern which are the most suitable and effectual means for attaining it. This wisdom and knowledge, then, are called riches of salva-tion—"The riches of salvation, wisdom, and knowledge"— because it is these which give everlasting salvation. Without

[1] Eccles. v. 12.   [2] Eccles. iv. 8.   [3] Prov. x. 4.

them thou hast lost it. Examine the depths of thy heart, and see whether these riches are there, and if not, strive to obtain them both by the exertions necessary for the acquisition of these riches, and also by praying for them without ceasing to God, since all thy care and trouble are of no use without God's blessing on them : " The blessing of the Lord maketh men rich."[1] Pray to Him continually to give thee wisdom and knowledge : the wisdom of desiring to labour for the true end only ; the knowledge, showing also in what way to labour.

III. Consider thirdly, that it is of little avail to be rich if thou hast not the means of guarding the riches thou hast acquired. If thou leavest them exposed to thieves, thou wilt run the risk of losing in a single day what thou hast gathered together with much difficulty in the course of years. Therefore, as the miser has his treasury, that is, the coffer in which he keeps under lock and key all the gold he has collected, so, too, must the just man have his. And what is this? It is the holy fear of God : " The fear of the Lord is his treasure." For it is this holy fear of God which is the guardian of the wisdom and the knowledge which are his riches. It guards them from men and from devils and from their own irregular appetites. (1) It guards them from men ; because he who is more afraid of displeasing God than men will not suffer them to turn him away from his end, nor to hold him back from the use of the means which conduce to that end : " It is better for me to fall into your hands without doing this thing, than to sin in the sight of the Lord."[2] (2) It guards them from the devils ; because he who is more afraid of the anger of God than of the fury of all his infernal enemies, closes his ears immediately to all those temptations which seek to turn him away from his end by alluring the concupiscible part to the love of temporal goods, or by discouraging the irascible part from the energetic use of every means for attaining that end : " But he answered without delay, that he would rather be sent into the other world."[3] (3) It guards them from the irregular appetites, which are like thieves in the house to the just man ; because, fearing the loss of God more than that of everything on earth, he is always on the alert not to yield to them when, either by force or cunning, they are preparing to rob him : " They that fear the Lord will seek after the things that are well pleasing to Him,"[4] not to themselves. Do not rely, then,

[1] Prov. x. 22.
[2] Daniel xiii. 23.　　[3] 2 Mach. vi. 23.　　[4] Ecclus. ii. 19.

on any riches of wisdom or knowledge that thou mayest possess, if thou dost not guard them in this treasury. Indeed, as the man who has most to lose has the most need of care in guarding it, so he who has most wisdom and knowledge has the greatest need of the fear of God.

———

SIXTEENTH DAY.

If thou shalt not watch, I will come to thee as a thief, and thou shalt not know at what hour I will come to thee (Apoc. iii. 3).

I. Consider first, how good a thief He is, Who here urges thee to be on the watch. Certainly He has no desire to take thee unawares; for if it were so, He would doubtless invite thee to sleep. Do not wonder at this; for the speaker here is no other than thy Jesus, Who, in His great love for thee threatens thee with every evil for thy good. Observe, therefore, that He does not positively say, "I will come to thee as a thief," but "If thou shalt not watch, I will come to thee as a thief." So, then, if thou shouldst be so unhappy as to see Him surprise thee suddenly at the hour of thy death, as a thief, the fault will be thine, not His. His object in giving thee notice that He will come to thee when thou least expectest it is that thou mayest expect Him continually.

II. Consider secondly, that when our Lord had once declared that if thou art not on the watch, He will come to thee, at thy death, like a thief, that is, suddenly, unexpectedly, undreamt of—"If thou shalt not watch, I will come to thee as a thief"—it might appear superfluous to add directly afterwards, that thou shalt not know the hour of His coming—"Thou shalt not know at what hour I will come to thee"—seeing that that has been declared sufficiently in the preceding words. But this is a mistake; it is not superfluous. And the reason is, that even when thou art not aware of the approach of a thief at midnight on account of the sleep in which thou art wrapped, it is possible that others may be so for thee, and may awake thee in time. In such a case the thief comes to thee "as a thief," and yet it cannot be said that thou knowest not the hour of his coming to thee, because there are persons who let thee know it. But in the case of which our Lord is here speaking, it will not be so; He will come "as a thief,"

who is not expected; and there is no one who will be able to
let thee know of the time of his coming: "Thou shalt not
know at what hour I will come to thee." His arrival will be
at a time unknown not only to thyself, but to all the physicians
who are attending thee, to all thy friends, relations, and
acquaintances; so that no one will be able to say to thee,
The thief is coming. And how many there are who die by
accident, so suddenly that they are known to be dead before
they are known to be dying! Our Lord warns thee that this
will be thy case, too, one day, if thou sleepest in sin. For this
is the punishment of one who does not awake, though he has
been warned to do so not once only, but again and again—
sudden death! "The man that with a stiff neck despised him
that reproveth him shall suddenly be destroyed."[1]

III. Consider thirdly, that even if thou art awake, waiting
for thy Lord, thou mayest still think that He will come to thee
like a thief in thy last hour, because He will come to take
from thee everything thou possessest, wealth, glory, greatness,
friends, relations, country, comforts, and thy body itself. But
this is the case only when thou art living in a state of attachment to these goods. For if, before He comes to take them
from thee, thou hast endeavoured to detach thyself entirely
from them, at least in affection, all that thou wilt have to do
at that hour will be to restore them readily to Him Who gave,
or rather lent them to thee. And, therefore, He will come to
thee, not as a thief, to take from thee what is thine, but as a
Master, to require from thee that which was only lent by Him
for thy use. He will come as a thief if thou hast an inordinate
affection for these goods. I say "as a thief," because, in
taking what is His own, He is not a thief, but "as a thief,"
because it will seem to thee that He is taking from thee what
is thine. Be, then, always prepared in mind to restore to thy
Lord what thou now possessest it is true, but only for a time.
And to this end watch over thy heart, call aloud to it, arouse
it, so that this poor heart of thine may never err so as to love
as its own possession what is only borrowed; then, neither in
this sense will thy Lord deal with thee at thy death like a thief,
but like a benefactor, because He will take from thee the less,
and give thee the greater; He will take away earth, and give
thee Heaven; He will take away what is transient, and give
thee what is eternal: "He shall appear to them that expect
Him unto salvation."[2]

[1] Prov. xxix. 1.   [2] Hebrews ix. 28.

SEVENTEENTH DAY.

They are laid in Hell like sheep: death shall feed upon them
(Psalm xlviii. 15).

I. Consider first, the great multitude of the damned:
"They are laid in Hell like sheep." They go down thither
in bands: "Gather them together as sheep for the slaughter."[1]
Neither is this to be wondered at, because, since the greater
part of men lead a bad life, it is but reasonable to expect them
to die a bad death. What wilt thou say if ever, which God
forbid, thou shouldst be in that multitude of the lost? Wilt
thou find any consolation in having so many companions in
damnation? What comfort can it be to a sheep to be taken
to the shambles, not alone, but with many others? "Thou
hast multiplied the nation, and hast not increased the joy."[2]

II. Consider secondly, that the same sinners who now bear
themselves so audaciously towards God, that it seems as though
they would, like untamed beasts, shake off the yoke of all His
just commandments, will be at the Last Day so wretched and
powerless, as to be incapable of making even the slightest
resistance to the sentence of their damnation, though they
would fain do so. And this, too, is what the Psalmist means
to express when he says, "They are laid in Hell like sheep."
See how easily a herd-boy takes a large flock of sheep to the
shambles. Even so will the Divine justice send down to Hell
the vast multitude of the reprobate, making those wretched
ones go thither of themselves without a word of appeal.

III. Consider thirdly, that the folly of sinners is such, that
the greater part of them are lost because they will not depart
from the common custom. It is the general excuse, "Every-
body does so;" and thus because they are unable to conquer
a base feeling of human respect, there are countless multitudes
who daily let themselves be allured by companions to games,
dances, riotous feasts, and yet more disgraceful places, going
"to dumb idols, according as you were led."[3] And this, too,
is the Psalmist's meaning when he says of all such persons,
"They are laid in Hell like sheep." Hast thou noticed what
the shepherd does when he sees his flock unwilling to go over
a ditch? He takes one sheep, and forces it to jump over,
then all the rest follow immediately. This is what the devil
does; he incites some one to introduce an evil practice, and

[1] Jerem. xii. 3.   [2] Isaias ix. 3.   [3] 1 Cor. xii. 2.

at once every one imitates him blindly, just like sheep. But do not thou follow the crowd if thou wouldst not perish with them : "Thou shalt not follow the multitude to do evil."[1]

IV. Consider fourthly, that the multitude of those who are every day lost, by their own will, being so great, Hell, vast as it is, will scarcely be able to contain them, when not only their souls, but their bodies are there. Therefore the Psalmist, who foresaw in spirit the manner in which this will be done, says that they will be herded together like sheep : "They are laid in Hell like sheep." Thou knowest how sheep are packed together when the fold is too small. So will it be with the reprobate. Hence thou mayest infer what will be the oppression, the strain, the despairing struggles amongst them ; how some will not be able to endure the weight which stifles them, nor others the pressure of those narrow bounds. See, then, how vain it is for those wretched ones to expect comfort from the multitude of their companions in misery, when it will, on the contrary, be one of their most intolerable torments.

V. Consider fifthly, that this oppression would of itself be enough to cause the death of the damned, if they were in a condition to be subject to it. But as these unhappy beings will not be able to die a second time, they will experience all the suffering belonging to death without any of the relief. And therefore, in conclusion, the Psalmist says that death will consume them gradually, so that it will destroy without killing them : "Death shall feed upon them." The word translated "feed" (*depascere*), describes the way in which cattle bite and pull and tear the grass in a meadow where they are feeding, while still leaving the roots uninjured. This is what death will do, as though it had found its sweetest pasture in the damned : "Death shall feed upon them." It will consume them, but never so as to make an end of them. By death also is here intended every kind of torment capable under other conditions of inducing death, unless, indeed, thou preferrest to understand, as many do, by the word "death," the devil, he being called Death, as having been the author of death, just as Christ is called the Life, because He is the author of life : "And behold a pale horse, and he that sat upon him : his name was Death, and Hell followed him."[2] But whatever this death may be, is it not madness to think so little about escaping from it, that such vast numbers choose rather to follow after it? "Hell followed him."

[1] Exodus xxiii. 2.   [2] Apoc. vi. 8.

EIGHTEENTH DAY.

Whosoever are led by the Spirit of God, they are the sons of God
(Romans viii. 14).

I. Consider first, the mark which the Apostle here gives
thee whereby to know certainly those who are the sons of God.
He says that they are excited to good by the Holy Spirit, and
that as by a higher power which is their master: "Whosoever
are led by the Spirit of God, they are the sons of God." All
the just are moved, guided, governed by the Spirit of God, but
not to all can the word be applied which is translated "led"
(*aguntur*), because all do not allow themselves to be acted on
by Him with the readiness implied by this word. And, therefore, observe that the Apostle does not here say, "Whosoever
are the sons of God, they are led by the Spirit of God," but
"Whosoever are led by the Spirit of God, they are the sons
of God." Those who are thus led are they who are at once
known to be what they are by the prompt obedience which
they render to their Father. How is it with thee as to this
unresisting following God's guidance in all things? Art thou
pliant to His inspirations, or art thou hard, obstinate, and
wilful? If thou art influenced by the spirit of fear in following
them, like the generality of the just, it is a sign that thou art
indocile, and therefore at the most thou art only moved in the
ordinary way, not in that which is described by the word
aguntur. If thou art influenced by the spirit of love, as is
the case with the nobler sort of the just, it is a sign that thou
art docile, and in that case it may be said of thee that thou
art moved not in the ordinary way only, but in that which is
described by *aguntur*. Thou art very evidently a son.

II. Consider secondly, that, at the first glance, these words
may give rise in thy mind to a false notion, namely, that God
compels the just by His grace to do right; but they really
prove the contrary: "Whosoever are led by the Spirit of God,
they are the sons of God;" then, it is perfectly certain that
those who "are led by the Spirit of God" do not thereby lose
a particle of their liberty, for then they would act not as sons
but as slaves. This word *aguntur* in no way means "they are
forced" or "compelled," but "they are borne:" "borne," that
is, as by the strongest natural inclination, making them answer
readily to the impulse that is given. It is the same word
which is used in the passage, "Jesus, being full of the Holy

Ghost, was led (*agebatur*) by the Spirit into the desert."[1] It is not said, "He went," but "was led, or borne," because when the Holy Ghost fills the sails of a man's heart, he not only goes whither he is called by Him, but flies like a ship before the wind. Thou oughtest, therefore, to remember that when God concurs with secondary causes to make them operate, He does so according to what is fit in each case. And so, with those which are necessary, such as plants, trees, or animals, He concurs so as to make them operate necessarily, because this is in conformity with their nature. With those which are free, as with men, He concurs so as to make them operate freely, because this is in conformity with theirs : "God dealeth with you as with His sons."[2] And this is why St. Augustine so well says on this text, that the sons of God are certainly moved by the Holy Spirit, but that "they are moved in order that they may move" (*aguntur ut agant*), just as it is with vessels which are moved by a gentle breeze. They "are moved," but so "that they may move" at the same time, for the breeze only invites them to motion and makes it easy to them. It invites them by the fair weather that it brings, and it makes motion easy by taking a share of the work, but still it does not force them to go on against their will, as a violent tempest would. Whenever the sailors wish to slacken sail and pause, the gentle breeze does not make any resistance, at least any violent resistance. So, too, is it with the Spirit of God : "Oh, how good and sweet is Thy Spirit, O Lord, in all things : "[3] "good," because He always moves men to what is good ; "sweet," because He moves without forcing them. He moves them by enlightening their understanding. And this is, as it were, to invite them by the fair weather He produces in their minds : "Thou gavest them Thy good Spirit to teach them."[4] And He moves them by strengthening their wills, which is by His doing together with them whatever they do, nay, doing far more than they do : "The Spirit of the Lord was their leader."[5] But, though this is to move, it is not to compel them. Do thou rather infer from this that, if in thy case the Holy Spirit does not "move so that thou mayest move," the fault is thine, who allowest that gentle breeze to blow in vain, like those Corinthians of whom it is written, "We, helping, do exhort you that you receive not the grace of God in vain."[6] Neither do thou say

[1] St. Luke iv. 1.   [2] Hebrews xii. 7.   [3] Wisdom xii. 1.
[4] Esdras ix. 20.   [5] Psalm lxiii. 14.   [6] 2 Cor. vi. 1.

that He does not breathe to help thee; for this, too, is thy fault. Call upon Him with all thy heart, and He will breathe on thee. This is the difference between the earthly and the heavenly wind. The former is often vainly asked for by sailors; the latter comes most readily when it is asked for: "I called upon God, and the spirit of wisdom came upon me."[1]

III. Consider thirdly, that there are three degrees of perfection in doing good: to do it in the right way, to do it promptly, and to do it gladly. In the first men are said to be just, in the second spiritual, in the third blessed on earth. And, therefore, the first displays in them the virtues, the second the gifts, and the third the beatitudes. If, then, thou wouldst have a still stronger assurance as to who are the undoubted sons of God, see who they are who show these three degrees of perfection in their works, by doing them, not only in the right way, but promptly, and not promptly only, but gladly. This also was what the Apostle intended to express when he said, "Whosoever are led by the Spirit of God, they are the sons of God." By saying they "are led" (*aguntur*), he has shown that the sons of God are not guided by caprice, like those "that follow their own spirit,"[2] but are entirely guided by the light of reason, subordinate and subject to that of faith: "His justices I have not put away from me."[3] And accordingly they are called just, because they possess both natural and supernatural justice. By saying "by the Spirit," he has shown that they do not move slothfully in doing what is right, as they do who are acted on by a slothful motive power, such as is that of the body, but promptly, as they do who are acted on by an agile, prompt, speedy, and strong motive power, such as is that of the Spirit, and accordingly they are called spiritual, because they are swift to do good: "Whither the impulse of the Spirit was to go, thither they went."[4] And by saying "of God," he has also shown that the spirit which moves them to act is not a gloomy or troubled spirit, but full of delight, as is the Spirit of God: "My Spirit is sweet above honey."[5] And accordingly they are called blessed on the earth, because it is not merely from the words of others that they know how sweet it is to converse with God, but from their own experience: "How sweet are Thy words to my palate, more than honey to my mouth."[6] Enter into

---

[1] Wisdom vii. 7.   [2] Ezech. xiii. 3.   [3] Psalm xvii. 23.
[4] Ezech. i. 12.   [5] Ecclus. xxiv. 27.   [6] Psalm cxviii. 103.

thyself, and examine whether there are these undoubted marks
of sonship in thy daily actions. It is but too possible that
only the first of the three is barely discernible.

IV. Consider fourthly, that if thou hast not these marks,
thou must set about acquiring them. And how is this to be
done? It is the virtues which must dispose thee to act
rightly; both those which belong to man as man, namely, the
moral virtues, and those which belong to him as partaker of
the Divine nature, namely, the theological virtues, and these
thou hast to strengthen by repeated acts of them, thereby
adding acquired habits to those which are called infused: "I
will be employed (or "exercised," *exercebor*) in Thy commandments."[1] It is the gifts, which are called the gifts of the Holy
Spirit, that dispose thee to act promptly; not, indeed, that
they make thee do any different actions from those which are
inspired by the virtues which have just been spoken of, but
they enable thee to perform them with the utmost freedom,
even making thee capable of instantly recognizing and following
Divine inspirations, especially in certain difficult and critical
cases, in which the light of reason alone would be insufficient:
"Thy good Spirit shall lead me into the right land."[2] And
what is it that disposes thee to act gladly? It is to do thy
actions for the love of God, desiring nothing of Him but
Himself only: "What have I in Heaven, and besides Thee
what do I desire upon earth?"[3] For this is what makes thee,
after all, blessed in poverty, in persecutions, in sorrow, and in
all else which was borne by Christ, and which is so opposed
to the world's teaching, namely, that thou art suffering it all
for God, to please God, to give glory to God, and because
thou wilt never depart in anything from the will of God. So
long as thou doest thy actions for any lower end, however
worthy, thou wilt be good, but not blessed on earth. Thou
wilt become blessed when thy actions are done purely for the
love of God: "How great is the multitude of Thy sweetness,
O Lord, which Thou hast hidden for them that fear Thee!"[4]
Thou hast revealed it to them that love Thee, and "hidden"
it "for them that fear Thee," that is, those that fear Thee, not
with a chaste but a servile fear.

---

[1] Psalm cxviii. 78.
[2] Psalm cxlii. 10. [3] Psalm lxxii. 25. [4] Psalm xxx. 20.

NINETEENTH DAY.

He who causeth a sinner to be converted from the error of his way, shall save his soul from death, and shall cover a multitude of sins (St. James v. 20).

I. Consider first, how terrible an evil sin is; it is "the error of the way," a going out of the way. But what way is this? It is the way that leads to Heaven; and this is why the evil is so terrible. For if thou hast gone out of the way that leads to thy native place on earth, still thou mayest chance to come to some city whose inhabitants are kind, courteous, and hospitable, who, stranger as thou art, will give thee a hearty reception. But if thou goest out of the way leading to thy heavenly home thou art lost; there is no other way but that to Hell, and what a barbarous land is that! "A man that shall wander out of the way of doctrine shall abide in the company of the giants."[1] The spies sent by the Israelites to make the circuit of the land of Canaan were so afraid when they came into a country of giants, that they returned in terror, saying, "There we saw certain monsters of the sons of Enac, of the giant kind, in comparison of whom we seemed like locusts."[2] What, then, will it be to find oneself in Hell, dwelling with the devils, who are, indeed, monstrous giants in the fury, ferocity, and arrogance which have been in them ever since they were bold enough to make war on God? Yet there it is that every one who leaves the right way must come, whether he "wander out of the way of doctrine" in faith or in morals. How is it with thee in this respect? Art thou unhappily out of the right way? If so, stop; and think, as is reasonable, about saving thy own soul before occupying thyself with others. Do not meditate further on the passage here set forth by St. James, for it is not intended for thee. How canst thou encourage others to turn back into the right way, whilst thou art going on in the wrong one? "Thou that teachest another teachest not thyself."[3] First think about getting back into the right road thyself, ceasing at least from giving bad example, as thou hast hitherto done, and then urge others also to return: "He that heareth, let him say, Come."[4]

II. Consider secondly, that as thou canst not expect to turn others back from the evil way so long as thou art walking in it thyself, thou mayest expect it when thou art in the right

[1] Prov. xxi. 16.
[2] Numbers xiii. 34.		[3] Romans ii. 21.		[4] Apoc. xxii. 17.

way, and therefore thou shouldst strive to do so. And who can tell the good that thou wilt then do? Thou wilt save thy neighbour's soul from death; and from what a death! A two-fold death, which robs the soul of a two-fold life, the life of grace and the life of glory. And do not regard the fact that the evil of this death is not visible to the eyes of thy imagination; it is enough that it is visible to those of faith: "She that liveth in pleasures is dead while she liveth."[1] If thou wouldst understand what is the state of the soul without God, Who is its life, consider the state of the body without the soul. When it has lost the soul the body has no longer the power of motion in any part; it has neither colour, nor beauty, nor strength, nor existence, and gradually it decays so as to infect the air, making even those once dearest to it fly from it. This too, nay, far worse than this, is the state of the soul which has lost its God; and the body is unconscious of its misfortune in having lost the soul, whereas the poor soul, on the contrary, which has lost God, if it is not immediately conscious of it, will be so when it awakes, as it were, from the sleep in which it is now wrapped. It will then see what it is to be dead as regards the loss of God, and immortal only so far as is necessary in order to feel the ruin, the misery, the fury, the anguish, the utter despair born of so great a loss. Dost thou not see how great a thing it is to save the soul of thy neighbour from such a death? It is something far different from being a saviour like Othniel, or Josue, or Gideon, and others like them, who, by their valour, preserved the life of their countrymen's bodies; it is to be a saviour after the very pattern of Jesus Christ, Who, by His words, gave life to the soul: "Saviours shall come into Mount Sion."[2] The other saviours stood, as it were, at the foot of Mount Sion to guard it for Jesus, Who was to come and set up His glorious throne there, that is, the pulpit: "But I am appointed King by Him over Sion, His holy mountain, preaching His commandment."[3] These saviours have gone up to preach together with Jesus: "For we are God's coadjutors."[4]

III. Consider thirdly, how, although mere charity ought to be sufficient to urge thee to help those who have gone astray, and to recall them from the way which leads to so terrible a death, God has, nevertheless, willed that thy charity shall not be without its reward. And, therefore, He tells us

---

[1] 1 Timothy v. 6.
[2] Abdias i. 21.　　[3] Psalm ii. 6.　　[4] 1 Cor. iii. 9.

that "he who causeth a sinner to be converted from the error
of his way" shall not only save the soul of his neighbour
"from death," but shall also cover the many sins which he
has himself committed : he "shall cover a multitude of sins."
I say, which he has himself committed, because although our
version only has *peccatorum*, without the addition of *suorum*,
it is the general opinion of sacred interpreters that the pronoun
is understood ; and many bishops have even expressed it when
quoting this text in their pastorals to urge men to join them
in working for the salvation of souls. And is not this a
surpassing reward? Herein are fulfilled the words of Job,
"The blessing of him that was ready to perish came upon
me,"[1] because the good which thou doest to thy neighbour
when he is on the point of perishing returns to thyself. It is
true, indeed, that when it is here said, "He shall cover a
multitude of sins," the sins spoken of may be either past or
present. Past sins are "covered" as to the debt which
remained to be paid for them in Purgatory, and present sins
are "covered" as to the guilt also. For if they are mortal
God is frequently moved by this act of charity to give grace
to detest them and to repent of them, and so to obtain their
remission directly ; and if they are venial, He is moved by this
act to remit them on the instant : "Before all things, have a
constant mutual charity among yourselves, for charity covereth
a multitude of sins."[2] At all events, thou mayest hope that
God will not punish thee for them with those spiritual punishments which are so much to be dreaded. For thou knowest
that when thy venial sins are very numerous, God, even if He
is not so angry as to turn away His face from thee, does at
least deprive thee of a thousand favours which He would
otherwise bestow on thee, either by giving thee more efficacious
aids in loving Him, or by preserving thee from temptations,
or by protecting thee from trouble, or by visiting thee in time
of prayer. Now, the act of charity which thou art performing
by helping thy neighbour seems to have the effect of rendering
God blind, as it were, to the venial sins that are in thee, and
making Him treat thee incomparably better than thou wouldst
deserve but for that act of charity. And this is what the
Apostle seems to wish thee to understand when he says, "He
who causeth a sinner to be converted from the error of his
way, shall save his soul from death, and shall cover a multitude
of sins." It may, indeed, be said that the just man (whose

---

[1] Job xxix. 13.   [2] 1 St. Peter iv. 8.

characteristic it is to devote himself to the salvation of others) covers "a multitude of his sins," because he will mend of them, thus at all events diminishing their number, their "multitude," by means of the abundant grace which he will obtain of God for his sanctification, so that, even if he has some slight sins, they will not be many. And this is the true covering of sins, that which is obtained from God by virtue of sanctifying grace: "Thou hast covered all their sins."[1] For there is a great difference between the way in which we cover our sins by acts of charity and that in which God covers them by sanctifying grace. We cover them by acts of charity to our neighbour, as it were, with a scarlet cloth, which does indeed hide the wounds, so that they do not excite horror, but which leaves them there; God covers them by sanctifying grace as with a life-giving ointment, which not only covers, but heals the wounds: "Blessed are they whose iniquities are forgiven, and whose sins are covered."[2] And this, too, thou wilt obtain if thou makest it thy business to bring back from their errors either those who are in danger, or who have fallen.

IV. Consider fourthly, that the proximate, and, so to speak, the direct method of bringing others back from their errors, is undoubtedly that of preaching, rebuking, counselling, warning, and, above all, of giving good example. There is, however, another, which is remote and indirect, and it is that of praying for those who are employed in the exercise of the first method. And therefore, as thou seest, the Apostle does not merely say, "He who converteth a sinner," but "He who causeth a sinner to be converted from the error of his way, shall save his soul from death, and shall cover a multitude of sins;" for all cannot equally devote themselves to bring back those in error to right faith or right morals, but all may at least help those who are bringing them back, just as those persons do who from the shore are watching sailors, casting to the shipwrecked planks, poles, and ropes, and who pray to God to prosper their zeal: "For the rest, brethren, pray for us, that the word of God may run and be glorified even as among you."[3] And why canst thou not pray for these wanderers themselves, and win their return from God? This is the surest method, if not also the most meritorious. For the man who treats with sinners of the conversion of sinners often spends his labour in vain, but he who does so with God always obtains his end. What excuse hast thou, therefore, if,

[1] Psalm lxxxiv. 3. [2] Psalm xxxi. 1. [3] 2 Thess. iii. 1.

not being able to cross mountains in order to bring back vast numbers of wanderers who are rushing to ruin, thou dost not beg of God to open their eyes to perceive their danger before the night comes when the time will have passed in which it was possible to turn back, and nothing but destruction is before them? "Pray one for another that you may be saved: for the continual prayer of a just man availeth much."[1]

TWENTIETH DAY.

Drop down dew, ye heavens, from above, and let the clouds rain the Just: let the earth be opened, and bud forth a Saviour (Isaias xlv. 8).

I. Consider first, that this blessed earth here spoken of is, according to the strictest interpretation, no other than our Lady, that inviolate Virgin and immaculate Earth, from which, without human agency, there sprang that Divine Bud, desired by Isaias so long before, when he exclaimed: "Let the earth be opened, and bud forth a Saviour." This being so, it will at once strike thee as strange that this form of expression should be used. For, if the Earth here spoken of was so inviolate as to remain as much closed during as before child-birth, how can the Prophet pray that it might be opened? "Let the earth be opened." But observe to Whom it was to be opened. To Him Who was able to come forth, and yet leave it inviolate. Wonder, then, no longer. A window is said to be open to the light when the shutters are removed and only the panes remain, although it is all the time closed to water, air, and to all the animals which may go about trying to get in. And why is it said to be open to light? Because the light is able to penetrate notwithstanding. Therefore, because the Divine Incarnate Word was able to pass through Mary's virginal womb, as light passes through crystal without injuring it—"For Sion's sake I will not hold my peace, till her Just One come forth as brightness "[2]—it might well be said to be opened for Him when He left it, since it is well known that everything which can be penetrated by the power of any one is said to be open to Him. Observe, therefore, that the first thing the Prophet asks is not that the earth may be opened, but that the heavens should rain—"Drop down

[1] St. James v. 16.   [2] Isaias lxii. 1.

dew, ye heavens, from above, and let the clouds rain the Just;
let the earth be opened, and bud forth a Saviour"—because
the favourable influences must first come from Heaven, and
then the earth must give her fruit: "The Lord will give
goodness, and our earth shall yield her fruits."[1] Thy heart is
closed earth, not because it is virgin, but because it is barren
and dry, yielding no fruit of devotion. Wouldst thou know
what is the real reason of this? It is that thou hardly ever
lookest up to heaven: "He prayed; and the heaven gave
rain, and the earth brought forth her fruit."[2]

II. Consider secondly, that Christ was to be conceived by
Mary purely through the operation of the Holy Ghost, and
therefore the Prophet here exclaims with uplifted eyes: "Drop
down dew, ye heavens, from above, and let the clouds rain
the Just." He invites the Divine Spirit to descend into Mary's
virginal womb, and to make it fruitful, so as in due time, like
chosen earth, to bring forth the blessed Fruit of the Incarnate
Word, by which we were to be saved. And if thou askest
why the Incarnation of the Word is compared to dew rather
than to any other rain, it is because the silence of His coming
was in proportion to the greatness of the salvation He brought.
Mary was seen to have conceived before it was possible to
know the manner of the conception: "Before they came
together, she was found with child of the Holy Ghost."[3]
Therefore it is that all will not equally experience the good
effects of this dew; but it is just as it was with that which fell
on Gideon's fleece: the first night the fleece was wet and not
the floor round it; and the second night the floor was wet and
not the fleece. So the first to benefit by Christ's coming were
the Jews, the rest of the world remaining dry; and afterwards
the rest of the world received the benefit and the Jews remained
dry: "To you it behoved us first to speak the word of God,
but because you reject it . . . we turn to the Gentiles."[4] Do
thou return lively thanks to God that thou art in a place
where this dew has descended most abundantly? but if as
yet thou hast gathered no fruit in consequence, what does
this show? It shows that thy heart is not of earth, but of
stone.

III. Consider thirdly, that Jesus is here, by antonomasia,
called "the Just"—"Drop down dew, ye heavens, from above,
and let the clouds rain the Just"—for to Him alone can that

---

[1] Psalm lxxxiv. 13.
[2] St. James v. 18. [3] St. Matt. i. 18. [4] Acts xiii. 46.

name be given. Every one who is holy can be called just, but not "the Just," for when we call any one just, it implies accidental justice, so to speak; but when we say "the Just," essential justice is implied. And there never has been essential justice save in Christ, Who, for that reason, is called Justice itself: "Who is made unto us . . . Justice."[1] In all others justice has been accidental, because it was possible for it to exist and not to exist in them; whereas it was essential in Christ, because in Him it was not possible for it not to exist; and if it is in others by grace only, it is in Him by nature. See, therefore, how well the Prophet here says: "Drop down dew, ye heavens, from above, and let the clouds rain the Just," because there were many just men at that time on the earth, but not "the Just." "The Just" had yet to come: "They have slain them who foretold of the coming of the Just One."[2] And when He came, whence could He come save from Heaven? Hence it is that, as there are in Christ two Natures, the Human and the Divine, it is with reference especially to the latter that the Prophet said, "Drop down dew, ye heavens, from above, and let the clouds rain the Just;" and with reference to the former, "Let the earth be opened, and bud forth a Saviour," because, if Christ was not just only, but "the Just," this was so because of the Divine Nature, to which sanctity is essential—"One is good, God"— and if Christ was a Saviour, this was not only by the Divine but by the Human Nature, which gave Him power, as our Head, to infuse salvation into us, in the same way that Adam, as our head, had infused perdition into us. In receiving Him as a Saviour, thou hast certainly to reverence, to thank, and to love Him; but in receiving Him as "the Just," this is not enough: thou hast also to imitate Him. Indeed, why shouldst thou not imitate Him as Saviour also, if so great a grace is given thee? But take heed, for the first glory which is here given to Him is that of being Just, and then that of Saviour; how, then, canst thou presume to reverse the order?

IV. Consider fourthly, how great is the salvation which this Saviour comes to bring thee. It is as great as the evils from which He comes to set thee free. There are two of these: the evil of guilt, and the evil of penalty. But in these two fatal species, how many separate evils are contained! Think on it thyself, if thou hast courage to go through the catalogue. And yet from all these this life-giving Bud is to

[1] 1 Cor. i. 30. [2] Acts vii. 52.

save thee: "I will raise up for them a Bud of renown (*nominatum*), that is, "named" or foretold for so many ages, "and they shall be no more consumed with famine in the land, through the want of all good, "neither shall they bear any more the reproach of the Gentiles."[1] See, therefore, that this great Saviour is not compared to a precious metal hidden in the earth, but to a bud, which springs from it spontaneously—"Let the earth be opened, and bud forth a Saviour"—in order to show thee that thou art not obliged to labour to find Him, but that He is to come of His own free choice to find thee, so great is His desire for thy happiness. How easily, then, mayest thou derive all possible good from Him by approaching to receive Him! Is it not very easy for thee to gather a bud from the earth? Just as easily mayest thou receive thy Saviour from Mary's arms, if thou approachest Him full of sorrow to lay before Him thy miseries, and to beg Him devoutly to deliver thee from them. "The earth shall yield her increase," or "bud" (*germen*), that is, Jesus; and His people "shall be in their land without fear" of their infernal enemies: "and they shall know that I am the Lord when I have broken the bonds of their yoke," that is, sin, "and shall have delivered them out of the hand of those that rule over them,"[2] that is, of their irregular appetites.

TWENTY-FIRST DAY.

ST. THOMAS THE APOSTLE.

Blessed are they that have not seen, and have believed (St. John xx. 29).

I. Consider first, that beatitude is, as it were, the centre in which the heart is at rest. And therefore thou canst not understand how it is that he who believes and does not see is here called "blessed" by Christ. For the more any one believes, the more does he desire to see that which he believes, as it is written, "Abraham desired that he might see My day;" and therefore such a one is not at rest. He is at rest who sees that which, believing, he desired to see, for then desire is turned into joy, as it was in Abraham: "He saw it, and was glad;"[3] therefore it is he who sees, not he who believes, that

---

[1] Ezech. xxxiv. 29. [2] Ezech. xxxiv. 27. [3] St. John viii. 56.

is blessed. But here thou hast to remember that, as has been
said more than once already, that there are two kinds of
beatitude : one in possession (*in re*), and another in hope
(*in spe*); one in fruit, the other in flower; one perfect, the other
imperfect. And therefore, though he who believes is certainly
not blessed *in re*, because he does not yet see that which he
believes, he is, at all events, blessed *in spe*, because by believing
it he is preparing himself to see it, which was Abraham's case.
He who sees is blessed *in re :* " Blessed are the eyes that see
the things which you see ; "[1] but this beatitude is reserved for
us in the next life, in the land of ripened fruits.   In this life,
when flowers are but opening, we must be content to enjoy
the beatitude *in spe*, which, although imperfect, is nevertheless
called beatitude, because a good which is hoped for with great
confidence is already half possessed.   And thou knowest that
the Apostle speaks of joy as accompanying hope, although it
belongs properly to a present good : " Rejoicing in hope."[2]
And why is this?   Because the hope of a true believer is so
certain, that if it does not give Paradise, it at least gives a
foretaste of it.   This, then, is the reason why Christ said,
" Blessed are they that have not seen, and have believed."
The reason is that sight is the reward answering to faith ;
and who can more surely promise himself sight than he who
believes, if he believes aright?   " Blessed are they that have
not seen, and have believed," is said just as it is said, " Blessed
are the poor, the meek, the merciful," and the rest, because
of the certainty which they all have of the reward answering
to these great virtues, if only they are faithful in practising
them.

II. Consider secondly, that if the beatitude proper to this
life is not to see but to believe, thou mayest think it better
for thee to care nothing about knowing how right, how
good, how beautiful, how worthy of belief that is which thou
believest, but to believe it blindly in thy prayer, without
meditating on it or entering into it, as though everything
which is added to sight were taken from faith.   But dost thou
not think that all the rest of God's servants understood, like
thyself, that the beatitude proper to this life is not to see but
to believe?   And yet all, or nearly all, of them have always
done their utmost to understand rightly what they believed :
" I am Thy servant, give me understanding that I may know
Thy testimonies ; "[3] not only "that I may believe," but " that

[1] St. Luke x. 23.   [2] Romans xii. 12.   [3] Psalm cxviii. 125.

I may know." If thy argument were valid, then it would be necessary, in order to increase the merit of the faithful, to leave in the Church two things only : ignorance and faith. And what would better please her enemies, who have always been attacked and destroyed by faith, certainly, but by faith allied with knowledge ? It is necessary, too, to consider to whom Christ was speaking when He said, "Blessed are they that have not seen, and have believed." It was to the doubting Thomas. It is one thing to seek reasons for believing, and another to believe, and, because of that belief, to seek all the more eagerly reasons for understanding how right, and good, and beautiful, and increasingly worthy of belief that which we believe is. The first of these two things is what Christ condemned in Thomas, and in him in all who will not believe without seeing : "Except I shall see I will not believe ;" the second is what has been always done by almost all the servants of God. They have vied with each other in seeking reasons to prove the truths that they believed, to elucidate and to establish them, as gold is proved by being tested. But it was not want of faith that was their motive, but love for the faith. This is what thou, too, oughtest to do in the state thou art in, praying God to grant to thee, also, in thy prayer that bright light which beams from His countenance : "Make Thy face to shine upon Thy servant, and teach me Thy justifications."[1] Therefore it is that the gift of understanding answers to faith, that he who believes may also seek to understand how right it is to believe.

III. Consider thirdly, that the devil deceives thee by leading thee to imagine that the merit of thy faith is lessened by so many reasons for it. This would indeed be the case if thy faith were greater or less in proportion to the force with which these reasons impress thy mind. But thou art bound to believe "above all things," as one who believes in God ; that is, to believe in such a manner that thy faith is unaltered when all reasons are dim to thee, and thou art left in darkness. "Evening, and morning, and at noon I will" alike "speak and declare"—that is, I will "speak" the great things that God hath done for my soul—and I will "declare" those that He hath promised. Besides, dost thou suppose that the merit of the faith of St. Gregory, St. Ambrose, St. Augustine, and all the other holy Doctors, was lessened by the great light that they had ? Rather, it was increased. For every one

[1] Psalm lxviii. 135

who well understands what he believes is naturally inclined
to love it the more. Therefore, if in such a case faith has
less merit on the one hand, it has more on the other—less by
reason of the easiness, more by reason of the love. And
thou knowest that faith is the more worthy in proportion
to the greatness of the charity which, as it were, gives it life.
And what more enkindles charity than a very bright light?
" O house of Jacob, come ye, and let us walk in the light of
the Lord."[1]   Do not, then, abuse the words of our Lord to
St. Thomas, " Blessed are they that have not seen, and have
believed," by condemning him who not only believes in his
prayer, but seeks also to understand, for it was not to such
a one that He addressed them.   He addressed them to the
man who refuses to believe anything but what he understands.
The words which apply to him who seeks to understand as
well as to believe are rather, " Blessed are the eyes that see
the things which you see."[2]   For what greater beatitude can
there be on earth than that of resembling, as it were, the
blessed in Heaven?   " We are happy, O Israel, because the
things that are pleasing to God are made known to us."[3]   If,
therefore, God does not give thee that kind of beatitude which
is almost *in re*, do thou rest fully satisfied with that which is
only *in spe*.   But if He does give it thee, thank Him for it.

IV. Consider fourthly, that God knows very well what is
best for thee.   Therefore, if thy state is one in which thou
art not capable of understanding what thou believest, because
of the habitual dimness of thy mind either from ignorance
or infirmity, or because God chooses thee to be in darkness
in order to try thee, then thou shouldst apply to thyself these
words, " Blessed are they that have not seen and have
believed," as though they were especially addressed to thee.
This is a great grace granted to us by our God.   It is His
will that the faith required of us should not consist in understanding, but in acquiescing in the truths He has revealed.
If it consisted in understanding them, what would become of
all those Christians who have neither means, nor talents, nor
time to do so.   It is enough for one who does not understand
them to submit his mind to what has been believed by all
those holy Doctors who have understood them, and God is
satisfied : " The oxen were ploughing, and the asses feeding
beside them."[4]   Humble thyself, therefore, by applying these

---

[1] Isaias ii. 5.
[2] St. Luke x. 23.   [3] Baruch iv. 4.   [4] Job i. 14.

words to thyself, and think that if it is for these learned men to spend their labour in continually cultivating, digging, ploughing, and preparing the fields of the Church to receive the seed which God then casts into it, it is enough for thee not to suffer thy thoughts to wander away from them, although thou art at rest all the time that they are toiling so hard. Is it not a great advantage for thee that when thou art incapable of understanding, God only requires of thee to believe? Whenever, therefore, the devil troubles thee with temptations against faith by setting before thee the difficulty of the mystery to which thou givest thy assent, do thou confound him by saying immediately, " Blessed are they that have not seen and have believed." So shalt thou put him to flight. And dost thou not perceive that the very fact of thy not understanding what God says is a reason the more for gladly believing it? " Behold, God is great, exceeding our knowledge."[1] How could a God be great Whose power and knowledge and wisdom and Providence in governing the world did not surpass the human understanding? There are so many marks to prove that God is the Author of the Christian religion, if only thou art attentive to them, that to doubt it would be the height of folly. Seek, then, no further; think only that thou must believe. Not that thou mayest not think very often about these remarks; thou art free to think of them, only do not make them a reason for believing, but for pitying the blindness of those who do not believe. And are they not, indeed, most pitiable? The city of salvation is always before their eyes, " seated on a mountain,"[2] and yet these miserable men are not ashamed to inquire of every one they meet where it is : " Many say, Who showeth us good things?"[3]

TWENTY-SECOND DAY.

Who shall give Thee to me for my Brother, sucking the breasts of my Mother, that I may find Thee without, and kiss Thee, and now no man may despise me? (Cant. viii. 1).

I. Consider first, that the object which the soul longs to attain in prayer is that closeness of embrace, that intimate union with her God, which is so often denoted in the Sacred Scriptures by the name of a chaste kiss. But all do not

[1] Job xxxvi. 26.   [2] St. Matt. v. 14.   [3] Psalm iv. 6.

receive this in the same way. Some, before attaining to
finding their God in prayer, have gradually to penetrate by
thought into the recesses of some one of those mysteries in
which, as it were, He is hidden, to meditate, to seek and
follow on His track, till at length God is moved to pity by the
labour they have endured, to admit them to Himself by means
of some word of unusual sweetness, some light of unwonted
brightness, making them feel the Divine presence in the
depths of their being and unite themselves to it.   These do,
doubtless, attain to the finding of God, but, as it were, in His
palace : they "find Him within."   And so the audience which
God grants them is like that which all sovereigns give on
state occasions, that is, after passing through a long succession
of apartments.   Others no sooner kneel down to pray than
they find God, so to speak, on the threshold : they "find Him
without ;" because, without any long preliminary reasoning,
at the very first elevation of the mind, they are at once united
with Him, their affections are present, the embrace is immediate, their tears are ready ; they have no labour to go through
in order to be admitted to the audience they so greatly desire.
This is the grace given to those who are raised by God to the
sublime gift of contemplation.   And it is this which the soul
mystically asks of God in these words : "Who shall give
to me . . . that I may find Thee without and kiss Thee?"
But now observe what soul it is that asks such great things.
It is the heavenly bride who, according to the Hebrew phrase-
ology, here calls the Bridegroom, not by the name of Spouse,
but of Brother, because they were of the same tribe.   And
yet even such a soul as this does not claim this favour as
due to her by right, but says, "Who shall give to me?"   And
dost thou, who hast scarcely risen out of the mire of thy sins,
venture to claim it for thyself, and disdaining the trouble
of meditation, desire instantly, by one act of faith recited
before thy prayer, to throw thyself into the arms of God, and
to enjoy Him in the sweetness of that contemplation which
is so delightful because it is found without seeking? Oh,
how greatly art thou deceived! Say first, "Who shall give
to me?"   Pray, knock, declare to God that thou art unworthy
of being honoured by a single glance of His ; and then, after
all, know that thou art still uncertain of receiving the gift of
prayer which thou desirest, because it is entirely gratuitous,
and although thou mayest hope for it, if thou strivest earnestly
to obtain it, thou canst never claim it.

II. Consider secondly, that a soul which has received such a grace, knows perfectly in the act of receiving it that no one can despise her—"Now no man may despise me"— And why? Because there is no creature that can venture to tempt her, by any offer, to separate herself from her God. What is the most contemptuous treatment that thy soul can possibly endure? It is that which the world inflicts on thee by inviting thee to follow after its vanities; it is that which the flesh inflicts on thee when it invites thee to indulge in its pleasures, luxuries, and amusements; it is that which the devil, thy chief enemy, inflicts on thee when he invites thee to imitate his ambition. Is not this an unheard-of contempt? When, therefore, the soul is united to God in the manner which has been described, who, she says, will ever be audacious enough to despise me so far as to tempt me to forsake that good to which I am joined? "Who shall separate us from the love of Christ?"[1] Riches, rank, pleasures, dignities are empty honours. Let who will have them; I value them not. She sees that she is now treated as a spouse by her Beloved, so tenderly does He caress her; and so she no longer dreads the idle words of those rivals who, before she was admitted to these heavenly nuptials, treated her with scorn, as though it were impossible for her ever to be so favoured. Now, in what state art thou? It may be that many of thy companions now despise thee when they see thee devote thyself so studiously to prayer, sometimes going so far as to inquire jestingly what degree of ecstatic contemplation thou hast reached by this time. Let them talk; for if by faithfully persevering in thy efforts, thou attainest to that which the bride here sighed after, thou wilt see that, even without ecstasies, the time will come when no one can scorn thee. What is there that people of the world will not endure to accomplish earthly nuptials? And wilt thou submit to nothing for the sake of those that are heavenly? But when will that time come? When, as soon as thou hast entered on thy prayer, thou art able to say to God with all thy heart, "Thou art here, and Thou art enough." This is the union which fears contempt from no one.

III. Consider thirdly, that the bride does not say here merely, "Who shall give to me," in any way, "that I may find Thee without, and kiss Thee, and now no man may despise me?" but she says more decidedly, "Who shall give

[1] Romans viii. 35.

Thee to me for my Brother, sucking the breasts of my Mother?" For, when she sees her Spouse on His lofty throne of glory, on which He is now reigning, it is as though she could not bring herself to hope for so close and sweet a union with Him as is expressed here by a "kiss." And so she thinks of Him as He was when an Infant in the arms of the Blessed Virgin Mary (whom, according to the usual custom of addressing the bridegroom's mother by the same title, the bride here calls her Mother), and, as such, she desires to clasp Him in her arms. And why is this, but that she may yet be able more freely to pour out her devout tenderness upon Him? And, therefore, Christ has appeared in this form to countless saints, perhaps more frequently than in any other, that they might enjoy Him with greater familiarity; for no one is deterred by any feeling of awe from clasping to his breast a sucking babe, from caressing, embracing, and covering him with kisses, as it is natural to do to so tender and sweet a creature. From this we may learn how greatly those persons were mistaken, who have asserted that it was an injury to the purity and perfection of contemplation to represent to one's imagination the Sacred Humanity of our Saviour, and that therefore we should always withdraw from all sensible ideas, banish every sort of form and figure, and fix the mind in a purely intellectual contemplation. But the bride, who is here speaking, is the model of a holy soul, and what is it she says? In the very act of desiring her Spouse to come to her in sublime contemplation, without any labour of meditation and of seeking Him, she desires to see Him as a Babe, and a sucking Babe on the breast of His Mother, as on a throne of grace: "Who shall give Thee to me for my Brother, sucking the breasts of my Mother, that I may find Thee without, and kiss Thee, and now no man may despise me?" One of the principal objects that God had in becoming Incarnate was that we might find it easier to unite ourselves to Him when we see Him made one of ourselves.

TWENTY-THIRD DAY.

Let us go with confidence to the throne of grace, that we may obtain mercy, and find grace in seasonable aid (Hebrews iv. 16).

I. Consider first, that Christ as a true King has two thrones, one of grace, the other of justice. He will sit upon the latter when He comes to judge us at the end of our life; He sits on the other so long as we live. One is future, therefore, the other present. He is now seated on the throne of grace to give to every one what is rightly asked of Him: "Ask, and you shall receive." And He will sit on the throne of justice to give that only which has been merited: "I will judge thee according to thy ways, . . . and My eye shall not spare thee, and I will show thee no pity."[1] What folly is thine, then, if thou dost not go to the throne of grace while thou mayest, but waitest till thou art summoned to that of justice. Therefore the Apostle says, "Let us go with confidence to the throne of grace;" because every culprit hastens at once to the throne of grace, whereas no one goes unsummoned to that of justice. Is it not a shame, then, that God should be obliged to invite thee to have recourse to Him, to lay all thy wants freely before Him? Thou art a culprit; but what then? If thou wert obliged to go to the throne of justice, then, as a culprit, thou wouldst have good reason to tremble at going, and to say to God, "Enter not into judgment with Thy servant;" but as it is to the throne of grace that thou hast to go, why, guilty as thou art, shouldst thou hesitate: "He shall give equal grace with grace,"[2] that is, He will give grace by redeeming thee equal to that which He gave by creating thee.

II. Consider secondly, what are the ends to obtain which we have to go to this throne. They are two: first, to gain pardon for the evil that we have done; secondly, to win grace proportionate to the good that we have to do. Therefore the Apostle says, "That we may obtain mercy, and find grace in seasonable aid." The pardon of evil is ascribed to mercy, which finds us in a state of misery so great as that of sin is, and releases us from it: "In My reconciliation have I had mercy on thee."[3] And therefore it is said with reference to this pardon: "That we may obtain mercy." The giving us strength to do good is attributed to grace: "We have grace,

---

[1] Ezech. vii. 3.   [2] Zach. iv. 7.   [3] Isaias lx. 10.

whereby let us serve, pleasing God, with fear and reverence:"[1] with "fear" as serving a Master; with "reverence" as serving a Father. And therefore it is said with reference to this strength, "That we may find grace in seasonable aid." We can obtain neither of these favours by the way of merit; not the remission of evil, that is, of sin, because so long as we are in sin we are not yet capable of meriting, being enemies of God: "The Highest hateth sinners;"[2] not the grace necessary for doing good, because although the end of merit, which is the glory of God proposed to us to merit, is a subject of merit, yet the beginning of that merit, which is grace, cannot be a subject of merit: "If by grace, it is not now by works, otherwise grace is no more grace."[3] What remains, therefore? That we should obtain it by earnest prayer: "Let us go with confidence to the throne of grace," that is, "with confidence of speech," as it is in another reading; because the obtaining a favour by way of supplication does not rest on the dignity of him who offers the petition, but on the kindness of him who receives it: "For it is not for our justifications that we present our prayers before Thy face, but for the multitude of Thy tender mercies."[4] Knowing, then, the importance of having recourse to this throne for two such great ends, how is it that thou dost not hasten to it? It is a plain sign that thou dost not value those ends if thou despisest the means.

III. Consider thirdly, what is the chief thing thou art to do in order to excite in thyself this confidence in asking of God with great liberty what thou needest for thy soul's good. Thou must penetrate deeply into the knowledge of thy nothingness: "Without Me you can do nothing."[5] It is certain that of thyself thou canst do absolutely "nothing;" thou canst not rise out of the evil into which thou art fallen, still less canst thou do the least good; and yet thou art under the strictest obligation to do that which of thyself thou art unable to do. What, then, dost thou fear? Canst thou imagine that when thou hast recourse to the goodness of thy God to beg Him to help and support thee, and to grant thee what thou hast need of in order to obey Him, He will not hear thee in due time? If, in such a case, God were not most ready to hear thee, He would both impose commandments on thee, and inspire thee with counsels beyond

---

[1] Hebrews xii. 28.   [2] Ecclus. xii. 3.
[3] Romans xi. 6.   [4] Daniel ix. 18.   [5] St. John xv. 5.

thy strength. And is it possible for thee to fear this from so good a God? "Let us go with confidence to the throne of grace;" for although God is not bound to give us anything independently of His own Divine promises (and therefore it is always true that what He gives us He gives of grace), yet He cannot fail to give it to us, not only by virtue of those promises, but of the commandments and counsels by which He obliges or invites us to serve Him truly. Speak boldly, then, "with confidence," and ask help of God. But what sort of help? that which He knows to be "seasonable." This is what is needed, and therefore it is this which thou hast always earnestly to ask of Him: "Let us go with confidence to the throne of grace, that we may obtain mercy, and find grace in seasonable aid;" not only "in seasonable time," that is, the time of this life, the only time when the tribunal of grace is open to us—"Behold, now is the acceptable time"— but also "in seasonable aid;" because it is not every kind of aid that is always equally seasonable for thee. That which is seasonable is that with which God foresees that thou wilt not neglect to correspond; and this is what thou art to ask continually of God, that thou mayest repent of evil and do good.

IV. Consider fourthly, that the knowledge that thou canst do nothing of thyself ought certainly to give thee great courage to hope in God, as has been said, and to beg seasonable aid of Him for everything which He now enjoins on thee, or only inspires thee to do. But what ought still more to give thee this courage is the knowledge that God commands thee to hope by a distinct precept: "Hope in thy God always."[1] So that, if thou dost not hope, no matter how great may be thy demerit or sin, thou offendest Him most grievously, and He counts thee at once among those who are the most odious of rebels, those, namely, who are guilty of high treason: "Woe unto them, for they have gone in the way of Cain."[2] What then wouldst thou have more? "Let us go with confidence to the throne of grace." If thy sovereign were to signify to thee that whenever thou shalt despair of his favour by saying, "My iniquity is greater than that I can deserve pardon,"[3] he will be so angry that he will regard thee and treat thee as a rebel, by banishing thee for ever out of his sight, wouldst thou ask for any further reason for trusting in him? Why, then, dost thou do so with regard to God? Has He, in

[1] Osee xii. 6. [2] S. Jude i. 11. [3] Genesis iv. 13.

Heaven, ever broken faith with any one? "Behold the generations of men, and know ye that no man hath hoped in the Lord, and hath been confounded."[1] If, then, thou hopest, canst thou imagine that thou wilt be the first to be confounded? It is enough that thou art one of those who hope, not of those who presume. And who are these last? They are those who expect to be saved without taking any trouble. Listen to what the Apostle says here: "Let us go with confidence to the throne of grace, that we may obtain mercy, and find grace in seasonable aid." If the whole favour has to consist in "seasonable aid," then we, too, have something to do on our side in order to be saved, otherwise we should be expecting not "aid," that is, help in the act we are doing, but exemption from doing anything, and this is granted to none: "Ought not Christ to have suffered these things, and so to enter into His glory?"[2] See, then, whenever thou really desirest aid, not exemption from good, how high thou hast to soar on the wings of hope. Thou hast to say to God that thou wilt hope in Him without doubting, because He enjoins this on thee; but that even though He were to cease to enjoin it, thou wouldst, nevertheless, continue to hope in Him as before, merely because of thy belief in His goodness. This is treating Him as the Master He is, gracious beyond thought: "Although He should kill me, I will trust in Him." So, too, art thou to say, if thou wouldst treat Him as He deserves; but, in order to show that, at the same time, thou wilt not cease to do what is right on thy side, thou must immediately add: "But yet I will reprove my ways in His sight, and He shall be my Saviour."[3]

TWENTY-FOURTH DAY.

Keep fidelity with a friend in his poverty, that in his prosperity also thou mayest rejoice (Ecclus. xxii. 28).

I. Consider first, that so long as a man is prosperous he cannot distinguish true from false friends, because both the one and the other alike pay him respect. In order to distinguish them it is necessary, although this is to his cost, for his fortunes to change, and for him to become poor instead of prosperous, when he least expected such a change: "A

[1] Ecclus. ii. 11.   [2] St. Luke xxiv. 26.   [3] Job xiii. 15.

friend is known in adversity."[1] Think, then, that this was one of the principal reasons why the King of glory, if we may use the expression, is now changing His state, and is about to descend from the height of His majesty to be born in a stable. He desires to prove clearly the fidelity of those who love Him. Oh, how many of those who worshipped Him so long as He was pouring riches on His people from His throne, will so despise Him when they see Him lying naked, cold, and weeping in the manger, as even to declare that they do not know Him! What wilt thou do? Thinkest thou that thou art firmly resolved to stay with Him? to cling to Him in this state of utter poverty? Blessed art thou if so it is. Thou mayest be quite certain that when the day comes on which He will sit down again on the throne which He has now left, none will be more liberally rewarded by Him in prosperity than those who have not forsaken Him in poverty : "Keep fidelity with a friend in his poverty, that in his prosperity also thou mayest rejoice."

II. Consider secondly, what it means to be faithful to Jesus in His poverty : "Keep fidelity with a friend in his poverty." It means to be glad to bear a like poverty with Him, and it means to be glad to relieve it. The first of these things is done by giving up all that belongs to us for Christ, and the second by keeping it indeed, but in order to distribute it piously amongst the poor. Thou mayest think, perhaps, that the second is more pleasing to Him because He has said to thee so expressly, "What you did to one of these My least brethren you did to Me;"[2] but thou art greatly mistaken : it is the first that is most pleasing to Him. Many persons have their understandings so blinded by attachment to their comforts that it seems to them more praiseworthy, wise, and salutary to relieve than to suffer the poverty of our Lord. This is not so ; who, thinkest thou, is preferred in the Gospels, Zaccheus, who became an almsgiver, and a most bountiful one, or Peter, James, John, and Andrew, who having nothing of their own but a boat, forsook that too for the love of God ? By leaving a small thing, the latter gained the dignity of the Apostolate ; while the former, by giving great things, did not gain, but, as St. Jerome remarks, he remained of low stature even after he had received and entertained our Lord in his own house. So much greater is the merit of him who begs with Christ than of him who, for Christ, relieves a great

[1] Ecclus. xii. 9.   [2] St. Matt. xxv. 40.

multitude of beggars. And no wonder, the former suffers
Christ's poverty together with Him ; the latter compassionates
it. Dost thou think it a higher merit to compassionate thy
neighbour's misery than to suffer it? The devil himself did
not seem to think so, who mocked at Job as at a man, virtuous
indeed, but not perfect, when he saw him make his palace an
asylum for the poor. It was when he saw him on the dunghill
instead of the palace, when there was no one to give him
shelter in his deep poverty, that the devil ceased to mock at
him. Do not flatter thyself, then, by thinking that it is better
for thee to spend thy property piously than to strip thyself of
it to follow thy Jesus, bare of all, like Him. But how if thou
shouldst know how to do neither the one nor the other ;
neither to strip thyself of thy goods to suffer with Christ, nor
even to spend them in compassionating Him? Most surely
thou canst not hope for His riches, if thou hast not shown
him any fidelity in His great poverty : "Keep fidelity with
a friend in his poverty, that in his prosperity also thou mayest
rejoice."

III. Consider thirdly, what will be the riches which Jesus
will vouchsafe at last to give thee if thou hast been a faithful
friend to Him in the poverty which He is now about to choose
for His state. It cannot be doubted that they will be two-fold,
temporal and eternal. Therefore, whether thou hast been
faithful to Him in that poverty by relieving it compassionately
or by enduring it, He will not only give thee Paradise, but
also that hundred-fold on earth which He has promised in
due proportion to him who has shared his wealth with Him,
and to him who has renounced it. Nevertheless, it would
seem that, in this passage, He particularly intended to speak
of the eternal riches. And therefore, He did not say simply,
"Keep fidelity with a friend in his poverty, that in his
prosperity thou mayest be made rich," but He said also "that
in his prosperity thou mayest rejoice." Is it not plain, therefore, that for thy joy, to be in these riches thou must needs
have reached the place where they are permanent? "His
soul shall dwell in good things."[1] For what joy canst thou
have in those riches which are liable to be lost any moment,
as earthly riches are? It can only be in those that are never
lost, in eternal riches, that thou canst really rejoice. And
consider now what an inestimable exchange this is; in relieving
or in suffering the poverty of thy Lord thou wilt have given

[1] Psalm xxiv. 13.

Him thy riches, which are worthless ; and in rewarding thee He will, on the other hand, give thee His, which are of infinite value.   This is what it is to be faithful in His poverty to such a King as Christ is.   If thou hast been faithful to an earthly king who has fallen into adversity, what will it be in his power to give thee when he is restored to his kingdom? At the most, he will give thee a small portion of it.   But if thou hast been faithful to Christ, He will make thee partaker with Him of His whole Kingdom.   Therefore it is not said, " Keep fidelity with a friend in his poverty, that of his prosperity also thou mayest rejoice," but "in his prosperity," that thou mayest know that His very Kingdom shall be all thine, as though thou wert co-heir together with Him.   See how plainly the next words tell thee this : " In the time of his trouble continue faithful to him, that thou mayest also be heir with him in his inheritance."

TWENTY-FIFTH DAY.

THE NATIVITY OF OUR LORD.

For the grace of God our Saviour hath appeared to all men, instructing us that, denying ungodliness and worldly desires, we should live soberly, justly, and godly in this world, looking for the blessed hope and coming of the glory of the great God and our Saviour Jesus Christ (Titus ii. 11—13).

I. Consider first, that the "grace" here spoken of is the intense love of Christ for us, a love which was certainly never merited by us, and therefore was wholly gratuitous of "grace." Now, this love was, as we all know, always the same in the Son of God, but it did not always appear.   It "appeared" most especially on this day, when for our sakes He condescended so far as to be seen lying on straw, clothed in human flesh, a naked trembling Babe, bathed in the tears which His eyes already began to shed for us.   And this is the meaning of the Apostle when he says, " The grace of God our Saviour hath appeared."   Till now this grace was entirely in Heaven : "O Lord, Thy mercy is in Heaven."[1]   Now, at length, it has come down from Heaven to earth, and therefore, whereas up to this time it was promised, prophesied, shadowed forth under various figures, it has on this day "appeared"

[1] Psalm xxxv. 6.

without a veil. How shameful would it be, then, if, on this very day when the love of Christ for thee appeared so clearly, thy love for Him were in no way to appear! But there is one way only in which love appears, the way of works: "In this we have known the charity of God, because He hath laid down His life for us."[1]

II. Consider secondly, that this love of God, our Saviour, is said to have "appeared to all men," although there were, and still are, so many who do not know it. The reason is because He has neglected nothing on His part to make Himself known. The sun appears on the horizon to all; if, therefore, many persons close their windows from his light, is this a reason for saying that he does not appear to them just as he does to others who do not close them? "The grace of God our Saviour hath appeared to all men," because it hath appeared to give light to all men. It is true that although this glorious Sun appeared to give light to all men, yet He has not given light to them all; and therefore, after having said, "The grace of God our Saviour hath appeared to all men," the Apostle immediately added, "instructing us," not "instructing all," but "us;" because all did not receive this light of instruction: "This is the judgment: because the light is come into the world, and men loved darkness rather than the light."[2] This Infant, Whom thou beholdest to-day lying on the straw, is come to enlighten thee. But if thou hast no desire to be enlightened, observe that the fault is not on His side. Oh, how many beams of truth is He engaged in shedding all around! They are as many as are the examples which as soon as He is born He begins to display to thee, so as to be to thee, not only God, but a Saviour-God; how far different from those false ancient gods, who could not save! "They pray to a god that cannot save."[3] If thou dost not fix thy eyes on these beams, the fault is thy own.

III. Consider thirdly, that these examples which Christ gave thee from the hour of His Birth till His Death consist in lessons how to order thyself in regard to thyself, to thy neighbour, and to God. And therefore, in regard to thyself, Christ has taught thee to live "soberly," that is, with moderation, so as, at the least, not to comply with thy desires unrestrainedly, but to moderate them in all things with temperance. In regard to thy neighbour, He has taught thee to live "justly," that is, according to the rules of reason,

[1] 1 St. John lil. 16.   [2] St. John iii. 19.   [3] Isaias xlv. 20.

which requires thee to behave to thy neighbour as thou wouldst like him to behave to thee. And, in regard to God, He has taught thee to live "godly," that is, like an obedient child. See how well Christ did all this from His Birth to His Death; and then reflect on thyself, and humble thyself if thou, on the contrary, doest it so ill. Thou wilt, perhaps, excuse thyself by saying that the world thou livest in is so wicked: "In this world." But this is the very thing which Christ set Himself to teach thee: how to live "soberly" amongst the dissolute, "justly" amongst the unjust, "godly" amongst the wicked, even as He did: "As a lily among thorns."

IV. Consider fourthly, that there are two main obstacles in the way of thus living "soberly" towards thyself, "justly" towards thy neighbour, and "godly" towards God, especially in so corrupt a world. One of these has its source in the understanding, the other in the will. Perverted maxims are the first, and unbridled desires the second. And therefore the Apostle bids thee, before all things, renounce both of these, "that, denying ungodliness and worldly desires, we should live soberly, justly, and godly in this world." Infidelity, as the Doctors here observe, is the greatest of all wickedness, and therefore it is this which must be renounced in the first place by humbly subjecting the understanding to all that the faith teaches; and this is renouncing perverted maxims: "Denying impiety." Then, even when infidelity is taken away, there still remains concupiscence, the parent of irregular appetites, to lead us into evil by reason of the corruption of our nature; and this, therefore, must be renounced in the second place: "And worldly desires." And these appetites are called "worldly," because they are the desires of things which pass away together with the world in which we live, desires of things which are temporal and transitory, and which must, at most, end with the world. And yet thou livest in a state of so much attachment to them as to despise eternal things for their sake. What blindness! Such desires, if they are very irregular, cannot fail to show that there are many remains of infidelity in thee. It is this which seduces thee: "Wickedness overthroweth the sinner."[1]

V. Consider fifthly, that as thou art greatly deterred from this sober, just, and godly life, which Christ comes into the world to teach thee by the infidelity of the mind, so, on

---

[1] Prov. xiii. 6.

the other hand, art thou wonderfully encouraged to it by the
continual thought of the beatitude prepared for thee in the
next life.	And therefore the Apostle says lastly, "Looking
for the blessed hope and coming of the glory of the great
God and our Saviour, Jesus Christ."	He does not say,
"Looking for the hoped for beatitude," but "the blessed
hope," to show us how certain the hope is which is founded
on the Divine promise, so certain, that it may be said that the
hope of the good is hardly to be distinguished from the good
that is hoped for.	It is true that this beatitude will not be
perfect till the Day of Judgment, because then the glory of
the body will be added to that of the soul ; and therefore the
Apostle not only says, "Looking for the blessed hope," but
adds, "And the coming of the glory of the great God and
our Saviour, Jesus Christ."	This God, Whom thou now
beholdest in swaddling-clothes weeping on the straw, seems
a little God, because He is made little.	But on that day, He
will not seem so : He will then appear as the great God He is
really in Himself, and therefore the Apostle here gives Him
the title of "great God."	"They shall see the Son of Man
coming in the clouds of heaven with much power and
majesty."	And so thou seest that, in the first Advent, He is
compared to dew : "Drop down dew, ye heavens ;"[1] and in
the second to lightning : "As lightning cometh out of the
east, and appeareth even into the west, so shall also the
coming of the Son of Man be."[2]	Meantime, what hast thou
to do ?	To look for this second coming with the solicitude
it merits : "My people shall long for My return."[3]	And do
not suppose that this second coming will be like the first.
The first was one of humiliation, the second will be one of
glory for Christ : "The coming of the great God and our
Saviour, Jesus Christ."	If therefore thou hast desired the
first, as being for thy benefit, still more shouldst thou desire
the second, as being for the glory of Christ.

[1] Isaias xlv. 8.		[2] St. Matt. xxiv. 27.		[3] Osee xi. 7.

TWENTY-SIXTH DAY.

ST. STEPHEN, THE PROTO-MARTYR.

Put ye on, therefore, as the elect of God, holy and beloved, the bowels of mercy, benignity, humility, modesty, patience; bearing with one another, and forgiving one another, if any have a complaint against another: even as the Lord hath forgiven you, so you also (Coloss. iii. 12, 13).

I. Consider first, that the Apostle means in this passage to urge the practice of those virtues by which the predestinate are most especially to be distinguished from the reprobate among the faithful. Wherefore he says to them, "Put ye on, as the elect of God, holy and beloved, the bowels of mercy," and the rest. He calls them "elect," because of their having been elected to glory; he calls them "holy," because of their being sanctified by grace; and he calls them "beloved," because of the love shown to them by God in both these gifts. As being such, he would have them "put on" all the virtues that are here named. But in order the better to understand their order, thou must first consider that there are two states in which men may be contemplated—prosperity and adversity. If thou contemplate them in the former, to see of what sort they are, both as regards others and themselves, thou wilt see that they are, for the most part, as regards others, cruel interiorly, and harsh exteriorly. And therefore the Apostle would have them "put on," towards others, "the bowels of mercy and benignity:" "the bowels of mercy," to correct interior hardness of heart; "benignity," to correct exterior harshness of behaviour. As regards themselves, they are usually interiorly vain, and exteriorly ostentatious; and therefore the Apostle would have them "put on," towards themselves, "humility and modesty:" "humility," to correct interior pride; "modesty," to correct exterior ostentation. Next, in the state of adversity, whether thou consider men with regard to themselves or to others, thou wilt see them, generally, impatient interiorly and resentful exteriorly. And, therefore, the Apostle would have them, in this state, "put on patience," and not only this, but all which he immediately goes on to express in the words, "bearing with one another," and the rest. And so "patience" is here opposed to the difficulty of interior endurance, and "bearing with one another," and the rest, is opposed to the quickness to show resentment exteriorly. And in this way the Apostle shows thee, indirectly, how differently the elect act from others, whether in the state of prosperity or adversity. How is it with thyself as to the virtues here mentioned? Reflect on this attentively; for these are the virtues which are here brought forward as marks of the most evident predestination—compassion, benignity, humility, modesty, patience, and the forgiveness of injuries. If these are wanting in thee, what great reason hast thou for fear!

II. Consider secondly, that this fear may be unduly increased in thee by this passage, because thou mayest say that not only "the bowels of mercy," but all the other virtues here mentioned by the Apostle, are those which are very much the result of a man's natural temperament, and that being naturally cruel, harsh, haughty, impatient, and extremely sensitive, thou canst not hope to be of the predestinate, because the marks here given are wanting in thee. But remember that this is just why the Apostle says so fitly "put ye on." Were the clothes of silk or wool, as the case may be, that thou wearest given thee by nature? Certainly not; by nature thou wert naked, but by industriously co-operating with the aids bestowed on thee by God, as Author of the natural order, thou art able to procure what is necessary to clothe and adorn the body. In the same way, by co-operating by thy exertions with the aids given thee by God, as Author of the supernatural order, thou must procure what is necessary to clothe and adorn the soul, still more than the body. Wilt thou be the first person who from being cruel has become merciful; from being harsh, gentle; from vain, humble; from ostentatious, modest; from resentful, patient? If this were not feasible, then the Apostle would not here say "put ye on." By saying "put ye on," he shows that he is speaking to the naked. Strive to do what is in thy power by overcoming nature through repeated acts of these virtues which are opposed to it in thee; and by this means thou wilt speedily possess the marks of the predestination which thou desirest; because to perform repeated acts of virtue is precisely to put on habits of virtue. What does the Apostle mean, thinkest thou, when he says, "Put ye on the bowels of mercy," and the rest? He means, perform acts of those virtues which I here enumerate to you, but let them be frequent, because acts that are performed at rare intervals do not generally suffice for the formation of habits; if, therefore, you do not think yourselves predestinate, act constantly as though you were so, and by this means you will become predestinate.

III. Consider thirdly, that as the Apostle said, "Put ye on the bowels of mercy," he might equally well have said "the bowels of love," but he chose rather to say "of mercy," in order to teach thee who the persons are to whom thou art to extend even thy interior charity, that is, to the unworthy. There are some persons with regard to whom there is no motive for doing or wishing them good but that of their exceeding misery, whether it be corporal or spiritual. Towards such persons the bowels of any kind of charity are not enough, there must be those which are here called "the bowels of mercy," those which, on this day, made the great proto-martyr, St. Stephen, so merciful towards those wretches who were stoning him. Most certainly he saw no merit in them to make him love them. On the contrary, he saw a multitude of reasons amply sufficient to prevent him from loving them, being, as they were, ungrateful, envious, brutal, and furious against him. What then? There being no other kind of charity by which he could have attained so far as not only to pray for them fervently, but to excuse them, he did so by the force of profound mercy. There are no bounds to these "bowels of mercy." They extend to all men. Do not be satisfied, then, with having bowels of ordinary love, but aim at having those of mercy also, which, as thou seest, are placed first among the marks of predestination : "Put ye on, as the elect of God, holy and beloved, the bowels of mercy," then follow the rest. Do thou, too, strive to give them the first place.

IV. Consider fourthly, that what made the first martyr so merciful to his persecutors was the example of Christ, which he had witnessed just before. This should be the chief motive with thee also. Therefore thou hearest how the Apostle encourages thee : "Even as the Lord hath forgiven you, so you also." Injuries are not forgiven, I grant thee, through the force of any kind of love, but through that of pure mercy. But see how Christ did even this, and in thy case also. Therefore, the redemption of the world is ascribed in the Sacred Scriptures to mercy rather than to any other sort of true love : "Through the bowels of the mercy of our God, in which the Orient from on high hath visited us."[1] "According to His mercy He saved us, by the laver of regeneration."[2]

[1] St. Luke i. 78.　　[2] Titus iii. 5.

"According to His great mercy He hath regenerated us into a lively hope."[1] If, therefore, by force of mercy, God went so far as to take human flesh and to die for thee (for thee, ungrateful one !) on the base tree of the Cross, canst not thou, by force of mercy, attain to doing or wishing some good to one who has injured thee even unjustly ? Thou canst, if only thou wilt : the grace is ready for thee. Happy wilt thou be if thou dost attain to this ; thou wilt gain the clearest possible earnest of salvation : "Blessed are the merciful, for they shall obtain mercy."[2]

TWENTY-SEVENTH DAY.

ST. JOHN THE EVANGELIST.

Will the eagle mount up at thy command, and make her nest in high places ? She abideth among the rocks, and dwelleth among cragged flints and stony hills where there is no access. From thence she looketh for the prey, and her eyes behold afar off. Her young ones shall suck up blood, and wheresoever the carcass shall be, she is immediately there (Job xxxix. 27—30).

I. Consider first, that all commentators here understand by the eagle the true contemplative, who is compared to the eagle because of its instinct. And what is this ? To fly high ? Not only so, but to delight in the highest mountain peaks. It is the same with the true contemplative ; the higher he soars, the greater is his enjoyment : "He makes his nest in high places ;" and the word translated "high" means "difficult," so that it is not merely in "high," but in "difficult" places that he delights. There are six degrees of contemplation. The first is simply in the imagination, and is that in which we contemplate visible creatures, admiring their multitude, variety, beauties, and the rest of their qualities which are represented to us by the senses only, and in which we praise God : "How great are Thy works, O Lord ! Thou hast made all things in wisdom."[3] The second is in the imagination aided by the reason, and in this we not only contemplate visible things, as has been said, but we still further endeavour, by the help of reason, to investigate their hidden properties, the end for which they were created, their differences, order, utility, and other characteristics, which do not appear at first

[1] 1 St. Peter i. 3.   [2] St. Matt. v. 7.   [3] Psalm ciii. 24.

sight: "Wonderful are Thy works, and my soul knoweth right well."[1] The third is in the reason aided by the imagination, and in this we rise from visible to the understanding of invisible things: "The invisible things of Him are clearly seen, being understood by the things that are made."[2] Neither do we merely reason from creatures to the Creator, which is, as it were, to make a ladder of them; but, further, in the properties which we see, for example, in water, in plants, in the sun and stars, we see reflected, as it were, the properties of grace when it sanctifies, of inspirations when they enter the heart, of lights when they aid, of Christ when He gives every kind of good to the world; and this is to make of them, as it were, a mirror: "Ask now the beasts, and they shall teach thee, and the birds of the air, and they shall tell thee."[3] The fourth is in the reason aided by the reason; and in this the reason, withdrawing itself as far as possible from the influence of the senses, remains fixed in the contemplation of purely spiritual truths; those which it understands it considers directly, in themselves only; those which it does not understand it infers from others like it which it does understand—for instance, from the pleasure which human sciences afford it infers that which the Beatific Vision will give: "He created in them the science of the spirit."[4] The fifth is above, but not contrary to reason, and it is that in which we contemplate those truths which the reason cannot entirely grasp of itself, but which it has no difficulty in approving, nay, in delighting in, when once they have been revealed to it. Such truths are the simplicity of the Divine Essence, its immensity, infinity, and the rest of its prerogatives, which are indeed superior but not contrary to reason, and which are revealed to us by faith: "Hear, O Israel, the Lord our God is one Lord."[5] The sixth is not only above reason, but treads it under foot, and therefore it contains those truths of the faith which regard the Trinity of the Divine Persons, and other mysteries of the like nature, which the reason is naturally disposed to reject, and yet which, when enlightened by God, it not only does not reject, but delights in beyond all others, rejoicing to see itself at one and the same time conquered and strengthened: "Behold, God is great, exceeding our knowledge."[6] The first two degrees are concerned with sensible, the two next with comprehensible, the two last with incomprehensible things;

---

[1] Psalm cxxxviii. 14.   [2] Romans i. 20.   [3] Job xii. 7.
[4] Ecclus. xvii. 6.   [5] Deut. vi. 4.   [6] Job xxxvi. 26.

and so the first are easy, the second high, and the third difficult. And it is in these last that the eagle loves to make her nest. Because the mind of the contemplative passes over the hills, alights on the mountains, but makes her nest on the highest peaks, "in difficult (*arduis*) places," that is, she dwells by preference on those truths which faith has discovered, and delights, at one time, in seeing how conformable to reason they are ; at another, how far they surpass it. Now, when thou hearest of this noble instinct which has been bestowed on the eagle, thou art at once able to understand what it is that thou hast to do, if God should ever be pleased to give thee so high a vocation. And here observe that if among all holy contemplatives there is no eagle reputed more glorious than St. John the Evangelist, the reason is that none have soared higher in their first flight. He began his at the point where others end theirs : " In the beginning was the Word."

II. Consider secondly, that although the eagle chooses the heights, it is not every height that she chooses, but those that are of stone—"She abideth among the rocks "—because the reason why the true contemplative delights in the secrets revealed to us by faith is not simply because they are very sublime, but because they are of faith, that is, solid, firm, sure, and incontestable. This is the sublimity which is most acceptable to him : " The fortifications of rocks shall be his highness."[1] And it is to be observed that the mysteries revealed to us by faith are divided into two classes : those which regard the Divinity, and those which regard the Humanity of our Lord. And therefore, as thou seest, there are also two kinds of high rocks among which the eagle makes his chosen dwelling. Some are inaccessible from their height, others not only from their height but their ruggedness : " She dwelleth among cragged flints and stony rocks, where there is no access." By the "rocks where there is no access" are signified the mysteries of the Divinity, which do, indeed, alarm the understanding of those who are weaker among the faithful, although if they cannot understand they at least admire them : "We cannot find Him worthily ; He is great in strength, and in judgment, and in justice, and He is ineffable. Therefore man shall fear Him, and all that seem to themselves to be wise shall not behold Him."[2] By the rugged rocks are signified the mysteries of the Sacred Humanity, which are to so many a continual occasion of

[1] Isaias xxxiii. 16.   [2] Job xxxvii. 23, 24.

stumbling—"They stumbled at the stumbling-stone"[1]—while the proud deride, because they do not understand them : "But we preach Christ crucified, unto the Jews, indeed, a stumbling-block, and unto the Gentiles foolishness."[2]   The true contemplative, like the eagle, makes his nest alike "among cragged flints " and "rocks where there is no access."   He does indeed make it first "among cragged flints," because he first dwells long upon the mysteries of the Humanity, and afterwards, "among rocks where there is no access," because he passes on to those of the Divinity.   But in process of time he goes from one to the other indifferently, as the full-grown eagle does, finding in them all an equally firm abiding-place.   In the "rocks where there is no access," she has an unbroken field for gazing on the sun in the zenith of his splendour, and in the "cragged flints" she has a refuge from the winds and storm and rain when night comes on.   Do thou also learn for thy profit, that as in the mysteries of the Divinity thou almost rivallest the blessed in seeing God, thou hast, in those of the Humanity, a special refuge from the tempests to which even eagles are sometimes suddenly exposed on their mountain-tops. There come times of desolation, of sadness, of weariness, and trial.   What oughtest thou then to do ?   Flee to the wounds of Christ, Who was pierced for thee : "He shall go into the clefts of rocks, and into the holes of stones, from the face of the fear of the Lord."[3]

III. Consider thirdly, that the eagle "dwelleth among rocks where there is no access," to escape thither from the troubling of men, and that she "dwelleth among cragged flints," to be also free from the inroads of animals, especially fierce ones, with which, except in cases of grave necessity, she avoids a contest.   Thou, too, wilt find the same two-fold advantage from dwelling like the eagle now in "rocks where there is no access," and again "among cragged flints."   When thou wouldst fain flee from the wearisomeness of intercourse with men, hasten to the mountains and set thyself to contemplate the ineffable joys of those who constantly behold the unveiled face of God, and thou wilt despise the society of all whom thou hast left below : "Our conversation is in Heaven."[4] When thou desirest to escape the insidious attacks of the devils, go to the rugged rocks, enter into the mysteries of the poverty, the contempt, the disfigurement, the torments of

[1] Romans ix. 32.
[2] 1 Cor. i. 23.   [3] Isaias ii. 21.   [4] Philipp. iii. 20.

Christ; for then it is that the devils will have least courage to approach thee.

IV. Consider fourthly, that the true contemplative is not so enamoured of the delight he takes in solitude as not to think of leaving it when there is the opportunity of rescuing some soul from sin. On the contrary, this is the food on which he is nourished. "The food of the just is the conversion of sinners," as St. Gregory observes on this text; therefore he discerns this food even from the loftiest heights, and as he has eyes that can discern from afar the misery of those sinners who are not only dead to God, but putrifying in their vices —his "eyes behold afar off"—so too, urged by his great zeal as by the keenest hunger, he flies swift and straight to make prey of them: "Wheresoever the carcass shall be, he is immediately there." This is to act like a noble eagle: to think not only of contemplation, but of the chase. So, too, in many different ways did St. John the Evangelist. Therefore, if thou admirest him when thou seest him on the mountain-top, fixing his gaze on the orb of the sun, like an eagle in her solitude, no less shouldst thou admire him when thou seest him in extreme old age hastening down ravines and precipices with no other object but that of reaching and, like an eagle, making prey of a depraved young man: "As an eagle flying to the prey."[1] This is the truly beautiful life, the mixed life, which is a combination of the active and contemplative lives. Such a one "makes his nest in high places," and yet "wheresoever the carcass shall be, he is immediately there."

V. Consider fifthly, that this does not certainly belong to mere beginners. And therefore, although the full-grown contemplative passes, like the eagle, from contemplation to action, and from action to contemplation, he does not allow his disciples to do so all at once. He makes them at first give themselves chiefly to silence, solitude, and prayer, and although from time to time he allows them a taste of the chase, it is never more than this. Therefore it is said that his "young ones shall suck up blood." It is much if, in the beginning of their course, they gradually accustom their palate to the great delight which is given by snatching a soul from sin in spite of the devil. The time will come when they will go on from "sucking up blood" to having their breast and feathers all steeped in it after the hot chase in which they will have been engaged, to snatch from the talons of the devil

[1] Job ix. 26.

the putrifying corpse which they beheld from the high places on the brink of destruction. But till that time comes it is enough to imbue them with the desire of the blood which is so sweet to the taste. This is what the eagle whom we celebrate to-day did with his dear children: "Because He hath laid down His life for us, we ought to lay down our life for the brethren."[1] He invited them, like a generous eagle, to "suck up blood," at least in desire.

VI. Consider sixthly, that as the sacred commentators concur in understanding the eagle in this passage to signify the contemplative, so, too, they observe that there is no rule by which any one can be raised to the state of contemplation: "Will the eagle mount up at thy command?" God alone must raise us to these heights: "I will lift thee up above the high places of the earth."[2] There will be days when even the eagle is weary, and feels neither strength nor spirit for his usual flights. What must he then do? He must humbly await the command of his Lord, which will restore his vigour. Meanwhile, if he cannot fly to the highest peaks, let him stay on those that are lower; and if he cannot even reach those, let him remain on the hills, for God will have even the eagle understand that if he has in contemplation two such powerful wings as knowledge and love, it is not of himself that he has them: "There were given to the woman two wings of a great eagle, that she might fly into the desert."[3] Neither do thou say that it is written of the just who have a special trust in God, that "they shall take wings as eagles,"[4] for although it is true that "they shall take" them, it is only when they are offered to them by God. And it is intended by this to show us the difference between those who trust greatly in God and those who do not trust in Him, because the former, when the day comes that they are able to act like eagles by following the lofty flights to which God invites them by knowledge and love, will not remain on the low ground like the latter through pusillanimity, fear, or attachment to former habits. And besides, although all those who "hope in the Lord take wings as eagles" for themselves, they cannot fasten them to the shoulders of others; they take them for themselves, not for others. And, therefore, it is necessary to wait till God plainly offers that which is a gift before going on to exercise it. And although a thing which

---

[1] 1 St. John iii. 16.
[2] Isaias lviii. 14.   [3] Apoc. xii. 14.   [4] Isaias xl. 31.

is a gift may sometimes be rightly asked of God, nevertheless we find in the Sacred Scriptures that the wings of a dove have been asked of God: "Who will give me wings like a dove, and I will fly and be at rest?"[1] but not those of an eagle. For her swift wings are given to the dove that she may fly to the upper regions of the air as high as is necessary to place herself in safety, whereas they are given to the eagle that she may rule those regions.

TWENTY-EIGHTH DAY.

THE HOLY INNOCENTS.

Behold, they whose judgment was not to drink of the cup shall certainly drink: and shalt thou come off as innocent? Thou shalt not come off as innocent, but drinking thou shalt drink (Jerem. xlix. 12).

I. Consider first, how cowardly thou art if thou shrinkest from a few crosses and trials which God sends thee that thou mayest win Heaven. These innocent babes, almost as soon as they were born, had to win it by suffering a most cruel death, being slain by the sword and dashed to pieces before the eyes of their mothers. And dost thou expect to reach it by suffering nothing? Oh, how greatly art thou deceived! "Behold, they whose judgment was not to drink of the cup shall certainly drink: and shalt thou come off as innocent? Thou shalt not come off as innocent, but drinking thou shalt drink." It is said that to drink of the cup was not "their judgment," because these infants, not being yet endowed with reason, were not only as yet incapable of suffering such a penalty, but even of being placed at the bar, so undoubted was their innocence. And yet they drank of the cup, as though they had been criminals, and criminals deserving death, for they drank it to the very dregs: *bibentes biberunt.* And dost thou, who art guilty, complain if it is thy lot to drink a few drops of it? Thou studiest how to take thy pleasure now in all things, to laugh and joke and dance and follow thy fancy; but what thou dost not suffer here thou wilt suffer hereafter: "Rejoice and be glad, O daughter of Edom, that dwellest in the land of Hus; to thee also shall the cup

[1] Psalm liv. 7.

come, and thou shalt be made drunk and naked,"[1] that is, "thou shalt be made drunk" with all that bitter draught of which thou now refusest to taste a drop; and "thou shalt be made naked" of all those pleasures or glories or dignities which now make up thy delight.

II. Consider secondly, that by this "cup" is meant avenging justice, according as it is written, "In the hand of the Lord there is a cup of strong wine full of mixture,"[2] and therefore observe its qualities. It is "a cup," because this justice is exercised by God in measure, that is, according to the quantity or the quality of the crimes that have to be punished : "Thou wilt give us for our drink tears in measure."[3] It is of pure or "strong wine" (*vini meri*), because as such wine has the power of so taking away a man's strength as not to leave him any longer master of himself either interiorly or exteriorly, so, too, has Divine justice. So that whereas a man in his senses may very often defend himself from human justice, escape from it, ward it off, this is impossible with Divine justice. Whoever is in its power must be as helpless as a drunken man : "Take the cup of wine of this fury at My hand ; and thou shalt make all the nations to drink thereof, unto which I shall send thee ; and they shall drink and be troubled and be mad because of the sword which I shall send among them."[4] It is pure wine, but not all of one kind ; it is wine "full of mixture," because Divine justice is bound by no law to one single kind of punishment, as human justice is ; it is made up of many kinds : "Fire and brimstone and storms of winds shall be the portion of their cup."[5] It is "in the hand of the Lord," because He is free to exercise His justice whenever He pleases ; there is no appointed time or place for Him, as there is for human judges. He does what He will : "He hath poured it out from this to that, but the dregs thereof are not emptied." Wilt thou refuse to drink of this cup when thy Lord offers it to thee in this life ? Take heed, for if the Innocents have to drink it merely because they are of Adam's race, much more sinners, that is, those who are laden with so many sins which they have themselves committed : "All the sinners of the earth shall drink." How canst thou, then, be the only one out of so many to go unpunished ? "Thou shalt not come off as innocent, but drinking thou shalt drink," that is, if thou wilt not drink this cup out of love, thou wilt have to

[1] Lament. iv. 21.   [2] Psalm lxxiv. 9.
[3] Psalm lxxix. 6.   [4] Jerem. xxv. 15, 16.   [5] Psalm x. 7.

drink it whether thou choosest or not : " And if they refuse to take the cup at thy hand to drink, thou shalt say to them : Thus saith the Lord of hosts, Drinking you shall drink."[1]

III. Consider thirdly, that it seems so hard to thee to be sometimes persecuted or punished when thou art innocent, that thou art bold enough to say that thou wouldst feel it less if thou wert guilty. But is not this the greatest of errors? Wouldst thou, then, prefer to drink the cup of God's justice as Herod did, who, out of rage at seeing himself slowly devoured by worms, attempted to kill himself with a knife, rather than to drink it as the infants did whom Herod slew? The evil most to be dreaded in the world is not punishment but guilt, therefore does God declare the punishment that men may avoid the guilt. Is it possible that thou preferrest the former with, rather than without, the latter? Not so; be willing for God to suffer thee to be both persecuted and punished although innocent. The time will come when He will do thee justice. Dost thou not see that in a few years the cup passed from the Innocents to Herod? So, too, mayest thou believe that it will be in thy case: "Behold, I have taken out of thy hand the cup of dead sleep, the dregs of the cup of My indignation; thou shalt not drink it again any more."[2]

TWENTY-NINTH DAY.

God, Who, at sundry times, and in divers manners spoke in times past to the fathers by the Prophets, last of all in these days hath spoken to us by His Son, Whom He hath appointed Heir of all things, by Whom also He made the world (Hebrews i. 1).

I. Consider first, that it may strike thee as strange that the practice of the New is so different from what it was under the Old Law. In the latter it was not lawful only, but praiseworthy, to desire to receive supernaturally an answer from God concerning what He was about to do : " Lord God, whereby may I know?"[3] So much so, that those who neglected to do so were often rebuked : "They consulted not the mouth of the Lord;"[4] "They have not asked at My mouth."[5] People then went purposely to the Prophets to inquire about the

---

[1] Jerem. xxv. 28. [2] Isaias li. 22.
[3] Genesis xv. 8. [4] Josue ix. 14. [5] Isaias xxx. 2.

least things : "Come, let us go to the seer."[1] Nor was it
only prognostications, but sometimes also visions, apparitions,
assurances that they were then at liberty to ask, nay, they were
invited to do so : "Ask thee a sign of the Lord thy God either
unto the depth of Hell or unto the height above."[2] Now, on
the contrary, nothing of the sort is allowed : "The Jews require
signs."[3] Not only would a man not be praised among Christians
who should act thus, but he would be blamed ; and the only
course which is now approved is to commend to God all the
works that we undertake, but not to desire to know the event
before the time. I do not deny that this may seem wonderful
to thee ; but it is because thou dost not yet fully understand
how great a benefit God has done us in giving us Christ. In
giving Him to us, He has already told us all that He could
tell : "A short word shall the Lord make upon the earth."[4]
And the reason is that everything which God formerly told
His people when He so often spoke to them by His Prophets,
and in so many other ways, was directed to the one end of
foretelling Christ : "The end of the Law is Christ, unto justice
to every one that believeth."[5] And although He frequently
gave answers concerning other matters, such as, whether they
were to march, to fight, and the like, still these very things
were all figures of what was to be done in future days by
Christ or by His disciples : "All these things happened to
them in figure."[6] And, therefore, it was right to ask of God
the certain, clear, accurate rule of everything that was to be
done, for He alone could know how the type was to be
arranged so as not to contradict the antitype. Now, the
antitype has appeared : "God, Who, at sundry times and in
divers manners, spoke in times past to the fathers by the
Prophets, last of all in these days hath spoken to us by His
Son." And, therefore, now that the types are at an end, all
that remains is to contemplate the antitype, to listen to what
He said, and to see how He acted when He came into the
world. If we do this, we shall know how we ought to behave
in all our actions. Of what use, then, is it to ask further?
The painter, who has the original before him, has no need to
inquire of the master how to employ the brush, he has only to
look at the original and copy it.

II. Consider secondly, how great an advantage, in this
case, our times have over those which were of old. "In times

[1] 1 Kings ix. 9.   [2] Isaias vii. 11.   [3] 1 Cor. i. 22.
[4] Romans ix. 28.   [5] Romans x. 4.   [6] 1 Cor. x. 11.

past," that is, under the Old Law, God did indeed speak, but
only to a few persons, "to the fathers," that is, to the nation
of the Jews only.   Now, "in these days," He has spoken to
that nation and to all others: "I appeared openly to them
that asked not after Me."[1]   Therefore, it is added that God
has spoken for the last time, "last of all," because, after this
time, He will speak no more, so that whereas formerly one law
followed the other, as being imperfect, one prophecy another,
one truth another; now, under the Gospel law, everything is
said in perfection: "It is consummated."   God spoke "in
times past" to servants by the mouth of servants, "by the
prophets:" now, "in these days," He has spoken to servants
by the mouth of His Son.   And, therefore, the Prophets spoke
precisely as servants, always saying: "Thus saith the Lord."
The Son spoke as a Master: "I say unto you."   And whereas
they spoke obscurely, as becomes servants, who are never
thoroughly informed of the secret—"The servant knoweth not
what his lord doth"[2]—He spoke with wonderful clearness,
because He, as the Son, knew everything: "The only-begotten
Son, Who is in the bosom of the Father, He hath declared
Him."   "In times past" God spoke "at sundry times and in
divers manners," as is the wont of one who does not tell
everything at once.   "In these days" He has always spoken
in the same way, a shorter, indeed, but far nobler way.   Dost
thou, then, deserve any excuse if thou dost not perceive the
priceless boon that God has bestowed on thee in causing thee
to be born not "in times past," but "in these days," in which
we live, "upon whom the ends of the world are come."[3]   All,
therefore, that thou hast to do now is to keep thy eyes and
ears turned to Christ.   Watch Him, to learn how He bore
Himself; listen to Him, to understand what He said: and
how much wilt thou immediately learn which is for thy soul's
good!   As to that of the body, do not concern thyself to
know anything, in the way that was formerly done.   For the
reason why many things of this sort were praiseworthy in the
Jews, in order to preserve their mortal life, was that they were
in a state of continual expectation of living to see Christ.
Therefore the dying Ezechias lamented in these words: "I
sought for the residue of my years; I said, I shall not see the
Lord God in the land of the living."[4]   And therefore, Simeon,
who did live to see Him, said: "Now Thou dost dismiss Thy

[1] Romans x. 20.
[2] St. John xv. 15.   [3] 1 Cor. x. 11.   [4] Isaias xxxviii. 11.

servant, O Lord, . . . because mine eyes have seen Thy salvation."[1] So that now this regard for the life of the body is no longer praiseworthy, for the only way of seeing Christ now is to die. Therefore, thou oughtest to give but little thought to thy body; think about thy soul, and thou shalt speedily know all that thou desirest concerning it by inquiring not of the servants, but of the Son of thy King.

III. Consider thirdly, how greatly that man errs who, out of a desire to think of God alone, turns away his thoughts habitually from what was done by Christ. He must be our oracle in everything we do on earth: "This is My Beloved Son, in Whom I am well pleased: hear ye Him."[2] How, then, can there ever be a time when we have no longer to converse with Him? To converse face to face with God, without a veil, is reserved for us in Heaven; on earth we are obliged to do this under a veil. Do we not know that in Christ "dwelleth all the fulness of the Godhead," even "corporally"?[3] Why, then, seek further? The more the splendour of the Divinity is tempered, the more suited is it to our weakness. Learn, then, from Him as Man the example thou hast to imitate; adore in Him as God the Infinity and Incomprehensibility that thou hast to believe. Therefore is it that, in order to set Him before thee as He is, both God and Man, the Apostle after saying, "God, Who, at sundry times and in divers places, spoke to the fathers in times past by the Prophets, last of all, in these days, hath spoken to us by His Son," immediately adds, "Whom He hath appointed Heir of all things, by Whom also He made the world." When he says, "Whom He hath appointed Heir of all things," he is speaking of Him in His Human Nature; when He says, "By Whom He made the world," he is speaking of Him in His Divine Nature. In His Human Nature, Christ is appointed by the Father Heir of all Divine goods, as being those of His Father, and so Heir of all nations, angels, archangels, and of all spirits subject to God in Hell as well as in Heaven: "Ask of Me, and I will give thee the Gentiles for Thy inheritance."[4] And therefore it is of Christ, according to this nature, that the Apostle here says: "Whom God hath appointed Heir of all things." Next, in His Divine Nature, He is the Creator of ages, and consequently of all besides; for the word translated "world" (*sæcula*) means also the ages of time. Now, the most difficult thing to conceive as having been created is, according to every school of philosophers, time, so much does it bear the appearance of having always existed. So that, what can there be that He Who made time has not made? Therefore, as thou seest, the Apostle does not here say "by Whom He made the ages," but "by Whom also," in order by this to show that the power of this Divine Son extended so far as even to create time. Neither do thou object that it is not said "Who made the ages," but "by Whom God made the ages," because the preposition "by" does not imply inferiority of power in any of the Divine Persons, but order. It is meant that the ages were made by Him, as in the archetypal pattern which in the mind of God is consubstantial with the Maker. Besides, if it was He "by Whom" His Almighty Father "made the ages," it follows that not His Father only, but He also was before all ages: "God is our King before ages."[1] Do thou, when contemplating Him as the Maker of ages, humble thyself reverently before His mighty power; and contemplating Him as universal Heir of every good thing which God can possibly give, and an Heir, not merely chosen, but appointed by Him, that is, certainly, unalterably, and immoveably Heir, understand that there can be no inheritance in store for thee but that which is thine by the grace of Christ: "That being justified by His grace, we may be heirs according to hope of life everlasting."[2]

THIRTIETH DAY.

Thy testimonies are wonderful ; therefore my soul hath sought them
(Psalm cxviii. 129).

I. Consider first, that if the Sacred Scriptures are full of meaning so wonderful as far to surpass the capacity of our minds, it may seem to thee that the holy King David ought to have been satisfied merely with believing them, and not to have set himself also to search into them. Yet he did not do so. Nay, it is for this very reason that he says he did search diligently into these meanings, because they were so wonderful: "Thy testimonies are wonderful; therefore my soul hath sought (*scrutata est*) them." And the reason is

[1] Psalm lxxiii. 12.　　[2] Titus iii. 7.

that, although this searching is to be abhorred as presumption
when it has its source in want of faith in those wonderful
sayings, it is not merely not to be abhorred, but greatly to be
commended when it proceeds from love to them: "They
received the word with all eagerness, daily searching the
Scriptures, whether these things were so."[1] And what sage
ever publishes his books that people may be satisfied with
believing what is in them? He publishes them that those
who are capable not only of believing but of understanding
their contents may read, study, consider them, and ponder the
value of every word: "The words of the wise shall be weighed
in a balance."[2] Why, then, shouldst thou suppose that a
God of infinite wisdom has acted differently in the Sacred
Books which He dictated to His servants by His own
mouth? He did this that we might be continually engaged
in fathoming and searching into them as into a rich mine:
"Search the Scriptures."[3] What, then, if thou shouldst disdain
to do God this honour? When thou art able to admire His
wisdom (which is the highest and final degree of contemplation, in which the soul may be said to go out of herself in
a state of astounded ecstasy: "I have heard Thy hearing and
was afraid"[4]), thou oughtest not to be satisfied with merely
believing it, which is the lowest degree of contemplation with
which it begins: "He that cometh to God must believe."[5]

II. Consider secondly, that there are two meanings of the
Sacred Scriptures, the literal and the spiritual or mystical
meaning, and both of these are most wonderful: "Thy
testimonies are wonderful." The literal is the direct meaning
produced by the force of the words, that which is, as it were,
the body containing the spiritual meaning, and therefore
which always lies at the top, on the surface, and as it were
outside. The spiritual is the indirect meaning, which is inside,
as the spirit is inside the body, for which reason it is called
spiritual: "I saw in the right hand of Him that sat upon the
throne a book written within and without:"[6] "within," as to
the spiritual; "without," as to the literal meaning. And here,
in order to admire the Divine Word, observe that the wisdom
of God is so great that by words He can mean things, as we
ourselves do when, for example, by Jerusalem we mean the
city which was the metropolis of Palestine; and by the things
signified by those words He can, at the same time, mean

[1] Acts xvii. 11. [2] Ecclus. xxi. 28. [3] St. John v. 39.
[4] Habacuc iii. 1. [5] Hebrews xi. 6. [6] Apoc. v. 1.

several other things according to His pleasure, which, although we may sometimes be able to do, we cannot do *ad infinitum*, as He can to Whose Divine mind there are no limits. And so, by Jerusalem, He has at the same time been able to mean other things which are not visible on the outer rind of the word, but which are to be found in its kernel by one who knows how to look closely into it.

Now, as God has had no other end in the Sacred Scriptures but to reveal to us what we ought to believe, to hope, and to do, according to the rules of the charity which is due to Him, so, too, are there three things to which He has alluded in the spiritual sense. These are (1) the Church Militant, which was to be set up on earth by Christ as its supreme Head, and under this meaning is especially included that which we are to believe; (2) the Church Triumphant, which was to be set up in Heaven by Christ, and under this meaning is especially included that which we must hope; (3) the faithful soul, which was to be espoused by Christ, and under this meaning is included especially that which we are or are not to do, according to the many precepts which are all collected together for us in that of charity. Hence it is that the spiritual meaning branches into three meanings, which are like three species subordinate to one genus, namely, the allegorical, the anagogical, and the moral, or, as some call it, the tropological meaning. The allegorical meaning belongs to the Church Militant, of which the Old Law was a figure. The anagogical belongs to the Church Triumphant, of which the Old Law was not, strictly speaking, a figure, but a shadow. The moral belongs to our soul. And so, by this word Jerusalem, which I have chosen for thee as an example, God doubtless always intended to signify the chief city of Palestine, but, still further, sometimes the Church Militant, sometimes the Church Triumphant, sometimes the faithful soul, and sometimes all three together, which is certainly a very wonderful manner of speaking: "Thy testimonies are wonderful." This is seen in the Psalms—"Praise the Lord, O Jerusalem "—because all the benefits which, according to the letter, David there prophesied would be bestowed by God upon the city of Jerusalem, when it should be rebuilt by Nehemias after the Babylonian captivity, he meant to signify far more, according to the spiritual meaning, with regard to the Church Militant and Triumphant, and to the devout soul which becomes, by means of lofty contemplation, in another sense, the Vision of Peace.

Seeing, then, that thou hast in the Divine Scriptures a language so admirable, how is it possible for thee not to be so charmed with it as to exclaim to God in thy turn, "Thy testimonies are wonderful, therefore my soul hath sought them"? Turn from idle romances, the authors of which, invent as they may, have never been able to succeed in imagining tales to equal in beauty the truths written by God in His Book: "The wicked have told me fables, but not as Thy Law."[1] And if, during the past year, thou hast paid thy God this homage of dwelling diligently on His words, then resolve to renew the practice during the one which is coming: "My eyes to Thee have prevented the morning, that I might meditate on Thy words."[2]

III. Consider thirdly, that the spiritual meanings of Scripture are also called mystical, and for this reason: because, although they are contained in the literal meaning (as the spirit is contained in the body), yet they do not always appear at first sight like the spirit, which shows itself at once by the motions of the body, and even from the expression, the manner, and the complexion. They have to be searched after with care, as being not only hidden but abstruse, like all mysteries. Therefore the holy King David said to God, "Thy testimonies are wonderful, therefore my soul hath sought them," not only "hath considered," but "sought them," because he did not presume to be able to penetrate into the words of God at once without trouble; he studied them, pondered them, made deep research into them, in order to discover all the hidden meaning that they might contain. It is true, indeed, that he did all this with a view to his soul's good, and therefore he said, "My soul hath sought them," not "my understanding" only, but "my soul," in order to include in one word the understanding and the will: "My soul hath desired thee in the night."[3] If thou settest thyself in thy prayer to fathom the meaning of the Scriptures in order to feed thy understanding only, thou art not doing what is right; thou oughtest to do this in order to prepare the food of the understanding for the profit of the will, which should at the same time be inflamed either to believe more firmly, to hope more strongly, or to love more ardently that which God reveals to thee in the depths of His Word. And this is the true gift of understanding, that which is not merely applied to speculative but to practical purposes: "Give me under-

[1] Psalm cxviii. 85.   [2] Psalm cxviii. 148.   [3] Isaias xxvi. 9

standing, and I will search Thy Law."[1] So that, by virtue of
this gift, not only hast thou to consider the Divine meanings
of Scripture in order to understand what they contain, but also
what they require of thee as the rule of all thy operations. If,
then, thou dost not possess this gift, at least not in any considerable degree, this is the reason: thou dost not put in
practice what God has frequently revealed to thee by its
means: "A good understanding to all that do it."[2]

THIRTY-FIRST DAY.

*For of Him, and by Him, and in Him are all things: to Him be glory
for ever. Amen* (Romans xi. 36).

I. Consider first, that the power, wisdom, and goodness of
the Three Divine Persons are the same in each: if it were not
so, it would follow that there was not in them One only God,
which is contrary to the teaching of faith: "There are Three
Who give testimony in Heaven, the Father, the Word, and the
Holy Spirit. And these Three are One."[3] The Divine power
is the efficient cause from which creatures receive their being,
and therefore it is attributed to the Father, as the Principle
whence they all proceed. Wisdom is the exemplary cause by
which they receive their form, and therefore it is attributed to
the Son, Who proceeds from the Father as His Image, but
a substantial Image, representing all the beauty which it is
possible for God to impart to the things which can be created
by Him. Goodness is the final cause from which they receive
their order, and therefore it is attributed to the Holy Spirit, as
proceeding from the Father and the Son by the way of love,
that is, as the motive power which gives to all things the being
of which they are capable, and preserves it. Understanding
this, it will be easy for thee to understand the deep meaning
of these few words, "For of Him, and by Him, and in Him
are all things: to Him be glory for ever," which thou hast to
consider on this last day of the year, in order to give glory to
God for every possible good. They mean: "For all things
are of Him, as powerful; by Him, as wise; in Him, as good:
to Him be glory for ever." By the words "of Him, by Him,
and in Him" is to be understood the Trinity of the Divine

[1] Psalm cxviii. 34.   [2] Psalm cx. 10.   [3] 1 St. John v. 7.

Persons; by the words "to Him" is to be understood the Unity of the Essence; and this being the same in all, it follows that the glory due to the Father, to the Son, and to the Holy Ghost, for what is done by Them for the good of men, is not different in each, but that one and the same glory is due to All, as to One and the same God—"to Him be glory," that is, to the God Who is powerful, and therefore "of Him are all things;" Who is wise, and therefore are "by Him;" Who is good, and therefore they are "in Him." Occupy thyself here in considering this beautiful union which is in the Three Divine Persons, in working each by each for thy good; and, knowing how greatly thou art bound to Them, animate thyself to give thyself wholly to Their service; so that whatever thou hast in thy power, whatever thou knowest, whatever thou willest, all may be for God, not dividing thy heart, but resolving to give it all to Him only: "With all thy strength love Him that made thee."[1]

II. Consider secondly, that it is not said, *De ipso omnia,* but *ex ipso;* because although it is true that whatever is *de ipso* is also *ex ipso,* it is not true that whatever is *ex ipso* is also *de ipso.* The Son is of one substance with the Father, and therefore of Him it is said that He is not only *ex ipso,* but also *de ipso,* "God of God" (*Deus de Deo*). Not so creatures; and therefore of them it is not said that they are *de ipso,* but only *ex ipso,* "all things of God" (*omnia ex Deo*).[2] Now, by this "all" the Apostle here meant created things also. And therefore it is that he said *ex quo* and not *de quo,* because the preposition *ex* is not used to express a consubstantial cause, as the preposition *de* is: "I will crop off a tender twig from (*de*) the top of the branches thereof, and I will plant it on a mountain high and eminent."[3] Do thou here take pleasure in observing the difference between thee and the Son of God. He is not only *ex ipso Deo,* but *de ipso;* whereas thou art only *ex ipso.* Although, indeed, since this Divine Son has exalted thee by grace to a participation of that nature which He has in common with the Father, He has made thee have thy being not only from, but in some sort of Him, in so sublime a way, that thou too art become a son of God: "He gave them power to be made the sons of God."[4] Neither do thou say that Christ is the Son of God by nature, and thou by adoption only; for, in the first place, dost thou not think that

---

[1] Ecclus. vii. 32.
[2] 1 Cor. xi. 12.
[3] Ezech. xvii. 22.
[4] St. John i. 12.

it is the greatest possible honour to be adopted by God as His own son? If it is thought so great a thing to be adopted by an earthly king, what must it be to be adopted by God? Consider next, that Divine adoption is very different from human. The latter makes the adopted son partaker of his father's inheritance, but not of his nature; the former makes him partaker of this also : "That you may be made partakers of the Divine Nature,"[1] as St. Peter said to all the just. Although, indeed, in Christ this nature is a substantial form, subsisting of itself in His Person, as in a being both human and Divine, and therefore it can be said that Christ is "God of God" (*Deus de Deo*). In the just this nature is an accidental form, which finds them already subsisting in their complete being of pure humanity. And therefore, although it may be said in a certain sense that they are sons of God—"I have said, You are gods and all of you the sons of the Most High"[2] —still they are only sons "born of (*ex*) God.[3] "Whosoever is born of (*ex*) God committeth not sin;"[4] "Whosoever is born of (*ex*) God overcometh the world;"[5] "Every one who doth justice is born of (*ex*) Him."[6] And just as that which gives to man his first natural being, and is the first interior principle of the natural motions by which he is moved, is, in the natural order, his nature; so that which gives to man his first supernatural being, and is the first interior principle of the supernatural motions by which he is moved, may be asserted to be his nature in the supernatural order. And this is sanctifying grace. Is it possible, then, that, possessing so exalted a dignity, thou shouldst ever so despise it as to become the devil's slave instead of the son of God?

III. Consider thirdly, that as all things are "of Him," so also are all things "by Him." But as this preposition "by" may occasion thee to fall into error, observe that in Divine things the Father truly does all things by the Son : "All things were made by Him."[7] But what does this mean? Is it that the Father gives power to the Son to do what He does, as He gives it to the kings of the earth, speaking of whom He says justly: "By Me kings reign"?[8] But the Father has all His power from Himself. The meaning is, that the Father acts, as it were, by means of the Son, but in a most sublime manner, because, in communicating to Him His Essence, He also

---

[1] 2 St. Peter i. 4.   [2] Psalm lxxxi. 6.   [3] St. John i. 13.
[4] 1 St. John iii. 9.   [5] 1 St. John v. 4.   [6] St. John ii. 29.
[7] St. John i. 3.   [8] Prov. viii. 15.

communicates the power of working; but not an instrumental power, either less than or different from His own, such as is that which He communicates to His ministers, but the very same, with no difference, except that the Father has it from Himself, and the Son from the Father. Hence, the Son also is the primary agent in all things, just as the Father is never a secondary agent: "For what things soever He doth, these the Son also doth in like manner,"[1] to the confusion of those who would assert that He is inferior to the Father. Nevertheless, it is said that "the Father maketh by the Son," and it is not said that "the Son maketh by the Father," because as the order of the Divine Persons cannot be regulated by Their power, that being the same in all the Three, it is regulated by Their relations to each other according to Their origin, which is different. The Son is, with regard to the Father, the direct Reason of all things which can be made by Him, as by the Chief Artificer; He is His Art, but an Art which is essential, intimate, innate, and consubstantial. Therefore, as we do not say that the art works by the artificer, but that the artificer works by the art, so neither is it said that the Son works by the Father, but the Father by the Son. Learn, therefore, from seeing that God cannot but work by an Infinite wisdom, which is His Art, as proper to Himself as His very Essence, not only to love Him in His dispositions and to admire Him in His decrees, but also to reverence Him in the depth of His judgments, which cannot be discerned by thy reason: "Who can say, Why dost Thou so?"[2]

IV. Consider fourthly, that as all things are "of Him" and "by Him," so also are they "in Him." This preposition "in" here signifies the power of containing, and so it may be attributed not to the Third Person only, but to the others also, because all things are contained in the Father, as the Efficient Cause, and in the Son, as the Exemplary Cause; but when applied to the Holy Ghost, as it seems to be in our text, it signifies the Moving Cause, which is that on which is founded the creation and also the preservation of all things. And this Moving Cause is, beyond a doubt, no other than Divine Love—"I have loved thee with an everlasting love"[3] —for God does not love things because they are, as we do, but He gives them their being because He loves them. It is the Divine Goodness, then, which, as it first gave existence to all created things, so too does not allow them to return to

---

[1] St. John v. 19.   [2] Job ix. 12.   [3] Jerem. xxxi. 3.

their original nothingness, for which reason they are all said to have their continued subsistence in Him : "They abounded with delight in Thy great goodness."[1] But every one knows that Goodness, as an attribute relating to the will, is ascribed to the Holy Ghost, the Primary Love. And therefore it is particularly said of Him, that "in Him are all things." Furthermore, the Holy Ghost is the bond uniting the Father to the Son and the Son to the Father ; and therefore He is, as it were, the support of whatever is done by Them, according to that saying : "Small things increase by union, and the greatest decay through disunion." Do thou, therefore, consider that union is productive of the same good in human as in Divine things. And therefore, when thou violatest charity in the community in which thou art living, know that thou art bringing upon it utter destruction, so far as thy power extends, because, although power and wisdom are necessary for its preservation, they are not sufficient of themselves, there must be union also. And whence can this union come but from the mutual charity of those who are united ?

V. Consider fifthly, that thou seest, therefore, that in Divine things (*in divinis*) unity is attributed to the Father, equality to the Son, and connection to the Holy Ghost. Unity is attributed to the Father, because unity presupposes no other before itself ; and so it represents to us the First Principle, that is, a Sovereign Power, receiving existence from none and giving it to all : "One God, the Father, of Whom are all things."[2] Equality is attributed to the Son : "He thought it not robbery to be equal with God,"[3] because there must be equality at least between Two. And although, undoubtedly, all the Three Divine Persons are, and are spoken of, as equal to each other, yet still, the First Person cannot constitute equality, because equality cannot be in unity alone ; and the Third Person finds equality already existing. Therefore, it is particularly attributed to the Second Person, as being the first to constitute it, that is, it is attributed to that Person to Whom also Wisdom is attributed, because it belongs to wisdom to equalize things. Connection is attributed to the Holy Ghost, because connection presupposes and joins extremes. And this connection is, as thou seest, altogether founded on mutual love, that is, the love of the Son to the Father, and of the Father to the Son. And to this love goodness also is ascribed, because this it is which makes the

---

[1] 2 Esdras ix. 25.   [2] 1 Cor. viii. 6.   [3] Philipp. ii. 6.

Father and the Son to agree in diffusing so many of their goods even outside of Themselves, so that the Father does nothing *ad extra* without the Son, neither does the Son do anything without the Father, but the operation of both is one, as Their power is. What, then, will be the consequence if mutual love is severed in a community. There can be no hope of any more good either within or without.

VI. Consider, sixthly, that when it is said, "Of Him, by Him, and in Him are all things," thou art to understand by "all things" all those things which have any sort of existence ; but this must be a real existence, and therefore thou art not to suppose that there is any kind of allusion to sins, for they have no existence, but an improper, unreal, and abusive existence, which is, in fact, only a want of perfection. Observe that, in every sin whatsoever, there is a want of each of one of those three Divine perfections, which give existence to everything. There is want of power, because to sin is not an act of strength, but of weakness. Strength consists in the subjection of the irregular appetites, so that they are compelled to obey reason. There is want of wisdom, because to sin is not an act of knowledge, but of ignorance, if, indeed, it may not rather be called blindness. There is want of goodness, because sin not only does not make man good, but it makes him exceedingly evil to himself and to others. Is it not quite certain, therefore, that "all things" in this passage cannot possibly include sins? This is the reason why sins are called "nothing." "Correct me, O Lord, but yet with judgment, and not in Thy fury, lest Thou bring me to nothing."[1] It is true, that if they are nothing, it is the most terrible nothingness possible, because the existence which they leave man is just enough for him to wish one day that he had none. Wilt not thou, too, pray to thy God not to bring thee to such a nothingness as this, that is, to a state in which He refuses thee those special or superabundant aids which thou hast ceased to merit through thy tepidity ; because no sooner are these withdrawn from thee, than it is impossible for thee not to sin: "They shall come to nothing, like water running down."[2]

VII. Consider lastly, how legitimate a conclusion this is: "Of Him, by Him, and in Him are all things ;" therefore, no one ought to ascribe glory to himself for anything, but to give it to God : "To Him be glory for ever." And when is it that thou ascribest to thyself the glory of any good thing that thou

[1] Jerem. x. 24.   [2] Psalm lvii. 8.

mayest have done? It is when thou regardest it with self-complacency, or self-praise, as though thou hadst done it of thyself. This is the greatest robbery of God that thou canst commit: for it is to rob Him of that glory which can only be due to Him. Glory is, in its nature, common to others besides God for the good that they do—"Glory to every one that worketh good"[1]—but with this difference, that the glory which is given to others can never be given to them as belonging to them, but only to them as working by the power of God: "He that glorieth, let him glory in the Lord."[2] It is only the glory which is given to God that can be given to Him as His without restriction. And yet how often dost thou make thyself the end of thy glory, thinking of thyself as if thou wert the primary agent in the good that thou doest! Say, on the contrary, "For of Him, and by Him, and in Him are all things: to Him be glory for ever." And why "for ever"? Because the glory which is given to every one ought always to be proportionate to his merit. But is it not, then, certain that the glory which ought to be given to God is an infinite glory? For, seeing that the power, the art, the love with which He operates in the very least thing are infinite, it follows that He merits infinitely to be glorified for each thing. Therefore, there being no glory that can be given to God by creatures which is infinite in intensity, it is just that that which is given should at least be infinite in extension, that is, for ages after ages, without end: "For of Him, and by Him, and in Him are all things: to Him be glory for ever," that is, not "any glory," but "all glory:" the glory of the heart, the glory of the lips, and the glory of action. Amen.

[1] Romans ii. 10.   [2] 2 Cor. x. 17.

www.ingramcontent.com/pod-product-compliance
Lightning Source LLC
Chambersburg PA
CBHW020848210326

41598CB00018B/1613